CONCRETE
TECHNOLOGY
Second Edition

A.R. SANTHAKUMAR

Former Dean and Chairman
Faculty of Civil Engineering
Anna University

Former Emeritus Professor
Department of Civil Engineering
Indian Institute of Technology Madras

OXFORD
UNIVERSITY PRESS

OXFORD

UNIVERSITY PRESS

Oxford University Press is a department of the University of Oxford.
It furthers the University's objective of excellence in research, scholarship,
and education by publishing worldwide. Oxford is a registered trade mark of
Oxford University Press in the UK and in certain other countries.

Published in India by
Oxford University Press
Ground Floor, 2/11, Ansari Road, Daryaganj, New Delhi 110002, India

ISBN-13: 978-0-19-945852-3
ISBN-10: 0-19-945852-9

Typeset in Times New Roman
by E-Edit Infotech Private Limited (Santype), Chennai
Printed in India by Repro India Ltd

Cover image: Atmosphere1 / Shutterstock

Preface

Concrete is the most widely used building material. It is versatile, has desirable engineering properties, can be moulded into any shape, and, more importantly, is produced with cost-effective materials. Although recent developments in plastics and other lighter materials have resulted in the replacement of concrete in some applications, the use of concrete worldwide has increased phenomenally, especially in infrastructure projects. In fact, these developments have complimented and improved the performance and use of concrete in structures.

The knowledge of concrete's controlled production, maintenance, testing, and repair is vital for a discerning designer to ensure its optimal use. The large number of failures of structures has underlined the need for a better understanding of the behaviour of concrete, especially in challenging environmental conditions. This, in turn, warrants the need to have a sound knowledge of the selection of materials, mix proportioning, and quality control methods.

About the Book

The book is intended to help students and professional engineers gain a broader understanding of concrete as a construction material. It discusses the subject in its historical perspective together with a chronology of the developments relating to concrete as a material and its applications and technologies. It explains the fundamental concepts of concrete construction in simple language and covers the aspects of material science, mix proportioning, and construction. This book combines sound theory with adequate practical examples based on empirical observations both from the laboratory and the field.

New to the Second Edition

The first edition of the book was published more than 10 years ago and since then there have been revisions of codes of practices both in India and abroad as well as advances in concrete technology practices. This edition, in addition to revision of standard codes, includes recent developments in the field of concrete technology. Some of the prominent changes in this edition are as follows:

- A new chapter on rheological models of concrete (chapter 6) focuses on the flow characteristics of fresh concrete
- Exclusive chapters on the properties of fresh and hardened concrete (chapters 4 and 7)
- A new chapter provides improved treatment of the different stages of manufacture of concrete (chapter 5)
- Updated and revised content with new sections on reactive powder concrete, pervious concrete, geopolymer concrete, and manufactured (M) sand
- Improved pedagogy with numerous additional illustrations and multiple choice questions at the end of each chapter

Content and Coverage

The text explains the theoretical and practical aspects of the subject exhaustively. Incorporating the provisions of the code, IS: 456-2000, it includes the latest developments in the field of concrete construction.

The book begins with a historical account of concrete and developments in this field in Chapter 1. Chapters 2 and 3 discuss the properties of constituent materials and chemical and mineral admixtures. Chapters 4 to 9 discuss the properties of fresh and hardened concrete, different stages of manufacture of concrete, rheological models of concrete, and mix proportioning to produce concrete to suit specific requirements. Chapter 10 is devoted to a discussion on the various types of steel reinforcement used in concrete structures. The effect of corrosion and durability of concrete are discussed in Chapters 11 and 12, respectively.

Research in material technology has led to the development of lightweight concrete for reducing foundation loads, high and ultrahigh-strength concrete for applications in tall buildings and bridges, high-performance concrete for special performance requirements, special concretes such as polymer concrete for high durability, and steel-fibre-reinforced concrete for preventing cracks in concrete. These have been discussed in detail in Chapters 13 to 17. Various aspects of the production of ready mix concrete and its use are covered in Chapter 18. The construction of structures such as dams and large bridge piers, where the dimensions of the volume of concrete necessitate taking extra care of excessive amounts of heat of hydration, needs special mass concrete. The issues involved in mass concreting are discussed in Chapter 19.

Reinforcement detailing and formwork erection are an essential part of concrete construction at the site. The ultimate goal of achieving excellent infrastructure depends on the people on the job, who need to know the intricacies of such detailing aspects as tolerances and accuracy of fabrication. It is well known that the best of designs using the most sophisticated computer programs will not make the structure behave well unless the concrete construction and the reinforcement detailing practice conform to the prescribed specification. This important aspect has been discussed in Chapter 20. Chapter 21 discusses structural concrete block masonry, and Chapter 22 deals with the quality control issues involved in the production of concrete and construction of concrete structures. Recognizing the importance of the rehabilitation of ailing structures, Chapters 23 and 24 discuss repair materials and technologies. Concrete structures exposed to aggressive environmental conditions are discussed in Chapter 25, and Chapter 26 deals with underwater concrete structures.

Concrete testing is gaining importance from the point of view of the need to monitor and evaluate the existing concrete structures for their safety and upkeep. Important testing techniques and procedures are discussed in Chapter 27. Chapters 28 to 31 focus on special materials in construction, concreting machinery and equipment, performance and maintenance of concrete structures, and future trends in concrete technology, respectively.

Acknowledgements

While writing the book, references have been made to many previous works on the subject. As an acknowledgement, these works have been included in *Bibliography*. I acknowledge with reverence my professors, P. Purushothaman, S.R. Srinivasan at the College of Engineering, Guindy, Chennai, and Thomas Paulay and R. Park at the University of Canterbury, Christchurch, New Zealand. I owe my learning of the subject to them. I thank my colleagues at the Indian Institute of Technology Madras for their constructive suggestions.

I am indebted to my parents, R. Pushpavathi and A.M. Ramalingam, who have been an abiding source of inspiration for me. I owe heartfelt thanks to my wife Vanaja for her unwavering love and support, without which this book would not have been possible.

Finally, I acknowledge the support and guidance provided by the editorial team at the Oxford University Press India and thank the Press for giving me an opportunity to author this book for the benefit of both students and professionals.

This book presents the latest scientific advances in concrete technology and addresses all the variability of concrete to enable a comprehensive treatment of interrelated issues. I hope that this work will be welcomed by the profession as an authoritative resource on concrete construction. Comments and suggestions for the improvement of the book are welcome and can be sent at santhaar@gmail.com.

A.R. SANTHAKUMAR

Contents

CHAPTER 1

Concrete: Past, Present, and Future

Concrete is weak compared to steel. It is also brittle. Yet, it is the most widely used building material. This is because of its versatility, and because it has desirable engineering properties. It can be made on-site using easily available materials and can be moulded into any shape and the surface can be textured and coloured for aesthetic purposes. Most importantly, it can be produced with cost-effective materials.

Can we use wood to build a dam or a furnace? Can we use steel to lay a pavement? Or, can we use asphalt to construct a building frame? The answer is obviously 'No'. However, we can use concrete to build all these structures elegantly. In the words of J.W. Kelly, '*It is used to support, to enclose, to surface and to fill.*'

Concrete possesses very good water-resistant properties and, hence, can be used in intake towers for drawing water, dams and water tanks for storing, and canal linings for transporting water. Structural elements are often built with reinforced concrete. Examples are pile foundations, footings, beams, floors, walls, columns, and roofs.

Concrete is a strong and tough material. Reinforced concrete resists cyclones, earthquakes, blasts, and fires much better than timber and steel if designed properly. Flying fragments pose a great hazard to people during cyclones and tornadoes. Walls built with concrete have sufficient strength and mass to resist wind-driven debris compared to wood or sheet steel walls.

When compared to wood or steel, concrete has an inherent fire-resistant property. It regains its properties on cooling when the temperatures reached and the duration of the fire are not abnormally high.

Compared to many other engineering materials, such as steel and rubber, concrete requires less energy input for its manufacture. Currently, a large number of mineral admixtures, which are waste products of other industries, are being beneficially used in making quality concrete. Thus, from the consideration of energy and resource conservation and sustainability of the environment, concrete is the most preferred material.

In India, reinforced concrete has been used extensively for the construction of houses, buildings, roads, bridges, and dams. The advantages of concrete are well known to engineers and architects. However, the use of concrete for low-cost housing and rural housing has not been extensive. There are many areas where there could be a considerable increase in the use of cement-based products. As compared to burnt clay bricks, which use the fertile topsoil depriving it for agricultural purpose, precast products such as concrete lintels, joists, and concrete framework for doors and windows provide better and cost-effective solutions for building affordable houses. Such houses require less maintenance compared to those made from other materials, making concrete competitive when lifecycle cost is considered.

1.1 Historical Background

Though concrete is an extensively used material today, there was a time when it was still unknown. In those days, structures were built using burnt clay bricks, stone, or steel. The Pyramids of Giza, Egypt (West 1985), were built around 2540 BC, that is, nearly 45 centuries ago. The Great Wall of China (Calder 1983), completed in about 230 BC, is 20 centuries old. However, the Romans made a number of structures with concrete. The Pantheon in Rome, Italy (Gunnar 1997), was constructed nearly 19 centuries ago. It is a 6-m-thick mass concrete covered by a 43.3 m diameter concrete dome roof. The fall of the Roman Empire led to the loss of information about concrete technology. The Egyptians, Romans, Chinese, Mayans, and Indians (National Geographic Society 1986) were using various kinds of natural cements at the beginning of 1700 AD. These cements contained some form of lime-sand mortar.

Around 1756, John Smeaton, trying to build the Eddystone Lighthouse (Gunnar 1997), conducted a series of experiments to find a substance that will set underwater. He used pozzolana for his construction and published the results of his experiments in a paper in 1791. Until the mid-19th century, concrete was made using lime. It is actually the invention of the steam locomotive that created a need for concrete for building rail bridges in Europe. In 1824, Joseph Aspdin (Mehta 1986) patented the material that he called *Portland cement*. The name was chosen because the concrete produced out of this cement resembled a well-known quality of stone mined on the isle of Portland in Dorset, England. However, it was I.C. Johnson (Evans 1998) who in 1845 noticed that sintered material produced superior cement. He produced this superior cement for the first time in his factory. However, lime still continued to be used in combination with Portland cement to provide better workability to the fresh mixture. Thus, the cement we use today has been in production for the last 150 years (Table 1.1).

Table 1.1 History of development of concrete structures

Year	Structures/Techniques
2540 BC	The Pyramid of Khufu (Cheops) is built in Giza, Egypt.
100 AD	The Pantheon is built in Rome, Italy.
1757	Leonhard Euler publishes the equations for elastic buckling of columns.
1773	Coulomb becomes the first to understand the theory of beams ($\sigma = My/I$), but his results were not widely known.
1816	A concrete bridge is constructed over the Dordogne River in Souillac, France.
1824	Joseph Aspdin patents a material that he calls Portland cement.
1850	The invention of reinforced concrete is credited to Joseph-Louis Lambot of France who built a rowboat, which was exhibited in the Paris exposition of 1854.
1853	Francois Coignet of France builds a 6-m-long roof using reinforced concrete.
1889	The Eiffel tower is completed, supported on massive concrete foundations.
1908	Concrete stress distribution is proposed.
1924	The world's production of Portland cement goes up to 50 million tonnes/year. An increasing number of bridges are being made of concrete to provide roads for the cars that were being mass produced on assembly lines.
1928	Patent by Eugène Freyssinet for prestressed concrete.
1950	Fly ash, a waste material extracted from the flue gas of coal-fired power plants, begins to be widely used to reduce the amount of cement required in concrete.
1976	It is discovered that the strength of concrete can be increased by adding micro-silica. Silica fume is obtained as a result of the environmental treatment of fumes from silicon furnaces.
1994	Reactive powder concretes with strength of 700 MPa and more are developed by Pierre Richard in France.
1997	The Confederation bridge in Canada is completed, with span of 250 m and a total length of 12.9 km. Concrete of strength up to 90 MPa is manufactured.

Since then, the developments related to concrete across the world have manifested themselves in the different concrete technologies used at construction sites today. Changes in material technology have led to new ways of manufacturing and delivering concrete. On-site mixing is slowly giving way to ready-mixed concrete. The ongoing developments in concrete-pumping techniques have presented new ways of transporting and placing concrete. From the humble beginning of Portland cement, today researchers are working on improving the strength of concrete to 700 MPa (reactive powder concrete), which was first developed by Pierre Richard in France.

India has been building noteworthy structures for thousands of years. There is enough evidence of the use of bricks in India nearly 3000 years back during the Indus Valley Civilization. The excavations of Mohenjo Daro and Harappa indicate this. Just as in Europe, the railway network in India too used bridges made of steel and concrete for crossing rivers. Prestressed concrete was used for the first time to build garages for military tanks in 1940. The first prestressed concrete bridge was constructed in 1949 on the Assam rail link, followed by the Palar road bridge near Chennai in 1952 with 23 spans of 28.4 m each. Concrete has been the main material for building high dams (Bhakra Dam, Bilaspur, 226 m high), long bridges (Ganga Bridge, Patna, 5575 m long; second Thane creek bridge, Mumbai, 1.837 km long; six-lane second Hooghly cable-stayed bridge, Kolkata, 837 m long), and monumental structures (Bahai Temple, New Delhi).

1.2 Components of Concrete

Hardened concrete can be considered to have three distinct phases: (a) the hardened cement paste (HCP) or matrix, (b) the aggregate, and (c) the interfacial or transition zone (TZ) between the HCP and the aggregate. For optimum performance, all the three phases should be considered explicitly. Figure 1.1 shows the cut and polished surface of concrete. In this figure, the first two phases can be identified easily.

When examined under a powerful electron microscope, the structure of HCP in the vicinity of large aggregates is observed to be very different from the structure of bulk

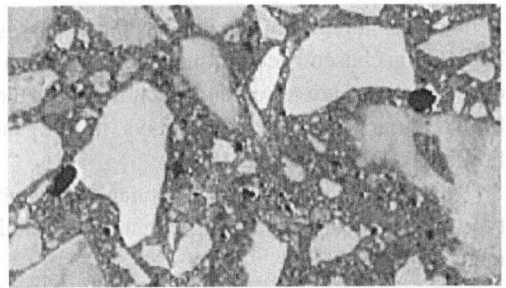

Fig. 1.1 Cut and polished surface of concrete

paste. The zone of 10–50 mm around the large aggregate is seen to be a weak porous transition zone. The aforementioned three phases can also have sub-phases. For instance, the aggregate phase consists of several minerals and voids. In addition, the structure of concrete undergoes changes with time and location (space). It is affected by temperature, humidity, and time. A brief description of the individual components of concrete is essential to understand the behaviour of these phases.

The HCP is about 30–40% of the volume of concrete. It is a product of Portland cement and water. The quantity of water is about 7–15% and cement about 14–21% of the volume. Their relative quantities affect the strength, and the strength is mainly controlled by the water–cement ratio.

The aggregates constitute 60–70% of the volume and comprise both fine and coarse aggregates. The aggregates are inert fillers. They contribute to concrete's weight and deformation characteristics. Figure 1.2 shows the constituents of concrete. Concrete also contains air, which is a part of the paste phase. It comes from two sources. Nearly 2% of the volume of concrete is entrapped air, out of which about 1–2% is sometimes deliberately introduced as entrained air using an air-entraining admixture. The entrained air in concrete makes it resist freeze–thaw cycles better, making it durable. However, it certainly causes a reduction in strength and density.

The important properties of concrete are (a) air content, (b) fluidity, (c) strength, (d) setting time, and (e) durability. Admixtures are added to modify one or more of these properties in the fresh or hardened state of concrete.

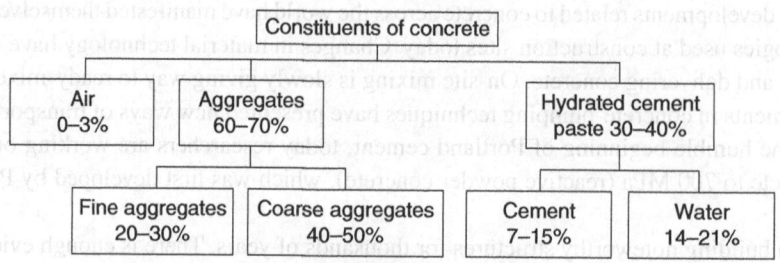

Fig. 1.2 Constituents of concrete

1.3 Strength Development

The transformation of fresh concrete to hardened concrete takes place in the following three stages:

1. **Fresh stage**   In this stage, concrete is plastic. It is workable and capable of being moulded.
2. **Transition stage**   In this stage, the workability of concrete reduces and the process of setting begins. The excess water evaporates along with heat of evolution, and its strength slowly develops.
3. **Hardened stage**   In the final stage, concrete becomes stiff and gains enough strength to support a load. Therefore, it has sufficient load-carrying capacity as per design.

The strength on the 28th day of this process is taken as the reference strength for hardened concrete. The 28th day is chosen because this gives an even four weeks' time ($7 \times 4 = 28$). For assessing the strength of concrete, cubes are tested on the same day of the week. That is, cubes made on a Monday are tested on the Monday four weeks later. In 28 days, most of the strength of normal concrete made with standard ordinary Portland cement is developed.

The strength development of concrete depends on its age, water–cement ratio (w/c), air content, cement type, aggregate type, paste aggregate bond, and the curing and environmental conditions in which it was made.

1.4 Different Types of Concrete

Concrete can be classified into various categories depending on the density and strength recommended by IS: 456-2000.

Based on density, concrete is classified as lightweight, normal-weight, and heavyweight concrete. The densities of these concretes are given in Table 1.2.

Table 1.2 Classification of concrete based on density

Classification	Density (kN/m³)
Lightweight concrete	18
Normal-weight concrete	24
Heavyweight concrete	32

The aggregates used in making concrete contribute mainly to its density. Normal-weight concrete is produced using natural sand and crushed stone (granite). For lightweight concrete, either lightweight aggregates, such as pumice, or pyro-processed and bloated aggregates, are used. These concretes are used for applications in which the load of gravity is to be reduced, e.g., for reducing the load on foundations. Heavyweight concrete is produced using high-density aggregates such as hematite or scrap steel pieces. These concretes are used for radiation shielding or increasing the weight of a structure for stability purposes.

The standard code of practice for plain and reinforced concrete, IS: 456-2000, has classified concrete on the basis of strength. Table 1.3 shows the three main categories based on strength.

Table 1.3 Classification of concrete based on strength

Classification	Maximum strength (MPa)	Type
Ordinary concrete	<20	Low-strength
Standard concrete	20–40	Medium-strength
High-strength concrete	40–80	High-strength

Table 1.4 Typical proportions of materials used in concretes of different grades

Material	Ordinary		Standard		High-strength	
	kg/m³	% vol.	kg/m³	% vol.	kg/m³	% vol.
Cement	255	8.1	356	11.3	510†	16.2
Water	178	17.6	178	19.7	178	17.7
Fine aggregate	800	29.9	848	31.7	889	32.6
Coarse aggregate	1170	43.7	1032	38.5	872	32.5
Cement paste						
Mass %	18		22		28	
Volume %	25		30		35	
w/c ratio by mass	0.69		0.50		0.35	
Strength*	15		30		55	

*Strength refers to the 28-day cube compressive strength as per IS specification.
†The cement content may be reduced using suitable admixtures.
Note: 100 kg/m³ = 1 kN/m³.

The typical proportions of constituent materials used to produce concretes of different types are shown in Table 1.4. These values are based on tests conducted with available aggregates and the standard OPC 43 grade cement. These values are only for comparison, not for field use. For field use, proper mix design, which has been explained in a later chapter, should be implemented and used based on field trial mixes. These values are based on the commercial mix designs used by the Building Technology Centre, Anna University.

There are many other special concretes such as fibre-reinforced concrete, latex-modified concrete, and roller-compacted concrete. The properties of these concretes are discussed in later chapters.

1.5 Units of Measurement

Most of the original developments in concrete technology occurred in Europe and the USA. The system of measurement used for centuries in the UK, the Commonwealth, and the USA has been the Imperial (British) system, based on the foot-pound-second (FPS) system. During the French Revolution, a system known as the metric system, based on the centimetre-gram-second (CGS) system, was introduced in France. Most countries have now switched over to the modernized form of the metric system known as the Systeme International d' Units (SI) or the International System of Units. It was agreed to adopt this system at the General Conference on Weights and Measures in 1960 by as many as 30 participating countries.

Today, however, the USA is the only country in the world using English units—feet (ft) for length, pounds (lb) for weight, and pounds per square inch (psi) for stress. Though in India we have generally adopted SI units, the FPS system is still in use in the construction industry; for example, a wall is still referred to as a 9" wall. This is primarily due to our brick size being in FPS units (9" × 4.5" × 3"). To enable our construction industry to participate in multinational activity, it is necessary to adopt the accepted SI system of measurement.

Table 1.5 Multiples of 10 and prefixes used in SI measurement

Multiplication factor expressed as a number	Multiplication factor expressed as a power of 10	Prefix	SI prefix symbol
10	10^1	deca	da
100	10^2	hecto	h
1000	10^3	kilo	k
1,000,000	10^6	Mega	M
1,000,000,000	10^9	Giga	G

Table 1.6 Sub-multiples of 10 and prefixes used in SI measurement

Sub-multiplication factor expressed as a number	Sub-multiplication factor expressed as a power of 10	Prefix	SI prefix symbol
0.1	10^{-1}	deci	d
0.01	10^{-2}	centi	c
0.001	10^{-3}	milli	m
0.000001	10^{-6}	micro	μ
0.000000001	10^{-9}	nano	n

Since both these systems are prevalent in the field of concrete construction, one must be conversant with English as well as SI units until a complete change to SI units takes place. That is, when we manufacture metric bricks, metric sieves, and think and act metric.

The SI system is based on the metre, the kilogram, and the second. Multiplying or dividing the basic quantities by multiples of 10 provides larger or smaller units. Each quantity thus formed has a prefix. Kilo denotes thousand times, Mega denotes million times, micro denotes a millionth part, and milli denotes a thousandth part. A complete list of the symbols and names of these prefixes expressed as multiplication and sub-multiplication factors of 10 and its powers is given in Tables 1.5 and 1.6, respectively.

In this book, we will be extensively using units for length, force, and pressure. The unit of force is Newton (N) and that of pressure is Pascal (Pa). One newton is the force required to accelerate 1 kg of mass by $1 \, \text{m/s}^2$. A stress of $1 \, \text{N/m}^2$ is expressed as 1 Pa. In order to represent very large or very small quantities, accepted prefixes are used for units. A list of commonly required conversion factors is given in Table 1.7.

1.6 New Developments and Future Trends

The many changes that have taken place in concrete technology over the last decade have led to the development of high-strength, high-performance concretes. Although the chemical composition of ordinary Portland cement has remained more or less the same, the proportions of the elements have been modified and the manufacturing techniques have been refined. This has led to the development of cements with higher strengths. The development of high-strength cements has made it possible to achieve economy since lesser quantities of high-strength cement are needed to provide much stronger and durable structures. New supplementary cementitious materials and pozzolans such as ground blast-furnace slag, metakaolin, fly ash, and silica fume are being increasingly used as additives along with new varieties of superplasticizers and viscosity modifying agents. These have helped in producing very efficient high-performance concretes.

The aforementioned developments in construction technology have led to the introduction of newer methods for efficient mixing, transportation, and placing of wet concrete. These include computer-controlled mixing, concrete pumping, and improvements in shotcreting and formwork technologies.

Table 1.7 Conversion table

To convert	Into	Multiply by	To convert	Into	Multiply by
Length			Fluid ounces (fl. oz)	Millilitres (mL)	28.41
Inches (in.)	Millimetres (mm)	25.4	Pints (p)	Litres (L)	0.568
Inches (in.)	Centimetres (cm)	2.54	Gallons (G)	Litres (L)	4.55
Feet (ft)	Metres (m)	0.3048	*Mass/weight*		
Yards (yd)	Metres (m)	0.9144	Ounces (oz)	Grams (g)	28.35
Miles	Kilometres (km)	1.6093	Pound-mass (lb)	Kilogram (kg)	0.45359
Area			Tons	Tonnes (t)	1.016
Square inches (sq. in.)	Square centimetres (sq. cm)	6.4516	Pound-force (lb-f)	Newtons (N)	4.448
Square feet (sq. ft)	Square metres (sq. m)	0.093	Kilogram force (kg-f)	Newtons (N)	9.807
Square yards (sq. yd)	Square metres (sq. m)	0.836	*Density*		
Acres (ac)	Hectares (ha)	0.405	Pounds per cubic yard	Kilograms per cubic metre	0.5933
Square miles (sq. m)	Square kilometres (sq. km)	2.58999	(lb/cu. yd)	(kg/m^3)	
Volume			*Pressure*		
Cubic inches (cu. in.)	Cubic centimetres (cm^3)	16.387	Kips per square inch (ksi)	Megapascal (MPa or N/mm^2)	6.895
Cubic ft (cu. ft)	Cubic metres (m^3)	0.0283	*Temperature*		
Cubic yards (cu. yd)	Cubic metres (m^3)	0.7646	Degrees Fahrenheit (°F)	Degrees centigrade (°C)	(°F − 32)/1.8

Today, concrete can be designed for high strength, high performance, or both. Specially desired characteristics such as low heat of hydration, low shrinkage, and high durability in a hostile environment can be designed and achieved. There has also been progress due to the availability of ready-mixed concrete as well as changes in specification, leading towards desired performance rather than desired strength.

Exercises

Review Questions

1. List three reasons for the use of concrete as a building material.
2. Trace the historical development of concrete technology.
3. Why are different units used for measuring strength?
4. What is a transition zone?
5. How is concrete classified according to strength in IS: 456-2000?
6. What are the constituents of concrete? State their relative proportions.
7. How are lightweight and heavyweight concrete produced?
8. How can the use of concrete in rural India be increased?
9. Distinguish between different types of concrete and their behaviours.
10. List some of the important concrete structures built in India, describing their significance.
11. List a few landmark structures built in concrete which may be described as milestones in the development of concrete technology around the world.
12. What are the stages of transformation of fresh concrete to hardened concrete?
13. Why is concrete tested on the 28th day after casting?

Multiple Choice Questions

1. How many distinct phases does the hardened concrete exhibit?
   a. One
   b. Two
   c. Three
   d. Four

2. In the hardened concrete the weakest portion is
   - a. Aggregate
   - b. Hardened cement paste
   - c. Transition zone
   - d. All of these

3. The density of concrete is largely influenced by
   - a. Aggregates
   - b. Water–cement ratio
   - c. Cement
   - d. Admixtures

4. Admixtures are added to concrete to
   - a. Modify the properties
   - b. Act as a filler
   - c. Reduce the cost
   - d. Act as replacement for fine aggregate

5. Before 19th century the binder used to make concrete was
   - a. Portland cement
   - b. Lime
   - c. Admixture
   - d. Surki

6. Portland cement was patented by
   - a. I.C. Johnson
   - b. Coloumb
   - c. John Smeaton
   - d. Joseph Aspdin

7. The first prestressed concrete bridge was built in India
   - a. Near Chennai
   - b. In Assam
   - c. Near Delhi
   - d. Near Mumbai

8. FPS system is still used in
   - a. UK
   - b. India
   - c. USA
   - d. China

9. The strength of concrete is based on the cube crushing strength on the
   - a. 30th day
   - b. 7th day
   - c. 14th day
   - d. 28th day

10. As per IS: 456-2000, standard concrete should have a strength range of
    - a. <20 MPa
    - b. 20 to 40 MPa
    - c. 40 to 89 MPa
    - d. >80 MPa

11. The weight of a cubic metre of normal concrete is about
    - a. 24 kN
    - b. 18 kN
    - c. 30 kN
    - d. 6 kN

12. In SI units, compressive strength of concrete is expressed in
    - a. psi
    - b. ksc
    - c. MPa
    - d. ksm

13. The major constituent of concrete is
    - a. Cement
    - b. Water
    - c. Air
    - d. Aggregates

14. Concrete in a structural member has to pass through
    - a. Fresh stage
    - b. Transition stage
    - c. Hardened stage
    - d. All of these

15. Water–cement ratio used to produce high-strength concrete is of the order of
    - a. 0.69
    - b. 0.50
    - c. 0.35
    - d. 0.20

Answers to Multiple Choice Questions

1. c	5. b	9. d	13. d
2. c	6. d	10. b	14. d
3. a	7. b	11. a	15. c
4. a	8. c	12. c	

CHAPTER 2

Constituent Materials

Several materials are used to manufacture good quality concrete. It is important to know the properties of cement, aggregate, and water, as they impart strength and durability to concrete. Of all the materials that influence the behaviour of concrete, cement is the most important constituent, because it is used to bind sand and aggregates and it resists atmospheric action. In this chapter, we will study the properties of all the major raw materials used to make concrete. The properties of chemical and mineral admixtures used to modify the properties of concrete are described more exhaustively in the next chapter.

2.1 Cement

The main raw material used for the production of cement is clinker. Clinker is an artificial rock made by heating limestone and other raw materials in specific quantities to a very high temperature in a specially made kiln. Portland cement is hydraulic cement made by finely pulverizing the clinker produced by calcining to incipient fusion a mixture of argillaceous and calcareous materials. It is the fine grey powder that is the most important ingredient of concrete; hence the name *cement concrete*. Cement undergoes a chemical reaction with water and sets and hardens when in contact with air or underwater. The global production of cement is approximately 1.5 billion tonnes per year.

Portland cement is a general term used to describe hydraulic cement. The typical raw materials used for making cement are limestone ($CaCO_3$), sand (SiO_2), stale clay (SiO_2, Al_2O_3, or Fe_2O_3), and iron ore (Fe_2O_3). Thus the chemical components of cement are calcium (Ca), silicon (Si), aluminium (Al), and iron (Fe). The calcareous component, lime (CaO), is derived from one of the following: limestone, chalk, marble, lime sand shell deposit, lime sludge. The argillaceous component (SiO_2, Al_2O_3, or Fe_2O_3) is derived from one of the following: clay, shale, tuff ash, slate, glass. Table 2.1 shows the raw materials used to derive the chemical components of cement.

Table 2.1 Raw materials used to derive the chemical components of cement

Calcium (Ca)	Silica (SiO_2)	Alumina (Al_2O_3)	Iron oxide (Fe_2O_3)
Limestone	Sand	Clay	Iron ore
Shale	Clay	Shale	Clay
Marl	Shale	Slag	Mill scale
	Marl		

2.1.1 Manufacturing Process

The cement manufacturing process can be divided into three distinct stages as follows:

1. Raw material acquisition
2. Clinker production in a kiln
3. Cement grinding, packing, and distribution

The manufacturing plant is generally located near the source of calcium; the required silica is transported to the plant. It is important to process the raw materials in order to obtain the correct composition and properties of cement. The components are heated in a kiln, leading to the formation of new compounds and producing clinker.

Raw material acquisition

The stages in the acquisition of raw materials include rock mining from a quarry, crushing in more than one stage, and storing with other raw materials (Fig. 2.1). After analysis, the raw materials are proportioned, ground to fine powder, and blended. Some cement plants add water to the material during grinding, then blend and store it as slurry (Fig. 2.2).

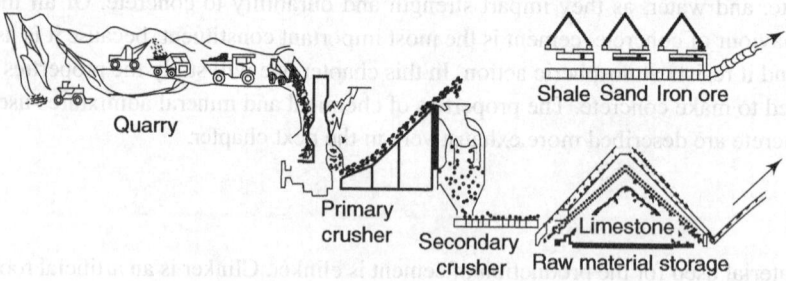

Fig. 2.1 Stages in the acquisition of raw materials

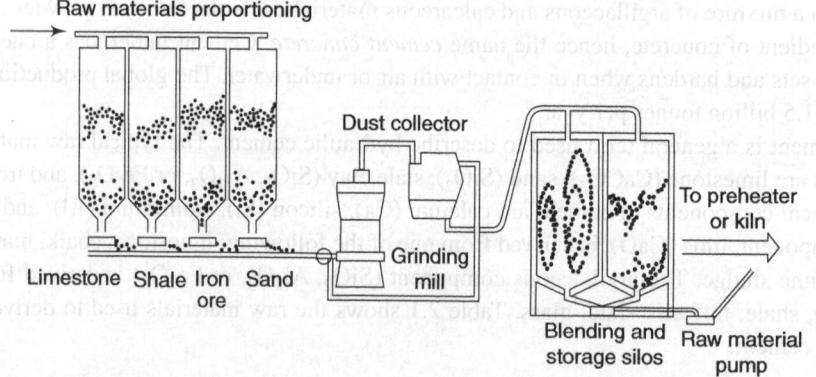

Fig. 2.2 Proportioning, grinding, and storing of raw materials

Clinker production

A large quantity of coal is required to make cement. The processes used to manufacture cement are the following:

a. The wet process
b. The dry process (nearly 75% of cement is produced using this process)
c. The preheater/precalciner process

Clinker is produced in a kiln. The first rotary kiln to produce Portland cement was patented in 1885 by Frederick Ransome. Subsequently, Thomas A. Edison developed the long kiln (50 m in length compared to the then existing 20–30 m length).

The chemistry of the wet process is the easiest to control. This process can be adopted with ease when the raw materials are wet. However, the fuel requirement (fuel required to evaporate the slurry water) of the wet process is large. Obviously, the dry process requires less fuel, and the preheater or precalciner process further enhances fuel efficiency and permits a high production rate.

A cement kiln, a huge inclined rotating furnace, is a continuous stream process vessel (Fig. 2.3) in which the feed and fuel are held in a dynamic balance. As the raw materials—limestone, clay, and shale—move toward the 1500°C flame, the chemical reaction transforms them into clinker. The five distinct processes occurring in the kiln are shown in Fig. 2.4. Thus the production of cement is an energy-intensive process. Clinker production requires high temperatures, which are generated by the combustion of coal. Alternative fuels, such as used tyres, engine oil, and spent solvents, can also be used in the kiln. This allows conservation of fossil fuels and reduces CO_2 emissions into the atmosphere.

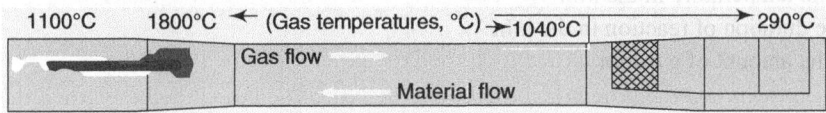

Fig. 2.3 Kiln—a continuous stream process vessel

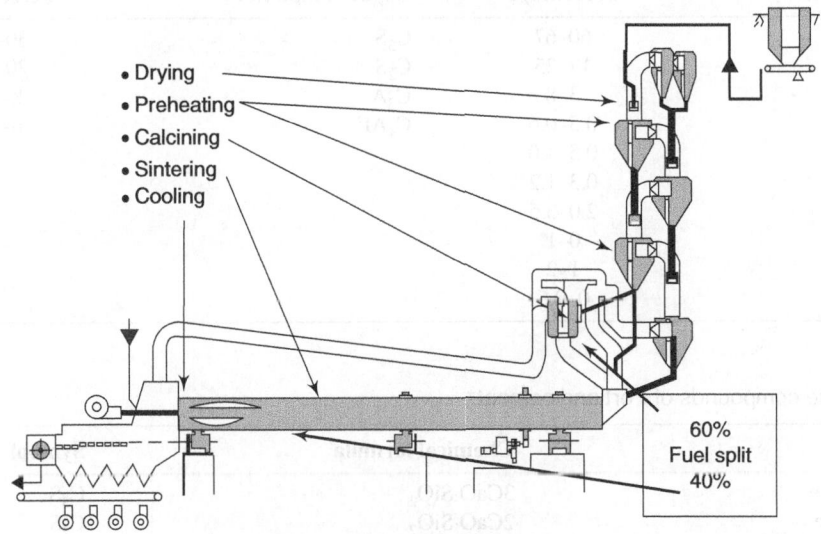

Fig. 2.4 Five processes performed in a kiln

Cement grinding and distribution

Clinker is stored at the construction site until it is required for grinding into cement. It is ground with gypsum and supplementary cementitious materials to form fine cement powder. Gypsum is added to regulate the setting time. Thus,

Pulverized clinker + gypsum = Portland cement

Mineral additions and supplementary cementitious materials include limestone, granulated slag, and fly ash. These materials reduce the requirement of clinker in cement. Hence, they reduce the use of natural materials and the emission of CO_2 in clinker production. It is estimated that 755 kg of CO_2 is released for every tonne of cement produced.

Cement products are distributed via road, rail, or sea transport. They are delivered to the user either in bulk containers or in standard 50 kg cement bags.

2.1.2 Chemical Composition of Portland Cement

Table 2.2 gives the oxide and Bogue composition of Portland cement. The *insoluble residue*, determined by dissolving cement in hydrochloric acid, is a measure of the adulteration of cement caused by the impurities in gypsum. The loss on ignition shows the extent of carbonation and hydration of free lime and free magnesia because of the exposure of cement to the atmosphere. Portland cement also consists of the Bogue compounds shown in Table 2.3; the last column shows the abbreviations used for these compounds. The typical percentages of the Bogue compounds found in cement are shown in Table 2.2. We can control the amount of such compounds produced using the following methods:

a. Changing the raw material feed
b. Changing the temperature in the kiln
c. Altering the duration of reaction in the kiln
d. Adjusting the amount of gypsum

Table 2.2 Oxide and Bogue compound compositions of Portland cement

Oxide compound	Percentage	Bogue compound	Percentage
CaO	60–67	C_3S	30–50
SiO_2	17–25	C_2S	20–45
Al_2O_3	3–8	C_3A	8–12
Fe_2O_3	0.5–0.6	C_4AF	6–10
MgO	0.5–4.0		
SO_3	0.3–1.2		
K_2O/Na_2O	2.0–3.5		
Others	0–1		
Loss on ignition	1–2		
Insoluble residue	0–0.5		

Table 2.3 Bogue compounds of Portland cement*

Compound	Chemical formula	Symbol
Tricalcium silicate	$3CaO \cdot SiO_2$	C_3S
Dicalcium silicate	$2CaO \cdot SiO_2$	C_2S
Tricalcium aluminate	$3CaO \cdot Al_2O_3$	C_3A
Tetracalcium alumino ferrite	$4CaO \cdot Al_2O_3 \cdot Fe_2O_3$	C_4AF

*The relative percentages of these compounds are shown in Table 2.2.

2.1.3 Hydration Chemistry of Cement

The reaction of cement when mixed with water is called *hydration*. Both C_3S and C_2S make up nearly 75% of cement. The hydration of these compounds is responsible for the setting and hardening of cement. In the presence of water, the silicates and aluminates listed in Table 2.2 form products of hydration, which result in a hard mass over a period of time. This hard mass is known as *hydrated cement paste*. The hydration surface reaction starts immediately once cement comes in contact with water. It is an exothermic reaction. The cement grains become smaller as the reaction proceeds, which produces hydration products. The hydration continues as long as heat and moisture are available. All four Bogue compounds along with gypsum are involved in the hydration reaction and only a very small amount of water is needed for it.

Hydration of C_3S

The chemical reaction of C_3S with water can be expressed as

$$C_3S + water \rightarrow C-S-H + C-H + heat$$

where C–S–H is calcium silicate hydrate and C–H is calcium hydrate. Typical C–S–H and C–H crystals are shown in Fig. 2.5.

C–S–H Calcium silicate hydrate constitutes 50–60% of the solids in the paste. It forms a continuous binding matrix. It is amorphous and fibrous and hence has a large surface area. It is an important factor for the strength development of cement paste.

C–H Calcium hydrate makes up about 20% of the solids in the paste. It exists in the form of thick, crystalline hexagonal plates and is embedded in the C–S–H matrix. Its growth fills the pore spaces. It does not significantly contribute to strength. Its leaching causes white patches and efflorescence.

The rate of heat generated by C_3S is shown in Fig. 2.6.

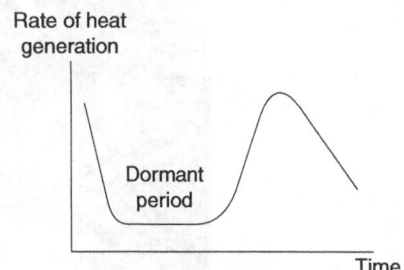

Fig. 2.5 Typical C–S–H and C–H crystals

Fig. 2.6 Rate of heat generated by C_3S

Hydration of C_2S

The hydration of C_2S is similar to the hydration of C_3S. The same products are generated. However, C_2S reacts slowly and hence generates less heat. It contributes to strength development at later stages.

Hydration of C_3A

This hydration reaction produces a substance called *ettringite* as follows:

$$C_3A + gypsum + water \rightarrow ettringite + heat$$
$$C_3A + ettringite + water \rightarrow monosulphoaluminate$$

If the amount of gypsum is too little, C_3A will react fast and can cause a 'flash set'. On the other hand, too much gypsum will delay setting and cause undue expansion. As shown in Fig. 2.7, ettringite is a crystalline and needle-like substance. It constitutes about 10–20% of the solid content. It is a long, slender, and prismatic crystal and is stable only in the presence of gypsum. It plays a minor role in strength development but contributes considerably to durability. Monosulphoaluminate is a stable hydration product. It is fairly crystalline. Figure 2.8 shows thin irregular plates clustered like a flower. Hence it fills the pores and can re-form ettringite in the presence of sulphate ions.

Fig. 2.7 Ettringite—A needle-like crystal

Fig. 2.8 Irregular plate structure of monosulphoaluminate

Hydration of C_4AF

The hydration of C_4AF is similar to that of C_3A; the same products are formed. However, C_4AF reacts slowly and hence generates less heat and combines well with gypsum. A summary of the behaviour of Bogue compounds in cement hydration is given in Table 2.4.

Table 2.4 Summary of the behaviour of Bogue compounds in cement hydration

Compound	Reaction time	Strength development	Setting time	Heat of evaluation
C_3S	Medium	High	Low	High
C_2S	Slow	Low/high	Low	Low
C_3A	Fast	Low	High	Very high
C_4AF	Medium	Low	Medium	Medium

In summary, the hydration of cement occurs at the surface of the grain. All the compounds react simultaneously, and a compound reaction takes place. The smaller grains hydrate first and the larger grains become smaller while they hydrate. Some very large grains never hydrate completely. The hydration rate has the following hierarchy:

$$C_3A, \ C_3S, \ C_4AF, \ \text{and} \ C_2S$$

Ettringite is formed first, followed by C–H and C–S–H. There is no change in the total volume of the cement paste as a result of hydration. After hydration, the paste composition consists of both the solid and water phases as shown in Fig. 2.9. Thus, hydration is a chemical reaction. The Bogue compounds react to form C–S–H, C–H, ettringite, monosulphoaluminate, and heat. Calcium silicate contributes to strength development and setting is influenced by calcium aluminate.

The hydration reactions can be summarized as follows:

$$2C_3S + 6H \Rightarrow C - S - H + 3CH \ (120 \ \text{cal/g})$$
$$2C_2S + 4H \Rightarrow C - S - H + CH \ (62 \ \text{cal/g})$$
$$C_3A + 3CSH_2 + 26H \Rightarrow C_6AS_3H_{32} \ (300 \ \text{cal/g})$$
$$2C_3A + C_6AS_3H_{32} + 4H \Rightarrow 3C_4ASH_{12}$$
$$C_4AF + 10H + 2CH \Rightarrow C_6AFH_{12}$$

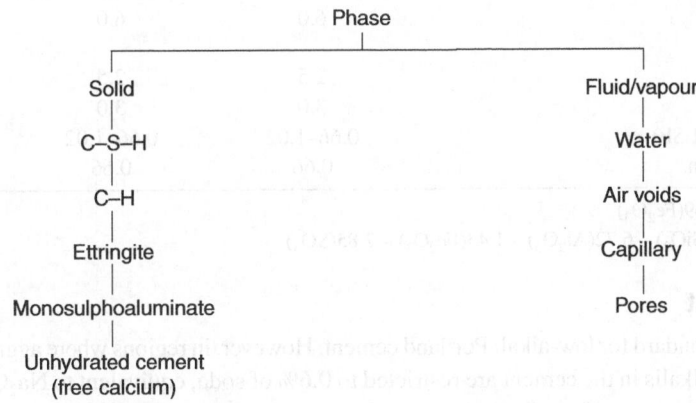

Fig. 2.9 Solid and fluid/vapour phases of the reaction products of hydration

2.1.4 Types of Cement

There are various types of cement in use. IS: 456-2000 (Indian Standard Code of Practice for Plain and Reinforced Concrete) permits the use of 10 different types of cement, which are described below:

Ordinary Portland cement

The Bureau of Indian Standards (BIS) has classified ordinary Portland cement (OPC) into three grades in order to produce different grades of concrete to meet the demands of the construction industry.

- *Grade 33* Ordinary Portland cement conforming to IS: 269-2013
- *Grade 43* Ordinary Portland cement conforming to IS: 8112-2013
- *Grade 53* Ordinary Portland cement conforming to IS: 12269-2013

The grade of cement indicates its mortar cube compressive strength in N/mm^2 at 28 days. The basic physical and chemical properties of the three grades of OPC are given in Table 2.5. Both grade 43 and grade 53 cement can be used for producing higher grades of concrete.

Table 2.5 Physical and chemical properties of various grades of ordinary Portland cement

	Grade 33	Grade 43	Grade 53
Physical properties			
Minimum compressive strength (MPa)			
3-day	16	23	27
7-day	22	33	37
28-day	33	43	53
Fineness			
Minimum specific surface (Blaine's air permeability method, m^2/kg)	225	225	225
Setting time (min)			
Initial, min.	30	30	30
Final, max.	600	600	600
Soundness, expansion			
Le Chatelier test (mm), max.	10.0	10.0	10.0
Autoclave test for MgO (%), max.	0.8	0.8	0.8
Chemical properties			
Loss on ignition (%), max.	5.0	5.0	4.0
Insoluble residue (%), max.	4.0	4.0	4.0
MgO (%), max.	6.0	6.0	6.0
SO$_3$ (%), maximum for			
C_3A^* > 5%	2.5	2.5	2.5
$C_3S^\dagger$ > 5%	3.0	3.0	3.0
Lime saturation factor (LSF)	0.66–1.02	0.66–1.02	0.8–1.02
Lime–alumina ratio, min.	0.66	0.66	0.66

$^*C_3A = 2.65(Al_2O_3) - 1.69(Fe_2O_3)$
$^\dagger C_3S = 4.07(CaO) - 7.6(SiO_2) - 6.72(Al_2O_3) - 1.43(Fe_2O_3) - 2.85(SO_3)$

Low-alkali cement

There is no separate standard for low-alkali Portland cement. However, in regions where aggregates have reactive silica or carbonates, alkalis in the cement are restricted to 0.6% of soda, equivalent to Na$_2$O + 0.658K$_2$O, with a view to avoid the occurrence of an alkali–aggregate reaction, leading to possible cracking and destruction of concrete. In such cases, instead of low-alkali Portland cement, blended cement can be used.

Blended cement

Portland cement containing a mineral additive becomes blended or composite cement. Blended cement is a hydraulic, cementitious product, similar to ordinary Portland cement, but has certain improved properties owing to the presence of the blending material in it.

The use of blended cement improves the properties of both fresh and hardened concrete. This can happen as a result of the extended hydration of the cement–pozzolana mixture, the reduced water demand, and the improved cohesion of the paste. Another important benefit is the improvement in durability resulting from the lower permeability and improved microstructure of concrete. This arises from the reduction in pore size and the refinement of the pore structure of the cement paste as well as from improvements in the properties of the 'interfacial zone' between the cement paste and the aggregate surface. This will be discussed in greater detail later. The basic physical and chemical properties of main types of blended cement are given in Table 2.6.

Table 2.6 Physical and chemical properties of blended cement (PPC and PBSC), SRC, and white cement

	PPC	PBSC	SRC	White cement
Physical properties				
Minimum compressive strength (MPa)				
3-day	16	16	10	14.4
7-day	22	22	16	19.8
28-day	33	33	33	29.7
Fineness				
Minimum specific surface (Blaine's air permeability, m²/kg)	300	225	225	225
Setting time (min)				
Initial, min.	30	30	30	30
Final, max.	600	600	600	600
Soundness, expansion				
Le Chatelier test (mm), max.	10.0	10.0	10.0	10.0
Autoclave test for MgO (%), max.	0.8	0.8	0.8	0.8
Additives (% by weight of cement)				
Fly ash	15–35	—	—	—
GGBS†	—	35–70	—	—
Degree of whiteness (%), min.	—	—	—	70
Chemical properties				
Loss on ignition (%), max.	5.0	5.0	5.0	—
Iron oxide (%), max.	—	—	—	1.0
Insoluble residue (%), max.	‡	4.0	4.0	2.0
Magnesia (%), max.	6.0	8.0	6.0	6.0
Sulphur (%), max. as	3.0	3.0	2.5	2.75§
Sulphuric anhydride (SO₃)				3.0*
Lime saturation factor (LSF)	—	—	0.66–1.02	0.66–1.02
Tricalcium aluminate (C₃A), (%) max.	—	—	5.0	—
Tetracalcium aluminoferrite + 2 tricalcium aluminate (C₄AF + 2C₃A), (%) max.	—	—	—	25.0

†GGBS denotes ground, granulated blast-furnace slag.
‡$x + [4.0(100 - x)/100]$, where x is the declared percentage of pozzolana in PPC.
§For $C_3A \leq 7\%$
*For $C_3A > 7\%$
Note: PPC denotes Portland pozzolana cement, PBSC denotes Portland blast-furnace slag cement, and SRC denotes sulphate-resisting cement.

Portland pozzolana cement (PPC)   This cement is manufactured either by grinding together Portland cement clinker, gypsum, and a pozzolana such as fly ash, or by uniformly blending Portland cement and fine pozzolana. The BIS has categorized PPC based on the pozzolana added to the mix. IS: 1489 (Part I) – 2015 gives provisions for fly ash based Portland pozzolana cement. The proportion of fly ash used as pozzolana can vary between 15% and 35% by weight of cement. The physical requirements of OPC and PPC are similar. Table 2.6 gives the physical and chemical requirements of PPC. Increased impermeability, lower heat of hydration, lower plastic shrinkage, reduced alkali–aggregate expansion, and improved resistance to aggressive chemical agents and corrosion are some of the major benefits obtained from the proper use of PPC. The use of PPC is, thus, desirable for enhancing durability. It results in improved long-term performance, especially for structures in aggressive environments. In mass concrete construction, PPC concretes have exhibited better behaviour with respect to cracking than OPC concretes because of their lower heat of hydration. The several uses of PPC include building construction, reinforced concrete structures, dams, foundations, machine-beds, plastering, and ornamental and other precast concrete products. High-strength PPC (with strengths of 43 and 53 N/mm^2) is preferred over OPC. However, the rate of strength gain may be slow if PPC is used in place of OPC. Hence, PPC concrete will require a longer curing period and longer setting and hardening time.

Portland blast-furnace slag cement (PBSC)   This cement is an intimately ground mixture of Portland cement clinker and granulated blast-furnace slag, either interground or ground separately and blended together. Granulated blast-furnace slag is a non-metallic product obtained by rapidly chilling or quenching molten slag, tapped from the blast furnaces of steel plants, in water. The slag constituent should be neither less than 35% nor more than 70% of the Portland cement (IS: 455-2015). PBSC generally has better fineness, lower heat of hydration, lower permeability, and better resistance to chemical attack and corrosion than OPC.

Portland slag cement (PSC)   This cement can be used for all construction jobs in place of ordinary Portland cement. However, its special properties render its adoption highly desirable for marine structures, municipal works such as sewers, structures involving large masses of concrete such as dams, retaining walls, bridge abutments, and structures exposed to sulphate-bearing soils such as foundations and concrete roads.

Sulphate-resisting Portland cement

Sulphate-resisting Portland cement (SRPC or SRC) counters what is commonly known as 'sulphate attack'. Soluble sulphate salts, such as sodium sulphate (Na_2SO_4) and calcium sulphate ($CaSO_4$), when present in ground water or soil penetrate through the pores of hardened concrete and chemically react with the tricalcium aluminate (C_3A) constituent of cement. The reaction product, ettringite, occupies a greater volume than the reacting compounds. This creates internal pressure, resulting in the cracking of the concrete, which eventually spalls and disintegrates.

SRPC has a low C_3A content, which helps in avoiding sulphate attack. The use of SRC is strongly recommended for structures in marshy lands, creek areas, coastal areas, sea water, and other areas where soluble sulphate salts are present beyond tolerable limits.

With respect to all other chemical and physical properties, SRPC is similar to ordinary Portland cement and, hence, can be used for all types of constructions where OPC, PPC, or PBSC is used. SRPC should conform to IS: 12330-1988 [Reaffirmed 2013], Indian Standard Specification for sulphate-resisting Portland cement (Table 2.6). Wherever sulphate concentration, measured in terms of SO_3 content, exceeds 300 ppm in ground water, or 0.2% in soil, the use of SRPC is recommended. When, besides sulphates, chlorides are present beyond permissible limits, it is preferable to use blended cement instead of SRPC.

Low-heat Portland cement

Low-heat Portland cement (LHC) has its chemical constituents proportioned in such a way that the heat liberated due to hydration is reduced. This makes it particularly suitable for use in massive structures such as dams, bridge abutments, and retaining walls. Its rate of gaining strength is lower than that of ordinary Portland cement. Hence, it requires a longer curing period. While using this cement, it is necessary to take adequate precautions with regard to the time of removal of the formwork. The Indian Standard Specification for Low Heat Portland Cement IS: 12600-1989 gives the physical and chemical requirements of LHC.

Hydrophobic cement

Adding water-repellent chemicals to OPC clinker at the grinding stage produces hydrophobic cement. This type of cement can be stored under humid and damp conditions for prolonged periods without losing much strength or forming lumps. The film formed around the cement grain breaks down when it is mixed vigorously with aggregates. Except for a slightly longer mixing time, there is no change in the procedure for making concrete with hydrophobic cement. IS: 8043-1991 gives the requirements and properties of hydrophobic cement.

Oil well cement

Oil well cement (OWC) is specially manufactured for use while drilling oil wells, to fill the space between the steel castings and the wall of the well. It sets in a controlled way under high-temperature and pressure conditions. This gives the slurry made from it sufficient time to reach the large depths normally associated with oil wells. However, once set, it develops strength rapidly and remains stable even at high temperatures.

Oil well cement conforming to the American Petroleum Institute Specifications for Class G high sulphate resistance (HSR) cement is now manufactured in India. It fulfils the specifications of general-purpose OWC and can be modified with suitable additives for application in deep oil wells. OWC is highly resistant to sulphate attack because of its intrinsic chemical composition.

White cement

Ordinary cement is grey in colour because of the presence of iron oxide in it. The raw materials in white cement are so chosen that the maximum iron oxide content is limited to 1% (Table 2.6). White cement is used primarily for decorative purposes. However, most white cements, which also fulfil the requirements of IS: 269-1989, can be used as a replacement for ordinary Portland cement in structural works. A variety of colours can be obtained by adding pigments to white cement (Table 2.7).

Table 2.7 Production of coloured finishes with white cement

Colours desired	Chemical names of colours	Approximate quantities required (kg/bag of cement)	
		Light shade	Medium shade
Grey	Lampblack* or	0.25	0.50
Blue black	Carbon black[†] or	0.25	0.50
Black	Black oxide[‡] of manganese or	0.50	1.00
	Mineral black	0.50	1.00
Blue	Ultramarine blue	2.30	4.10
Brownish red to dull brick red	Red oxide of iron	2.30	4.10
Bright red to vermilion	Mineral turkey red	2.30	4.10
Red sandstone to purplish red	Indian red	2.30	4.10
Brown to reddish brown	Metallic brown (oxide)	2.30	4.10
Buff, colonial tint, and yellow	Yellow ochre or	2.30	2.00
	Yellow oxide	1.00	2.00
Green	Chromium oxide or	2.30	4.10
	Greenish blue ultramarine	2.70	4.10

*Only good quality lampblack should be used.
[†]Carbon black is light in weight and requires very thorough mixing.
[‡]Black oxide or mineral black is probably most advantageous for general use. Use 5 kg of oxide for each bag of cement.

2.1.5 Cement Storage

Since cement is a very finely ground hygroscopic material (i.e., it readily absorbs moisture), necessary precautions should be taken to ensure that it is kept dry and free from moisture. The storage shed should have a proper floor, raised at least 150 mm above ground level, and should be provided with airtight doors and windows. Cement stored for a long time tends to deteriorate; an indicative rate of its deterioration is given in Table 2.8(a). A comparison of the rate of hardening of concrete made with fresh and stored cement is given in Table 2.8(b).

It is a good practice to move cement in and out of the storage shed using the 'first-in-first-out' scheme. The drainage systems on the roof of the shed and around the storage should be well maintained, especially during monsoon. At the construction site, the cement bags should be kept on a raised platform and covered with tarpaulin.

Table 2.8(a) Expected loss of strength of cement due to long storage periods

Period of storage of cement	Minimum expected reduction in strength at 28 days (%)*
Fresh	0
3 months	20
6 months	30
1 year	40
2 years	50

*The values are indicative. If cement is likely to have deteriorated during storage, it should be sent to a laboratory for testing the exact % reduction in strength.

Table 2.8(b) Rate* of hardening of concrete made with fresh and stored cement

Age at testing	Concrete made with cement stored in bags under normal conditions for 6 months (%)	Concrete made with fresh cement (%)
7 days	73	100
28 days	75	100
6 months	84	100

*The rates are indicative. If cement is likely to have deteriorated during storage, it should be sent to a laboratory for testing.

2.1.6 Cement Testing

Tests to which a particular cement type must be subjected to are stipulated in the concerned BIS specifications given in Tables 2.5 and 2.6. These tests are intended to quantitatively assess the properties. The properties examined include chemical composition and physical characteristics such as fineness, soundness, setting time, and strength of cement. Samples used for cement testing are selected from a mixture of approximately equal portions of cement selected from at least 12 different bags or packages when the cement is not loose, or 12 different positions in a heap or heaps when the cement is loose. This ensures a fairly average sample. The minimum weight of the final sample taken should be 5 kg. The selected sample is stored in an airtight container until testing.

Chemical composition

Loss on ignition   One gram of the sample is heated for 15 min in a platinum crucible (or for 1 h in a porcelain crucible) at a temperature of 900–1000°C. It is then cooled and weighed. Loss on ignition is restricted to 5% by weight.

Insoluble residue   A well-stirred mixture of 1 g of cement, 40 cm³ of water, and 10 cm³ of concentrated hydrochloric acid (specific gravity 1.18) is boiled for 10 min and filtered. The container is rinsed five times and the filter washed ten times with hot water. The residue is washed, filtered with hot water, and boiled for 10 min with Na_2CO_3 solution (2N). The solution is filtered again, through the same filter paper, with acid (2N) and finally with water till it is free from chlorides. The filter paper is dried, ignited, and weighed to give the weight of the insoluble residue. The insoluble residue is restricted to 2–4% depending on the type of cement by weight.

Lime and alumina   The percentage of lime to silica, alumina, and iron oxide when calculated using the formula

$$\frac{CaO - 0.7SO_3}{2.8SiO_2 + 1.2Al_2O_3 + 0.65Fe_2O_3}$$

is restricted to 1.02 and cannot be less than 0.66. An excess of free lime will lead to the unsoundness of cement.

Magnesia   If the percentage of free magnesia exceeds 6%, the cement becomes unsound.

The Bureau of Indian Standards has prescribed the following physical tests to control the quality of Portland cement with respect to finesse, setting time, soundness, and strength.

Finesse test

The finesse of cement has a bearing on the rate of hydration. Surface area of cement particles is more for finer cement. Therefore, finer cement gains strength faster but also evolves more heat at early stages.

a. Sieving method: The finesse of cement can be determined by dry sieving as recommended by IS: 4031 (Part 1)-1996. For this method, the standard IS 90 micron sieve is used. Based on sieving, the proportion of cement whose grain size is larger than the mesh size of the sieve is determined. The residue is weighed and the result is expressed as a percentage of the cement weight taken. The percentage residue should not exceed 10%.

b. Air permeability method (Blaine method): The finesse of cement is measured as specific surface. Specific surface of cement is expressed as surface area in square metres of all cement particles in 1 kilogram of cement. The higher the specific surface, the finer the cement and faster will be the hydration.

The test consists of measuring the time taken for a fixed quantity of air flow through a compacted cement bed of specified dimension and porosity. Under standard condition, the specific surface of cement is proportional to $\sqrt{t}$, where t is the time interval of a given quantity of air flow through the compacted cement bed. The number and size range of individual pores in the specified bed are determined by the cement particle size distribution. This also determines the time for specified air flow. Blaine's air permeability apparatus is shown in Fig. 2.10. The method is comparative rather than absolute. Hence, a reference specimen of known property is required

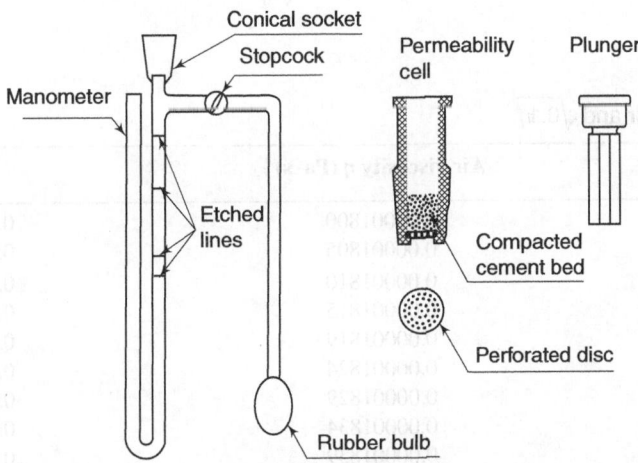

Fig. 2.10 Blaine's apparatus

to compare and determine the specific surface of actual sample. The reference sample is used to calibrate and determine the apparatus constant k.

A bed of cement is prepared in a special permeability cell having a porosity $e = 0.500$, the weight of cement taken being m_1. Therefore,

$$m_1 = 0.500\rho v$$

where ρ = Density in g/cm^3
   v = Volume of the cement bed in cm^3
The specific surface is given by

$$S = \frac{k\sqrt{e^3}\sqrt{t}}{\rho(1-e)\sqrt{0.1\eta}}$$

where k = Apparatus constant
   e = Porosity of cement bed
   t = Measured time
   ρ = Density of cement in g/cm^3
   η = Viscosity of air at test temperature
The viscosity of air at various temperatures is given in Table 2.9. The apparatus constant k is determined using reference cement and performing the test with the equation

$$k = \frac{S_0\rho_0(1-e)\sqrt{0.1\eta}}{\sqrt{e^3}\sqrt{t_0}}$$

where S_0 = Specific surface of reference cement
   ρ_0 = Density of reference cement in g/cm^3
   t_0 = Measured time for reference cement
   η = Air viscosity at test temperature in Pa·s
with a specified porosity of $e = 0.500$

Thus,

$$k = \frac{1.414 S_0\rho_0\sqrt{0.1\eta}}{\sqrt{t_0}}$$

Table 2.9 Viscosity of air and $\sqrt{0.1\eta}$

Temperature (°C)	Air viscosity η (Pa·s)	$\sqrt{0.1\eta}$
16	0.00001800	0.001342
17	0.00001805	0.001344
18	0.00001810	0.001345
19	0.00001815	0.001347
20	0.00001819	0.001349
21	0.00001824	0.001351
22	0.00001829	0.001353
23	0.00001834	0.001354
24	0.00001839	0.001356

Linear interpolation can be used for the determination of intermediate values.

Example 2.1

Given the following data, determine the specific surface of the given sample of cement:

Finesse value of reference sample cement = 225,000 mm^2/g
Density of reference sample = 3.15 g/cm^3
Density of given sample of cement = 3.10 g/cm^3
Temperature at test = 24°C
Time taken for reference sample = 18,000 s
Time taken for actual sample = 20,000 s

Solution
Apparatus constant

$$k = \frac{1.414 S_0 \rho_0 \sqrt{0.1\eta}}{\sqrt{t_0}}$$

$$= \frac{1.414 \times 225,000 \times 3.15 \times 0.001356}{\sqrt{18,000}}$$

$$= 10.128$$

Specific surface of the sample of cement:

$$S = \frac{k\sqrt{e^3}\sqrt{t}}{\rho(1-e)\sqrt{0.1\eta}}$$

$$= \frac{10.128 \times \sqrt{0.5^3} \times \sqrt{20,000}}{3.10(1-0.5) \times 0.001356}$$

$$= 240936.34 \text{ mm}^2/\text{g}$$

Setting time

To enable concrete to be transported, placed, and compactly laid in position, the initial setting of cement should be sufficient. Once the concrete has been laid, it should harden so that the structure can be put to use as soon as possible. The initial setting of cement is that stage in the process of hardening after which any cracks appearing do not generally reunite. The final setting is that stage when it has attained sufficient strength and hardness, with which the material develops the ability to carry load. The Vicat apparatus (Fig. 2.11) with a plunger or needle (Fig. 2.12) is used to determine the setting time for cement.

Standard consistency test

Cement paste of standard consistency has a specified resistance to penetration by a standard plunger. The water required for such a paste is determined by trial penetrations of pastes with different water contents. Content of water is

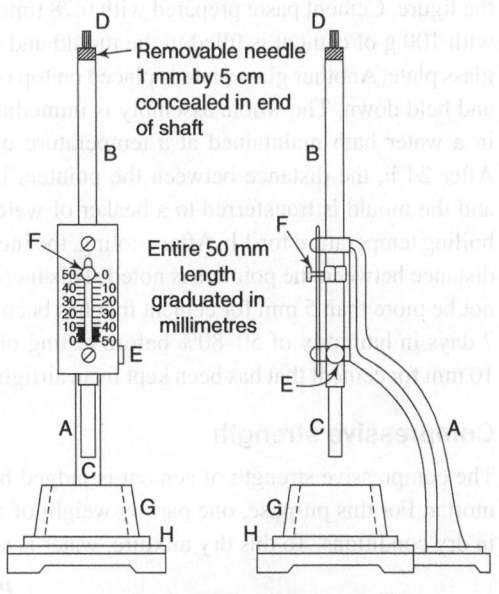

Removable needle 1 mm by 5 cm concealed in end of shaft

Entire 50 mm length graduated in millimetres

Fig. 2.11 Vicat apparatus

expressed as a percentage by mass of the cement. Vicat apparatus with the plunger C (shown in Fig. 2.12) is used for the test. The plunger shall be of non-corrodible metal in the form of a right cylinder of 50 ± 1 mm effective length and of 10.00 ± 0.05 mm diameter. The total mass of moving parts shall be 300 ± 1 g. The test is repeated with pastes containing different water contents until one is found to produce a distance between the plunger and the base plate of 6 ± 1 mm. The water content of that paste is recorded to the nearest 0.5% as the water for standard consistency.

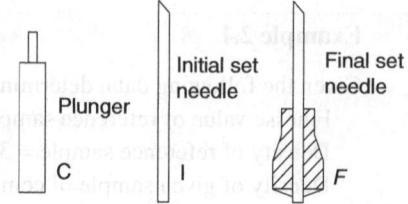

Fig. 2.12 Penetration needles of the Vicat apparatus

A paste of 300 g of cement made with 0.85 times the amount of water required for a paste of normal consistency is filled in the mould. The lower end of the rod of the Vicat apparatus is fitted with a needle I, 1 mm in cross section (Fig. 2.12). This needle is brought in contact with the surface of the paste and released. The initial setting is said to have taken place when the needle fails to penetrate beyond a point 5 mm above the glass plate kept at the base. The time taken from the instant water is added to the cement to the moment the needle fails to penetrate as described above is the initial setting time. It should not be less than 30 min for ordinary Portland cement.

For determining the final setting time, the 1-mm square needle is replaced by the needle F shown in Fig. 2.12. This needle has an annular attachment around the 1-mm square needle, projecting 0.5 mm below it. It is brought in contact with the paste in the mould and released instantly. The final setting is considered to have taken place when the attachment fails to make any impression on the surface whereas the square needle makes one. The time taken from the instant water is added to the cement to the moment the circular attachment fails to make an impression on the surface of the cement paste is known as the final setting time. For ordinary Portland cement, the final setting time should not be more than 10 h. The test is performed in an air-conditioned room with 90% humidity and a temperature of 25–29°C.

Soundness

It is essential for concrete not to undergo large changes in volume after setting. Change in volume is termed *unsoundness* and can cause cracks, undue expansion, and disintegration of concrete.

This test is carried out with the help of Le Chatelier's apparatus (shown in Fig. 2.13). It consists of a split brass cylinder to which two pointers A are attached, one on each side of the split. The dimensions are shown in the figure. Cement paste prepared with 0.78 times the water required to prepare a paste of normal consistency with 100 g of cement is filled in the mould and placed on a glass plate. Another glass plate is placed on top of the mould and held down. The whole assembly is immediately placed in a water bath maintained at a temperature of 27–32°C. After 24 h, the distance between the pointers is measured and the mould is transferred to a beaker of water heated to boiling temperature for 1 h. After cooling, the increase in the distance between the pointers is noted. This increase should not be more than 5 mm for cement that has been aerated for 7 days in humidity of 50–80% before testing or more than 10 mm for cement that has been kept in an airtight container.

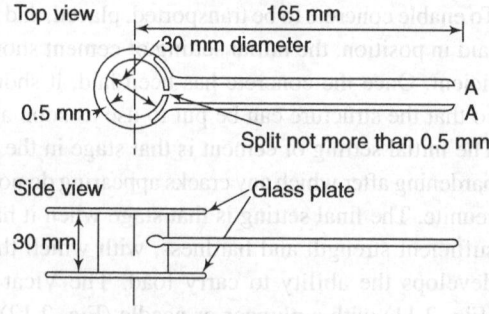

Fig. 2.13 Le Chatelier's apparatus

Compressive strength

The compressive strength of cement is judged by determining the compressive strength of cement and sand mortar. For this purpose, one part by weight of cement is mixed with three parts by weight of standard sand in dry conditions. To this dry mixture, water is added as per the formula

$$P = P_n / 4 + 3.0$$

where P is the percentage of water by weight of the dry materials and P_n is the percentage of water required for making a cement paste of normal consistency. Cement, sand, and water are thoroughly mixed to give a paste of uniform colour; however, these should not be mixed for less than 3 min or more than 4 min. Cubes having 7.086-cm sides are cast using this paste and kept in an atmosphere of 90% humidity and 25–29°C temperature for 24 h. These are then removed from the moulds and kept submerged in clean water until the time of the test.

Three cubes each are tested in a compression-testing machine after 3 days and then 7 days. The compressive strength of ordinary Portland cement of Grade 33 should not be less than the following values: 16 N/mm² after 3 days and 22 N/mm² after 7 days. A detailed specification based on the grade of cement is given in Table 2.5.

Tensile strength

The tensile strength of cement and sand mortar is tested to judge the tensile strength of cement. To do so, briquettes of standard dimensions as shown in Fig. 2.14 are prepared. These briquettes have a uniform thickness of 1 in. (25.4 mm) and a minimum cross-sectional area of 645 mm² at the throat section. The test has been devised to have a tensile load on 1 in.² of the cross section, and hence all dimensions of the specimen are in FPS units. For preparing briquettes, one part by weight of cement and three parts by weight of sand are mixed with the quantity of water determined from the following formula:

$$P = 0.2P_n + 2.5$$

Cement, sand, and water are mixed thoroughly to get a uniform colour. The mix is then filled in the moulds. The moulds are beaten down with a standard spatula (Fig. 2.15) till water appears on the surface. These are then turned upside down and again a small heap of mortar is placed on them and beaten down. The surfaces are smoothened with the blade of a trowel. The briquettes are taken out of the moulds after keeping them at 25–29°C with 90% humidity for 24 h. Six such specimens are tested in briquette-testing machine fixtures (Fig. 2.16) after 3 days and then 7 days. The tensile strength for good Portland cement should not be less than 2 N/mm² after 3 days and not less than 2.5 N/mm² after 7 days. Though this test has currently been withdrawn by the BIS, it is useful to determine the tensile strength of cement. This test is covered under the American Society for Testing Machine Standard (ASTM) C307.

2.2 Aggregates

Aggregates are the major ingredients of concrete. They constitute 70–75% of the total volume, provide a rigid skeleton

A battery of three moulds for casting briquettes

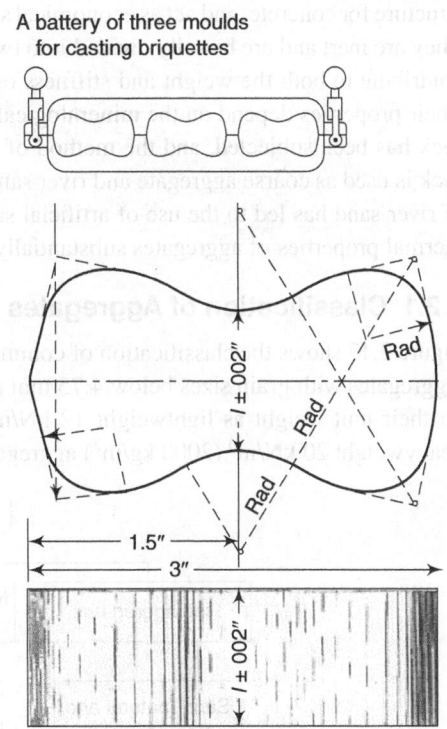

Fig. 2.14 Standard dimensions of a briquette

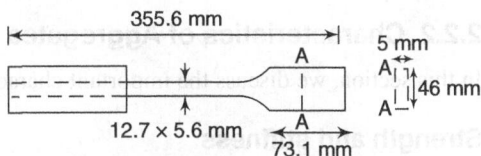

Fig. 2.15 Standard spatula

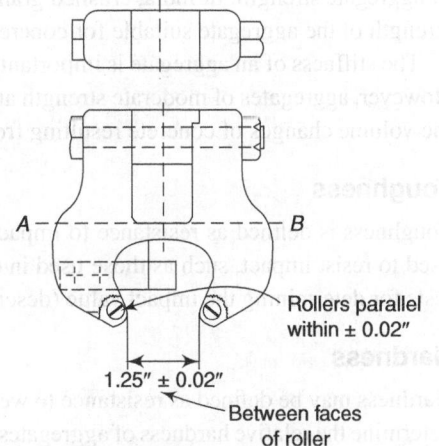

Fig. 2.16 Fixture for briquette test in a tensile strength test

structure for concrete, and act as economical space fillers. IS: 383-1970 defines the requirements of aggregates. They are inert and are broadly divided into two categories, i.e., fine and coarse, depending on their size. They contribute to both the weight and stiffness of concrete. Generally, coarse aggregates are derived from rock. Their properties depend on the mineralogical composition of rock, the environmental exposure to which the rock has been subjected, and the method of crushing employed to get the different sizes. In India, crushed rock is used as coarse aggregate and river sand is preferred for fine aggregate. Of late, the lack of availability of river sand has led to the use of artificial sands, especially in southern states. The physical, chemical, and thermal properties of aggregates substantially influence the performance of concrete.

2.2.1 Classification of Aggregates

Figure 2.17 shows the classification of common aggregates. Aggregates are classified as either fine or coarse. Aggregates with grain sizes below 4.75 mm are termed fine aggregates. Aggregates are also classified based on their unit weight as lightweight 12 kN/m^3 (1200 kg/m^3), normal-weight 15 kN/m^3 (1500 kg/m^3), and heavyweight 20 kN/m^3 (2000 kg/m^3) aggregates.

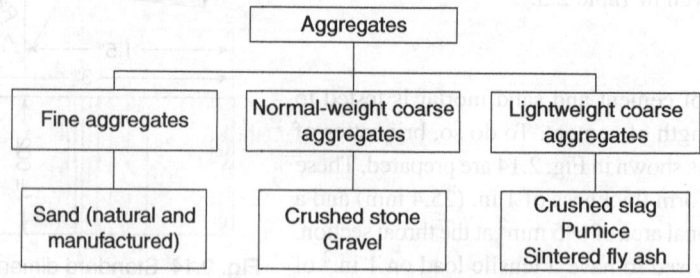

Fig. 2.17 Classification of common aggregates

2.2.2 Characteristics of Aggregates

In this section, we discuss the important characteristics of aggregates.

Strength and stiffness

The strength of concrete in general cannot exceed the strength of the aggregates that constitute it. However, it is not possible to directly test an aggregate for its strength. The aggregate crushing value is used as an index of aggregate strength. In India, crushed granite is used as an aggregate. A good average value of the crushing strength of the aggregate suitable for concrete is 80–100 MPa.

The stiffness of an aggregate is important for maintaining the dimensional stability of concrete under load. However, aggregates of moderate strength and modulus of elasticity can advantageously be used to withstand the volume changes of concrete resulting from thermal or expansive causes.

Toughness

Toughness is defined as resistance to impact. A test for toughness is generally done for aggregates that are used to resist impact, such as those used in concrete employed to build aircraft runways. There are standard tests for determining the impact value (described later in Section 2.2.4), which is used to quantify toughness.

Hardness

Hardness may be defined as resistance to wear or abrasion. The test described later in Section 2.2.4 is used to determine the relative hardness of aggregates. This property is important for aggregates used in pavements and industrial floors. The requisite material properties of aggregates used for general roller-compacted concrete (RCC) work are summarized in Table 2.10.

Table 2.10 Material properties of aggregates used for general roller-compacted concrete (RCC) work

Properties	Limiting values (%)	
	For wearing surfaces	For surface other than wearing surfaces
Crushing value	30	45
Impact value	30	45
Abrasion value (Los Angeles)	30	50
Soundness (average loss of weight after five cycles)		
Fine aggregates	10*	15†
Coarse aggregates	12*	18†

* When tested with Na_2SO_4
† When tested with $MgSO_4$

Cleanliness

The bonding of cement paste takes place at the surface of the aggregates. Hence, the aggregates should be clean, in order to ensure good paste–aggregate bonds. The permissible levels of impurities in aggregates are given in Table 2.11.

Table 2.11 Permissible levels of impurities in aggregates

Deleterious substances	Fine aggregates*		Coarse aggregates*	
	Uncrushed	Crushed	Uncrushed	Crushed
Coal and lignite	1.00	1.00	1.00	1.00
Clay lumps	1.00	1.00	1.00	1.00
Material finer than 75 mm	3.00	3.00	3.00	3.00
Shale	1.0	—	—	—
Total of percentages of all deleterious materials†	5.0	2.00	5.00	5.00

*Percentage by weight, max.
†Mica is excluded.

Shape and texture

Aggregates are round or angular in shape. Angular aggregates possessing well-defined edges are obtained by crushing rocks such as granite. These aggregates obtained from poor laminated rocks are classified as flaky if their thickness is small compared to their width. Such aggregates should be avoided. Rounded aggregates require less cement paste and water for a given workability.

The surface texture of an aggregate is an index of its smoothness or otherwise. An aggregate with rough texture will provide a better aggregate–cement bond and is, hence, preferred. The shape and surface texture of aggregates influence important properties such as paste requirement, strength, and workability of concrete as shown in Table 2.12.

Table 2.12 Influence of shape and texture on paste requirement, strength, and workability

Type of aggregate	Paste requirement	Strength	Workability
Rounded	Low	Low	High
Angular	High	High	Low
Rough	High	High	Low
Smooth	Low	Low	High

Physical soundness

The physical soundness of an aggregate ensures protection against mechanical actions. These mechanical actions are induced by volume changes due to environmental conditions such as freezing or thawing, wetting or drying, and swelling. The deterioration of concrete due to freeze–thaw cycle is shown in Fig. 2.18. This causes a portion of the aggregate in the concrete to pop out (see Fig. 2.19). According to IS: 383-1970, the average loss of weight after 10 cycles should not exceed 12% and 18% when tested with sodium sulphate and magnesium sulphate, respectively.

Chemical soundness

An aggregate is said to be chemically unsound if it has been damaged due to a chemical reaction. A typical example of chemical unsoundness is the reaction of aggregates with alkali. The reaction between the active silica constituents of aggregate and the alkalis of cement leads to disruption of concrete, as shown in Fig. 2.20. As a result of this reaction, an alkali–silicate gel is formed. This alters the borders of the aggregate and the paste. An internal pressure is developed, leading to expansion cracking and, subsequently, disruption of the cement paste.

The reactivity of aggregate depends on particle size, porosity, alkali content, and fineness of cement. The rate of the alkali–aggregate reaction is also affected by the presence of water in the paste. The reaction is accelerated under wetting and drying conditions. Temperature in the range 10–40°C accelerates the reaction.

Expansion due to an alkali–aggregate reaction can be controlled by

a. using sound aggregates,
b. using a mixture of good and suspect aggregates,
c. controlling the permeability of concrete by using a low water–cement ratio, a mineral admixture, or a suitably good finish,
d. using special low-alkali cement, and
e. using a good surface sealant.

Gradation

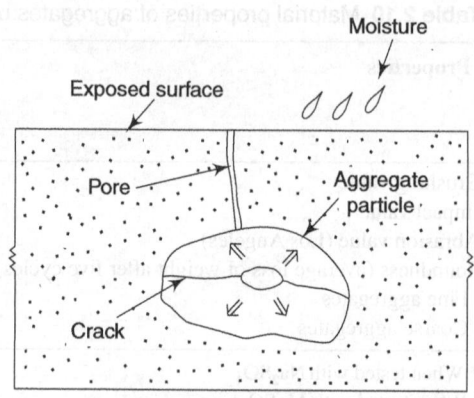

Fig. 2.18 Deterioration of concrete due to freeze–thaw cycle

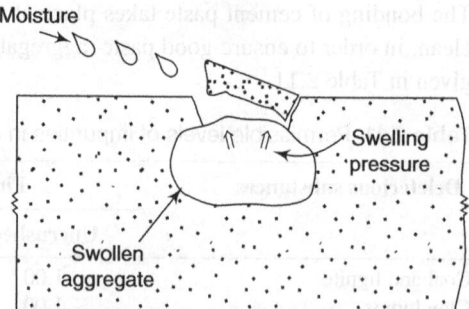

Fig. 2.19 Popping out of a portion of concrete

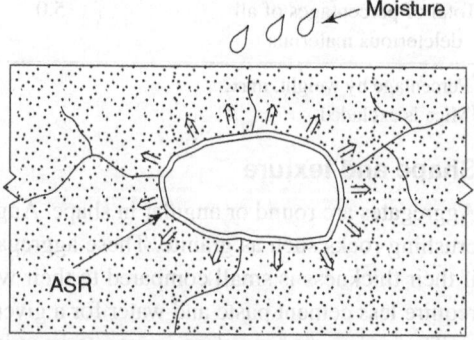

ASR: alkali–silica reaction
Fig. 2.20 Alkali–aggregate reaction

Gradation refers to the particle size distribution of aggregates. The gradation of coarse aggregates plays an important role in workability and paste requirements. The gradation of fine aggregates affects the workability and finishability of concrete. Grading is a very important property of the aggregate used for making concrete, in view of its effect on the packing of particles, resulting in the reduction of voids. This in turn influences the water demand and cement content of concrete.

Consistency of grading within and among consignments of aggregates is thus vital in order to ensure uniform quality of concrete.

Grading is described in terms of the cumulative percentage of weights passing a particular IS sieve. The typical grading limits for coarse and fine aggregates are given in Tables 2.13 and 2.14, respectively. Figure 2.21

Table 2.13 Typical grading limits for coarse aggregates

Sieve designation (mm)	Percentage passing for single-sized aggregates of nominal size (%)					
	63 mm	40 mm	20 mm	16 mm	12.5 mm	10 mm
80	100	—	—	—	—	—
63	85–100	100	—	—	—	—
40	0–30	85–100	100	—	—	—
20	0–5	0.20	85–100	100	—	—
16	—	—	—	85–100	100	—
12.5	—	—	—	—	85–100	100
10	0–5	0–5	0.20	0–30	0–45	85–100
4.75	—	—	0–5	0–5	0–10	0.20
2.36	—	—	—	—	—	0.5

Table 2.14 Typical grading limits for fine aggregates

Sieve designation	Percentage passing for single-sized aggregates of nominal size (%)			
	Grading zone I	Grading zone II	Grading zone III	Grading zone IV
10 mm	100	100	100	100
4.75 mm	90–100	90–100	90–100	95–100
2.36 mm	60–95	75–100	85–100	95–100
1.18 mm	30–70	55–90	75–100	90–100
600 µm	15–34	35–59	60–79	80–100
300 µm	5–20	8–30	12–40	15–50
150 µm	0–10	0–10	0–10	0–15

shows how standard grading and actual grading vary. Coarse aggregates are either graded (contain particles having more than one size) or single-sized (i.e., are mainly retained on one sieve).

The fineness modulus (FM) is a gross measure of aggregate gradation and is associated with fine aggregates. This is incorporated in the mix proportioning process. It is defined as the sum of the cumulative percentage weight retained on a standard set of sieves divided by 100. Fine aggregate, based on the fineness modulus, is classified as fine, medium, or coarse:

- Fine FM between 2.3 and 2.6
- Medium FM between 2.6 and 2.9
- Coarse FM between 2.9 and 3.2

Figure 2.21 shows the typical relationship between the percentage passing and particle size for a well-graded aggregate.

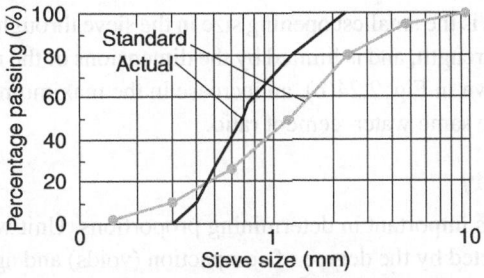

Fig. 2.21 Comparison of standard and actual grading

The various types of gradation are discussed below.

Uniform grading As shown in Fig. 2.22, in uniform grading, all particles are of the same size. Note that this produces a large volume of voids irrespective of particle size. Hence the paste requirement for this concrete is high.

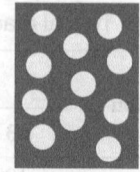

Fig. 2.22 Uniform grading

Continuous grading Continuous grading incorporates a combination of particles of many sizes (see Fig. 2.23). Hence, it minimizes the volume of voids but increases the particle surface area. This is the preferred gradation.

Gap gradation This involves grading in which one or more sizes are omitted. This type of concrete is generally used for architectural or aesthetic purposes.

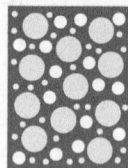

Paste requirement variation The paste requirement decreases as the uniformity of gradation decreases and the continuity of gradation increases. The left-most sample in Fig. 2.24(a) has a particle size of 31 mm and requires 250 mL of paste. The sample in the centre has two particle sizes, 31 mm and

Fig. 2.23 Continuous grading

10 mm, and requires 215 mL of paste. The sample on the right-hand side has a continuous grading of 31 mm down and requires the least quantity of paste, 200 mL.

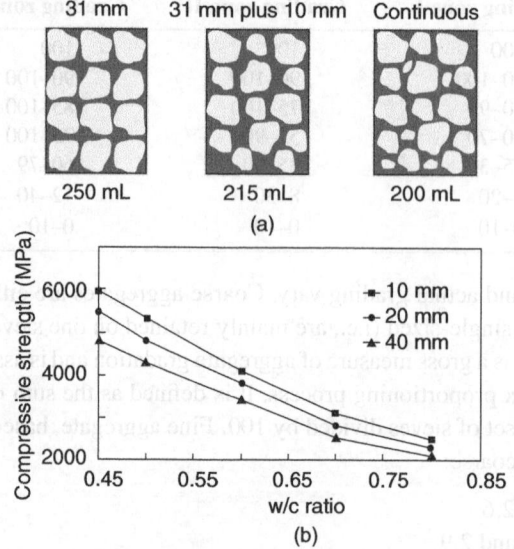

Fig. 2.24 (a) Paste requirement variation with type of aggregates; (b) Influence of maximum size of aggregates on compressive strength

Maximum size of aggregate

The maximum size of aggregate is the smallest opening size in the sieve through which all aggregates can pass. This influences the maximum strength, and is limited by the dimensions of the members and the congestion of the rebars in a member. As shown in Fig. 2.24(b), an increase in the maximum size of an aggregate increases the compressive strength for the same water–cement ratio.

Unit weight/specific gravity

The unit weight of aggregates is important in determining proportions. Unit weight refers to bulk density or mass per unit volume. It is affected by the degree of compaction (voids) and aggregate moisture (presence of water). The typical dry unit weight is 1600 kg/m^3.

Specific gravity refers to the relative (as compared to water) density of a unit volume of aggregate:

$$\text{Specific gravity} = \frac{\text{Density of material}}{\text{Density of water}}$$

The difference between specific gravity and unit weight is related to the voids within and surrounding the aggregates. Apparent specific gravity (ASG) is computed considering solids only. Bulk specific gravity (BSG) accounts for 1–2% of the void space within an aggregate particle. Bulk specific gravity is also a function of the moisture present in the aggregate. Figure 2.25 shows the assumptions made while computing unit weight, BSG, and ASG. Thus, BSG (dry) accounts for air-filled voids and BSG (saturated surface dry) accounts for water-filled voids. The typical value of specific gravity is 2.65 for coarse aggregates. The typical values of bulk density and specific gravity and the percentage of voids present are summarized in Table 2.15.

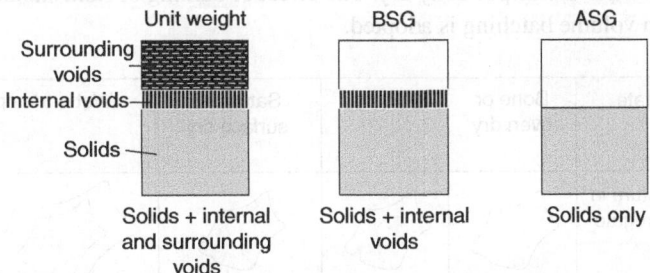

Fig. 2.25 Unit weight, bulk specific gravity (BSG), and apparent specific gravity (ASG)

Table 2.15 Bulk density, specific gravity, and percentage of voids in aggregates

Material	
	Bulk density (kg/L)
River sand	
Fine	1.44
Medium	1.52
Coarse	1.60
Beach or river shingle	1.60
Broken stone	1.60
Stone screenings	1.44
Broken granite	1.68
	Specific gravity
Trap	2.9
Granite	2.8
Gravel	2.66
Sand	2.65
	Voids (%), average
River sand	
Fine	43
Coarse	35
Mixed and moist	38
Mixed and dry	30
Broken stone, graded	
25 mm max. size	46
50 mm max. size	45
63 mm max. size	41
Stone screenings	48

Water absorption and moisture effects

All aggregates have pores in which water can be held for a long time. The different moisture conditions in which aggregates exist are shown in Fig. 2.26. Absorption and surface moisture determination is important in order to maintain the desired water–cement ratio in the mix design. Table 2.16 shows the approximate water absorption percentages of different aggregates. Very light and porous aggregates may absorb as much as 25% by weight.

Bulking of sand

Moist sand has more volume than dry sand. This is called *bulking*. It is caused by a thin film of water surrounding it, thereby increasing the voids. The volume of sand goes on increasing with the increase in moisture content up to a certain limit. Thereafter, the film breaks down and the moisture starts filling the pores. A further increase in the moisture decreases the volume, and finally at about 10–20% of the moisture content, the sand has the same volume as when it was perfectly dry. The effect of bulking of sand should be considered in mix design, especially when volume batching is adopted.

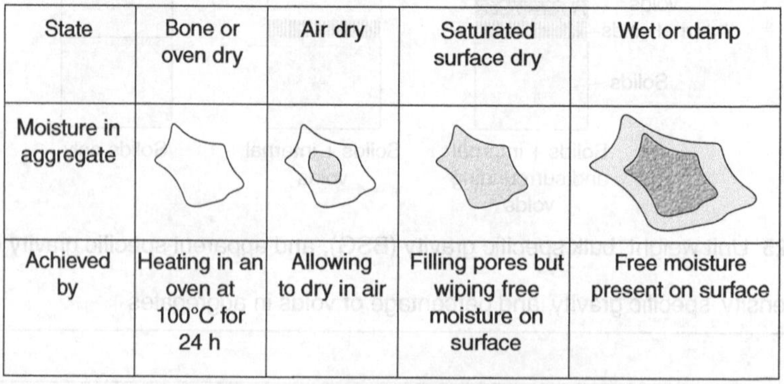

State	Bone or oven dry	Air dry	Saturated surface dry	Wet or damp
Moisture in aggregate				
Achieved by	Heating in an oven at 100°C for 24 h	Allowing to dry in air	Filling pores but wiping free moisture on surface	Free moisture present on surface

Fig. 2.26 Moisture conditions of aggregates

Table 2.16 Approximate water absorption by different aggregates

Aggregate	Water absorption (%)
Average sand	1.0
Pebbles and crushed limestone	1.0
Trap rock and granite	0.5
Porous sandstone	7.0

2.2.3 Alternatives to Aggregates

Regular concrete makes use of river sand as fine aggregate. Various types of stones or natural gravels are used as coarse aggregates. Many environmental protection groups object to the mining of river sand as it tends to lower the water table and causes deterioration of the quality of natural habitat. Hence, there is a growing interest to find replacement for sand using different recycled materials such as sintered fly ash, blast furnace slag, and other industrial waste products. Crushed recycled concrete itself has been used for this purpose. Though aggregates are considered inert, their physical properties are important in imparting desirable properties to concrete both in the fresh state and in the hardened state. The aggregates also must be economical. The physical properties of aggregates influence the fresh properties of concrete such as workability and dimensional stability and hardened properties of strength and durability.

The alternatives chosen should be clean, devoid of chemical impurities or clay and fine particles which will affect hydration of cement.

Recycled aggregates used in place of sand have lower specific gravity and higher water absorption. It has been found that concrete made with recycled aggregates has lower compressive strength and modulus of elasticity.

Field testing to validate the use of particular waste aggregate is necessary to ensure that required strength and durability is attained. In addition, recycled aggregates tend to have higher water absorption and a lower specific gravity than conventional aggregates.

All types of waste aggregates including broken glass bottles can be used in concrete applications. Glass aggregates in concrete can lead to alkali–silica reaction between the cement and the glass aggregate. This may lead to cracking and disintegration and decreased long-term durability.

Lack of reliable and authentic data on alternative aggregates hinders their widespread use. Further research is needed before these can be confidently used in structural concrete applications.

2.2.4 Tests on Aggregates

Aggregates are tested for strength, hardness, abrasion, particle shape and texture, porosity and water absorption, deleterious constituents, alkali–aggregate reaction, etc., as described below.

Crushing strength

A sample of parent source rock is cut and dressed to have specimen dimensions of 25 mm diameter and 25 mm height with a tolerance of ± 0.5 mm. The sample should be representative of the quarry. The end faces should be at right angles to the cylindrical axis. Direct compression is applied to the ends of the cylinder at a rate of about 5 tonnes/min. The load required to produce the crushing of the specimen is observed. The crushing strength is given by the ratio of the crushing load to the cross-sectional area over which the load acts. This test measures the quality of the rock rather than the quality of the aggregate. It can also be used to assess the nature of the source of the aggregate.

Crushing value

The apparatus used for this test, shown in Fig. 2.27, consists of an assembly of an open-ended cylinder (15 cm in diameter) and a base plate, inside which a plunger works. The material to be tested should pass through a 12.5-mm sieve and get retained on a 10-mm sieve. The sample is filled adopting standard filling procedure to ensure good compaction. The whole assembly is placed in a compression-testing machine and subjected to a load of 40 tonnes at a rate of 4 tonnes/min. After the load is released, the crushed aggregate is sieved through a

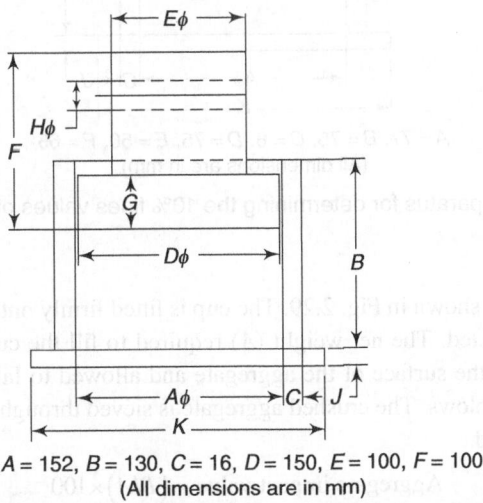

$A = 152, B = 130, C = 16, D = 150, E = 100, F = 100$
(All dimensions are in mm)

Fig. 2.27 Apparatus used to determine the crushing value of aggregates

2.36-mm sieve. The ratio of the material passing through the sieve to the total weight is called the crushing value. The permissible crushing values for different applications are 45% for concrete for non-wearing surfaces and 30% for concrete for wearing surfaces. Certain weaker aggregates may give crushing values of 25–30%. In such cases, this test is not reliable because the material gets crushed before the full load is applied and gets compacted.

Ten per cent fines value

The material to be tested should pass through a 12.5-mm sieve and retained on a 10-mm sieve. The aggregate is placed in the apparatus for crushing strength (see Fig. 2.28) and load is applied on it to cause a total penetration of the plunger in 10 min of about 15 mm for rounded aggregates, 20 mm for natural crushed aggregates, and 24 mm for honeycombed aggregates.

After reaching the required penetration, the material is sieved through a 2.36-mm sieve. The fines passing through the sieve are weighed and expressed as a percentage of the original sample. Normally this percentage is 7.5–12.5%. If no further load is applied to produce fines within 7.5–12.5%, the load required to produce 10% fines is given by

$$L_{10} = \frac{14\,X}{Y+4}$$

where X is the load in tonnes and Y is the mean percentage of fines at X tonnes.

The load required to produce 10% fines from 12.5-mm to 10-mm-sized particles has been observed, but no data is available to correlate the 10% fines value with the crushing value discussed earlier. BS: 882-1965 prescribes a minimum value of 10 tonnes for aggregates to be used in wearing surfaces and 5 tonnes for other concretes as 10% fines values.

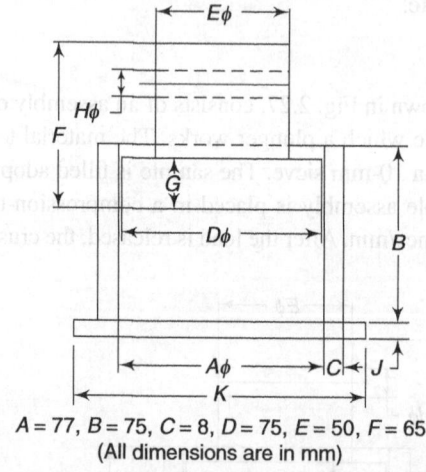

A = 77, B = 75, C = 8, D = 75, E = 50, F = 65
(All dimensions are in mm)

Fig. 2.28 Apparatus for determining the 10% fines values of aggregates

Impact test

The impact testing apparatus is shown in Fig. 2.29. The cup is fitted firmly onto the base plate. The aggregate is filled in the cup and compacted. The net weight (A) required to fill the cup is determined. The hammer is raised up to 380 mm above the surface of the aggregate and allowed to fall freely on the aggregate. The sample is subjected to 15 such blows. The crushed aggregate is sieved through a 2.36-mm sieve. The fraction (B) passing the sieve is weighed.

$$\text{Aggregate impact value} = (B/A) \times 100$$

This test is considered as an alternative to the crushing value test.

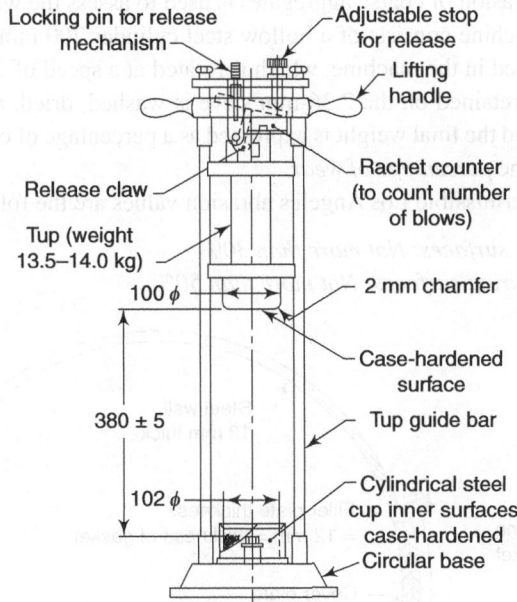

Locking pin for release mechanism

Adjustable stop for release

Lifting handle

Release claw

Tup (weight 13.5–14.0 kg)

Rachet counter (to count number of blows)

100 φ

2 mm chamfer

Case-hardened surface

380 ± 5

Tup guide bar

102 φ

Cylindrical steel cup inner surfaces case-hardened

Circular base

Fig. 2.29 Apparatus for impact testing of aggregates (all dimensions are in mm)

Test for hardness and abrasion resistance

Aggregates can break during handling, mixing, etc. Therefore, it is necessary for them to have good abrasion resistance, so that the damage in ready-mixed concrete due to prolonged mixing is minimal. Aggregates get degraded and result in excess fines. This affects the workability of concrete. The degradation of the fineness modulus with increase in mixing time is shown in Fig. 2.30.

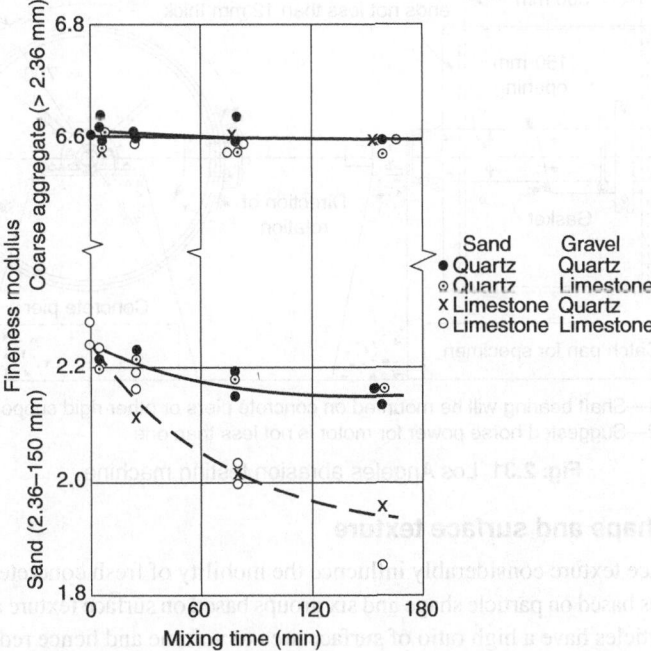

	Sand	Gravel
●	Quartz	Quartz
⊙	Quartz	Limestone
×	Limestone	Quartz
○	Limestone	Limestone

Fig. 2.30 Degradation in fineness modulus of coarse aggregates with increase in mixing time

Los Angeles test for the abrasion of coarse aggregates is used to assess the wear resistance. The apparatus is shown in Fig. 2.31. The machine consists of a hollow steel cylinder 700 mm in diameter and 500 mm in length. The test sample is placed in the machine, which is rotated at a speed of 20–33 rpm. After completing 500 revolutions, the material retained on the 2.36-mm sieve is washed, dried, and weighed. The difference between the original weight and the final weight is expressed as a percentage of original weight of the sample. This percentage is known as the *percentage of wear*.

As per IS: 383-1970, the permissible Los Angeles abrasion values are the following:

- *Concrete used for wearing surfaces: Not more than 30%*
- *Concrete used for non-wearing surfaces: Not more than 50%*

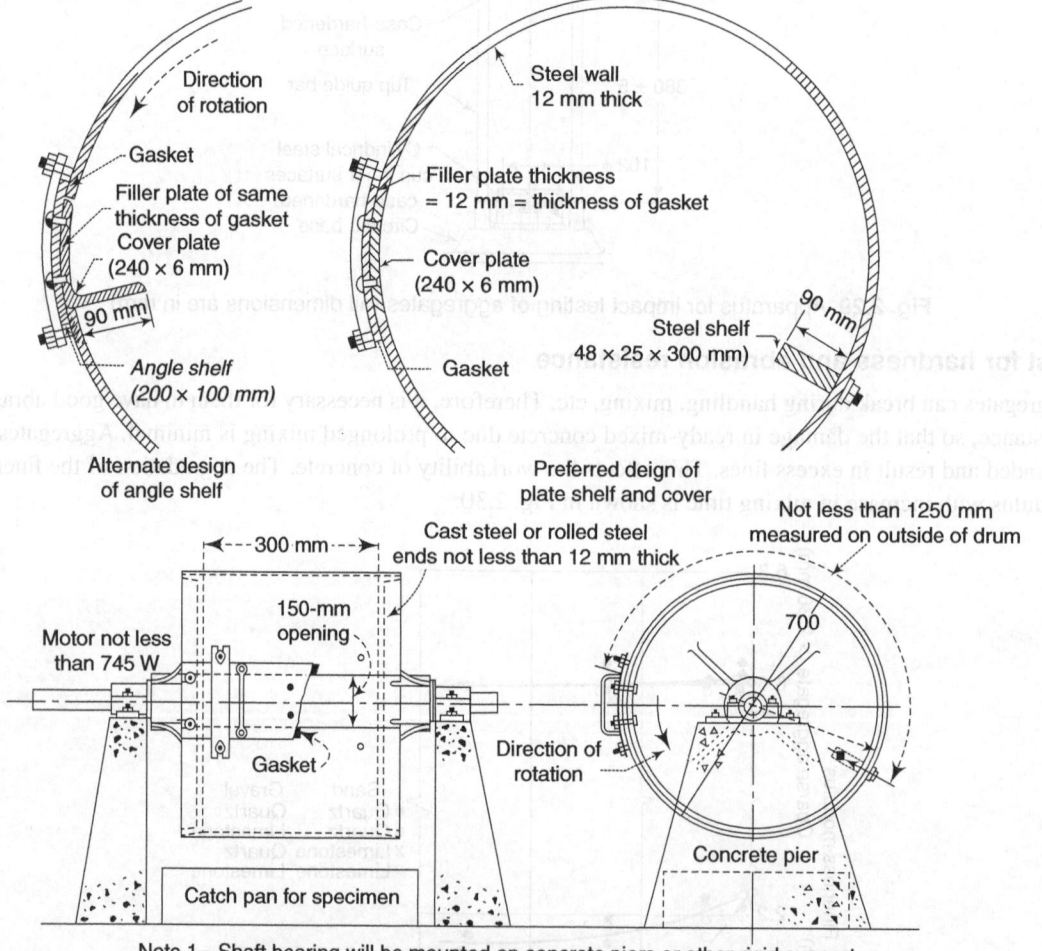

Fig. 2.31 Los Angeles abrasion testing machine

Tests for particle shape and surface texture

Particle shape and surface texture considerably influence the mobility of fresh concrete. As per IS: 383-1970, four groups of aggregates based on particle shape and six groups based on surface texture are listed in Table 2.17. Elongated and flaky particles have a high ratio of surface area to volume and hence reduce the workability of the mix. This also affects the durability of concrete.

Table 2.17 Classification of aggregates based on particle shape and surface texture

Classification	Description	Example
Classification based on particle shape		
Rounded	Fully water-worn or completely shaped by attrition	River or seashore gravel; desert, seashore, and wind-blown sand
Irregular or partly rounded	Naturally irregular or partly shaped by attrition and having rounded edges	Pit sand and gravel, dug flints or rocks
Angular	Possessing well-defined edges formed at the intersection of roughly plane faces	Crushed rocks of all types, talus, screens
Flaky	Usually angular; thickness is small relative to the width and/or length	Laminated rocks
Classification based on surface texture		
Glassy	Conchoidal fracture	Black flint
Smooth	Water-worn or smooth due to fracture of laminated or fine-grained rock	Gravel, chert, slate, marble, certain rhyolite
Granular	Fracture showing more or less uniform rounded grains	Sandstone, volite
Rough	Rough fracture of fine- and medium-grained rock containing no easily visible crystalline constituents	Basalt, felsite, porphyry, limestone
Crystalline	Easily visible crystalline constituents	Granite, gabbro, gneiss
Honeycombed and porous	Visible pores and cavities	Brick, pumice, foamed slag, clinkers, expanded clay

Flakiness index

A particle is said to be *flaky* when its thickness is less than 0.6 times the mean size fraction to which the particle belongs. A 20-mm-sized aggregate passes through a 20-mm sieve and is retained in a 16-mm sieve. Hence the average size of the aggregate is $(20 + 16)/2 = 18$ mm. The limiting thickness for flakiness is $0.6 \times 18 = 10.8$ mm. Hence, if one of the dimensions of the aggregate is less than 10.8 mm, it is considered flaky. IS: 2386(Part I)-1963 recommends the following test for flakiness index.

To determine the flakiness index, a sample is taken and passed through a 63-mm to 6.3-mm sieve and the particles collected separately. Each particle is passed through an aperture of thickness 0.6 times the mean sieve size. If a particle passes through the aperture, it is flaky. All the flaky particles are collected and weighed. The following formula is then applied:

$$\text{Flakiness index} = \frac{\text{Weight of particles passing through the gauge}}{\text{Weight of the sample}}$$

A minimum of 200 pieces of any fraction must be tested. A flakiness index not greater than 25% is suggested for coarse aggregates.

Elongation index

The elongation index can be obtained using a test gauge called the *length gauge* using the formula

$$\text{Elongation index} = \frac{\text{Weight of particles passing through the length gauge}}{\text{Weight of the sample taken}}$$

A particle is said to be elongated if one of its dimensions is larger than 1.8 times the mean size. Therefore, taking the same example of a 20-mm down aggregate, the limiting length for elongated aggregate is $1.8 \times 18 = 32.4$ mm. Hence, if one of the dimensions of the aggregate is larger than 32.4 mm, the aggregate will be considered elongated. It is advisable to keep the flakiness index below 25% and the elongation index below 30% to obtain good quality concrete.

Surface texture

Surface texture is a measure of the smoothness and roughness of the aggregate. IS: 383-1970 classifies surface characteristics into five groups, namely, glassy, smooth, granular, crystalline, and honeycombed and porous. This grouping is broad and is based on the visual examination of the specimen. The rough and porous texture is preferred. This improves the bond of the cement paste and hence the compressive and flexural strengths by 20%.

Sieve analysis

IS: 2386(Part I)-1963 recommends the sieve analysis. This test consists of the simple operation of dividing aggregates into fractions, each consisting of particles of the same size. The sieves used for the test have square openings. Sieves are described by the sizes of their openings as 80 mm, 63 mm, 50 mm, 40 mm, 25 mm, 20 mm, 16 mm, 12.5 mm, 10 mm, 6.3 mm, 4.75 mm, 3.35 mm, 2.36 mm, 1.70 mm, 1.18 mm, 850 μm, 600 μm, 425 μm, 300 μm, 212 μm, 150 μm, and 75 μm. All the sieves are mounted in frames one above the other in ascending order. The sieves used for coarse aggregates are of sizes 80 mm, 40 mm, 20 mm, 10 mm, 4.75 mm, 2.36 mm, 1.18 mm, 600 μm, 300 μm, and 150 μm. All the sieves are mounted on a sieve shaker. Aggregate of known quantity is placed over the top sieve, and after sieving through the test sieves, the residue in each sieve is weighed. The percentage of weight retained to the total weight is calculated, from which the percentage passing is determined. The requirements of percentage passing for coarse and fine aggregates are specified by IS: 383-1970.

Fineness modulus

The fineness modulus is an empirical number representing the cumulative percentage of weights retained on various sieves. The larger the fineness modulus, the coarser is the aggregate. The sum of the cumulative percentages of weight retained divided by 100 is known as the fineness modulus of aggregate. The sieve sizes used are No. 100 (150 μm), No. 50 (300 μm), No. 30 (600 μm), No. 16 (1.18 mm), No. 8 (2.36 mm), and No. 4 (4.75 mm), and 3/8 in. (9.5 mm), 3/4 in. (19.0 mm), 1-1/2 in. (38.1 mm), and larger, increasing in the ratio of 2 to 1. For fine aggregates this value ranges from 2 to 3 and for 20-mm coarse aggregates this value ranges from 6 to 7. The fineness modulus is between 4 and 4.2 for mixed (all in) aggregates. The fineness modulus can also be considered as the weighted average of the sieve on which the material is retained, the sieves being counted from the finest. This gives an indication of the probable behaviour of the concrete mix.

Example 2.2

In a sieve analysis, a sample of dry aggregate of weight 25 kg is sieved and separated through a series of sieves with progressively smaller openings. Once separated, the weight of particle retained on each sieve is measured. The table below shows the data obtained. Perform a fineness modulus analysis, draw the grading curve, and comment on the result obtained.

Data

Total weight of aggregates taken = 25 kg

Sieve size (mm)	Weight retained (kg)
38	0
19	0
9.5	0
4.75	1
2.36	4.25
1.18	6.25
0.6	7
0.3	3.75
0.125	1.5

Solution

Sieve size (mm)	Weight retained	Cumulative weight retained	% Retained	% Passing
38	0	0	0	100
19	0	0	0	100
9.5	0	0	0	100
4.75	1	1	4	96
2.36	4.25	5.25	21	79
1.18	6.25	11.5	46	54
0.6	7	18.5	74	26
0.3	3.75	22.25	89	11
0.125	1.5	23.75	95	5
	Total		329	

Fineness modulus = 329/100 = 3.29

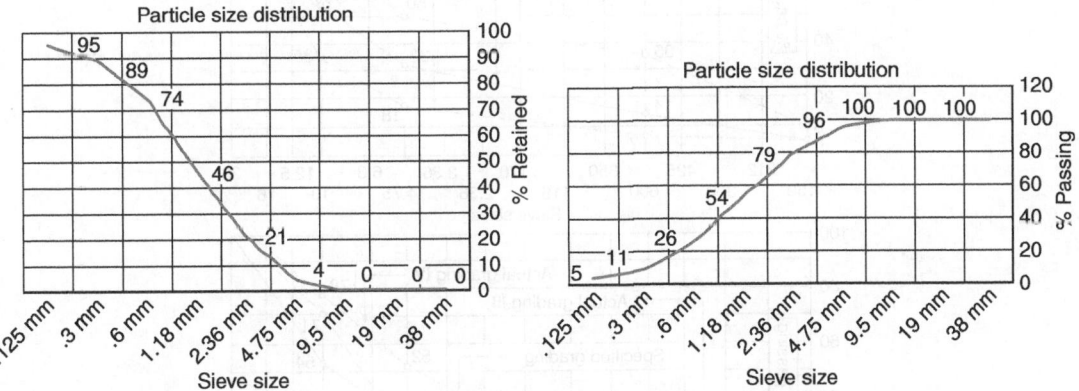

Particle size distribution curves for a typical fine aggregate

The result of the sieve analysis represented graphically is known as the grading curve. If the grading curve is uniform and linear, it indicates that the aggregates are evenly distributed. A study of the particle size distribution graph is important. A steep slope indicates the presence of a particular size and a flat slope indicates a lesser proportion or even a missing size fraction. This information has been used in a later chapter for concrete mix proportioning.

Interpretation of grading curves

Figures 2.32 and 2.33 show graphical representation of particle size distribution. The specified grading is also shown in these figures. In Fig. 2.34, grading I lies above the specified grading. This shows that the actual grading is finer than the specified one. Grading II is coarser than grading I as well as that specified. Grading III shows a flat gradient between 600 μm and 4.75 mm. This indicates a missing particle size (between 4.75 mm and 600 μm). Mixes made with such aggregates tend to segregate. This type of grading is called *gap grading*. Grading IV indicates that a slope between 600 μm and 4.75 mm is in excess (steep). This leads to a harsh mix that is difficult to trowel and finish.

Porosity and absorption

Porosity and water absorption of aggregates influence the workability of fresh concrete, the bond between aggregate and cement paste, and the durability of concrete. Porous aggregates become the reservoir of the free moisture inside the aggregate. Pores may also absorb water added to the mixture and result in partial hydration of cement due to insufficient moisture.

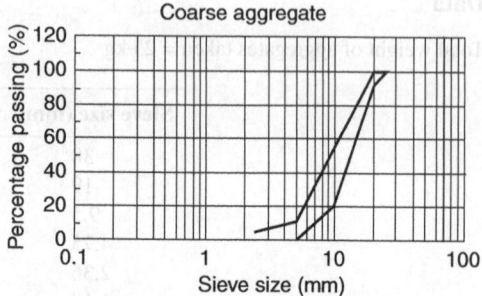

Fig. 2.32 Particle size distribution of coarse aggregates

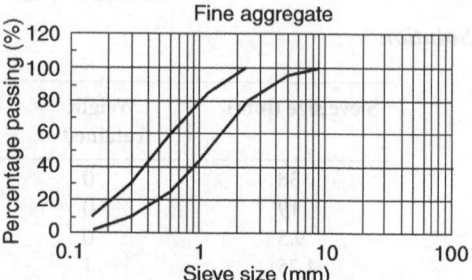

Fig. 2.33 Particle size distribution of fine aggregates

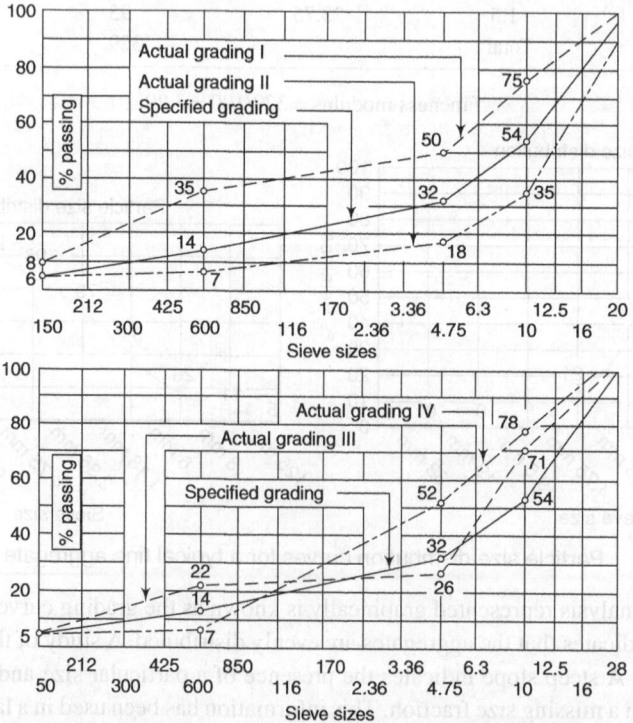

Fig. 2.34 Gradings I–IV relative to the specified grading

IS: 2386(Part III)-1963 recommends the following test for water absorption and specific gravity. The sample is screened by a 10-mm sieve and thoroughly washed. A wide-mouthed glass jar of 1.5 L capacity is filled with distilled water and the sample immersed in it. The entrapped air is removed by gentle agitation. The vessel is topped up with distilled water and weighed (A). The vessel is then emptied and filled with distilled water and weighed (B). The surface of the aggregate is dried with cloth and exposed to the atmosphere away from direct sunlight. The dried aggregate is weighed (C). The aggregate is then placed in an oven and dried at 100°C for 24 h and then weighed (D).

- True specific gravity = $D/[C - (A - B)]$
- Apparent specific gravity = $D/[D - (A - B)]$
- Water absorption (% by weight) = $100 (C - D)/D$

Tests for deleterious constituents

Several impurities can be considered as deleterious constituents, which affect the strength, workability, and long-term performance of concrete. These are grouped as follows:

a. Coatings around aggregates which interfere with their interfacial bond with cement
b. Fine particles which increase the specific surface area and hence affect workability
c. Soft, friable materials which form a weak link in the composite material
d. Materials which react chemically with cement and interfere with its hydration

Iron pyrites, coal, mica, shale, clay, alkalis, seashells, and organic impurities are examples of undesirable impurities in concrete.

IS: 383-1963 has prescribed limits on permissible quantities of impurities. There exists a simple field method for determining the silt content in sand which is illustrated in Example 2.3.

Example 2.3

For the purpose of field determination of silt content in aggregates, the following experiment is conducted. Common salt solution is placed in a 250 mL jar. Sand is then added till the water level reaches 100 mL mark. More salt solution is added till the level reaches 150 mL. After agitation the solution is allowed to stand still for 3 h. The depth of solid layer and the depth of silt layer (Fig. 2.35) are measured.

Data

Total depth of solid layer = 50 mm
Depth of silt layer = 1.5 mm

Solution

Percentage of silt = $(1.5/50) \times 100 = 3\%$
Percentage of silt should be less than 6%.

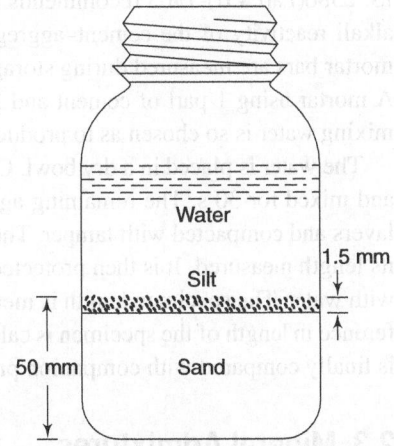

Fig. 2.35 Silt determination in fine aggregates

Test for bulking characteristics of sand

The test for the bulking characteristics of sand is performed by taking two measuring jars and adding moisture to sand in steps of 5%. It is seen that bulking increases gradually with increase in moisture content up to a certain point. After reaching a maximum, the increase in moisture content tends to decrease the total volume.

Finally, the volume tends to reach its lowest value. Figure 2.36 shows a plot of the increase in volume of sand with increase in moisture content. From the graph, the following parameters can be obtained: A, the initial percentage of moisture content, B, the initial percentage of bulking of sand, C, the maximum percentage of bulking, and D, the percentage of moisture content at maximum bulking.

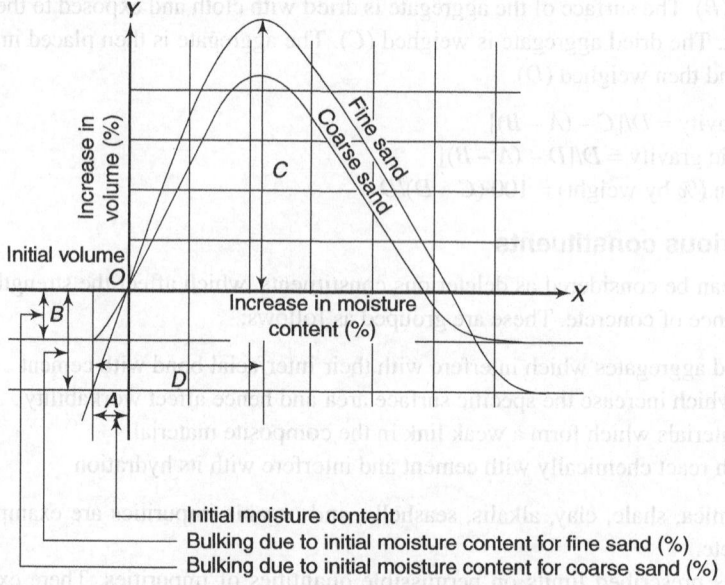

Fig. 2.36 Bulking characteristics of sand

Alkali–aggregate reaction

IS: 2386(Part VII)-1963 recommends the mortar bar method of testing for the determination of the potential alkali reactivity of the cement–aggregate combination. The expansions developed by the combinations in mortar bars are measured during storage under prescribed conditions of testing at different intervals of time. A mortar using 1 part of cement and 2.25 parts of graded aggregate by weight is prepared. The amount of mixing water is so chosen as to produce a flow of 105–120% in a standard flow test.

The water is placed in a dry bowl. Cement is added to it and mixed for 30 s. Half of the aggregate is added and mixed for 30 s. The remaining aggregates are added and mixed for 90 min. The mould is filled in two layers and compacted with tamper. The specimen is left in the mould for 24 h, after which it is removed and its length measured. It is then protected against loss of moisture and placed in a container, but not in contact with water. The specimen length is measured periodically at intervals of 1, 2, 3, 6, 9, and 12 months. The difference in length of the specimen is calculated. The expansion of the prisms observed in the sample aggregate is finally compared with companion prisms made with known aggregates having no alkali reaction.

2.3 Mineral Admixtures

Mineral admixtures are finely divided siliceous materials which are added to mixtures in relatively large quantities. They are classified as follows:

a. Reactive minerals which are either pozzolanic or cementitious or both. The example of a pozzolanic admixture is low-calcium fly ash. The example of a cementitious admixture is ground, granulated blast-furnace slag. High-calcium fly ash is both pozzolanic and cementitious.
b. Inert mineral fillers which have no pozzolanic or cementitious properties. They are generally added as fillers, e.g., silica fume.

Materials belonging to the first group are added as replacements for cement. They react with calcium hydroxide in the hydrated cement paste and form complex compounds which reduce permeability and improve the ultimate strength and durability, besides improving the economy of the mix. Some cements like PPC and PBSC, discussed earlier, contain these mineral admixtures. They are used both on-site and in ready-mixed concrete. IS: 456-2000 recommends the use of the following mineral admixtures: fly ash, silica fume, rice husk ash, metakaolin, and ground, granulated blast-furnace slag.

Table 2.18 summarizes the requirements of fly ash for being used as pozzolana in concrete. Its use in modern concrete has gained acceptance and it is being extensively used for producing high-performance concretes. The properties and uses of mineral admixtures are discussed in later chapters. The role of chemical admixtures is discussed in the next chapter.

Table 2.18 Requirements of fly ash for being used as pozzolana

Characteristic	Requirement of grade I fly ash
Physical requirements	
Fineness: minimum specific surface by Blaine's permeability method (m²/kg)	320
Lime reactivity, average compressive strength (N/mm²), min.	4.5
Minimum compressive strength at 28 days (N/mm²), min.	Not less than 80% of the strength of the corresponding plain cement mortar cube
Drying shrinkage (%), max.	0.15
Soundness expansion autoclave test (%), max.	0.8
Chemical requirements	
Silicon dioxide (SiO_2) + aluminium oxide (Al_2O_3) + iron oxide (Fe_2O_3) (% by mass), min.	70.0
Silicon dioxide (SiO_2) (% by mass), min.	35.0
Magnesium oxide (MgO) (% by mass), max.	5.0
Total sulphur in the form of sulphur trioxide (SO_3) (% by mass), max.	2.75
Available alkalis as sodium oxide (Na_2O) (% by mass), max.	1.5
Loss on ignition (% by mass), max.	12.0

2.4 Water

Water is the next most important ingredient after cement for making concrete. It is also the least expensive. Careless use of water can lead to poor quality concrete. Therefore, a detailed study of the quantity and quality of water required for making good quality concrete is essential. The purpose of water in concrete is threefold.

1. It distributes the cement evenly.
2. It reacts with cement chemically and produces calcium silicate hydrate (C–S–H) gel.
3. It provides for workability, i.e., it lubricates the mix.

There are two sources of water in concrete: (a) intentionally added water, known as mix water, and (b) aggregate moisture, which can either add water to the mixture or absorb water from the mixture.

2.4.1 Quantity of Water

The reaction products of hydration consist of 20–30% calcium hydroxide, which is crystalline. These crystalline products are surrounded by calcium silicates and aluminates, which are in colloidal form. The combination of $Ca(OH_2)$ and C–S–H is called *gel*. When this gel hardens, it imparts adequate strength to the cement paste and thus to concrete. Figure 2.37 shows a schematic representation of cement gel hydration with water–cement ratios 0.2, 0.3, and 0.5. Note that a small amount of water is needed to hydrate cement. Additional water is required to lubricate the mix. Too much water can lead to the creation of capillary pores.

The quantity of water is the most important parameter and is controlled by the water–cement ratio. As the quantity of water in a mix goes up, the following effects are noticed: the strength decreases, durability decreases, workability increases, cohesion decreases, and the economy may increase at the expense of quality and reliability.

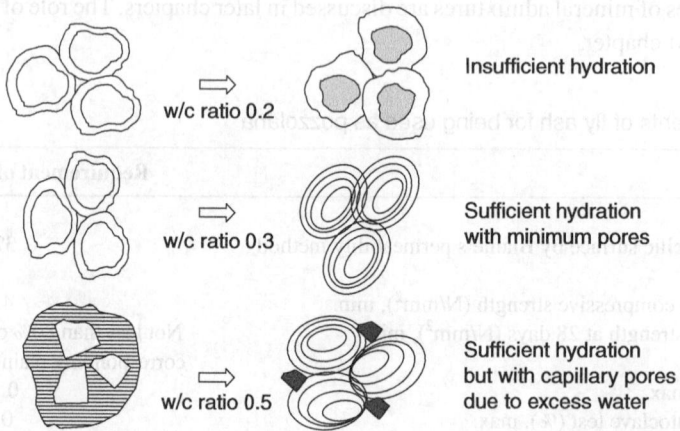

Fig. 2.37 Schematic representation of insufficient, sufficient, and excess water for hydration

2.4.2 Water–Cement Ratio

Duff Abrams in 1919 established the law now well known as Abrams' law: *With given concrete materials and conditions of testing, the quantity of mixing water used alone determines the strength of concrete as long as the mix is workable*.

For a set of concrete ingredients and conditions of curing and testing, the strength of hardened concrete is inversely proportional to water–cement ratio. In other words, the strength decreases as the water–cement ratio increases. Abrams established this law which determines the strength of concrete based on water–cement ratio, provided the mix is workable. The limitation of this law is that the effect of entrapped air on strength is ignored. The above law can be numerically expressed as:

$$S = \frac{A}{B^x} \text{ psi}$$

where x = Water–cement ratio by volume

S = Strength at 28 days in psi

A and B are constants depending on the type of cement used

A varies in the range of 7000 to 14,000

B varies in the range of 6 to 12

The above equation in SI units can be expressed as

$$S = \frac{A \times 6.895 \times 10^{-3} \text{ MPa}}{B^x}$$

A typical relationship as per this law is shown in Fig. 2.38. The insert gives the influence of workability on strength.

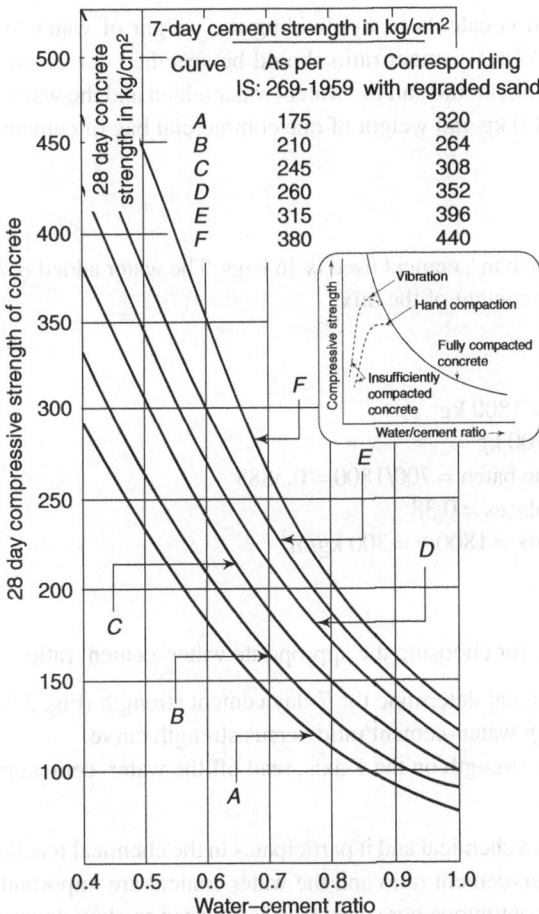

Fig. 2.38 Water–cement ratio versus strength

Example 2.4

In the expression for Abrams' law $S = \dfrac{A}{B^x}(6.895 \times 10^{-3})$ MPa, the constants A and B for a type of cement used are 8000 and 10, respectively. Estimate the strength of concrete if the water–cement ratio adopted by volume is 0.40.

Solution

Strength, $S = (8000/10^{0.40}) \times 6.895 \times 10^{-3} = 21.63$ MPa

---

This law, though developed by Abrams in the USA, has now been adopted by many countries, which have developed their own charts using the cement available in their countries. The current trend is to use the water–cement ratio as the ratio of water to cement by weight rather than by volume for a given mix.

$$\text{w/c} = \frac{\text{water (kg)}}{\text{cement (kg)}}$$

Thus, the water–cement ratio is calculated by dividing the weight of water by the weight of cement in a specified batch of concrete. Water–cement ratio should be specified correct to the nearest 2 decimals. For example, as w/c = 0.52, it should include all the water (water added and the water available in the aggregates). Weight of 1 litre of water is 1.0 kg and weight of one commercial bag of cement is 50 kg.

Example 2.5

In a concrete batch of size 6 m^3, cement used is 36 bags. The water added is 700 L. Determine the water–cement ratio and cement content of the mix.

Solution

Cement used = 36 × 50 = 1800 kg
Water added = 700 L = 700 kg
Water–cement ratio of the batch = 700/1800 = 0.3888
Corrected to 2 decimal places = 0.38
Cement content of the mix = 1800/6 = 300 kg/m^3

The following steps are taken for choosing the appropriate water–cement ratio:

1. Verify the type of cement and determine the 7-day cement strength (Fig. 2.38).
2. Choose the corresponding water–cement ratio versus strength curve.
3. For the required concrete strength on the *Y* axis, read off the water–cement ratio from the corresponding curve.

Note that water is also used as a chemical and it participates in the chemical reaction of hydration. It is imperative to note that both the water–cement ratio and the water content are important parameters influencing the properties of concrete. A discontinuous pore structure is required to slow down and stop the infiltration and movement of moisture/water.

Example 2.6

Find the water–cement ratio for concrete of strength M25 if the cement to be used has a 7-day strength of 245 kg/cm^2.

Solution

Referring to Fig. 2.38, from curve *C*, the water–cement ratio to be adopted is 0.52.

We should understand that increasing water content in a batch dilutes the effect of cement paste, increasing its volume, reducing its density, and lowering its strength. In addition to decreasing the strength, concrete placed with a higher water–cement ratio will have higher permeability, decreased resistance to weathering, increased cracking, and decreased durability because of increased volume changes during alternate wetting and drying.

Concrete in the fresh state placed with a high water–cement ratio suffers from the risk of segregation and increased tendency to bleed. Tables 2.19(a) and 2.19(b) show the recommended maximum values of water–cement ratios for different exposure conditions needed in practice based on IS: 456-2000 and ACI 318-05, respectively.

Table 2.19(a) Recommended values of maximum water–cement ratios for different environmental conditions

S. No.	Environmental exposure condition	Water–cement ratio (max)	Minimum grade of concrete
1	Mild (Protected against weather)	0.55	M20
2	Moderate (Exposed to condensation and rain)	0.50	M25
3	Severe (Exposed to severe rain and alternate wetting and drying)	0.45	M30
4	Very severe (Exposed to sea water spray, corrosive fumes, or freezing conditions)	0.45	M35
5	Extreme (Tidal zone members or exposed to aggressive chemicals)	0.40	M40

Table 2.19(b) Recommended values of maximum water–cement ratios for different exposure conditions

S. No.	Exposure condition	Maximum water–cement ratio	Minimum strength
1	Low permeability	0.50	35
2	Freeze–thaw conditions	0.45	40
3	Chlorides, deicing chemicals, salt water spray or sea water requiring corrosion protection	0.40	45

2.4.3 Sources of Water Used for Concreting

a. Batch water: This is the main source of water for concrete mixing. Water discharged into the mixer based on calculated water–cement ratio from municipal/corporation supply, or water pumped from underground sources like well or borewell or recycled water piped and let in through flow meters, etc., constitutes the main batch water.

b. Ice: To control the heat of hydration during hot weather, ice may be used while mixing water. By the time mixing is over, ice should be completely dissolved.

c. Transit addition of water in the truck: Planned addition of water during transit by truck operator, provided water–cement ratio and other conditions are met, is sometimes done to regulate the workability.

d. Free moisture on aggregates: This depends on the saturation and condition of the aggregates and this must be considered as additional water for water–cement ratio consideration. Tests may be required to assess the quantity of this water.

e. Recycled water coming from wash water of mixer and other waste water.

Generally potable water fit for drinking which has no taste or odour can be used as mixing batch water for concrete applications. However, some water that is fit for drinking may not be fit for concreting. The quality of water required for concreting is discussed later in Section 2.4.5.

2.4.4 Indirect Water—Bulking Quantity Test

Volume increase of sand is caused by moisture present in sand. Finer the sand, the bulkier it is. The sand we use normally contains moisture, which is the indirect water that goes into the mix when concrete is made by batching the ingredients. This indirect moisture is to be accounted for ensuring quality concrete. If we do not take this moisture into account, it will make the water–cement ratio more and the concrete weak. Bulking depends on the percentage of moisture content and the particle size of sand. Both moisture present in the sand and increase in volume can be assessed by conducting a bulking quantity test.

The extent of bulking can be estimated by a simple bulking quantity field test which is described below:

1. Fill a sample of moist fine aggregate (sand) into a measuring cylinder. Note down the level, say h_1.
2. Pour water into a measuring cylinder and completely cover the sand with water and shake it. Since the volume of the saturated sand is the same as that of the dry sand, the saturated sand completely offsets the bulking effect. Note down the level of sand, say h_2.
3. Subtract the final level h_2 from the initial level h_1 (i.e., h_1-h_2), which shows the bulking of sand under test.
4. Calculate the percentage of bulking using the formula given below:

$$\text{Percentage of bulking} = \left[\frac{(h_1 - h_2)}{h_2}\right] \times 100$$

From Fig. 2.36 one can determine the quantity of moisture present in the aggregates which should be found and deducted from the total water to be added based on water–cement ratio calculations.

2.4.5 Water Quality

The quality of water used must be checked for ensuring good quality concrete. Water used for mixing and curing should be free from oil, acid and alkali, salts, and organic material. It should be of potable quality and generally purer than that required for drinking. (The human body has a better threshold for salts and sugar than concrete!) Whenever there is uncertainty of quality, water should be tested before use.

2.4.6 Impurities in Water

Impurities in water can be of three types: chemical, physical, or biological.

Chlorides   Chlorides can cause corrosion of the steel reinforcement and can accelerate setting. The water used may be contaminated with chlorides because of it being sea water, the presence of admixtures and de-icing salts, or deliberate chlorination for disinfection.

Sulphates   Sulphates can lead to the reformation of ettringite as well as reduction of long-term strength levels.

Organic matter   The effects of organic matter on concrete are varied. If algae are present in water, it should not be used because it will affect setting and strength development.

Sugar   Sugar retards the setting time. Too much sugar may 'kill' the concrete (i.e., it will not set).

Waste water   It is best not to use waste water. Alternatively it can be used after proper testing and treatment.

Table 2.20 gives the typical limits of impurities in water as per IS: 456-2000.

Table 2.20 Impurities in mixing and curing water

Solids	Permissible limits, max. (mg/L)
Organic	200
Inorganic	3000
Sulphates (as SO_3)	400
Chlorides (as Cl)	
For plain concrete	2000
For reinforced concrete	500
Suspended matter	2000

2.4.7 Aggregate Moisture

As pointed out earlier, aggregates have moisture within their voids and on their surfaces. This moisture must be accounted for when proportioning concrete mix. All moisture contributes to the weight of aggregates. Surface moisture contributes to the amount of mixing water. Table 2.21 gives the approximate quantities of surface water carried by aggregates.

Table 2.21 Surface water carried by aggregates

Aggregate	Approximate quantity of surface water	
	Percentage by mass	L/m^2
Very wet sand	7.5	120
Moderately wet sand	5.0	80
Moist sand	2.5	40
Moist gravel or crushed rock	1.25–2.5	20–40

2.4.8 Water Quality Tests

Regular testing of water is important to ensure its suitability for concreting. Certain chemicals which are sometimes found in water sources may cause long-term durability problems that take years to develop.

Determination of acids and alkalis The hardness of water can be determined by titration. The sample of water is titrated against acidic and basic solutions of prescribed normality with prescribed indicators as follows. To neutralize a 200-mL sample of water using phenolphthalein as an indicator, not more than 2 mL of 0.1N NaOH is required. The end of the titration is reached when the colourless phenolphthalein turns pink. If the end point is not reached, it means that the water contains more acid than is permissible and is hence unfit for use. To neutralize a 200-mL sample of water using methyl orange as an indicator, not more than 10 mL of 0.1N HCl is required. The end point is reached when the red methyl orange turns yellow. If the end point is not reached, it means that the water contains more alkalis than is permissible and is hence unfit for use.

Determination of total solids The total quantity of solids in water is determined by evaporating and drying a measured sample in an oven at 105°C for 1 h. The residue after evaporation and drying is weighed and expressed in mg/L. The average 28-day compressive strength of at least three 15-cm cubes prepared with the water proposed to be used should not be less than 90% of the average strength of three similar concrete cubes prepared with distilled water. The initial setting time of cement with the proposed water to be used should not be less than 30 min and should not differ by more than 3 min from the initial setting time of the same cement with distilled water. The pH value of hydrogen ion concentration is a measure of the acidity or alkalinity of water. If it is more than 7, then the solution is alkaline. The pH value of water should not be less than 6. IS: 456 2000 states that water found satisfactory for mixing concrete can also be used for curing concrete. However, it should not produce any objectionable stain on the surface of concrete. The presence of tannic acid or iron compounds in curing water is objectionable, as these react chemically with water.

Exercises

Review Questions

1. What are the raw materials used for the production of cement?
2. What is clinker and how is it produced?
3. Describe the Bogue chemical compound composition of Portland cement.

4. Describe the hydration reaction of important Bogue compounds indicating the products of hydration.
5. Describe the role played by gypsum in the hydration reaction of cement.
6. List the various types of cement indicating their use for different applications.
7. What are the important chemical tests conducted on cement to determine its quality?
8. What is soundness of cement and how is it tested?
9. How are aggregates classified?
10. Define fineness modulus. Give the practical range of fineness modulus values for coarse and fine aggregates.
11. Write explanatory notes on (a) uniform grading, (b) gap grading, and (c) continuous grading.
12. What are the effects of the shape and texture of aggregates on the strength and workability of concrete?
13. What are the different moisture states in which aggregates exist?
14. What is the effect of the maximum size of aggregate on concrete strength?
15. List the various tests conducted on coarse aggregate indicating the property being tested.
16. Describe the importance of sieve analysis in determining particle size distribution.
17. Distinguish between the true and apparent specific gravity of aggregates.
18. Describe the importance of the quality of water used for concreting.
19. How does increasing the quantity of water influence the properties of fresh and hardened concrete?
20. Describe a test to determine the initial moisture content of fine aggregates in a construction site.

Numerical Problems

1. For the type of cement used, the constants A and B in Abrams' law are 10,000 and 7, respectively. Find the strength of concrete if the water–cement ratio used is 0.45. What is the percentage by which the strength will increase if the water–cement ratio is lowered to 0.35?
2. Find the specific surface of cement with the following data obtained from Blaine's apparatus:
   Fineness value of reference cement = 225,000 cm²/g
   Density of reference cement = 3.10
   Density of given sample of cement = 3.15
   Time taken for air to go through the reference sample = 4 h
   Time taken for the air to go through the given sample = 4 h 30 min
   Test temperature = 24°C
3. A batch of 5 m³ of concrete was made with the following:
   Cement used = 32 bags
   Water used = 490 L
   Find the water–cement ratio adopted and the cement content of the mix.
4. Find the fineness modulus of the coarse aggregates for the following data.
   The weights retained on the following set of sieves are:

S. no.	Sieve size	Weight retained (g)
1	38.1 mm	36
2	19.0 mm	405
3	9.5 mm	378
4	4.75 mm	81
5	2.36 mm	0
6	1.18 mm	0
7	600 μm	0
8	300 μm	0
9	150 μm	0

The weight of aggregates taken is 900 g.

5. Find the fineness modulus of the aggregates for the following data:

S. no.	Sieve size	Weight retained (g)
1	38.1 mm	27
2	19.0 mm	261
3	9.5 mm	243
4	4.75 mm	63
5	2.36 mm	54
6	1.18 mm	81
7	600 μm	90
8	300 μm	45
9	150 μm	18

The weight of aggregates taken is 900 g.

Comment on the results obtained.

Multiple Choice Questions

1. The main raw material for the manufacture of cement is
   a. Bauxite
   b. Limestone
   c. Fly ash
   d. Gypsum

2. The weight of a standard cement bag is
   a. 25 kg
   b. 50 kg
   c. 100 kg
   d. 75 kg

3. Di-calcium silicate is a
   a. Bogue compound
   b. Gel
   c. Crystal
   d. Inert filler

4. The major oxide of cement is
   a. Calcium oxide (CaO)
   b. Silicon dioxide (SiO_2)
   c. Aluminium oxide (Al_2O_3)
   d. Magnesium oxide (MgO)

5. If the amount of gypsum is more, then
   a. Concrete will set fast
   b. Concrete will have a flash set
   c. Concrete will set normally
   d. Concrete will take long time to set

6. Hydration reaction of cement
   a. Will absorb heat
   b. Will evolve heat
   c. Will require pressure for the reaction
   d. Is neither exothermic nor endothermic

7. The grade of cement indicates
   a. Rate at which cement reacts
   b. Permeability class of cement
   c. Durability class of cement
   d. 7-day strength of mortar cube produced by cement

8. Use of blended cement is preferred because of
   a. Improved cohesion of the paste
   b. Lower permeability
   c. Improved durability
   d. All of the above

9. Le-Chatelier's apparatus is used to test cement for
   a. Setting time
   b. Consistency
   c. Specific gravity
   d. Soundness

10. Initial setting time for cement should be
    a. >120 min
    b. >60 min
    c. >30 min
    d. >600 min

11. Toughness of aggregates is a measure of
    a. Durability
    b. Strength
    c. Resistance to impact
    d. Resistance to creep

12. Uniform grading of coarse aggregates
    a. Produces compactness
    b. Produces large volume of voids
    c. Has no effect on voids
    d. Produces better finish

13. In the test to assess the cleanliness of fine aggregates it was found that the depth of solids was 50 mm and depth of silt was 4 mm. The percentage of silt is
    a. 4
    b. 2
    c. 8
    d. 6

14. Bulking of sand is
    a. Increase in volume of sand because of addition of water
    b. Decrease in volume of sand because of addition of water
    c. Saturating sand with water
    d. Drying sand in an oven to drive out all moisture

15. Abrams' law relates water–cement ratio with
    a. Density
    b. Strength
    c. Durability
    d. Stiffness of mix

16. Strength of concrete S is related to the water–cement ratio x by which of the following relationships in which A and B are constants
    a. $S = A/B^x$
    b. $S = A \cdot B^x$
    c. $S = A + B^x$
    d. $S = A - B^x$

17. For RCC structures the permissible limit for chloride in mg/L is
    a. 2000
    b. 1500
    c. 500
    d. 250

18. The larger the finesse modulus of sand
    a. The finer the aggregate
    b. The coarser the aggregate
    c. The graded the aggregate
    d. The gap graded the aggregate

19. Aggregate grading represents
    a. Particle size distribution of aggregates
    b. Quality of aggregates
    c. Impurities in the aggregates
    d. Voids in the aggregates

20. Alkali–aggregate reaction can be controlled by
    a. Using sound aggregates
    b. Controlling permeability of concrete
    c. Using low water–cement ratio
    d. All of the above

Answers to Multiple Choice Questions

1. b	6. b	11. c	16. a
2. b	7. d	12. b	17. c
3. a	8. d	13. c	18. b
4. a	9. d	14. a	19. a
5. d	10. c	15. b	20. d

CHAPTER 3

Chemical and Mineral Admixtures

Admixtures are materials used to modify the properties of fresh and hardened concrete. They are classified as chemical and mineral admixtures. Chemical admixtures are used in the construction industry for building strong, durable, and waterproof structures. They are mainly used for the following four purposes:

1. Some chemicals are mixed with concrete ingredients and spread throughout the body of concrete to favourably modify the moulding and setting properties of the concrete mix. Such chemicals are generally known as *chemical admixtures*. Admixtures are added to concrete to give it certain desirable properties in either the fresh or the hardened state. Most admixtures result in modifying more than one intended property.
2. Some chemicals are applied on the surface of concrete to protect it during or after its setting.
3. Some chemicals are applied on the surface of moulds used to form concrete to effect easy mould-releasing operation.
4. Some chemicals are applied to bond or repair broken or chipped concrete.

Mineral admixtures are siliceous materials which have fine particle size. They are added to concrete either as a filler or to improve certain desired properties such as durability. Mineral admixtures are classified as either pozzolanic or cementitious. They are either natural materials like metakaolin or by-products of industries such as fly ash.

Table 3.1 lists various types of chemical admixtures used in concrete and the purpose for which they are used. These durability enhancing mineral admixtures or supplementary cementing materials are used for imparting long service life to reinforced concrete structures. In this chapter, we will discuss the role of these chemical and mineral admixtures in concreting practices and products.

3.1 Accelerators

Accelerators reduce the setting time and generally produce early removal of forms and early setting of concrete repair and patch work. They are helpful in cold weather concreting. The most common accelerator for plain concrete work is calcium chloride ($CaCl_2$). Its quantity in the concrete mix is limited to 1–2% by weight of cement. The presence of $CaCl_2$ can cause corrosion of embedded steel. It reduces resistance against sulphate attack and may cause an alkali–aggregate reaction. For prestressed and reinforced concrete, $CaCl_2$ cannot be used. Instead calcium formate is preferred as an accelerating admixture for such concretes. The properties and types of accelerating admixtures are shown in Fig. 3.1.

Table 3.1 Types of chemical admixtures

Type of admixture	Effect produced on concrete
Accelerators	Accelerating and water-reducing
Retarders (water-reducing)	Retarding and water-reducing
Plasticizers (air-entraining)	Air-entraining, water-reducing, and increasing workability
Waterproofers	Damp proofing
	Permeability reducing
Pumping superplasticizers	Flowing concrete, increasing workability, inducing flowability
Pigments	Colouring
Viscosity modifiers	
Miscellaneous	
Coating	
Sealants	
Grouts	
Fungicides	
Resins and polymeric substances	
Minerals	
Gas-forming	
Air de-entraining	
Alkali–aggregate reducing	

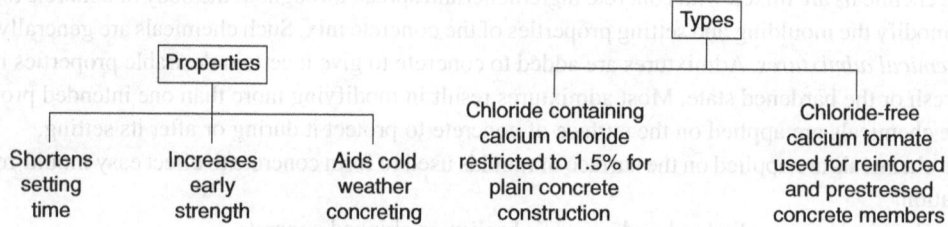

Fig. 3.1 Properties and types of accelerating admixtures

3.2 Retarders

Retarders increase the setting time of the concrete mix and reduce the water–cement ratio. Usually up to 10% water reduction can be achieved. A wide range of water-reducing and set-retarding admixtures are used in ready mixed concrete. Usually, these chemicals are derived from lignosulphonic acids and their salts, hydroxylated carboxylic acid and their salts, and sulphonated melamine or naphthalene formaldehyde. They have a detergent-like property, and work on the principle that water-reducing agents migrate to the surface of water, as shown in Fig. 3.2. This increases the surface activity and hence imparts a soapy property to the mix and delays setting.

Both accelerators and retarders affect cement hydration, setting time, and strength gain. The effect of retarders and accelerators on hydration of cement is shown in Fig. 3.3.

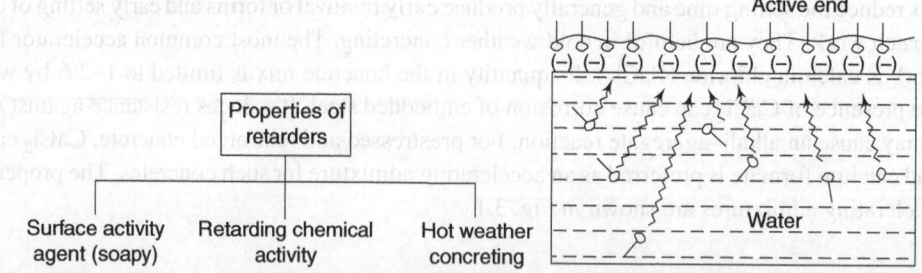

Fig. 3.2 Migration of water-reducing agents to the surface of water, thus producing soapy property

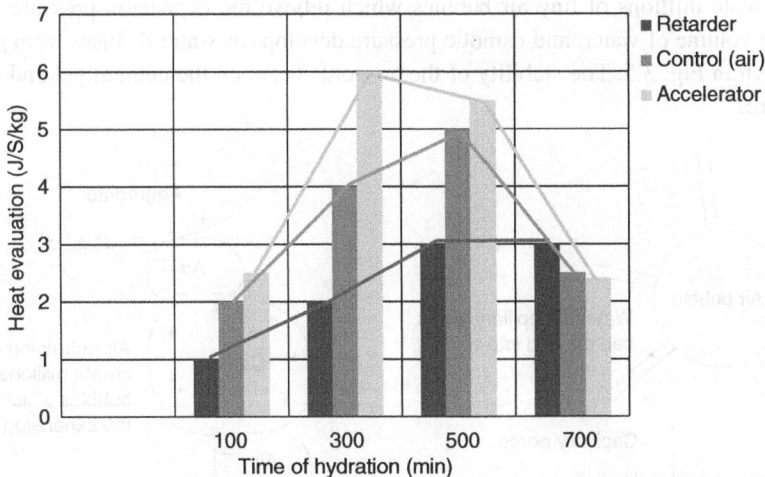

Fig. 3.3 Effect of accelerators and retarders on hydration of cement

3.3 Plasticizers

A plasticizer is defined as an admixture added to wet concrete mix to impart adequate workability properties. As shown in Fig. 3.4, plasticizers can be of the following three types:

1. Finely divided minerals
2. Air-entraining agents
3. Synthetic derivatives

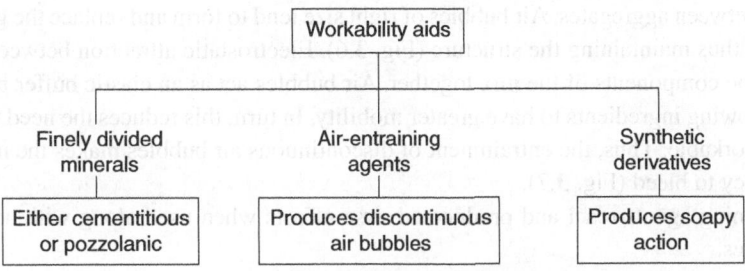

Fig. 3.4 Types of plasticizers

Their properties are described in the following sections.

3.3.1 Finely Divided Minerals

They are either cementitious or pozzolanic. Natural cements, hydraulic lime, and slag cements belong to the former category, whereas fly ash and heat-treated clays belong to the latter. They are used as workability aids, and help in reducing bleeding by way of adding finer particles to the mix.

3.3.2 Air-entraining Agents

These admixtures help in protecting concrete subjected to repeated freeze–thaw cycles. Concrete with entrained air has higher workability and cohesiveness. Air-entraining agents not only reduce segregation and bleeding, but also ensure durability against frost.

Air-entraining agents are derived from synthetic detergents, salts of sulphonated lignin, fatty acids, organic salts of sulphonated hydrocarbons, or salts of wood resins.

These agents create millions of tiny air bubbles which relieve the expansion pressure. They result in a 9% increase in the volume of water, and osmotic pressure develops as water diffuses from gel pores into the capillaries as shown in Fig. 3.5. The stability of the air voids between the cement gel and aggregate is also shown in this figure.

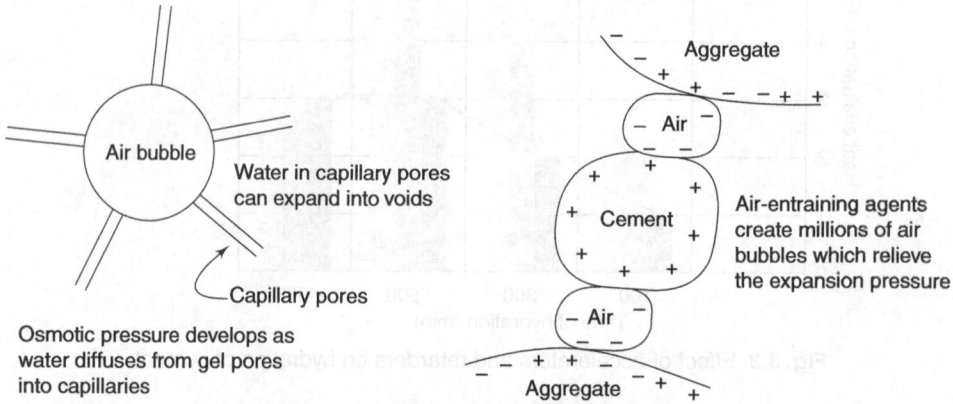

Fig. 3.5 Activity of air-entraining agents

Cohesion problems occur in a concrete mix when the internal structure is insufficient to hold the mix together. Internal structure stability is due to both interparticle elastic-static attraction and because of physical support between particles. If some fractions in the grading of the sand are missing, then full physical support of the particles by each other may not occur. There can also be insufficient attraction between particles. These lead to cohesion problems and encourage segregation.

1% to 2% additional air over what is normally present will improve cohesion. Air will increase the volume allowing support between aggregates. Air bubbles of right size tend to form and replace the gaps in the grading of fine aggregates, thus maintaining the structure (Fig. 3.6). Electrostatic attraction between the bubbles and aggregates holds the components of the mix together. Air bubbles act as an elastic buffer between aggregate and cement gel allowing ingredients to have greater mobility. In turn, this reduces the need to add more water to make the mix workable. Thus, the entrainment of discontinuous air bubbles makes the mix more fluid but reduces the tendency to bleed (Fig. 3.7).

Some air-entraining agents react and produce adverse effects when used along with accelerating or set-retarding admixtures.

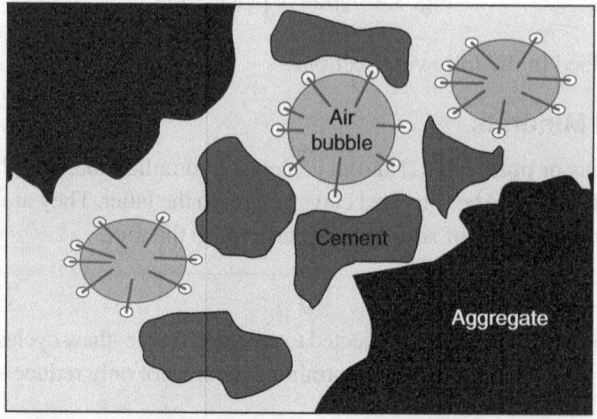

Fig. 3.6 Role of air bubbles in concrete

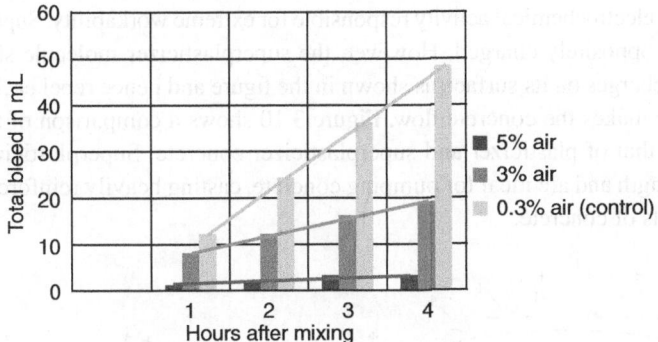

Fig. 3.7 Effect of air entrainment on bleed

3.3.3 Synthetic Derivatives

Synthetic derivatives are surface-active agents which introduce soapy action into the mix. These are primarily added to increase workability. The best example of a synthetic derivative is benzene sulphonate. Chemically, they comprise the same chemicals as found in retarders, and hence they also generally retard the setting time.

The effects of surface-active agents on the mix are shown in Fig. 3.8. As can be seen, the water entrapped in the floc is released, workability is improved, and entrapped air is released.

These derivatives may react differently with different types of cement. Hence, a careful study of the type of cement is required before choosing a particular synthetic derivative.

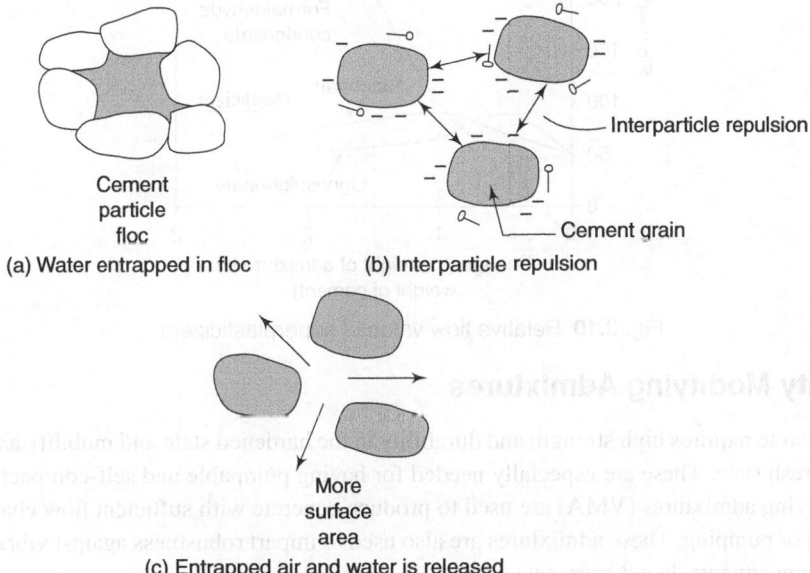

Fig. 3.8 Surface-active agent

3.4 Superplasticizers

Superplasticizers produce extreme workability and thus flowing concrete. They achieve reduction in the water content without loss of workability. Their use generally leads to an overall reduction in the cost.

Figure 3.9 shows the electrochemical activity responsible for extreme workability. Superplasticizer molecules and cement grains are oppositely charged. However, the superplasticizer molecule sits in between cement grains, exhibiting like charges on its surface as shown in the figure and hence repel each other. This increases the mobility and hence makes the concrete flow. Figure 3.10 shows a comparison of the flow characteristic of fresh concrete with that of plasticizer and superplasticizer concrete. Superplasticizers enable savings in cement for a given strength and are ideal for pumping concrete, casting heavily reinforced concrete members, and the precast elements of concrete.

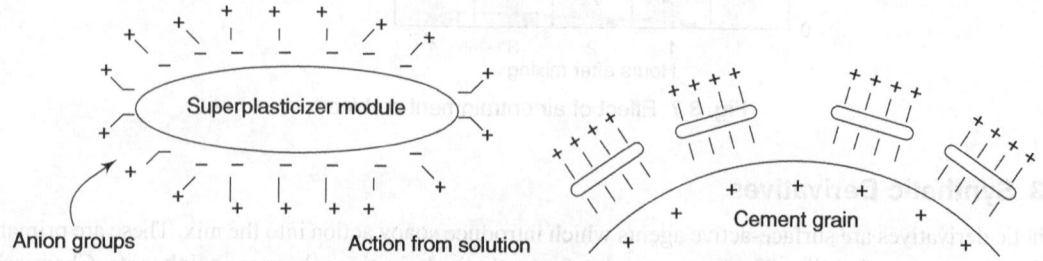

Fig. 3.9 Electrochemical activity of superplasticizers

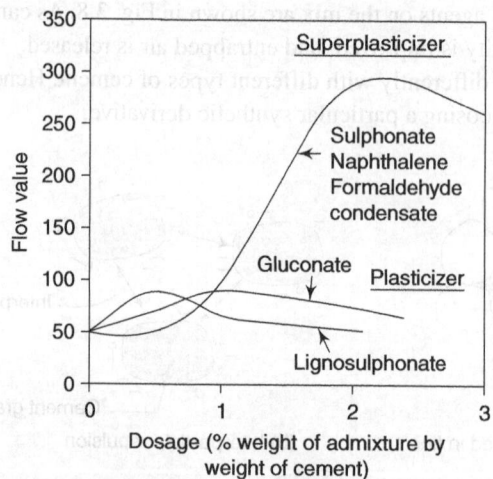

Fig. 3.10 Relative flow value of superplasticizers

3.5 Viscosity Modifying Admixtures

Present-day concrete requires high strength and durability in the hardened state and mobility and flow characteristics in the fresh state. These are especially needed for having pumpable and self-compacting properties. Viscosity modifying admixtures (VMA) are used to produce concrete with sufficient flow characteristics for self-compacting or pumping. These admixtures are also used to impart robustness against vibration or impact so that the mix constituents do not segregate.

Viscosity modifying admixtures are also referred to as stabilizers, viscosity enhancing admixtures (VEA), and water retaining admixtures (WRA). They are used to modify the rheological properties of cement paste. The rheology of concrete in the fresh state can be physically quantified based on yield point and plastic viscosity. Yield point is the force required to move the fresh concrete from rest. It is dependent on workability, and thus is based on slump. Plastic viscosity is the resistance of fresh concrete to flow under external pressure. Viscosity is developed by friction between various components of the mix. The speed of flow can be related to viscosity and is assessed by the time taken to flow during slump.

Figure 3.11(a) and (b) depict the slow-speed, high-viscosity and fast-speed, low-viscosity flow determined by slump, respectively. Addition of VMA increases the plastic viscosity. Admixtures are used along with VMA to have a desired yield point.

These VMAs are based on poly-carboxylate technology. They are available either as powder or in the form of dispersed fluid for easy dosing. The dosage varies based on application, and the normal range of dosage is 0.1–1.5% by weight of cement.

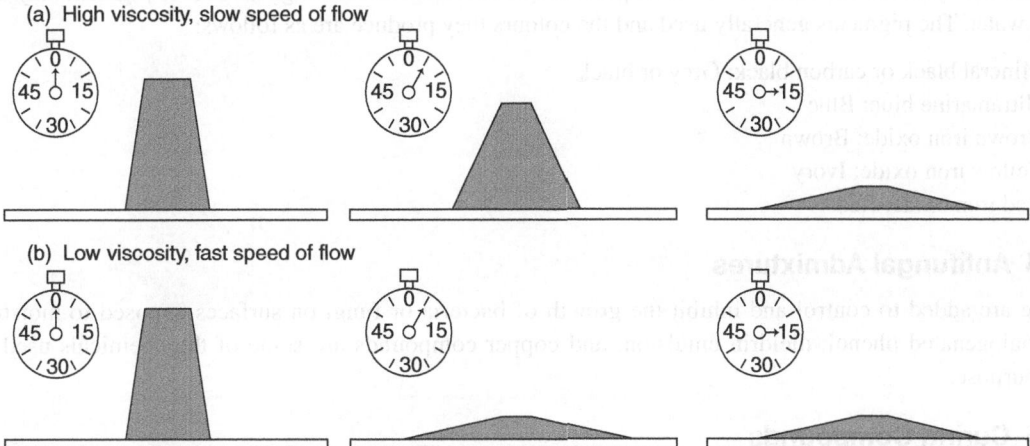

Fig. 3.11 Plastic viscosity (*Source*: http://www.efnarc.org/pdf/Guidelines%20for%20VMA%20 (document%20180).pdf. Reproduced with permission from EFNARC in association with EFCA)

3.6 Waterproofers

These chemicals are added to concrete or mortar at the time of preparation to make the structure waterproof. They react with the lime contained in cement to form inorganic salts which block the pores and capillaries, thereby reducing moisture penetration.

Waterproofers may be obtained in powder or liquid form and consist of pore-filling or water-repellent materials. The chief materials in the pore filling class are alkaline silicates, notably silicates of soda, aluminium, zinc sulphates, and chlorides of aluminium and calcium. These are all chemically active and accelerate the setting time of concrete but may encourage corrosion of embedded steel. The chemically inactive pore-filling materials are finely ground chalk, fuller's earth, and talc. They act as workability aids and result in an increase in density. They are used in conjunction with calcium soaps. Materials in the water-repellent class are soda and potash soaps, to which lime, alkali, or silicates are sometimes added. Chemically inactive materials in the water-repellent class are calcium soaps, resins, vegetable oils, fats, waxes, bitumen, and coal tar.

3.7 Miscellaneous Admixtures

Several other chemicals are also used in concrete for achieving specific results. The important ones are as follows.

3.7.1 Gas-forming and Expansive Chemicals

These chemicals are used to produce lightweight concrete as well as to cause expansion on application, such as in grouts for anchor bolts. They are of the non-shrinking type. The principal chemicals used are hydrogen peroxide, metallic aluminium, and activated carbon. Sometimes bentonite clays and natural gum are also used. These find application in deep well sealing.

3.7.2 Corrosion-inhibiting Chemicals

These chemicals resist corrosion of reinforcement. Generally, alkalinity of concrete is an adequate protection for steel. However, in adverse environments, sodium benzoate, calcium lignosulphonate, and sodium nitrate have shown good results.

3.7.3 Pigments

Natural and synthetic materials are used for producing various colours. A large amount of pigment may need extra water. The pigments generally used and the colours they produce are as follows:

- Mineral black or carbon black: Grey or black
- Ultramarine blue: Blue
- Brown iron oxide: Brown
- Yellow iron oxide: Ivory
- Red iron oxide: Red

3.7.4 Antifungal Admixtures

These are added to control and inhibit the growth of bacteria or fungi on surfaces exposed to moisture. Polyhalogenated phenol, dieldrin emulsion, and copper compounds are some of the chemicals used for this purpose.

3.7.5 Curing Compounds

These admixtures are either wax-based or resin-based. When coated on freshly laid concrete, they form a temporary film over the damp surface. This prevents or slows down water evaporation and allows sufficient moisture retention in concrete for curing. This chemical curing method is used when ordinary wetting at regular intervals is not possible, especially in hot and dry climates, or in instances like slipform shuttering work for chimneys, cooling tower constructions, etc.

3.7.6 Sealants

These are used to seal joints. They are formulated from synthetic rubbers or polysulphides. The choice of a sealant depends on the location of the joint, its movement capability, and the function the sealant is expected to perform.

3.7.7 Floorings

These are usually toppings based on metallic or non-metallic aggregates, which are mixed with cement and placed over a freshly laid concrete base. These compounds, in a highly viscous liquid form, mixed with recommended fillers at the site, are formulated using resins and polymers such as epoxy, acrylic, polyurethane, or polysulphide. They give excellent wear resistance.

3.7.8 Floor Coatings

Floor coatings based on silicon fluorides harden concrete, making it more resistant to impact, abrasion, and chemical attack. On the other hand, polymeric floor coatings, such as epoxy, acrylic, urethane, and polysulphide, are used for making the floor surface dust-free, non-magnetic, and chemical- or fungus-resistant.

3.7.9 Guniting Aids

These help in the early setting of the concrete mix and create a good bond between new and old surfaces, in the process reducing loss on rebound during the guniting application. Their effectiveness depends on the efficiency of controlled water addition.

3.8 Mineral Admixtures

IS: 456-2000 permits the use of mineral admixtures for modifying the properties of concrete. The use of industrial waste products such as fly ash, which has both pozzolanic and cementitious properties, leads to cost and energy saving. The following minerals can be added to concrete either as admixtures or as a part of cement.

3.8.1 Fly Ash

Fly ash is solid fine-grained material resulting from combustion of pulverized coal or lignite in thermal plants. It is estimated that 110 million tonnes of fly ash is being generated in India annually. The land required for disposal of this waste is huge. Based on the collection method, fly ash is classified as follows:

1. Bottom ash: Collected from bottom of boilers
2. Pond ash: Collected from electrostatic precipitators but stored in ponds mixed with water in a wet condition
3. Dry fly ash: Separated from end precipitator and stored in a dry state in silos

Fly ash can be grouped under either high-calcium (15–35%) or low-calcium (< 10%) type depending on its CaO content. The unburnt carbon content in fly ash should be less than 5%.

Electron microscope photographs show that the particles in fly ash occur as solid spheres of silica glass, the particle size varying from <1 μm to 100 μm. The majority of particles are of 20 μm size. 10% to 15% of the particles should have a size more than 45 μm. The surface area of the particles is in the range of 300–400 m^2/kg. The use of fly ash is recommended from the point of view of the durability it imparts to concrete, in addition to economy and energy-saving considerations.

The chemical and physical properties of fly ash as recommended by IS: 3812-2013 are shown in Tables 3.2 and 3.3.

Table 3.2 Chemical properties of fly ash

Characteristic	Max/Min	
	Siliceous fly ash	Calcareous fly ash
Silicon dioxide (SiO_2) plus aluminium oxide (Al_2O_3) plus iron oxide (Fe_2O_3) in percent by mass (*min*)	70	50
Silicon dioxide (SiO_2) in percent by mass (*min*)	35	25
Magnesium oxide (MgO) in percent by mass (*max*)	5.0	5.0
Total sulphur as sulphur trioxide (SO_3) in percent by mass (*max*)	5.0	5.0
Available alkalies as equivalent sodium oxide (Na_2O) in percent by mass (*max*)	1.5	1.5
Total chlorides in percent by mass (*max*)	0.05	0.05
Loss on ignition in percent by mass (*max*)	7.0	7.0

Table 3.3 Physical properties of fly ash

Characteristic	Max/Min
Fineness—Specific surface in m^2/kg by Blaine's permeability method (*min*)	200
Particles retained on 45 micron IS sieve (wet sieving) in percent (*max*)	50
Soundness by autoclave test—Expansion of specimen in percent (*max*)	0.8

3.8.2 Silica Fume

Silica fume, a very fine non-crystalline SiO_2, is a by-product of ferro-silicon industry. It is made at a temperature of approximately 2000°C. Its size is about 0.1 μm (20–25 m^2/g). Compared to cement, the particle size of silica fume is 2 orders finer. It acts as an excellent pore-filling material. It can be used in proportion of 5–10% of the cement content in a mix. Important properties of a typical silica fume sample tested are given in Table 3.4.

Table 3.4 Physical and chemical properties of silica fume

Physical properties			Chemical properties		
	Test value	Max		Test value	Max/Min
Moisture content	0.80%	3%	SiO_2	93%	Min 85%
Fineness >45 micron	1.60%	10%	SO_3	0.60%	Max 1%
Specific gravity	2.18		Loss on ignition	3.50%	Max 6%
Pozzolanic activity @ 7 days	27	31			
Soundness	<2%	0.05%			

3.8.3 Rice Husk Ash

It is a waste product of rice mills. Each tonne of paddy can generate 0.2 tonne of husk. This is a highly reactive pozzolanic admixture. It is produced by controlled combustion of husk-retaining silica in the non-crystalline form, with cellular structure. The fineness is of the order of 50–250 m^2/g (0.2 to 2 μm).

3.8.4 Metakaolin

Metakaolin is obtained by calcination of pure or refined clay at temperatures of 650–850°C and by grinding it subsequently to achieve a fineness of 700–900 m^2/kg. It is a highly reactive pozzolana.

3.8.5 Ground Granulated Blast Furnace Slag

The chemical composition of ground granulated blast furnace slag (GGBS) indicates the presence of silica glass which contains calcium, magnesium, and aluminium. It has a cementitious character. Before use, it is necessary to dry and grind it to a particle size of less than 45 μm (~500 m^2/kg). It is seen that a particle of size less than 10 μm contributes to the early strength. A particle size of 10–45 μm contributes to the later age strength. Sizes greater than 45 μm do not generally hydrate. Important properties of a typical sample of GGBS are shown in Table 3.5.

The use of mineral admixtures is beneficial in reducing thermal cracking, enhancement of strength, impermeability, pore refinement, and durability against chemical attack.

Table 3.5 Physical and chemical properties of GGBS

Physical properties			Chemical properties		
	Code requirement	Typical value		Code requirement	Typical value
Density (kg/m^3)		2880	Loss on ignition (%)	<3%	1.1%
Fineness (sp. surface) (cm^2/g)	Not less than 2750	4490	Sulphate content (SO_3) (%)	<2.5%	0.20%
Standard consistency (%)		29	Chloride ion content (%)	<0.1%	0.02%
Initial setting time (min)	220		Magnesia content (%)	<14%	10.30%
Soundness (mm)	0		Alkali content (%)	Not specified	0.70%
Flexural strength (28 days) (MPa)	<9	8	Sulphur content (%)	<2.0%	0.90%
Comp. strength (28 days) (MPa)	>30	40	Moisture content (%)	<1.0%	0.10%

Exercises

Review Questions

1. Classify the various concrete chemicals based on their use.
2. Why are chloride-based accelerators not used in prestressed concrete structures?
3. Distinguish between plasticizers and superplasticizers.
4. List the different types of workability aids.
5. How does a surface-active agent increase workability?
6. Why do superplasticizers perform better than surface-active agents?
7. What method will you adopt to cure concrete in areas of water shortage?
8. What are the different chemicals used to obtain the desired colours on a concrete surface?
9. How are mineral admixtures classified?
10. Distinguish between pozzolanic and/or cementitious admixtures.

Multiple Choice Questions

1. Admixtures are used in concrete
   a. To modify the fresh properties of concrete only
   b. To modify the hardened properties of concrete only
   c. To modify either fresh and/or hardened properties of concrete
   d. To externally apply rendering only

2. Calcium chloride is not preferred as an accelerating admixture because
   a. It is expensive
   b. It does not accelerate the hydration properly
   c. It causes corrosion of embedded steel
   d. It adversely affects strength of concrete

3. Retarders are used in ready-mixed concrete
   a. To increase the setting time
   b. To reduce the setting time
   c. To reduce the water–cement ratio
   d. To increase the setting time and at the same time effect reduction in water–cement ratio

4. Plasticizers are used in concrete
   a. To impart desirable workability
   b. To increase the strength
   c. To increase the resistance to water penetration
   d. To arrest early cracking

5. Air-entraining agents are used in concrete as admixtures
   a. To relieve the bursting pressure during wetting and drying
   b. To introduce discontinuous air bubbles
   c. To aid workability
   d. To effect all the above

6. The mineral admixture which is not derived from the by-product of an industry but made from natural source is
   a. Fly ash
   b. Metakaolin
   c. Silica fume
   d. GGBS

7. The mineral admixture which has the finest particle size is
   a. Fly ash
   b. Blast furnace slag
   c. Silica fume
   d. Rice husk ash

8. Waste materials like fly ash when added to concrete as admixture
   a. Increase permeability
   b. Increase cracking
   c. Increase bleeding
   d. Increase durability

9. The particle size of silica fume is about
   a. Same size as cement
   b. Half the particle size of cement
   c. One hundredth the particle size of cement
   d. One tenth the particle size of cement

10. Metakaolin is
    a. An industrial waste product
    b. A filler for blocking pores
    c. A highly reactive pozzolana
    d. An inert material

Answers to Multiple Choice Questions

1. c
2. c
3. d

4. a
5. d
6. b

7. c
8. d
9. c

10. c

CHAPTER 4

Properties of Fresh Concrete

This chapter concentrates principally on the properties of concrete in its fresh state. Though the fresh state is transient, its condition seriously affects the behavioural properties of the final product. Poor compaction and improper curing can lead to porous concrete with low strength and high permeability.

4.1 Workability

ASTM: C125-93 defines workability as the 'property determining the effort required to manipulate a freshly mixed quantity of concrete with minimum loss of homogeneity'. In essence, workability is defined as the ease of placement with resistance to segregation. It is indeed difficult to define workability without in some way involving the type of construction (such as congestion of reinforcement). The definition of workability given in ACI: 116R-90 is 'the property of freshly mixed concrete or mortar which determines the ease and homogeneity with which it can be mixed, placed, consolidated and finished'. Therefore, the workability of concrete is associated with terms such as flowability, mobility, stability, resistance to segregation, and pumpability. Each of these terms has a specific meaning but all of them are related to workability.

Workability is necessary to compact concrete to the maximum possible density. We can express the strength of partially compacted concrete compared to the strength of fully compacted concrete in terms of the strength ratio:

$$R_{\text{strength}} = \frac{f_{\text{ck}} \text{ of partially compacted concrete}}{f_{\text{ck}} \text{ of fully compacted concrete}}$$

Similarly, we can express the density of partially compacted concrete compared to the density of fully compacted concrete in terms of the density ratio:

$$R_{\text{density}} = \frac{\gamma \text{ of partially compacted concrete}}{\gamma \text{ of fully compacted concrete}}$$

As the density reduces, voids generally increase. Figure 4.1 shows that a reduction in the density of concrete from a density ratio of 0.8 to a density ratio of 0.6 reduces the strength ratio from 0.65 to 0.45. This indirectly shows that strength reduces with increase in voids. The presence of 5% voids may reduce the strength by as

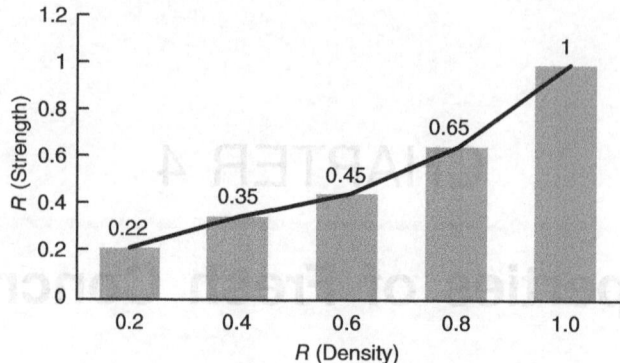

Fig. 4.1 Relationship between R(Strength) and R(Density)

much as 30%. This is in agreement with Feret's expression, relating the strength to the sum of the volume of water and air in hardened cement paste, as shown below:

$$f_c = K \left(\frac{c}{c+w+a} \right)^2$$

where f_c is the strength of concrete, c is the volume of cement, w is the volume of water, a is the volume of air, and K is a constant.

Voids are caused by entrapped air as well as the removal of excess water by evaporation (due to a high water–cement ratio). They are also caused by water trapped under large aggregates and the congestion of reinforcements. Air voids (bubbles) are accidentally introduced air and are more easily expelled from wet mixes than from dry ones. At optimum moisture content, one can achieve maximum density and through it arrive at maximum strength.

4.1.1 Factors Affecting Workability

Several factors affect the workability of concrete. They relate to the properties of aggregate, cement, water, and entrapped air. In Chapter 2 we enumerated the factors affecting workability, especially with respect to aggregates. The properties of aggregates that affect workability include their maximum size, grading type, shape, and texture. For a given workability, there exists a unique aggregate–cement ratio which gives the maximum strength. The effect of low values of the aggregate–cement ratio is to increase workability. For aggregate–cement ratios of the order of 2, the workability is adequate with a slump of about 100 mm.

Among the factors affecting workability, water content expressed in L/m³ is most important. IS: 456-2000 specifies the maximum water content permitted for different environments with respect to durability. In Chapter 2, we saw how the quality of aggregates affects the requirement of water in a mix. For normal concrete without air entrainment, and for the same workability, the water content decreases with increase in the maximum size of aggregates. It is also obvious that the water requirement decreases with decrease in slump. Hence, to produce durable concrete, one must use minimum water, which will happen if we select the maximum possible size of aggregates, with a good compacting effort requiring smaller slump values. When air is entrained, the water content can be reduced in accordance with the relation shown in Fig. 4.2.

Though the maximum size of aggregate is preferred from the point of view of the minimum cement content, the workability is governed by the maximum size, grading, shape, and texture, as discussed in Chapter 2. Therefore, the ratio of coarse aggregates to fine aggregates, which is governed by grading, and water–cement ratio have to be considered together to produce a good cohesive mix. For a given workability, there is a unique value of the coarse–fine aggregate ratio which requires the lowest water content.

Thus, there are three factors which affect the workability of a mix: (1) water–cement ratio, (2) aggregate–cement ratio, and (3) water content. Though the first two factors are independent, the third is interdependent with the other two factors. For example, if the water content is kept constant while reducing the aggregate–cement ratio, the water–cement ratio decreases, though the workability is not unduly reduced.

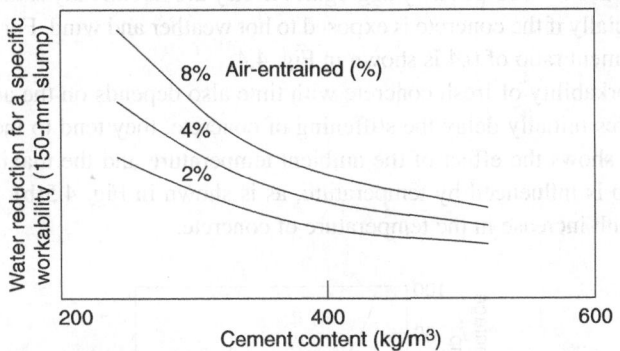

Fig. 4.2 Effect of air entrainment on water content

Effect of environmental conditions

The workability of the concrete mix is also affected by the temperature of concrete and, therefore, by the ambient temperature. On a hot day it becomes necessary to increase the water content of the concrete mix in order to maintain the desired workability. The amount of mixing water required to bring about certain changes in the workability also increases with temperature.

Effect of time

The time required for depositing concrete and compacting is referred to as placing time. Fresh concrete loses workability with the increase in placing time mainly because of the loss of moisture due to evaporation. A part of the mixing water is absorbed by the aggregate or lost by evaporation in the presence of sun and wind, and a part of it is utilized in the chemical reaction of hydration of cement. The loss of workability varies with the type of cement, the concrete mix proportions, the initial workability, and the temperature of concrete. On an average, a 125-mm slump concrete may slump only by about 50 mm during the first one hour. The workability in terms of the compacting factor decreases by about 0.10 during the period of one hour from the time of mixing. The decrease in workability with time after mixing may be more pronounced in concrete with plasticizers. The effect of placing time on workability is illustrated in Fig. 4.3.

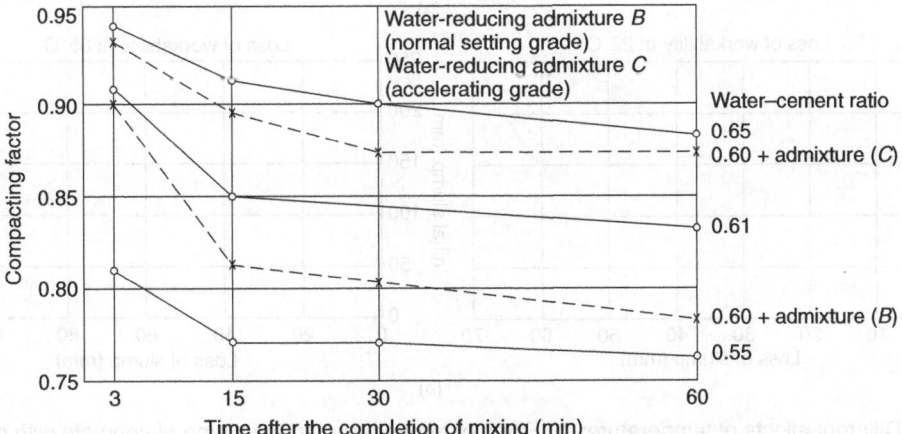

Fig. 4.3 Effect of placing time on workability

Effect of temperature

Freshly mixed concrete stiffens with the passage of time. This is different from the hardening of the mix. As time passes, water is lost due to absorption by aggregates if they are not already saturated. Some water is lost due to evaporation, especially if the concrete is exposed to hot weather and wind. For example, the slump loss with time for a water–cement ratio of 0.4 is shown in Fig. 4.4.

The change in the workability of fresh concrete with time also depends on the admixtures used. Though water-reducing admixtures initially delay the stiffening of concrete, they tend to increase the loss of slump with time. Figure 4.5(a) shows the effect of the ambient temperature and the maximum size of aggregates on slump. Loss of slump is influenced by temperature, as is shown in Fig. 4.5(b). Note that loss of slump significantly increases with increase in the temperature of concrete.

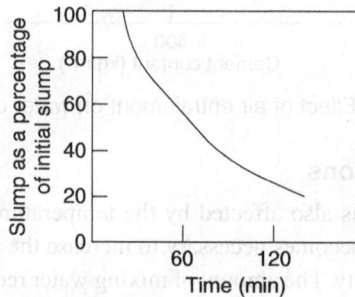

Fig. 4.4 Loss of slump with time elapsed since mixing (*Source*: Meyer and Perenchio 1980. Reproduced with permission from Portland Cement Association)

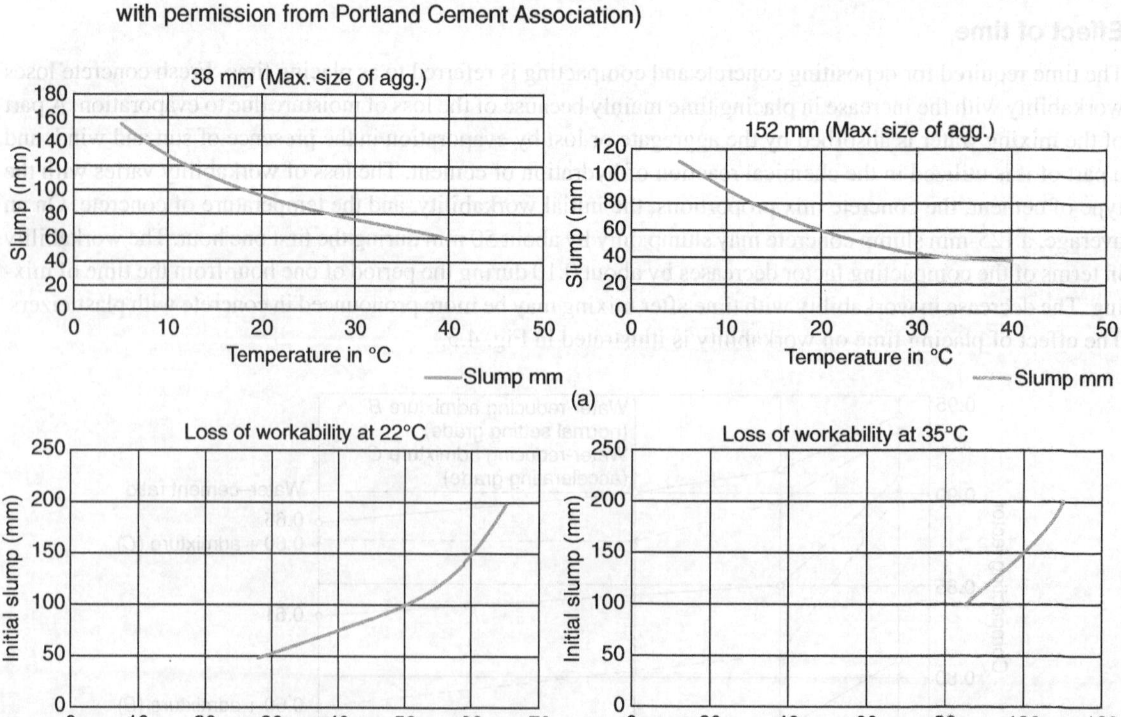

Fig. 4.5 Different effects of temperature: (a) Effect of temperature on the slump of concrete with different max. size of aggregates (b) Effect of temperature on the loss of workability

4.1.2 Measurement of Workability

There is no definite test, in the strict sense, to measure the effort required to manipulate fresh concrete. However, the mobility or flowability of concrete is measured in quantitative terms using the tests described below.

Slump test

This test is used as a simple site test. The mould for this test is the 300-mm-high frustum of a cone. It is placed upright as shown in Fig. 4.6 and filled with three layers of concrete. Each layer is tamped 25 times with a 16-mm-diameter steel rod with a rounded nose. The top is struck off level and the mould is held firmly against the slab base. After this, the mould is lifted gently, causing the unsupported cone to slump, as shown in Fig. 4.7. The decrease in height of the cone is designated the *slump value* and is measured correct to 5 mm. The test has to be carried out carefully and the sides of the cone can be moistened to avoid friction.

The recommended slump values are shown in Table 4.1. The slump test has certain limitations. Slump occurs under self-weight and hence does not reflect ease of compaction. In construction work, concrete is vibrated dynamically. The slump test does not reflect this. One can probably say that slump reflects the 'yielding' of concrete under its own weight. In spite of the aforementioned drawback, the slump test is a very useful site test and can be used to monitor batch-to-batch feeding into a mixer machine and the resulting variation in workability. The three different ways in which slump occurs are shown in Fig. 4.8. An even (true) slump shows a cohesive mix, a shear slump indicates a harsh mix, and a floppy (collapse) slump indicates a flowing mix. Zero slump indicates stiff mix.

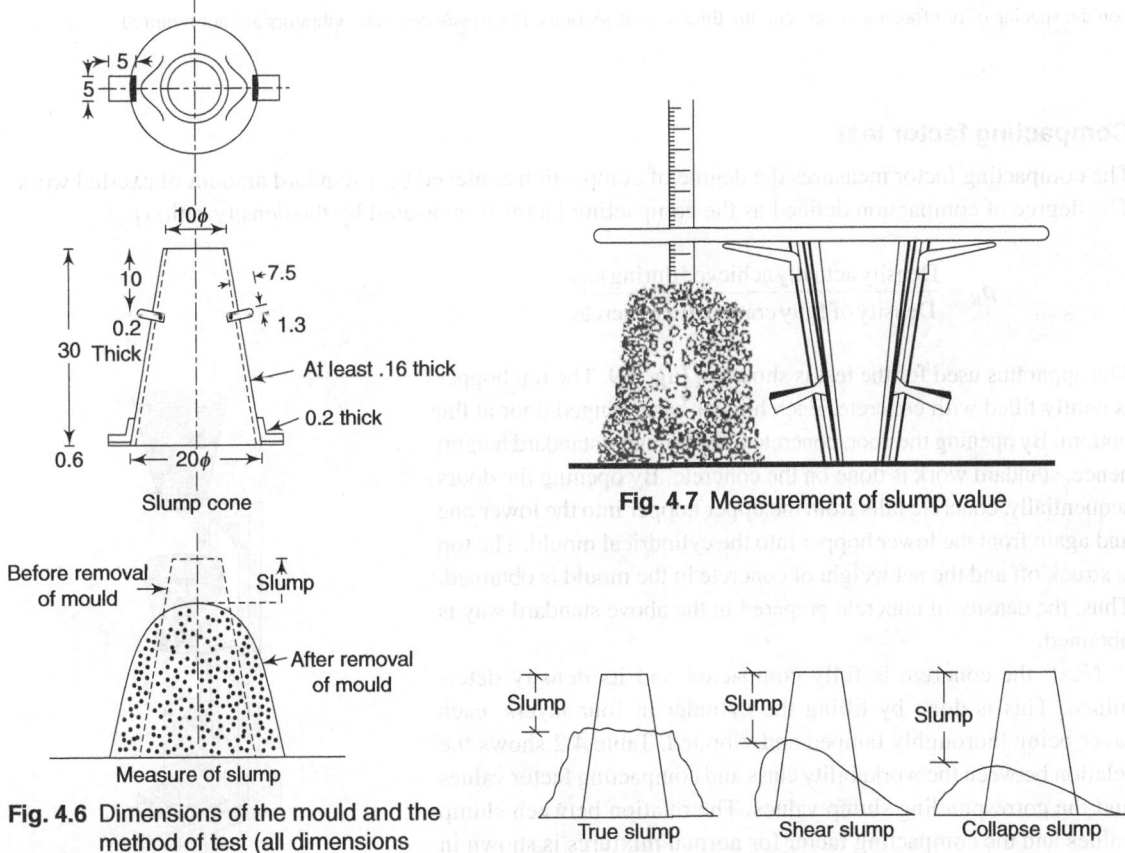

Fig. 4.6 Dimensions of the mould and the method of test (all dimensions in cm)

Fig. 4.7 Measurement of slump value

True slump   Shear slump   Collapse slump

Fig. 4.8 Types of slump

Table 4.1 Recommended slump values

Placing conditions*	Degree of workability	Slump (mm)
Blinding concrete; shallow sections; pavements using pavers	Very low	Too small to measure
Mass concrete; lightly reinforced sections in slabs, beams, walls, columns; floors; hand placed pavements; canal lining; strip footings	Low	25–75
Heavily reinforced sections in slabs, beams, walls, columns; slipform work; pumped concrete	Medium	50–100
		75–100
Trench fill; in situ piling	High	100–150
Tremie concrete	Very high	Too large to measure

*For most placing conditions, internal vibrators (needle vibrators) are suitable. The diameter of the needle is determined based on the spacing of reinforcement bars and the thickness of sections. For *tremie* concrete, vibrators are not required.

Compacting factor test

The compacting factor measures the degree of compaction achieved by a standard amount of exerted work. The degree of compaction defined as the compacting factor is measured by the density ratio (ρ_R):

$$\rho_R = \frac{\text{Density actually achieved during test}}{\text{Density of fully compacted concrete}}$$

The apparatus used for the test is shown in Fig. 4.9. The top hopper is gently filled with concrete. Each hopper has a hinged door at the bottom. By opening the door, concrete falls through a standard height; hence, standard work is done on the concrete. By opening the doors sequentially, concrete falls from the upper hopper into the lower one and again from the lower hopper into the cylindrical mould. The top is struck off and the net weight of concrete in the mould is obtained. Thus, the density of concrete prepared in the above standard way is obtained.

Next, the concrete is fully compacted and its density determined. This is done by filling the cylinder in four layers, each layer being thoroughly tamped and vibrated. Table 4.2 shows the relation between the workability class and compacting factor values and the corresponding slump values. The relation between slump values and the compacting factor for normal mixtures is shown in Fig. 4.10.

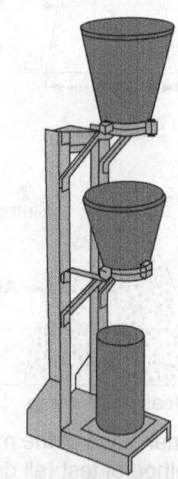

Fig. 4.9 Compacting factor apparatus

Table 4.2 Compacting factor values and the corresponding slump values

Workability class	Compacting factor	Corresponding slump (mm)
Very low	0.78	Not specified
Low	0.85	25–75
Medium	0.92	50–100
		75–100
High	0.96	100–150
Very high	—	Not specified

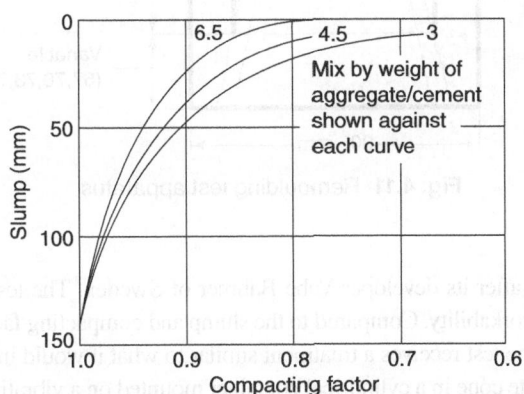

Fig. 4.10 Relation between slump values and compacting factor

Flow test

The flow test indicates the susceptibility of concrete to segregation, and gives a satisfactory performance for concretes of consistencies for which the slump test can also be used. The test consists of moulding a fresh concrete cone on the platform of the flow table and giving it 25 jolts, each having a drop of magnitude 12.5 mm. The spread of the concrete is measured in terms of the increase in diameter of the concrete heap and is expressed as the measure of the flow or consistency of the concrete. The flow is expressed by measuring the spread of concrete with respect to the original diameter.

Remoulding test

The remoulding test was developed by Powers (1932). The test set-up is shown in Fig. 4.11. A standard slump cone is placed inside a cylinder of diameter 305 mm and height 203 mm. The cylinder is mounted rigidly on the flow table described earlier, adjusted to give a 6.3-mm drop. There is a 210-mm-diameter, 127-mm-high inner ring inside the main cylinder. The distance between the bottom of the ring and the bottom of the cone can be adjusted from 67 mm to 76 mm.

The slump cone is filled as per the procedure described earlier. It is then removed and a disc-shaped rider weighing 1.95 kg is placed on top. The table is now jolted at the rate of one jolt per second until the bottom of the rider is 81 mm above the base plate. At this point, it can be observed that the shape of the concrete has changed from that of the frustum of a cone to that of a cylinder. The effort required to remould a cone into a cylinder is expressed in terms of the number of jolts.

The remoulding test is a valuable laboratory test, especially for assessing the remoulding characteristics of the mix, which are directly related to workability.

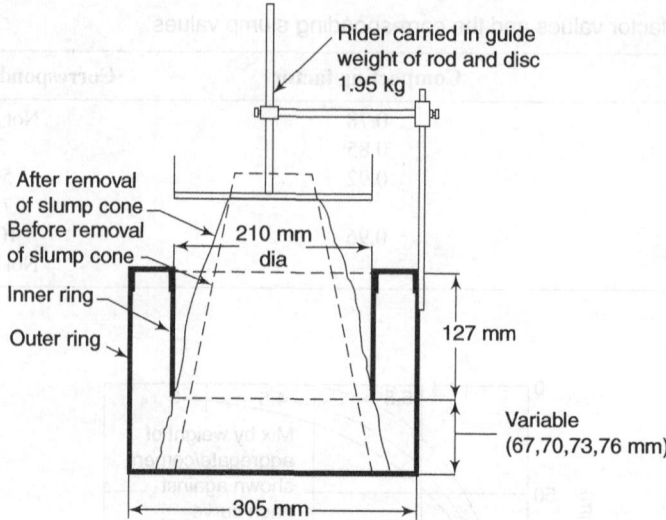

Fig. 4.11 Remoulding test apparatus

Vebe test

The Vebe test has been named after its developer Vebe Bahrner of Sweden. The test is suitable for stiff concrete mixes having low or very low workability. Compared to the slump and compacting factor tests, the Vebe test has the advantage that the concrete in the test receives a treatment similar to what it would in actual practice. The test consists of moulding a fresh concrete cone in a cylindrical container mounted on a vibrating table (Fig. 4.12). When the concrete cone is subjected to vibration using a standard vibrator, it starts to occupy the cylindrical container by way of getting remoulded. The remoulding is considered complete when the concrete surface becomes horizontal. The time (in seconds) required for the complete remoulding is considered as a measure of workability and is expressed as the number of Vebe seconds. The end point of the test, when the concrete surface becomes horizontal, has to be ascertained visually. This introduces a source of error, which is more pronounced in concrete mixes of high workability. Consequently, a low Vebe time is recorded. For concrete with slump in excess of 125 mm, the remoulding is so quick that the time cannot be measured accurately. The test is therefore not suitable for concrete with higher workabilities, i.e., slumps of 100 mm or more. An approximate relation between slump and Vebe time is given in Fig. 4.13.

The test gives satisfactory results for mixes with the Vebe time varying between 3 and 30 s. Both the Vebe and remoulding tests determine the time required for compaction and hence are related to the work done.

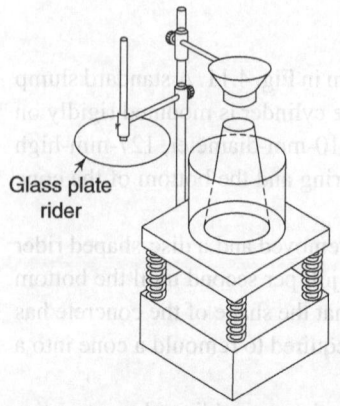

Fig. 4.12 Vebe apparatus

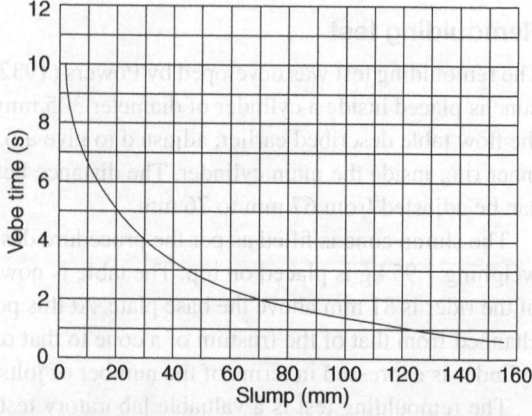

Fig. 4.13 Relation between slump and Vebe time

Ball penetration test

This is a simple field test. The Kelly ball is hemispherical in shape, weighing 13.6 kg and having a diameter of 152 mm, as shown in Fig. 4.14. The apparatus was developed by J.W. Kelly. When the ball is placed on concrete with thickness not less than 200 mm, it sinks due to its own weight and indicates the consistency of the mix for control purposes. It is difficult to directly correlate slump values with Kelly ball penetration values. However, for a specific mix, the relation is shown in Fig. 4.15.

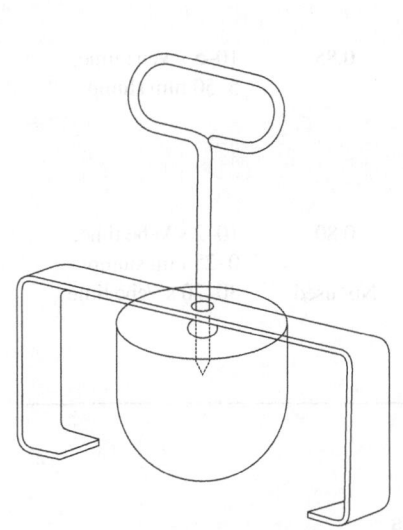

Fig. 4.14 The Kelly ball

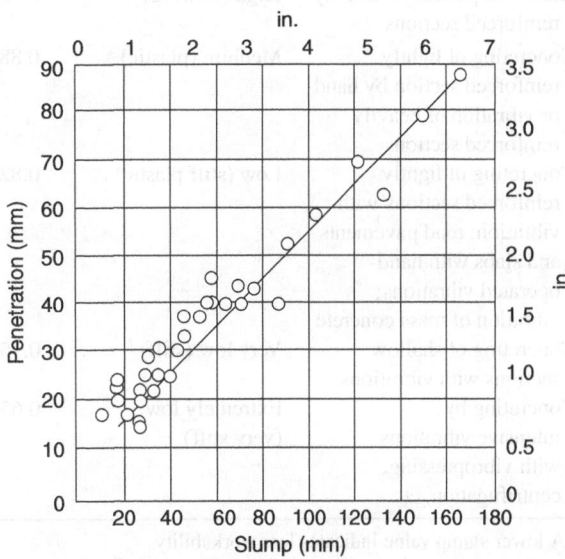

Fig. 4.15 Relation between Kelly ball penetration values and slump values (*Source*: Kelly and Polivka 1955. Reproduced with permission from American Concrete Institute)

As each of the above tests measures only a particular aspect of workability, there is no rigid correlation between the workabilities of concrete as measured by different test methods. In the absence of definite correlations between the different measures of workability under different conditions, it has been recommended that, for a given concrete, the appropriate test should just be used rather than correlating it with other tests. However, Table 4.3 gives the range of expected values of workability measured by different test methods for comparable concretes.

The major drawbacks of tests on workability are as follows:

a. The tests are quite arbitrary and empirical as far as the measurement of workability is concerned because each of these tests is a single-point test measuring a single quantity, which at times may classify two concretes as identical, which actually may behave quite differently on the job.
b. The results from these tests depend on minor variations in the techniques used to carry out the tests, i.e., they are operator-sensitive.
c. None of the tests is capable of dealing with concrete having a large range of workabilities. For example, the slump test is quite incapable of differentiating between concretes of very high workabilities (collapse slump) and those of very low workabilities (zero slump). Table 4.4 lists the test methods appropriate for concretes of various workabilities.

However, with all their limitations, empirical tests have made concrete mix design possible. There is a strong need for the development of a new rational test based on the rheological properties of concrete.

Table 4.3 Suggested values of workability of fresh concrete for different placing conditions

Placing condition	Degree of workability	Value of workability			
		Compacting factor, maximum size of aggregate			Vebe time slump for 20-mm aggregate
		10 mm	20 mm	40 mm	
Hand compaction of heavily reinforced sections	High (flowing)	0.95	0.95	0.95	125–150 mm slump*
Concreting of lightly reinforced section by hand or vibration of heavily reinforced sections	Medium (plastic)	0.88	0.90	0.92	5–2 s Vebe time,** 25–75 mm slump
Concreting of lightly reinforced sections with vibration; road pavements and slabs with hand-operated vibrations; vibration of mass concrete	Low (stiff plastic)	0.82	0.84	0.85	10–5 s Vebe time, 5–50 mm slump
Concreting of shallow sections with vibrations	Very low (stiff)	0.75	0.78	0.80	10–2 s Vebe time, 0–25 mm slump
Concreting by intensive vibrations with vibropressing, centrifugation, etc.	Extremely low (very stiff)	0.65	0.68	Not used	30–20 s Vebe time

*A lower slump value indicates less workability.
**A lower Vebe time indicates more workability.

Table 4.4 Test methods appropriate for concretes of various workabilities

Workability	Method
Very low	Vebe test
Low	Vebe test, compacting factor test
Medium	Compacting factor test, slump test
High	Compacting factor test, slump test, flow test
Very high	Flow test

4.2 Compactability

Compactability is the ease with which concrete can be compacted. In other words, it is the amount of internal work required to produce complete compaction. This property of the mix depends upon many factors such as the amount, fineness, and chemical composition of cement, the amount of water, the grading and shape of fine aggregates, and the presence or absence of entrained air and admixtures. The addition of certain admixtures and entrained air greatly increases compactability without increasing the slump.

4.3 Mobility

Mobility is the ease with which concrete can flow into the formwork around steel, forming adequate bonds, i.e., the ability to be moulded. This measure depends on the type of formwork, the arrangement of steel in the mould, the method adopted in moulding, the time lag between mixing and pouring, and the nature of work.

The main concrete mixers are of either tilting or pan type. Tilting-type mixers are preferable for mixes having low workabilities and large-sized aggregates. Non-tilting type of mixers are generally stationary. These are used in mixing plants for the production of large quantities of concrete. Their discharge rates are slow and particularly suitable for the stiff and cohesive mixes used for precast concrete. Another type of mixer used mainly for the construction of roads is the dual-drum mixer which operates in series. In a dual-drum mixer, a part of the mixing is carried out in one drum and then discharged to the second drum for final mixing and discharging. Special types of mixers are also available for shotcreting as well as prepacked aggregate concretes.

Generally, the value of average strength of concrete goes up with mixing time (Fig. 4.16). The coefficient of variation of the strength of concrete produced reduces with increase in mixing time (Fig. 4.17). Depending on the capacity of the mixer, an optimum mixing time is recommended.

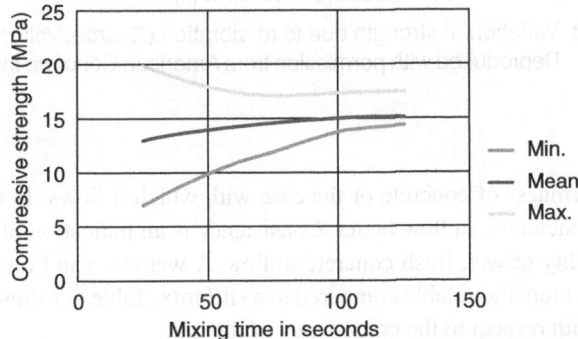

Fig. 4.16 Relation between compressive strength and mixing time

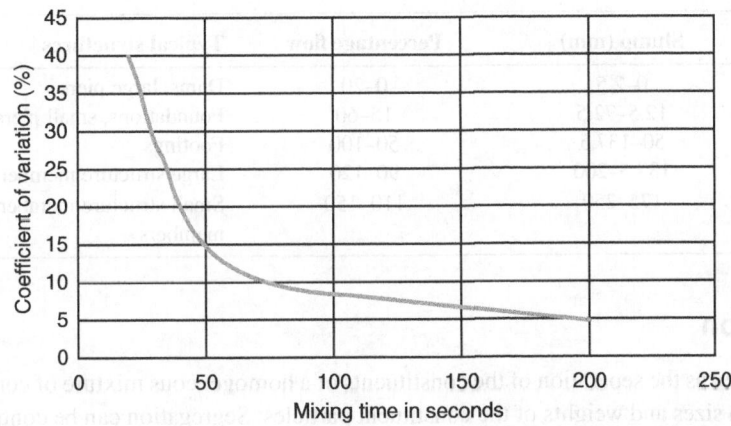

Fig. 4.17 Relation between the coefficient of variation of strength and mixing time

4.4 Stability

The ability of concrete to remain a stable, homogeneous, coherent mass without segregation both during handling (static stability) as well as during vibrations (dynamic stability) is termed stability. The traditional method of producing compaction in fresh concrete is hand poking. This is not a very efficient technique. The common types of vibrators used are internal vibrators, external vibrators, table vibrators, and special vibrators such as surface vibrators either mechanically or electrically operated, and vibrating rollers used for thin slabs. Generally, concrete must be vibrated as soon as it is placed in the formwork. However, in certain cases, to ensure a good bond between the layer of concrete already laid and fresh concrete, it is vibrated again. This is called re-vibration. The variation of strength due to re-vibration is shown in Fig. 4.18. With the aforementioned factors in mind, the different tests to be conducted on fresh as well as hardened concrete will be discussed later.

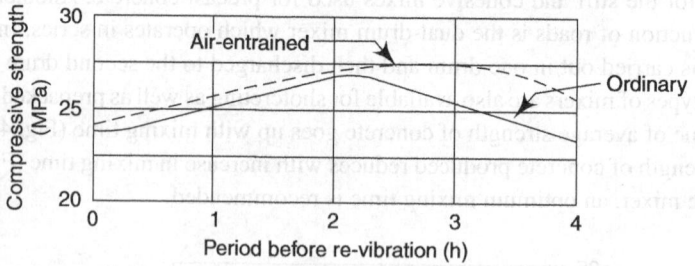

Fig. 4.18 Variation of strength due to re-vibration (*Source*: Vollick 1958. Reproduced with permission from American Concrete Institute)

4.5 Consistency

Consistency refers to the firmness of concrete or the ease with which it flows. In turn, it refers to the mean degree of wetness as wet concretes can flow better. Consistency is an indication of workability. It shows the relative mobility, or the ability of wet, fresh concrete to flow. A wet mix can flow better compared to a dry mix. Similarly, a soft mix is more mouldable compared to a stiff mix. Table 4.5 shows the types of mixes used for different applications with respect to the consistency of the mix.

Table 4.5 Consistency of mixes used for different applications

Consistency	Slump (mm)	Percentage flow	Typical structures
Dry	0–2.5	0–20	Dams, large piers
Stiff	12.5–72.5	15–60	Foundations, small piers
Medium	50–137.5	50–100	Footings
Wet	137.5–200	90–120	Large structure members, beams, slabs
Sloppy	175–250	110–150	Small structure members, thin slabs, small members

4.6 Segregation

Segregation is defined as the separation of the constituents of a homogeneous mixture of concrete. It is caused by the differences in sizes and weights of the constituent particles. Segregation can be controlled by properly choosing the grading of aggregates and by carefully handling wet mixes.

In relatively lean and dry mixes, segregation can be caused by the coarser particles separating out because they travel farther along the slope or settle to a greater extent than finer particles. The second form of segregation occurs in very wet mixes in which the cement–water paste separates from the mix.

Segregation can also be caused by poor handling, such as dropping wet concrete from a considerable height, passing long chutes along a slope, and discharging concrete carelessly against some firm obstruction. It may also be caused by the vibration of concrete. Though vibration provides a useful means of compaction, over-vibration leads to segregation. This can happen when vibration is allowed to continue for too long. It leads to the separation of coarse aggregates from the mix. These aggregates settle at the bottom, and the cement–water paste moves to the top in the form of laitance (scum). This laitance is different from bleed water (discussed in the next section).

Segregation is difficult to measure. However, its occurrence is easily detected. The flow test can indicate the susceptibility of a mix that is likely to segregate. In dry mixes, heavier particles move away and occupy the edges of the flow table. In wet mixes, the cement–water paste tends to move away from the middle, and the centre of the flow table is left only with coarser particles.

The proneness of concrete to segregation may be found by a practical field test described here. A cylinder mould is filled with concrete and vibrated for 10 min. After this, when the mould is stripped, one can observe the distribution of particles to determine the proneness of the mix to segregation.

4.7 Bleeding

Bleeding is also known as 'water gain'. It is the accumulation of water at the surface, which accompanies the sedimentation of freshly mixed concrete. This happens due to the inability of the solid constituents of the mix to hold all the mixing water, and they settle *downwards* due to gravity and the water moves *upwards*.

Bleeding is expressed quantitatively as the total settlement per unit height of concrete or as the percentage of mixing water. In extreme cases, this can be nearly 20%. Bleeding is a function of (a) air velocity, (b) temperature, and (c) humidity. If the rate of bleeding is roughly equal to the rate of evaporation as shown in Fig. 4.19, then bleeding will not cause any problem. If the rate of bleeding is less than the rate of evaporation, then the surface becomes dry, because of which cracks appear on it. The restraint at the bottom encourages such cracks.

Figure 4.20 shows the effect of bleeding in a concrete structure. Bleed water collects under the longitudinal reinforcement. When this water evaporates, voids get formed, leading to a weak bond between the reinforcement and the surrounding concrete. Bleeding thus establishes a path for water to permeate through, increasing the permeability of concrete.

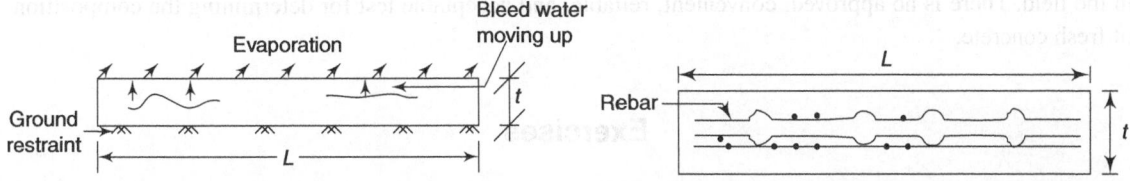

Fig. 4.19 Evaporation and bleeding	**Fig. 4.20** Effect of bleeding in a concrete structure

The evaporation of water from the surface of concrete depends on (a) the relative humidity of the surrounding air, (b) the ambient temperature, and (c) the velocity of wind. Figure 4.21 shows the effect of relative humidity on the loss of water from the surface. Figure 4.22 shows the effect of temperature on the loss of water from the surface.

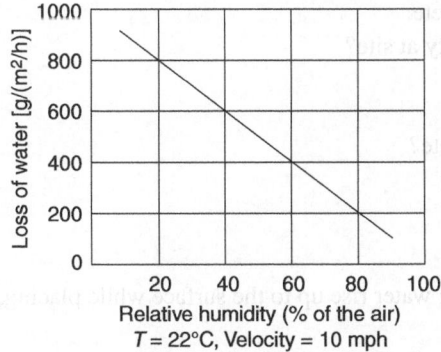

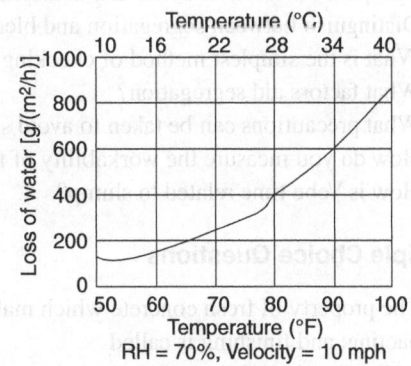

Fig. 4.21 Effect of relative humidity on loss of water from a concrete surface	**Fig. 4.22** Effect of temperature on loss of water from a concrete surface

4.7.1 Test for Bleeding

The ASTM: C232 test is used for quantifying bleeding. A sample of concrete is placed and consolidated in a container 254 mm in diameter and 279 mm in height. The bleed water accumulated on the surface is withdrawn at 10-min intervals during the first 40 min and thereafter at 30-min intervals. Bleeding is expressed in terms of the amount of accumulated water, as a percentage of the mixing water in the test sample.

Bleeding of fresh concrete can be controlled in the following ways:

a. Bleeding decreases with increase in the fineness of cement.
b. Bleeding can be controlled by moderating C_3A content. More C_3A content leads to less bleeding.
c. Adding an accelerator as an admixture controls bleeding, as the concrete sets faster.
d. The rate of bleeding decreases with decrease in temperature.
e. Rich mixes are less prone to bleeding than lean mixes.
f. Mineral admixtures reduce bleeding.

Initial bleeding continues steadily but decreases with the stiffening of the mix. It stops when the mix gets stiffened sufficiently enough to terminate the process of sedimentation. A weak layer results if the bleed water is used and remixed at the top surface. This can be avoided by delaying the commencement of the finishing operation. Overworking the surface should be avoided; this causes laitance at the top.

4.8 Analysis of Fresh Concrete

The two most important parameters of fresh concrete are cement content and water–cement ratio. A few test methods based on the physical and chemical parameters have been suggested by ASTM as well as BIS to determine these two parameters. However, there seems to be no reliable method that can be practically applied in the field. There is no approved, convenient, reliable, and acceptable test for determining the composition of fresh concrete.

Exercises

Review Questions

1. Discuss the efficacy of the various definitions put forward for workability of fresh concrete.
2. What is the relationship between the strength and density of concrete?
3. What are the various factors which affect the workability of concrete?
4. Compare the relative merits and demerits of various workability tests.
5. Distinguish between segregation and bleeding of concrete.
6. What is the simplest method of checking the workability at site?
7. What factors aid segregation?
8. What precautions can be taken to avoid segregation?
9. How do you measure the workability of flowing concrete?
10. How is Vebe time related to slump?

Multiple Choice Questions

1. The property of fresh concrete which makes the mixing water rise up to the surface while placing, compacting, and finishing is called
   a. Segregation
   b. Bleeding
   c. Bulking
   d. Slumping

2. The property of fresh concrete in which ingredients like coarse aggregates and fine aggregates separate while placing or compacting is called
   a. Segregation
   b. Bleeding
   c. Bulking
   d. Slumping

3. Workability of concrete is directly proportional to
   a. Aggregate–cement ratio
   b. Time for transit
   c. Grading of aggregate
   d. Air in the mix

4. Workability of concrete is inversely proportional to
   a. Time of transit
   b. Water–cement ratio
   c. Air in the mix
   d. Size of coarse aggregates

5. Air entrainment in concrete increases
   a. Strength
   b. Workability
   c. Shrinkage
   d. Unit weight

6. Use of pozzolana cement instead of ordinary Portland cement
   a. Decreases workability
   b. Increases bleeding
   c. Increases shrinkage
   d. Increases strength

7. In order to obtain best workable concrete, the shape of aggregates should be
   a. Rounded
   b. Elongated
   c. Flaky
   d. Angular

8. The field test to check ordinary workable concrete for construction is
   a. Vicat needle test
   b. Sieve analysis
   c. Compaction factor test
   d. Slump cone test

9. Bleeding can be prevented by
   a. Controlling water content
   b. Using finely ground cement
   c. Controlling compaction
   d. All of the above

10. Bleeding of concrete is generally due to
   a. Excess of water
   b. Too much finishing
   c. Large size coarse aggregates
   d. All of the above

Answers to Multiple Choice Questions

1. b	4. a	7. a	10. d
2. a	5. b	8. d	
3. c	6. c	9. d	

CHAPTER 5

Stages of Manufacture of Concrete

The materials that are used to make good concrete can also be used to make poor concrete. If there is poor control over the process of making concrete, we get only bad-quality concrete. However, with the same ingredients, if good care is taken to effect control over the process of concreting, only then can we get good-quality concrete. Thus, for good-quality concrete both quality materials and good control over the process of concreting are vital. The various stages in the manufacture of concrete are as follows:

a. Batching
b. Mixing
c. Transporting
d. Placing
e. Compacting
f. Curing
g. Finishing

This is schematically shown in Fig. 5.1. It should be borne in mind that quality control is most important during all stages of manufacture of concrete.

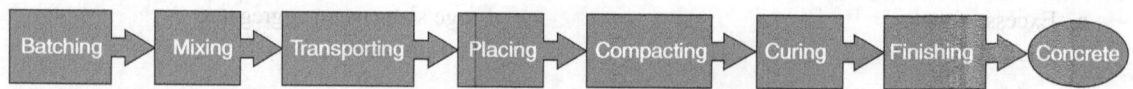

Fig. 5.1 Stages of manufacture of concrete

5.1 Batching Concrete

Batching is the process of measuring specified quantities of cement, aggregate, water, and admixtures as per the mix proportions for a specified grade of concrete. The two methods of batching are as follows:

a. Volume batching
b. Weigh batching

5.1.1 Volume Batching

Volume batching is not a good method because of the inaccuracies it introduces in the measurement of granular materials. Loose sand in a moist condition occupies more volume than dry compacted sand due to the phenomenon of *bulking*. Hence, the effect of bulking must be considered while measuring sand. Despite drawbacks, for less important non-engineered small works, this method is adopted because of its ease in application. However, it is unscientific and hence not recommended for important works.

The quantity of the two variables that should be considered in volume batching are as follows:

a. Relative proportion of the ingredients of concrete in terms of volume
b. Water–cement ratio

Table 5.1 shows the recommended values of the relative proportion of ingredients by volume for different grades of concrete.

Table 5.1 Relative proportion of ingredients generally used in volume batching

Grade of concrete	Relative proportion of ingredients by volume			Specified as
	Cement	Sand	Coarse aggregate	
M10	1	3	6	1:3:6
M15	1	2	4	1:2:4
M20	1	1.5	3	1:1.5:3
M25	1	1	2	1:1:2

Water–cement ratio used must be as per IS: 456-2000 based on environmental condition.

The typical gauge box used to measure the volume of ingredients is shown in Fig. 5.2. The steps involved in the volume batching process are given below:

Step 1 Cement is first measured by weight. One bag of cement weighs 50 kg. This has a volume of 35 L or 0.035 m³. Cement should not be measured by the volume because its unit weight will vary depending on the type of compaction during filling of the cement bags.

Step 2 The gauge box shown in Fig. 5.2 is used for measuring the required volume of coarse aggregate and sand. Generally the volume of the box is 35 L, that is, the volume of one bag of cement.

Step 3 The moisture present in the sand must be accounted for during batching. The quantity of water is found by multiplying the water–cement ratio by the weight of cement. From this quantity, the water present in the aggregate is deducted to arrive at the quantity of water to be added to the mix. Water is measured in litres. However, it can also be measured by weight (in kg) as its unit weight is 1.

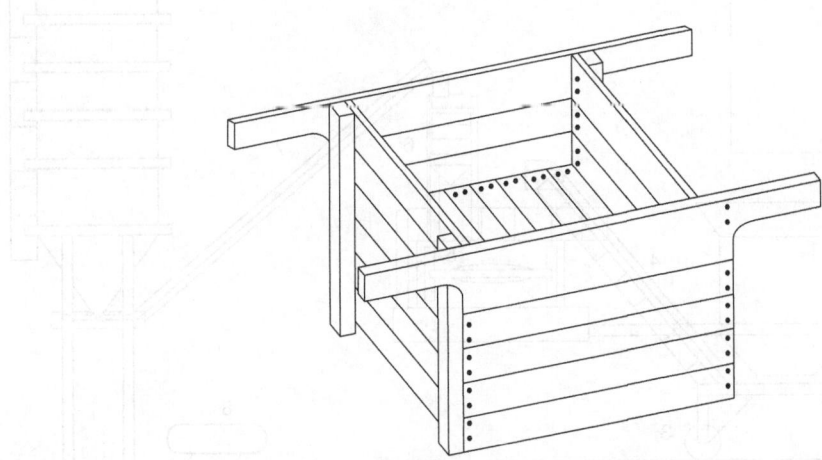

Fig. 5.2 Typical gauge box

5.1.2 Weigh Batching

Weigh batching is the correct method of measuring the materials that are used to make concrete. Use of weigh batching system facilitates accuracy, flexibility, and simplicity. For large works, a weigh batching plant is used. A typical batching plant consists of the following components:

a. Aggregate bins for storage of aggregates
b. Feeding mechanism such as scrapers, conveyors, and hoist
c. Mixers for mixing the materials
d. Measuring system such as balances, scales, and electronic weighing systems
e. Storage tank for water and water measuring system
f. Dispenser for chemical admixtures
g. Control room

Figure 5.3 shows a typical modern batching plant. Aggregates are stored in silos and in bins protecting them from rain, dust, and wind. Scrapers and conveyors are used to feed the mixers. Two types of plants are used. It may be either continuous or cyclic. In continuous type, feeding is done continuously and automatically, whereas in cyclic type, loading and discharge are done cycle after cycle and, generally, manually.

IS: 456-2000 stipulates the measuring accuracy of ingredients. Cement is measured for ±2% accuracy and other materials such as aggregates, water, and admixtures are measured for ±3% accuracy. The ingredients enter one side of the mixture machine and get discharged from the other side.

Measurements of weight are accurately made by a system of mechanical levers and load cells. The discharge is done in modern machines automatically by weigh hopper gates operated by compressed air cylinders. Presetting a desired batch weight is possible using punched cards or digit switches and microprocessor computers.

Microwave moisture gauges are employed to determine the moisture in aggregates and in most plants a flow meter is used to measure and regulate water quantity. In some plants, water is also weighed.

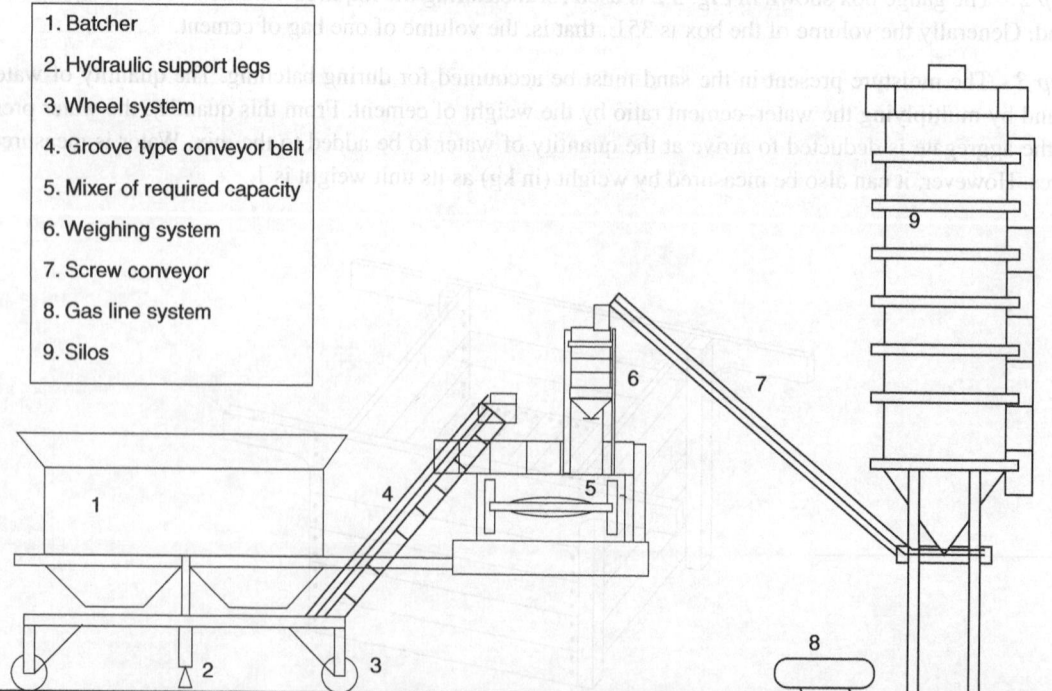

1. Batcher
2. Hydraulic support legs
3. Wheel system
4. Groove type conveyor belt
5. Mixer of required capacity
6. Weighing system
7. Screw conveyor
8. Gas line system
9. Silos

Fig. 5.3 Typical concrete batching plant

IS: 4925-2004 'Concrete Batching and Mixing Plant Specification' specifies the tolerances in batching depending on the type of material (see Table 5.2).

The minimum weight of any material for which the aforementioned tolerances apply is given by the following equation:

$$W = \frac{0.3 \times R}{T}$$

where

W = Minimum weight (in kg)

R = Scale capacity (in kg)

T = Weight tolerance (in %)

Proper batching ensures better quality. Variations in strength and other properties result from variations in proportions. If the properties of concrete, such as strength, density, and slump, exhibit only little variation, we say the concrete is of uniform quality. On the other hand, if the variation of a property, for example, strength variation, is large, it is not a uniform concrete. If the slump variation is large, it is not a uniform concrete. Every property, such as air content and aggregate content, should show only little variability and should be as uniform as possible.

Table 5.2 Tolerances in weight during batching materials for concrete

S. no.	Type of material	Tolerance permitted
1	Cement and other cementitious material	±1%
2	Sand, aggregates	±2%
3	Water	±1%
4	Admixture	±3%

5.2 Mixing Concrete

Thorough mixing is essential for the production of a uniform quality of concrete. The method chosen for mixing should be capable of effectively mixing concrete material containing the largest specified aggregates to produce a uniform mixture for the practically required slump.

5.2.1 Methods of Mixing

Concrete is mixed either by hand mixing or by machine mixing, based on the quality and quantity of concrete required. Normally for mass concrete, where good quality of concrete is required, a mechanical mixer is used.

Hand Mixing

Mixing by hand is employed only for specific cases where quality is not of much importance, either because of the unimportant nature of the work or because the quantity of concrete required is less. Hand mixing generally does not produce uniform concrete and hence should not be normally used, unless it is for very small domestic works. The steps involved in hand mixing are given below:

Step 1 Before mixing, aggregates are washed with water to remove impurities such as dirt, dust, or any other unwanted material.

Step 2 The required quantity of sand is measured and is spread evenly on a platform.

Step 3 The required quantity of cement is evenly spread on the previously spread sand.

Step 4 The sand and cement are mixed carefully using a shovel, and the mix is turned over and over again, until a uniformly coloured mix is obtained.

Step 5 This sand–cement mix is uniformly spread and a measured amount of coarse aggregate is spread on this mix to obtain a uniform thickness.

Step 6 The whole mass should be mixed a number of times, at least three times by shovelling, and turning over, and by twisting from centre to side, then back to the centre, and again to the sides.

Step 7 A pit is made in the middle of the mixed heap.

Step 8 Three-quarters of the total quantity of water required is added while the materials are turned in towards the centre with spades. The remaining water is added by a water-can, slowly turning the whole mixture over and over again. This operation is continued until a uniform colour and consistency is obtained throughout the heap.

Step 9 5% extra cement is added (than that specified for machine mixing) when hand-mixed cement concrete is produced.

Mechanical Mixing

Mechanical mixers can be divided into two main types: batch mixers and continuous mixers. Batch mixers produce concrete batch by batch, one batch at a time. The operation is intermittent. The raw material is loaded at one end and the concrete is discharged at the other end. This constitutes a cycle of operation which is repeated until enough quantity of concrete is produced. Continuous mixers produce concrete at a specified rate. The raw materials are continuously entered at one end and mixed concrete exits from the delivery end.

The following are the two types of batch mixers which are commonly employed:

a. Horizontal or inclined axis (Drum mixers) (see Fig. 5.4)
b. Vertical axis (Pan mixers) (see Fig. 5.5)

The drum mixers have a drum, with fixed blades, rotating around its axis. The components of the drum mixer are shown in Fig. 5.4.

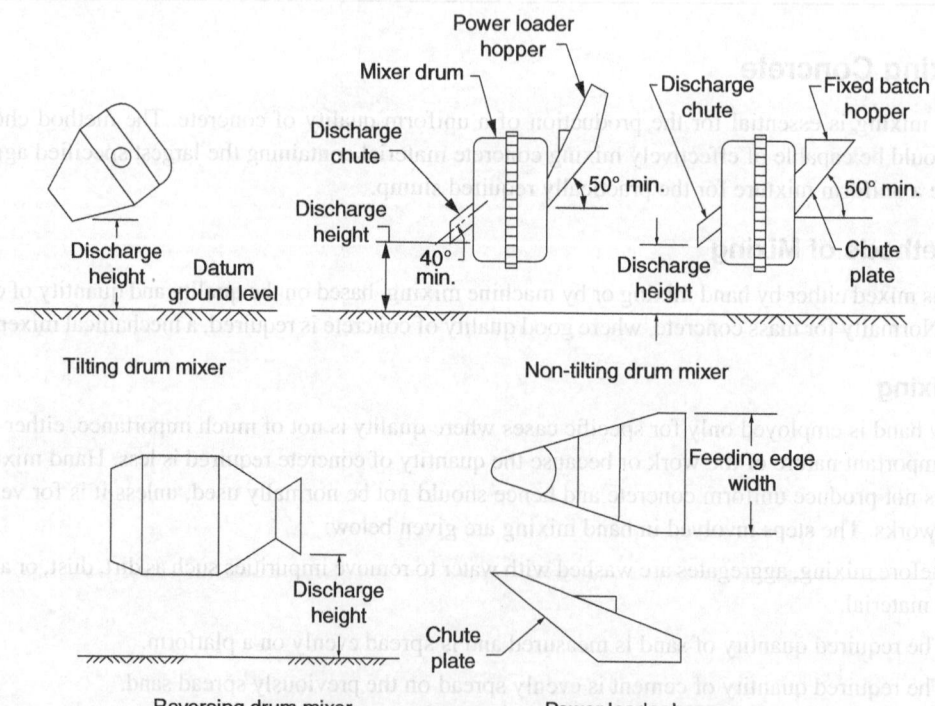

Fig. 5.4 Components of a drum mixer

A typical pan-type mixer is shown in Fig. 5.5. Pan mixers, on the other hand, may have either the blades or the pan rotating around the vertical axis.

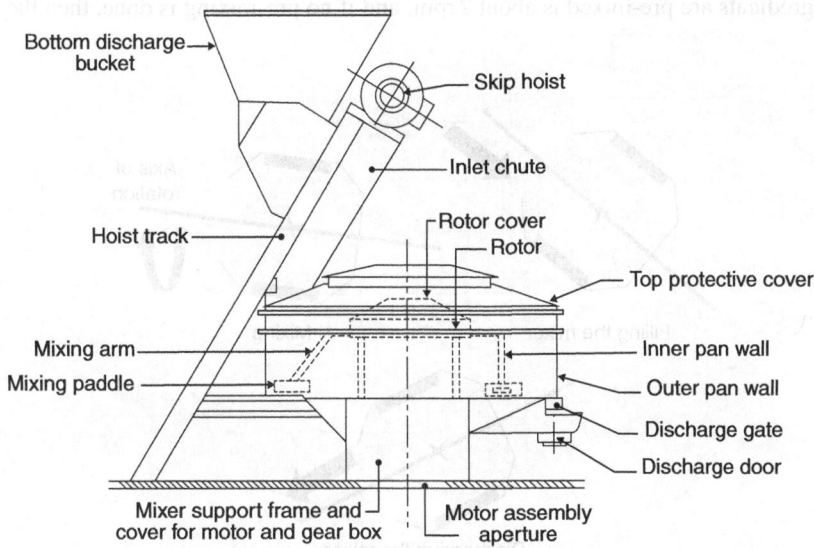

Fig. 5.5 Typical pan-type concrete mixer

Working of Drum Mixer

The blades of the drum mixer are attached to a movable drum. The material is input into the opening on the left side. When the blades rotate, they drag and lift the material and make it fall. This operation ensures thorough mixing. For good mixing, the rotation speed is controlled. The following are the three types of drum mixers:

a. Tilting drum   b. Non-tilting drum   c. Reversing drum

The tilting type, shown in Fig. 5.6, is commonly used because of its convenience. In the non-tilting type (Fig. 5.7), material is input into the opening on the left and after mixing, discharged from the right opening. In the reversing type, the same opening is used for input and for also discharging the material. The drum rotates clockwise for mixing and anticlockwise for discharging.

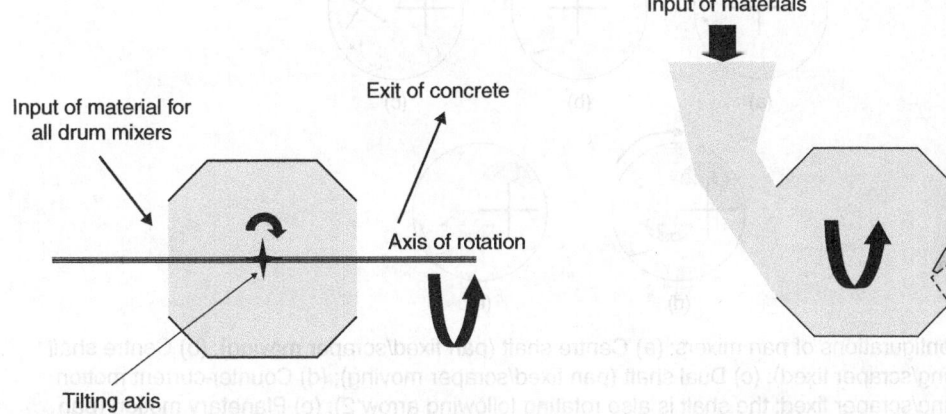

Fig. 5.6 Typical cross section of a drum   **Fig. 5.7** Cross section of a non-tilting mixer

Three operations, namely filling the mixer, mixing, and discharging the mixed concrete, are performed to get thoroughly mixed good fresh concrete. These operations are shown schematically in Fig. 5.8.

Mostly tilting drum mixers are used in trucks employed in ready mixed concrete production. The speed of mixing if the ingredients are pre-mixed is about 2 rpm, and if no pre-mixing is done, then the rate of mixing is 15 rpm.

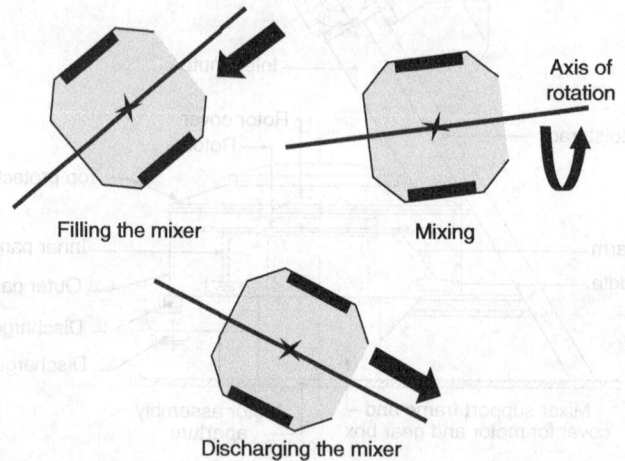

Fig. 5.8 Operations performed to get good thoroughly mixed fresh concrete using a tilting drum mixer

Working of Pan Mixer

In pan mixers, either the pan or the blade, or both the pan and the blade rotate. Different configurations of blades are shown in Fig. 5.9.

In the configurations shown in Fig. 5.9(a) and (b), the axis of rotation of the pan and blade coincide. However in Fig. 5.9(c) and (d), there are two rotations—one having dual shaft and the other having counter-current motion. Figure 5.9(e) shows a pan mixer in which both the pan and the scraper are rotating. To discharge the mixer the trap door at the bottom is utilized.

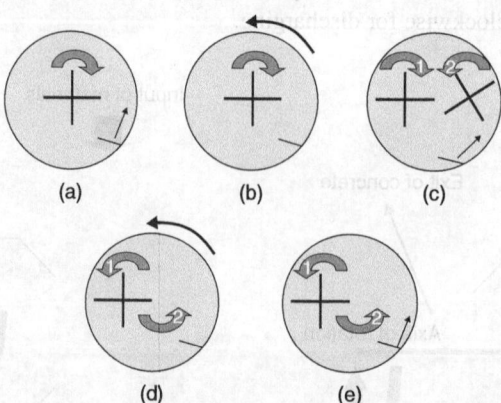

Fig. 5.9 Various configurations of pan mixers: (a) Centre shaft (pan fixed/scraper moving); (b) Centre shaft (pan rotating/scraper fixed); (c) Dual shaft (pan fixed/scraper moving); (d) Counter-current motion (pan rotating/scraper fixed; the shaft is also rotating following arrow 2); (e) Planetary motion (pan fixed/scraper moving; the shaft is also rotating following arrow 2)

Continuous Mixers

They are usually of non-tilting type. They have screw-type blades rotating in the middle of the drum. The drum is tilted 15° downward towards the discharge end. A major use is for low-slump concrete employed in road works.

5.2.2 Mixing Time

The mixing method includes loading of material, mixing time which imparts a particular energy to the ingredients, and the discharge method. The efficiency of mixing depends on the duration of loading, mixing, and discharge. A typical schedule of loading, mixing, and discharge is shown in Fig. 5.10.

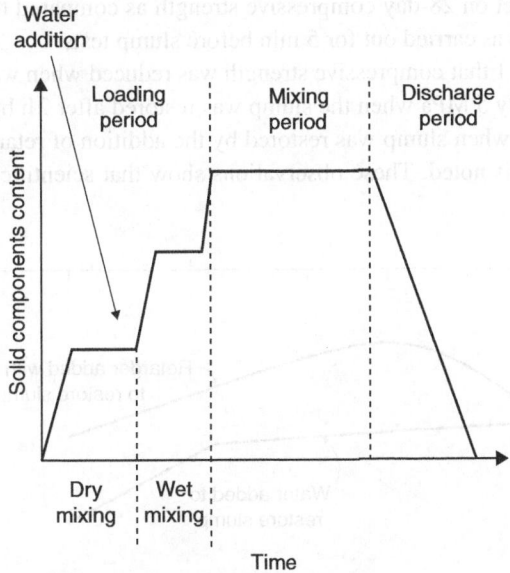

Fig. 5.10 Mixing schedule

5.2.3 Perfomance Attributes of Mixed Concrete

The concrete produced should exhibit uniform properties of all the signficant variables. The properties that are considered for uniformity are the following:

a. Workability in terms of slump
b. Density of concrete
c. Air content
d. Compressive strength

The efficiency of mixing is determined by the uniformity of concrete produced. Thus what is required is homogeneous concrete. A direct measure of homgeneity depends on concrete composition, such as distribution of various constituents, including air content present in various samples taken during the concrete discharge.

5.2.4 Re-tempering

Re-tempering is remixing of wet concrete that has stiffened due to loss of workability primarily caused due to delay in placing the concrete after mixing it. The delay in the delivery of concrete from the mixing plant can occur due to time-consuming transportation, long distance, and sometimes traffic hold-up. Generally, concrete that has stiffened is rejected by the engineer at site. However, from economic and environmental considerations, such a rejection should be justified; if not, stiffened concrete should be reused. To reuse the stiffened concrete, usually specific quantity of cement and water are added.

BIS does not allow remixing of partially hardened mortar or concrete. Therefore, if this concrete is to be reused, a detailed quality study must be undertaken to know the pros and cons of re-tempering. If the codes are not followed, additional water must be added to the concrete to restore the slump and provide sufficient workability for proper placing and compaction. The disadvantage is that the additional water means a higher water–cement ratio and lower strength and durability.

Analysis of re-tempered concrete shows that a very wet mix having a delay of 1 h gains a strength increment of 2–15% when proper re-tempering is applied with additional cement and water or suitable admixtures. However, further delay renders a decrease in strength.

It is well known that retarding admixtures are capable of delaying the hardening of concrete to the extent that it can be vibrated or finished much later than normal without any detriment to strength or durability.

Figure 5.11 shows the effect on 28-day compressive strength as compared to the addition of water alone to restore the slump. Mixing was carried out for 5 min before slump tests.

It can be seen from Fig. 5.11 that compressive strength was reduced when water was added at any period. However, it increased by nearly 5 MPa when the slump was restored after 2 h by the addition of the retarder. At 4 h strength also increased when slump was restored by the addition of retarder and water. After 6 h only a slight reduction in strength is noted. These observations show that scientific re-tempering is possible for field applications.

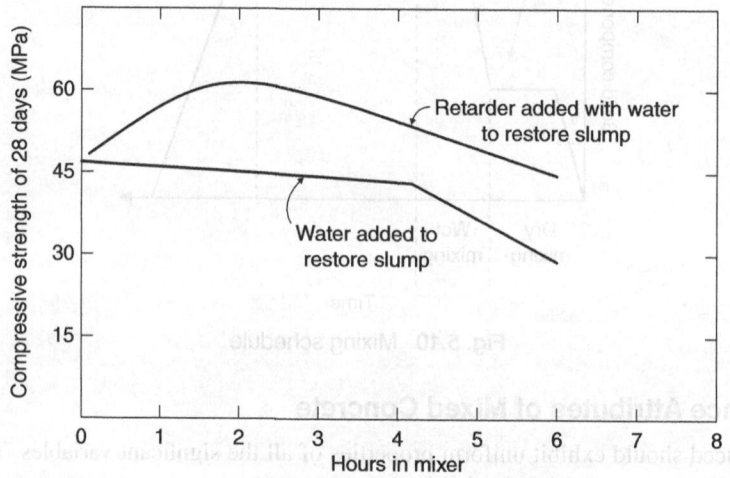

Fig. 5.11 Effect of using water/retarder on compressive strength for restoring the slump

5.3 Transporting Concrete

Concrete is produced using one of the following two methods:

a. *Ready mixed concrete*: Concrete is produced away from the site of construction and is transported in ready-mix trucks and delivered at site, requiring no further treatment before placing in forms.

b. *Site mixed concrete*: In site mixing, the plastic concrete from the mixer is transported to the forms before placing.

The various issues related to transportation of ready mixed concrete are discussed in Chapter 18 on ready mixed concrete. Here in this chapter the aspects related to transportation of site mixed concrete are discussed. The main objective while transporting concrete is to ensure that parameters, such as water–cement ratio, slump, consistency, air content, and uniformity, are not changed before reaching the forms for placing. The process of transportation should not lead to either loss of slump or segregation.

The following methods are used for transporting concrete:

a. *Direct discharge into forms by short chutes*: Short chutes in a semi-circular shape stiffened at intervals are simple and economical to use. Free fall of concrete from a height of more than 2 m must be avoided.
b. *Barrows*: Manual wheel barrows of approximately 80 kg capacity can be used for long horizontal distances. For major works, power barrows of 800 kg capacity, up to 300 m hauls are used.
c. *Dumpers and trucks*: These are used for horizontal long hauls. Because of jolting, especially if the terrain is rough, the concrete during transit has the risk of segregation.
d. *Elevating towers and hoist*: In multi-storied buildings, elevating towers are used for lifting concrete buckets. The lifted concrete is then distributed by either chutes or barrows. This type of transportation can be used where high lifts are required.
e. *Monorail system*: In tunnels and in dam sites, a single track is laid to carry a monorail power wagon which moves at a speed of 80 m/min. This type of transportation can be used for covering long distances.
f. *Cranes and cableways*: When concreting is to be done in a large project covering mountains and valleys, cranes and cableways are used to provide three-dimensional transport enabling both horizontal and vertical movement. Depending on the site condition, the type of crane can be chosen. It may be a derrick, crawler, or wheel mounted.
g. *Belt conveyor*: It can be used when hauling concrete over long distances. It is not very much recommended because of its vulnerability to segregation. The initial setting-up cost is also high. Discharge can be as high as 115 m³/h.
h. *Concrete bucket and skip*: These are common equipment. Figure 5.12 shows a typical concrete bucket. The capacity of the skip varies from about 0.2 m³ to 10 m³.

Fig. 5.12 Concrete bucket and skip

The discharge of concrete is controlled by the shape of the gate and proper flow.

a. *Tremie*: This pipe is used to transport concrete under water. It consists of a pipe and a funnel at the top. The tremie is first filled with bentonite slurry and then displaced by wet concrete. The use of tremie is further discussed in Chapter 26.
b. *Concrete pumps and pneumatic placers*: A concrete pump is used for transporting concrete by pumping. There are two types of concrete pumps:
   • In the first type, a concrete pump is attached to a truck. It is known as a trailer-mounted boom concrete pump because it uses a remote-controlled articulating robotic arm (called a *boom*) to accurately place the concrete. Boom pumps are mostly used in larger construction projects as they are capable of pumping at very high volumes and because of the labour-saving nature of the placing boom. They are a revolutionary alternative to truck-mounted concrete pumps.
   • In this type, a concrete pump is either mounted on a truck or a trailer and is known as a truck-mounted concrete pump. It is also referred to as a *line pump* or trailer-mounted concrete pump depending on where it is mounted. This pump requires flexible concrete placing hoses attached to the outlet. These hoses are linked and lead to wherever the concrete is to be placed.
c. *Boom conveyors*: Using boom conveyors concrete can be placed at locations like the raft slab of a basement, as shown in Fig. 5.13. Concrete with a flowing character can be efficiently placed at even inconvenient locations with ease.

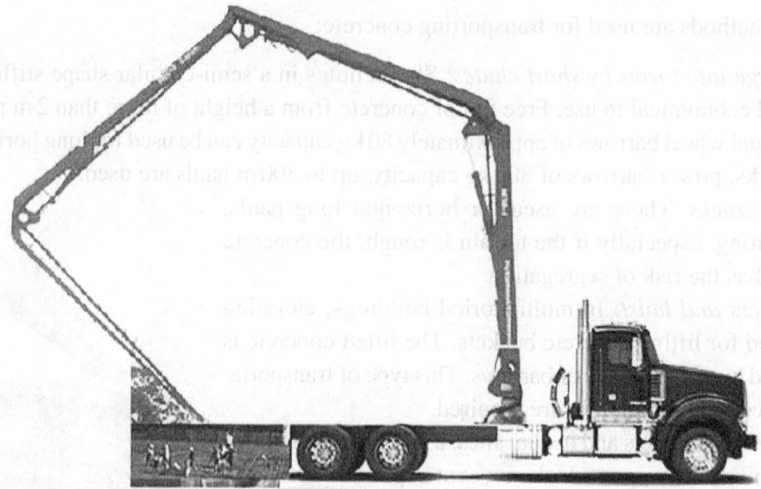

Fig. 5.13 Typical boom placer in operation

5.4 Placing Concrete

The placing operation consists of equipment, layout, proposed procedures, and methods. It should be planned and no concrete should be placed until the formwork is inspected and found suitable for placement. The equipment for conveying concrete should be practically capable of continuous flow of concrete during depositing. The drop in height during placement should not be such that it encourages segregation.

The concrete should be placed in its final position before the cement reaches its initial setting. The placed concrete must be compacted in its final position within 30 min of leaving the mixer. After final compaction, the placed concrete should not be disturbed.

While placing, concrete should be deposited, as close as possible, to its final position. Concrete should not be moved or re-handled or caused to flow in a manner which may encourage disturbance to the final position. In particular, the following aspects may cause weakness in the strength of the finished concrete:

a. Segregation where concrete is dropped, as in columns and walls
b. Loss of materials
c. Displacement of reinforcement
d. Dislocation of shuttering or embedded inserts
e. Cracking and loss of strength

5.4.1 Methods of Placing

The following methods are employed for placing concrete:

a. *Free fall method*: Concrete is placed by free fall from top to bottom. Concreting should first be done at the centre; it should then flow to the edges. For most structures, this type of placement is adopted. To avoid segregation, the free fall height should be restricted.
b. *Pump line method*: In this method, concrete is pumped through a pipe. The pipe is usually flexible and can be positioned to fill the formwork uniformly.
c. *Tremie method*: Gravity-fed concrete into the tremie is generally adopted for underwater concreting. Concrete is introduced into the hole and fed by a pump into the tremie. The concrete falls due to gravity and is placed continuously until the pipe is full.

5.4.2 Striking Forms

The formwork and false work stability are both important and should be checked carefully. Generally, the following three checks are made:

a. Structural strength of members and connections
b. Lateral stability
c. Overall stability

The time at which striking is permitted is usually related to the strength of concrete. The determination and proper assessment of compressive strength is important. This involves many factors such as water–cement ratio, type of cement used, and temperature. The main criterion for removal of formwork from walls and columns is the strength achieved such that mechanical damage is avoided during striking.

5.5 Compacting Concrete

Compaction is intended to expel the entrapped air so that coarse and fine aggregate particles in wet concrete mass are packed together without any voids. This action is intended to increase the density of concrete. In addition, good compaction helps to achieve the following:

a. Increases the strength of concrete
b. Increases the bond between the concrete and reinforcement
c. Increases the abrasion resistance
d. Decreases permeability
e. Aids in reducing shrinkage and creep

Proper compaction makes the concrete flow to the nooks and corners of the formwork, avoiding formation of honeycombs. This helps in obtaining a good finish, especially for vertical surfaces. Normal concrete, when it is placed in the formwork, contains 5–25% of air voids. By following a good compaction procedure, this can be reduced to 1–3%. Figure 5.14 shows the relationship between strength and voids present for a typical M25 concrete. It can be seen that the presence of 10% air voids reduces the strength by more than 50%.

Figure 5.15 shows the two stages of compaction, namely (a) the initial liquefaction of concrete when the aggregates are made to slump and fill the formwork and (b) the final expulsion of entrapped air. The first stage takes about 3–5 s, and the second stage takes 7–15 s when the entrapped air bubbles tend to come to the surface. The operation of compaction should be continued until air bubbles no longer come to the surface.

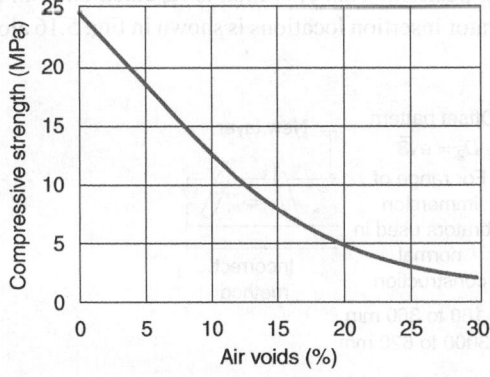

Fig. 5.14 Loss of strength due to presence of air voids

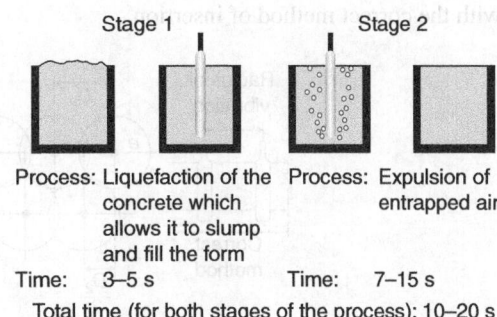

Fig. 5.15 Two-stage process of compaction

5.5.1 Methods of Compaction

The following are the two methods of compaction:

a. Hand compaction
b. Mechanical compaction

Hand Compaction

Hand compaction is used for minor unimportant structures because of its inefficiency. The mix design used should suit the hand compaction with respect to the slump. The following three approaches are used for hand compaction:

a. *Roding*: A 2 m-long rod of 16 mm diameter is used to poke the concrete at the corners of the formwork to expel entrapped voids. The thickness of layers used for enabling roding is about 10–15 cm.
b. *Ramming*: This method is used for compacting on the ground in foundations, etc., for plain concrete, and is generally not used for RCC.
c. *Tamping*: The top surface of slabs is generally beaten down by wooden rammers of thickness 10 cm with a handle. This is intended to achieve both leveling and compaction simultaneously.

Mechanical Compaction

Mechanical vibration causes expulsion of air and enables compaction of concrete. Various types of mechanical vibrators are used for this purpose. The following are the main types used:

a. Immersion vibrator (internal or needle vibrator)
b. Surface vibrator
c. Form vibrator
d. Table vibrator
e. Platform vibrator

Internal Needle Vibrators

These are the most common type of vibrators used at a majority of construction sites. They consist of a tubular housing within which an eccentric weight rotates and thus induces vibration on the surrounding mass. The frequency range adopted is 3000–5000 rpm. The diameter of the needle varies from 20 to 75 mm. They are available in lengths of 25–90 cm. Depending on the diameter of the needle and frequency of vibration, the zone of influence varies from 100 to 600 mm.

Immersion vibrators are driven by a flexible shaft connected to a motor, or an electric, petrol, or diesel motor situated within the tubular housing or by compressed air. The motor-driven vibrator is more common at construction sites in India.

Needle vibrators are inserted vertically into fresh concrete as quickly as possible and held stationary until the air bubbles coming out cease to rise to the surface. This usually takes about 20 s. The vibrator should be slowly withdrawn and reinserted vertically in an adjoining position. This operation is repeated until all the concrete is well compacted. A typical pattern of needle vibrator insertion locations is shown in Fig. 5.16 along with the correct method of insertion.

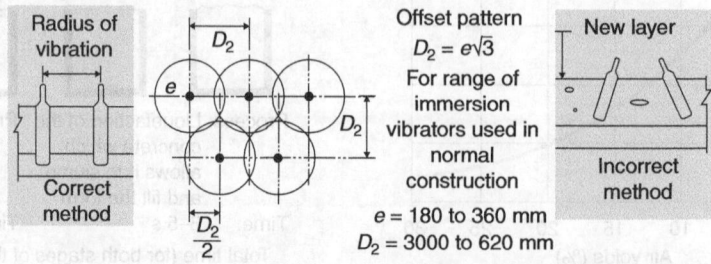

Fig. 5.16 Correct method of using a needle vibrator

Surface Vibrators

Surface vibrators are applied to the top surface of concrete, and act downwards from there. They are used to compact large slabs, industrial floors, pavements, etc. They also aid in leveling and finishing the surface of the slab or pavement. Generally, these are not effective beyond 15–20 cm.

A *vibrating beam screed* is the most common type of surface vibrator. It consists of a metal beam on which a vibrating unit is attached. In general, the centrally mounted vibrating unit has a span of 6 m. A typical unit is shown in Fig. 5.17. These units are pulled by hand to effect compaction. As can be seen, surface vibrators act from the top towards the bottom and hence, as we go down, the efficiency of compaction decreases. Thus, surface vibrators are effective for slabs up to 150 mm.

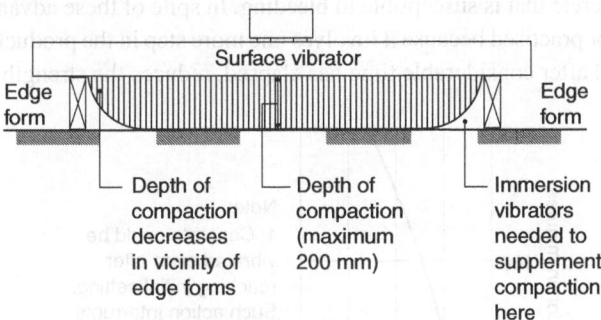

Fig. 5.17 Typical surface vibrator

Form Vibrators

Form vibrators are also called *external vibrators*. These are used in members where the reinforcements are congested. The vibrators are clamped to formwork and when the form vibrates the concrete gets compacted. For this application, the formwork should also be designed to resist the forces imposed on it by the attached vibrator.

Table Vibrators

This type is usually adopted in laboratories where small specimens are put on a vibrating table. Standard cubes and cylinders cast as companion specimens in the lab are generally subjected to compaction by table vibrators.

Platform Vibrators

These are similar to table vibrators but are of larger size. They are used in precast industries to vibrate large panels such as walls and hollow-core slab units.

5.5.2 Under Vibration

Due to insufficient vibration, concrete is not compacted well enough. This is known as under-vibration. It reduces the strength and durability of concrete, and also affects the surface finish. Under-vibration is also the normal cause of honeycombs in concrete. It can be avoided by proper supervision of duration of vibration and spacing between insertion of needles to avoid shadow areas being not vibrated.

5.5.3 Over Vibration

Excessive vibration causes segregation of lager particles from finer ones. It is characterized by excessive thickness of mortar at the surface of the concrete. The surface concrete will also have a frothy appearance. This problem usually occurs in poorly proportioned mixes. Controlling the water–cement ratio and the duration of vibration are a few measures that are adopted to overcome this problem.

5.5.4 Re-vibration

Generally, concrete is vibrated as soon as it is placed in the formwork. However, in certain cases, to ensure a good bond between the first layer of concrete already laid and fresh concrete, it is vibrated again. This is called re-vibration. The variation of strength due to re-vibration is shown in Fig. 5.18.

It is necessary to vibrate concrete when it is fresh, before the initial setting. However, when concrete is laid in layers, the bottom layer stiffens before the top layer is laid. To ensure a good bond between the layers, the top of the bottom layer can be re-vibrated if it can regain its plastic state. Based on experimental results, it has now been established that concrete can be re-vibrated.

The improvement in strength because of re-vibration results from the expulsion of trapped water and is found to be highest in concrete that is susceptible to bleeding. In spite of these advantages, re-vibration is not ordinarily recommended or practised because it involves one more step in the production of concrete. Careless re-vibration, when applied after considerable time has elapsed, reduces the strength of concrete.

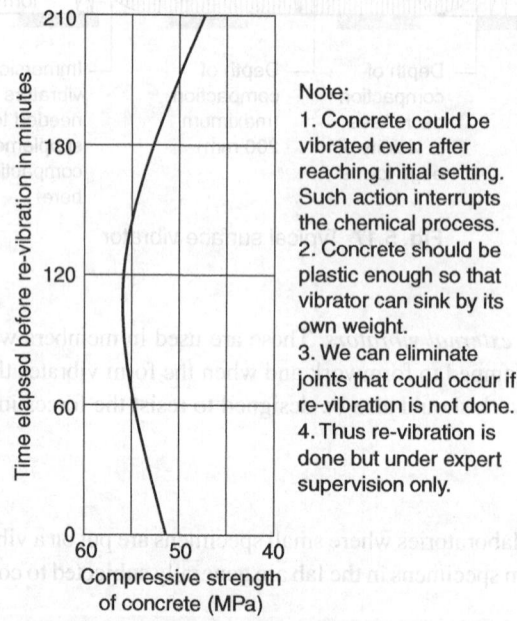

Note:
1. Concrete could be vibrated even after reaching initial setting. Such action interrupts the chemical process.
2. Concrete should be plastic enough so that vibrator can sink by its own weight.
3. We can eliminate joints that could occur if re-vibration is not done.
4. Thus re-vibration is done but under expert supervision only.

Fig. 5.18 Effect of re-vibration to eliminate cold joints permitted only under expert supervision

5.6 Curing Concrete

Curing refers to maintaining satisfactory moisture content and temperature in fresh concrete in order to achieve the desired strength and hardness.

5.6.1 Importance of Curing

Drying removes the water needed for hydration. Without adequate hydration, concrete tends to be weak. Temperature is an important parameter to consider for proper curing. In outdoor concreting, temperature, humidity, wind velocity, etc., contribute to the evaporation of water. Properly cured concrete has better durability and better surface hardness, and is less permeable.

Powers (1932) has shown that hydration is significantly less when the relative humidity in the pores drops below 80%. Lerch (1957) conducted tests to study the effects of air temperature, wind velocity, and the temperature of concrete on the loss of water available for hydration. Figures 5.19–5.21 present the findings.

Figure 5.22 shows the effect of loss of moisture on the reduction of humidity in the pores.

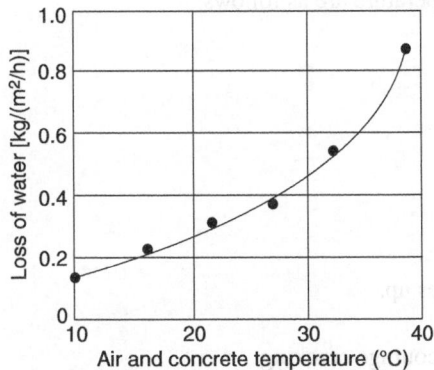

Fig. 5.19 Effect of air temperature on loss of
water from concrete

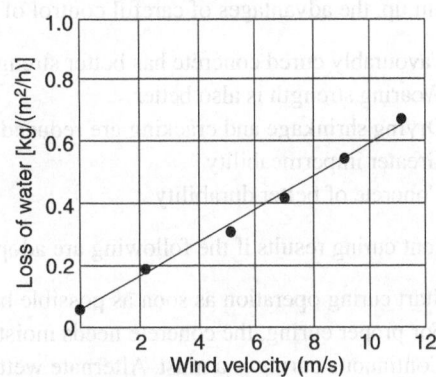

Fig. 5.20 Effect of wind velocity on loss of water
from concrete

The prevention of loss of moisture from concrete is important not only from the point of view of strength development but also from the point of view of prevention of plastic shrinkage, decrease in permeability, and improvement of resistance to abrasion. The loss in the 28-day strength seems to be directly related to the loss of moisture during the first three days. From Fig. 5.23, it is evident that a 5% loss in moisture leads to nearly 75% loss in strength. Hence, continuous curing during the first three days is a must. Intermittent curing seems to be even worse than no curing at all.

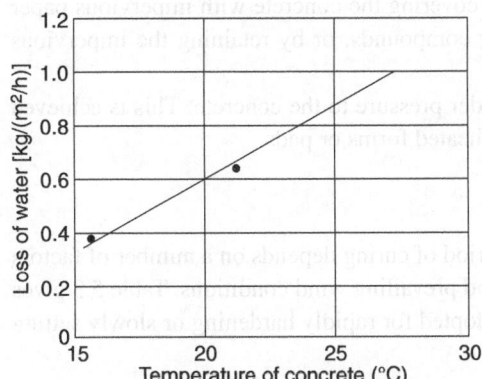

Fig. 5.21 Effect of temperature of concrete
on loss of water from concrete

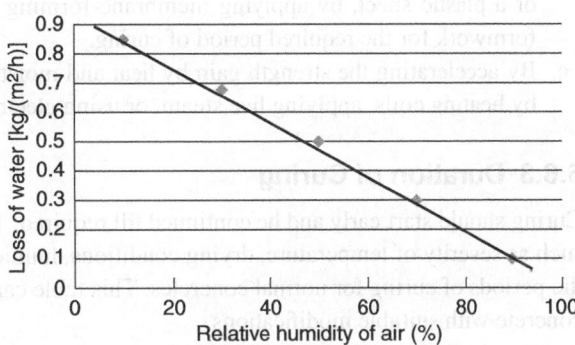

Fig. 5.22 Effect of relative humidity on loss of water
from concrete

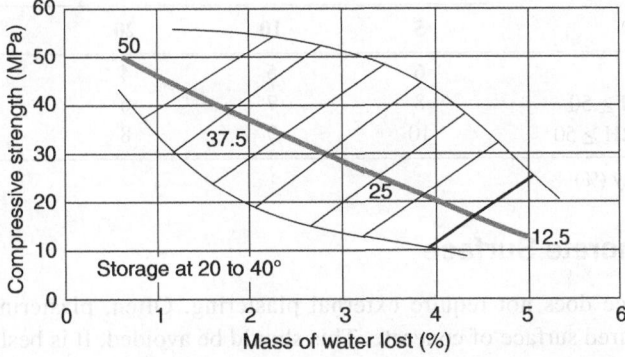

Fig. 5.23 Relationship between compressive strength and loss of water during first three days

To sum up, the advantages of careful control of moisture and temperature are as follows:

a. Favourably cured concrete has better strength
b. Wearing strength is also better
c. Drying shrinkage and cracking are reduced
d. Greater impermeability
e. Concrete of better durability

Efficient curing results if the following are adopted:

a. Start curing operation as soon as possible before concrete dries up.
b. For proper curing, the concrete needs moisture.
c. Continuous curing is a must. Alternate wetting and drying encourage cracking.
d. The ideal curing temperature is 23°C.
e. Cure concrete for at least 10 days.

5.6.2 Methods of Curing

Concrete is kept moist by the following methods:

a. By maintaining the mixing water in concrete without evaporation during the early strength development period. Ponding, immersion, spraying, fogging, and providing saturated wet covering prevent evaporation and maintain a cool temperature which is beneficial in hot weather.
b. By reducing the loss of mixing water from the surface by covering the concrete with impervious paper or a plastic sheet, by applying membrane-forming curing compounds, or by retaining the impervious formwork for the required period of curing.
c. By accelerating the strength gain by heat and moisture under pressure to the concrete. This is achieved by heating coils, applying live steam, or using electrically heated forms or pads.

5.6.3 Duration of Curing

Curing should start early and be continued till required. The period of curing depends on a number of factors such as severity of temperature, drying conditions, humidity, and prevailing wind conditions. Table 5.3 gives the periods of curing for normal concretes. This table can be adopted for rapidly hardening or slowly setting concrete with suitable modifications.

Table 5.3 Minimum curing period in days

	Temperature (°C)				
	5	10	20	30	40
No sun, RH* ≥ 80	6	5	4	3	3
Medium sun and winds, RH ≥ 50	8	7	6	5	4
Strong sun or high winds, RH ≥ 50	10	9	8	7	6

*RH denotes relative humidity (%)

5.7 Finishing Concrete Surface

A well-concreted surface does not require external plastering. Often, plastering is done to hide the defects in the manufactured surface of concrete. This should be avoided. It is best to have exposed form finish.

5.7.1 Different Types of Finishes

The following three types of rubbed finishes can be provided:

a. *Smooth rubbed finish*: This is applied within a day after removal of the formwork. The wetted surface is rubbed with Carborundum abrasive unit until the desired finish is attained. No external grout is used.

b. *Grout cleaned finish*: This is achieved by the application of grout. The grout is made using one part of Portland cement and 1.5 parts of fine sand mixed with enough water to achieve the consistency of a thick paint. White cement is added to match the surrounding concrete. The grout is scrubbed into the voids, and then the surface is rubbed and kept damp for 36 h.

c. *Cork floated finish*: This requires a stiff grout to be applied to the wetted surface. Proportions of sand and cement are 1:1, with white cement included as needed for colour matching. The grout is compressed into the voids by grinding with a slow-speed grinder, and a cork float is used to produce the final finish with a swirling motion.

5.8 Autogenous Healing

Fine cracks appear in concrete in the initial stages due to shrinkage and temperature. If these cracks are allowed to close without lateral displacement under moist conditions, they tend to heal completely due to hydration of the cement that had not hydrated until now. This phenomenon is called autogenous healing of concrete. The width of cracks that can undergo this type of healing is estimated to be 0.1–0.2 mm. Moisture conditions and the application of minute pressure help autogenous healing (Fig. 5.24). Cracks having smaller widths heal faster (within about a week).

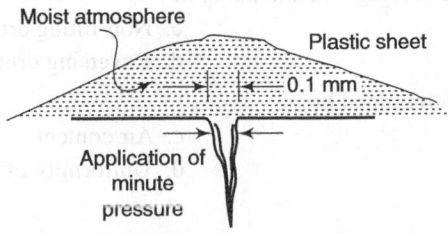

Fig. 5.24 Autogenous healing encouraged by favourable forces

Exercises

Review Questions

1. What are the stages of manufacture of concrete?
2. Why is weigh batching preferred for important works? Explain.
3. What are the disadvantages of volume batching?
4. List the components of a typical weigh batching plant.
5. What are the methods of mixing concrete?
6. What are the two common types of mixers used?
7. Distinguish between continuous mixers and batch mixers.
8. What performance attributes of the mixed concrete should one examine for uniformity of concrete produced?
9. What is re-tempering of concrete? Discuss.

10. What is the mixing schedule and how does it affect the production of concrete?
11. What are the various methods of transporting concrete?
12. What aspects cause weakness in concrete during placing?
13. What is meant by striking form?
14. What are the disadvantages of poor compaction?
15. What are the various methods of compaction?
16. List the different types of vibrators used for concreting.
17. What is meant by over-vibration?
18. Under what circumstance is concrete re-vibrated?
19. What are the advantages of good and efficient curing?
20. What is autogenous healing? Explain.

Multiple Choice Questions

1. Weigh batching is preferred because
   a. It is easier to weigh the materials
   b. Volume of sand is uncertain
   c. All cement bags weigh 50 kg
   d. Accurate balances are available

2. In weigh batching, aggregates are measured to an accuracy of
   a. 1%
   b. 2%
   c. 3%
   d. 4%

3. Tilting drum mixers are preferred
   a. For easy loading and unloading
   b. For efficient operation
   c. Because they are easily available
   d. Because they can be mounted on trucks

4. The same opening is used for loading and unloading in
   a. Tilting drum mixer
   b. Pan-type mixer
   c. Non-tilting drum mixer
   d. Reversing drum mixer

5. The efficiency of mixing is determined by
   a. Water–cement ratio
   b. Density
   c. Air content
   d. Uniformity of all properties

6. Re-tempering is
   a. Remixing of concrete which has stiffened to increase the workability
   b. Addition of water to concrete
   c. Addition of admixture to concrete
   d. Checking of the slump of concrete

7. Segregation occurs during transport of concrete because
   a. The concrete is jolted
   b. The concrete could be dropped from a height
   c. The mix becomes harsh because of loss of moisture
   d. All of these

8. Which equipment is used for moving concrete vertically and horizontally
   a. Monorail
   b. Trucks
   c. Lift
   d. Crane and cable way

9. Concrete can be placed in difficult locations in tall buildings and structures using a
   a. Bucket
   b. Dumper
   c. Concrete pump
   d. Boom conveyor

10. Striking form can be done only after examining
    a. Duration after concreting
    b. Type of concrete used
    c. Strength gain attained by concrete
    d. All of these

11. Compaction is intended to
    a. Remove air voids
    b. Increase the strength
    c. Make the concrete impermeable
    d. Increase durability

12. Curing should be started
    a. Immediately after concreting, before the concrete dries up
    b. 24 h after concreting
    c. 48 h after concreting
    d. 36 h after concreting

Answers to Multiple Choice Questions

1. b   4. d   7. d   10. d
2. b   5. d   8. d   11. a
3. a   6. a   9. d   12. a

CHAPTER 6

Rheological Models of Fresh Concrete

6.1 Introduction

Wet concrete mix is a system in which cement paste and water bind aggregates consisting of sand and gravel or crushed stone into a homogeneous mass. The flow characteristics of wet concrete depend on variables involving the properties of aggregates and other ingredients of the mix. Rheology is the science of flow of materials. Rheological models of fresh concrete give us information about the ability of fresh concrete mix to flow under its own weight and fill the moulds without voids. With respect to fresh concrete, we use the following terms to specify the flow characteristics:

a. Workability
b. Flowability
c. Flow under its own weight
d. Stability
e. Compactability
f. Pumpability
g. Fillability
h. Finishability

With the advent of more fluid concretes (pumpable concrete, self-levelling concrete), it has become necessary to measure the flow properties of concrete in definite quantifiable terms. The flow characteristics of concrete are complicated, unlike simple fluids, and depend on the relationship between stress, strain, rate of strain, and time.

In reinforced concrete, the ability of concrete to flow through reinforcing bars depends on the yield stress of fresh concrete and the dimensions of the formwork. These are identified as continuum properties. As a general rule, the largest aggregate size should be smaller than 1/5 times the smallest dimension of form, namely the thickness of the wall. The mix design of concrete involving flow characteristics should consider spacing of reinforcing bars in the structure to be cast. These dimensions are identified as having finite size effects. The flow of concrete depends on both continuum and discrete (finite size) effects. Ideally, for good flow, design should consider both rebar structure and proper size of aggregates in the concrete mix design.

6.2 Simple Flow Test

Figure 6.1 shows a simple slump flow test. It is an attempt to evaluate the flow properties of concrete. Here slump and slump flow are used to describe concrete flow.

In addition, concrete consistency classes can be assessed based on flow table spread and compactability by compaction factor test. The slump measures vertical settlement of fresh concrete, whereas the flow table spread measures horizontal increase in diameter when the cone is lifted. Compactability indicates how well the concrete can be compacted without voids. The concrete consistency classes give us an idea of flow characteristics from the perspective of producing a good, homogeneous concrete mass which can be compacted to have maximum density.

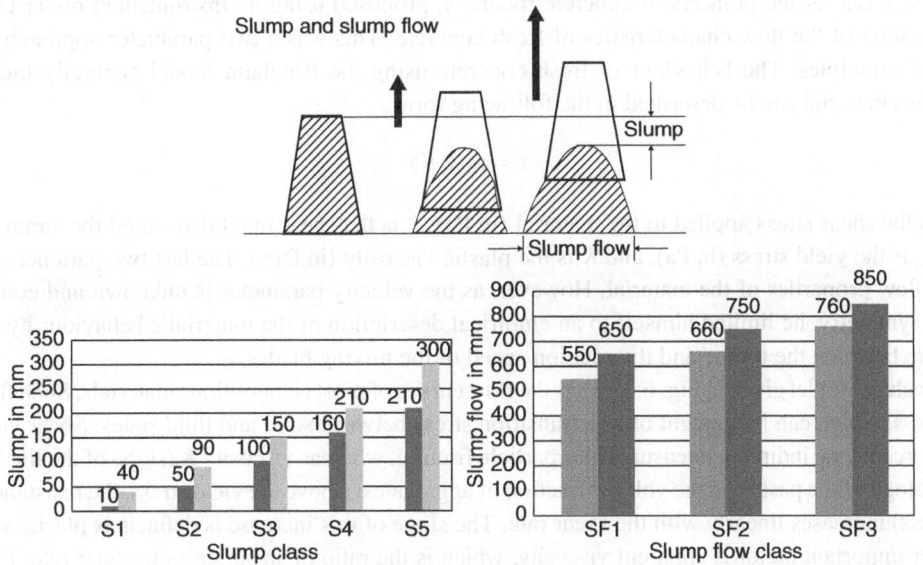

Fig. 6.1 Slump and slump flow

6.3 Rheological Models

Rheology is the science of flow and deformation of material. Flow or deformation of wet material is caused by forces applied on the material. Thus, it is logical to study the connection between applied forces and the deformation or the deformation rate.

In rheology, we consider dynamic deformations rather than static deformations, as we do in mechanics. Thus, shear strain is replaced by shear strain rate. By a simple approximation, we can say the shear strain rate is average velocity of flow divided by flow thickness (Fig. 6.2). Based on the flow situation, the shear strain rate varies as shown by the indicative values in Table 6.1.

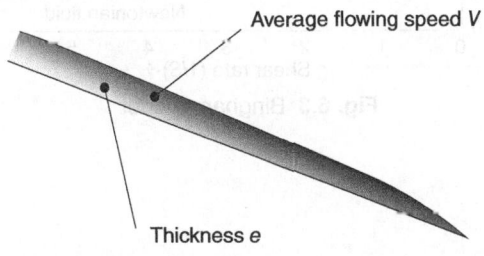

Shear strain rate = V/e

Fig. 6.2 Definition of shear strain rate for shear flow

Table 6.1 Shear strain rates for different activities (flow situation)

Activity (flow situation)	Maximum shear strain rate* (1/S)
Hand mixing	10–60
Truck mixing	10
Pumping	20–40
Casting	10

*Values based on material available in literature.

GH Tattersall, one of the pioneers of concrete rheology, proposed using an instrumented mixer to obtain a characterization of the flow characteristics of fresh concrete. This was a two-parameter approach based on rheological principles. The behaviour of fresh concrete using the Bingham model generally intended for cementitious material can be described in the following form:

$$\tau = \tau_0 + K\dot{\gamma} \tag{6.1}$$

where τ is the shear stress applied to the material (in Pa), $\dot{\gamma}$ is the shear rate (also called the strain gradient) (in s^{-1}), τ_0 is the yield stress (in Pa), and K is the plastic viscosity (in Pa·s). The last two parameters characterize the flow properties of the material. However, as the velocity parameter is unknown and complex due to lack of symmetry, he limited himself to an empirical description of the material's behaviour by using the relationship between the torque and the rotation speed of the mixing blades.

The Bingham model given in Fig. 6.3 shows the flow curves of most cementitious materials. Such fluids have a yield stress, which can be thought of as a transition stress between solid and fluid states. Shear yield stress (y-axis intercept), τ_0, indirectly measures interparticle friction, whereas viscosity K (slope of the line) depends on the rheology of the paste and the volume fraction of aggregates. Above the yield stress, the resistance to flow (shear stress) increases linearly with the shear rate. The slope of this increase is defined as plastic viscosity.

Another important factor is apparent viscosity, which is the ratio of shear stress to shear rate. Fluids that exhibit yield stress have infinite apparent viscosity at zero shear rates, as can be seen in Fig. 6.4. This explains why slow flows occur in particular at the corners of the formwork. They are mainly governed by yield stress.

Though Bingham model is the most commonly adopted, there are other rheological models depending on the type of fluid whose flow is being characterized. A few important models are shown in Fig. 6.5.

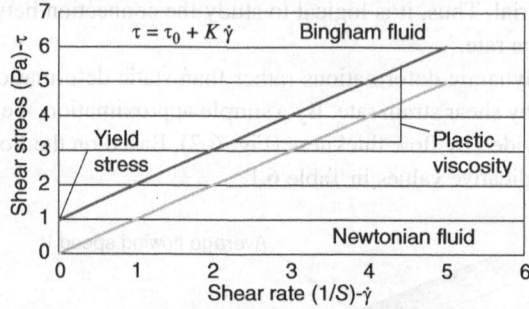

Fig. 6.3 Bingham model

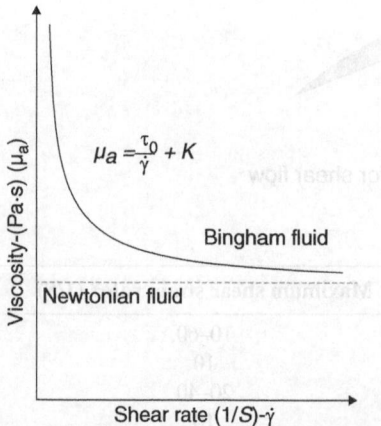

Fig. 6.4 Variation of viscosity with shear rate

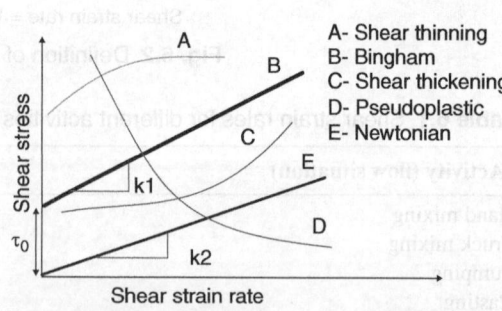

A- Shear thinning
B- Bingham
C- Shear thickening
D- Pseudoplastic
E- Newtonian

Fig. 6.5 Various flow models

6.4 Schematic Differences in Flow Curves of Different Types of Concrete

The shear yield strength of self-compacting concretes is of the order of 0–60 Pa. This is low when compared to traditional concrete. In addition, we should account for the variability of plastic viscosity. Hence a few more models relating shear stress and shear strain rate have been in use. These models are shown in Fig. 6.5. In model A, stress increases with strain rate which is known as shear thinning. B is the Bingham model which shows that stress increases linearly with strain rate starting from a residual shear stress. C shows shear thickening behaviour which is the opposite of shear thinning. Unlike Bingham model, some others accommodate non-linear behaviour between stress and strain rate. Model D shows the effect of viscosity modifying agents such as polysaccharides. Note that in this model there is a drop in shear stress at high strain rates. This behaviour is known as pseudoplastic and is exhibited by certain gels having fluid property when they are stirred but becomes thixotropic (semi-solid) while standing without disturbance.

Figure 6.6 schematically shows the difference in flow curves of different types of concrete. It is noteworthy to see that self-compacting concrete has a lower yield stress. Similarly, we can also note that high-strength concrete has higher plastic viscosity.

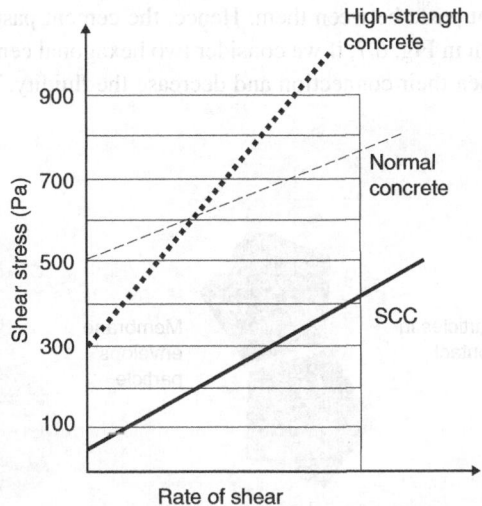

Fig. 6.6 Difference in flow curves of different types of concrete

6.5 Rheological Test Methods

Most of the concrete rheometers apply a shear rate and measure the shear stress response of the concrete. The rheological test methods for concrete tend to fall into one of the following categories based on the mode by which the concrete is forced to flow:

Confined flow Concrete flows under its own weight through an orifice. The flow can be due to applied pressure. Flow cone, filling ability devices, and flow test through an opening are methods which come under this category.

Free flow Concrete flows freely under its own weight or is poked and inserted by a rod or a plunger by only gravitational force. Slump and penetrating rod come in this category.

Vibration flow Concrete flows under the influence of vibration which is applied through a vibrating table. The Vee Bee test belongs to this category.

Rotational rheometer Concrete is sheared between two parallel surfaces, one or both of which are rotating. Generally, it has been observed that rheometers give reliable results when the yield values of concrete (and also the ratio of yield stress to plastic viscosity) are low.

Measurements from rheometers provide useful information regarding the nature of the flow of concrete. Such measurements help in visualizing superplasticizers—viscosity modifying agent systems used in self-compacting concretes. It helps in solving compatibility issues related to different types of materials used in self-compacting concretes. Despite these, it is currently not used in concrete technology field practice.

6.6 Factors Affecting Rheological Properties

A number of factors affect the flow characteristics of concrete. The effect of mixing, cohesion, strain gradient between aggregates, use of admixtures, and VMA are a few factors which affect the flow. These are briefly discussed here.

Mixing of Concrete

From a practical standpoint, a system must be deagglomerated in a way that is representative of each application considered. When mixing concrete, it is important to understand that rigid aggregates concentrate the strain gradient in the cement paste between them. Hence, the cement paste is subjected to much higher stresses than expected. As shown in Fig. 6.7, if we consider two hexagonal cement particles in the mix left to hydrate, hydration will strengthen their connection and decrease the fluidity. This would happen mostly for low-intensity mixing.

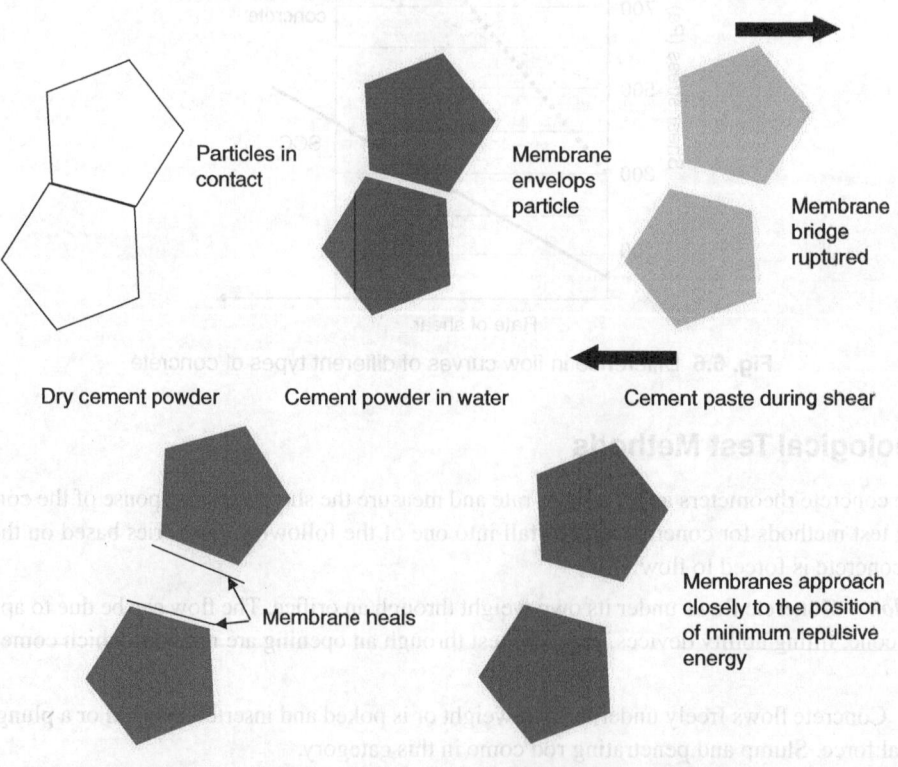

Fig. 6.7 Effect of mixing

On the other hand, in high-intensity mixing, first the particles hydrate, and then they come into contact. Hence there is no connection. Under this situation fluidity is improved. The intensity of mixing destroys the bond and attraction between the particles.

Effect of cohesion

Yield stress in concrete is the result of attractive forces between fine particles in the system which include, in addition to cement, slag, fly ash, silica fume, etc. These attractive forces result in a cohesive system. Due to shear, few of these bonds get broken (Fig. 6.8). When enough bonds are broken, the particles move farther apart. In reality, this signals transition from solid to viscous/liquid state.

Effect of water and superplasticizers

Addition of water reduces both yield stress and plastic viscosity as shown in Fig. 6.9(a). However, addition of superplasticizers has an effect only on yield stress (see Fig. 6.9(b)).

Other factors

There are a number of other factors which influence the flow of concrete such as heat of hydration and entrained air. These effects are discussed in relevant chapters.

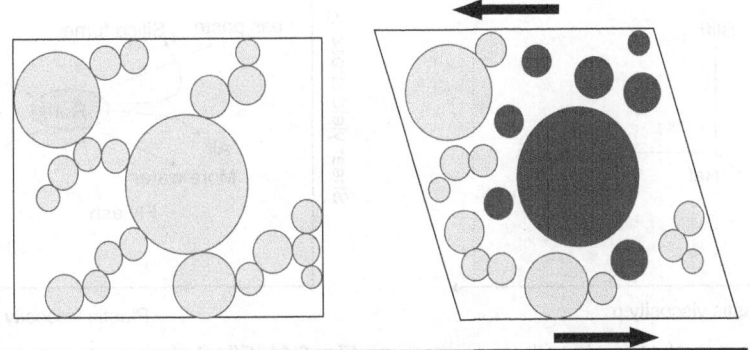

Fig. 6.8 Solid to liquid transition due to yield stress

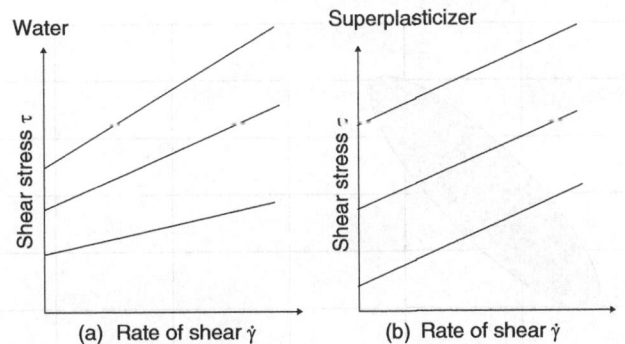

Fig. 6.9 Effect of water and superplasticizer on flow characteristics

6.7 Effect of Rheological Properties on Different Types of Concrete

In this section we examine the influence of rheological properties on different types of concrete.

Different types of concrete

Figure 6.10 shows the three-dimensional relationship between different types of concrete and the rheological parameters. For example, compared to reference concrete, a wet concrete can be produced by decreasing both yield stress and plastic viscosity, whereas a stiffer concrete can be produced by increasing the yield stress. Initially, addition of silica fume decreases viscosity (fine particle content increases the flow), whereas higher dosage increases the yield stress as well as viscosity.

Different additives

The effect of air, water, and other mineral admixtures on rheological parameters of concrete is shown in Fig. 6.11.

Rheological properties of different concretes

Shear yield stress of self-compacting concrete is in the range of 0–50 Pa, whereas for normal concrete it is high in the range of 100–300 Pa as shown in Fig. 6.12. Plastic viscosity of normal concrete is in the range of 0–40 Pa·s. However, plastic viscosity for self-compacting concrete is fairly high having a range of 50–90 Pa·s.

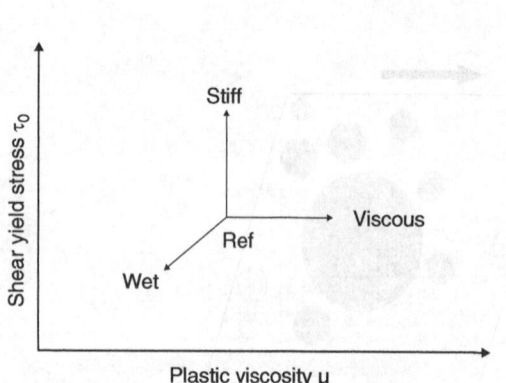

Fig.6.10 Different types of concrete with respect to flow characteristics

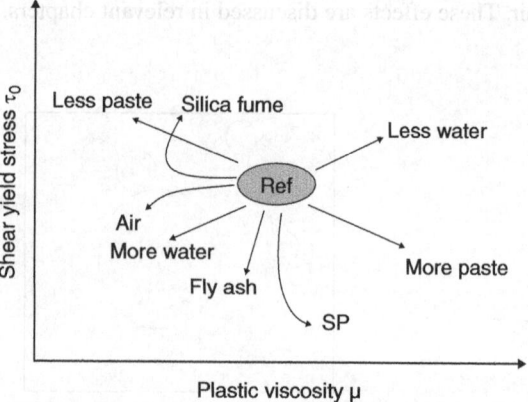

Fig. 6.11 Effect of air, water, and mineral admixtures on flow characteristics of concrete

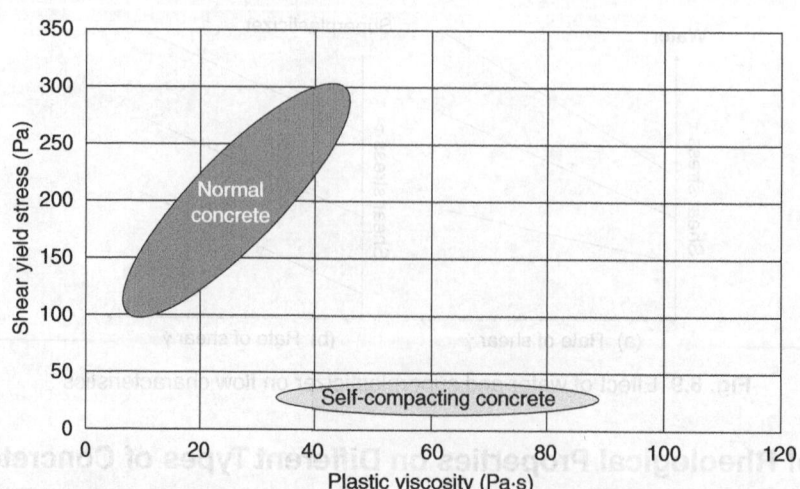

Fig. 6.12 Rheological properties of normal and self-compacting concrete

Control of segregation

Segregation can be controlled by the addition of viscosity modifying agents such as water-soluble polysaccharides which exhibit a pseudoplastic rheological behaviour. Such materials show a decrease in shear stress at high shear rates. As the viscosity of water with such VMAs is low, the concrete flows without hindrance. When the concrete comes to rest (this is when segregation can occur), the viscosity of the concrete mix due to the presence of VMA increases, which helps in keeping the mixture stable. Thus stability, mobility, and compactability are all achieved simultaneously.

Exercises

Review Questions

1. List the flow properties of fresh concrete.
2. What are the continuum and discrete effects that affect RCC when concreting?
3. How is a simple flow test used to classify slump and flow characteristics?
4. Define shear stress, shear rate, and yield stress with respect to Bingham model.
5. How is shear rate connected with mixing, pumping, and casting?
6. How is viscosity related to shear stress?
7. List two flow models used in rheological studies.
8. What are the schematic differences in the flow curves of different types of concrete?
9. List the rheological test methods describing their relative use.
10. What are the factors that affect rheological properties of concrete?
11. Discuss the practical applications of rheological methods in concrete technology.
12. Compare the rheological properties of normal concrete with self-compacting concrete.

Multiple Choice Questions

1. Rheology of fresh concrete deals with
   a. Strength of concrete
   b. Young's modulus of concrete
   c. Load carrying capacity of concrete
   d. Flow characteristics of fresh concrete

2. Shear rate has the unit of
   a. Distance
   b. Time
   c. Distance/Time
   d. 1/Time

3. The rheology of fresh concrete does not include
   a. Stability of fresh concrete
   b. Stress through fresh concrete mass
   c. Water–cement ratio
   d. Compatibility of concrete

4. The flow properties of fresh concrete mainly depend on
   a. Deformation characteristics of fresh concrete
   b. Richness of mix
   c. Strength of concrete
   d. Modulus of concrete

5. Shear yield stress is the lowest for
   a. Normal concrete
   b. High-strength concrete
   c. Ultrahigh-strength concrete
   d. Self-compacting concrete

6. The difference between Bingham fluid and Newtonian fluid is
   a. Yield stress is non-zero in the former
   b. Yield stress is non-zero in the latter
   c. Plastic viscosity is zero in the latter
   d. Plastic viscosity is zero in the former

7. Low-intensity mixing causes
   a. Defloculation
   b. Floculation
   c. Breaking of the bond between crystals
   d. Good flow

8. The model used for simulating fresh concrete is
   a. Bingham model
   b. Newtonian model
   c. Vicat model
   d. Navier–Stoke model

9. Addition of water to the concrete mix
   a. Reduces plastic viscosity
   b. Increases plastic viscosity
   c. Does not change plastic viscosity
   d. Does not affect flow

10. Plastic viscosity is high for
    a. Normal concrete
    b. High-strength concrete
    c. Self-levelling concrete
    d. Concrete without fines

Answers to Multiple Choice Questions

1. d   4. a   7. b   10. c
2. d   5. d   8. a
3. c   6. a   9. a

CHAPTER 7

Properties of Hardened Concrete

The quality of concrete is determined by its mechanical properties as well as its ability to resist deterioration. The mechanical properties can be broadly divided into short-term and long-term properties (Fig. 7.1). The long-term properties include the behaviour of concrete under creep, shrinkage, and fatigue and durability characteristics, i.e., characteristics that help it withstand environmental forces, such as impermeability, abrasion, fire resistance, and freeze–thaw resistance. As far as the short-term (instantaneous) properties are concerned, the strength of concrete is the most important characteristic, as it has a strong relationship with quality. The compressive strength of concrete is widely used for specifying the concrete being used in structural drawings and at construction sites. Strength as a parameter is used for controlling as well as evaluating almost all other properties of concrete because of its relationship with durability and dimensional stability. Durability properties of concrete including resistance to fire are discussed in a later chapter.

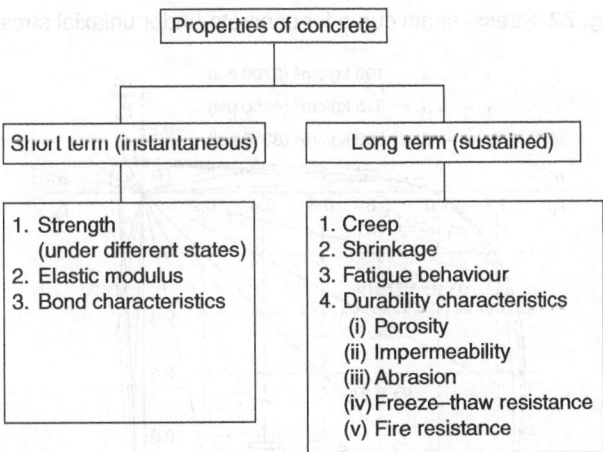

Fig. 7.1 Properties of concrete

7.1 Strength under Uniaxial and Multiaxial Stresses

In India and in many European countries, the short-term properties of concrete such as elastic modulus, tensile and compressive strength, shear strength, and stress–strain characteristics are expressed in terms of the uniaxial cube compressive strength of a 15 cm × 15 cm × 15 cm cube, moist cured for 28 days. This compressive strength is used as the design basis; it is also used for establishing the type of concrete by routine testing, which involves making cubes at the construction site and testing them at the end of 28 days. The cube compressive strength is represented by σ_{cu}. By conducting tests on a cube, only strength parameters can be obtained. In some countries, such as the USA, the compressive strength of concrete is assessed by testing $6'' \times 12''$ (150 mm × 300 mm) cylinders. Testing cylinders enables one to obtain the stress–strain properties of concrete under uniaxial compression. The cylinder compressive strength is represented by f_c'.

Figure 7.2 shows a typical stress–strain curve for concrete under uniaxial compression. Normal-strength concrete reaches peak stress (called the *maximum compressive cylinder strength*) at a strain (ε) of about 0.002. The failure load of the cylinder occurs at a strain of about 0.0035 or more. The portion of the stress–strain curve beyond maximum stress and up to the failure load is known as the *falling branch*. A gradual falling branch is preferred over a steep falling branch. This depicts a ductile failure. A steep falling branch exhibits a brittle (or sudden) explosive failure.

The tensile strength of concrete is low, and it is susceptible to sudden failure under tension. Hence, while designing structures, the tensile strength of concrete is ignored.

In structures, concrete is rarely subjected to uniaxial stress. In deep beams and walls, it is subjected to biaxial stress. The inset in Fig. 7.3 shows the state of biaxial stress for a typical concrete element. The strength

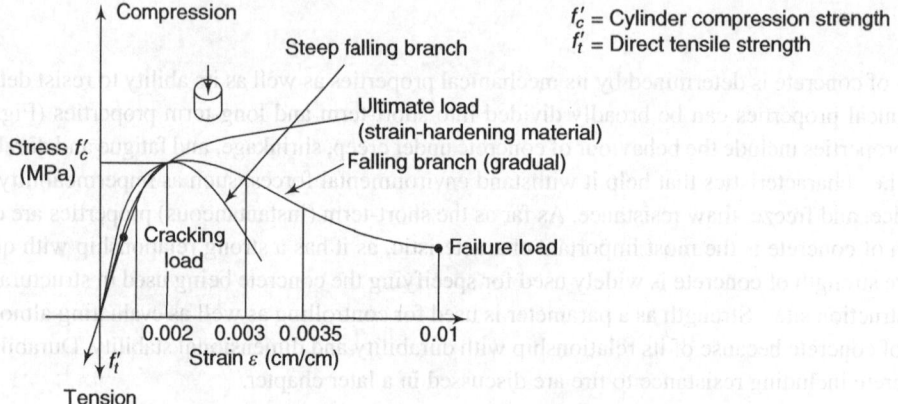

Fig. 7.2 Stress–strain curve for concrete under uniaxial stress

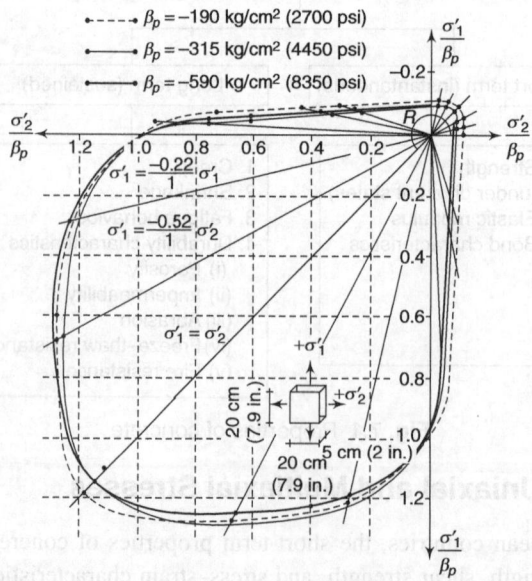

Fig. 7.3 Strength envelopes for concrete under biaxial loading (*Source*: Kupfer et al. 1969. Reproduced with permission from American Concrete Institute)

envelope of concrete under biaxial in-plane stress developed by Kupfer et al. (1969) is shown in Fig. 7.3. Note that the compressive failure stress increases under biaxial stress. Under compression-tension, the compressive strength decreases linearly with increasing tension, whereas under biaxial tension, the strength seems to be independent of biaxial loading.

7.2 Failure Modes

The failure mode of concrete under stress depends on the following parameters:
(i) the state of stress, (ii) the type of test, and (iii) the effect of loading type.

7.2.1 State of Stress

Under uniaxial tension, cracks initiate and grow rapidly. The pre-existing cracks at the interface between aggregate and mortar join up with the new cracks formed due to loading, leading to brittle failure.

Cracks start appearing in normal concrete under uniaxial compression when it is subjected to about 50% of the ultimate load. At this stage, a stable system of cracks already exists at the interface between coarse aggregates and mortar. At higher stresses, cracks start appearing in the mortar matrix. Finally the interface cracks reach the cracks in the mortar matrix at about 80–85% of the ultimate load. This is followed by significant stiffness loss and failure.

7.2.2 Type of Test

Figure 7.4 shows the failure cracks in a cylinder and a cube. Note the difference between the orientations of the cracks. The cylinder initially bulges and forms near-vertical cracks. The cracks in the cube occur at an inclination of about 30° to 45°. The influence of the loading platen surface offering shear resistance to the top portion of the cube causes the cracks to occur at a much steeper angle (45°). Such steep cracks do not occur in the cylinder.

Cylinder strength will be less than the cube strength. This is due to the type of failure explained earlier. It is reasonable to assume the following relationship between cube and cylinder strength.

$$f_c' = (0.85 \text{ to } 0.9)\,\sigma_{cu}$$

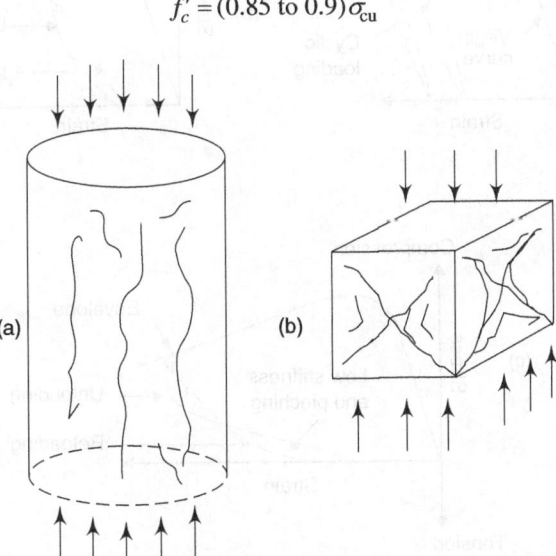

(a)   (b)

Fig. 7.4 Typical failure cracks in concrete (in compression) in a cylinder (a) and a cube (b)

7.2.3 Effect of Loading Type

Sustained load causes concrete to fail at a lower ultimate load compared to specimens tested under short-term loading in laboratories. This occurs due to progressive micro-cracking under sustained load. Figure 7.5 shows the relationship between short-term and sustained (long-term) loading strengths. Note from the failure limit that the long-term strength is only 80% of the short-term strength.

An increase in the loading rate (represented by a decrease in loading duration *t*) leads to improved strength. Thus, the strength of concrete increases with the rate at which load is applied.

Plain concrete subjected to repeated loading exhibits both strength and stiffness degradation as shown in Fig. 7.6. The unloading curve—after peak load has been reached—shows non-linearity. The reloading curve

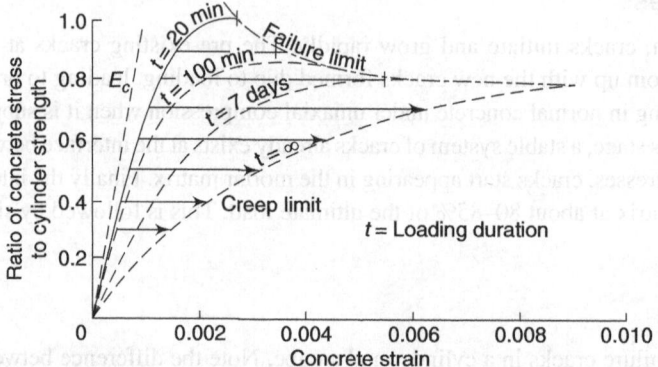

Fig. 7.5 Relationship between short-term and long-term strength (*Source*: Rusch 1960. Reproduced with permission from American Concrete Institute)

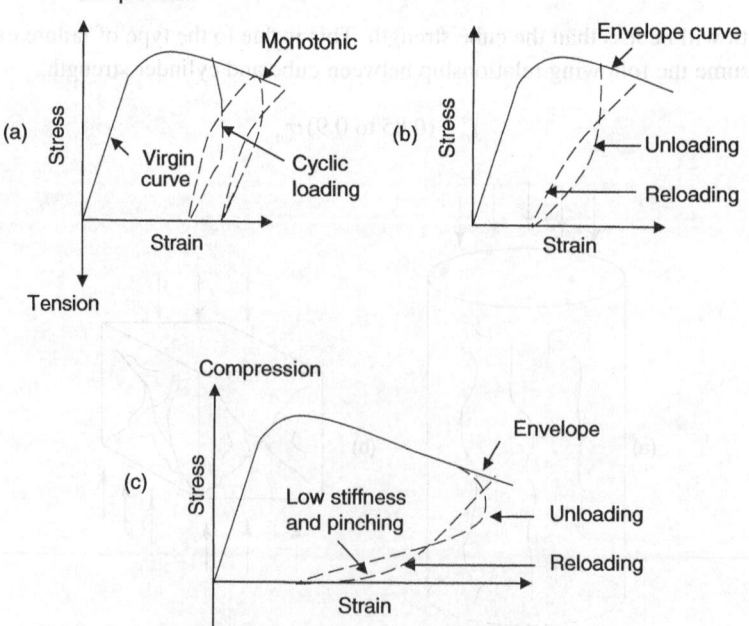

Fig. 7.6 Effect of cyclic load on plain concrete behaviour: (a) monotonic and cyclic loading, (b) unloading and reloading at low strain levels, and (c) unloading and reloading at high strain levels

shows low stiffness at low loads. These properties become important while designing structures which are subjected to seismic loads.

7.3 Tensile Strength

Tensile strength is an important property of hardened concrete. The tensile strength of concrete is very low compared to its compressive strength. Generally tensile strength of concrete is determined by indirect methods because of the difficulty in applying direct tension to concrete. However, sophisticated equipment are available now using which we will be able to perform direct tension test. We will discuss about these special tests later in the chapter on testing. Herein, we will consider normal tensile strength evaluation of concrete using simple laboratory procedures. It should be kept in mind that these approximate methods generally give higher values compared to direct uniaxial tensile strength. The following two types of tests are common:

- Splitting tensile strength test
- Flexural tensile test

7.3.1 Splitting Tensile Strength

Splitting tensile strength test determines the tensile strength in an indirect way. The compression load P is applied at diametrically opposite points of a cylinder as shown in Fig. 7.7. Note the uniform tension produced across the diameter joining load points which splits the cylinder into two halves. Using elastic solution maximum tensile stress on the vertical diameter is calculated as

$$f_t = \frac{2P}{\pi DL}$$

where

P = Compressive load at failure

D = Diameter of cylinder

L = Length of cylinder

The above test result is referred to as "splitting tensile strength" of concrete. It is about 1/10th of the compressive strength.

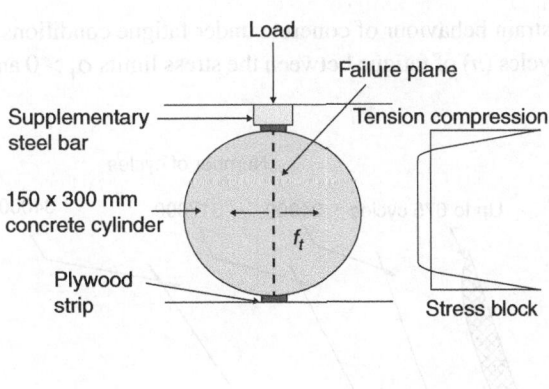

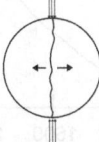

Fig. 7.7 Splitting tensile strength

7.3.2 Flexural Tensile Strength—Modulus of Rupture

The flexural test is performed on a 150 mm × 150 mm × 500 mm beam specimen, subjected to middle third loading as shown in Fig. 7.8. The flexural tensile stress is known as *modulus of rupture*.

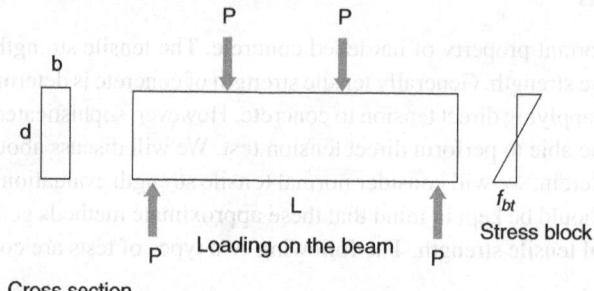

Cross section

Loading on the beam

Stress block

Fig. 7.8 Flexural tensile strength

The beam specimen is loaded by the two-point loading, and the failure tensile stress f_{bt} "modulus of rupture" is calculated as follows:

$$f_{bt} = \frac{PL}{bd^2}$$

where
f_{bt} = Modulus of rupture
P = Load at failure
L = Span between supports
b = Width of the beam
d = Depth of the beam

During test the failure should occur within the middle third loading.

7.4 Fatigue

Figure 7.9 shows the stress–strain behaviour of concrete under fatigue conditions. The specimen to be tested is subjected to a number of cycles (*n*) of fatigue between the stress limits $\sigma_1 > 0$ and $\sigma_2 > \sigma_1$. The stress range

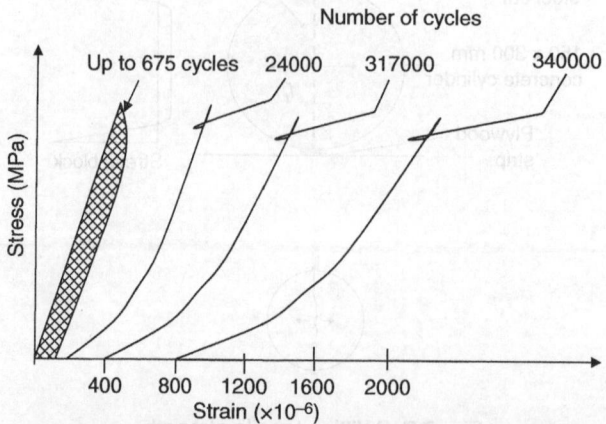

Fig. 7.9 Effect of fatigue loading on the behaviour of concrete

for the test is σ_1 to σ_2. After a specified number of cycles, the fatigue-damaged concrete is tested to determine the changes in the stress–strain properties caused by fatigue. The number of cycles imposed is shown on the curve itself. The stress–strain curve changes with the number of load repetitions. Initially it exhibits normal behaviour (concave downwards) known as *strain-softening behaviour*. With the increase in the number of cycles, the behaviour changes to *strain hardening*, i.e., concave upwards.

The changes in fatigue behaviour can be identified to occur in three phases.

- *Phase I*: Strain increases rapidly with increase in the number of cycles.
- *Phase II*: This is the stable phase; the strain increases linearly.
- *Phase III*: This stage is associated with the effect of instability; the strain increases progressively till fatigue failure occurs.

The area under the hysteresis loop (shown shaded in Fig. 7.9) is of interest because it represents energy dissipated. The strain at failure due to fatigue can be quite large (0.004 cm/cm).

The modified Goodman diagram (Fig. 7.10) for compressive fatigue can be used to assess the vulnerability of concrete to fatigue damage. It is seen that if the static sustained load is high, the range of stress a member can withstand for a given number of cycles is less. The modified Goodman diagram for flexural fatigue of concrete is shown in Fig. 7.11. The fatigue strength in flexure for 10 million cycles is only 55% of the static strength. Bond deterioration due to fatigue is important in RCC structures subjected to cyclic and reversed cyclic loading. This type of load occurs during earthquakes. The effect of fatigue on bonds is expressed in terms of the cumulative slip that occurs at the steel–concrete interface.

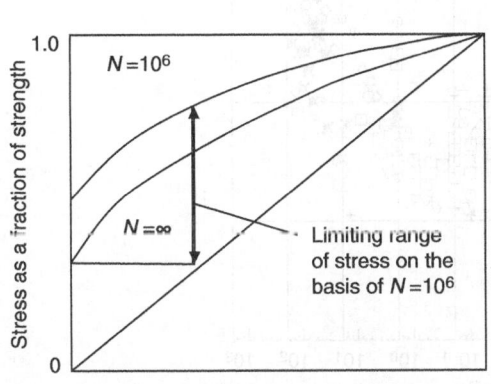

Fig. 7.10 Limiting stress range for compression fatigue

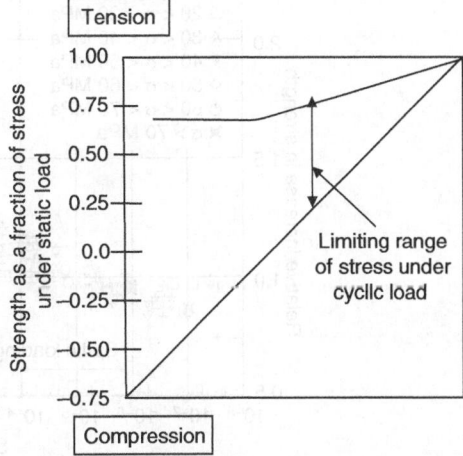

Fig. 7.11 Limiting range of stress under flexural fatigue

7.5 Impact Strength

The impact strength of concrete is important when it is subjected to sudden (impact) load or repeated impact load as witnessed in forge hammer foundations. Runway concrete pavements are also subjected to repeated impacts due to landing and take-off of aircraft. The impact strength of concrete increases with increase in compressive strength.

Figure 7.12 shows the relation between impact strength and compressive strength of concrete made with different types of aggregates. Angular and surface rough aggregates (broken granite) exhibit better impact strength. The rate of loading has a profound effect on strength. Figure 7.13 shows the relation between the percentage increase over static strength and the rate of loading. Impact at a loading rate 2–3 orders high increases the strength by about 50–60%. The effect of strain rate on the percentage increase in strength is shown in Fig. 7.14.

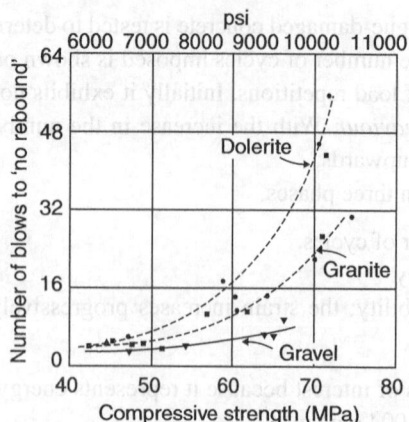

Fig. 7.12 Relation between impact strength and compressive strength for concrete made with different types of aggregates (*Source*: Green 1964. Reproduced with permission from ICE Publishing)

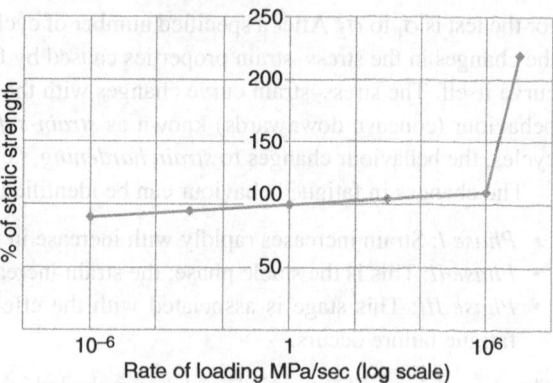

Fig. 7.13 Relation between compressive strength and rate of loading up to impact level

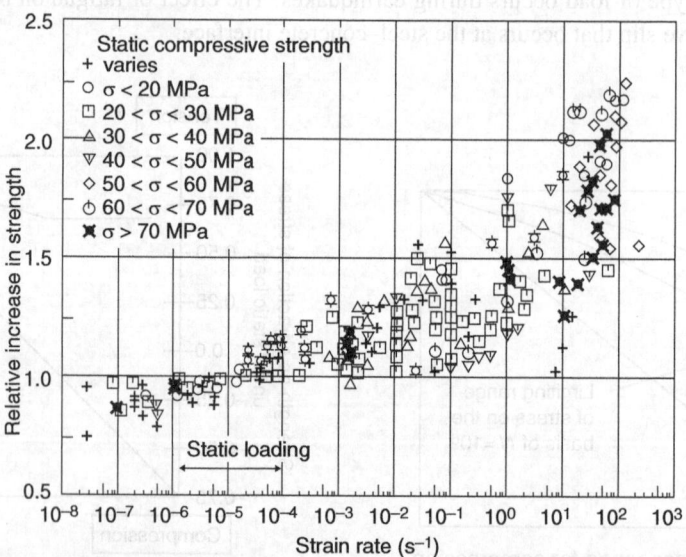

Fig. 7.14 Relation between relative increase in compressive strength and the strain rate for concretes of different strengths (*Source*: Bischoff and Perry 1991)

7.6 Abrasion Resistance

Abrasion resistance of concrete is desired for applications involving flow of water in spillways of dams and when bridge deck top surface is used as a wearing surface. It is considered as one of the prime requirements of durability. Deterioration of concrete surface may occur due to abrasion by sliding, scraping, percussion or action of abrasive materials carried by water. Abrasion loss under physical effects suffered by concrete pavements (roads and air-fields), industrial floors, railway platforms, dockyards, and footpaths should be evaluated and minimized.

In such cases, it becomes difficult to assess the abrasion resistance of concrete because the damage caused by the abrasive action in each case varies. However, evaluation of relative resistance of concrete surfaces is possible.

The test method based on IS: 9284-1979 'Reaffirmed 2002 Method of Test for Abrasion Resistance of Concrete' can be used to quantify the resistance. The surface of the concrete cubes is subjected to impingement of an abrasive charge (Fig. 7.15). As a result, abrasion of the concrete surface of the cubes occurs, and resulting loss in mass of the cubes is taken as the abrasion loss of concrete. The maximum loss suffered after test should be within the limits specified in Table 7.1.

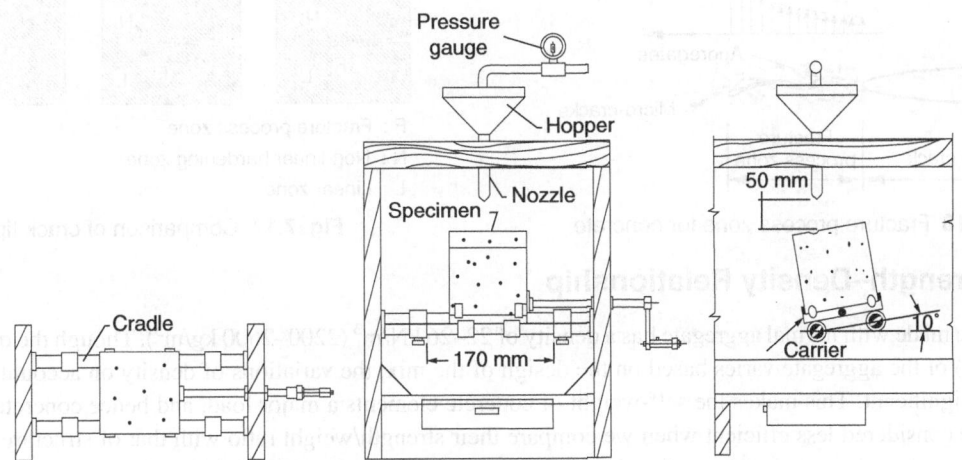

Fig. 7.15 Pneumatic sand blasting cabinet showing the cradle

Table 7.1 Suggested values of abrasion loss

S. No	Surfacing category	Maximum value of loss
1	Concrete pavement	0.16
2	Factory floor	0.16
3	Dockyard	0.16
4	Railway platform	0.24
5	Footpath	0.4

7.7 Fracture Properties of Concrete

Formation of cracks requires certain amount of energy which agrees with fracture mechanics concept. Finite element analysis based on strength criteria is not applicable near cracked zones because the results vary with mesh size. Limit analysis based on plasticity theory cannot be applied for brittle type of failure like punching shear and unreinforced shear failures. Size effects can be addressed by fracture mechanics. Concrete with tension softening type of failure can be tackled by fracture mechanics. Hence, there is justification for using fracture mechanics for concrete.

In classical mechanics critical stress does not depend on structure size. Plain concrete shows strong size effect.

At the crack tip linear elastic fracture mechanics allows the stress to approach infinity. Infinite stress cannot develop in real materials. A certain range of inelastic zone must develop around the crack tip. This zone (Fig. 7.16) is considered as fracture process zone. It includes micro-cracks. Cohesive pressure should still exist, and the behaviour is non-linear.

In concrete, as compared to metals, strain softening instead of strain hardening dominates the behaviour. In a ductile material, the fracture process zone is usually small (Fig. 7.17), whereas in concrete fracture process zone is large. Hence concrete crack tip is not defined clearly.

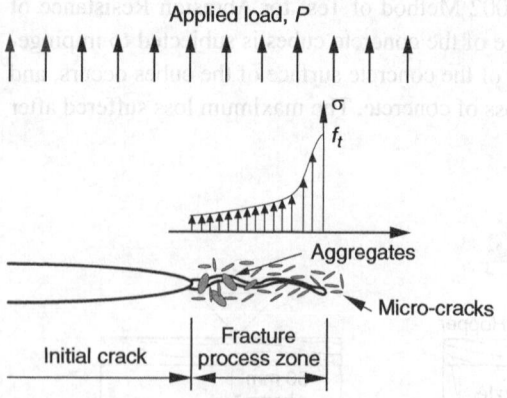

Applied load, *P*

Aggregates

Micro-cracks

Initial crack | Fracture process zone

Fig. 7.16 Fracture process zone for concrete

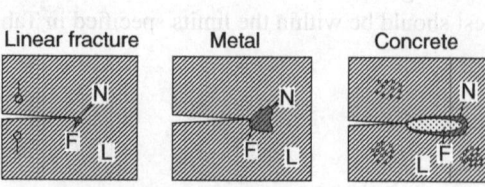

Linear fracture | Metal | Concrete

F : Fracture process zone
N : Non-linear hardening zone
L : Linear zone

Fig. 7.17 Comparison of crack tip

7.8 Strength–Density Relationship

Concrete made with normal aggregate has a density of 22–26 kN/m^3 (2200–2600 kg/m^3). Though the quantity (volume) of the aggregate varies based on the design of the mix, the variations of density on account of this are not significant. This makes the self-weight of concrete elements a major load, and hence concrete structures are considered less efficient when we compare their strength/weight ratio with that of structures made of either steel or aluminium. It must then be fully recognized that high self-weight is a drawback as far as the use of normal-weight structural concrete is concerned.

Figure 7.18 traces the improvement in the structural efficiency of concrete since the introduction of light-weight concrete. Considering the strength/density ratio, the efficiency has improved by more than five times.

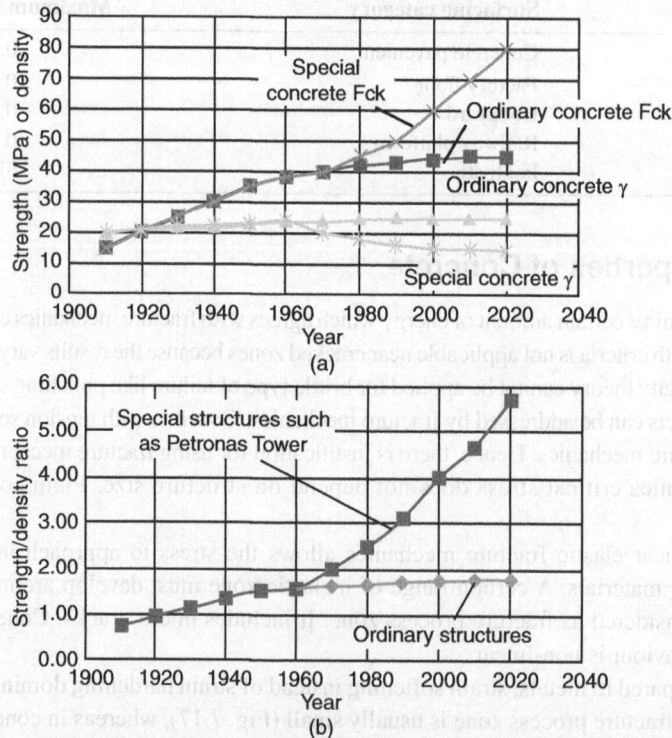

Fig. 7.18 (a) Variation of strength and density of concrete over the years; (b) Increase in structural efficiency of concrete over the years

Concrete is considered a two-phase material—aggregate enveloped by mortar as shown in Fig. 7.19. One way to reduce the weight of concrete is to reduce the density of the mortar phase. This can be done by omitting fine aggregates to produce 'no-fines concrete'. No-fines concrete contains coarse aggregates surrounded by an approximately 1.3-mm-thick layer of cement paste. There exist, therefore, large voids, which reduce the strength of this concrete. This also means that there will not be any capillary movement of water. The variation of the compressive strength of no-fines concrete with respect to its density is shown in Fig. 7.20.

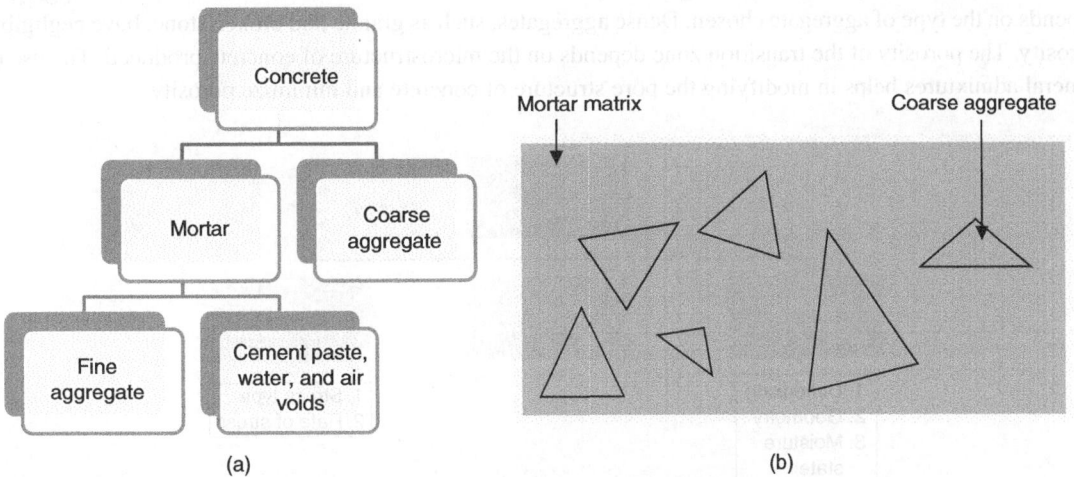

Fig. 7.19 (a) Concrete as a two-phase material; (b) Coarse aggregate inclusion in continuous mortar matrix

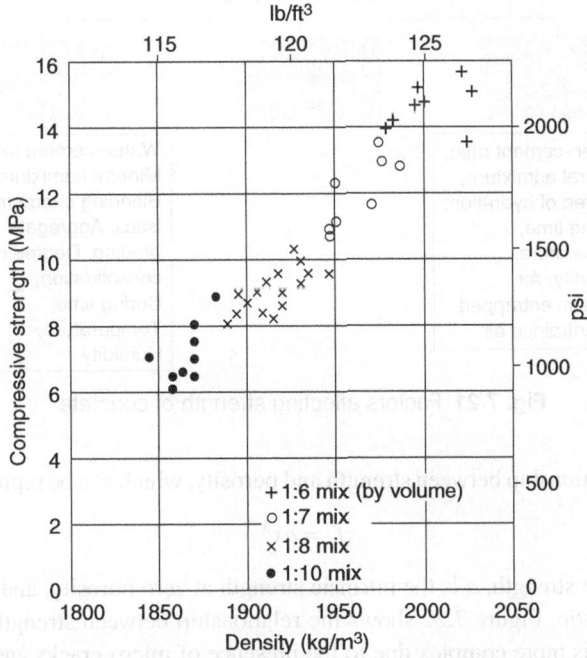

Fig. 7.20 Compressive strength of 28-day-old no-fines concrete as a function of its density at the time of testing (*Source*: McIntosh et al. 1956)

7.9 Parameters Affecting Strength

Figure 7.21 shows the various parameters that affect the strength of concrete. For simplicity, these parameters can be divided into three categories: (a) specimen parameters, (b) material parameters, and (c) loading parameters.

Specimen parameters   The dimension, moisture state, and shape of a specimen influence the strength.

Material parameters   Mix proportions are varied in order to produce concrete of different strengths. Indeed the number of parameters that contribute to strength are many. The most important of them is porosity. Porosity in concrete can result from either the matrix, aggregate, or the interfacial transition zone. The matrix porosity is influenced by the water–cement ratio, degree of hydration of cement, and the air content. Porosity of aggregate depends on the type of aggregate chosen. Dense aggregates, such as granite and broken stone, have negligible porosity. The porosity of the transition zone depends on the microstructure of concrete produced. The use of mineral admixtures helps in modifying the pore structure of concrete and minimize porosity.

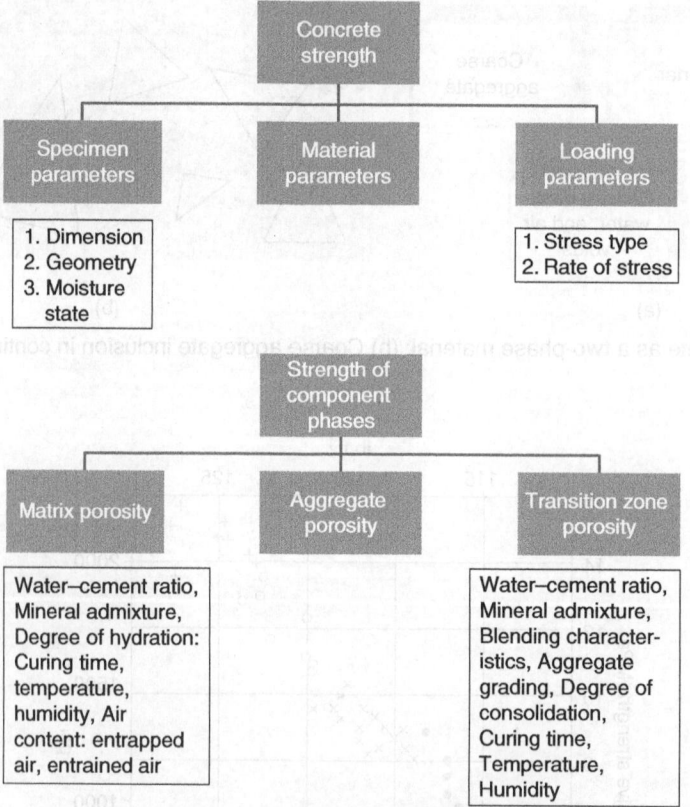

Fig. 7.21 Factors affecting strength of concrete

There exists an inverse relationship between strength and porosity, which can be represented by the expression

$$f_c = ax^3$$

where f_c is the compressive strength, a is the intrinsic strength at zero porosity, and x is the solid/space ratio known as the *gel–space ratio*. Figure 7.22 shows the relationship between strength and gel–space ratio. In concrete, this relationship is more complex due to the presence of micro-cracks and a weak transition zone, which will be discussed in detail in the next chapter.

Loading parameters The rate of loading also influences the strength. Concrete exhibits higher strengths at a faster rate of loading.

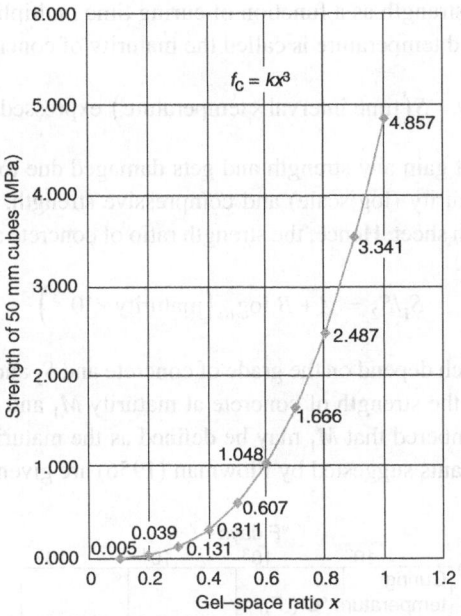

Fig. 7.22 Porosity strength relationship for solids

7.10 Effect of Age on the Strength of Concrete

Though the strength of concrete traditionally refers to the 28-day strength of concrete, concrete gains strength even after 28 days. The rate of gain in strength for concrete with different water–cement ratios is different. This rate is lower in concrete with higher water–cement ratio as shown in Fig. 7.23.

The strength versus time curve for concrete becomes important if we want to put a structure to early use. In prestressed concrete work, the strength at an early age should be known if we want to transfer the prestress at an early age. This property also becomes important for the early removal of formwork. The long-term strength is more than the 28-day strength because hydration takes place even after a very long period, especially if the concrete is in a moist atmosphere. However, IS: 456-2000 code does not recommend taking advantage of the long-term strength even if the structure is loaded only at a later time. This conservative approach takes care of the durability problems with age in concrete, which may offset the gain in strength with age beyond 28 days.

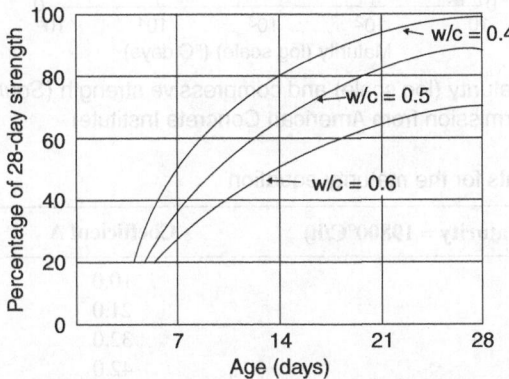

Fig. 7.23 Relative gain in strength with time for concrete made with different water–cement ratios

7.11 Maturity of Concrete

The strength of concrete increases with hydration. The rate of hydration of cement increases with temperature. Hence, it is possible to express strength as a function of curing time multiplied by temperature. The product of time interval for hydration and temperature is called the maturity of concrete. Thus,

$$\text{Maturity} = S\left(\text{time interval} \times \text{temperature}\right) \text{ expressed in } °C/h$$

Below $-12°C$, concrete does not gain any strength and gets damaged due to the action of frost. Figure 7.24 shows the relation between maturity (log scale) and compressive strength. Note that this relation is linear when plotted on a semi-log graph sheet. Hence, the strength ratio of concrete as it matures can be expressed as

$$S_1/S_2 = A + B \log_{10}\left(\text{maturity} \times 10^{-3}\right)$$

where A and B are constants which depend on the grade of concrete and S_1 and S_2 are the strengths of concrete at any two maturities, i.e., S_1 is the strength of concrete at maturity M_1 and S_2 is the strength of concrete at maturity M_2. It should be remembered that M_1 may be defined as the maturity of concrete used at 18°C for 28 days. The values of the constants suggested by Plowman (1956) are given in Table 7.2.

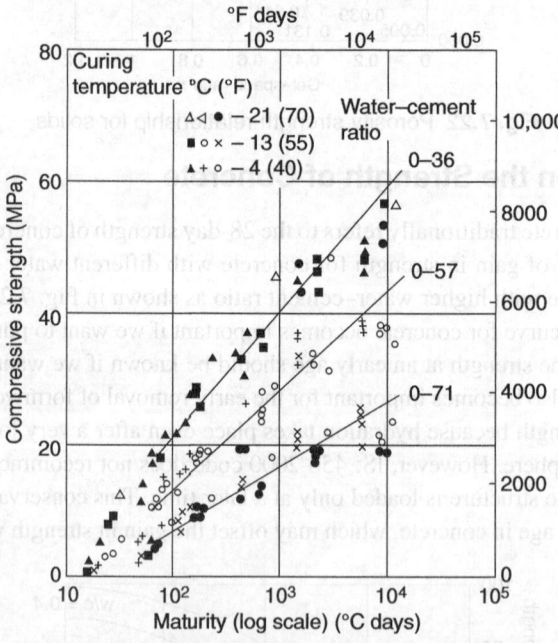

Fig. 7.24 Relation between maturity (log scale) and compressive strength (*Source*: Lew and Reichard 1978. Reproduced with permission from American Concrete Institute)

Table 7.2 Plowman's coefficients for the maturity equation

Strength at 28 days at 18°C (maturity = 19800°C/h)	Coefficient A	Coefficient B
<17	10.0	68.0
17–35	21.0	61.0
35–52	32.0	54.0
52–69	42.0	46.5

The strength–maturity relation is uncertain at low-strength maturity values. It is quite useful in determining the time required for the removal of the formwork of concrete. It is also useful for fixing the temperature of steam curing in order to obtain a particular level of maturity in a precast component before it is sent out of the factory for use.

The maturity concept can also be used to correlate the strength of concrete in a structure with the strength 'predictions' of the compression specimen, especially if the compression specimen is cured at a different temperature in the laboratory.

7.12 Stiffening Time

The reaction between cement and water is the primary cause for the setting of concrete. The phenomena of stiffening, setting, and hardening are physical manifestations of the progressive hydration reaction of cement.

The initial and final setting times of cement are points defined arbitrarily by the Vicat test, which determines the onset of solidification in fresh cement paste. Similarly, the setting of concrete is defined as the onset of solidification in a fresh concrete mixture. Both the initial and final setting times of concrete are also arbitrarily defined by test methods such as the penetration resistance method (ASTM: C403 and IS: 8142). The initial and final setting times, as measured by the penetration resistance method, are two distinct milestones in the process of solidification of the mixture. These are purely functional points in the sense that the former defines the limit of hardening and the latter defines the beginning of development of mechanical strength. They do not have to coincide exactly with the periods marking the end or the complete loss of workability and the beginning of mechanical strength development. Instead, the initial set represents approximately the time at which fresh concrete can no longer be properly mixed, placed, and compacted; the final set represents approximately the time after which strength begins to develop at a significant rate. Obviously, the knowledge of the changes in concrete characteristics as defined by the initial and final setting times can be of considerable value in scheduling concrete construction operations.

Briefly, in the penetration resistance test, the initial and final setting times are defined as times at which the penetration resistances are 3.5 MPa (500 psi) and 27.5 MPa (4000 psi), respectively. It is to be noted that these arbitrarily chosen points do not indicate the strength of concrete. In fact at 3.5 MPa concrete has no compressive strength, while at 27.5 MPa the compressive strength of concrete may only be about 0.7 MPa. The principal factors controlling the setting times of concrete are cement composition, water–cement ratio, temperature, and admixtures. Cements that are quick setting, false setting, or flash setting tend to produce concrete with corresponding characteristics. In the case of cement, the water–cement ratio obviously affects the setting time. The laboratory setting time data of cement paste does not coincide with the actual setting times of concrete made with the same cement, because the water–cement ratios in the two cases are usually different. In general, the higher the water–cement ratio, the longer is the setting time.

Exercises

Review Questions

1. Which properties relate to the long-term performance of concrete structures?
2. Why is compressive strength used as a reference in design and execution?
3. Why are the failure modes of a cylinder and a cube different?
4. Describe the effect of the rate of loading on concrete strength.
5. What is the importance of lightweight, high-strength concrete?
6. What are the various parameters affecting the strength of concrete?
7. What do you understand by the 'falling branch' of the concrete stress–strain curve? What is its importance?

8. Why should we apply fracture mechanics to concrete?
9. How does concrete behave under multi-axial stress?
10. What is the effect of repeated loading on concrete?
11. What is the difference between 'split tensile strength' and 'flexural tensile strength'?
12. What is the difference between strain hardening and strain softening behaviour?

Multiple Choice Questions

1. The stress–strain curve for concrete is obtained by subjecting a concrete cylinder to
   a. Uniform rate of stress
   b. Uniform rate of strain
   c. Constant stress
   d. Constant strain

2. Compared to static tests, dynamic test of concrete results in
   a. Higher value of elastic constants
   b. Lower value of elastic constants
   c. Same value of elastic constants
   d. No influence

3. The inelastic behaviour of concrete is due to
   a. Use of large aggregates
   b. Creep
   c. Higher water–cement ratio
   d. Propagation of cracks

4. Thermal conductivity of concrete
   a. Decreases with density
   b. Increases with density
   c. Is not affected by density
   d. Depends on the type of cement

5. Thermal coefficient of expansion for concrete is about
   a. 3×10^{-8}
   b. 3×10^{-6}
   c. 3×10^{-4}
   d. 3×10^{-3}

6. Concrete attains its full strength in
   a. 7 days
   b. 14 days
   c. 28 days
   d. 60 days

7. The strength of concrete is decreased by
   a. Vibration
   b. Impact
   c. Fatigue
   d. All of the above

8. The tensile strength of concrete is about _____ % of compressive strength
   a. 10
   b. 20
   c. 30
   d. 40

9. The permissible stress of concrete subjected to fatigue should be
   a. 25%
   b. 50%
   c. 75%
   d. 90%

10. The standard size of the cube for compression strength test is
    a. 50 mm
    b. 100 mm
    c. 150 mm
    d. 200 mm

Answers to Multiple Choice Questions

1. b	4. b	7. d	10. c
2. a	5. c	8. a	
3. d	6. c	9. b	

CHAPTER 8

Dimensional Stability of Concrete

Concrete is generally considered brittle. It deforms under load, exhibiting elastic behaviour before cracking. In addition, it is subjected to shrinkage during the drying process and creeps under sustained loading, and is susceptible to thermal effects. Hence, a quantitative knowledge of its deformation behaviour under instantaneous as well as sustained loading effects such as creep and shrinkage is necessary for an understanding of the modelling of its behaviour for building structures such as bridges and tall buildings. The deformation of concrete under short-term and long-term loads is examined in this chapter.

8.1 Dimensional Stability of Concrete

When water evaporates from the surface of freshly laid concrete at a rate faster than the bleed water replacement, the surface of the concrete dries and shrinks. As against free shrinkage, the surface of concrete is restrained by the layer of concrete below it. Hence tensile stresses develop in the surface layer of the weak, stiffening plastic concrete, resulting in shallow cracks of varying depths. This phenomenon is known as *plastic shrinkage cracking*.

The stress–strain behaviour of concrete depends on a number of variables, such as the properties of the materials with which concrete is made, and the loading parameters, such as the rate of loading and the age at testing. Figure 8.1 shows the typical stress–strain behaviour of cement paste, aggregate, and concrete. The variation (non-linearity) of the stress–strain curve for concrete can principally be attributed to the cracking of the cement paste. The cement paste as well as the aggregates exhibit linear stress–strain properties.

Figure 8.2 shows the sequence of behaviour when concrete is subjected to increasing uniaxial compression. Internal cracking develops through a number of stages depending on the level of applied stress. The cracks start at the aggregate–paste interface known as the *transition zone* and spread into the cement mortar phase as the level of stress increases. When the ultimate load is reached, the cracks get interconnected and contribute to the attainment of the failure stress, leading to ultimate collapse. Thus, the following stages of cracking behaviour are clearly witnessed:

Stage 1 Even before the load is applied, micro-cracks exist in the interface (transition zone). The number and width of cracks increase in the mortar phase with increase in the level of stress. The severity of cracking depends on the bleeding characteristics, the strength of the transition zone, and the curing history. At low load (about 25% of the ultimate load) the transition zone remains stable.

Stage 2 Above 25% and up to about 50% of the ultimate stress, the level of stress increases. At 50–60% of the ultimate stress, cracks begin to appear in the mortar matrix.

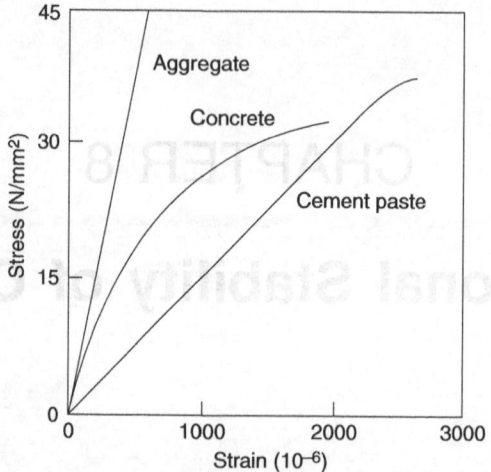

Fig. 8.1 Typical stress–strain behaviour of cement paste, aggregate, and concrete (*Source*: Hsu 1971. Reproduced with permission from American Concrete Institute)

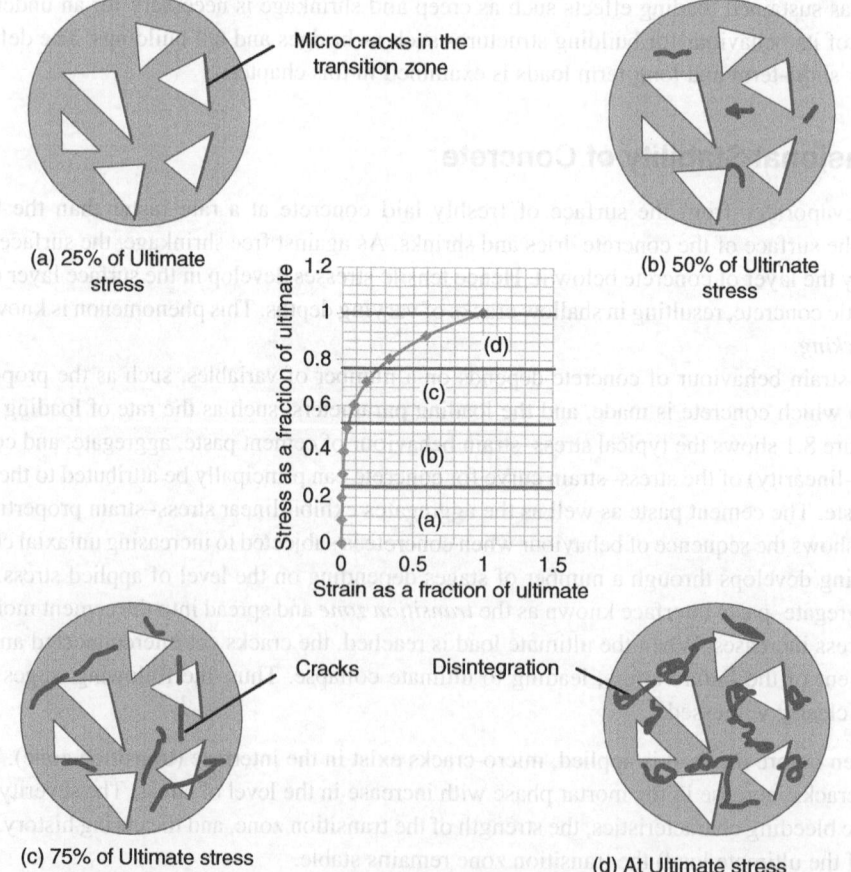

Fig. 8.2 Stress–strain behaviour of concrete under stages of loading and pictorial representation of cracking and disintegration

Stage 3 The cracks start becoming unstable at about 75% of the ultimate stress. As the strain rate increases, the cracks in the mortar phase also increase.

Stage 4 Above 75% of the ultimate stress, the cracks in the matrix join up with the cracks in the transition zone and failure becomes imminent. Upon reaching the ultimate load, the stress decreases with increase in strain. Thus, concrete has an almost linear ascending strain-softening characteristic up to ultimate strain and a falling stress–strain characteristic as shown in Fig. 8.3.

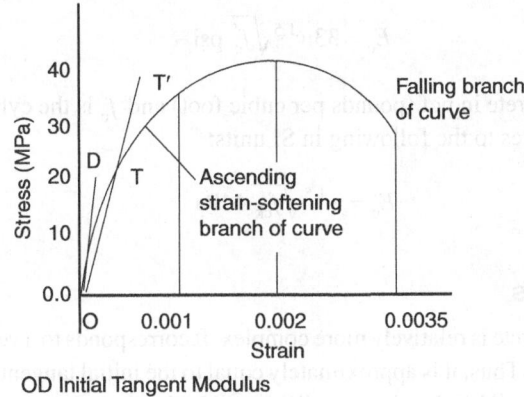

OD Initial Tangent Modulus
OT Secant Modulus @ 20 MPa

Fig. 8.3 Typical stress–strain curve for concrete

8.2 Modulus of Elasticity of Concrete

Though the stress–strain characteristics of concrete are non-linear from the beginning, the initial tangent *OD* to the stress–strain curve is regarded as the initial tangent modulus. The modulus varies with the load level, as well as with the rate of loading. At any load level *T*, other than at *O*, the modulus is estimated based on secant line *TT'* at the corresponding load level. Thus, there is a need to properly define the modulus, accounting for these variations.

8.2.1 Static Modulus

The modulus of elasticity of concrete is generally related to compressive strength. The relationship depends on the aggregate type, mix proportion, rate of loading, curing conditions, and method of measurement. The modulus of concrete under static loading conditions is generally known as its static modulus. The value of the static modulus *E* for concrete is determined on the basis of the uniaxial stress–strain curve obtained from a standard test cylinder (15 cm diameter, 30 cm height). The non-linearity of the stress–strain curve leads to the following three definitions of static modulus:

 i. Tangent modulus evaluated as the slope of the tangent to the stress–strain curve at any point. Thus, the initial tangent modulus refers to the shape of the tangent at zero stress, shown as line *OD* in Fig. 8.3.

 ii. Modulus evaluated as the slope of the line drawn from the origin to a point on the stress–strain curve corresponding to 45% of the ultimate stress, shown as line *OT* in Fig. 8.3.

iii. Chord modulus evaluated as the slope of the line drawn between any two points on the stress–strain curve or as defined by a prescribed standard. Static chord modulus of elasticity is defined by ASTM C469 as the ratio of the difference of the stress at 40% of the ultimate strength and the stress at 50 millionth (50×10^{-6}) of the strain to the difference in strain corresponding to the stress at 40% of the ultimate strength and 50 millionth strain (50×10^{-6}). IS: 456-2000 gives the following empirical formula for static modulus:

$$E = 5000\sqrt{f_{ck}} \text{ MPa} \qquad (8.1)$$

where f_{ck} is the characteristic compressive strength of concrete in Megapascals. Though the unit weight of concrete influences the modulus IS code, Eq. (8.1) does not reflect it. The ACI code suggests the following formula:

$$E_c = 33w^{1.5}\sqrt{f_c'} \text{ psi} \qquad (8.2)$$

where w is the weight of concrete in pcf (pounds per cubic foot) and f_c' is the cylinder strength of concrete in psi. The above equation reduces to the following in SI units:

$$E_c = w^{1.5}\sqrt{f_{ck}} \text{ MPa} \qquad (8.3)$$

8.2.2 Dynamic Modulus

The dynamic modulus of concrete is relatively more complex. It corresponds to a very small instantaneous strain due to suddenly applied stress. Thus, it is approximately equal to the initial tangent modulus but is considerably larger than the static secant modulus. For low, medium, and high-strength concrete, the dynamic modulus is generally 40%, 30%, and 20%, respectively, higher than the static modulus (Mehta 1986). Figure 8.4 shows a typical test set-up for determining the dynamic modulus of concrete.

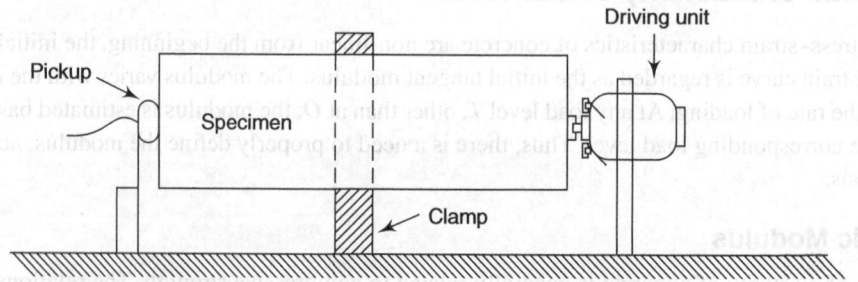

Fig. 8.4 Test arrangement for the determination of dynamic modulus

8.3 Factors Affecting the Modulus of Elasticity of Concrete

Since concrete is a multiphase solid, no direct relationship can exist between its density and modulus as in single-phase solids such as metals. The modulus is influenced by density, porosity, mix proportion, modulus of elasticity of the ingredients, and the characteristics of the transition zone. These parameters determine the elastic behaviour of concrete.

8.3.1 Effect of Aggregates

Dense aggregate leads to a high value of modulus of elasticity (E) for concrete. A larger proportion of coarse aggregate leads to a high value of E as well. Table 8.1 summarizes the values of E for various ingredients as well as concrete. Note that a very large value of E of the aggregate will lead to an elastic mismatch among the aggregate, mortar, and cracks in the transition zone.

8.3.2 Effect of Hydrated Cement Paste

The elastic modulus of cement paste is determined by its porosity. The water–cement ratio, air content, admixture dosage, and degree of cement hydration control the porosity of cement paste. The modulus of elasticity of concrete can be represented based on the following simple equation:

$$E_c = E_a g + E_p(1 - g) \tag{8.4}$$

where E_c is the modulus of elasticity of concrete, E_a is the modulus of elasticity of aggregate, E_p is the modulus of elasticity of cement paste, g is the volume fraction of the aggregate, and $1 - g$ is the volume fraction of the cement paste.

8.3.3 Effect of the Transition Zone

As discussed in Section 8.1, void space and micro-cracks influence the stress–strain behaviour. The existing cracks in the transition zone and the orientation of C–H crystals as well as existing void spaces make the transition zone weak. This causes the elastic modulus to drop gradually with increasing loads.

Table 8.1 Modulus of elasticity of aggregate, cement paste, and concrete

Description	E (10^5 MPa)
Granite	1.4
Sandstone	0.2–0.5
Expanded shale	0.07–0.21
Hydrated cement paste	0.07
Concrete	0.1–0.2

8.4 Poisson's Ratio

Poisson's ratio is defined as the ratio of lateral strain to axial strain under uniaxial loading within the elastic range. For concrete, the value of Poisson's ratio generally lies between 0.15 and 0.20. In the inelastic range, which occurs in concrete even within service loads, volume dilation results from internal micro-cracking. This leads to an apparent increase in Poisson's ratio under increasing axial strain.

Figure 8.5 shows a comparison of lateral strain for low and medium-strength concrete along with the corresponding axial strains. Observe that medium-strength concrete exhibits less volume dilation than low-strength concrete, implying less internal micro-cracking. This may mean that the effectiveness of hoop steel for confinement will be less for high-strength concrete.

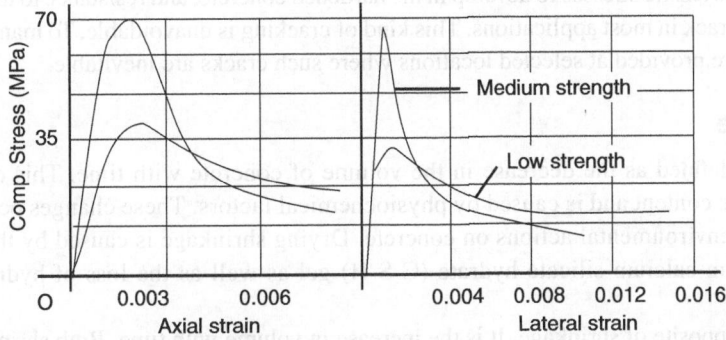

Fig. 8.5 Axial stress versus axial strain and lateral strain for low and medium-strength concrete

8.5 Mechanics of Setting and Hardening

We now discuss the main physical mechanism of setting and hardening of concrete. Before setting starts, the constituents of concrete are in a suspension. The downward settlement of heavier particles, such as the coarse aggregate, takes place due to gravity. A clear form of bleed water appears at the top. This phenomenon is generally known as *sedimentation*.

The volume of hydrates formed due to the reaction of cement and water is less than the initial volume of the ingredients by 8–12%. Due to this, a potential shrinkage of 3–4% exists in the cement paste. This is known as *Le Chatelier's contraction*. The reaction between cement and water is highly exothermic, liberating heat. This heat of hydration starts getting liberated early upon the addition of water to cement and continues during strength gain, which elevates the temperature of concrete.

At the beginning of the setting process, the solid grains exist in a connected liquid phase. Hydration starts from the surfaces of the cement grains. The hydration product is known as gel. The cement grain gets hydrated and forms a thick crust, which in fact delays the hydration of the inner core of the cement grain. The formation of hydrates around the grains makes the crystals coalesce, and in less than an hour concrete changes (sets) into a continuous solid. Ordinarily, the hydration of concrete never ends. The hydrate layer (crust) formed becomes thicker but the cores of the cement grains never get completely hydrated. Therefore, only a fraction of hydration is completed during the setting process.

Just after setting, the porous network is completely filled with water, and *self-desiccation* takes place. The liquid phase reduces rapidly. In high-strength concrete, the high rate of self-desiccation sometimes leads to early cracking. This becomes more pronounced if a mineral admixture such as silica fume is used.

The post-setting process results in the following:

- Internal growth of the skeleton (hardening): the set layers join up and form a skeleton; hydration continues in the pores bridged by the skeleton of the hydrated gel.
- Reduction of water content in the pore space (self-desiccation): This is similar to drying and causes the skeleton of the hydrated mass to shrink. This is followed by desiccation. After demoulding, the drying process begins from the surface at ambient temperature. This phase is very short if the water content is reduced to just that required for hydration (as in hardened cement paste).

8.6 Shrinkage, Creep, and Thermal Effects

Concrete suffers from shrinkage and creep, which are time-dependent deformations. These along with cracking are the greatest concerns for designers because their effects are difficult to predict.

Almost all concrete is mixed with more water than required for the hydration of cement paste. Much of this excess water evaporates, causing concrete to shrink. Resistance to this shrinkage provided by sub-surface reinforcement causes tensile stresses to develop in the hardened concrete, and resistance to the drying shrinkage causes concrete to crack in most applications. This kind of cracking is unavoidable. To manage such cracking, contraction joints are provided at selected locations where such cracks are inevitable.

8.6.1 Shrinkage

Shrinkage can be defined as the decrease in the volume of concrete with time. This decrease is due to changes in moisture content and is caused by physiochemical factors. These changes occur without stress and are caused by environmental actions on concrete. Drying shrinkage is caused by the loss of surface-adsorbed water from calcium-silicate-hydrate (C-S-H) gel as well as the loss of hydrostatic tension in small capillaries.

Swelling is the opposite of shrinkage. It is the increase in volume with time. Both shrinkage and swelling are expressed as dimensionless strain (mm/mm) under standard conditions of temperature and humidity.

Shrinkage is mainly caused by the deformation of the paste, though the stiffness of the aggregate affects it. Figure 8.6 shows the typical variation of shrinkage strain with time. It also shows the effect of rewetting. There are a number of interdependent factors which affect shrinkage. This makes it difficult to correctly predict the effect of shrinkage without proper testing.

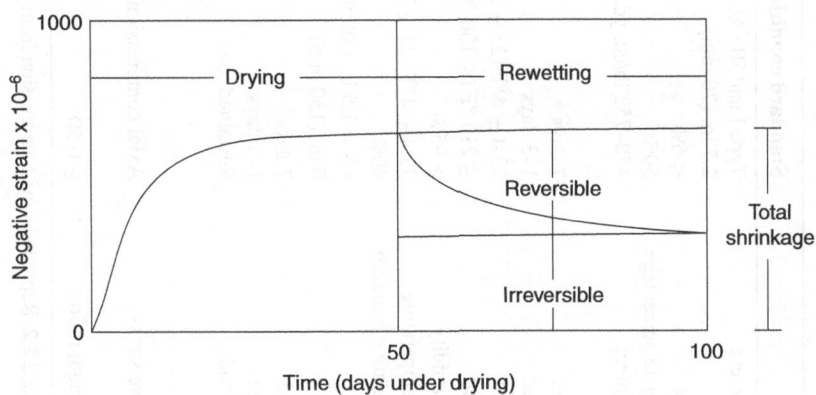

Fig. 8.6 Variation of shrinkage strain with time

Factors affecting drying shrinkage

Table 8.2 summarizes the factors affecting concrete creep and shrinkage. These are described below:

Material and mix proportion and environmental conditions   A higher water–cement ratio leads to higher shrinkage. As water–cement ratio increases, the paste strength and stiffness decrease, and as the water content increases, shrinkage also increases. In addition, shrinkage decreases with increase in the member size as well as with higher relative humidity at the site.

Aggregate type and content   Aggregates restrain the extent of shrinkage of the paste. Hence, concrete with larger aggregate content exhibits smaller shrinkage. Similarly, concrete with a higher modulus aggregates or aggregates having a larger surface roughness resists shrinkage more.

Cement type and content   Rapid-hardening cement gains strength rapidly and hence shrinks more than ordinary cement due to the increase in fineness and higher water demand. Shrinkage-compensating cements minimize or eliminate shrinkage cracks. In general, the shrinkage of concrete, S_c, can be expressed as:

$$S_c = S_p(1-g)^n \tag{8.5}$$

where S_p is the shrinkage of cement paste, $1 - g$ is the volume of cement paste, and $n = 1.2$–1.7 based on aggregate-related properties. IS: 456-2000 recommends (in the absence of test data) an approximate value of 0.003 for total shrinkage strain. Note that the magnitude of shrinkage strain is roughly as much as the magnitude of maximum concrete strain attained on loading near collapse.

8.6.2 Creep

When a viscoelastic material is subjected to stress for a particular duration, its strain changes over time. This time-dependent increase/decrease in the strain of hardened concrete subjected to a sustained stress is termed *creep*. It is obtained by deducting the following from the total measured strain in a laboratory-loaded specimen under constant humidity: (a) instantaneous strain usually considered as elastic strain, (b) thermal strain due to change in temperature, and (c) shrinkage strain of the specimen. Thus, creep in concrete is a post-elastic

Table 8.2 Factors affecting concrete creep and shrinkage, and the variables considered in the recommended prediction method

	Factors	Variables	Standard conditions
Concrete composition (creep and shrinkage)	Cement paste content	Type of cement	Type I and III (ACI)
	Water–cement ratio	Slump	2.7 in. (70 mm)
	Mix proportion	Air content	≤ 6%
	Aggregate characteristics	Fine aggregate percentage	50%
	Degree of compaction	Cement content	470–752 lb/cu. yd. (279–446 kg/m²)
Initial curing	Length of initial curing	Moist-cured	7 days
		Steam-cured	1–3 days
	Curing temperature	Moist-cured	73.4 ± 4 °F (23 ± 2 °C)
		Steam-cured	≤ 212 °F (≤ 100 °C)
	Curing humidity	Relative humidity	≥ 95%
Member geometry and environment (creep and shrinkage)	Concrete temperature	Concrete temperature	73.4 ± 4 °F (23 ± 2 °C)
	Concrete water content	Ambient relative humidity	40%
	Geometry (size and shape)	Volume–surface ratio (v/s) or minimum thickness	v/s = 1.5 in. (38 mm)
			6 in. (150 mm)
Loading (only creep)	Concrete age of load Application	Moist-cured	7 days
		Steam-cured	1–3 days
	Duration of loading period	Sustained load	Sustained load
	Duration of unloading period (number of load cycles)		
	Type of stress and distribution across the section	Compressive stress	Axial compression
	Stress-strength ratio	Stress–strength ratio	≥ 0.50

Source: 'Prediction of Creep, Shrinkage, and Temperature Effects in Concrete Structures'. ACI 209R-92. Table 2.2.2. Reproduced with permission from American Concrete Institute

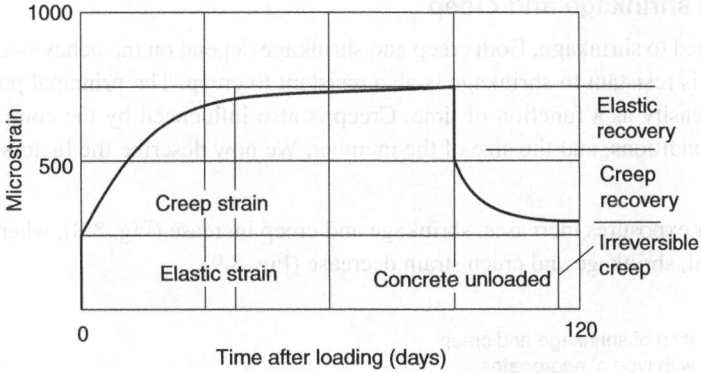

Fig. 8.7 Creep deformation with respect to time

phenomenon. Figure 8.7 shows creep deformations with time. It shows the effect of loading and unloading. The total creep can be divided into recoverable and irrecoverable creep.

With respect to loading and restraining conditions and environmental effects, the various definitions of creep are as follows:

True or basic creep   Creep with no loss of moisture (theoretical creep under 100% relative humidity)

Specific creep   Creep strain per unit applied stress:

$$\text{Specific creep} = \frac{\varepsilon_{cr}}{\sigma} \tag{8.6}$$

Drying creep   The additional creep due to a specimen undergoing drying in addition to applied stress

Creep coefficient   The ratio of creep strain to elastic strain:

$$\text{Creep coefficient} = \frac{\varepsilon_{cr}}{\varepsilon_{el}} \tag{8.7}$$

The creep of concrete is principally affected by the creep of cement paste

$$C_c = C_p (1-g)^\alpha \tag{8.8}$$

where C_c is the creep of concrete, C_p is the creep of cement paste, and g is the aggregate content, where α is a function of the volume and the stiffness of the paste and aggregate respectively:

$$\alpha = f(V, V_a, E, E_a) \tag{8.9}$$

where V and V_a represent volume of paste and aggregate, E and E_a represent shiffness of paste and aggregate.

IS: 456-2000 recommends the ultimate creep coefficients given in Table 8.3, depending on the duration of loading.

Table 8.3 Variation of creep coefficients with duration of loading

Duration of loading	Creep coefficient
7 days	2.2
28 days	1.6
1 year	1.1

Factors affecting shrinkage and creep

Creep is closely related to shrinkage. Both creep and shrinkage depend on the behaviour of hydrated cement paste. Concrete that is resistant to shrinkage is also resistant to creep. The principal parameter influencing creep is the load intensity as a function of time. Creep is also influenced by the composition of concrete, the environmental conditions, and the size of the member. We now describe the factors affecting creep and shrinkage.

a. As the time after exposures increases, shrinkage and creep increase (Fig. 8.8), whereas with increase in aggregate content, shrinkage and creep strain decrease (Fig. 8.9).

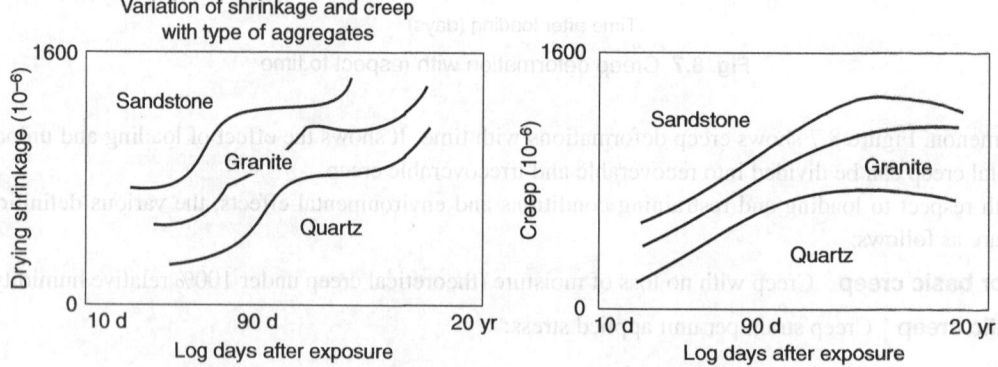

Fig. 8.8 Variation of shrinkage and creep with type of aggregates

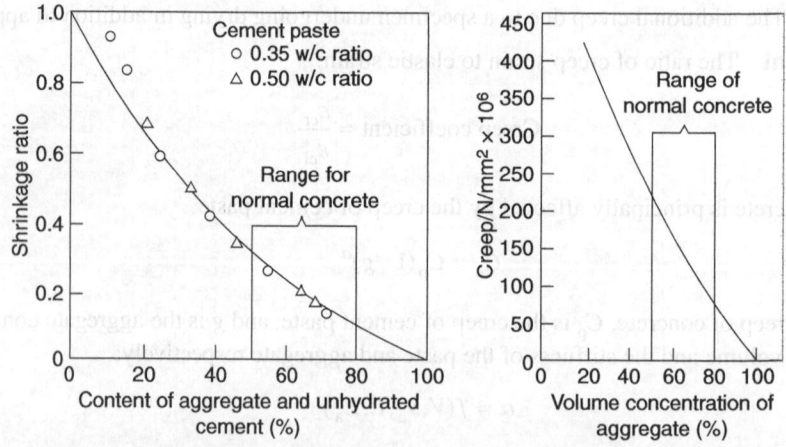

Fig. 8.9 Variation of creep and shrinkage with the aggregate content (*Source*: ACI Monograph 6 1973. Reproduced with permission from American Concrete Institute)

b. For constant cement content, an increase in the water–cement ratio increases both the drying shrinkage and the creep (Fig. 8.10).

c. For a constant water–cement ratio, an increase in the cement content reduces the creep but increases the shrinkage (Fig. 8.11). This is the only case in which the opposite effects are experienced by shrinkage and creep.

d. Humidity is the most important environmental parameter affecting both creep and shrinkage. A decrease in the relative humidity of air decreases both shrinkage and creep (Fig. 8.12).

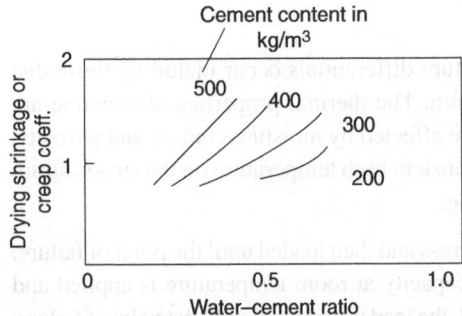

Fig. 8.10 Variation in drying shrinkage and creep coefficient with water–cement ratio

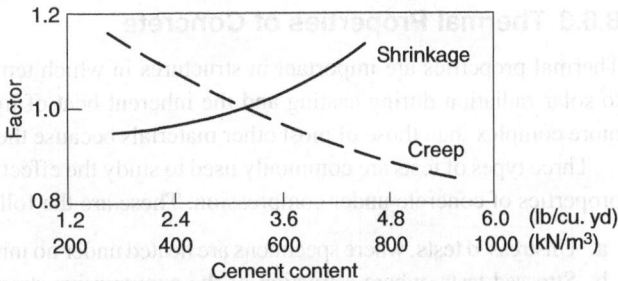

Fig. 8.11 Variation of creep and shrinkage with the cement content in the water–cement ratio (*Source*: Jones et al. 1971. Reproduced with permission from American Concrete Institute)

e. The higher the temperature, the larger the creep and shrinkage, provided there is no variation in the curing condition.

f. The duration of loading affects creep but not shrinkage. This is because the effect of strength on loading at a given stress level creep is lower in specimens loaded at a later time (Fig. 8.13).

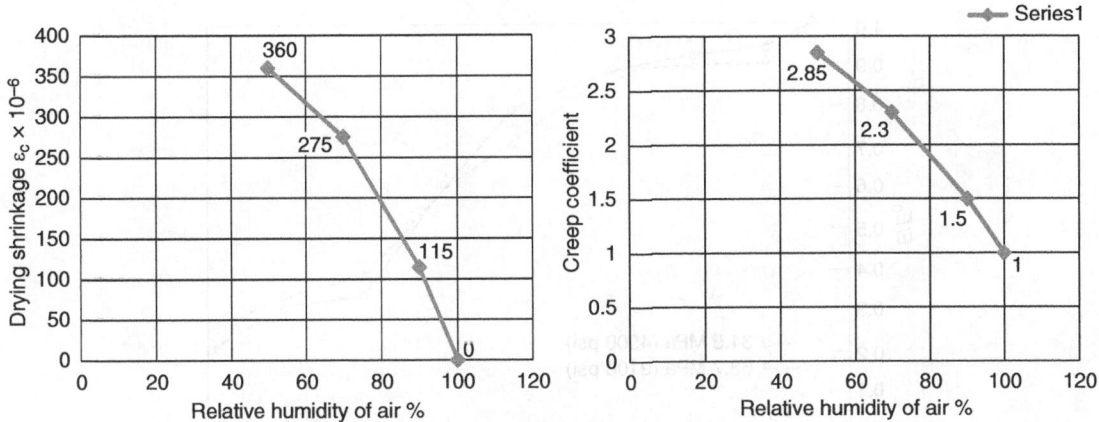

Fig. 8.12 Variation of creep and shrinkage with relative humidity

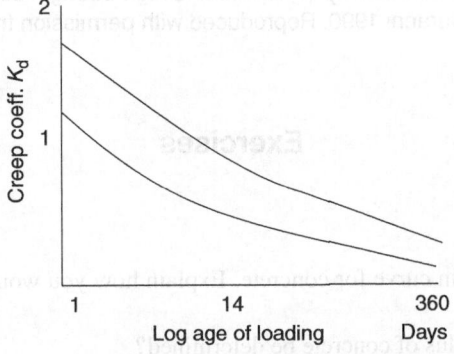

Fig. 8.13 Variation of creep coefficient with age of loading

8.6.3 Thermal Properties of Concrete

Thermal properties are important in structures in which temperature differentials occur including those due to solar radiation during casting and the inherent heat of hydration. The thermal properties of concrete are more complex than those of most other materials because these are affected by moisture content and porosity.

Three types of tests are commonly used to study the effect of transient high temperature on the stress–strain properties of concrete under compression. These are the following:

a. Unstressed tests, where specimens are heated under no initial stress and then loaded until the point of failure.
b. Stressed tests, where a fraction of the compressive strength capacity at room temperature is applied and sustained during heating. When the target temperature is reached, the load is increased until the point of failure.
c. Residual unstressed tests, where the specimens are heated without any load, cooled to room temperature, and then loaded until the point of failure.

For normal-strength concrete, when the exposure temperature is greater than 450°C, the residual unstressed strength decreases. When tested under pre-compressed load, the loss in strength is less, as compressive load helps to delay the spread of cracks. The moisture content plays a significant role in reducing the strength due to increase in temperature. The variations of the moduli of elasticity of normal and high-strength concrete with temperature are shown in Fig. 8.14. The moduli decrease by about 5–15% when exposed to temperature in the range of 100–300°C. At 800°C, the value of E reduces to 20–25% of its value at room temperature.

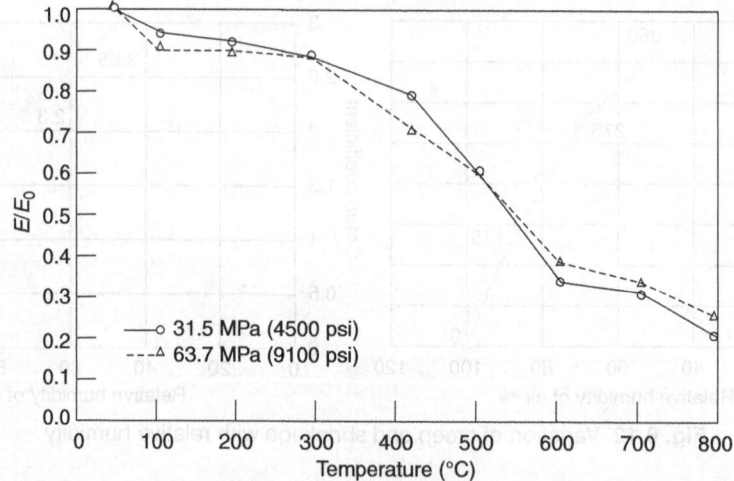

Fig. 8.14 Variations in the moduli of elasticity of normal and high-strength concrete with temperature (*Source*: Castillo and Duranni 1990. Reproduced with permission from American Concrete Institute)

Exercises

Review Questions

1. Draw the typical stress–strain curve for concrete. Explain how you would determine the various elastic moduli for concrete.
2. How can the dynamic modulus of concrete be determined?
3. Which factors affect the cracking of concrete?

4. What is the peculiarity of Poisson's ratio for concrete?
5. Distinguish between the setting and hardening of concrete.
6. Define shrinkage and creep.
7. Why are shrinkage and creep treated together?
8. What are the factors that affect the shrinkage and creep of concrete?
9. How do strength and Young's modulus vary with temperature for concrete? Comment.

Multiple Choice Questions

1. Bleeding of concrete occurs when
   a. Finer particles go down to the bottom
   b. Coarser particles get separated
   c. Cement paste rises to the surface
   d. Honeycombs are formed

2. Bleeding is
   a. Good as it helps finishability
   b. Good as it helps workability
   c. Good as it reduces voids
   d. Not good because it forms a weak layer at the top

3. The separation of coarse aggregate from mortar is called
   a. Bleeding
   b. Segregation
   c. Slumping
   d. Creeping

4. Segregation in concrete results in
   a. Impermeability
   b. Increase in density
   c. Decrease in workability
   d. Honeycombs in concrete

5. Shrinkage in concrete is due to
   a. Loss of moisture due to evaporation
   b. Large size coarse aggregate
   c. Too much of fine aggregate
   d. Poor workability

6. Creep of concrete in column results in
   a. Loss of moisture
   b. Loss of load sustained
   c. Loss of impermeability
   d. Cracking

7. Which of the following has the lowest density
   a. Concrete
   b. Steel
   c. Bricks
   d. RCC

8. The unit weight of RCC is generally taken as
   a. 24 kN/m^3
   b. 22 kN/m^3
   c. 25 kN/m^3
   d. 28 kN/m^3

9. Loss in modulus of elasticity when temperature of concrete is raised by 400°C is
   a. 15%
   b. 40%
   c. 60%
   d. 70%

10. An increase in cement content
    a. Will reduce creep
    b. Will increase creep
    c. Will have no influence on creep
    d. Will reduce shrinkage

Answers to Multiple choice Questions

1. c   4. d   7. c   10. a
2. d   5. a   8. c
3. b   6. b   9. a

CHAPTER 9

Proportioning Concrete Mix

Concrete is a mixture of Portland cement, water, and coarse and fine aggregates. It consolidates into a hard mass because of a chemical reaction called hydration between cement and water. The coarse aggregate (CA) in the mixture acts as a filler, whereas fine aggregate (FA) fills up the voids in the CA. Cement with water acts as the binder. The mobility (workability) of the mixture is provided by the cement paste, fines with water, and admixtures.

Concrete mix proportioning is governed by the properties required in the fresh (plastic) as well as the hardened state. The properties of plastic concrete are important for proper placing and compaction. The strength and durability of the final structure is provided by hardened concrete. The two are related primarily to the water–cement ratio. The ease of placement is governed by the workability. For good workability, one requires a larger water–cement ratio. However, this affects the strength and durability.

Proportioning a concrete mix for a given purpose is thus the art of obtaining a suitable ratio of the various ingredients of concrete with the required properties at the lowest cost.

This can be achieved by a suitable choice of materials and their proportions. A properly proportioned concrete mix with certain requirements of workability, strength, and durability should have the minimum possible cement content to make the mix most economical.

This chapter deals with proportioning mixes for medium-strength concretes—M40 and below. Proportioning ingredients for high-strength concrete is discussed in Chapter 14.

It should, however, be observed that precise relations have not yet been established between the desirable properties of concrete and the specific characteristics of the mix such as the water–cement ratio, aggregate–cement ratio, and grading. In addition, elusive qualities such as aggregate–particle shape and texture play important roles in mix proportioning. Hence, concrete mix proportioning cannot be converted into a fully automatic process or a computer input–output sequence.

Proportioning a concrete mix is more of an art than a pure science and methods are still evolving. The properties used as data, based on sample testing, backed up by experience and knowledge, are essential. Nevertheless, they are used only in the first stage of proportioning a mix. Invariably, no mix proportion is final without making trial mixes in the field. Based on the observations made on trial mixes, the mix proportions are adjusted and refinements carried out to get optimum proportions. Proportioning is still very much a trial-and-error problem. Proportioning arrived at in the laboratory is really a means of providing, at best, a starting point for field refinements for final use.

9.1 Approach to Concrete Mix Design

The essential functional properties of hardened concrete, such as strength, durability, and appearance, can be satisfied only if the workability and cohesiveness of fresh concrete are suitable for the given working conditions. The standard procedures for judging or measuring these properties are given in IS: 456-2000. All the essential properties including permeability can be achieved if the concrete mix has adequate cement content, optimum water content, and can be worked into and placed in forms that will give the maximum density under the available effort to compact without resulting in the separation or segregation of the aggregate during handling and placing. Apart from the essential requirement that concrete should be properly compacted to obtain the maximum strength and durability, the water–cement ratio rule is fundamental to the production of good concrete, whether in the field or in the laboratory.

The mobility or workability of fresh concrete and the strength of hardened concrete are the two most essential properties. These can be observed and measured, and can give an idea of the overall performance of the concrete mix in its wet state and its long-term strength when it hardens. Workability can be measured by the slump value, the compacting factor, or the Vebe test. The compressive strength is usually measured on 150-mm cubes cast from concrete produced in the laboratory or in the field using a standard procedure. The strength test of the cubes, however, does not truly represent the actual strength of concrete in the structure, owing to the variation in the degree of compaction, curing conditions at the site, etc. In addition, the strength of the structure also depends on the conditions of the concrete in the actual application, the shape and size of the structural member, and the manner in which the member is placed in relation to other units. However, owing to convenience, the characteristic strength f_{ck} of concrete (defined later), as obtained from cubes or cylinders, has come to be accepted as the yardstick for the performance of hardened concrete.

Conventional nominal-mix proportions such as 1:1:2, 1:2:4, and 1:3:6 have no significance for strength or durability. In these, the quantity of fine aggregates is fixed irrespective of the cement content and the maximum size of aggregate. Hence, the variations in the quality of concrete produced are inevitable. Therefore, nominal-mix proportions, arbitrarily fixed, have no significance in modern concrete technology and should never be used.

9.2 Principles of Mix Proportioning

The data required for proportioning a concrete mix are the following:

a. The environmental exposure conditions for different concrete structures according to IS: 456-2000 are listed in Table 9.1.

Table 9.1 Environmental exposure conditions

Environment	Exposure conditions for concrete surfaces
Mild	• Protected from weather or other aggressive conditions
Moderate	• Sheltered from severe rain or freezing while wet • Exposed to condensation and rain • Continuous contact with water • In contact with or buried under non-aggressive soil/ground water • Sheltered from saturated salt air in coastal areas
Severe	• Exposed to severe rain, alternate wetting and drying or occasional freezing while wet or severe condensation • Completely immersed in sea water • Exposed to coastal environment
Very severe	• Exposed to sea water spray, corrosive fumes, or severe freezing conditions while wet • In contact with or buried under aggressive subsoil/ground water
Extreme	• Surfaces of members in tidal zones • Members in direct contact with liquid/solid aggressive chemicals

b. The grades of concrete, e.g., M20 and M25, which denote the characteristic strengths f_{ck} of 20 and 25 N/mm², respectively, and the standard deviation S according to IS: 456-2000, are reproduced in Tables 9.2 and 9.3, respectively.

Table 9.2 Grades of concrete

Group	Grade designation	Specified characteristic compressive strength of a 150-mm cube at 28 days (N/mm²)
Ordinary concrete	M10*	10
	M15	15
	M20	20
Standard concrete	M25	25
	M30	30
	M35	35
	M40	40
	M45	45
	M50	50
	M55	55
High-strength concrete	M60**	60
	M65	65
	M70	70
	M75	75
	M80	80

*M refers to the mix and the number refers to the specified compressive strength of a 150-mm cube at 28 days, expressed in N/mm².

**For concrete of compressive strength greater than M55, the design parameters given in IS: 456-2000 may not be applicable. This has been discussed in Chapter 14 on high-strength concrete.

Table 9.3 Assumed standard deviations

Grade of concrete	Assumed standard deviation* (N/mm²)
M10	3.5
M15	
M20	4.0
M25	
M30	5.0
M35	
M40	
M45	
M50	

*The values correspond to proper storage of cement, weigh batching of all materials, controlled addition of water, regular checking of all materials, aggregate grading, controlling moisture content, and periodical checking of workability and strength. (Wherever there is a deviation from these conditions the values should be increased by 1 N/mm².)

c. The type of cement, e.g., ordinary Portland, Portland pozzolana, and Portland slag, from a total of 10 cements, according to IS: 456-2000 and the relevant properties of these cements are summarized in Table 9.4(a). The end uses of these cements are briefly described in Table 9.4(b).

Table 9.4 Important properties and uses of various types of cement

(a) Relevant properties of cements

Type of cement	IS code	Fineness, min. (m²/kg)	Setting time (min.)		Soundness		Compressive strength (MPa)	
			Initial	Final	mm*	%**	7 days	28 days
OPC33	269-1989	225	30	600	10	0.5	22	33
OPC43	8112-1989	225	30	600	10	0.5	33	43
OPC53	12269-1987	225	30	600	10	0.5	37	53
PPC (fly ash based)	1489-1991 (Part I)	300	30	600	10	0.8	22	33
PSC (slag)	455-1989	225	30	600	10	0.8	22	53
SRC	12330-1988	225	30	600	10	0.8	16	33
White cement	8042-1989	225	30	600	10	0.8	19.8	29.7
RHC	8041-1990	325	30	600	10	0.5	At 3 days: 27	—
Low-heat PC	12600-1989	320	60	600	10	0.5	16	35

*Le Chatelier method expansion in mm
**Autoclave method expansion in %

(b) End uses of various cements

Type of cement	End use
Ordinary Portland cement grade 33	Used for general civil construction works under normal conditions; normally not used where high-grade concrete is required due to limitations of strength
Ordinary Portland cement grade 43	Used for general construction work due to non-availability of grade 33
Ordinary Portland cement grade 53	Used in roller-compacted concrete (RCC) and prestressed concrete members of higher grades, cement grouts, instant plugging mortars, etc., where initial high strength is the required criterion
Portland slag cement (PSC)	Provides better protection against chloride and sulphate attacks. Preferred over ordinary Portland cement (OPC) for use in constructions where the structures are susceptible to sulphate and chloride attacks, e.g., marine structures or structures near the sea, sewage disposal treatment works, and water treatment plants
Portland Pozzolana cement (PPC)	Makes concrete more impermeable and dense as compared to OPC. The long-term strength gain of PPC is higher compared to OPC. PPC produces less heat of hydration and offers a greater resistance to the attack of aggressive environments than OPC. Can be used for most types of construction
Sulphate-resisting Portland cement (SRC)	Amount of C_3A is restricted to lower than 5% and $C_3A + C_4AF$ to lower than 25%. Can be used wherever OPC/PPC/PSC is used, but is advantageous for foundations, piles, basements, underground structures, and sewage and water treatment plants. Gives better resistance to sulphate environments
Low-heat Portland cement	Particularly suitable for making concrete for dams and many other types of water-retaining structures, bridge abutments, massive retaining walls, piers, thick slabs, etc. Used where control of heat of hydration is a necessity
Rapidly hardening cement (RHC)	Used for repair and rehabilitation works and where speed of construction and early completion are required due to limitations of time. Preferred also for underwater works

(contd)

(contd)

Type of cement	End use
Hydrophobic Portland cement	Manufactured under special requirement for high rainfall areas to improve the shelf life of cement. The cement particles are given a chemical coating during manufacturing, which makes them water-repelling. Not affected by high humidity and can be stored for longer periods without significant loss of their setting property compared to OPC
White cement	Made from raw materials containing very little iron oxide and manganese oxide. Limited quantities of certain chemicals which improve the whiteness of cement are added during manufacture. This type of cement is generally meant for non-structural works. It is used for special purposes such as in the manufacture of mosaic tiles and wall paintings and for introducing architectural effects

d. The type and size of aggregates, e.g., natural sand, crushed stone, and gravel, and their sources of supply.

e. The nominal maximum size of aggregates, e.g., 40, 20, and 10 mm according to IS: 383-1970 and IS: 456-2000, reproduced in Table 9.5.

f. Maximum and minimum cement content (kg/m³), according to IS: 456-2002, reproduced in Table 9.6(a) for 20-mm aggregates and Table 9.6(b) adjustments for other sizes.

g. The maximum free water–cement ratio by weight, required to be employed owing to considerations of strength and/or durability for different exposures, and in order to meet workability and other requirements [Table 9.6(a)].

Table 9.5 Maximum size of aggregates

Size (mm)	Condition*
40	Used where there is no restriction to the flow of concrete, such as in large member sizes and in mass concrete work
20	Used for most work with usual member sizes
10	Used for members with thin sections, for closely spaced and heavily reinforced members, and for members having small cover (15 mm)

*The maximum size should be as large as the condition will permit.

Table 9.6 Minimum cement content

(a) Minimum cement content*, maximum water–cement ratio, and minimum grade of concrete for different exposures with normal-weight aggregates of 20 mm nominal maximum size

Exposure	Plain concrete			Reinforced concrete		
	Minimum cement content (kg/m³)	Maximum free water– cement ratio	Minimum grade of concrete	Minimum cement content (kg/m³)	Maximum free water– cement ratio	Minimum grade of concrete
Mild	220	0.60	**	300	0.55	M20
Moderate	240	0.60	M15	300	0.50	M25
Severe	250	0.50	M20	320	0.45	M30
Very severe	260	0.45	M20	340	0.45	M35
Extreme	280	0.40	M25	360	0.40	M40

*The cement contents prescribed in this table are irrespective of the grades of cement and inclusive of mineral admixture additions. Additions such as fly ash or ground and granulated blast-furnace slag may be taken into account in concrete composition with respect to the cement content and water–cement ratio if the suitability is established, and as long as the maximum amounts taken into account do not exceed the limit of pozzolana and slag specified in IS: 1489 (Part 1) and IS: 455, respectively. The maximum cement content is restricted to 450 kg/m³.

**The minimum grade for plain concrete under mild exposure conditions is not specified.

(b) Adjustments in minimum cement contents for aggregates having nominal maximum sizes other than 20 mm

Nominal maximum size aggregate (mm)	Adjustment to minimum cement contents in Table 9.6(a) (kg/m³)
10	+ 40
20	0
40	−30

h. The degree of workability of concrete based on placing conditions [Tables 9.7(a) and (b)].
i. Air content, inclusive of entrained air (Table 9.8).
j. The type of admixture used, if any.
k. The maximum/minimum density of concrete.
l. The maximum/minimum temperature of fresh concrete.
m. The type of mixing and curing water, e.g., fresh potable water or ground water.

Table 9.7 Degree of workability of concrete

(a) Measured in accordance with IS: 1199

Placing conditions	Degree of workability	Slump (mm)
Blinding concrete; shallow sections; pavements using pavers	Very low	Not specified
Mass concrete; lightly reinforced sections in slabs, beams, walls, and columns; floors; hand-placed pavements; canal lining; strip footings	Low	25–75
Heavily reinforced sections in slabs, beams, walls, columns; slipform work; pumped concrete	Medium	50–100 75–100
Trench fill	High	100–150
In situ piling; tremie concrete	Very high	Not specified

(b) For different placing conditions along with appropriate test values

Workability class				
Extremely low	**Very low**	**Low**	**Medium**	**High**
Sections subjected to extremely intensive or prolonged vibration; pressure may also be required	Small sections subjected to intensive vibration, and large sections to normal vibration	Simply reinforced sections with vibration, and large sections without vibration	Simply reinforced sections without vibration, and heavily reinforced sections with vibration	Heavily reinforced sections without vibration

Nominal maximum size of aggregate (mm)														
10	20	40	10	20	40	10	20	40	10	20	40	10	20	40
Slump in mm														
0	0	0	0	0–10	0–25	0–5	10–25	25–50	5–25	25–50	50–100	25–100	50–125	99–175
Vebe time														
>20						5–10			3–5			2–3		
Compacting factor														
0.65	0.68		0.68	0.78	0.78	0.83	0.85	0.85	0.9	0.92	0.92	0.95	0.95	0.95

Table 9.8 Mean total air content in fresh concrete

Nominal maximum size aggregate (mm)	Entrained air (%)
20	5 ± 1
40	4 ± 1

n. The source of water and the type of impurities present in it (Table 9.9). The mix should be designed to obtain a grade of concrete having the desired durability (Tables 9.1 and 9.2) and workability [Tables 9.7(a) and (b)], and a characteristic strength f_{ck} not less than the values (shown in Table 9.10) satisfying the compliance requirement.

Table 9.9 Impurities present in water

Content	Standard tested against	Permissible limit, max. (mg/L)
Organic	IS: 3025(Part 18)	200
Inorganic	IS: 3025(Part 18)	3000
Sulphates (as SO_3)	IS: 3025(Part 24)	400
Chlorides (as Cl)	IS: 3025(Part 32)	2000
		For concrete not containing embedded steel and 500 for reinforced concrete work
Suspended matter	IS: 3025(Part 17)	2000

Table 9.10 Characteristic compressive strength compliance requirement

Specified grade	Mean of the group of four non-overlapping consecutive test results (N/mm²)	Individual test results (N/mm²)
M15	$\geq f_{ck} + 0.825 \times$ established* standard deviation (rounded off to the nearest 0.5 N/mm²) Or	$\geq f_{ck} - 3$
M20 or above	$f_{ck} + 3$, whichever is greater $\geq f_{ck} + 0.825 \times$ established standard deviation (rounded off to the nearest 0.5 N/mm²) Or $f_{ck} + 4$, whichever is greater	$\geq f_{ck} - 4$

*In the absence of an established value for standard deviation, the values given in Table 9.3 may be assumed. An attempt should be made to obtain the results of 30 samples as early as possible to establish the correct value of standard deviation pertaining to the concerned site condition.

9.3 Properties of Concrete Related to Mix Design

Durability and strength are important properties of concrete in the hardened state. To achieve these desirable properties, fresh concrete should be workable so that it can be compacted with the least effort. Thus, the idea of mix proportioning is to ensure proper workability of concrete in the fresh state at the optimum water–cement ratio to give the hardened concrete required strength to withstand loads and adequate durability performance over the design life of the structure against aggressive environment.

9.3.1 Durability

Durability may be defined as resistance to weathering action due to environmental conditions such as changes in temperature and humidity, chemical attack, abrasion, frost, and fire. Durability problems occur in concrete due to the following factors: (a) chemical impurities in cement and water, (b) combination of high-alkali cement and reactive aggregate, (c) changes in moisture content during service life, and (d) changes in temperature. In porous concrete, durability is greatly reduced because of the dissolution of soluble constituents, permeation, as well as the disintegration of concrete due to water freezing in its pores and voids. The durability of concrete depends essentially on its pore structure. The volume of the hardened paste is less than the combined volume of the cement and water. Hence, voids and pores are bound to exist in cement paste. The number and size of pores naturally increases with increase in the quantity of water used for preparing the concrete mix [Fig. 9.1 (a–c)]. At a water–cement ratio of 0.6 with 100% hydration of cement, the capillary porosity defined by volume fraction $P = 0.3$ and the corresponding solid/space ratio $1 - P = 0.7$ leads to a paste strength of about 85 N/mm^2 and results in a permeability of about 20×10^{-12} cm/s. The larger the water–cement ratio employed, more permeable the resulting concrete becomes because of increase in capillary porosity. Note that an increase in the water–cement ratio leads to an increase in capillary porosity, decrease in strength, and an increase in permeability. Therefore, from considerations of durability, the water–cement ratio has to be restricted to a maximum prescribed value. Thus, achieving strength alone is not adequate for constructing a durable structure with a prescribed service life; the capillary porosity, which leads to permeability, should be limited.

The extent of corrosion of steel in reinforced concrete structures is influenced by the thickness and quality of cover provided. With strong, dense aggregates, an impermeable cover can be obtained by adopting a lower water–cement ratio and by ensuring thorough compaction during concreting. This should be followed by proper curing procedures to enable adequate hydration. For guidance, the maximum water–cement ratio and minimum cement content required to ensure durability of concrete under various conditions of exposure (Table 9.1) are given in Tables 9.6(a) and (b). Table 9.11(a) summarizes the type of cement, the minimum cement content, and

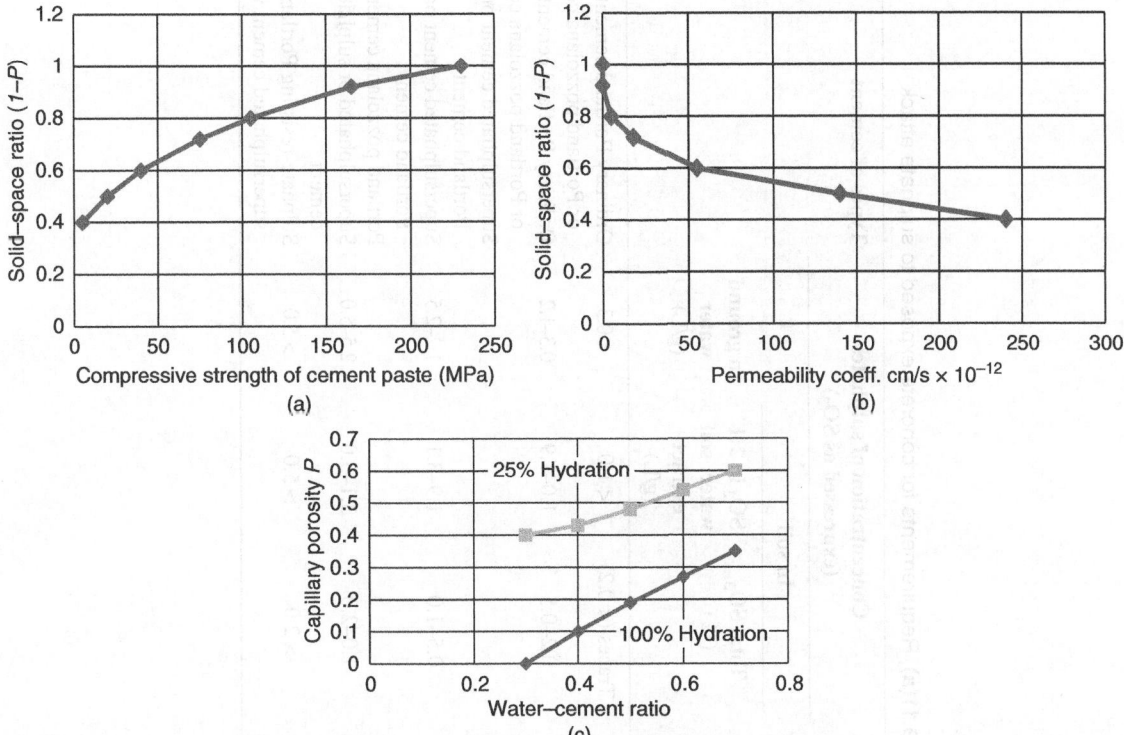

Note: P defines the ratio of voids to the total volume and 1 – *P* defines the ratio of solids to the total volume.

Fig. 9.1 Effect of water–cement ratio and degree of hydration on permeability

Table 9.11(a) Requirements for concrete exposed to sulphate attack

Class	Concentration of sulphates (expressed as SO_3)			Types of cement	Dense, fully compacted concrete made with 20 mm nominal maximum size aggregates	
	In soil		In ground water (g/L)		Minimum cement content (kg/m³)	Maximum free water–cement ratio
	Total SO_3 (%)	SO_3 in 2:1 water: soil extract (g/L)				
1	Traces (< 0.2)	< 1.0	< 0.3	Ordinary Portland cement or Portland slag cement or Portland pozzolana cement	280	0.55
2	0.2–0.5	1.0–1.9	0.3–1.2	Ordinary Portland cement or Portland slag cement or Portland pozzolana cement	330	0.50
				Supersulphated cement or sulphate-resisting Portland cement	310	0.50
3	0.5–1.0	1.9–3.1	1.2–2.5	Supersulphated cement or sulphate-resisting Portland cement	330	0.50
4	1.0–2.0	3.1–5.0	2.5–5.0	Portland pozzolana cement or Portland slag cement	350	0.45
				Supersulphated or sulphate-resisting Portland cement	370	0.45
5	> 2.0	> 5.0	> 5.0	Sulphate-resisting Portland cement or supersulphated cement with protective coatings	400	0.40

the maximum free water–cement ratio to be used for different levels of sulphate concentration. Table 9.11(b) gives the limits of chloride content in concrete according to IS: 456-2000.

Table 9.11(b) Limits for chloride contaminants in concrete

Type of concrete	Maximum total acid-soluble chloride content (kg/m^3 of concrete)
Concrete containing metal aggregates and concrete steam cured at elevated temperature and prestressed concrete	0.4
Reinforced concrete or plain concrete containing embedded metal inserts	0.6
Concrete not containing embedded metal inserts or any material requiring protection from chloride	3.0

9.3.2 Workability

Workability is defined as the amount of useful internal work required to produce full compaction of concrete. Workable concrete does not mean concrete that flows. A workable concrete is one that is cohesive, flows well, and can, without segregation, be fully compacted with the envisaged type of compaction (type of vibration). This requirement enables proper handling and placing of fresh concrete with thorough compaction around the entire reinforcement, and ensures the complete filling of the leak-proof mould or form erected for this specific purpose. The workability of concrete depends on the following factors:

a. Type of aggregate—rounded, angular, flaky, etc.
b. Grading of CAs and FAs, i.e., poorly graded, well graded, or containing an excess of any size fraction
c. Quantity of cement paste in the mix
d. Consistency of the paste

Consequently, workability is governed by the aggregate–cement ratio and water–cement ratio for a given type and grading of aggregate. For a majority of normal concretes with given aggregates, workability depends on the water content (L/m^3) of concrete. Workability is associated with the following properties:

Ease of flow

The ease of flow depends on the method employed to overcome internal friction through proper lubrication of the surface of the aggregates. This can be done by

a. increasing the quantity of water (not recommended since it will reduce the strength because of the increased water–cement ratio) and
b. reducing the surface area of the aggregates (larger aggregates lead to smaller surface areas; for a given workability, the water demand is less for larger maximum size aggregates; this is illustrated in Fig. 9.2).

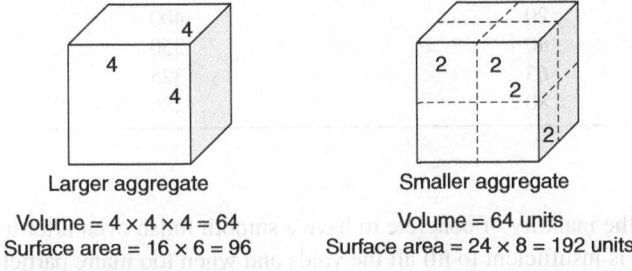

Larger aggregate
Volume = 4 × 4 × 4 = 64
Surface area = 16 × 6 = 96

Smaller aggregate
Volume = 64 units
Surface area = 24 × 8 = 192 units

Fig. 9.2 Effect of the maximum size of aggregates on workability

Note that for small aggregates, the surface area doubles and hence the flow gets retarded. This technique is good for leaner mixes and ordinary concrete. For high-strength concrete, smaller aggregates are preferred due to the requirement of the particle packing mechanism. Hence, coarse grading gives a smaller surface area than fine grading (smaller aggregates). Angular or flat particles result in a larger surface area compared to cubical particles as shown in Fig. 9.3.

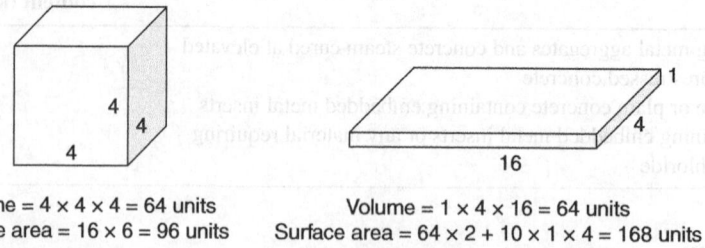

Volume = 4 × 4 × 4 = 64 units
Surface area = 16 × 6 = 96 units

Volume = 1 × 4 × 16 = 64 units
Surface area = 64 × 2 + 10 × 1 × 4 = 168 units

Fig. 9.3 Effect of the shape of aggregates on surface area

Hence, the shape of aggregates plays an important role in workability. The FA–CA ratio can be adjusted to get the most favourable workability for all FA + CA aggregates. Table 9.12 guides the selection of the FA–CA ratio for various zones of sands in order to induce good flow.

Table 9.12 Guide to selection of FA–CA ratio[*]

Maximum size of CA (mm)	Zone I	Zone II	Zone III	Zone IV
10	1:1	1:1.5	1:2	1:3
20	1:1.5	1:2	1:3	1:3.5
40	1:2	1:3	1:3.5	—

[*]These values are for initial guidance only.

Prevention of segregation (cohesiveness)

During concreting, the separation of CA from mortar is called segregation. The risk of segregation exists either when there is too much water and insufficient fines in the mix, the mix is subjected to shock, the formwork is leaking, or when fresh concrete is under water. Table 9.13 gives the recommended minimum quantity of fines required to ensure cohesiveness.

Table 9.13 Recommended minimum quantity of fines to prevent segregation

Nominal size of CA (mm)	Quantity of fines (kg/m^3 of compacted concrete)
10	525
20	400
40	350
63	325
80	280

Harshness of mix

Harshness is defined as the inability of concrete to have a smooth finish even after trowelling. This happens when the cement mortar is insufficient to fill all the voids and when too many particles are large or have the same size. Too little mortar makes the CA rise to the surface of the concrete being cast.

Bleeding

Bleeding is the phenomenon of the development of a layer of water at the top surface of freshly placed concrete. A little excess amount of trowelling will cause the water and cement to rise to the surface, leaving behind the CA at the bottom. This problem is caused by too much floating or over-vibration.

The degrees of workability suggested for different placing conditions and the appropriate slump values as recommended by IS: 456-2000 are given in Table 9.7(a). The table does not consider the maximum size of aggregate as a variable. Table 9.7(b) further classifies workability limits in terms of the compacting factor, slump value, and Vebe time and is more useful for mix proportioning.

It may be noted that Table 9.7(b) gives the workability of a particular mix as measured by each of the test procedures—compacting factor, slump, and Vebe time. These data have been provided by the Cement and Concrete Association (C&CA), London, for guidance. The results may not coincide with all the values suggested, as they vary according to specific site conditions. This is because certain properties of concrete, e.g., the particle shape of the CA, may have a greater effect on one of the test procedures than on the others. However, it is worthwhile to make these broad comparisons, especially for very high and very low workability, since the slump values in these ranges are not reliable. The values given in Table 9.7(b) refer to ordinary concretes of normal density (2400 kg/m^3).

9.3.3 Strength

The strength of concrete depends on many parameters such as quality and quantity of cement and water, grading of aggregate, batching, mixing, placing, compaction, and curing. However, the water–cement ratio is the most important parameter influencing the strength of concrete. Since the quantity of water controls the workability for a given set of materials (cement and aggregates), it is possible to achieve different workabilities by altering the water content but maintaining the water–cement ratio. This will ensure the attainment of the required strength.

The strength of hardened paste depends on the chemical composition and the fineness of cement. Any addition of water in excess of that required for adequate hydration of the cement results finally in the increased porosity of the paste. When concrete is not fully compacted, voids may remain in the mass, and this has a pronounced effect on the strength of concrete. One per cent air voids reduce the strength of concrete by 4–5%.

When sound aggregates are used, the strength of concrete depends solely on the water–cement ratio for a particular type of cement, age, method of testing, and curing. According to Abrams' Law, compressive strength can be expressed as:

$$F = \frac{A_1}{B_1^x}$$

or

$$\log F = \log A_1 - x \log B_1 \tag{9.1}$$

where F is the compressive strength of concrete, A_1 and B_1 are constants depending on the age of concrete, quality of cement, and the type and quality of aggregate, and x is the water–cement ratio by weight. Abrams' Law will hold good for sound aggregates and fully compacted concretes. The constants A_1 and B_1 are determined experimentally for the materials proposed to be used in the concrete. Lean and rich mixtures give the same strength for the same water–cement ratio. Professor Abram had also made the important discovery that the strength of concrete is mainly governed by the quantity of water used to make it, and is independent of the aggregate–cement ratio, provided the concrete is workable.

Uncertainties affecting strength

Though Abrams' Law is conceptually unfailing, other variables such as the maximum size, grading of aggregates, and strength gain rate of different types of cements in the initial periods of hydration (0–10 days) also affect the magnitude of the compressive strength attainable ultimately after full hydration. These uncertainties are discussed in this section:

a. The water–cement ratio may be selected from a chart giving the relation between compressive strength and free water–cement ratio for fully compacted, dense-aggregate concrete [Fig. 9.4(a)–(b)] made with Indian cement. However, emphasis is laid here on the limitation arising out of correlating the 7-day strength of cement with the 28-day strength of concrete. IS: 269-1976 pertaining to ordinary Portland cement specifies the

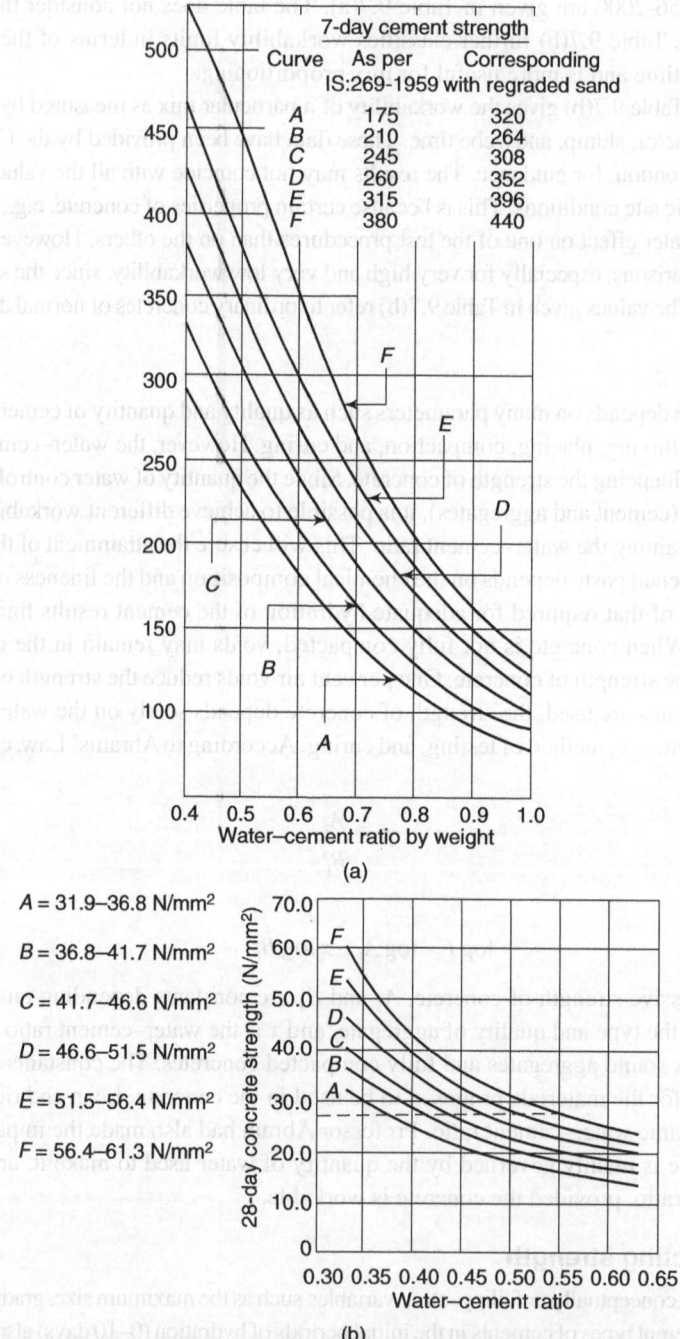

	7-day cement strength		
Curve	As per IS:269-1959		Corresponding with regraded sand
A	175		320
B	210		264
C	245		308
D	260		352
E	315		396
F	380		440

(a)

A = 31.9–36.8 N/mm²

B = 36.8–41.7 N/mm²

C = 41.7–46.6 N/mm²

D = 46.6–51.5 N/mm²

E = 51.5–56.4 N/mm²

F = 56.4–61.3 N/mm²

(b)

Fig. 9.4 Design curves for (a) the 7-day compressive strength of cement and (b) the 28-day strength of cement

compressive strength of cement at 3 days and 7 days. The strength gain varies for different cements, i.e., no two cements will give the same 28-day strength when tested in accordance with the relevant IS specifications.

b. It may be pertinent to note that the water–cement ratio law is subject to other variables such as the maximum size as well as grading of aggregates. The work done by Cordon and Gillespie (1963) has shown that depending on the maximum size of the aggregate, the cement content for a specific strength is altered, since at the same water–cement ratio, different ranges of strength are attainable for different sizes of aggregate. The recommended grading zones for 40, 20, and 10 mm maximum size of aggregate are shown in Figs 9.5, 9.6, and 9.7, respectively.

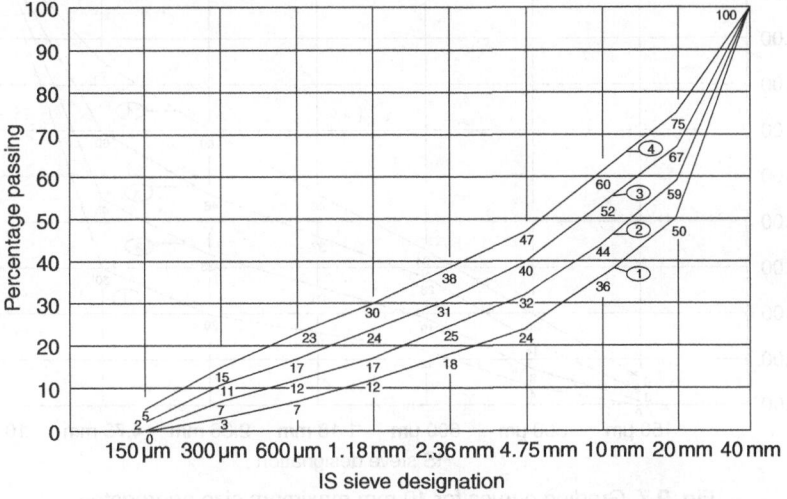

Fig. 9.5 Grading curves for 40 mm maximum size aggregate

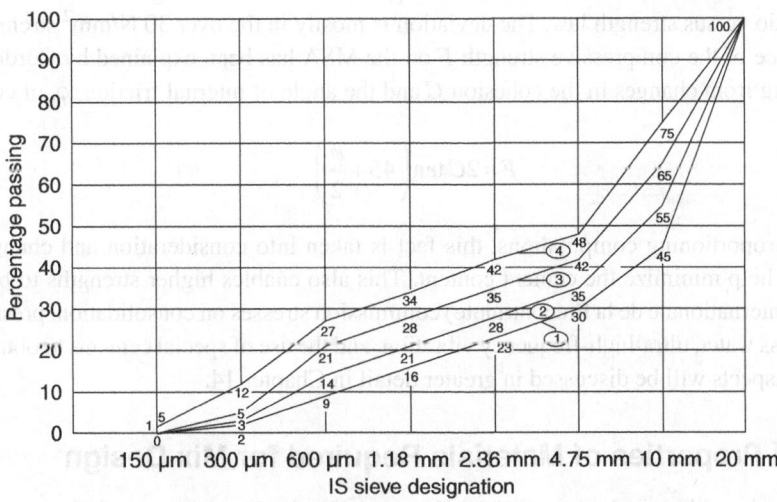

Fig. 9.6 Grading curves for 20 mm maximum size aggregate

High-strength concrete

Prior to 1960, it was generally understood that as the maximum size of aggregate (MSA) is increased, better strength can be ensured per unit cement spent, primarily because of the reduced sand and water requirements made possible by the lowered surface area. While this premise is largely true, it is now fairly well established that for compressive strength over M30 the MSA shifts to a lower size. For instance, the cement contents for two levels of strength could be as follows:

Compressive strength (N/mm²)	Cement content (kg/m³)		
	20 mm MSA	40 mm MSA	75 mm MSA
25	300	265	230
35	370	440	Not feasible

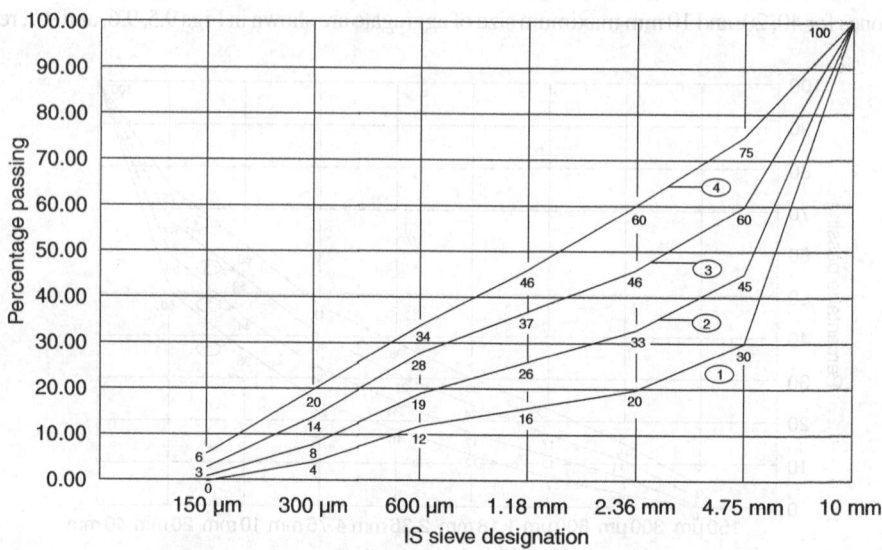

Fig. 9.7 Grading curves for 10 mm maximum size aggregate

For the same water–cement ratio, lower the MSA, higher the concrete strength. This defies the well-known water–cement ratio versus strength law. The deviation is mostly in the over 30 N/mm² strength zone.

The dependence of the compressive strength F on the MSA has been explained by Cordon and Gillespie (1963) as resulting from changes in the cohesion C and the angle of internal friction, ϕ, of concrete. Thus,

$$F = 2C \tan\left(45 + \frac{\phi}{2}\right) \tag{9.2}$$

In modern mix-proportioning computations, this fact is taken into consideration and changes in the MSA recommended to help minimize the cement content. This also enables higher strengths to be achieved. The FIP (Fédération Internationale de la Précontrainte) commission stresses on consolidation, pressure application, expulsion of excess water, ultra-high-frequency vibration, and the use of special cements to obtain high-strength concrete. These aspects will be discussed in greater detail in Chapter 14.

9.4 Physical Properties of Materials Required for Mix Design

The physical properties of the constituent materials to be assessed before commencing mix design are discussed here.

9.4.1 Cement

The physical property of cement to be assessed is the minimum compressive strength (N/mm²) of the particular type of Portland cement used at different ages according to the relevant IS specification (IS: 269-1976, IS: 1489-1976, and IS: 455-2000, respectively, for ordinary Portland, Portland pozzolana, and Portland slag cements). Table 9.4 summarizes these strength properties for various cements. The most common types of cement used are ordinary Portland cement (OPC), Portland pozzolana cement (PPC), and Portland blast-furnace slag cement (PBSC). Hydration occurs as long as the relative humidity in the pores is above 0.85, provided sufficient water is available for the chemical reactions. On an average, 1 g of cement requires only 0.253 g of water for complete hydration. As hydration proceeds, the ingress of water by diffusion through the deposit of hydration products around the original cement grain becomes more and more difficult, and the rate of hydration decreases continuously. In mature pastes, the particles of C–S–H form an interlocking network, which is a gel having a specific surface and 'capillary' pores according to their sizes (Fig. 9.8). Strength-giving properties and phenomena such as creep and shrinkage are attributable to the porous structure of the gel, and the strength is attributable to the bond afforded by the enormous surface area (Fig. 9.9).

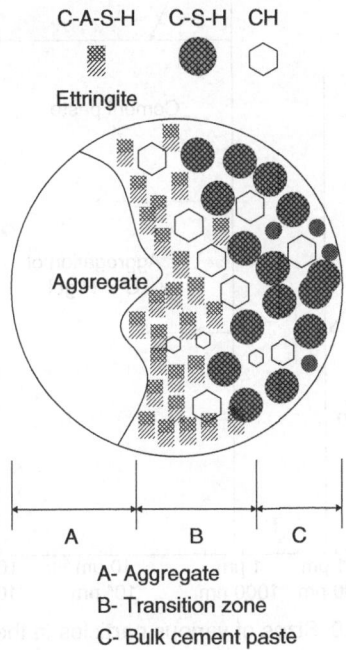

Fig. 9.8 C–S–H gel and transition zone

9.4.2 Aggregate

The physical properties of aggregate to be assessed are its size to be used, particle shape, colour, surface texture, density (heavyweight or lightweight), impurities, and its effect on the durability of concrete, conforming to IS: 383-1970. The products of cement hydration completely surround and bind together the aggregate particles into a solid hardened mass. The strength of concrete is a 'chain system' governed by the weakest link, be it the cement paste link, the aggregate, or the aggregate–cement paste interface (Fig. 9.10). The strength at the aggregate–mortar interface (transition zone) is perhaps more critical; hence the shape, size, and texture of the CA must be carefully selected. The aggregate should consist of clean, hard, strong, and durable particles free from chemicals or coatings of clay or other fine materials that will hinder their bonds with the cement paste and reduce the transition zone strength.

Very sharp and rough aggregate particles or flat and elongated particles require more fine material to produce a workable concrete. Accordingly, the water and, therefore, cement requirements increase. Excellent concrete is made by crushed stone, but the particles should be roughly uniform and cubical in shape. Natural, rounded aggregates having smooth surfaces are better from the point of view of workability, but their bonds with mortar may be weaker. However, their use need not be precluded and the deficiency can be very well taken care of in mix proportioning. The maximum size of the aggregate governs the strength and workability of concrete. As explained earlier for mixes up to strength M30, a larger maximum size of aggregates gives better strength results.

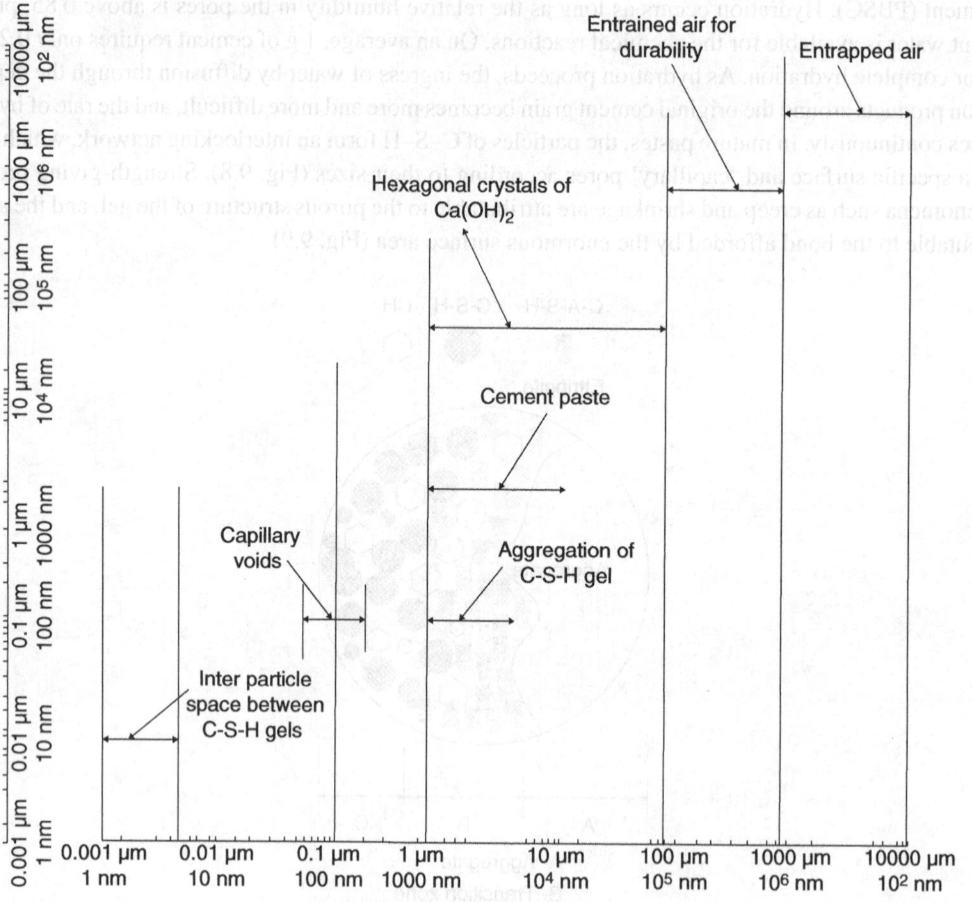

Entrained air for durability

Entrapped air

Hexagonal crystals of Ca(OH)₂

Cement paste

Capillary voids

Aggregation of C-S-H gel

Inter particle space between C-S-H gels

	0.001 μm	0.01 μm	0.1 μm	1 μm	10 μm	100 μm	1000 μm	10000 μm
	1 nm	10 nm	100 nm	1000 nm	10^4 nm	10^5 nm	10^6 nm	10^2 nm

Fig. 9.9 Sizes of various particles in the gel

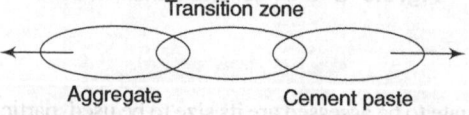

Transition zone

Aggregate   Cement paste

Fig. 9.10 Strength 'chain system' of concrete

A statement of fact valid for all mix proportions is—the smaller the maximum size of CA, the greater the proportion of FA needed for concretes of identical cement content and workability. Also, lower the cement content of the mix, larger the proportion of FA needed; more angular the CA, greater the proportion of FA required. The relative proportions of CA to FA and their distribution into different sized fractions are given by sieve analysis. The results are plotted as grading curves. Depending upon the maximum size of aggregate, the workability required, the concentration of reinforcement, the cross-section of the member, etc., standard 'ideal' combined grading curves specifying the limits of the various fractions are recommended for guidance. In general, all the fractions of both types of aggregates should be present in the desired proportions. For example, experience has shown that excess of fine sand results in higher water and cement content, while excess of coarse sand gives harsh, unworkable mixes. For strength less than M30 or when small-sized CA is used, a finer combined grading should be preferred, while for richer mixes a coarser combined grading is more desirable and economical.

For any mix design, there is no single overall grading which is suitable for all sets of conditions. A finer grading should be used for leaner mixes, more workable mixes, and for concrete made with poorly shaped aggregate particles, i.e., angular, flaky, or elongated particles with rough texture. A larger proportion of fine particles should also be employed in mixes where the maximum size of the aggregate is small. This can be observed from the ranges of grading shown in Figs 9.5–9.7. Likewise, no single grading is the best for any given type of mix. A finer grading should be used for conditions which tend to promote segregation, e.g., transporting in jolting or vibrating containers over long distances, by pneumatic placers, or on a series of conveyor belts; discharging down-inclined chutes into a heap; dropping through a considerable height or over reinforcement; and placing in formwork which is not mortar-tight. When concrete is to be placed underwater or pumped, there should be enough FA to provide a more plastic and cohesive mix than may otherwise be required. The higher the workability of concrete, the greater is the proportion of FA needed.

9.4.3 Water

The tolerable concentrations of some impurities in mixing and curing water as specified in IS: 456-2000 are summarized in Table 9.9.

9.4.4 Admixtures

The physical properties, type, and the method of use of admixtures must be assessed.

9.4.5 Importance of Finer Fractions

The selection of a suitable grading requires experience and sound judgement. The final proportions of fine and coarse aggregates should be based on the behaviour of the concrete when trial mixes are made under full-scale field conditions. It is safer to use a high, rather than a low, proportion of fines even though it may increase the richness of the mix slightly. This ensures protection against segregation, harshness and, therefore, avoids poor compaction. For instance, using irregular gravel CA of 20 mm maximum size, natural sand, and normal methods of transport, the overall grading should be somewhere near curves 2 and 3 of Fig. 9.6 for medium workability.

The amount of FA passing the 300-μm and 150-μm IS sieves influences the workability, finish, surface texture, and water gain. For concrete to have a smooth finish and proper cohesiveness, the minimum amount of FA passing the 300-μm sieve should be at least 15% and that passing the 150-μm sieve at least 3–4% of the total quantity of FA. The presence of adequate fines is more important in wetter (sloppy) mixes than in stiffer mixes, and in leaner mixes than in richer mixes.

The specific gravity of aggregate is also relevant to mix proportioning. For heavyweight concrete of density exceeding 3200 kg/m^3, materials such as crushed barytes, limonites, iron ores, and steel punchings or small castings with specific gravities of 4.25, 4.75, 4–5, and 7.8, respectively, are used as aggregates. Natural aggregates usually have specific gravities between 2.6 and 2.7. Hence, allowances should be made for the difference between the specific gravities of these materials and those of ordinary aggregate, which is assumed to be 2.6. If the specific gravity of FA is higher than that of CA, the mix will tend to segregate; however, it is better for both the aggregates to have nearly the same specific gravity. Aggregates that conform to the requirements of IS: 383-1970 do not have a large deficiency or excess of any size, give a smooth grading curve, and produce the most satisfactory concrete.

9.5 Variability of Test Results

The standard deviation of a given grade of concrete can be calculated from the results of individual tests of concrete cubes, using the formula

$$S = \sqrt{\frac{\Sigma \Delta^2}{n-1}}$$

(9.3)

where Δ is the deviation of the individual test strength from the average strength of n specimens and n is the number of specimen test results. If at least 30 test results for a particular grade of concrete on site with the same materials and equipment are not available, the standard deviation S, irrespective of the degree of control, may be assumed from Table 9.3.

9.6 Acceptance Criteria for Concrete

Brief excerpts from the acceptance criteria for structural concrete suggested in clause 16 of IS: 456-2000 are reproduced below (refer to Table 9.10).

The concrete shall be deemed to comply with the strength requirements when both of the following conditions are met:

a. The mean strength determined from any group of four consecutive test results complies with the following limits:

$$\text{Mean of test results} > f_{ck} + 0.825\,S$$
$$> f_{ck} + 4\,\text{N/mm}^2$$

b. Individual test result $> f_{ck} - 4\,\text{N/mm}^2$

Thus, with a view to obtain the specified minimum characteristic strength f_{ck} of concrete in the field, the mix should be designed for somewhat higher (preliminary) test strength depending on the degree of quality control and the tolerance level adopted.

9.6.1 Determining the Laboratory Design Strength of Concrete

The method of determining the laboratory design strength follows from the acceptance criteria for concrete suggested in IS: 456-2000. The target strength can be calculated based on the equation

$$\text{Target strength} = f_{ck} + 1.65\,S \tag{9.4}$$

The target laboratory strength can also be obtained using the formula

$$\sigma_c = \frac{f_{ck}}{1 - t(\gamma/100)} \tag{9.5}$$

where f_{ck} is the required compressive strength in the field, σ_c is the target laboratory strength, and t is the dimensionless factor for the specified tolerance level, given by

$$t = 1.65, 1.58, \text{ or } 1.54$$

where the number of samples tested is 10, 20, or 30, respectively, permitting 1 in 15 test results to fall below σ_c. γ is the coefficient of variation, taken as 10%, 15%, and 20%, respectively, for a very good, good, and fair degree of quality control in the field.

9.6.2 Quality Control of Concrete

The importance of quality control of concrete in deriving optimum benefit from the materials employed is increasingly being realized. Normally, specifications stipulate the 28-day compressive strength requirements. Though useful in establishing a criterion, the limitations of compressive strength data must be considered. Test specimens indicate the potential rather than the actual strength of concrete in a structure, while poor workmanship in the placing and curing of concrete may cause strength reduction which will not be reflected in the cube test. To place too much reliance on too few tests is, therefore, incorrect.

Most specifications require a minimum strength to be exceeded, but statistical analysis indicates that such specifications are difficult to comply with. The probable strength (f_{ck}) is known only after 28 days of casting. It would, therefore, be more realistic to take a calculated risk based on the probabilities arrived at by statistical methods and allow a certain percentage of 'lows'. By permitting a reasonable number of lows, the engineer can hold the builder to a standard that is possible to scrutinize. Otherwise, the site supervisor should be able to test the 3-day strength and predict the 28-day strength for enforcement of quality. Herein, the accelerated boiling water method of testing gains significance.

It may be mentioned that the engineer-in-charge may permit a higher percentage of 'lows' for mass concrete, foundation concrete, and concrete used in hydraulic structures. Many irrigation and power structures are designed with 20% lows.

9.7 Methods used for Proportioning Concrete Mixes

The main methods used for concrete mix proportioning are the following:

a. Trial mixes
b. Nominal mixes
c. Road Research Laboratory (RRL) method
d. Department of Environment (DOE) method
e. Minimum voids method
f. Maximum density method
g. Fineness modulus method
h. American Concrete Institute (ACI) method
i. Bureau of Indian Standards (IS) method

We describe the trial mixes method first as it is an essential component of all other methods. It enables field tests to be conducted on the mix and adjustments to be made thereafter.

9.7.1 Trial Mix

Concrete mix proportioning methods are based on assumptions regarding the type and quality of materials used. Therefore, it must always be verified that the materials (cement, aggregates, plasticizers, etc.) used will behave according to the assumptions. Therefore, trial mixes are made. The subsequent feedback from trial mixes forms an important part of the mix proportioning process. Depending on how much the results of the trial mix differ from the assumed mix proportion values, the following alternatives are used:

- Employing the trial mix proportions directly for manufacturing concrete at the site
- Modifying the trial mix proportions based on judgement and then using the revised mix proportions for making concrete at the site
- Preparing further trial mixes to include changes in the major variables based on the feedback generated from the previous mix proportions

This method assumes that (a) the stipulation of an appropriate water–cement ratio will result in a concrete of the desired characteristic strength, provided full compaction can be achieved, (b) the CA is graded, and (c) the FAs and CAs are stacked separately. To determine the free water–cement ratio corresponding to the given characteristic strength of concrete, Fig. 9.4(a) or 9.4(b) can be used.

The cost of a concrete mix is governed primarily by the cement content. The objective of a satisfactory concrete mix at minimum cost can, therefore, be achieved by a mix design process that gives the aggregate combination requiring the least amount of cement at a water–cement ratio fixed by the combined requirements of strength as well as workability.

Example 9.1

Design a plain concrete mix for a residential building assuming the following data:

Building is in a mild environmental zone

Compressive strength of cement at 7 days: 43 N/mm^2

CA: crushed stone

Maximum size of aggregate: 40 mm

Workability of the mix: medium

Slump of concrete required: 50 mm

Characteristic strength of concrete at 28 days: $f_{ck} = 20$ N/mm^2

Solution

Step 1 Find the water–cement ratio to achieve $f_{ck} = 20$ N/mm^2 at 28 days, assuming 30 samples and standard deviation $S = 4.6$ N/mm^2. Using the formula

$$f_{av} = f_{ck} + \left(1.65 - \frac{1.65}{\sqrt{N}}\right) S \tag{9.6}$$

we have

$$f_{av} = 20 + \left(1.65 - \frac{1.65}{\sqrt{30}}\right) \times 4.6 = 26.2 \, \text{N/mm}^2$$

From Fig. 9.4(a) or 9.4(b), the free water–cement ratio for 26.2 N/mm^2 compressive strength at 28 days is 0.5 with cement of strength 43 N/mm^2. The cement content for a very mild environment is 220 kg/m^3 [Table 9.6(a)].

Step 2 Take 0.9 kg cement and 0.45 kg water and make a slurry. Take 4.5 kg of surface-dry sand, add it gradually to the slurry, and mix thoroughly until the resultant mortar

  i. stands vertically when a trowel is inserted into it, turned through 90°, and lifted away vertically
 ii. becomes self-healing, i.e., vertical cuts about 20 mm deep and 25 mm apart made with a trowel edge on top of the mortar face heal by themselves—a condition wherein the cement slurry rises up to the bottom of the cuts

These conditions can be achieved with a few trials. Now, weigh the remaining sand and deduce the quantity of sand added to the slurry. Suppose the quantity of sand added to cement is found to be 2.88 kg. The sand–cement ratio, therefore, works out to about 3.2 by weight.

Step 3 Use the nomograph shown in Fig. 9.11 for the sand-reduction factor. Enter the nomograph from the left at the point indicating a sand–cement ratio of 3.2, connect it with the point on the straight line representing a water–cement ratio of 0.5, and produce it to meet the right-hand line. Thus, a sand-reduction factor of about 1.4 can be read off. The quantity of sand required for a 50-kg (35-L) bag of cement would be $(50 \times 3.2)/1.4 = 114$ kg. If the sand is assumed to have a density of about 1550 kg/m^3, the volume of sand required is $114/1550 = 0.073$ m^3 = 73 L.

Step 4 Prepare a trial mix as follows. Take a quarter bag of cement, 6.25 kg water (i.e., 25 kg or 25 L for one bag of cement, as water weighs 1 kg/L), 28.50 kg surface-dry sand, and 68 kg surface-dry CA. Mix cement, sand, and water with half the coarse aggregate (34 kg of CA). Add progressively further quantities of CA until the slump of the mix is estimated to be 50 mm. Weigh the remaining CA and calculate the final weights. The weight of CA added is 20.5 kg. The total weight of CA is $34 + 20.5 = 54.5$ kg. If the above adjustment takes too long, repeat the trial mix with the final weights of the materials and recheck. Let the quantity of CA used for a quarter bag of cement be 54.5 kg. Therefore, for one bag of cement the aggregate weight is $4 \times 54.5 = 218$ kg. If the density of CA is taken to be 2600 kg/m³, the quantity required would be $54.5/2600 = 0.02$ m³ $= 20$ L (i.e., 80 L per bag of cement). Thus, the final proportions by weight and by volume are as follows:

Material	By weight (kg)	Bulk density kg/m³	By volume (L)	Proportion by volume	Proportion by weight
Cement	50 (1 bag)	1429	35	1	1
Sand	114	1550	73	2.0	2.28
CA	218	2600	80	2.28	4.36
Water	25	1000	25	0.5 (w/c)	0.5 (w/c)

Recheck the workability of the mix and the compressive strength of the concrete cubes at 28 days.

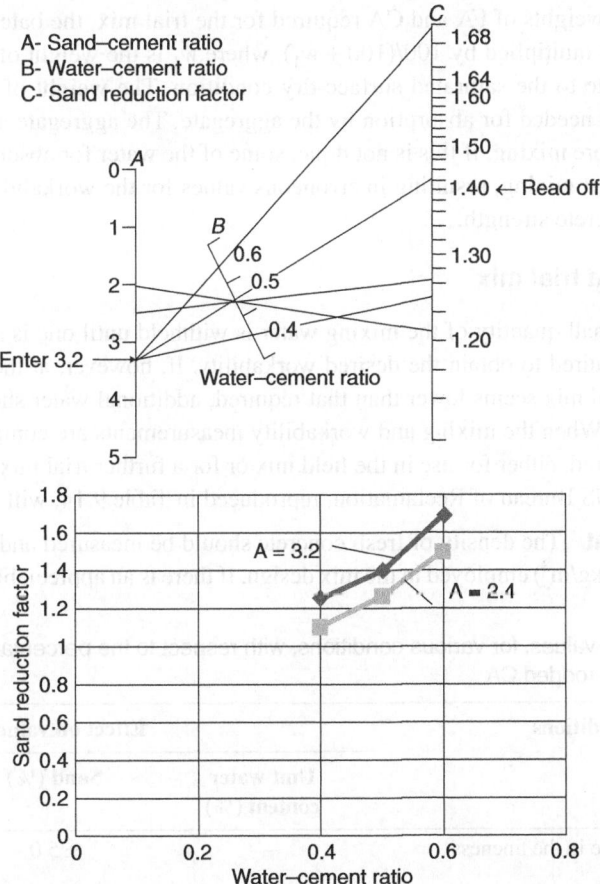

Fig. 9.11 Estimating sand reduction factor using a nomograph

Correction for wet sand

The mortar trial in step 2 should be done after the sand is surface dried and the percentage of water contained in it is determined. For the trial mix in step 4, the quantity of wet sand added should be increased by the weight of the water it contains. Also, the weight of water in the trial mix should be reduced by an amount equal to the weight of the water present in the sand.

Calculations of the quantity of constituents

Mix proportions indicate the weights in kilograms of different materials required to make 1 m³ of compacted concrete. Six cubes of 150-mm sides can be cast from a 0.05 m³ mix; thereafter it is possible to carry out the slump, Vebe, and density measurements separately. Therefore, the trial mix for a total volume of 0.05 m³ can be made for assessing the suitability of the mix. Thus, the individual batch weights (in kilograms) for the 0.05 m³ trial mix will be equal to the appropriate contents for 1 m³ multiplied by 0.05.

Adjustments for weight of aggregate and water

For aggregate to be batched in a surface-wet condition, the quantity of free water required should be determined in order to make adjustments to the weights of aggregate and water to be added to the mix. For aggregate to be batched in a dry condition, the batch weights of the aggregate should be reduced and the weight of mixing water increased to allow for the absorption of some of the mixing water by the dry aggregate. To derive the weights of FA and CA required for the trial mix, the batch weights obtained from the mix design should be multiplied by $100/(100 + w_1)$, where w_1 is the weight of water (per cent) needed to bring the dry aggregate to the saturated surface-dry condition. The weight of mixing water should be increased by the quantity needed for absorption by the aggregate. The aggregate also should be brought to a saturated condition before mixing; if this is not done, some of the water for absorption will act effectively as free water at the time of mixing, resulting in erroneous values for the workability of the mix and a possible decrease in the concrete strength.

Tests conducted on a trial mix

Vebe test   Initially a small quantity of the mixing water is withheld until one is satisfied that the quantity withheld is definitely required to obtain the desired workability. If, however, at the designed water content the workability of the trial mix seems lower than that required, additional water should be added to achieve the required workability. When the mixing and workability measurements are completed, a change of water content may still be required, either for use in the field mix or for a further trial mix. For this adjustment, the values suggested by the US Bureau of Reclamation, reproduced in Table 9.14, will be useful.

Determination of weight   The density of fresh concrete should be measured and the resultant value compared with the density (in kg/m³) employed in the mix design. If there is an appreciable difference, the contents

Table 9.14 Adjustment of values, for various conditions, with respect to the percentages of unit water content, sand, and dry-rodded CA

Changes in stipulated conditions	Effect on values of		
	Unit water content (%)	Sand (%)	Dry-rodded CA (%)
Each 0.1 increase or decrease in the fineness modulus of sand	—	± 5.0	± 1.0
Each 25-mm increase or decrease in slump	± 3.0	—	—
Each 1% increase or decrease in air content	± 3.0	± 0.5 to 1.0	—
Each 0.05 increase or decrease in the water–cement ratio	—	± 1.0	—
Each 1% increase or decrease in sand content	± 1.0	—	± 2.00
For angular CA	+ 7.0 to 10.0	+ 3.0 to 5.0	—
For low-slump concrete as in rigid pavements	− 3.0	− 3.0	+ 6.0

Note: If aggregates are proportioned by the percentage of sand method, use the first and second columns under 'Effect on values of'; if they are proportioned by the dry-rodded CA method, then use the first and third columns.

(in kg/m^3) or the unit proportions of the trial mix will differ from those given in the initial mix design. The initial design values should then be corrected by the ratio of the measured density to the assumed density so as to obtain the actual weights (in m^3) of the materials in the trial mix.

Strength test The strength test results should be compared with the average strength and adjustments made, if required, to the water–cement ratio by employing the curves given in Fig. 9.4(a) or 9.4(b). Minor adjustments may be made to the mix proportions for use in field mixes without preparing further trial mixes. However, if large adjustments to the water–cement ratio are required, then it would be advisable to prepare a second trial mix, adopting the revised proportions, and recalculate the batch weights based on the updated values.

9.7.2 Nominal Mix

The nominal mix method is recommended by IS: 456-2000 for concrete of strength M20 or lower. However, it is not advisable to use this method for structural applications. The proportions of materials for nominal mixes are given in Table 9.15.

Table 9.15 Proportions for nominal mix concrete

Grade of concrete	Total quantity of dry aggregates by mass per 50 kg of cement, to be taken as the sum of individual masses of FAs and CAs (kg), max.	Proportion of FA to CA (by mass)[*]	Quantity of water per 50 kg of cement, max. (mL)
M5	800	Generally 1:2 but	60
M7.5	625	subject to an upper	45
M10	480	limit of 1:1.5 and	34
M15	330	a lower limit of	32
M20	200	1:2.5	30

[*]The proportion of FAs to CAs should be adjusted from the upper limit to the lower limit progressively as the grading of FAs becomes finer and the maximum size of CA becomes larger. Normally, graded CA is used.

For example, for an average grading of FA (that is, zone II of table 9.4 of IS: 383), the proportions will be 1:1.5, 1:2, and 1:2.5 for the maximum sizes of aggregate 10, 20, and 40 mm, respectively.

The following important points must be noted from Table 9.15.

a. The first column gives five grades of concrete, i.e., M5, M7.5, M10, M15, and M20. They roughly correspond to 1:4:8, 1:3:6, 1:2:4, 1:1.5:3, and 1:1:2 nominal mixes, respectively.

b. The second column prescribes the maximum limits of total aggregates (FAs and CAs) per bag of cement. For example, the maximum quantity of total dry aggregate for M15 is 330 kg, which is the sum of the weight of both FAs and CAs. If a given water–cement ratio results in a low slump, then the total quantity of aggregate per 50 kg of cement can be reduced. On the other hand, for high slump, the total quantity of aggregates should not be increased, because the table gives the upper limits of aggregates. However, the water content can be reduced, thereby decreasing the water–cement ratio. Note that what is prescribed is the maximum aggregate content and the maximum water–cement ratio. For example, for grade M15, water ≤ 32 L should be added. Figure 9.12 indicates that if the water–cement ratio increases, the cement–aggregate ratio decreases (from 1:3 to 1:4.5, and so on).

c. The third column prescribes the proportion of FA to CA, which should, however, be changed according to the zone of sand and MSA used. The previous section discussed the various issues involved at length. As a rough estimate, Table 9.12 can be used to set the FA–CA ratio. In every batch, the slump test must be conducted first. Then the FA–CA ratio should be varied till the desired minimum value of FA, slump, cohesiveness, and finish are obtained, with water–cement less than or equal to that prescribed in column 4 of Table 9.15.

d. The typical gradings for various MSAs are given in Tables 9.16–9.18.

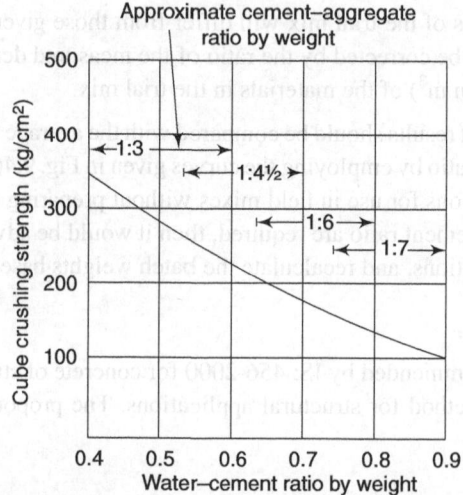

Fig. 9.12 Cement–aggregate ratio in relation to water–cement ratio and crushing strength of concrete

Table 9.16 Grading types for 40 mm MSA (Fig. 9.5)

Grading type	Percentage passing IS sieve								
	40 mm	20 mm	10 mm	4.75 mm	2.36 mm	1.18 mm	600 μm	300 μm	150 μm
1	100	50	36	24	18	12	7	3	0
2	100	59	44	32	25	17	12	7	0
3	100	67	52	40	31	24	17	11	2
4	100	75	60	47	38	30	23	15	5

Table 9.17 Grading types for 20 mm MSA (Fig. 9.6)

Grading type	Percentage passing IS sieve							
	20 mm	10 mm	4.75 mm	2.36 mm	1.18 mm	600 μm	300 μm	150 μm
1	100	45	30	23	16	9	2	0
2	100	55	35	28	21	14	3	0
3	100	65	42	35	28	21	5	0
4	100	75	48	42	34	27	12	1.5

Table 9.18 Grading types for 10 mm MSA (Fig. 9.7)

Grading type	Percentage passing IS sieve						
	10 mm	4.75 mm	2.36 mm	1.18 mm	600 μm	300 μm	150 μm
1	100	30	20	16	12	4	0
2	100	45	33	26	19	8	1
3	100	60	46	37	28	14	3
4	100	75	60	46	34	20	6

Adjustments for bulking and moisture content in aggregates Depending on the condition of the aggregates, the moisture has to be adjusted to ensure that the maximum value of the water–cement ratio is not exceeded. The tests described earlier can be used to determine the moisture content in aggregates.

Example 9.2

Using the following data in the field, determine the proportions of the various ingredients of M20 concrete.

Grade of concrete: M20

Maximum size of aggregate: 20 mm graded, moist

FA: Sieve analysis indicates zone III sand, which is moderately wet

Desired slump range: 25–50 mm

Capacity of the equipment available: 200 L, i.e., one bag of cement to be used per batch

Vibration: Mechanical

Solution

Step 1 Choose the maximum size of aggregates from the workability point of view for M20. Larger aggregates are preferable. Choose the initial proportions from the IS table for nominal mixes given below.

Particulars	Cement	Total dry aggregates	FA:CA aggregates	FA	CA	Water–cement	Water
Before adjustment	50 kg	200 kg	1:2	50	100	0.6	30

Aggregate quantities may need a correction of ±5.

Step 2 Conduct a test for the bulking of FAs and make adjustments for the bulkage. The bulkage is 20% based on the test. For the correction of bulkage, increase the volume of sand by 20% = (120/100) × 50 = 60 L. Adjust the water to be added to the mixture. Tests show that the surface moisture carried by aggregate is 80 L/1000 L for FA and 40 L/1000 L for CA. The correction for moisture is as follows:

For FAs,

$$\frac{80}{1000} \times 60 = 4.8\,L$$

For CAs,

$$\frac{40}{1000} \times 100 = 4\,L$$

Total moisture = 4.8 + 4.0 = 8.8 L

Step 3 Estimate the quantities after correction.

Particulars	Cement	Total dry aggregates	FA:CA	FA	CA	Water–cement	Water
After correction in first trial	50	200 kg	1:2	60	100	0.6	30 – 8.8 = 21.2 L

Step 4 Calculate the quantities required per batch according to the capacity of the mixer. In the present case, the mixer capacity is assumed to be 200 L. Hence the batch is worked out for 50 kg of cement.

Step 5 Using the available container, measure the quantities. The containers have capacities in multiples of 5 L. Therefore, for FA use the 30-L container twice and for CA use the 25-L container four times. Add 21.2 L of water.

Step 6 Feed all the material into the mixer. After mixing, discharge and conduct the slump test. In addition, observe the cohesiveness and harshness.

Cast the cubes and test the strength after 3, 7, and 28 days. Repeat steps 1–6 by slightly changing FA:CA until acceptable results are obtained from both the strength and workability tests.

The operations can be carried out as shown in the following table.

Operations involved in mix design

	Particulars	Cement	Total dry aggregates	FA:CA	FA	CA	Water–cement	Water
Step 1	Initial consumption	50	200	1:2	50	100	0.6	30
Step 2a	Correction for bulkage	20% of FA		$\frac{50}{100} \times 120 = 60\,L$				
Step 2b	Correction for moisture			in FA $= \frac{80}{1000} \times 60 = 4.8\,L$				
				in CA $= \frac{40}{1000} \times 100 = 4\,L$				
Step 3	Quantity after correction	50	200	1:2	60	100	0.6	21.2 (30–8.8)
Step 4	Quantity per bag of cement	50	200	1.2	60	100	0.6	21.2
Step 5	Selection of container	(a) FA 30 L – 2 = 60 L						
		(b) CA 25 L – 4 = 100 L						
Step 6	Mix the ingredients and prepare trial texts							

9.7.3 Road Research Laboratory Method

The Road Research Laboratory (RRL), UK, carried out an exhaustive study on concrete mix proportioning, the results of which have been published in *Road Note No. 4* (1950). The RRL method has found extensive application in UK, India, and elsewhere. The following parameters are involved.

Aggregates

Nominal maximum size of aggregate Three sizes, namely, 40, 20, and 10 mm, are used for normal reinforced concrete works. In certain special cases, 75–80 mm sizes are used.

Grading of aggregate For each of the MSAs that can be used in reinforced concrete works, four types of gradings are given in Figs 9.5–9.7.

Types of aggregate Rounded, irregular, and crushed rocks are used as aggregates.

Workability

Workability can be categorized into five groups, based on extremely low, very low, low, medium, and high slump values.

Strength

Average test strength This is based on the characteristic strength and standard deviation. Characteristic strength is defined as the strength of concrete below which not more than 5% of the test results are expected to fall. The strength is fixed based on this target.

The other parameters considered in mix proportioning are the type of cement and the durability requirements of concrete.

Tables 9.19(a)–(i) give aggregate–cement ratios for different water–cement ratios for designated different workability levels. These tables are based on tests conducted by several documented research works. After determining the water–cement ratio from strength and/or durability considerations, the workability acceptable at the site is determined. For this purpose, the aggregate–cement ratio is found from the tables and a trial mix is made; if this mix is unsatisfactory, a suitable adjustment is made in the aggregate–cement ratio and a new trial mix is made.

The water–cement ratio versus strength curves such as those shown in Fig. 9.4 are used to determine the free water–cement ratio required to obtain the given average strength f_{av} of concrete. The values given here are for fully compacted, dense-aggregate concrete of medium strength made with Indian Portland cements of average quality and cured at like temperatures.

The effective water content includes (a) the free water in the sand and CAs, and (b) the water added at the mixer, but excludes the water absorbed by the aggregate. When the quantity of water required in the mix has been estimated, the quantity of water to be added at the mixer is determined by subtracting the free water in the aggregate from the required water. The determination of free water in the aggregate, however, is a matter of site tests and control. Usually the type of aggregate to be selected is fixed, as only those that are economically feasible in the locality.

Table 9.19 Aggregate–cement ratios required to give the four degrees of workability with different water–cement ratios (by weight) and gradings (Figs 9.5–9.7)

(a) 40-mm rounded gravel aggregate

Workability grading* w/c ratio	Very low				Low				Medium				High			
	1	2	3	4	1	2	3	4	1	2	3	4	1	2	3	4
	Aggregate–cement ratio by weight†															
0.35	5.0	4.5	3.9	3.4	4.3	3.9	3.5	3.1	3.4	3.1	2.9	2.7				
0.40	7.0	6.5	5.7	4.9	5.9	5.6	5.0	4.4	4.7	4.6	4.3	3.8	4.1	4.0	3.9	3.5
0.45	8.9	8.6	7.7	6.5	7.6	7.4	6.7	5.8	6.0	6.1	5.7	5.0	5.2	5.3	5.0	4.6
0.50				8.0			8.2	7.2	7.5	7.6	7.1	6.3	6.3	6.5	6.2	5.7
0.55								8.4		8.9	8.1	7.3	$S^{\ddagger}$	7.7	7.4	6.7
0.60																7.6

*Refer to Fig. 9.5.
†These values have been obtained by extrapolation of other data and are not based directly on the results of trial mixes.
‡S indicates that the mix would segregate.

(b) 40-mm irregular gravel aggregate

Workability grading* w/c ratio	Very low				Low				Medium				High			
	1	2	3	4	1	2	3	4	1	2	3	4	1	2	3	4
	Aggregate–cement ratio by weight															
0.35	4.0	3.9	3.6	3.2	3.4	3.3	3.2	2.9								3.1
0.40	5.3	5.2	4.8	4.3	4.5	4.5	4.2	3.8	3.8	3.8	3.7	3.4	3.4	3.5	3.3	4.0
0.45	6.6	6.5	6.0	5.3	5.6	5.6	5.3	4.8	4.6	4.7	4.6	4.3	4.1	4.4	4.3	4.8
0.50	7.8	7.7	7.1	6.3	6.6	6.6	6.3	5.7	5.4	5.7	5.5	5.1	4.8	5.9	5.1	5.5
0.55			8.1	7.3	7.6	7.6	7.2	6.6	6.2	6.5	6.3	5.9	$S^{\dagger}$	S	5.9	6.3
0.60								7.4	7.0	7.3	7.1	6.6			6.7	6.9
0.65								8.8	7.8	8.1	7.8	7.3			7.3	7.4
0.70												7.9				8.0
0.75																

*Refer to Fig. 9.5.
†S indicates that the mix would segregate.

(c) 40-mm crushed rock aggregate

Workability grading[*] w/c ratio	Very low				Low				Medium				High			
	1	2	3	4	1	2	3	4	1	2	3	4	1	2	3	4
	Aggregate–cement ratio by weight[†]															
0.35	3.4	3.4	3.2	2.9	4.0	3.8	3.6	3.3								
0.40	4.9	4.6	4.2	3.8	4.9	4.7	4.4	4.2	3.3	3.3	3.2	3.0	3.1	3.1	2.9	2.7
0.45	6.0	5.7	5.2	4.7	5.8	5.6	5.3	5.0	4.1	4.1	3.9	3.8	3.7	3.8	3.7	3.4
0.50	7.2	6.8	6.2	5.6	6.6	6.4	6.1	5.8	4.8	4.8	4.7	4.6	4.4	4.5	4.5	4.2
0.55	8.1	7.7	7.1	6.4	7.4	7.2	6.9	6.6	5.5	5.5	5.4	5.3	S[‡]	5.2	5.2	4.8
0.60		8.6	8.0	7.2	8.1	7.9	7.6	7.3	6.1	6.2	6.1	6.0		S	5.9	5.6
0.65			8.8	7.9		8.5	8.3	7.9	S	6.9	6.8	6.6			6.5	6.2
0.70				8.6				8.5			7.5	7.3			7.1	6.8
0.75												7.8				7.4

[*]Refer to Fig. 9.5.
[†]These values have been obtained by extrapolation of other data and are not based directly on the results of trial mixes.
[‡]S indicates that the mix would segregate.

(d) 20-mm rounded gravel aggregate

Workability grading[*] w/c ratio	Very low				Low				Medium				High			
	1	2	3	4	1	2	3	4	1	2	3	4	1	2	3	4
	Aggregate–cement ratio by weight															
0.35	4.5	4.2	3.7	3.2	3.8	3.6	3.3	3.0	3.1	3.0	2.8	2.6				
0.40	6.6	6.1	5.4	4.5	5.3	5.1	4.6	4.1	4.2	4.2	3.9	3.6	3.7	3.8	3.6	3.3
0.45	8.1	7.6	6.7	5.8	6.9	6.6	5.9	5.1	5.3	5.3	5.0	4.6	4.6	4.8	4.5	4.1
0.50			8.0	7.0	8.2	8.0	7.0	6.0	6.3	6.3	6.0	5.5	5.5	5.7	5.4	4.8
0.55				8.1			8.2	6.9	7.3	7.3	7.0	6.3	6.3	6.5	6.1	5.5
0.60								7.7			8.0	7.1	S[†]	7.2	6.8	6.1
0.65								8.4				7.8		7.7	7.4	6.6
0.70															7.9	7.1
0.75																7.6

[*]Refer to Fig. 9.6.
[†]S indicates that the mix would segregate.

(e) 20-mm irregular gravel aggregate

Workability grading[*] w/c ratio	Very low				Low				Medium				High			
	1	2	3	4	1	2	3	4	1	2	3	4	1	2	3	4
	Aggregate–cement ratio by weight															
0.35	3.6	3.6	3.5	3.0	3.0	3.0	3.0	2.7								
0.40	4.9	4.8	4.6	4.1	3.9	3.9	3.9	3.5	3.3	3.4	3.4	3.2	3.1	3.2	3.2	2.9
0.45	6.0	5.8	5.5	5.0	4.8	4.8	4.7	4.3	4.0	4.1	4.1	3.9	S[†]	3.8	3.8	3.5
0.50	7.2	6.8	6.4	5.9	5.5	5.5	5.4	5.0	4.6	4.8	4.8	4.5		4.4	4.4	4.1
0.55	8.3	7.8	7.3	6.7	6.2	6.2	6.1	5.7	S	5.4	5.3	5.1		4.9	4.9	4.7
0.60	9.4	8.7	8.1	7.4	6.9	6.9	6.7	6.3		6.0	5.9	5.6		S	5.4	5.2
0.65				8.0	7.5	7.5	7.3	6.8		S	6.4	6.1			5.8	5.7
0.70					8.0	8.0	7.8	7.3			6.8	6.6			6.2	6.1
0.75								7.9			7.2	7.0			6.6	6.5
0.80											7.5	7.4			S	7.0

[*]Refer to Fig. 9.6.
[†]S indicates that the mix would segregate.

(f) 20-mm crushed rock aggregate

Workability grading* w/c ratio	Very low				Low				Medium				High			
	1	2	3	4	1	2	3	4	1	2	3	4	1	2	3	4
	Aggregate–cement ratio by weight															
0.40	4.5	4.1	3.8	3.5	3.5	3.5	3.2	3.0								
0.45	5.5	5.0	4.6	4.3	4.3	4.2	3.9	3.7	3.7	3.7	3.4	3.3	3.5	3.5	3.2	3.1
0.50	6.5	5.9	5.4	5.0	5.0	4.9	4.5	4.3	4.2	4.2	3.9	3.8	S†	3.9	3.8	3.5
0.55	7.2	6.6	6.0	5.7	5.7	5.5	5.0	4.8	4.7	4.7	4.5	4.3		S	4.3	4.0
0.60	7.8	7.2	6.6	6.3	6.3	6.0	5.6	5.3	S	5.2	4.9	4.8			4.7	4.5
0.65	8.3	7.7	7.2	6.9	6.9	6.5	6.1	5.8		5.7	5.4	5.2			5.2	4.9
0.70	8.7	8.2	7.7	7.5	7.4	7.0	6.6	6.3		6.2	5.8	5.7			5.5	5.3
0.75			8.2	8.0	7.9	7.5	7.0	6.7		S	6.2	6.1			5.8	5.7
0.80							7.4	7.2			6.6	6.5			6.1	6.0

*Refer to Fig. 9.6.
†S indicates that the mix would segregate.

(g) 10-mm rounded gravel aggregate

Workability grading* w/c ratio	Very low				Low				Medium				High			
	1	2	3	4	1	2	3	4	1	2	3	4	1	2	3	4
	Aggregate–cement ratio by weight															
0.40	5.6	5.0	4.2	3.2	4.5	3.9	3.3	2.6	3.9	3.5	3.0	2.4	3.5	3.2	2.8	2.3
0.45	7.2	6.4	5.3	4.1	5.5	4.9	4.1	3.2	4.7	4.3	3.7	3.0	4.2	3.9	3.4	2.9
0.50		7.8	6.4	4.9	6.5	5.8	4.9	3.8	5.4	5.0	4.3	3.5	4.8	4.5	4.0	3.4
0.55			7.5	5.7	7.4	6.7	5.7	4.4	6.1	5.7	4.9	4.0	5.3	5.1	4.5	3.9
0.60				6.5		7.5	6.4	5.0	6.7	6.3	5.5	4.5	5.8	5.6	5.0	4.3
0.65				7.2			7.1	5.6	7.3	6.9	6.1	5.0	S†	6.1	5.5	4.7
0.70							7.7	6.2	7.9	7.5	6.7	5.5		6.6	6.0	5.1
0.75								6.7			7.2	5.9		7.1	6.5	5.5
0.80								7.2			7.7	6.3		7.6	6.9	5.9

*Refer to Fig. 9.7.
†S indicates that the mix would segregate.

(h) 10-mm irregular gravel aggregate

Workability grading* w/c ratio	Very low				Low				Medium				High			
	1	2	3	4	1	2	3	4	1	2	3	4	1	2	3	4
	Aggregate–cement ratio by weight															
0.40	4.1	3.8	3.3	2.8	3.3	3.1	2.8	2.3								
0.45	5.1	4.8	4.3	3.6	4.1	3.9	3.5	3.0	3.5	3.4	3.2	2.8	3.2	3.1	3.0	2.7
0.50	6.1	5.8	5.2	4.4	4.8	4.6	4.2	3.7	4.2	4.1	3.8	3.4	S†	3.8	3.6	3.2
0.55	7.0	6.7	6.1	5.2	5.5	5.3	4.9	4.3	S	4.7	4.4	4.0		4.4	4.2	3.7
0.60	7.9	7.6	7.0	6.0	S	6.0	5.6	4.9		5.3	5.0	4.5		4.9	4.7	4.2
0.65			7.8	6.8		6.6	6.2	5.5		5.9	5.6	5.0		5.4	5.2	4.6
0.70						7.2	6.8	6.1		6.4	6.1	5.5		5.9	5.7	5.0
0.75						7.8	7.4	6.7		6.9	6.6	6.0		6.4	6.1	5.4
0.80							8.0	7.3		7.4	7.1	6.4		6.8	6.5	5.8

*Refer to Fig. 9.7.
†S indicates that the mix would segregate.

(i) 10-mm crushed rock aggregate

Workability grading* w/c ratio	Very low				Low				Medium				High			
	1	2	3	4	1	2	3	4	1	2	3	4	1	2	3	4
	Aggregate–cement ratio by weight															
0.40	3.7	3.3	2.8	2.0												
0.45	4.5	4.1	3.5	2.6	3.8	3.6	3.0	2.2	3.3	3.1	2.7	2.1				
0.50	5.2	4.9	4.2	3.2	4.4	4.2	3.6	2.7	3.8	3.7	3.2	2.6	S‡	3.2	2.9	2.4
0.55	5.9	5.6	4.9	3.8	4.9	4.8	4.2	3.2	S	4.2	3.7	3.0		3.7	3.4	2.8
0.60	6.6	6.3	5.5	4.3	S	5.3	4.7	3.7		4.7	4.2	3.4		4.2	3.8	3.2
0.65	7.3	7.0	6.1	4.8		5.8	5.2	4.2		5.1	4.6	3.8		4.6	4.2	3.6
0.70	7.9	7.6	6.7	5.3		6.3	5.7	4.6		5.6	5.1	4.2		5.0	4.6	4.0
0.75			7.3	5.8		6.8	6.2	5.0		6.0	5.5	4.6		5.4	5.0	4.4
0.80			7.8	6.3		7.2	6.6	5.5		6.4	5.9	5.0		5.8	5.4	4.7

*Refer to Fig. 9.7.

†S indicates that the mix would segregate.

The water–cement ratio and, hence, the cement content in a concrete mix is only one factor among others required to establish concrete strength. There is more than one way to arrive at an economical concrete mix. For instance, to produce high-strength concrete, one should consider (a) the most efficient maximum size of aggregate, (b) the surface texture of the aggregate available, and (c) the particle shape of aggregate. When the paste quality is such that the bond strength governs the concrete strength, it may prove more economical to increase concrete strength by decreasing the MSA or by using smaller CA particles. We will discuss this aspect further in the chapter on high-strength concrete.

According to IS: 456, the nominal maximum size of CA should be the largest one economically available but not greater than one-fourth the minimum thickness of a member, provided that the concrete can easily be laid all around the reinforcement and can properly fill the corners of the formwork. Generally, for normal reinforced concrete constructions, aggregates of 20 mm nominal maximum size are considered satisfactory. For heavily reinforced concrete members, e.g., ribs of main beams, the nominal maximum size of the CA could be 5 mm less than the minimum cover to the reinforcement, or 5 mm less than the minimum clear distance between the main bars, whichever is smaller.

From Figs 9.5–9.7 and the corresponding Tables 9.16–9.18, it is observed that for 40, 20, and 10 mm MSA, grading type 4 is the finest and suitable for very wet or lean mixes with high workability and poorly shaped aggregates, whereas grading type 1 is the coarsest and suitable for dry mixes. For convenience, the overall gradings in Figs 9.5–9.7 are denoted with reference to the four gradings for each of the three nominal MSAs. They have been so numbered that the higher numbers refer to gradings which represent a higher proportion of fine particles. These are not ideal curves—they represent the gradings used in the tests from which the data were obtained.

Adjustment for different extents of water absorption of aggregates   It is important for the calculation of the water–cement ratio of the mix to account for both the water added at the mixer and the water used in the aggregate if it is damp or moist. However, water in the damp aggregate occurs partly on the surface of the particles and is partly absorbed into the pores, where it is not readily available to affect the properties of concrete. There is no generally accepted method to account for the water in a mix absorbed by the aggregate, although extreme conditions are denoted by the terms 'total water–cement ratio', used when the absorbed water has been included, and 'free water–cement ratio', used when the absorbed water has been excluded. In general practice, "saturated surface-dry condition" refers to aggregates that have their open pores filled by water.

The values of aggregate absorption used in the investigations undertaken by C&CA are shown in Table 9.20. For Indian conditions, the approximate quantities of moisture absorbed (by weight) by various aggregates may be assumed to be the following:

Average sand	1.0%
Pebbles and crushed limestone	1.0%
Trap rock and granite	0.5%
Porous sandstone	7.0%
Very light and porous aggregate	25.0%

It should be noted that the values suggested are only approximate estimates. For field application, it is recommended that these be obtained based on tests. The tables of *Road Note No. 4* (1950) are based on the 'free water–cement ratios'.

The C&CA tabulation shows that the absorption is higher for smaller sizes of CA, and may exceed 2% for continuously graded natural gravel. Based on a thumb rule, an aggregate absorption of 2% for grade M20 concrete can change the water–cement ratio and affect the 28-day strength of concrete by as much as 2 N/mm^2 and make a difference from the 'medium' to 'very low' category in the degree of workability.

Table 9.20 Percentage of water absorption in different types of aggregates

Type of aggregates	Rounded aggregate		Irregular aggregate		Crushed granite	
Time	10 min	24 h	10 min	24 h	10 min	24 h
Particle size	Percentage of water absorption					
600 to 300 μm	1.7	1.9	0.4	0.5	1.1	1.2
1.18 mm to 600 μm	2	2.2	1.1	1.3	1.2	1.3
2.36 to 1.18 mm	1.9	2.2	2.0	2.4	1.2	1.3
4.75 to 2.36 mm	1.6	2.0	2.8	3.4	1.2	1.3
10 to 4.75 mm	1.2	1.5	3.2	3.6	0.6	0.6
20 to 10 mm	0.5	0.8	1.4	1.9	0.3	0.4

If the aggregates to be used for any particular work have absorption values differing appreciably from those of the materials on which Tables 9.19(a–i) are based, an approximate adjustment can be made to the total water–cement ratio after the design has been prepared (Table 9.20). As a rough guide, the total water–cement ratio should be increased by the product of half the increase in the water absorption of the available material over the reference material and the aggregate–cement ratio by weight. However, it should be noted that the problems which arise from the uncertainty of the expression of the water–cement ratio are important in comparing the results of tests made under different conditions and where the values of the water–cement ratio are specified.

Summarizing, concrete mix design with the RRL method involves mainly the following steps:

Step 1   Fix the average compressive strength f_{av} from the given characteristic strength of concrete, assuming a suitable standard deviation S.

Step 2   Determine the water–cement ratio from the considerations of strength and/or durability of concrete.

Step 3   Choose the workability to suit the conditions at the site.

Step 4   Determine the gradings of the aggregate based on sieve analysis.

Step 5 Mix the proportions of the aggregates of various sizes so that the combined grading matches reasonably with the grading type given in the grading curves.

Step 6 Finally, choose an aggregate–cement ratio from the appropriate table.

Step 7 Make the trial mix and check it for cohesiveness.

Example 9.3

Design a concrete mix, intended for a reinforced concrete footbridge, from the following data:
Characteristic strength f_{ck} at 28 days: 20 N/mm^2
Cement to be used: Ordinary Portland
7-day strength: 33 N/mm^2
Workability: Low
CA: Crushed rock available in nominal maximum sizes of 20 mm and 40 mm
FA: Natural sand
Specific gravity: 2.8 for CA, 2.7 for FA

The gradings of aggregate are given in Table 9.21.

Table 9.21 Individual grading (percentage passing)

IS sieve designation	FA	CA	
		20 mm	**40 mm**
40 mm	100	100	100
20 mm	100	95	21
10 mm	100	30	12
4.76 mm	99	8	4
2.36 mm	75	5	1
1.18 mm	53	—	—
600 μm	35	—	—
300 μm	13	—	—
150 μm	3	—	—

Solution

Step 1 To determine the average compressive strength f_{av}, assume f_{ck} = 20 N/mm^2 and S = 4.0 N/mm^2 (Table 9.10).

$$f_{target} = f_{ck} + 1.65S = 20 + 1.65 \times 4$$
$$= 20 + 6.6 = 26.6 (\text{or } f_{ck} + 5 = 25)$$

Adopt 26.6 N/mm^2.

Step 2 Referring to Fig. 9.1(a) or (b), the water–cement ratio is approximately 0.5; the minimum cement content is 300 kg/m^3 for moderate exposure [Table 9.6(a)].

Step 3 To suit the site conditions, the workability is low.

Step 4 Determine the proportions of aggregates so that their combined grading conforms to, say, type-1 grading, to begin with, and type-2 grading preferably. The individual gradings of the selected aggregates are shown in Table 9.21. Let $X:Y:Z$ = FA:20-mm aggregate:40-mm aggregate. Note that X is FA; Y is 20 mm CA, and Z is 40 mm CA. Assuming that 50% of the combined aggregate passes the 20-mm IS sieve,

$$X + 0.95(Y) + 0.21(Z) = 0.5(X + Y + Z)$$

Assuming that 24% of the combined aggregate passes the 4.75-mm IS sieve,

$$0.99(X)+0.08(Y)+0.04(Z)=0.24(X+Y+Z)$$

Assume $X = 1$. Solving the two equations,

$$Y=0.86 \text{ and } Z=3.036$$

$$X:Y:Z=1:0.86:3.036$$

$$\text{Total} =1+0.86+3.036=4.896$$

The grading of the combined aggregate obtained by mixing the FAs and CAs in the above proportion is shown in Table 9.22. It shows that the combined grading obtained compares favourably with the type-1 grading in Fig. 9.6.

Table 9.22 Combined grading

IS sieve designation	Passing sieve			Sum	Percent passing	
	FA × 1	20-mm aggregate × 0.86	40-mm aggregate × 3.036		Sum (1 + 0.86 + 3.036) = 4.896	Type-1 grading
40 mm	100	86	304	490	96	100
20 mm	100	82	64	246	48	50
10 mm	100	26	36	162	32	36
4.75 mm	99	7	12	118	23	24
2.36 mm	75	4	3	82	16	18
1.18 mm	53	—	—	53	10	12
600 µm	35	—	—	35	7	7
300 µm	13	—	—	13	3	3
150 µm	3	—	—	3	1	0

Step 5 From Table 9.19, the aggregate–cement ratio is 6.6 for 40-mm irregular gravel aggregate, the water–cement ratio is 0.5, and the workability is low. The aggregate–cement ratio tables are based on specific gravities of 2.5 and 2.6, respectively, for the CA and FA and an adjustment is desirable for different specific gravities. It can be assumed that if the specific gravity of the FA and CA is 2.6–2.7 and 2.5–2.6, respectively, the workability relations will hold good, but if it is very different from these values, a correction will have to be made. In this event, adjust the gradings and the mix to obtain the correct proportions by volume. This correction is not often likely to be required, but will certainly be necessary with heavy aggregates used for making heavy concrete. For weight-batched concrete, calculate the quantities of the various constituents by weight after they have been proportioned by volume. Assume the specific gravity of cement to be 3.12. For example, the average specific gravity of aggregate is

$$2.6\times\frac{1}{4.896}+2.5\times\frac{3.896}{4.896}=2.52$$

The average specific gravity of the actual aggregate is

$$2.7\times\frac{1}{4.896}+2.8\times\frac{3.896}{4.896}=2.779$$

The aggregate–cement ratio for irregular aggregate is

$$6.6\times\frac{2.779}{2.52}=7.278$$

The mix proportion by weight has a water–cement ratio of 0.5.

$$\text{Sand: } \frac{1}{4.896} \times 6.778 = 1.38$$

$$\text{20-mm aggregate: } \frac{0.86}{4.896} \times 6.778 = 1.19$$

$$\text{40-mm aggregate: } \frac{3.086}{4.896} \times 6.778 = 4.2$$

Cement: FA: 20-mm CA: 40-mm CA = 1 : 1.38 : 1.19 : 4.2 water–cement ratio = 0.5

Step 6 As the mix appears to be under-sanded, make a second trial mix to obtain a reasonably good concrete mix of, say, medium workability, using the type-2 grading curve. The yield of concrete per 50-kg bag of cement is shown in Table 9.23.

Table 9.23 Yield of concrete per 50-kg bag of cement

Material	Weight (kg)	Specific gravity	Volume (m³)
Cement	50	3.120	16.03×10^{-3}
FA	76.5	2.750	27.82×10^{-3}
20-mm CA	54.5	2.880	18.92×10^{-3}
40-mm CA	213.5	2.880	74.13×10^{-3}
Water	25	1.000	25.00×10^{-3}
Total			161.90×10^{-3}
Entrapped air at 1.5%			2.43×10^{-3}
Total	419.5		164.33×10^{-3}

Yield of concrete per 50-kg bag of cement = 164.33 L

$$\text{Unit weight or density of concrete} = \frac{419.5}{164.33 \times 10^{-3}} = 2550 \, \text{kg/m}^3$$

Cement content

The cement content is determined by dividing the density of fresh concrete by the sum of the aggregate–cement ratio by weight, the water–cement ratio by weight, and 1 to represent cement. Thus, assuming an average aggregate–cement ratio of 6.0, the cement content = 2550/(6.0 + 0.5 + 1) = 340 kg/m³ > 335 kg/m³, which is the minimum required to meet the durability requirements, and no change in the mix proportions needs be made to satisfy this criterion.

Example 9.4

Design a suitable mix for a reinforced concrete footbridge, assuming the following data:
Cement used: Ordinary Portland
7-day strength: Between 33 and 43 N/mm²
CA: Crushed rock (granite)
FA: Natural sand of specified grading
Maximum size of CA: 40 mm
Characteristic strength of concrete: 20 N/mm² (200 kg/cm²) at 28 days
Grading of CA: 10 mm size graded down and 40 mm size graded down
Size of section: Average structural member of 500 mm minimum horizontal dimension

Type of concrete: Plain concrete in severe environment
Compaction: Needle vibrators used for compaction
Workability: Medium
Grading of aggregate required: Type 2 from Fig 9.6

Solution

Step 1 Average strength of concrete and water–cement ratio: As in the previous example, $f_{av} = 26.2$ N/mm². Figure 9.4(a) gives a water–cement ratio of about 0.60.

Step 2 Aggregate–cement ratio: Refer to Table 9.19(b) for 40 mm nominal size irregular gravel aggregate, while the aggregates available are crushed rock and natural sand. In the absence of recorded data for such combinations of crushed rock and natural sand aggregates, it may be mentioned that the relevant data required for design would lie somewhere near those for crushed rock aggregates given in Table 9.19(c). In addition, the effect of the shape of sand particles on the workability of the mix is far greater than that of CA. Hence, it is judicious to employ Table 9.19(b), giving data for irregular gravel aggregate, which would be fairly close to the case wherein crushed rock and natural sand are used. Thus, the aggregate–cement ratio by weight is adopted as 7.3 using grading curves 2 and 3 of Figs 9.5 and 9.6 respectively read with Table 9.19.

Step 3 Choosing proportions for aggregates: The grading does not appear to be very important. However, grading 2 is adopted, as it will contain a smaller percentage of fine aggregate. The given FA (natural sand) and CA (40 mm) + CA (20 mm) can be combined in the proportion 30:26:44 by weight to provide a grading approximating to curve 2.

Step 4 Batching proportions by weight: Cement:sand:10-mm down CA:40-mm down CA

$$=1:7.3\times\frac{30}{100}:7.3\times\frac{26}{100}:7.3\times\frac{44}{100}$$

i.e., 1:2.2:1.9:3.2 by weight.

Step 5 Batching proportions by volume: As the proportions of aggregate by weight have to be converted into their equivalents by volume, the bulk densities of the aggregates should be determined first. For the given aggregates, laboratory tests give, say, the following bulk densities:

$$\text{Natural sand} = 1630 \text{ kg/m}^3$$
$$\text{10-mm down CA} = 1520 \text{ kg/m}^3$$
$$\text{40-mm down CA} = 1470 \text{ kg/m}^3$$

Hence, the batch proportions by volume using one 50-kg bag of cement are

$$\text{Cement} = 50 \text{ kg} = 35 \text{ L}$$

$$\text{Sand} = \frac{2.2\times50}{1630}\times1000 = 67.5 \text{ L}$$

$$\text{CA 10-mm down} = \frac{1.9\times50}{1520}\times1000 = 62.5 \text{ L}$$

$$\text{CA 40-mm down} = \frac{3.2\times50}{1470}\times1000 = 108.5 \text{ L}$$

$$\text{Water} = 0.6\times50 = 30 \text{ kg} = 30 \text{ L}$$

That is, cement:sand:10-mm CA:40-mm CA = 35:67.5:62.5:108.5 = 1:1.95:1.8:3.1 by volume. *Note that the proportions by weight and those by volume are different.*

Step 6 Cement content: From a laboratory test, the bulk density of concrete = 2400 kg/m³. The proportions by weight are 1:2.2:1.9:3.2.

Weight of concrete per 50-kg bag batch $=(1+2.2+1.9+3.2+0.6)\times50=8.9\times50=445$ kg

$$\text{Volume of concrete per 50-kg bag batch}=\frac{445}{2400}=0.185\ m^3$$

Number of 50-kg bags of cement per m^3 of concrete $= 1/0.185 = 5.4$

$$\text{Total weight of bags}=5.4\times50=270\,\text{kg}$$

Hence, cement content $= 270$ kg/m^3. Check, as in the previous example:

$$\frac{2400}{7.3+0.6+1}=\frac{2400}{8.9}=270\,\text{kg/m}^3$$

which practically tallies with the minimum cement content of 260 kg/m^3 derived from Table 9.6(a) for plain concrete in a very severe exposure condition.

9.7.4 Department of Environment Method

This British method of concrete mix design, popularly referred to as the Department of Environment (DOE) method, is used in UK and other parts of the world and has a long established record. The method originates from *Road Note No. 4* described in the previous section. In 1975, the note was replaced by *Design of Normal Concrete Mixes*, published by the British DOE. In 1988, this publication was issued in a revised and updated edition to allow for changes in the various British standards.

The DOE method utilizes British test data obtained at the Building Establishment, the Transport and Road Research Establishment, and the British Cement Association. The aggregates used in the test conformed to BS 882 and the cement to BS 12 or BS 4027. This method can be used in India with minor modifications given below.

Mix design stages

In the DOE method, the mix design is determined in the following five stages:

Stage I   Determine the free water–cement ratio required for strength

1. Either use a specified margin or calculate a margin for a given proportion of defectives and statistical standard deviation (Table 9.10).
2. Obtain the target mean strength by adding the margin to the required characteristic strength (recommended to use for $f_{av} = f_{ck} + 5$).
3. If air entrainment is specified, calculate an artificially raised modified target mean strength.
4. Either accept a specified free water–cement ratio or obtain the maximum free water–cement ratio, which will provide the target mean strength for concrete made from the given CA type and cement with the given properties.

Stage II   Determine the free water content required for workability

1. Either use a specified free water content or obtain the minimum free water content, which will provide the desired workability for concrete made with the given FA type, CA type, and maximum size of CA.
2. If the free water content has been determined for workability, adjust the required free water content if air entrainment is specified, and adjust further if a water-reducing admixture is specified.

Stage III   Determine the required cement content

1. Obtain the minimum cement content, which is required for strength, by dividing the free water content obtained in Stage II by the free water–cement ratio obtained in Stage I.
2. Check the minimum cement required for strength against the maximum cement content permitted, and give a warning if the former exceeds the latter. (This calls for the use of mineral admixtures.)

3. Check the minimum cement content required for strength against the minimum cement content allowable for durability, and adopt whichever is greater as the cement content of the mix.

4. Divide the free water content by the cement content used in the mix to obtain a modified free water–cement ratio.

Stage IV   Determine the total aggregate content

1. Obtain the overall aggregate density.
2. Obtain the fractional volume of the aggregate by subtracting the proportional volumes of free water and cement from a unit volume.
3. Calculate the FA and CA contents from the total aggregate content obtained in this stage (step 2) and the percentage of FA.

Stage V   Determine the FA content

1. Either use a specified value of the percentage of FA or obtain this percentage, which will provide the desired workability for concrete made with the given grading of FA, maximum size of CA, and the free water–cement ratio obtained in Stage III.
2. Calculate the FA and CA contents from the total aggregate content obtained in Stage IV and the percentage of FA.

Other requirements

Sometimes the maximum cement content is specified to limit the heat of hydration and/or shrinkage, and sometimes limits are set for concrete density. The standard grading curves in Figs 9.5–9.7 can be used to estimate the mix proportions for the aggregates available in India.

9.7.5 Minimum Voids Method

The voids in the CA have to be filled in by the cement paste. Some extra sand paste and cement paste are needed to provide mobility to the mix and allow additional voids to be created by the action of the sand particles on the CA, and likewise that of the cement particles on the sand.

The minimum voids method, based on an experimental approach, follows a procedure opposite to that used in the trial mix method. The proportion of CAs and FAs is first determined, and then the cement paste added to obtain concrete of desired consistency. The procedure is as follows:

1. Determine the dry-rodded unit weights of the CAs and the FAs.
2. Take the proportions of graded CAs and FAs as 30:70, 40:60, 50:50, 60:40, 70:30, etc., and mix them thoroughly. Determine the dry-rodded unit weights of the combined aggregates with different proportions of CAs and FAs.
3. Given the specific gravity of CAs and FAs, determine the percentages of voids corresponding to each of the unit weights.
4. Plot a curve with the percentage of voids along the *y*-axis and the proportion of CAs and FAs along the *x*-axis. Then determine the proportion of CAs and FAs giving the maximum density/minimum percentage of voids from the curve, and use this proportion for making a trial mix.
5. Take known weights of CAs and FAs in the required proportion and mix them.
6. Prepare separately a sufficient quantity of cement paste with the required water–cement ratio.
7. Add the paste gradually to the aggregate, mixing it continuously, until a mix of the desired consistency is obtained.

In the past, the minimum voids theory has provided an 'ideal' or desirable aggregate grading resulting in an elliptical grading curve. The theory assumed that the maximum quantity of solid particles that can be packed into a concrete mix would give the highest strength and, consequently, the best concrete. However, subsequent investigations revealed that an aggregate graded to provide the maximum density led to harsh, unworkable mixes; also, a certain excess of cement paste above that required to fill the voids in sand is necessary to give workable mixes. Likewise, an excess of cement mortar above that needed to fill the voids in the CA gave more workable concretes. Based on the fineness modulus and surface area theories, other criteria have been employed to define 'satisfactory' gradings.

Weymouth (1938)'s particle interference theory gives perhaps the most rational approach to perfect aggregate grading—the theory embodies a mathematical relation between successive sizes, so that the coarsest particles distributed in a mix are separated by a distance equal to the sum of the mean diameter of the next smaller size and the thickness of the cement films between the particles. To establish the absolute volume of each size of solid particles satisfying the criterion, the relation may be extended from the largest particles to the finest ones, including cement. A generally accepted criterion is that for any given aggregate there are several gradings which suit the required workability and cement and water contents, within reasonable limits, provided that a specific degree of uniformity of grading is maintained within these limits. The following example very briefly illustrates the method.

Example 9.5

Design a concrete mix wherein the CA and sand, respectively, contain 40% and 30% voids. Assume 100 L of CA.

Solution

$$\text{Sand required} = 40 + 40 \times \frac{10}{100} \text{ extra} = 40 + 4 = 44 \text{ L}$$

$$\text{Cement paste required} = \left(44 \times \frac{30}{100}\right) + \left(44 \times \frac{30}{100}\right) \times \left(\frac{15}{100}\right) \text{extra}$$

$$= 13.20 + 1.98 = 15.18 \approx 15.2$$

Thus, dry cement paste = 15.2 × 1.2 = 18.24 L, approximately. Therefore, the proportions by volume of cement:sand:CA are 18.24 L:44 L:100 L, i.e., 1:2.41:5.48 by volume.

9.7.6 Maximum Density Method

The maximum density method uses a theoretical approach to decide the grading of aggregates, by using Fuller (1907)'s grading curves. Fuller's empirical expression for arriving at the proportion of different grain sizes, starting with the largest aggregate size and ending with the smallest size including cement particles, is

$$P_d = 100 \left(\frac{d}{D}\right)^{1/2} \tag{9.7}$$

where D is the MSA, d is the size of the particle (so that the percentage of all particles smaller than d is required), and P_d is the percentage of particle sizes smaller than d in the total mixture. Therefore, if 40 mm MSA is assumed, the percentage of particles smaller than 20, 10, and 4.75 mm ($\approx$ 5 mm) are

$$P_{20} = 100 \left(\frac{20}{40}\right)^{1/2} = 70.7\%$$

$$P_{10} = 100 \left(\frac{10}{40}\right)^{1/2} = 50.0\%$$

$$P_5 = 100 \left(\frac{5}{40}\right)^{1/2} = 35.35\%$$

Hence, the total quantity of CA is 100% – 35.35% = 64.65% of the total coarse mix, i.e., CA, FA, and cement. Using these values, the grading of CA is determined (Table 9.16). For comparison, the grading limits of CA specified in IS: 383-1970 are given in Table 9.24.

It may be seen from this table that the grading of CA obtained from Fuller's expression is a good mean between the limits specified in IS: 383-1970.

Table 9.24 Grading of 40-mm CA obtained using Fuller's expression compared with that specified in IS: 383-1970

IS sieve designation (mm)	Quantity between sieve	Quantity passing sieve	Percentage passing sieve	Percentage passing, for graded aggregate of nominal size 40 mm, according to IS: 383-1970
40	—	64.65	100	95–100
20	29.30	35.35	54.70	30–70
10	20.70	14.65	22.65	10–35
4.75	14.65	0.00	0.00	0–5
	64.65	—	—	—

The maximum density method assumes that the specific gravities of CA, FA, and cement are the same, which, however, is not correct. Therefore, an approximate weighted average for specific gravity may be assumed for using this method. An alternative approach would be to use solid volumes instead of weights for determining the proportions, and to finally convert these volumes into weights.

The preceding calculations show that for 40 mm MSA, the total quantity of FA together with cement is 35.35%. This means that for concrete with a given MSA, the aggregate–cement ratio is fixed, and consequently the workability for the required strength of concrete is also fixed. However, workability also depends on the means available for compaction. Thus, with this method, the requirements of both strength and workability cannot be simultaneously fulfilled. To overcome this, cement and sand may be so adjusted that the ratio of their combined proportions to the total solids remains constant, which is possible if the aggregate–cement ratio is fixed. The proportions are calculated for this aggregate–cement ratio, a trial mix is prepared with the required water–cement ratio, and the consistency of the mix is noted. If the mix is not satisfactory, further trial mixes can be made with different aggregate–cement ratios until a satisfactory mix is obtained.

Example 9.6

Determine the proportion of cement, FA, and CA assuming an aggregate–cement ratio of 6.2 by weight for $D = 40$ mm maximum size of CA. The specific gravities of cement and aggregate may be taken as 3.10 and 2.80, respectively.

Solution

Step 1   Aggregate–cement ratio by volume = $6.2 \times 3.10/2.8 = 6.86$.

Step 2   Assume unit solid volume of cement.

Step 3   Total volume of solids = $6.86 + 1.00 = 7.86$ units.

Step 4   Percentage of particles finer than 4.75 mm, or say 5 mm:

$$P_5 = 100\left(\frac{5}{20}\right)^{1/2} = 50\%$$

Step 5   Solids in CA = $(50/100) \times 7.86 = 3.93$ units. Solids in FA = $7.86 - 3.93 - 1.00 = 2.93$ units. The mix proportion by volume is 1:2.93:3.93 approximately.

Step 6   The mix proportion by weight is

$$1:2.93\times\frac{2.8}{3.1}:3.93\times\frac{2.8}{3.1}$$

i.e., 1:2.65:3.54 is recommended. Assume the loose unit weights of cement, FA, and CA to be 1440 kg/m³, 1650 kg/m³, and 1600 kg/m³. Incidentally, the mix proportion by volume is

$$1:2.65\times\frac{1440}{1650}:3.54\times\frac{1440}{1600}$$

i.e., 1:2.31:3.18.

9.7.7 Fineness Modulus Method

Developed by Professor Abram, this method uses a single parameter—the fineness modulus (FM)—to express the grading of aggregate. To determine the fineness modulus, the IS sieve designations used start from the smallest size of 150 μm; the larger designations are such that a particular size is twice that of the preceding designation. The cumulative percentages retained on each sieve are added and divided by 100, and the figure so obtained is called the fineness modulus. The ideal gradings of aggregate recommended by Professor Abram are given in Table 9.25, and it may be noted that they follow Fuller's gradings. The fineness moduli of these gradings are as follows:

$$\text{FM of 150-mm aggregate} = \frac{1100 - 334}{100} = 7.66$$

$$\text{FM of 75-mm aggregate} = \frac{1000 - 331}{100} = 6.69$$

$$\text{FM of 40-mm aggregate} = \frac{900 - 326}{100} = 5.74$$

$$\text{FM of 20-mm aggregate} = \frac{800 - 320}{100} = 4.80$$

Table 9.25 Ideal gradings of aggregate

MSA (mm)	Percentage passing IS sieve designation										
	150 mm	75 mm	40 mm	20 mm	10 mm	4.75 mm	2.36 mm	1.18 mm	600 μm	300 μm	150 μm
150	100	70	50	36	25	18	12.5	9	6.25	4.5	3
75	100	100	70	50	36	25	18	12.5	9	6.25	4.5
40	100	100	100	70	50	36	25	18	12.5	9	6.25
20	100	100	100	100	70	50	36	25	18	12.5	9

However, it is practically impossible to obtain aggregate conforming to the ideal gradings. Therefore, based on his own experience, Professor Abram suggested maximum and minimum grading limits as limits of fineness moduli (Table 9.26).

Table 9.26 Limits of fineness moduli

Aggregate type	Maximum size (mm)	Limits of fineness moduli	
		Maximum	Minimum
Fine aggregate	Not specified	3.5	2.0
Coarse aggregate	20	6.9	6.0
	40	7.5	6.9
	75	8.0	7.5
	150	8.5	8.0
All-in aggregates	20	5.1	4.7
	25	5.5	5.0
	40	5.9	5.4
	75	6.3	5.8
	150	7.0	6.5

By comparing the values of the fineness moduli of ideal gradings with those of the suggested limits, it is observed that the actual gradings are finer than the ideal gradings. This is reasonable as the ideal gradings tend to produce harsh mixes.

Aggregates stacked in different stockpiles can be suitably combined to provide a grading which has the required fineness modulus. Let A and B be the respective fineness moduli of the CAs and FAs. If C is the FM derived by combining the CAs and FAs (generally termed as 'all-in' aggregates), the proportion in which they have to be combined is determined as follows.

Let P_a be the fraction of aggregate with FM $= A$ and P_b be the fraction of aggregate with FM $= B$.

$$P_a + P_b = 1$$
$$C = P_a A + P_b B$$

Solving for P_a and P_b,

$$P_b = \frac{A - C}{A - B}$$

i.e., the percentage of $P_b = \dfrac{A-C}{A-B} \times 100$

Example 9.7

Determine the proportion of aggregates required to give a suitable grading based on their fineness moduli. Assume the following gradings of aggregates:

IS sieve designation (mm)	Aggregate size associated with percentage passing IS sieve (mm)		
	40	20	10
40	90	—	—
20	10	95	—
10	2	15	85
4.75	—	3	20

Solution

$$\text{FM of 40-mm aggregate} = \frac{900 - 102}{100} = 7.98$$

$$\text{FM of 20-mm aggregate} = \frac{800 - 113}{100} = 6.87$$

$$\text{FM of 10-mm aggregate} = \frac{700 - 105}{100} = 5.95$$

Combining the 20- and 10-mm aggregates to give an average FM of 6.45,

$$P_1 = \frac{6.87 - 6.45}{6.87 - 5.95} \times 100 = 46\%$$

Combining the 40- and 20-mm aggregates to give an average FM of 7.20,

$$P_2 = \frac{7.98 - 7.20}{7.98 - 6.45} \times 100 = 51\%$$

The proportion of 40-mm:20-mm:10-mm aggregate would be

$$(100 - 51) : 51 \times \frac{100 - 46}{100} : 51 \times \frac{46}{100}$$

i.e., 49:28:23. The actual grading of such a combination would be as follows:

IS sieve designation (mm)	Percentage passing sieve
40	$0.9 \times 49 + 1.0 \times 28 + 1.0 \times 23 = 95$
20	$0.1 \times 49 + 0.95 \times 28 + 1.0 \times 23 = 54$
10	$0.02 \times 49 + 0.15 \times 28 + 0.85 \times 23 = 25$
4.75	$0 + 0.03 \times 28 + 0.20 \times 23 = 6$
	180

$$\text{FM of combined aggregate} = \frac{900 - 180}{100} = 7.2$$

After the grading is fixed, trial mixes are prepared to make the final adjustments. The procedure involves the following step-by-step operations:

1. Fix the water–cement ratio from considerations of strength and durability.
2. Decide the MSA in accordance with the requirements of congestion and member size.
3. Determine the gradings of the various aggregates.
4. Decide the proportions of aggregates of different sizes.
5. Prepare a trial mix. Make final adjustments, if necessary, to obtain concrete of desired workability.

9.7.8 American Concrete Institute Method

In 1991, the American Concrete Institute (ACI) published guidelines for normal, heavyweight, and mass concrete mix design. The main features of the method of mix design, as described by ACI Committee 211, are presented below:

1. The required (target) average compressive strength (f'_{cr}) at 28 days for mix design is determined by adding up an empirical factor (k) to the design compressive strength (f'_c) similar to Eq. (9.4).
2. Slump, as a measure of workability, is selected depending upon the type of structure and complexity of the pouring conditions as per Table 9.7(b).
3. Water content is determined based on the nominal maximum size aggregate (NMSA), type of concrete (air-entrained or non-air entrained), and specified slump using Table 9.27. Then it is adjusted for the type of aggregates. Air content, as percentage of the concrete volume, is estimated depending upon the air-entrained or non air-entrained type of concrete, exposure conditions, and NMSA.
4. The water–cement ratio is selected based on the target strength [Fig. 9.4 and Table 9.6(a)] and the type of concrete (air-entrained or non air-entrained).
5. Cement content is calculated based on the water–cement ratio and the water content.
6. Coarse aggregate content, as dry-rodded bulk (percentage) of concrete unit volume, is determined based on the NMSA and the fineness modulus of sand using Table 9.28.
7. Once the water content, cement content, air content, and the coarse aggregate content per unit volume of the concrete is determined, the fine aggregate (F_{agg}) is calculated by subtracting the absolute volume of the known ingredients from the unit volume of the fresh concrete (in this case 1 m^3) as following:

$$F_{agg} = 1 - X$$

where X = sum of all other ingredients (air, water, cement, and coarse aggregates) in cubic metre calculated for 1 m^3 of concrete.

Table 9.27 Approximate mixing water and air content requirements for different slumps and nominal maximum sizes of aggregates (ACI:211.1-91)

Slump (mm)	Water.kg/m³ of concrete for indicated nominal sizes of aggregate							
	9.5*	12.5*	19*	25*	37.5*	50**	75†‡	150‡‡
Non-air-entrained concrete								
25 to 50	207	199	190	179	166	154	130	113
75 to 100	228	216	205	193	181	169	145	124
150 to 175	243	228	216	202	190	178	160	–
Approximate amount of entrapped air in non-air-entrained concrete, per cent	3	2.5	2	1.5	1	0.5	0.3	0.2
Air-entrained concrete								
25 to 50	181	175	168	160	150	142	122	107
75 to 100	202	193	184	175	165	157	133	119
150 to 175	216	205	197	184	174	166	154	–
Recommended average total air content, per cent for level of exposure:								
Mild exposure	4.5	4.0	3.5	3.0	2.5	2.0	1.5***††	1.0***††
Moderate exposure ‡	6.0	5.5	5.0	4.5	4.5	4.0	3.5***††	3.0***††
Extreme exposure ‡‡	7.5	7.0	6.0	6.0	5.5	5.0	4.5***††	4.0***††

* The quantities of mixing water given for air-entrained concrete are based on typical total air content requirements as shown for "moderate exposure". These quantities of mixing water are for use in computing cement contents for trial batches at 20 to 26°C. They are maximum for reasonably well-shaped angular aggregate grades within limits of accepted specifications. Rounded coarse aggregate will generally require 18kg less water for non-air-entrained and 16kg less for air-entrained concretes. The use of water-reducing chemical admixtures, ASTM C 484, may also reduce mixing water by 6 per cent or more. The volume of the liquid admixtures is included as part of the total volume of the mixing water.

†The slump value for concrete containing aggregate larger than 40 mm are based on slump tests made after removal of particles larger than 40 mm by wet-screening.

‡These quantities of mixing water are for use in computing cement factors for trial batches when 76 mm or 160 mm normal maximum size aggregate is used. They are average for reasonably well-shaped coarse aggregates. Well-graded from coarse to fine.

§Additional recommendations for air-content regarding tolerances are given in a number of ACI documents, including ACI 201, 346, 318, 301, and 302. ASTM C 94 for ready mixed concrete also gives air content limits. Consideration must be given to selecting an air content that will meet the needs of the job and also meet the applicable specifications.

** For concrete containing large aggregates which will be wet-screened over the 40 mm size prior to testing for air content, the percentage of air expected in the 40 mm minus material should be as tabulated in the 40 mm column. However, initial proportioning calculations should include the air content as a percent of the whole.

††When using large aggregate in low cement factor concrete, air entrainment need not be detrimental to strength. The mixing water requirements can be reduced sufficiently to improve the water–cement ratio and to thus compensate for the strength reducing effect of entrained air concrete. Generally, therefore, for these large nominal maximum sizes of aggregate, air contents recommended for extreme exposure should be considered even though there may be little or no exposure to moisture and freezing.

‡‡These values are based on the criteria that a small per cent air is needed in the mortar phase of the concrete. If the mortar volume will be substantially different from that determined in this recommended practice, it may be desirable to calculate the needed air content by taking small per cent of the actual mortar volume.

Reproduced with permission from American Concrete Institute, *Standard Practice for Selecting Proportions for Normal, Heavy Weight and Mass Concrete* (ACI 211.1-91) Reapproved 2002, Table A1.5.3.3.

8. Finally, water content is adjusted based on the absorption and the current moisture content of the coarse and fine aggregates, to account for deviation from saturated surface dry condition of the aggregates using Table 9.14. The mix proportions are arrived at using Table 9.29.

9. Scaling the batch up or down depending on requirement can be done based on trial mixes.

Table 9.28 Volume of coarse aggregate per unit volume of concrete

Nominal maximum size of aggregate, mm	Volume of dry-rodded coarse aggregate[*] per unit volume of concrete for different fineness modulus[†] of fine aggregate			
	2.4	2.6	2.8	3.00
9.5	0.50	0.48	0.46	0.44
12.5	0.59	0.57	0.55	0.53
19	0.66	0.64	0.62	0.60
25	0.71	0.69	0.67	0.65
37.5	0.75	0.73	0.71	0.69
50	0.78	0.76	0.74	0.72
75	0.82	0.80	0.78	0.76
150	0.87	0.85	0.83	0.81

[*]Volumes are based on aggregates in dry-rodded condition as described in ASTM C 29.

These volumes are selected from empirical relationships to produce concrete with a degree of workability suitable for usual reinforced construction. For less workable concrete such as required for concrete pavement construction, they may be increased by about 10%. For more workable concrete, such as may sometimes be required when placement is to be by pumping, they may be reduced up to 10%.

[†]See ASTM Method 136 for calculation of fineness modulus.

Reproduced with permission from American Concrete Institute, *Standard Practice for Selecting Proportions for Normal, Heavy Weight and Mass Concrete* (ACI 211.1 91), Table A1.5.3.6.

Table 9.29 First estimate of mass of fresh concrete

Nominal maximum size of aggregate, mm	First estimate of concrete unit mass, kg/m^{3*}	
	Non-air-entrained concrete	Air-entrained concrete
9.5	2280	2200
12.5	2310	2230
19	2345	2275
25	2380	2290
37.5	2410	2350
50	2445	2345
75	2490	2405
150	2530	2435

Reproduced with permission from American Concrete Institute, *Standard Practice for Selecting Proportions for Normal, Heavy weight and Mass Concrete* (ACI 211.1-91), Table A1.5.3.7.1.

9.7.9 Bureau of Indian Standards Method

Specification of a concrete mix

The objective of concrete mix design is to find the proportions in which concrete materials—cement, water, FA, and CA—should be combined in order to provide the specified strength, workability, and durability and possibly meet other requirements as listed in standards such as the IS: 456-2000. The specification of a concrete mix must therefore define the materials and the strength, workability, and durability to be attained.

Cement In mix design, the most important property of cement is its influence on the strength of concrete. The mix design allows the use of standard cements listed in Table 9.4.

Water In mix design, water is assumed not to contain impurities that affect its suitability for concrete (refer to Table 9.9).

Fine aggregate The type (crushed or uncrushed) and grading of FA are important in mix design and should be specified. The density should also be specified (Table 9.21).

Coarse aggregate The type (crushed or uncrushed) and maximum size of CA are important in mix design and should be specified. The density should also be specified (Figs 9.5–9.7).

Strength The required concrete strength, the age at testing, and the method of testing must be specified for a designed concrete mix. In mix design a target mean strength is aimed for. The target mean strength is generally derived from a characteristic strength or 'grade of concrete', to which a 'margin' is added to allow for statistical variation. This is usually based on quality control, effected in the field (refer to Fig. 9.4).

Workability The required workability and the method of testing must be specified for a concrete mix [refer to Table 9.7(a)].

Durability Tests for durability are not normally specified for concrete. Instead conditions are set for the mix, which are intended to provide the required durability. The conditions may include (a) a maximum free water–cement ratio, (b) a minimum cement content, and (c) a given percentage of entrained air or a given total air content (refer to Tables 9.6 and 9.8).

Example 9.8

An actual design of a concrete mix intended for reinforced concrete work for a thermal power station is summarized here. The characteristic strength of the concrete required is M25. Ennore sand is used and the sieve analysis has been supplied beforehand. The maximum size of CA used is 20 mm and low workability is assumed. The solution illustrates mix design for durability.

Solution

For the sake of brevity, only the key results are highlighted here. The detailed working calculations are similar to those given for the reinforced concrete footbridge example given under the RRL method and hence not detailed here. Nevertheless, the example clarifies concrete mix design based on the IS method for severe exposure conditions.

Step 1 Judging from the conditions at the site, the water–cement ratio should be as low as possible, and should not in any case exceed 0.45 for very severe exposure [see Table 9.6(a)]. So, adopt a water–cement ratio of 0.45 by weight. This would entail higher cement content, but such a rich mix is necessary for the durability of the concrete.

Step 2 Using Ennore sand, on the basis of the sieve analysis of the aggregate, sand:CA 10-mm down:CA 20-mm down = 39:20:41.

Step 3 For a water–cement ratio of 0.45, the aggregate–cement ratio interpolated from Table 9.19 for grading curve no. 1 is 5.4 by weight.

Step 4 Mix proportions by weight = cement:sand:CA 10-mm down:CA 20-mm down = 1:2.1:1.1:2.2.

Step 5 Cement content is approximately 320 kg/m^3, which satisfies the severe exposure condition requirement of Table 9.6(a), for reinforced concrete.

Steps of mix design

(Adopted from IS 10262 : 2009 *Concrete Mix Proportioning Guidelines*)

Step 1   Selection of water–cement ratio

Different cements, supplementary cementitious materials, and aggregates of different maximum size, grading, surface texture, shape, and other characteristics have influence on strength. Hence, they may produce concretes of different compressive strength for the same free water–cement ratio. Therefore, the relationship between strength and free water–cement ratio should preferably be established for the actual materials used. In the absence of such data, the preliminary free water–cement ratio (by mass) corresponding to the target strength at 28 days may be selected from the established relationship. The free water–cement ratio selected according to strength requirement should be checked against the limiting water–cement ratio for the requirements of durability and the lower of the two values adopted.

Step 2   Selection of water content

The water content of concrete is influenced by a number of factors, such as aggregate size, aggregate shape, aggregate texture, workability, water–cement ratio, cement and other supplementary cementitious material type and content, chemical admixture, and environmental conditions. An increase in aggregate size, a reduction in water–cement ratio and slump, and use of rounded aggregate and water-reducing admixtures will reduce the water demand. On the other hand, increased temperature, cement content, slump, water–cement ratio, aggregate angularity, and a decrease in the proportion of the coarse aggregate to fine aggregate will increase water demand. The quantity of maximum mixing water per unit volume of concrete may be determined from Table 9.30. The water content in the table is for angular coarse aggregate and for 25 to 50 mm slump range.

Step 3   Calculation of cementitious material content

The cement and supplementary cementitious material content per unit volume of concrete may be calculated from the free water–cement ratio and the quantity of water per unit volume of concrete. The cementitious material content so calculated should be checked against the minimum content for the requirements of durability and greater of the two values adopted. The maximum cement content shall be in accordance with IS: 456.

Step 4   Estimation of coarse aggregate proportion

Aggregates of essentially the same nominal maximum size, type, and grading will produce concrete of satisfactory workability when a given volume of coarse aggregate per unit volume of total aggregate is used. Approximate values for this aggregate volume are given in Table 9.31 for a water–cement ratio of 0.5, which may be suitably adjusted for other water–cement ratios. It can be seen that for equal workability, the volume of coarse aggregate in a unit volume of concrete is dependent only on its nominal maximum size and grading zone of fine aggregate. The correction to the values given in Tables 9.30 and 9.31 is given in Table 9.32.

Table 9.30 Maximum water content per cubic metre of concrete for nominal maximum size of aggregate

Nominal maximum size of aggregate, mm	Maximum water content[*] kg
10	208
20	186
40	165

[*]Water content corresponding to saturated surface dry aggregate.

Note: These quantities of mixing water are for use in computing material for trial batches.

Table 9.31 Volume of coarse aggregate per unit volume of total aggregate for different zones of fine aggregate

Nominal мaximum size of aggregate, mm	Volume of coarse aggregate* per unit volume of total aggregate for different zones of fine aggregate			
	Zone IV	Zone III	Zone II	Zone I
10	0.50	0.48	0.46	0.44
20	0.66	0.64	0.62	0.60
40	0.75	0.73	0.71	0.69

*Volumes are based on aggregates in saturated surface dry condition.

Table 9.32 Corrections to the values given in Tables 9.30 and 9.31

Change in conditions other than those given in Table 9.30	Correction for water content	Correction for sand content in total aggregate (%)
Sand conforming to Zone I, Zone III, or Zone IV	0	+1.5 for Zone I
		−1.5 for Zone III
		−3.0 for Zone IV
Increase or decrease in compacting factor value by 0.1 (for workability)	± 3%	0
Each 0.05 increase or decrease in water–cement ratio	0	± 1%
For rounded aggregates (gravel)	−15 kg/m^3	−7%

Step 5 Combination of different coarse aggregate fractions

The coarse aggregate used should conform to IS: 383. Coarse aggregates of different sizes may be combined in suitable proportions so as to result in an overall grading conforming to IS: 383 for particular nominal maximum size of aggregate. The estimation of entrapped air can be made based on Table 9.33.

Table 9.33 Approximate air content

Nominal size of aggregate (mm)	Percentage of entrapped air
10	3.0
20	2.0
40	1.0

Step 6 Estimation of fine aggregate proportion

Coarse and fine aggregate quantities are determined by finding the absolute volume of cementitious material, water and the chemical admixture; by dividing their mass by their respective specific gravity, multiplying by 1/1000 and subtracting the result of their summation from unit volume. The values so obtained are divided into coarse and fine aggregate fractions by volume in accordance with coarse aggregate proportion.

Step 7 Trial mixes

The calculated mix proportions shall be checked by means of trial batches. Workability of the trial mix shall be measured. The mix shall be carefully observed for freedom from segregation and bleeding and its finishing properties.

Example 9.9

For the parameters given below design an M30 grade concrete.

Characteristic strength required at site: 30 N/mm^3 (M30)

MSA: 20 mm

Shape of CA: Angular

Degree of workability: 0.85

Degree of quality control: Fair

Degree of exposure: Severe

Data on material:

Cement used: Grade 53 conforming to IS: 12269-1987

Specific gravity of cement = 3.15

CA: 20 and 12.5 mm mixed in the ratio 60:40

Sand: Conforming to zone II

Sieve analysis of sand:

IS sieve designation	Cumulative percentage		Passing specification for zone II
	Retained	Passing	
4.75 mm	—	100	90–100
2.36 mm	10	80	75–100
1.18 mm	30	70	55–90
600 μm	64	36	35–59
300 μm	92	8	8–30
150 μm	98	2	0–10

Sieve analysis of 20-mm CA:

IS sieve designation (mm)	Cumulative percentage		Passing specification for graded single-size aggregate
	Retained	Passing	
20	25	75	85–100
12.5	90	10	—
10	100	—	0–20
4.75	100	—	0–5
2.36	—	—	—

This table shows that there are no particles smaller than 10 mm.

Sieve analysis of 12.5-mm CA:

IS sieve designation (mm)	Cumulative percentage		Passing specification for graded single-size aggregate
	Retained	Passing	
20	—	100	100
12.5	15	85	85–100
10	60	40	0–45
4.75	100	—	0–10
2.36	—	—	—

Sieve analysis of 20- and 12.5-mm CA combined in the ratio 60:40:

IS sieve designation (mm)	Cumulative percentage		Passing specification for graded single-size aggregate
	Retained	**Passing**	
20	10	90	85–100
12.5	48	52	—
10	77	23	0–20
4.75	100	—	0–5
2.36	—	—	—

The sample conforms to the requirements of 20-mm graded single-size aggregate.

Specific gravity of CA: 2.67
Specific gravity of sand: 2.6
Free surface moisture:
    CA: Nil
    Sand: 2%
Water absorption:
    CA: 0.50%
    Concrete is pumpable

Solution

Step 1   Pick up the standard deviation, SD = 5 (Table 9.3).

Step 2   Target strength $= f_{ck} + (1.65 \times SD) = 30 + (1.65 \times 5) = 38.25$ N/mm^2.

Step 3   For the required target strength, determine the water–cement ratio. The water–cement ratio required is 0.49.

Step 4   Compare 0.49 with the value given in Table 9.6. From the table, the maximum water–cement ratio for severe exposure is 0.45. So, adopt a water–cement ratio of 0.45.

Step 5   Select the water and sand content from Table 9.30 for 20 mm nominal maximum size of CA, sand conforming to Zone II, for 25–50 mm slump water content per cubic metre of concrete is 186 kg. Estimated water content at 3% for every 25 mm slump for 100 mm slump is – 186 + (6/100) × 186 = 197. As superplasticizer is used water content can be reduced by 20%. Based on trials 25% reduction was achieved.

    Therefore water content = 197 × 0.75 = 147.75

Step 6   Having determined the water–cement ratio as 0.45 and the water content as 147.75 L, we calculate the cement content as 147.75/0.45 – 328.33 kg.

Step 7   Check for minimum cement content according to Table 9.6 for durability criteria. The minimum cement content for severe exposure is 320 kg/m^3. Hence the worked-out cement content satisfies the required condition, and we adopt the cement content of 329 kg. This is also less than the maximum permissible limit of 450 kg/m^2.

Step 8   Volume of coarse aggregate per unit volume of total aggregates corresponding to 20 MSA and Zone II aggregate (Table 9.31) is 0.62 for water–cement ratio 0.5.

    For water–cement ratio 0.45 which is 0.05 lower the proportion of CA is increased by 0.01
    (0.01 for every .05 change)

    Corrected proportion of volume of CA for water–cement ratio of 0.45 is 0.62 + 0.01 = 0.63

    Concrete is pumpable and hence the value is reduced by 10%

Hence volume proportion of CA $= 0.9 \times 0.63 = 0.567$

And volume proportion of FA $= 1 - 0.567 = 0.433$

Step 9   Mix calculation

a. Volume of concrete $= 1 \text{ m}^3$

b. Volume of cement $=$ (Mass of cement/Sp. Gravity of cement)/1000

$$= (329/3.15)/1000 = 0.104$$

c. Volume of water $=$ (Mass of water/Sp. Gravity of water)/1000

$$= (147.75/1)/1000 = 0.148$$

d. Volume of chemical admixture

Superplasticizer of 2% by mass of cement

$$= 329 \times 2/100 = 6.58$$

Volume of chemical admixture $=$ (Mass of admixture/Sp. Gravity of SP)/1000

$$= (6.58/1.145)/1000 = 0.006 \text{ m}^3$$

e. Volume of all in aggregate $= 1 - (0.104 + 0.148 + 0.006) = 0.742$

Step 10

Mass of CA $=$ Volume of all in aggregates $\times$ Proportion of CA $\times$ Specific gravity of CA $\times 1000 = 0.742 \times 0.567 \times 2.74 \times 1000 = 1152$

Mass of FA $=$ Volume of all in aggregates $\times$ Proportion of FA $\times$ SP of FA $\times 1000 = 0.742 \times (1-0.567) \times 2.74 \times 1000 = 881.51$

Proportion by weight:

Cement	Water	Sand	20-mm CA	12-mm CA
329	147.75	881.57	691	(1152–691) = 461

Proportion by ratio:

Cement	Water	Sand	20-mm CA	12-mm CA
1	0.45	2.67	2.1	1.4

9.8 Important Factors Affecting the Compressive Strength of Concrete

Water–cement ratio

The water–cement ratio is the main factor affecting the compressive strength of concrete at all ages. Every increase of 0.01 in the water–cement ratio decreases the strength by 1–1.5 N/mm^2.

Quality of cement and chemical constituents

The composition of Portland cement differs from factory to factory. The early strength of cement is related to its tricalcium silicate (C_3S) content—the higher the C_3S content relative to the C_2S content, more quickly the strength is gained after mixing. The finer the grinding of cement, the greater is the early strength developed. Typical examples are rapidly hardening and high-strength ordinary Portland cements. The age–strength relation is also related to the quality and type of curing. In general, this relation indicates that the ultimate strength will be same even though the early strengths may differ due to a change in chemical composition. If the 28-day compressive strength is assumed to be 100%, the probable strength at 90 days would be nearly 30% more. Later age strengths are especially significant when Portland pozzolana cement is used.

Apart from the differences in the strength of concrete due to the use of different types of cement, there is also some variation in concrete strength due to the variation in the strength of any particular type of cement. The reason for this variation is the various sources used for that cement, as also the normal variations in production from any given project over a period of time. For example, assuming a typical case at the age of 28 days, the standard deviation of the strength of ordinary Portland cement from different sources may be approximately 5 N/mm^2, whereas the corresponding value for the same type of cement from a single source may be only about 3 N/mm^2.

It may be noted that the curves relating compressive strength to the water–cement ratio [Fig. 9.4(a)] pertain to average cements of specified strengths at 7 days; it is quite likely that the cement available at a given site may give average strengths which are higher or lower than those specified. If the specified minimum strength of cement is to be realized in practice, the average free water–cement ratio for mix design may be multiplied by the factors suggested below to allow for the varying characteristics of cement.

Proportion of cement giving required strength	2 out of 3	4 out of 5	9 out of 10
Factor	0.98	0.95	0.92

However, if laboratory or full-scale trial mixes are envisaged with a consignment of cement expected to be actually used at the site, the above adjustments need not be made because the results of the trial mix will provide the changes required to account for these factors.

Storage of cement

The quality of cement stored in bags in godowns gradually deteriorates due to hydration. The losses in strength for different periods of storage are 15% in 3 months, 30% in 6 months, and 50% in a year.

Aggregate

The efficiency of cement increases with the use of aggregates with larger maximum sizes, especially for leaner mixes. Well-graded natural gravel or natural sand gives practically the same strength as well-graded crushed aggregate and natural sand concrete.

Water

Every extra litre of water added over the desired quantity in a 50-kg bag batch alters the water–cement ratio by 0.02–0.03, and affects the strength of concrete by 1–2 N/mm^2. The accuracy of batching water should be maintained within 1%.

Moisture in aggregate

The moisture in sand can cause a variation of as much as 20% in the quantities of materials batched by volume. For example, with a moisture content of 5–6%, sand weighing 1600 kg/m^3 would bulk and actually weigh 1280 kg/m^3, which should be allowed for in volume batching. In addition, the workability and yield are also affected by bulking. These factors have to be necessarily considered.

Slump

A 3% change in water content can cause a variation of 25 mm in slump, and this would generally affect the concrete strength by 2 N/mm^2.

Degree of compaction

The presence of 1% voids in the mix reduces the strength of concrete by 5%. Thus, with improper compaction and 5% voids, a well-proportioned concrete of strength 20 N/mm^2 would actually exhibit a strength of 15 N mm^2 only.

Temperature at the time of moulding cubes

If cubes are moulded or cured in water at a temperature higher than the normal 20–25°C, they will show increased strengths. Concrete moulded initially at a higher temperature may give high early strength; however, long-term strengths are known to be superior when concrete is moulded at about 20°C or lower, but not lower than 5°C.

Curing efficiency

Concrete develops its full strength (100%) when kept wet continuously for 28 days. The drying of concrete at some intermediate stage will, in the long run, lower its strength below that of a fully wet-cured specimen (typical values recorded were 93% to nearly 59% of the full strength).

Curing temperature

At very low temperatures, concrete gets frozen and may not hydrate. At temperatures above 10°C, a very small change in temperature causes a variation of 0.2–0.4 N/mm^2 in the strength of concrete.

Moisture content at the time of cube tests

Concrete cubes must be tested in a moist (surface dry) condition as soon as possible after removal from a curing tank. Oven-dried cubes are likely to show higher strengths by up to 80%.

Direction and rate of loading

Cubes are generally tested at right angles to the direction in which they are cast (filled), failing which there may be a 10% decrease in strength. The rate of application of load is usually 14 (N/mm^2)/min. Cubes tested at very high strain rates (speeds) may show increased strengths by up to 80%.

Duration of loading

Constant or continuous loading of a structure may cause fatigue and reduce the strength of concrete to 50% of its ultimate strength. Under a sustained load, failure may occur at a strength of 90% for a 1-hour loading, 80% for a 2-month loading, and 70% for a 30-year loading.

Degree of lateral restraint

If the cubes are supported laterally, increased strengths are likely to be obtained. Triaxial compression tests have revealed that a 28-N/mm^2 concrete can take an axial compressive load of about four to five times the normal load with a significant lateral restraint.

Compression machine and operation factors

Due to the continuous operation of a compression machine, its gauges become unreliable and need to be recalibrated. Errors may be introduced even while centring the cubes or due to the wear and tear of the compression plates. The mere friction of the spherical seat can affect the strength by nearly 20%. A 0.75-mm misalignment in the machine can reduce the strength of concrete by 9%.

Exercises

Review Questions

1. The process of mix proportioning cannot be computerized. Why?
2. What measures can be taken to make mix proportioning economical and scientific?
3. What are the defects of the currently used method of mix proportioning in India? How can it be made more scientific?

4. Explain the importance of grading? Theoretical grading, which produces maximum density, cannot be used in practice. Why?
5. Explain how you will account for the moisture present in sand while mix proportioning.
6. List the methods used for mix proportioning indicating the drawbacks of each method.
7. Explain the importance of the maximum size of aggregate for normal-strength concrete mix design.
8. Describe the significant variables affecting the workability of concrete.
9. Why is it desirable to use the minimum quantity of water in mix proportioning?
10. What tests are necessary to check the adoptability of a particular mix proportion for field use?

Multiple Choice Questions

1. The entrained air in concrete
   a. Increases workability
   b. Decreases workability
   c. Increases strength
   d. None of these

2. Separation of ingredients from concrete during transportation is called
   a. Bleeding
   b. Creep
   c. Segregation
   d. Shrinkage

3. For making impermeable concrete the process entails
   a. Thorough mixing
   b. Proper compaction
   c. Proper curing
   d. All of these

4. After casting an ordinary cement concrete on drying, it
   a. Shrinks
   b. Expands
   c. Remains unchanged
   d. Can expand or shrink

5. Approximate ratio of strength of cement concrete at 3 months to that at 28 days of curing is
   a. 1.15
   b. 1.30
   c. 1.0
   d. 0.75

6. Workability of concrete can be improved by
   a. More sand
   b. More cement
   c. More fine aggregate
   d. Fineness of coarse aggregate

7. Fineness modulus of fine aggregate is between
   a. 2–3.5
   b. 3.5–5.0
   c. 5–6.0
   d. 6–7.5

8. If slump of concrete mix is 250 mm, then its workability is
   a. Low
   b. Medium
   c. High
   d. Very high

9. If compaction factor is 0.95, then workability of concrete is
   a. Very low
   b. Low
   c. Medium
   d. High

10. Split tensile strength of concrete can be determined by
    a. Brazilian cylinder test
    b. Vicat apparatus
    c. Cube test
    d. Briquette test

Answers to Multiple Choice Questions

1. a	4. a	7. a	10. a
2. c	5. c	8. d	
3. d	6. b	9. d	

CHAPTER 10

Concrete Reinforcement

Reinforcement in a concrete structure accounts for one-third of the cost of the structure. The fabrication of bars, tendons, and meshes used for reinforcement adds almost 20–30% to the basic cost of steel. Therefore, a civil engineer should have knowledge of the fabrication of these reinforcing elements and their placement in the formwork. Steel is the predominant reinforcement material used in concrete structures. It should be well embedded in concrete to impart tensile strength.

The best of structural designs can be ruined if the design and drawing are not carried out in the field as per the design assumptions and expectations. This highlights the need for the proper fabrication and effective placement of reinforcement during construction. Proper reinforced concrete construction depends upon the workforce on the job—those who understand and design the structures, those who know the characteristics and limitations of the materials that are handled at the site, and those who fabricate the reinforcements at the site.

10.1 Reinforcement in Concrete

Concrete is strong under compression but weak under tension and shear. Whenever concrete members are likely to be subjected to tension, they are reinforced with steel. Bars, mats, or cages of steel are used as reinforcement. These can be classified as active or passive. The active ones are those used for prestressing and the passive ones are those subjected to stress upon the application of loads. Concrete structures use reinforcements of different types: reinforcing bars—plain or ribbed, welded mesh, fibre reinforcement, and prestressing wires and strands. Steel is used as reinforcement in concrete owing to its following qualities:

- High tensile strength
- Good bond with concrete
- Nearly same coefficient of expansion as that of concrete
- Adequate quality control possible during the manufacture of steel

10.2 Purpose and Location of Steel in Concrete

Depending on the type of force resisted by them in finished elements [Fig. 10.1(a)–(c)] or structures, reinforcements fall into the following categories:

1. Main or effective reinforcements, which withstand tensile forces produced by both live and dead loads (indicated as 1 and 3 in Fig. 10.1). While the bars indicated as 1 carry tensile forces, the bars indicated as 2 supplement concrete by carrying compression forces.

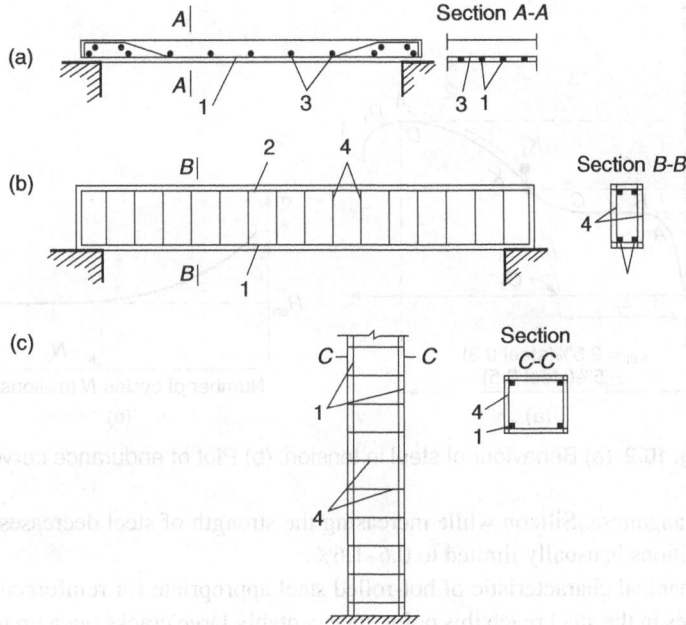

Fig. 10.1 Purpose and location of steel

2. Secondary reinforcements, which may be subdivided into the following:
    a. Web reinforcements, bent-up bars, and end mesh withstand shearing forces and diagonal tension. They also tie the entire block of reinforcement together (indicated as 4 in Fig. 10.1). In beams these are called 'stirrups' and in columns, 'laterals'. Generally they are not provided for slabs.
    b. Distribution bars spread the loads evenly among the main reinforcing bars. They also take care of the temperature and shrinkage forces (indicated as 3 in Fig. 10.1).
    c. Anchorages are used for keeping all the reinforcing bars in place during the assembly of the cage and mats. These bars aid in anchoring the bars in the ends.

Steel is incorporated at different locations in concrete, according to structural behaviour. Longitudinal steel bars are located near the bottom in a simple beam to prevent the failure of the concrete under tension. To resist shear, specially shaped bars called *stirrups* are used and placed vertically across the beam. These are more closely placed near the supports and their density decreases towards mid-span. Continuous beams require tension steel at the bottom, between the supports, and at the top surface of the supports, in cantilevers, and on overhangs. The surfaces of concrete members are lightly reinforced to prevent cracks due to the initial shrinkage of concrete and the temperature effects from season to season. It is a general practice to provide this reinforcement at right angles to the main reinforcement, or in the form of a mesh in an unreinforced member.

10.3 Mechanical Properties of Reinforcement Steel

Hot-rolled as well as *cold-worked* steels are employed as reinforcement. Hot-rolled steels, called *mild* or *soft*, have a *yield region* on their stress–strain curves and exhibit considerable elongation e_{ur} at failure [indicated as 1 in Fig. 10.2(a)].

The most important factor affecting the mechanical properties of steel is its chemical composition. The ultimate tensile strength of pure iron (ferrite) is comparatively low. An increase in the strength of steel and a reduction of the elongation at failure is achieved by introducing carbon and alloying additions (manganese, silicon, chromium, etc.) into its composition. Carbon, however, reduces the ductility and weldability of steel, and is therefore limited to 0.2–0.3%. A marked increase in strength without a large reduction of ductility is

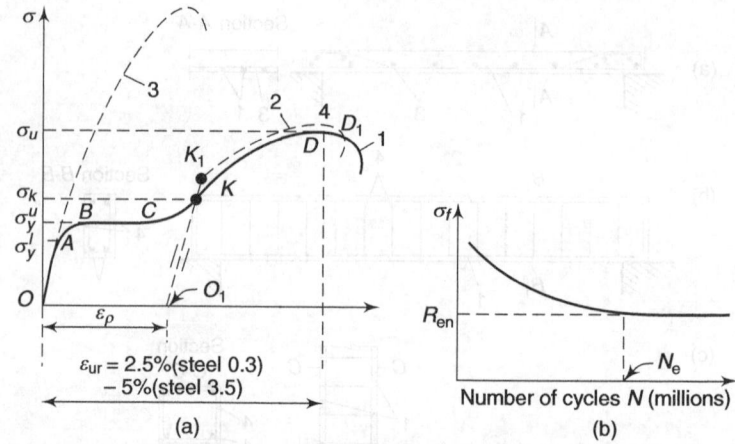

Fig. 10.2 (a) Behaviour of steel in tension; (b) Plot of endurance curve

obtained by adding manganese. Silicon while increasing the strength of steel decreases its weldability. The extent of alloying additions is usually limited to 0.6–1.6%.

The principal mechanical characteristic of hot-rolled steel appropriate for reinforced concrete is its *yield point*. When the stresses in the steel reach this point, unacceptably large cracks open up in the tension zone of the concrete, and then, as a result of considerable deflection of the member, the concrete in the compression zone crushes and fails. Thus, the ultimate strength of steel σ_u is not reached and not utilized in the load-carrying capacity of the member. For this reason, the minimum yield point of hot-rolled reinforcement steel is considered for strength calculations, and a material having inferior strength is rejected based on the yield strength (σ_y) test.

To ensure a more complete utilization of the properties of hot-rolled steel, it is artificially strengthened by working on it in a cold state. The resulting cold-worked metal has higher strength characteristics, which makes the use of steel in buildings economical. In essence, the strengthening of steel by cold-working consists of the following [see Fig. 10.2(a)]: A tensile stress $\sigma_K > \sigma_y$ is induced in a steel bar beyond the yield point on the strain-hardening part of the stress–strain curve. This corresponds to the strengthening or hardening of the metal up to $\sigma_K > \sigma_y$. After the load is removed, the unloading curve takes the form of a straight line, and the bar receives a permanent strain OO_1. With repeated loading, the new stress–strain curve begins to correspond to the unloading line, KO_1, since the plastic strains in the steel have already been taken up, and KO_1 remains parallel to the section OA, reflecting the elastic behaviour of the material. The bend of the curve—the beginning of the new yield area—now, however, occurs at a higher stress σ_K.

Hot-rolled steel is hardened by extrusion either to this stress $\sigma_K > \sigma_y$ or to prescribed strains exceeding those at the yield point, and this is known as *strain hardening* or *work hardening*. With time, owing to the so-called ageing of steel, the new yield point rises slightly (point K_1), a small yield area appears, and the ultimate strength (point D_1) increases a little.

Under repeated loading, fatigue failure of the reinforcement is possible at a reduced tensile strength. Fatigue failure occurs suddenly, without warning, and is brittle in nature (without yield).

To investigate the resistance of reinforcement to alternating stresses, experimental results are used to plot an *endurance* curve [Fig. 10.2(b)]. The number N of stress cycles is plotted along the x axis and the maximum value of the alternating stress σ_f of the reinforcement is plotted along the y axis. With a reduction in σ_f, the number of cycles required for failure increases. The endurance limit of steel is the point at which the endurance curve becomes a horizontal line, beginning at N_e million cycles. The strength of reinforcement under repeated loading is taken as its *endurance limit* R_{en} (the stress corresponding to N_e cycles, where the curve becomes a horizontal straight line). Experiments have shown that the minimum value of R_{en} depends on the cycle stress ratio $\sigma_{min}/\sigma_{max}$. When the load is repeated, the maximum and minimum stresses sustained by the material are

σ_{max} and σ_{min}, respectively. For example, when $\sigma_{min}/\sigma_{max} = -1$ (a symmetrical cycle), $R_{en} = 0.33 \sigma_y$; when this ratio is equal to zero, $R_{en} = 0.5\sigma_y$.

Depending on the method of cold working (extrusion, flattening, drawing, twisting, etc.), the properties of cold-worked steel approach more or less those of hard-drawn steel, which have no clearly marked yield point and which exhibit brittle failure (i.e., failure occurring before reaching an elongation of less than 3–4%).

Hard steels include heat-hardened steels produced in the form of bars or high-strength wire (without hardening by cold-working). The heat treatment consists in hardening the steel by heating up to 800°C followed by rapid quenching in oil and then tempering in a lead bath at about 500°C. For such steel, the service strength can be taken as being equal to the minimum rejection value of the increased yield strength (with extrusion hardening), or the minimum rejection value of the ultimate tensile strength (with cold-working by drawing and flattening).

The cold-worked steel is heated to a high temperature above 350–400°C to effect the annealing of steel. This causes a loss of the increased strength. For this reason, reinforced concrete members in which such steel is used should be reliably protected against the action of high temperatures at all stages of construction and service.

The use of steel hardened by extrusion and cold-worked by flattening as reinforcement in members subjected to repeated loads is not recommended (owing to the possibility of the steel losing its strain hardening).

10.4 Classification and Identification of Bars

The ribbed bars used as concrete reinforcement are made in following strength grades:

- Mild steel grades of Fe 250
- High-strength deformed steel bars of grades

> Fe 415, Fe 415D, 415S
>
> Fe 500, Fe 500D, 500S
>
> Fe 550, Fe 550D
>
> Fe 600

The figures following the symbol Fe indicate the specified minimum 0.2% proof stress or yield stress in MPa. The letter D and S following the strength grade indicates the categories with same specified minimum 0.2% proof stress/yield stress but with enhanced additional requirements to make them behave more ductile or more seismic resistant compared to others. Beyond the limit 600 N/mm^2 strength, the bars are more effective as tendons for prestressing. Hot-rolled and cold-worked bars are used for ribbed reinforcement.

With the availability of several types of bars, it might be difficult to identify the bar with its make and grade. While the make of the bar can be distinguished by the surface deformation, the grade marking by colour code is difficult. Ribbed bars have either two or four longitudinal ribs with cross ribs in between. The typical rib pattern is shown in Fig. 10.3. Bars of less than 12 mm diameter are produced in the two-rib design, whereas

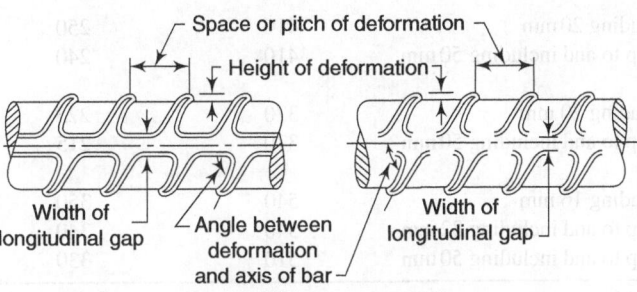

Fig. 10.3 Typical rib pattern

bars of over 12 mm diameter are produced in four-rib designs. Bars having 12 mm diameter are produced in both two-rib and four-rib designs. Grade Fe 500 bars can be identified by the presence of the asterisk (*) after every 300 mm; other bars do not have any such marks (Fig. 10.4).

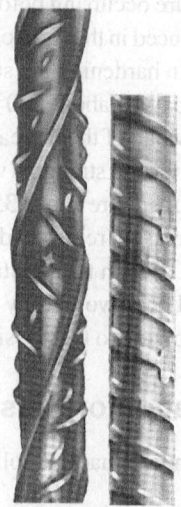

Fig. 10.4 Fe 500 with asterisks marked

In 1967–68, India adopted ribbed, cold twisted (CTD) bars with a guaranteed minimum proof strength of 415 N/mm². This grade, Fe 415, replaced plain mild steel bars, resulting in large savings in both the consumption of steel and the overall cost of construction. With a view to effect further economy, manufacturers produced weldable steel. The basic structural requirements of Fe 415 and Fe 500 CTD bars from the point of view of structural engineers are reviewed here. For efficient application in reinforced concrete, the reinforcing bars have to satisfy the following basic requirements: tensile strength, bond strength, ductility, bendability, weldability, fatigue strength, corrosion resistance, and fire resistance. Both mild steel bars and CTD bars have become obsolete and are not used in modern construction.

The properties of mild and medium tensile bars are presented in Table 10.1(a).

Table 10.1(a) Mechanical properties of mild and medium tensile bars

Type and nominal size of bar	Ultimate tensile stress (min)	Yield stress (min)	Elongation* per cent (min)
Mild steel grade I			
For bars up to and including 20 mm	410	250	23
For bars over 20 mm, up to and including 50 mm	410	240	23
Mild steel grade II			
For bars up to and including 20 mm	370	225	23
For bars over 20 mm, up to and including 50 mm	370	215	23
Medium tensile steel			
For bars up to and including 16 mm	540	350	20
For bars over 16 mm, up to and including 32 mm	540	340	20
For bars over 32 mm, up to and including 50 mm	510	330	20

*Elongation on a gauge length $5.65 \sqrt{A_0}$ where A_0 is the cross-sectional area of the test piece.

Various codes of practice as per Indian Standards used for concrete reinforcement are summarized in Table 10.1(b). High-strength deformed bars for concrete reinforcement are produced in India by cold twisting, controlled cooling, and/or micro-alloying. Though mild steel and CTD bars are permitted to be used as reinforcement, practically 90% of bars used in India are produced using Thermo Mechanical Treatment (TMT) process. The TMT process is described in Section 10.4.2. The mechanical properties of high-strength deformed bars are given in Table 10.1(c).

The chemical composition of the reinforcing bars of various grades is summarized in Table 10.2.

Table 10.1(b) Code of practice for concrete reinforcement

IS code	Title
IS: 432	Mild steel and medium tensile steel bars (Part I) and hard-drawn steel wire (Part II)
IS: 1786	High-strength deformed steel bars and wires for concrete reinforcement specification
IS: 2502	Bending and fixing of bars for concrete reinforcement
IS: 5525	Recommendations for detailing of reinforcement in reinforced concrete works
IS: 6461 (Part III)	Glossary of terms relating to cement concrete: Part 3 concrete reinforcement
IS: 1566	Hard-drawn steel wire fabric for concrete reinforcement
IS: 2751	Welding of mild steel bars used for reinforced concrete construction
IS: 2090	High tensile steel bars used in prestressed concrete
IS: 6003	Indented wire for prestressed concrete. Part I: Cold-drawn stress relieved wire; Part II: As-drawn wire
IS: 6006	Uncoated stress relieved strand for prestressed concrete

Table 10.1(c) Mechanical properties of high-strength deformed bars

Property	Fe 415	Fe 415D	Fe 415S	Fe 500	Fe 500D	Fe 500S	Fe 550	Fe 550D	Fe 600
0.2% proof stress/ yield stress, *Min*, N/mm²	415.0	415.0	415.0	500.0	500.0	500.0	550.0	550.0	600.0
0.2% proof stress/ yield stress, *Max*, N/mm²	—	—	540.0	—	—	625.0	—	—	—
TS/YS ratio*, N/mm²	≥1.10, but TS not less than 485.0 N/mm²	≥1.12, but TS not less than 500.0 N/mm²	1.25	≥1.08, but TS not less than 545.0 N/mm²	≥1.10, but TS not less than 565.0 N/mm²	1.25	≥1.06, but TS not less than 585 N/mm²	≥1.08, but TS not less than 600.0 N/mm²	≥1.06, but TS not less than 660 N/mm²
Elongation, per cent, *Min*, on gauge length 5.65√A, where *A* is the cross-sectional area of the test piece	14.5	18.0	20.0	12.0	16.0	18.0	10.0	14.5	10.0
Total elongation at maximum force, per cent, *Min*, on gauge length 5.65√A, where *A* is the cross-sectional area of the test piece†	—	5	10	—	5	8	—	5	—

* TS/YS ratio refers to ratio of tensile strength to the 0.2% proof stress or yield stress of the test piece.
† Test, wherever specified by the purchaser.

Table 10.2 Maximum permissible percentage of constituents in the chemical composition

Constituent	Max. Per cent								
	Fe 415	Fe 415D	Fe 415S	Fe 500	Fe 500D	Fe 500S	Fe 550	Fe 550D	Fe 600
Carbon	0.30	0.25	0.25	0.30	0.25	0.25	0.30	0.25	0.30
Sulphur	0.060	0.045	0.045	0.055	0.040	0.040	0.055	0.040	0.040
Phosphorus	0.060	0.045	0.045	0.055	0.040	0.040	0.050	0.040	0.040
Sulphur and phosphorus	0.110	0.085	0.085	0.105	0.075	0.075	0.100	0.075	0.075

Permissible stresses in steel reinforcement

As specified in IS: 456-2000, permissible stresses in steel reinforcement shall not exceed the values specified in Table 10.3. In flexural members the value of T (given in Table 10.3) is applicable at the centroid of the tensile reinforcement subject to the condition that when more than one layer of tensile reinforcement is provided, the stress at the centroid of the outermost layer shall not exceed by more than 10% the value given in Table 10.3.

Table 10.3 Permissible stresses in steel reinforcement

Type of stress in steel reinforcement	Permissible stresses in N/mm^2		
	Mild steel bars conforming to grade 1 of IS 432 (Part 1)	Medium tensile steel conforming to IS 432 (Part 1)	High yield strength deformed bars conforming to IS 1786 (Grade Fe 415)
Tension (σ_x or σ_{xv})			
(a) Up to and including 20 mm	140	Half the guaranteed yield stress subject to a maximum of 190	230
(b) Over 20 mm	130		230
Compression in column bars (σ_x)	130	130	190
Compression in bars in a beam or slab when the compressive resistance of the concrete is taken into account	The calculated compressive stress in the surrounding concrete multiplied by 1.5 times the modular ratio or σ_x, whichever is lower		
Compression in bars in a beam or slab where the compressive resistance of the concrete is not taken into account:			
(a) Up to and including 20 mm	130		190
(b) Over 20 mm		Half the guaranteed yield stress subject to a maximum of 190	190

Tensile strength

The yield strengths of high-strength bars are 415, 500, 550, and 600 N/mm^2, which are well above those of plain mild steel yield strength of 250 N/mm^2. The ultimate tensile strength of the bar is at least 10–15% more than that of the bar's yield strength. This potential increase in the yield strength has made these bars highly useful for structural engineers. There is also a stipulation that tensile strength/yield strength ratio is as specified in Table 10.1(c). Note that maximum limit for yield stress/proof stress has been specified for Fe 415S and Fe 500S seismic resistant bars as 540 and 625 N/mm².

Bond strength

The bond between concrete and steel depends directly on the deformations over the bars. An optimized deformation profile has been arrived at for deformed bars to give the best results. A general criterion has been recommended for the profile of such steels by several national standards including Indian Standard IS: 1786.

The slipping characteristics observed during pull-out tests on deformed bars have indicated that at a slip of 0.1 mm, the bar develops a bond of strength almost three to four times that of a plain round bar. In addition, the overlapping oblique lugs and the continuous longitudinal ribs provide a continuous bond.

Ductility

Ductility of reinforcement is an important criterion for the desired performance of a reinforced concrete member especially during earthquake loading. The elongation at rupture over a standard gauge length is accepted to be an index for ductility quality. The percentage elongation is a very important criterion for earthquake-resistant construction, and this should never be less than 14.5% for any steel. For ductile bars (Fe 415D and Fe 500D) this limit has been raised to 18% and 16% respectively. For seismic resistant bar the limit has further been raised to 20% for Fe 415S and 18% for Fe 500S. In addition for both D and S bars elongation limits have been specified for uniform elongation (elongation up to maximum force level). For 415D and 500D it is 5%, whereas for 415S and 500S it is 10% and 8%, respectively.

Bendability

This is an important requirement for reinforcing bars—in view of the cranking, kinking, and 90° bending of the bars, which are required frequently during fabrication. This in turn is directly related to the ductility of the material and the deformation characteristics.

If the ribs and lugs on the bar cause a serious notch effect, the bendability is greatly reduced. In the case of CTD bars, the geometry of ribs and lugs is designed in order to have a uniform streamline effect and symmetry incorporated during the production process. The lugs gradually merge into the core and do not run into the longitudinal ribs to make stress-raising corners. Sharp corners between the lugs or ribs and the core, if any, are rounded up due to plastic flow during twisting. The resulting profile of the ribs and lugs in the bars after twisting, therefore, minimizes the notch effect. This absence of the notch effect and the closely controlled chemistry of steel give CTD bars their bendability. The bend test is performed as per IS: 1599. The test piece, when cold, is doubled over the mandrel by continuous pressure until the sides are parallel. The specimen is considered to have passed the test if there is no rupture or cracks visible to a person of normal vision on the bent portion.

Weldability

For guaranteed weldability, carbon equivalent should not exceed 0.42%. High-strength deformed bars are steels with a restricted range of carbon, equivalent to a maximum of 0.42%, and therefore can be welded without the need for special electrodes or special welding techniques. The only precaution to be observed while welding is to permit the heat to be dissipated after one pass of welding (one should be able to touch the weld) before depositing the next pass.

Fatigue strength

Reinforced concrete members subjected to alternating loads producing minimum and maximum stresses of sufficiently high intensity may fail due to the fatigue in steel. The maximum repetitive stress that can be sustained by the steel section without failure for 2 million load cycles is commonly accepted as a measure of fatigue strength. Generally, in reinforced concrete structures, the stress due to the dead load is at least half the total stress; therefore, fatigue need not be considered in the design. Fatigue should be considered only in special cases, when a lightweight structure is subjected to a relatively high repeating live load.

Corrosion resistance

Protection against corrosion of reinforcing bars in *in situ* structures results from the cement slurry coating that surrounds it. However, this layer of cement slurry is sensitive to mechanical and chemical aggression, and has to be protected by a sufficiently thick cover of dense concrete.

Reports of extensive surveys have established that corrosion takes place, practically in all cases, due to faulty design or construction, i.e., due to lack of adequate cover or due to poor porous concrete with a high water–cement ratio.

Large widths of cracks in the tensile zone of concrete structures can damage the protective slurry layer glue around the bars, and therefore their maximum width needs to be limited. Tests have shown that with a crack width of 0.2 mm no corrosion takes place normally and with a crack width of 0.1 mm no corrosion takes place even under the extreme conditions of aggressive atmosphere. The maximum widths of cracks in structures reinforced with high-strength deformed bars do not exceed 0.1 and 0.2 mm for working stresses of about 200 and 300 N/mm^2, respectively.

The initial rusting of these bars is known to improve their bond strength over brand new steel. The effect of corrosion of bars during storage on the behaviour of structures can be judged as follows:

- Storage over several months leads to corrosion, which is generally associated with a weight loss of not more than 1–3%. The weight reduction is beyond tolerance limits only in the case of exceptionally heavy corrosion, wherein a corresponding devaluation of the bar may be called for.
- The bond characteristics of these bars are not adversely affected by corrosion.
- Protection against the corrosion of a bar resulting from its embedment in concrete is unaffected whether or not the bar was rusted before embedment.

Fire resistance

Reinforced concrete structures subjected to fire get severely affected when the temperature of the reinforcing steel exceeds 500–600°C. Around this temperature, the proof strength of CTD steel bars reduces to about 250 N/mm^2, which is the level of actual stress for Fe 415, designated by either the working stress method or the limit state method at service load.

Some recent tests on both Fe 415 and Fe 500 and plain mild steel (MS) bars in samples from different countries have indicated that the behaviour of both the groups of bars is about the same as the temperature increases. The tests have also shown that

a. the retention of strength at elevated temperatures appears to vary inversely with the carbon content,
b. the strength of high-strength steel bars is as stable over a range of time of exposure as plain MS, and
c. the indicated mode of failure in such bars is not due to the recrystallization and reversion to lower strength of the parent material.

10.4.1 Fe 415/Fe 500 CTD Bars

Up to the 1960s, the Indian construction industry had been using 250 MPa, low-yield strength, MS plain bars for concrete reinforcement. Attempts to increase the yield strength by the conventional method of increasing the carbon content in steel led to reduced ductility, bendability, and weldability of the bars.

In the 1970s, CTD bars were introduced as a product of the cold-twisting technology to overcome this strength–ductility problem. In these bars, the carbon content is restricted to a low level to impart acceptable values of ductility, bendability, and weldability and the proof strength was increased from 250 MPa to a guaranteed value of 415 MPa by cold twisting. The bond strength was increased by an appropriate ribbing pattern. In addition, judicious micro-alloying was done for achieving the desired properties in both TISCON-50 and TISCON TMT-50.

10.4.2 Thermo-mechanically Treated (TMT) Bars

TMT bars are a recent technological advancement for the production of high-strength deformed steel bars for concrete reinforcement. In this process, higher strength is obtained by thermo-mechanical treatment, wherein the steel bars are intensively cooled immediately after rolling. Sudden reduction in temperature creates a hardened surface layer with the internal core still hot. While the atmosphere cools down further, tempering takes place by the heat from the core. This process of thermo-mechanical treatment improves the strength and ductility of the bars. Besides thermo-mechanical treatment, the carbon content in TMT bars is also brought down, leading to the following advantages: higher strength, superior ductility, weldability, bendability, thermal resistance with its residual strength being 60% of the room temperature strength even at 600°C, corrosion resistance, and cost competitiveness in relation to CTD bars.

The TMT process in principle entails inline water cooling of the hot-rolled bars as they emerge from the hot-rolling stand at a temperature of 1000–1100°C. Direct water quenching results in the reduction of the surface temperature to around 200°C and results in the formation of martensite at the surface layers; the core of the bar remains austenitic. As the bar emerges from the quenching zone, the thermal gradient across the bar section causes heat to flow from the hot austenite core towards the bar surface. This results in tempering of the surface martensite and an equalization of the surface and core temperatures takes place. During subsequent atmospheric cooling of the rolled bar on the cooling bed, the hot austenite core is gradually transformed into a ferrite-pearlite microstructure. The TMT rebars thus develop a composite microstructure: a ductile ferrite-pearlite core and a tough tempered martensite rim. The composite microstructure is primarily responsible for the combination of the apparently contradictory metallurgical properties of high strength and ductility.

The interesting feature of the TMT process relates to the flexibility of producing different strength grades from one base chemistry. This is achieved by controlling the process parameters such as water pressure, volume of water, number of water jackets for quenching, and equalization temperature.

These bars are made by the manufacturers under three grades conforming to minimum yield strengths of 415, 500, and 550 MPa. TMT rebars are made as per IS: 1786 (specification for high-strength deformed bars for concrete reinforcement). These grades are categorized in Table 10.4. Note that TMT bars offer better ductility (based on % elongation) compared to the requirement specified by IS:1786.

Table 10.4 Mechanical properties of TMT bars

	IS: 1786		Sail TMT			Tiscon TMT		RINL rebars	
Grade	Fe 415	Fe 415	Fe 500	Fe 550	TMT42	TMT50	Fe 415	Fe 500
Yield strength	415	415	500	550	450	530	460	540
Tensile strength	485	500	580	630	510	580	520	585
Elongation (%)	14.5	22	20	18	20	18	20	16
Bend test								
up to 22 mm	3d	2d	2d	2d	2d	3d	2d	3d
> 22 mm	4d	2d	2d	2d	2d	4d	—	—

10.4.3 Micro-alloyed Steel

Small quantities of metals such as titanium, vanadium, niobium, and other elements are used either separately or in combination to improve grain refinement and increase strength, toughness, and ductility of steel. The advantages of micro-alloyed steel are the following:

- *High ductility*: Generally, this type of steel exhibits higher total elongation, of 20% as against a require-ment of 14.5% (IS: 1788). Hence this type of steel is suitable for earthquake-resistant structures in severe seismic zones.
- *Higher toughness*: Toughness relates to the capacity of a material to absorb impact without fracture. The fine-grained structure of this type of steel increases its energy-absorption capacity and hence increases its toughness.
- *Uniformity across the section*: Micro-alloyed steels have nearly uniform properties throughout the cross-section compared to CTD and TMT bars.

Micro-alloy bars are being produced in India in the grades of Fe 415 and Fe 500 ranging from 6 mm to 25 mm in diameter.

10.4.4 Ductile Earthquake Resistant Bars

Ordinary CTD/TMT bars have a percentage elongation of 14.5% at failure. The bars with specification D are intended to be used in situations where high ductility is required. They perform well in situations where dynamic loading is expected such as in machine foundations. They are required to have larger elongation at failure as shown in Table 10.1(c). Their designation D denotes that they are more ductile than other normal bars. They should also satisfy the norm with respect to total elongation at maximum force as specified in this table. However, no maximum limit has been specified for their proof stress/yield stress. These bars are prefered to be used in Seismic Zone III.

The bars with specification S, in addition to having improved ductility, should also satisfy the stipulation with respect to maximum proof stress/yield stress limit 540 N/mm² for Fe 415S and 620 N/mm² for Fe 500S steel. These steels are intended to be used in Seismic Zones IV and V.

10.4.5 Corrosion Resistant Steel (CRS) Bars

Corrosion takes place when steel oxidizes because of excessive presence of moisture and oxygen in the air. Corrosion of steel is even more rapid in areas of high humidity, places close to the sea, or in the presence of saline or salty water. Gases emitted in industrial areas also increase corrosion of the metal. Chlorides present in the concrete also accelerate corrosion. Wherever there is salinity in air, along the coastline, in sea water, in ground water, or where there are acid particles in the air, corrosion problems occur. Corrosion attacks RCC structures like buildings, bridges, dams, industrial plants, and others. Such an attack leads to shortening of their service life.

CRS bars are intended to fight the deadly menace of corrosion. Normally to make CRS bars, a special alloy of steel with desired quantity of copper, nickel, and chromium are carefully added in a specified combination to the molten steel. Copper fills the pores and resists corrosion; nickel toughens the steel and increases the fatigue resistance; and chromium resists oxidation and arrests pitting corrosion.

By using such CRS steel bars one expects the structure to perform well for a longer period even in an aggressive environment. These bars have dual microstructures, i.e., the surface layer is tempered martensite while the inner shell is ductile ferrite-pearlite.

10.4.6 Coating of Bars for Corrosion Resistance

There are a number of coating systems employed each having its own merits. They are briefly described in this section.

Cement-polymer composite coating system (CPCC)

This is a new method developed by Central Electrochemical Research Institute, Karaikudi, Tamil Nadu. This system consists of application of one coat of rapid setting primer followed by a coat of cement polymer sealing product. The primer and sealing products have thermoplastic acrylic resin as basic raw material. Sealing product is formulated using resin mixed with cement as pigments. Both rapid setting primer and sealing coats are patented items. This system has been developed mainly as a factory/shop process.

Fusion bonded epoxy coating (FBEC)

The fusion bonded epoxy coating is a process where epoxy powder is applied by electrostatic spray on hot steel at a pre-set temperature. The powder, when in contact with the hot bar, melts, flows, gels, cures, cools, and produces a well-adhered continuous corrosion-resistant protective coating. This thermosetting is an irreversible process and provides the best protection to rebar against corrosion. It prevents attack of chloride ions on the metallic surface and occurrence of electro-chemical reaction initiating corrosion of steel. The epoxy coated rebars were first used in Pennsylvania state in 1973 and commercially produced in the USA since 1976.

Coating has to be done as per IS: 13620-1993 'Fusion Bonded Epoxy Coated Reinforcing Bars' specification.

Hot dip galvanizing

Galvanizing is the process of deposition of zinc over the surface of reinforcing bars. One of the methods to prevent the steel from corrosion is to galvanize the bars. Zinc coating offers protection in more than one way. Firstly, inter-metallic and metallic layers of zinc act as a physical barrier between the steel material and corrosion environment. In this case, corrosion resistance of the steel is due to the corrosion resistance of zinc. Further, wherever steel is exposed to corrosive environment due to the breakdown of the protective coating, steel is still protected by the selective dissolution of zinc. Galvanizing or the method of zinc coating can be achieved principally by electro-galvanizing and hot dip galvanizing methods. Hot dip galvanizing process is the most widely used method.

Epoxy-phenolic IPN coating

Central Building Research Institute (CBRI), Roorkee, has developed an epoxy-phenolic interpenetrating polymer network (IPN) system for protection of structural steels and has extended it to rebars. It is a polymer formulation consisting of more than two polymers cross-linked in the network form by synthesis. Polymers are interwoven to each other and continue to link together by permanent entanglements devoid of chemical linking. The use of such coated bars is not a substitute for good construction practices including adequate cover and other durability requirements specified in the relevant chapters in this book.

10.4.7 Fibre Reinforced Plastic (FRP) Bars

FRP bars are intended to be used as a reinforcement in areas where steel reinforcing has a limited lifespan due to the effects of corrosion. They are also used in situations where electrical or magnetic transparency is needed. In addition to reinforcing for new concrete construction, FRP bars are used to structurally strengthen existing concrete members.

A structural reinforcing bar is made from filaments or fibres held in a polymeric resin matrix binder. The FRP bar can be made from various types of fibres such as glass (GFRP) and carbon (CFRP). FRP bars have a surface treatment that facilitates a bond between the finished bar and the structural element into which it is placed.

The design and construction of structures with FRP bars can be made using ACI 440 1R 06 recommendations.

10.5 Probable Errors in Placement of Reinforcement and Their Effects

Proper reinforcement fabrication and its placement in forms are vital in concrete construction to achieve the design requirements. The following are some of the prevailing errors in the placement of reinforcement in the field practice:

- Improper anchorage of the main reinforcement at the intersection of the cantilever slab and beam
- Ineffective placement of reinforcement with respect to the depth of the member, which reduces the lever arm in the beam or slab
- Discontinuity of reinforcement at continuous edges of the slabs
- Deformations and dislocations of reinforcement during concreting due to construction loads (e.g., workers walking over the reinforcements)
- Inadequate lap lengths
- Cluster of reinforcing bars at splicing points
- Inadequate spacing of bars
- Insufficient or non-uniform cover for reinforcement, which leads to corrosion
- Improper use or absence of cover blocks
- Reduction in bar diameter due to corrosion and improper storage of the bars at the site
- Bars with larger diameters used due to the non-availability of bars with correct diameters
- Improper detailing and placing of stirrups

10.6 Reinforcement Fabrication

Reinforcement fabrication is the process of assembling reinforcing steel bars into mats or cages prior to concreting at the site. At present, fabrication including decoiling, cutting, and bending is primarily a manual operation in India. If crude fabrication techniques are employed, the resulting errors and wastage of materials will affect the performance and quality of the structure. Accuracy of fabrication and dimensional tolerances are important and should be ensured at the site of the job.

The fabrication of bars, tendons, and meshes adds almost 20–30% to the basic cost of steel. Fabrication of reinforcement involves the following steps:

- Studying the bar bending schedule and detailing
- Straightening bars from coils or U-bend shapes
- Cutting the bars to sizes
- Bending the bars to specified shapes
- Bundling, marking, and stacking
- Transportation and handling at site
- Coating and treating of bars, where special protective measures for corrosion protection are needed
- Splicing of bars or welding, as necessary
- Binding the bars to form a rigid cage

Some of these operations are described below.

10.6.1 Bar Bending Schedule

As fabrication and bending tend to be on-site operations, bar bending schedules are made to give complete information regarding the bars to be cut, bent, and bundled and to enable the location of the bars on-site to be readily determined. This eliminates the need to search for dimensions of the bent bars. While placing the reinforcement in the forms, there is a possibility of errors arising out of the use of different methods of dimensioning especially for a cranked bar. The recommendations given in IS: 5525 with respect to detailing of reinforcement in reinforced concrete works eliminate such errors. The tolerances in the fabrication of these bars are also specified to be ±25 mm in length. A typical bar bending schedule is shown in Table 10.5.

10.6.2 Handling and Storing of Bars

Methods for unloading reinforcing bars vary with the type of structure and the position and means of access to the site. The bars should be unloaded at a place where it is convenient to cut, bend, and store them. Bars delivered to the site can be unloaded by hand, as near as possible to the storage area, to avoid having to carry the bars through long distances.

Bars should be stacked over timber sleepers, such that different lengths and sizes can be identified easily. Bars can also be stacked vertically or in tiers in racks, if the site has limited space. When storing in open areas, lime or cement coats are normally applied over reinforcing bars to avoid rusting due to aggressive atmosphere.

10.6.3 Surface Preparation

The surface condition of reinforcing bars may affect the bar strength. The main factors affecting bonds are the presence of scale, rust, oil, and mud on the surfaces. Scales appear when bars are rolled. Scaling also results from the cooling of hot metal. Loose scales are normally removed when the bars are handled, prior to fabrication. Rust improves bonds because it increases the normal roughness of the surface. It is considered harmful only when the area of cross-section or the weight of the bars is reduced due to rusting.

Bars should be unloaded and stored in such a manner that they are free of mud. Any mud coating on the bars should be washed off before using the bars. Occasionally, bars get covered with machine oil or grease,

Table 10.5 Typical bar bending schedule

S. No.	Bar mark	Diameter (mm)	Shape of bending (dimensions in cm)	Length (m)	Quantity (m)	Total length	Weight (kg)
1	Bottom layer (straight) bar	20	⊢——655——⊣ 12⟋ ⟍12	6.79	4	27.16	@2.47/m = 67.08
2	First cranked bar	16	140.5 · · · 140.5 9.5 52–37 9.5 300 45°	7.04	1	7.04	
3	Second cranked bar	16	185.5 52–37 185.5 410	7.04	1	7.04	
4	Third cranked bar	16	130.5 52 37 130.5 520	7.14 Subtotal of S. No. 2 to 4	1	7.04 21.12	@1.58/m = 33.37
5	Top bar	10	⊢——655——⊣ 7.5 7.5	6.70	2	13.40 = 8.31	@0.62/m
6	Stirrup	8	18.4 43.3	1.43	54	77.22	@0.39/m = 30.12

Total weight = 138.88 kg

which must be removed before placing. This may be done with a fire torch or by wiping the surface with solvents. Otherwise, this leads to ineffective bond between steel and concrete. Finally, the reinforcement should be washed free of all impurities before placement.

10.6.4 Rust Prevention

Cement paste normally provides a highly alkaline environment that protects embedded steel against corrosion. Concrete, which is well compacted and well cured with a low water–cement ratio, has low permeability and hence minimizes the penetration of atmospheric moisture as well as other components such as oxygen, chloride ion, carbon dioxide, and water, which encourages corrosion of the steel bar.

In very aggressive environments, the bars may be coated with special materials developed for this purpose. Coating on reinforcing steel, therefore, serves as a means of isolating the steel from the surrounding environment. Common metallic coatings contain galvanizing zinc. High chloride concentration around the embedded steel corrodes the zinc coating, followed by corrosion of steel. Hence, this treatment is used for moderately aggressive environments. For highly corrosive atmospheres caused by chloride ions from the de-icing salts applied to protect against sodium chloride and calcium chloride, usually near seashores, epoxy coating is applied to protect steel reinforcing bars from corrosion. Such bars have acceptable bond and creep characteristics. The coat normally applied is 0.1 to 0.3 mm thick. The reinforcement is coated with epoxy in the factory itself, where the steel rods are manufactured. Such reinforcements are known as fusion-bonded epoxy coated steel. The coating should be done in accordance with the provisions of IS: 13620-1993 specification. Steel manufacturers also produce TMT bars with better corrosion resistance, termed corrosion-resistant steel (CRS). The performance of CRS bars is better in resisting corrosion compared to plain bars. However, the use of CRS bars will only delay the process of corrosion. It will not prevent corrosion once and for all.

10.6.5 Bar Straightening and Cutting

The preparation of reinforcement starts with straightening and cleaning the bars, and cutting them into required lengths. Mild steel bars, especially in the lower diameter range (6–10 mm), are produced in coils. To convert these coiled bars into straight pieces, they have to be run through a straightening machine. The standard practice is to shear the bars in a group, to a tolerance of 2.5 cm in length. Hand cutting tools are sufficient and are normally employed for smaller jobs. For bulk cutting of larger diameter bars, powered cutting machines are used (see Fig. 10.5). Gauging the length to be cut is a built-in feature of the machine. With increase in the

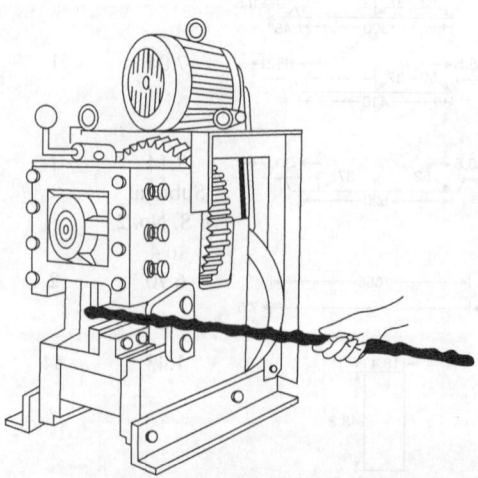

Fig. 10.5 Powered cutting machine

strength and diameter of the bars, the number of bars that can be cut together in a stroke decreases. Lighter machines can be run at the rate of 30–40 cuts per minute.

10.6.6 Bar Bending

Bars must be bent into shapes accurately in accordance with the dimensions given in the bar bending schedule. Any deviation in these dimensions can have an adverse effect on the strength of the structure, and the fabrication will not be correct if the bars are inaccurately bent. Special attachments are used for fabricating spirals, stirrups, and rings at high speeds along with powered bar bending machines (Fig. 10.6).

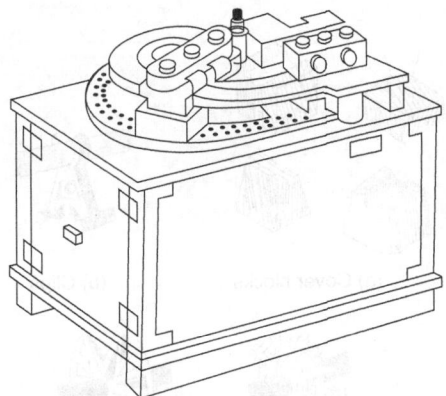

Fig. 10.6 Bar bending machine

10.7 Placing Reinforcing Bars

Whether fabricated at the site individually or by factory operation, reinforcement meshes and cages are handled and placed inside the formwork. The accuracy of placement depends on the dimensional accuracy of the forms and the accuracy of the grillwork. In placing and tying the reinforcement, special care has to be taken to obtain accuracy, and certain accessories are employed for this purpose. In handling fabricated bars, care should be taken to consider the following:

a. Shock loading and dropping of bars from heights could lead to brittle fracture.
b. Coating, if any, prescribed for the bars should be ensured to be free from scratches, lime, or cement wash patches for short-duration open storage. Fusion-bonded epoxy coating for durability in an aggressive atmosphere, zinc-based paints, or galvanization of bars, etc., can be adopted.
c. Each type of bar so identified in the bending schedule (Table 10.5) should be bundled and tagged together for identification and ease of placement.

10.7.1 Wire Ties

The bars are tied together at the intersections at sufficiently close intervals. This provides a rigid framework, in order to prevent the displacement or dislocation before and while concrete is being placed. For general work, a 16-gauge black soft-iron wire is most suitable. It is advisable to keep the ends of the ties reasonably short so that they do not touch the formwork, as this may lead to rust spots on the face of the concrete. For proper tying, standard binder clips and binder coils are available, which ensure the gripping action.

10.7.2 Cover and Spacing of Bars

To protect a bar or other reinforcement from corrosion, and to ensure that the combined action of the concrete and steel is effective, the bars are embedded at a certain minimum distance from the outer face of the concrete. This distance is called the *cover* of the concrete. The least cover to be provided is specified in the codes of practice. It is generally given in the specifications or the drawings. Depending on the element and specified thickness for cover, different sizes and shapes of prefabricated cover blocks are employed. Simple mortar blocks tied with set-in binding wire [Fig. 10.7(a)], asbestos cement-pressed blocks, moulded plastic cover blocks or clips [Fig. 10.7(b)], and even plastic-capped steel wire are employed; some selected shapes and sizes of cover blocks are commercially available. The shapes vary with end use—columns, beams, or slabs. Some of these plastic cover blocks are now manufactured and used in India. The various types of cover blocks available are summarized in Table 10.6.

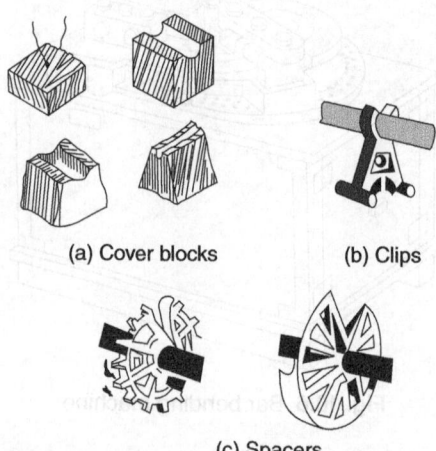

(a) Cover blocks   (b) Clips

(c) Spacers

Fig. 10.7 Cover blocks, clips, and spacers

Table 10.6 Types of cover blocks

Cover (thickness, mm)	Maximum bar size (mm)	Remarks
18–50 (clips)	Up to 36	• Used for positioning horizontal reinforcement • Lightweight, non-staining, and non-corrosive
25–50 (blocks)	Up to 36	• Supports heavy bars • Used in foundations and beams • Eliminates voids
18–75 (mesh chair)	Up to 36	• Supports reinforcing mesh • Used in floors, slabs, walls, pipes, and exposed aggregate panels • Accurate cover
15–30 (side spacer)	Up to 20	• Minimizes risk of dislodging when concrete is placed • No damage to formwork • Provides positive grip
18–40 (circular spacer)	Up to 30	• Provides all around cover to the reinforcement • Accurate, economical • Used in columns, walls, and other precast units
25–35 (grip spacer)	Up to 36	• Used for column reinforcement • Provides locking device • Firm grip and no spacer displacement

The spacing between two parallel bars must be such that it ensures that there is sufficient concrete around the bar. The prevailing codes, therefore, recommend that the horizontal distance between two parallel bars should not be less than the greater of the following dimensions: (a) the diameter of the bars or of the larger bar if the two bars are not of the same size; (b) 5 mm more than the nominal maximum size of the coarse aggregate. Proprietary bar spacers are now available [Fig. 10.7(c)], ensuring accurate spacing. Annular bar spacers, or similar other bracers which are readily available, can be conveniently slipped onto the bar from the sides.

10.7.3 Chairs

To support heavy reinforcement mesh form in one or more layers, chairs are employed as temporary supports during the progress of concreting (Fig. 10.8). These elements—also known as bolsters—are fabricated out of small wires (wire bars) in various shapes, supporting a single bar or a series of bars in one line. If the surface of the concrete is exposed to a severe atmosphere, the legs of the chairs are dip-galvanized or provided with plastic cups. In the supporting top layer of bars, reinforcing bars themselves are bent up in self-supporting shapes. They should also be adequately protected against corrosion.

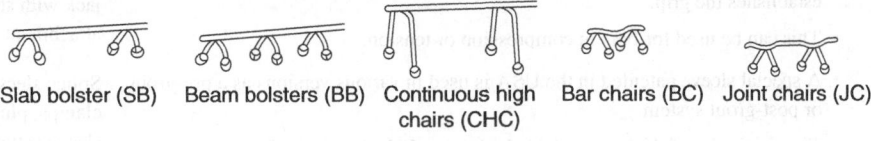

| Slab bolster (SB) | Beam bolsters (BB) | Continuous high chairs (CHC) | Bar chairs (BC) | Joint chairs (JC) |

Fig. 10.8 Types of chairs

10.7.4 Splicing of Bars

Splicing of bars is increasingly becoming a necessity because it is impossible to provide full-length continuous bars without lapping, when used as horizontal bars in walls, temperature bars in slabs, or vertical bars in high walls and piers. The location and kind of splices are normally shown in placing drawings. The locations of these splices depend on the recommendations of the code of practice (IS: 456-2000). No substitution in type and location should be made. The three general types of splices are: (a) lapped, (b) welded, and (c) mechanically coupled.

In most cases, lapped splices are more economical than the other types. Due to the close spacing of bars, it is not always possible to provide lap splices, especially on large-sized bars. The amount of lap might be sufficient to make some other type of splice more economical. Mechanical splices can be categorized as screwed couplers, friction clamps, pressed sleeves, and bonded sleeves (Fig. 10.9). The various types of mechanical splices and the equipment needed to install them are given in Table 10.7. The various types of welding adopted for making the sleeves are shown in Fig. 10.10.

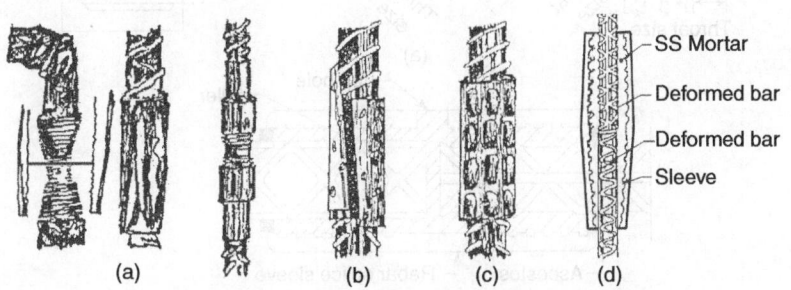

Fig. 10.9 Mechanical splices: (a) screwed couplers, (b) friction clamp, (c) pressed sleeve, (d) bonded sleeve

Table 10.7 Mechanical splices

Type of splice	Characteristics/arrangements required/application	Equipment
Screwed couplers [Fig. 10.9(a)]	• The ends of the bars to be jointed are provided with machined screw threads in the left or right direction. Plastic caps are provided to safeguard the threads which are removed before coupling the bars. • Effective in compression and tension. • Splicing can be prefabricated or done *in situ*.	Wrenches used to twist the bars
Friction clamps [Fig. 10.9(b)]	• Employs thin sheet sleeves of welded bars of small diameter arranged such that by tightening these two parts of the sleeves, friction develops between the bars to be spliced and the sleeve helps in positioning the bar and developing a bond between the deformations of the bars and the sleeves. • Holes provided in the sleeve help in the bonding through the concrete mortar. • Employed mostly for end bearing compression bars. • Unequal diameter bars can be spliced by adding a reducer sleeve.	Small tools such as the aligning rig and the hammer to drive the wedges
Pressed sleeves [Fig. 10.9(c)]	• A sleeve inserted onto the ends of the bar and squeezed by a machine establishes the grip. • This can be used for bars in compression or tension.	High-pressure hydraulic jack with squeezing attachment
Bonded sleeves [Fig. 10.9(d)]	• A special sleeve patented in the USA is used in various versions as a pre-grout or post-grout system. • The continuity of the bar is established through the bonding of a special grout injected into the sleeve. • Several advantages have been claimed by this system and it is very popular in many countries.	Splice sleeves, support clamps, plugs and seals, sleeve setters, grout mixer, and grout pump
Welded splices	• The bar ends are shaped to the specifications and welded with overfilling at joints. • Angles, plates, etc., are used to reinforce the strength at welds in order to ensure satisfactory performance [Fig. 10.10(a)]. • Patented weld splices, where the joint sleeves are filled with molten metal, are also used in Europe and other countries. • Usually, molten metal is used for micro-alloy, hot-rolled bars [Fig. 10.10(b)].	Welding machinery, other equipment depends on the type of bars; careful selection of electrodes and correct current during arc welding is important

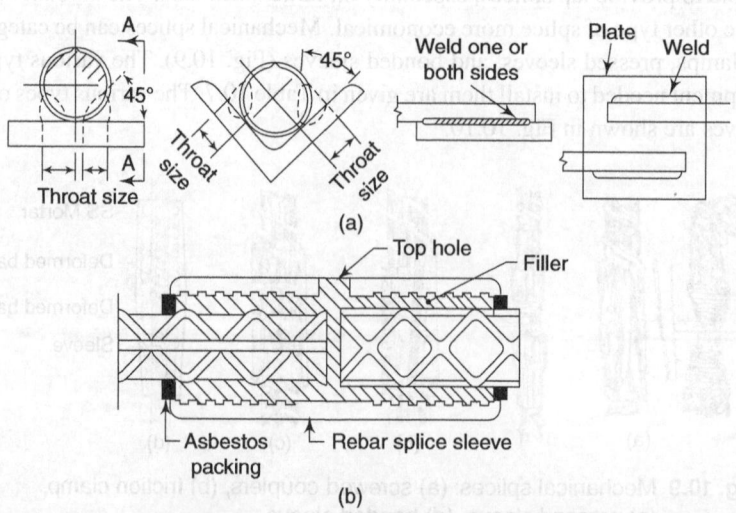

Fig. 10.10 (a) Welding bars with plates and angles; (b) Molten metal sleeve

10.7.5 Tolerances in Placing Bars

The strength of any concrete structure can be affected by improper positioning of the reinforcing bars. For example, the lowering of the top bars or raising of the bottom bars by 10 mm more than specified in a 15-cm slab could reduce its load-carrying capacity by 20%. Therefore, the placement and cover tolerances are emphasized in reinforcement placing as follows:

Depth of 20 cm or less: ± 10 mm depth

Depth greater than 20 cm: ± 15 mm depth

These reasonable tolerances are allowed in order to permit economical bar placing and do not apply to the fabrication of bars. There is no negative tolerance permitted for the cover.

10.7.6 Inserts

Inserts are shaped geometrically in efficient forms to ensure proper distribution of forces. This is also achieved by distributing reinforcing bars attached to the insert material. Some of the standard inserts are form ties, hanger attachments, structural connectors, and lifting inserts. The uses of different types of inserts in precast concrete construction are given in Table 10.8.

A typical hanger attachment is shown in Fig. 10.11.

Table 10.8 Inserts used in precast concrete construction

Insert	Use
Hanger attachments	Used in precast girders and *in situ* slab systems where the suspension of the formwork is necessary.
Structural connectors	Buried inside the elements and used for connecting two elements. Designed to transfer loads and moments. Examples: bracket fixtures, lift-slab collars, and panel bracings.
Lifting inserts	Used to lift precast elements such as wall panels, slabs, and beams. Different shapes and sizes are available. The inserts are galvanized or electroplated. In architectural panels, where complete corrosion resistance is required, stainless steel inserts are used. Single and multiple lifting inserts are used depending on the weight, dimensions, number, and location of openings in the panels. The strength of inserts is determined by using the classical theory of the shear cone failure mechanism.
Deflection inserts	These inserts are used for the proper alignment and positioning of prestressing strands/wires. Various designs are available, depending on the site requirements. One example is the use of steel support with rollers as a deflection insert.

10.8 Reinforcement Detailing

The application of sophisticated computer programs using very powerful computers underlines the fact that more attention is in general directed towards the perfection of analysis and calculation procedures than towards the seemingly more trivial problems of proper detailing. Many young engineers, well trained in the latest novelties of modern computer techniques, struggle to read the plans of their draftsmen, and counsel them about detailing problems. Past and present experiences have shown, however, that more damages occur due to shortcomings in the general construction and detailing rather than due to faulty calculation. This fact can be well observed at earthquake disasters, where properly detailed structures may show damage but rarely collapse.

In the first part of this section, the IS code provisions for slabs, beams, columns, and footings are described. An attempt is made to rationalize reinforcement detailing in structural drawings. The second part describes the art of detailing at joints and future trends for certain specific components which ensure ductile behaviour after the structure reaches the designed ultimate load.

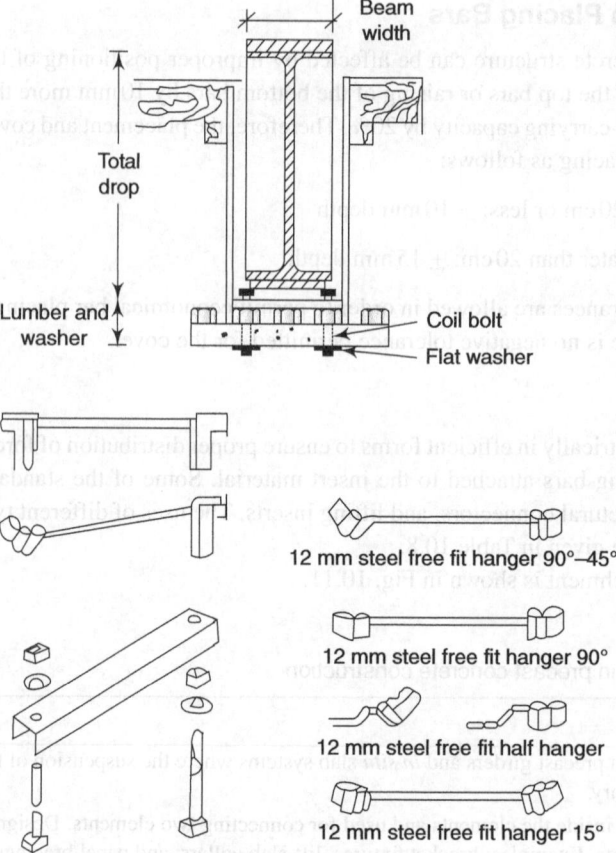

12 mm steel free fit hanger 90°–45°

12 mm steel free fit hanger 90°

12 mm steel free fit half hanger

12 mm steel free fit hanger 15°

Fig. 10.11 Typical hangers

10.8.1 Detailing Requirements as per IS Code

Here we study the provisions of the reinforcement detailing specified in the IS codes with some simple suggestions, along with the recommendations of the reinforcement detailing in the drawings provided in IS: 5526.

Slabs

The limits of the slab thicknesses as functions of the panel dimensions are intended to minimize the deflections caused by vertical vibrations. Another pertinent variable incorporated in the present code besides the limit on the depth–span ratio is the reinforcement percentage.

As per IS: 456-2000, the main bar of the slab should be of diameter 6–12 mm. The minimum reinforcement in either direction of the slab must be 0.15% and 0.12% of the cross-sectional area with mild steel and deformed bars, respectively. The necessity to minimize the cracks due to temperature, shrinkage, and effective dispersion of concentrated loads resulted in the specification of the maximum bar spacing of 460 mm in either direction of the slab. This provision is further made stringent by specifying a maximum bar spacing of three times the depth in the main direction and five times the effective depth in the transverse direction for better distribution of cracks. Further, some additional provisions are made in the codes regarding the curtailment of reinforcement.

Beams

Rectangular, T, and L beams are frequently used in reinforced concrete. The depth of the beams (h) varies over a wide range; depending on the load, it varies from 1/10 to 1/20 span. It is recommended that the depth must be taken in multiples of 25 mm for sizes up to 500 mm and multiples of 50 mm for greater sizes. The width of the section is selected from the range $0.33h$ to $0.50h$. The dimension of the breadth may also correspond to the design of the supporting member.

Longitudinal reinforcement

Deformed or plain bars with diameters of 10–40 mm are deployed for longitudinal reinforcement. They may be arranged in one or two rows.

Minimum longitudinal steel of

$$A_s (\text{mm})^2 = \frac{0.85bd}{f_y}$$

where
A_s = Area of steel
b = Breadth of beam
d = Depth of beam
f_y = Yield stress of steel

has been specified to avoid sudden failure due to the fracturing of steel. However, when high-strength concrete is used, the quantity of minimum longitudinal steel should be suitably increased. To avoid congestion of reinforcement, the maximum area of the reinforcement is restricted to 4% of the cross-sectional area.

Shear reinforcement

Transverse reinforcement is usually designed for excess shear, generally defined as the difference between the designed shear load and the diagonal cracking load of the beam without shear reinforcement. At least one stirrup should cross the inclined shear crack along the axis of the member. Along with the temperature and shrinkage requirements, this has necessitated the restriction of the maximum spacing of stirrups to $0.75d$ or 450 mm, whichever is less. The whole longitudinal reinforcement must be enclosed by stirrups to counter the risk of longitudinal dowel cracks, and it is for this reason that bent-up bars cannot be used alone. It is a good practice to use four-legged stirrups when the width of the beam exceeds 350 mm.

Side face reinforcement

With beam depth exceeding 750 mm, longitudinal bars are installed on the sides of the beams not more than 300 mm apart. The total area of these bars should be at least 0.1% of the cross-sectional area of the beam rib. The longitudinal spacer bars (hanger bars) used in beams should have a diameter of 12–16 mm.

Splicing

The continuity of bars may be provided with a lap length equal to twice the developmental length. Alternatively, splicing may be effected with welding according to the standard practice. Splicing may also be effected with the help of sleeves or couplers as explained in Section 10.7.4.

Columns

Columns are predominantly square, rectangular, or circular in cross section. The lateral dimensions of columns up to 500 mm are taken in multiples of 50 mm and of those exceeding 500 mm in multiples of 100 mm. A section less than 200 mm × 200 mm is not acceptable as a column.

Longitudinal reinforcement

Longitudinal reinforcement in columns consists of bars with diameters varying from 12 to 40 mm. IS: 456-2000 recommends a maximum steel percentage of 6% to avoid congestion. IS codes also stipulate a minimum value of reinforcement percentage (0.8%) to ensure adequate strength in the case of accidental eccentricity.

Transverse reinforcement

Transverse reinforcement is required in columns for both preventing buckling of main bars and resisting horizontal shear.

Footings

The thickness of the footings should at least be 150 mm. In two-way square footing, the reinforcement extending in each direction is distributed uniformly across the full width of the footing. The concrete in the footing just below the columns is confined, which helps in enhancing the bearing strength of the footing. The eventual failure of concrete in this zone may be due to transverse splitting. This splitting is avoided by extending the longitudinal reinforcements of the columns along with their lateral ties into the footing to a length equal to the development length in tension or compression.

Reinforcement detailing in structural drawings

Design decisions are conveyed to the construction site by means of detailed drawings. These drawings should clearly indicate the type of bars used in construction, their lengths, and their locations. Some simplified rules are recommended below for the production of detailed drawings:

a. A scale of 1:10 or 1:20 is recommended for drawing.
b. The beam, column, and slab lines should be drawn as thin lines, and the reinforcement should be shown in full in plans and elevations.
c. An indication of reinforcement with one or two typical bars should be shown in full in plans and elevations.
d. Bars should be fully detailed once only, this being on plan or elevation. In the other view, only the number should be indicated.
e. All dimensions should be in millimetres. These dimensions need not be followed with the letters 'mm'. If some dimensions are in metres, they must be followed with the letter 'm' or be written in millimetres not followed by any symbol, i.e., 2 m or 2000.
f. Bars should be named and their names should indicate the type of reinforcement, diameter of reinforcement, and the number, spacing, and location of bars.

Three types of bars are commonly used, namely, mild steel round bars, deformed bars (Fe 415), and high-strength deformed bars (Fe 500). These bars may be denoted by the symbols R, Y, and X, respectively. The diameter of the bar in millimetres is indicated after these symbols as R8, Y10, etc.

The amount of steel used is indicated along with the number of bars. The number of bars is indicated at the start of the name of the bars as 2-Y10 (2 bars of Fe 415 grade of 10 mm diameter). The spacing of the reinforcement in millimetres is indicated with the symbol as R8-200 (round mild steel bars of 8 mm diameter is provided at 200 mm spacing). It is a good practice to always indicate the number of bars along with the spacing as 20-R8-200, meaning that 20 members of 8-mm-diameter mild steel round bars are spaced at 200 mm centre to centre. The final number in the figure designates the bar type number referred in the bar schedule.

Detailed recommendation for beams and columns

A typical detailing of the bars for a beam is shown in Fig. 10.12. This detailing indicates to the fieldworker that there are five types of bars, out of which three are longitudinal reinforcements differing in diameter and length.

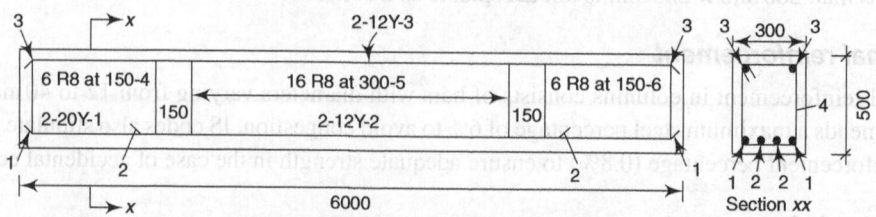

Note: Bar types 4 and 6 are similar. Hence there are five types of bars.

Fig. 10.12 Typical detailing of a beam

Detailing of slabs and walls

Slabs and walls should be detailed in the plan and in sufficient number sections to show the level of each layer of reinforcement. The following symbols may be attached along with spacing: T for 'top', B for 'bottom', EF for 'earth face', FF for 'far face', and E for 'either way'. The details of reinforcement for a footing and for a two-way restrained slab are shown, respectively, in Figs 10.13 and 10.14. For example, the detail '20 R8 200T-1' in a slab means that 20 mild steel reinforcements are spaced at 200 mm centre to centre at the top and that this reinforcement bar is numbered 1 in the bar schedule.

10.8.2 Specific Cases of Detailing of Reinforcement

This section describes the detailing of reinforcement in some of the important structural members. Detailing consists of the preparation of placing drawings, reinforcing bar details, and bar lists, which are used for the fabrication and placement of reinforcement in structures. The details presented here are restricted to situations that are frequently encountered in practice.

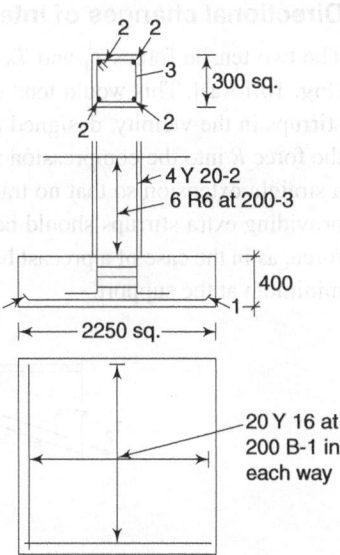

Fig. 10.13 Typical detailing of a column footing

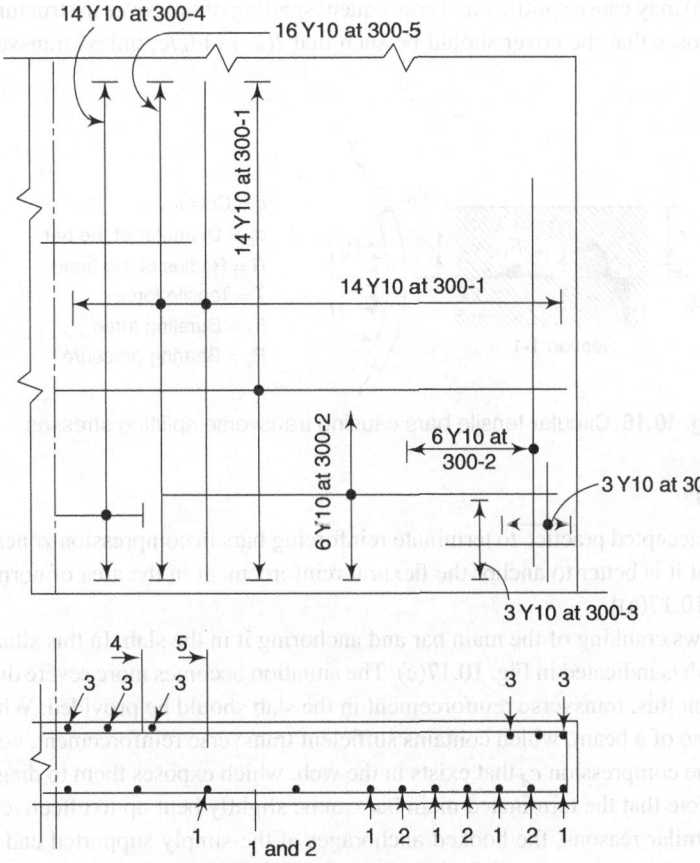

Typical reinforcement at the corner of a slab

Fig. 10.14 Typical detailing of a two-way restrained slab

Directional changes of internal forces

The two tensile forces T_1 and T_2 at the kink of a tensile bar are not unidirectional. Hence, a force R results [Fig. 10.15(a)]. This would tend to cause a splitting crack along the bar. When the change in angle is small, stirrups in the vicinity, designed to resist a conservative force of $1.5R$, would prevent cracking and transfer the force R into the compression zone. For larger angular changes, the reinforcement should be anchored by a straight extension so that no transverse force develops, as shown in Fig. 10.15(b). The same precaution of providing extra stirrups should be observed when abrupt changes occur in the direction of the compressive force, as in the case of a precast beam having varying depth with the maximum depth at the mid span and the minimum at the support.

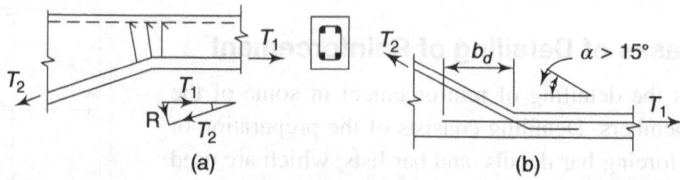

(a)   (b)

Fig. 10.15 Directional changes of tension force: (a) using ties when angular change is small, (b) main bars overlapping when inclination is large

Curved bars (Fig. 10.16) may cause splitting and consequent spalling of the cover in structures such as circular plates. Leonhardt proposes that the cover should be such that $R > 144d_b^2/c$, unless transverse reinforcement is provided.

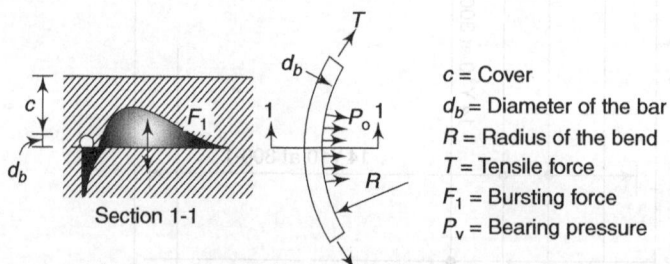

c = Cover
d_b = Diameter of the bar
R = Radius of the bend
T = Tensile force
F_1 = Bursting force
P_v = Bearing pressure

Section 1-1

Fig. 10.16 Circular tensile bars causing transverse splitting stresses

Detailing of beams

It has been a generally accepted practice to terminate reinforcing bars in compression zones. However, recent research has shown that it is better to anchor the flexural reinforcement in the area of normal pressure at the support reaction [Fig. 10.17(a)].

Figure 10.17(b) shows cranking of the main bar and anchoring it in the slab. In this situation, the possible crack location in the slab is indicated in Fig. 10.17(c). The situation becomes more severe due to end moments M in the slab. To prevent this, transverse reinforcement in the slab should be provided. When bars are terminated in the tension zone of a beam, which contains sufficient transverse reinforcement, considerable benefit may be derived from the compression c_d that exists in the web, which exposes them to diagonal pressures, as shown in Fig. 10.18. Note that the terminated main bar can be slightly bent up to effectively utilize this web compression c_d. For similar reasons, the hooked anchorages at the simply supported end of a beam can be much improved if the hooks are tilted or, preferably, made to lie in a near horizontal position. Splitting effects are largely counteracted by the normal pressure originating from the support reaction.

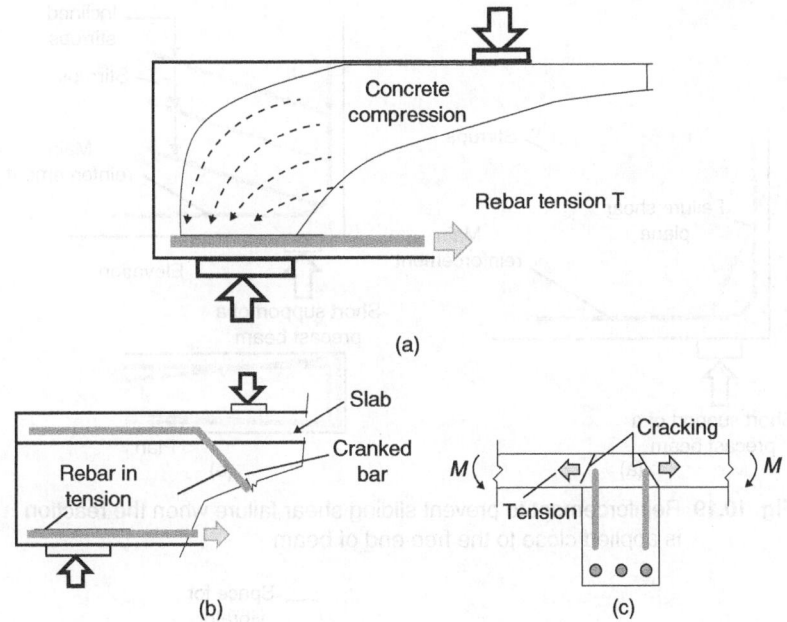

Fig. 10.17 Transverse pressure of tension at the anchorage of beam reinforcement

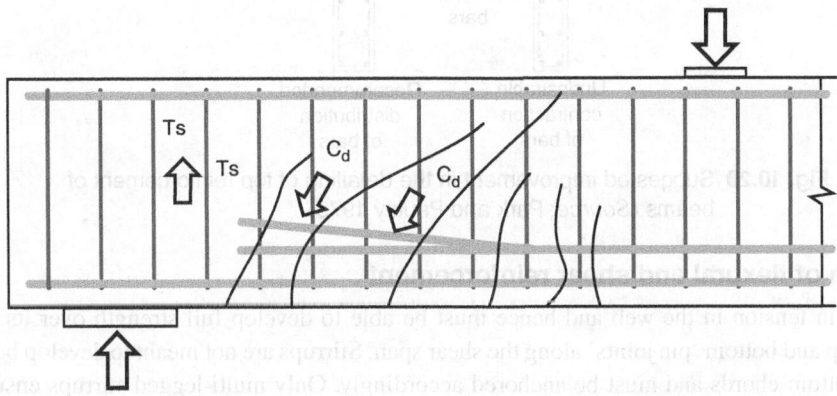

Fig. 10.18 Anchorage of flexural bars in the diagonal compression zone

In precast beams, the length available for end anchorage may be so short that only special devices such as bars, plates, and angles welded to the end of the reinforcement can ensure the development of the required bar strength. Owing to the fact that the support point is very often close to the free end of the beam, failure along a steep diagonal crack is a distinct possibility [Fig. 10.19(a)]. Additional inclined small diameter bars can be provided to ensure that no sliding failure occurs [Fig. 10.19(b)].

The spreading of the steel provided at the top for resisting negative support moment into the adjoining spans, preferably using small diameter bars, has the advantage of providing better access to the vibrators (Fig. 10.20) and a favourable distribution of compression in the flange.

In beams with a small shear span-depth ratio (*a/d*), arch action is the predominant mode of shear resistance after the onset of diagonal cracking. As a result, the flexural reinforcement is required to function as the tie for this arch. In such circumstances, it is better to carry the entire flexural reinforcement to the support and ensure that closely spaced stirrups are used near the support.

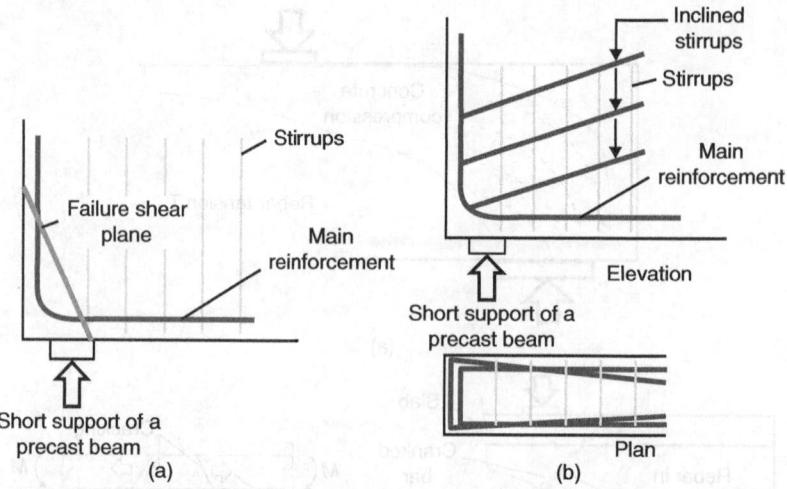

Fig. 10.19 Reinforcement to prevent sliding shear failure when the reaction is applied close to the free end of beam

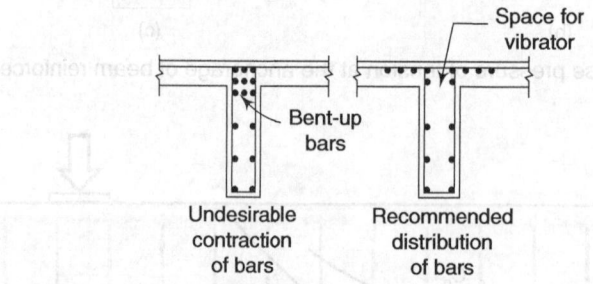

Fig. 10.20 Suggested improvement in the detailing of top reinforcement of beams (*Source*: Park and Paulay 1975)

Introduction of flexural and shear reinforcement

The stirrup is in tension in the web and hence must be able to develop full strength over its entire height between the top and bottom 'pin joints' along the shear span. Stirrups are not meant to develop bonds between the top and bottom chords and must be anchored accordingly. Only multi-legged stirrups ensure that bond forces develop at the right places (i.e., at each longitudinal bar). The undesirable concentration at diagonal compression in wide beams is represented in Fig. 10.21. The corner bars are rigid compared to middle main bars sitting on the flexible portion of the stirrup. This makes the compression c_d to arch towards the corner bars and cause stress concentrations.

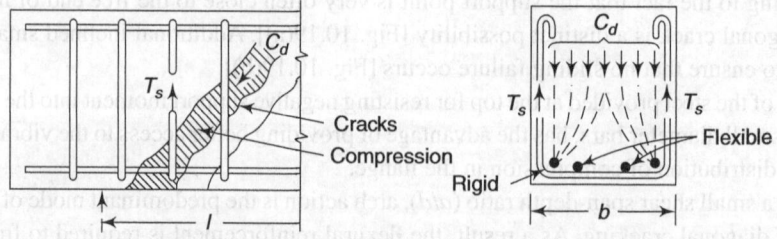

Fig. 10.21 Undesirable distribution of diagonal compression because of wide stirrups (*Source*: Park and Paulay 1975)

Bent-up diagonal bars [shown in Fig. 10.22(a)] are unsuitable as shear reinforcement due to the following reasons:

a. When widely spaced, they cause large stress concentrations in the bends, which can lead to splitting [Fig. 10.22(b)], particularly when the arrangement is unsymmetrical [Fig. 10.22(c)].
b. When bent-up bars are closely spaced (which would eliminate the undesirable effects), they deprive the flexural reinforcement of too many bars. Hence, at the point of cranking, there may be a chance of flexural cracking.
c. Compared to stirrups, they do not provide confinement for the concrete in the compression zone.
d. They generally lead to large crack widths.
e. They are more difficult to manufacture and handle on-site and are therefore relatively expensive.

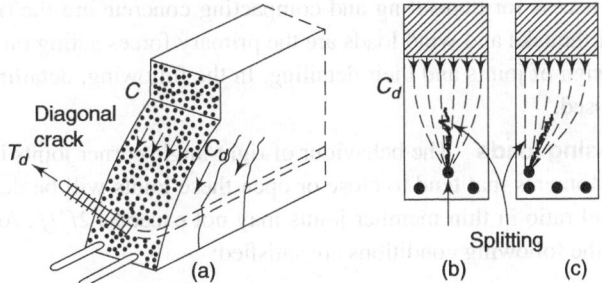

Fig. 10.22 Isolated bent-up bars not supporting compression forces satisfactorily
(*Source*: Park and Paulay 1975)

Detailing of supports and load points

When the reaction R_b is applied to the soffit of the beam as shown in Fig. 10.23(a), the critical section for shear is approximately at a distance d from the support. However, when the reaction R_t is applied from above, as shown in Fig. 10.23(b), the critical section is clearly at the face of the support. In the second case, additional precautions must be taken by the detailer to ensure that the reaction is 'guided' into the correct area of the supported beam or slab, which is suspended. It is important to note that the reaction for gravity loads always develops at the bottom of the beam, regardless of whether it is simply supported or continuous.

Additional stirrups (suspender stirrups) must be provided in the main beam at the intersection of the secondary beams to avoid horizontal splitting cracks along the flexural reinforcement of the supporting girder. When beams of equal depth meet, the bottom steel of the secondary beam should be above the bottom reinforcement of the supporting girder. In addition, a bar bent in the form of a horseshoe must be introduced to take the reaction from the bottom of the secondary beam to the top of the main beam.

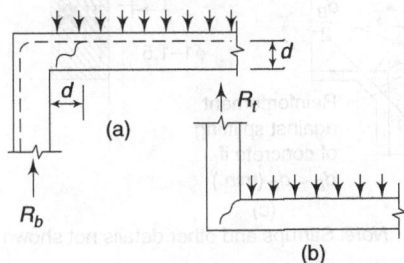

Fig. 10.23 Formation of diagonal cracks indicating a critical section when the
reaction is applied (a) from below or (b) from above a beam

Detailing of joints

The essential requirements for the satisfactory performance of a joint in a reinforced concrete structure can be summarized as follows:

a. A joint should exhibit a service load performance equal to that of the members it joins.
b. A joint should possess a strength that corresponds at least to the more adverse load combinations the adjoining members can possibly sustain, several times if necessary. That is, the joint should not fail under any condition.
c. The strength of the joint should not normally govern the strength of the structure, and its behaviour should not impede the development of the full strength of the adjoining member, making it one of the most important components in a frame.

Ease of construction and access for depositing and compacting concrete are the other prominent issues of joint design. Whenever earthquake and wind loads are the primary forces acting on a member, expert advice must be sought for the design of joints and their detailing. In the following, detailing of members subject to static loads alone is discussed.

Corner joints under closing loads   The behaviour of right-angle corner joints is fundamentally affected by the sense of loading. Moments that tend to close or open these joints will be dealt with separately. Tests reveal that the tension steel ratio in thin member joints may not exceed $1.2f_c'/f$. Adequate strength of such joints may be achieved if the following conditions are satisfied:

a. Tension steel is continuous around the corner (it should not be lapped within the joint).
b. Tension bars are bent to sufficient radius to prevent bearing or splitting failures under the bars. Nominal transverse bars placed under the bent bars are beneficial in this respect.
c. The amount of tension reinforcement is conservatively limited to $P = 6f_c'/f_y$, where f_c' and f_y are in the same unit. A typical detailing for achieving good performance is given in Fig. 10.24(a-c).

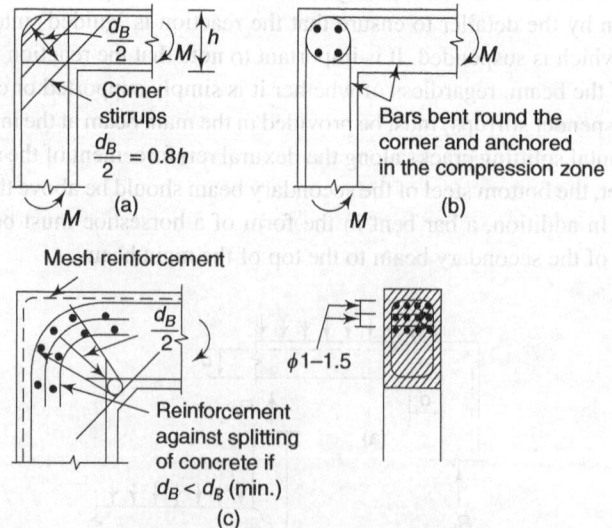

Note: Stirrups and other details not shown

Fig. 10.24 Detailing at the closing corners: (a) corner provided with stirrups, (b) main bars bent round and anchored in the compression zone, (c) corners provided with reinforcement against splitting of concrete

Corner joints under opening loads   When the applied moments tend to open a right-angled corner as shown in Fig. 10.25, the conditions are unfavourable. In these joints, secondary reinforcement to resist diagonal tension should be used as shown in Fig. 10.26(a-e).

Joints in multi-storey frames   Multi-storey buildings must be designed to withstand wind and/or earthquake loads. The behavior of joints under repeated loadings is a cause for concern.

Exterior joints   Considering stress resultants, it is apparent that diagonal tension and compression stresses (f_t and f_c) are induced in the panel zone of the joint. Diagonal tension may be high when the ultimate capacity of the adjoining member is developed, and this can lead to extensive diagonal cracking. The severity of diagonal tension cracking is influenced by the flexural steel content and the magnitude of axial compression load on the column. The important aspects to be considered are the following:

a. Bond performance of the bars anchored in the panel of the joint
b. Effect of repeated loading on internal cracking

The principles of detailing for opening and closing corners can be applied to a beam-column joint. The tensile reinforcement of a beam should be anchored effectively into the column, taking care that the radius of bend

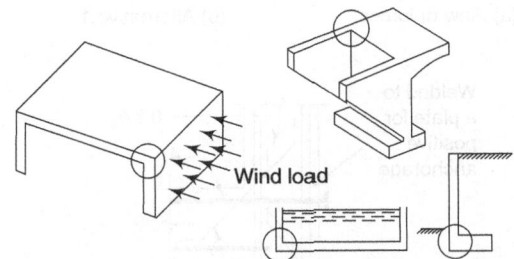

Fig. 10.25 Examples of continuous corners that may be subjected to opening loads

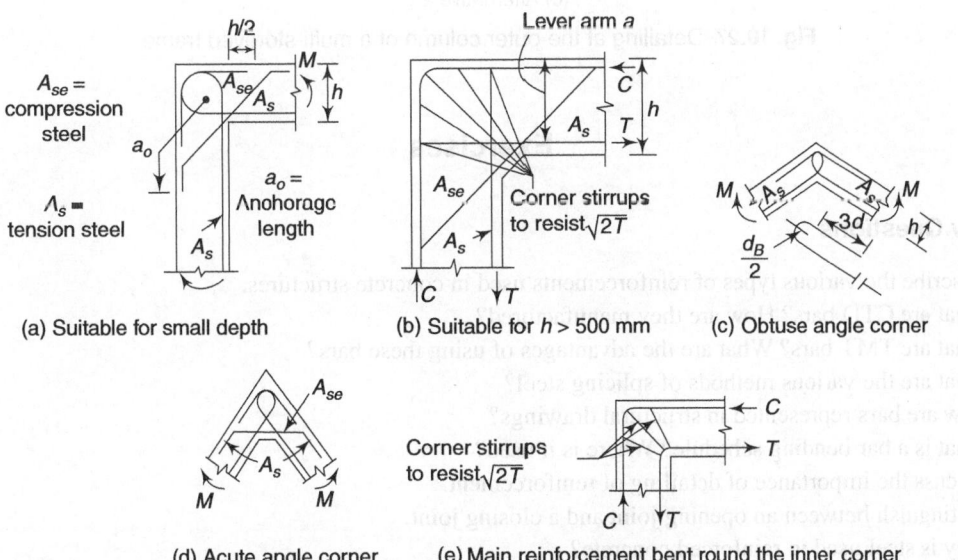

(a) Suitable for small depth   (b) Suitable for $h > 500$ mm   (c) Obtuse angle corner

(d) Acute angle corner   (e) Main reinforcement bent round the inner corner

Fig. 10.26 Detailing at the opening corners

is greater than the minimum required to prevent the splitting of concrete. Alternatively, the beam reinforcement can be provided with positive anchorage and additional bars in order to transfer the tensile forces from the reinforcement above the beam to that below the beam [Fig. 10.27(a)]. Inclined reinforcement is required at the upper corner in order to reduce cracking at the junction [Fig. 10.27(b) and (c)]. Excessive amount of beam steel is likely to create problems for anchorage.

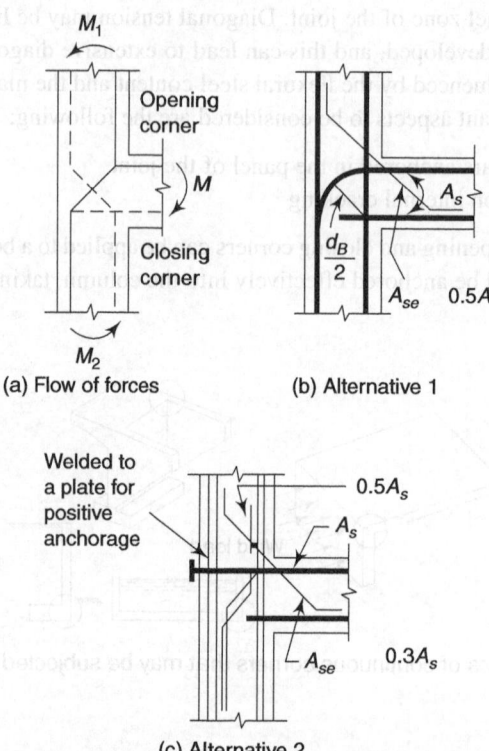

(a) Flow of forces   (b) Alternative 1

(c) Alternative 2

Fig. 10.27 Detailing at the outer column of a multi-storeyed frame

Exercises

Review Questions

1. Describe the various types of reinforcements used in concrete structures.
2. What are CTD bars? How are they manufactured?
3. What are TMT bars? What are the advantages of using these bars?
4. What are the various methods of splicing steel?
5. How are bars represented in structural drawings?
6. What is a bar bending schedule? Where is it used?
7. Discuss the importance of detailing of reinforcement.
8. Distinguish between an opening joint and a closing joint.
9. Why is steel used in reinforced concrete?
10. Distinguish between active and passive reinforcement.

Multiple Choice Questions

1. In RCC columns it is more economical to use
   a. Fe 250 steel
   b. Fe 415 steel
   c. Fe 500 steel
   d. Either Fe 415 or Fe 500 steel

2. In normal high-strength deformed bar without S or D classification, % elongation at failure should be
   a. 5%
   b. 8%
   c. 10%
   d. 14.5%

3. In high-strength deformed bars, ribs are introduced to improve
   a. Strength
   b. Fire resistance
   c. Corrosion resistance
   d. Bond between steel and concrete

4. To check bendability, after bending observations are made using
   a. A microscope
   b. A lens
   c. Naked eye with good vision
   d. A special optical gauge

5. In CTD bars, cold working is intended to
   a. Increase the yield stress
   b. Increase the ultimate stress
   c. Increase the ductility
   d. Increase the earthquake resistance

6. In Fe 500D, the letter D denotes that
   a. It is a special bar
   b. It is a ductile bar
   c. It is a seismic resistant bar
   d. It has better fire resistance

7. Cover blocks are used
   a. To ensure proper placement of bar in the formwork
   b. To ensure proper cover to steel
   c. To support the bar during concreting
   d. All of the above

8. The minimum % of high-strength steel that should be used in a slab is
   a. 0.12%
   b. 0.15%
   c. 0.20%
   d. 0.08%

9. The minimum % of steel to be used in a column is
   a. 1%
   b. 1.5%
   c. 0.3%
   d. 0.8%

10. Detailing of reinforcement is intended to
    a. Avoid wrong placement of steel
    b. Show the bar anchorages
    c. Show the location of the laps and bends
    d. All of the above

Answers to Multiple Choice Questions

1. c	4. c	7. d	10. d
2. d	5. a	8. a	
3. d	6. b	9. d	

CHAPTER 11

Corrosion

The main concerns in civil engineering are the climate and its effects, the environmental and atmospheric conditions, and the properties and behaviour of the metals and other materials used. Studies on the damage, deterioration, and durability of the materials used for construction attract the attention of civil engineers. Among the most pressing concerns for structural concrete durability is the corrosion of the steel reinforcement embedded in concrete.

Corrosion is defined as the destruction (or deterioration) of materials due to chemical (or electrochemical) reaction with the environment, and also the loss of steel due to the formation of rust. The corrosion of steel reinforcement is the depassivation of steel with reduction in concrete alkalinity through carbonation. Most materials undergo corrosion on exposure to natural environments (such as air, water, and soil) or other artificial environments (such as gases, liquids, and moisture). Hazards to human life and economic losses occur due to premature deterioration and destruction of steel in buildings, bridges, culverts, pipes, marine and offshore structures, towers, water supply and sanitary fittings, carpentry and electrical fittings, implants for the human body, etc.

Corrosion deteriorates concrete because the product of corrosion—ferric oxide (brown in colour)—occupies a greater volume (more than 2 to 10 times) than steel and exerts substantial bursting stresses on the surrounding concrete. The outward manifestations of rusting include staining, cracking, and spalling of concrete. The progress of the process of corrosion is generally a geometric progression with respect to time. Consequently, the cross-section of steel is reduced. With time, structural distress may occur either due to the loss of the bond between steel and concrete, or due to the cracking and spalling of concrete, or as a result of the reduced steel cross-sectional area. The latter effect can be of special concern in structures containing high-strength prestressing steel in which a small amount of metal loss could possibly induce a tendon failure. However, the process leading to the ultimate failure is slow, and normally gives years of warning to the maintenance engineer.

11.1 Corrosion of Reinforcement in Concrete

The corrosion of steel reinforcement causes an increase in the volume of oxidized compounds when compared with the volume of the base metal dissolved (Fig. 11.1). This increase in volume results in tensile forces leading to cracks in the concrete around the steel reinforcement and further accentuates the effect of the corrosive environment. Therefore, a more complete picture of the behaviour of the corroding steel in the reinforced concrete members would be desirable.

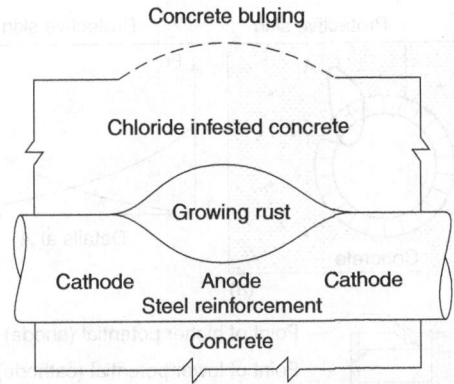

Fig. 11.1 Corrosion mechanism

Protection of reinforcement from corrosion is provided by the alkalinity of concrete, which leads to the passivation of steel. The formation of a protective skin around steel is shown in Fig. 11.2(a). The reserve of calcium hydroxide is very high, so there is no need to expect steel to corrode even when water penetrates to the reinforcement in the concrete. Owing to this, even the occurrence of small cracks (up to 0.1 mm in width) or blemishes in concrete need not necessarily lead to corrosion damage. However, environmental influences, and CO_2 in particular, reduce the pH value of concrete from 12.6 to 8.0, thus removing the passivating effect of alkalis. In conjunction with the existing humidity, a reduction in the pH value leads to the corrosion of reinforcement.

The presence of an electrical potential is a prerequisite for the occurrence of electrochemical corrosion [Fig. 11.2(b) and (c)]. The electrochemical potential may be created by any of the following:

- Differential aeration—difference in concentration of oxygen on the steel surface
- Differential ion concentration—metal ions, dissolved salts, and pH of concrete in the vicinity of steel may cause this
- Differential surface properties—Small blemishes on the surface of the reinforcement formed during rolling generally termed as mill scales or breaks in coatings, impurities in concrete, etc., may be responsible for this

The reinforcement in concrete has a passivating layer of gamma ferric oxide (Fe_2O_3). Any cracks that occur in the protective film are quickly repaired in the presence of sufficiently high hydroxyl ion concentration, forming ferrous hydroxide first and then cubic ferric oxide (Fe_3O_4) and gamma oxide (Fe_2O_3). The corrosion mechanism is shown in Fig. 11.1. There are two mechanisms by which the corrosion of steel can occur:

a. Direct ion oxidation
b. Electrochemical reaction

The latter is the more common mechanism. The reactions that take place are as follows:

$$Fe \rightarrow Fe^+ + 2e \ (\text{anodic reaction})$$
$$4e + O_2 + 2H_2O \rightarrow 4(OH) \ (\text{cathodic reaction})$$
$$Fe^{++} + 2(OH) \rightarrow Fe(OH)_2 \ (\text{ferrous hydroxide})$$
$$2Fe^{+++} + 6(OH) \rightarrow 2Fe(OH)_3 \ (\text{ferric hydroxide})$$
$$2Fe(OH) \rightarrow Fe_2O + H_2O$$

or

$$\text{At the anode}: Fe + 2(OH) \rightarrow Fe(OH)_2 + 2e$$
$$\text{At the cathode}: O_2 + 2H_2O + 4e \rightarrow 4(OH)$$

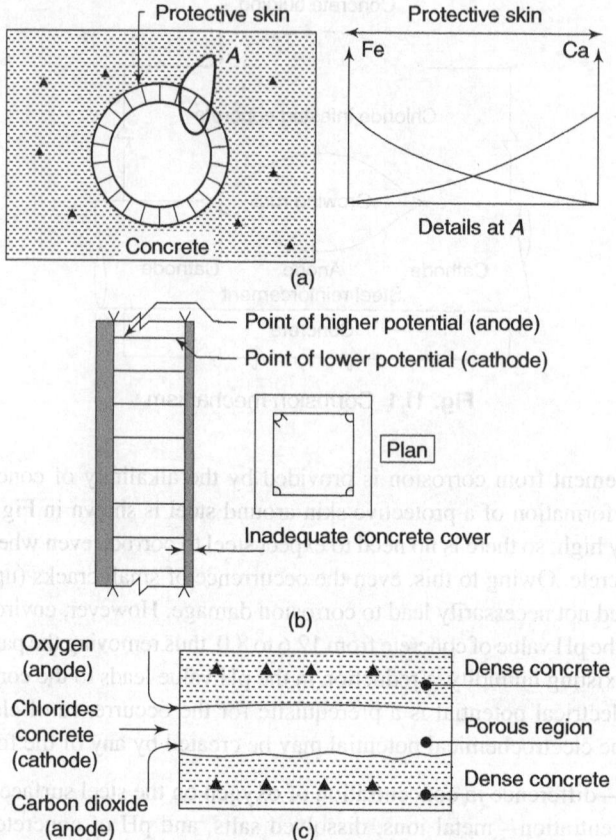

Fig. 11.2 (a) Formation of a protective skin around steel; (b) Typical case of potential difference in the ties of a column; (c) Typical case of potential difference in concrete

It is observed that for corrosion to occur and continue, oxygen and water are required. There is no corrosion below 30% relative humidity. At 70–80% relative humidity, the rate of corrosion is the highest. The chloride ion forms hydrochloric acid with water. This acid contaminates the concrete with chloride ions around the steel. The chloride ions present in the concrete surrounding the reinforcement ultimately destroy the passive protective film on the steel. Similarly, the formulation of $CaCO_3$ also destroys the passive environment by bringing the pH value down from 12.5 to 9.0 and less (Fig. 11.3). Due to this phenomenon, the steel is activated locally to form the anode, the remaining passive surface acting as the cathode. In such cases, localized corrosion in the form pitting occurs.

The schematic representation of corrosion due to chloride ions is shown in Fig. 11.4. The reactions that occur are:

$$Fe^{++} + Cl^- + H_2O \rightarrow Fe(OH)_2 + HCl$$

$$HCl \rightarrow Cl^- + H^+$$

The degree of protection against corrosion is provided by the pore fluid containing $Ca(OH)_2$. The pH of $Ca(OH)_2$ solution at 25°C is 12.53 at maximum solubility of $Ca(OH)_2$. The pH still remains at 11.27 when the concentration of $Ca(OH)_2$ is only 5.5% of the maximum.

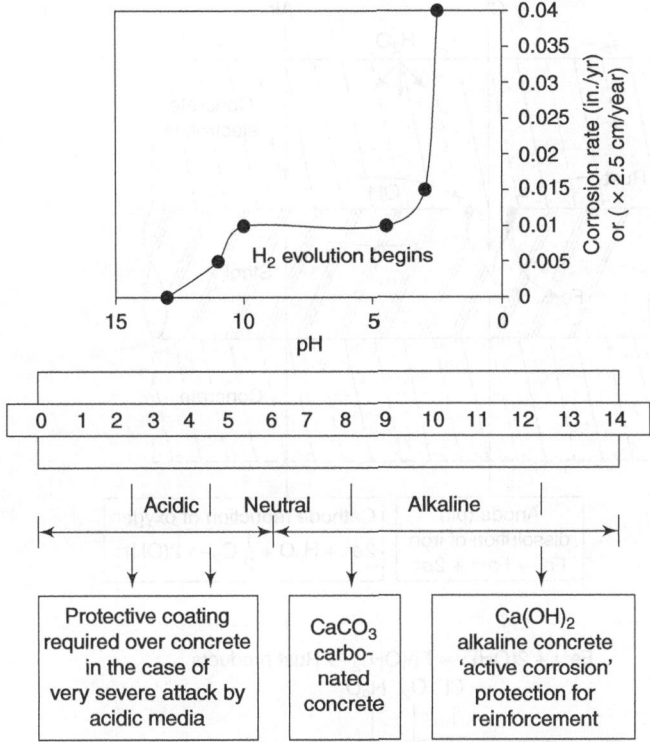

Fig. 11.3 Schematic representation of pH value

Pitting corrosion takes place due to chloride ions at a certain potential called the *pitting potential*. This potential is higher in dry concrete than at high humidity. As soon as a pit starts (Fig. 11.4), the potential in the neighbourhood of the reinforcement drops and a new pit is formed. Eventually, there may be a large-scale spread of corrosion, and general corrosion starts. Figure 11.5 shows the quantity of corrosion products versus time. Note that the deterioration during the propagation period is faster than that during the initiation period.

11.2 Factors Influencing Corrosion

The factors that generally influence corrosion of reinforcement in concrete structures are pH value, carbonation of concrete, chlorides and sulphates, moisture, oxygen, ambient temperature and relative humidity, severity of exposure, quality of construction materials, permeability of concrete, cover to reinforcement, initial curing conditions, formation of cracks, high carbon content in the reinforcement, high stress levels, inadequate grouting of prestressed tendons, rusted reinforcement prior to embedment, alkali–aggregate reaction, potential difference associated with liquid, contact with other metals, stray currents, and absence of periodical maintenance. The influences of some of the important factors are discussed in this section.

11.2.1 pH Value

The pH value of moist concrete is normally about 12.0, which is sufficient to passivate the reinforcement against corrosion. When it reduces to below 8.0, the carbonation of concrete takes place and, in turn, corrosion initiates. Figure 11.3 shows pictorially the relationship between the pH value, corrosion rate, and the acidic media, which encourages corrosion by destroying the protective layer around steel.

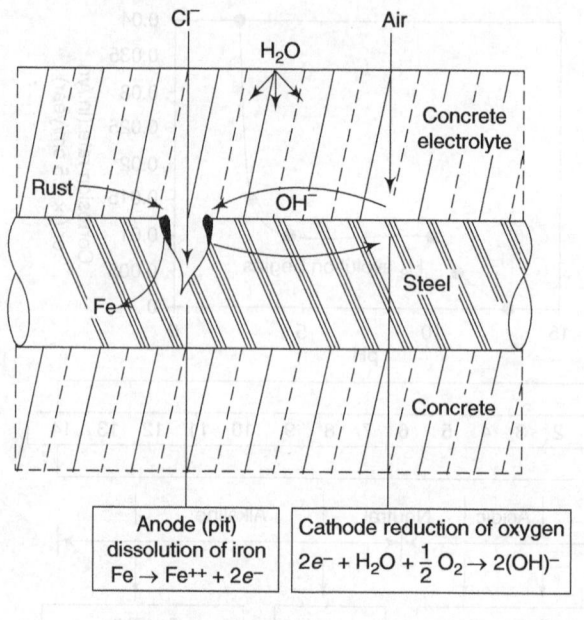

Fig. 11.4 Chloride-induced macro-cell corrosion of steel in concrete

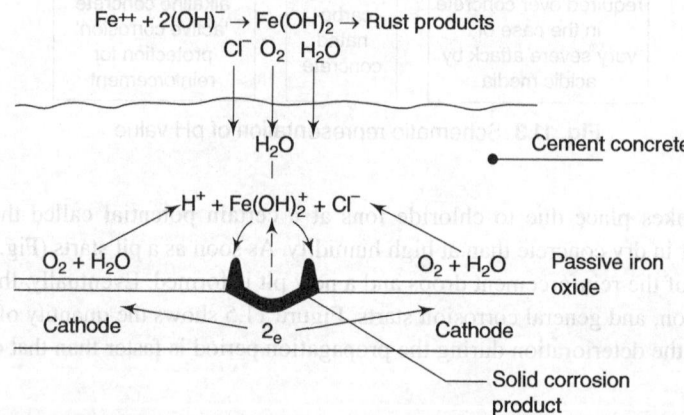

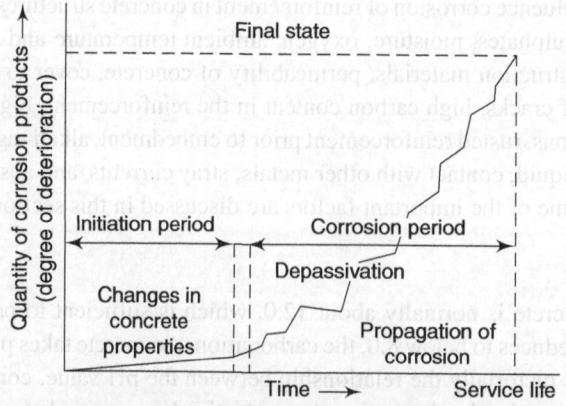

Fig. 11.5 Quantity of corrosion products versus time

11.2.2 Carbonation of Concrete

Carbonation is a process by which CO_2 converts free lime into $CaCO_3$ and water, thereby reducing the pH further. As the pH reaches about 9.0, the passivation of reinforcement is lost and corrosion starts based on the availability of oxygen and water. The penetration depth of this pH is generally called the *carbonation depth*. This process is shown in Fig. 11.6.

The rate of carbonation depends on the following parameters: (a) permeability of concrete, (b) CO_2 concentration in the air, (c) moisture in the gel and capillary pores, and (d) relative humidity of the atmosphere. As carbonation reaches the reinforcement, the passivating influence of concrete is lost and the reinforcement starts corroding in the presence of moisture and oxygen. Figure 11.7 shows carbonation time versus water–cement ratio, and Fig. 11.8 shows carbonation depth versus time. Note that carbonation time decreases with increase in water–cement ratio and with decrease in cover. Therefore, in order to increase the service life one should adopt a low water–cement ratio and adequate cover. Stirrups, which are closer to the outer surface, corrode faster than the main bar in a beam and hence it is advisable to provide excess steel for stirrups to avoid shear failure. The schematic process can be expressed as

$$Ca(OH)_2 + CO_2 + H_2O \rightarrow CaCO_3 + 2H_2O$$

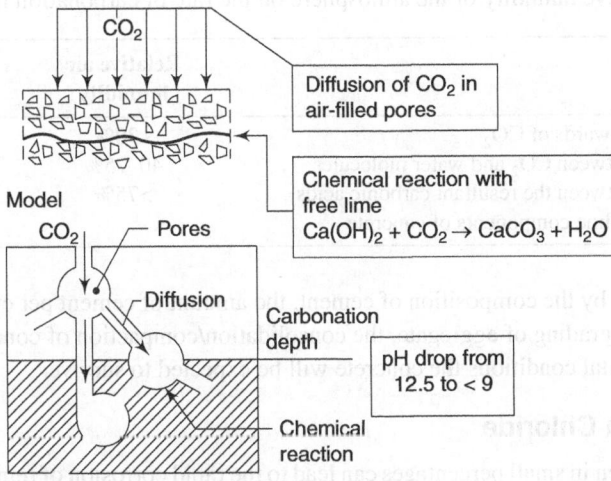

Fig. 11.6 Carbonation

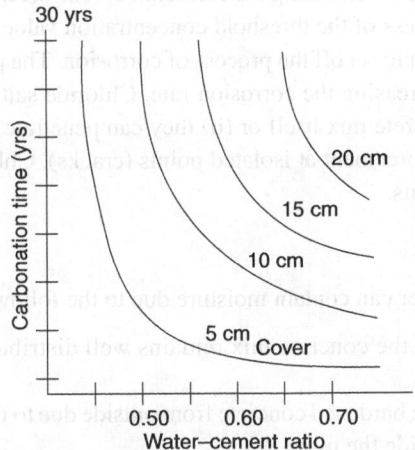

Fig. 11.7 Variation of carbonation time with water–cement ratio

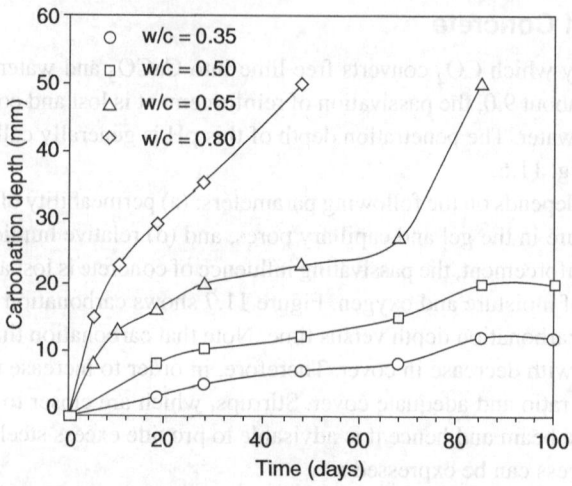

Fig. 11.8 Variation of carbonation depth with time

The influence of the relative humidity of the atmosphere on the rate of carbonation is shown below.

Phase	Process	Relative air humidity	Rate of carbonation
1	Diffusion inwards of CO_2	<30%	Low
2	Reaction between CO_2 and water molecules	40–75%	High
3	Reaction between the resultant carbonic acids and the alkaline components of concrete	>75%	Very high

Carbonation is controlled by the composition of cement, the amount of cement per cubic metre, the concrete mix design including the grading of aggregate, the consolidation/compaction of concrete, the curing of concrete, and the environmental conditions the concrete will be expected to survive.

11.2.3 Reaction with Chloride

The presence of $CaCl_2$ even in small percentages can lead to the rapid corrosion of reinforcement, as it reduces the electrical resistivity of concrete and helps to promote galvanic cell action. The presence of chlorides increases shrinkage cracks in concrete, further accentuating the corrosion of reinforcement in an aggressive environment. The ingress of chloride ions in excess of the threshold concentration value reduces the alkalinity of concrete and breaks down the protective film to set off the process of corrosion. The porosity and permeability increase the rate of penetration of ions, increasing the corrosion rate. Chloride salts can enter concrete in two ways: (i) they may be present in the concrete mix itself or (ii) they can penetrate into the hardened concrete where it is permeable and reach the reinforcement at isolated points (cracks). Chlorides induce pitting corrosion in steel reinforcement at isolated points.

11.2.4 Moisture

A reinforced concrete (RC) member can contain moisture due to the following reasons:

a. Water that is used for making the concrete mix remains well distributed and enclosed in the concrete mass.

b. Water that finds its way into the hardened concrete from outside due to the subsequent absorption of water which may circulate freely inside the mass.

11.2.5 Oxygen

The differential aeration cell, set up by the differential concentration of oxygen, sets off the process of corrosion. Thus, in RC members, the regions with the least oxygen concentration become anodic while those with larger oxygen concentration become cathodic.

11.2.6 Ambient Temperature and Relative Humidity

The rate of corrosion is directly proportional to the ambient temperature. The rate of carbonation increases with increase in temperature. The pH value limit below which corrosion is induced decreases with increase in ambient temperature. The rate of corrosion increases with increase in relative humidity, in the range 50–70%.

11.2.7 Severity of Exposure

The rate of corrosion is proportional to the severity of the exposure to the environment. Due to severe exposure, concrete in the cover region undergoes rapid deterioration and the reinforcement in turn loses its passivity and starts corroding.

11.2.8 Quality of Construction Material

The use of construction material contaminated with a significant level of chlorides/sulphates leads to the depassivation of reinforcement and sets off corrosion.

11.2.9 Quality of Concrete

The quality of concrete is one of the most common causes of early deterioration. Good quality concrete is dense and almost impermeable. This property prevents the easy access of external agents into the mass of concrete through the cover and thus to the reinforcement.

11.2.10 Cover to the Reinforcement

Lack of adequate cover contributes much to corrosion in an aggressive environment. A well-compacted and continuous, even if thin, cover of good quality concrete on reinforcement is sufficient to protect it from corrosion. The following are the required thicknesses of covers for various levels of exposure:

> For severe exposure: At least 50 mm thick
> For moderate exposure: At least 40 mm thick
> For mild exposure: At least 30 mm thick
> For normal exposure: At least 20 mm thick

11.2.11 Initial Curing Conditions

The process of corrosion depends on the permeability of the cover concrete and the initial curing conditions. If the curing is sufficient, then the impermeability of the cover concrete increases. High permeability of concrete to liquids and gases makes the alkalis in the hardened cement paste combine more or less completely with CO_2 and SO_2. This encourages corrosion of embedded steel.

11.2.12 Formation of Cracks

The formation of cracks is dangerous for protection against corrosion. Once concrete cracks, the external depassivating agents can penetrate deep into concrete and set off the process of corrosion.

Cracks running transversely to the reinforcement are less harmful than the longitudinal cracks along the reinforcement. Thus, in order to induce the process of corrosion and to keep it going, at least one of the following conditions must exist in any RC structure:

• Chloride ion concentration in excess of the threshold value at the interface of the reinforcement and concrete or sufficient advancement of the carbonation front to destroy the passivity of the ferric oxide surface layer on the reinforcement
• Adequate moisture in the concrete to facilitate the movement of chloride ions and provide a conduction path between the anodic and the cathodic areas on the steel
• Sufficient oxygen supply to the cathodic areas in order to maintain such areas in a depolarized condition
• Differences in electrochemical potentials at the surface of the reinforcement
• Low values of electrical resistivity of concrete
• Relative humidity in the range 50–70%
• Higher ambient temperature

11.3 Damages Caused by Corrosion

Corrosion of steel can be termed a 'creeping disaster' for a reinforced concrete structure. It begins with signs of weakness in the structure, progresses with time causing the structure to exhibit unsightly cracks, and finally causes the collapse of the structure. The following sections trace the events that lead to the collapse of a reinforced concrete structure due to corrosion.

11.3.1 Formation of White Patches

CO_2 reacts with $Ca(OH)_2$ in the cement paste to form $CaCO_3$. The free movement of water carries the unstable $CaCO_3$ towards the surface and forms white patches. This indicates the occurrence of carbonation.

11.3.2 Brown Patches along Reinforcement

When reinforcement starts corroding, a layer of ferric oxide is formed. This brown product resulting from corrosion may permeate from the steel surface along with moisture to the concrete's outer surface without cracking the concrete. This leads to brown discolouration on the surface of concrete.

11.3.3 Occurrence of Cracks

Increase in volume of corrosion products exerts considerable bursting pressure on the surrounding concrete, resulting in cracking. Hairline cracks on the concrete surface directly above the reinforcement and running parallel to it are positive visible indication that the embedded reinforcement is corroding. These cracks indicate that the expanding rust has grown enough to split the concrete.

11.3.4 Formation of Multiple Cracks

As corrosion progresses, multiple layers of rust are formed on the reinforcement, which in turn exert considerable pressure on the surrounding concrete, resulting in the widening of hairline cracks. In addition, a number of new hairline cracks are also formed. The bond between concrete and the reinforcement is considerably reduced. If this happens, there will be a hollow sound when the concrete is tapped at the surface with a light hammer.

11.3.5 Spalling of Cover Concrete

Due to the loss in bond between steel and concrete and the formation of multiple layers of cracking, the cover concrete starts peeling off. At this stage, the size of the bars is reduced.

11.3.6 Snapping of Bars

The continued reduction in the size of bars results in the snapping of the bars. This occurs in ties and stirrups. At this stage, the size of the main bars is also reduced.

11.3.7 Buckling of Bars and Bulging of Concrete

The spalling of the cover concrete and snapping of ties causes the main bars to buckle. This results in the bulging of concrete in that region. This is followed by the collapse of the structure. When the corrosion of reinforcement starts, the deterioration is usually slow but after the end of the initiation period, it advances very fast in geometric progression.

Corrosion can also cause structural failure due to reduction in cross-section of the main bar, and hence leads to reduced load-carrying capacity. It is possible to arrest the process of corrosion at any stage by altering the corrosive environment in the vicinity of the reinforcement.

11.4 Preventive Measures in New Construction

Preventive measures for controlling the corrosion of steel embedded in concrete focus on sound corrosion engineering principles directed towards the following aspects:

11.4.1 Design Factors

Corrosion engineering principles are based on the following design factors: low water–cement ratio, high-strength concrete, higher minimum cement content, thicker concrete cover, proper detailing of reinforcement, and moderate stress levels.

11.4.2 Construction Aspects

Adequate compaction of concrete, effective curing, production of impervious concrete, effective grouting of tendons, and periodical maintenance constitute sound engineering practices which must be implemented.

11.4.3 Reinforcement Protection

The reinforcement in new construction can be protected using the following:

a. Cement-based coatings
b. Galvanizing/zinc-based paints
c. Epoxy coating
d. Bitumen-based paints
e. Phosphate coatings

Coating for reinforcement bars must satisfy the following:

a. Ensure uniform coating on the deformed surface configuration of the bars
b. Be flexible enough to allow post-coated bending of bars
c. Be mechanically stable to sustain handling, transportation, and fabrication of reinforcement

d. Provide the facility of easy application
e. Resist corrosion

11.4.4 Surface Coating for Concrete

Any coating selected to be applied on the surface of concrete as a preventive measure against the corrosion of reinforcement must perform the following functions:

a. Control carbonation of concrete
b. Resist chloride penetration
c. Control moisture penetration of concrete
d. Provide supplementary protection in case of inadequate cover
e. Protect concrete from sulphate attack
f. Protect reinforcement from corrosion

In addition, any paint system used as coating for concrete should possess the following properties:

a. Adhesion to the surface of concrete
b. Alkali resistance
c. Abrasion resistance
d. Flexibility/viscosity
e. Weather resistance
f. Carbon dioxide diffusion resistance
g. Water vapour diffusion resistance
h. Water penetration resistance
i. Chloride/sulphate penetration resistance
j. Capable of breathing out moisture from inside

Based on these functional requirements and other properties, surface coating materials can be classified into bitumens, elastomers, polymers, silicones, silanes, and vegetable oils.

11.5 Tests for Existing Structures

Rehabilitation of RC structures that have been affected by corrosive environments can be achieved effectively if the factors influencing the durability of the structure and the extent of damage are investigated systematically and the various problems and causes afflicting the structure are isolated (see Fig. 11.9). The procedure for assessing the damage in such structures consists of the following steps:

11.5.1 Physical Inspection

Cracks appearing in concrete along the reinforcement, spalling of concrete, tilting of structures, excessive deflections, crack formation on the surface of concrete indicating carbonation, etc., can be relied upon to give sufficient data. During visual inspection, structures affected by corrosive environments should be differentiated from those showing deterioration due to structural inadequacy.

11.5.2 Inspection of Records

A study of the architectural and structural designs pertaining to the structure under investigation helps to isolate features which are more prone to corrosive environments. The structural drawings could help in arriving at the expected lifespan of the building apart from assessing the nature and extent of treatment required to rehabilitate the structure.

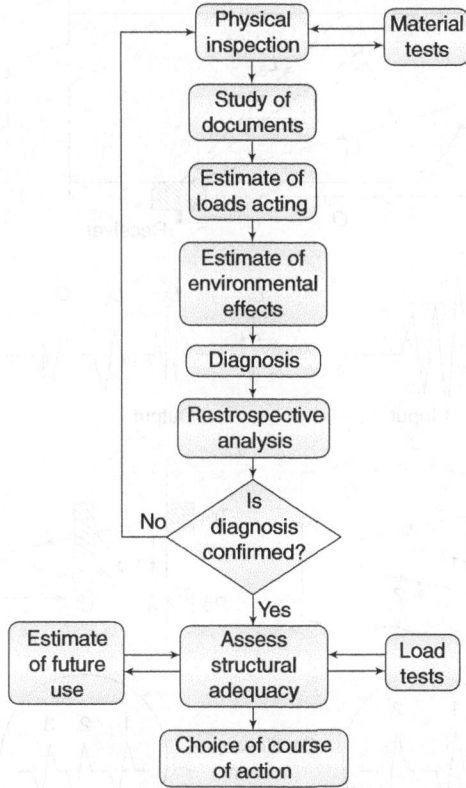

Fig. 11.9 Damage assessment procedure

11.5.3 Corrosion Monitoring Techniques

Non-destructive tests

Rebound hammer test   This test can be used to evaluate the compressive strength of the affected portion, and the result can be correlated with the designed strength of the elements of the structure.

Ultrasonic pulse-velocity test   This test can be used to investigate the extent and depth of deep-rooted cracks. In this test, an ultrasonic pulse is sent with the help of a transmitter and received on the opposite face with the help of a receiver. The measurement of the velocity of sound can indicate the path taken by the sound wave due to the presence of cracks in the structure and the location/depth of the crack can be monitored on a cathode ray oscilloscope. Figure 11.10(a) shows the effect of internal cracking on the display obtained in an oscilloscope screen. Note that in a cracked specimen, the number of peaks of the wave has increased from two to three. The effect of air voids or pores on the waveform is shown in Fig. 11.10(b). The damaged specimen has a spread-out waveform.

Potential measurement   Under favourable conditions, it is possible to use electrical measurements from the surface of the concrete to get an indication of the condition of steel inside. Electrical connections have to be made with a rebar by making a small hole in the concrete. However, in porous concrete such a hole may not be required, as wetting by the electrolyte itself may establish the required electrical contact with the rebar. The electrical connection with concrete is made by means of an electrolyte which wets both concrete

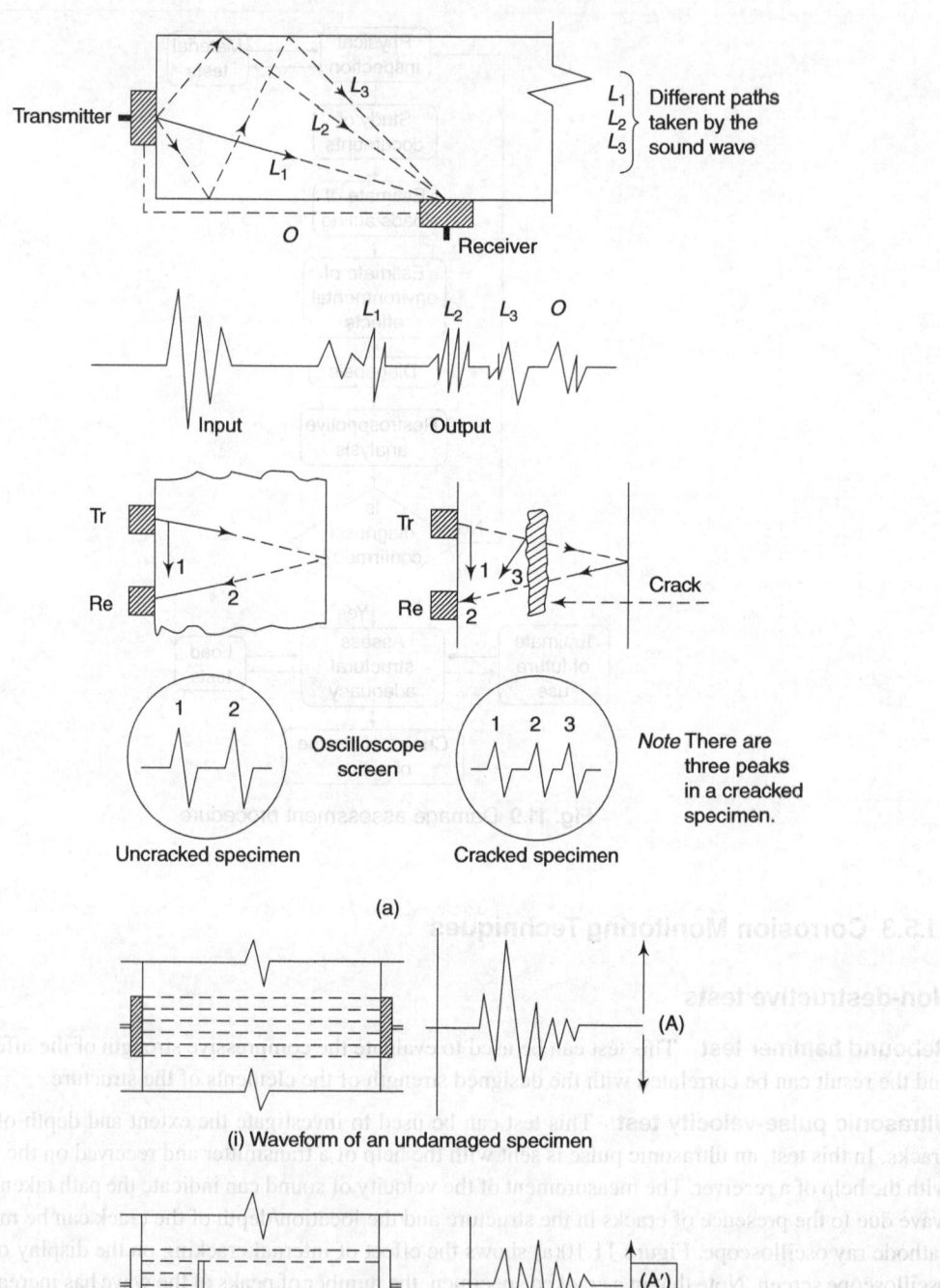

(a)

(i) Waveform of an undamaged specimen

(ii) Waveform of a damaged specimen

(b)

Fig. 11.10 Ultrasonic test on concrete: (a) effect of internal cracking and (b) effect of damage (crushing or voids)

and the conductor, to which another wire is attached. The electrolyte and conductor are made into a probe which can be held against the surface of concrete. Figure 11.11 shows the half-cell potential test set-up. The following three electrodes are used:

a. Saturated calomel electrode
b. Silver/silver chloride electrode
c. Copper/copper sulphate electrode

The probability of reinforcement corrosion is assessed based on open circuit potential (OCP) values based on half-cell potential measurements made using the equipment

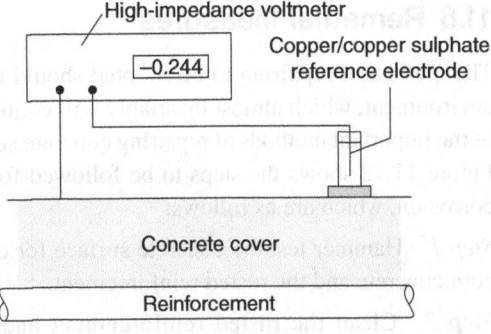

Fig. 11.11 Half-cell potential test set-up

shown in Fig. 11.11. The detection and assessment of corrosion in concrete structures is essential. When standard calomel electrode is used, an OCP value of < -500 indicates severe corrosion condition in the structure. A value between -500 and -350 indicates high risk, and a value > -350 but < -200 signifies intermediate and a value > -200 low corrosion risk.

The corrosion process of steel in concrete can be followed using several electrochemical techniques. Monitoring of OCP is the most typical procedure to the routine inspection of reinforced concrete structures. Its use and interpretation are detailed and explained in the ASTM C876 'Standard Test Method for Half-Cell Potential of Reinforcing Steel in Concrete' code.

However, in combination with resistivity measurements, contours of equipotential maps help in assessing corrosion. In this technique, the corrosion cell ratio is determined as follows:

$$\text{Corrosion cell ratio} = \frac{\text{Maximum difference in potential between the cathodic and anodic regions}}{\text{Average electrical resistivity in the anodic region}}$$

Active corrosion in the anodic region is probable in cases where this ratio is not less than 5.0.

Destructive tests

In destructive tests, core samples are obtained from the affected part with the help of core cutting machines, and the compressive strength is determined. The compressive strength of the core, after appropriate modifications to account for size effects, can be compared with the standard cube results of the originally laid concrete after allowing for the age factor.

Testing reinforcement

Specimens of exposed rebars can be taken and tested for metallurgical properties and tensile characteristics.

Carbonation test

The carbonation test is conducted to evaluate the depth up to which concrete has become virtually acidic. The following method is adopted:

1. A pH indicator solution of phenolphthalein in dilute alcohol is sprayed on a fresh concrete surface.
2. The pH indicator changes colour according to the alkalinity of the concrete.
3. As the pH value decreases from 10.0 to 8.2 and below, the indicator colour changes from dark pink to colourless, indicating the condition of concrete. Note that the carbonated concrete will be colourless.

11.6 Remedial Measures

The method of repairing to be adopted should be specifically suited to arrest further corrosive action of the environment, which almost invariably will continue to ravage the structure even after its rehabilitation. Some of the important methods of repairing concrete structures damaged by corrosive environments are given below. Figure 11.12 shows the steps to be followed for repairing a structural member that has deteriorated due to corrosion, which are as follows:

Step 1   Hammer test the concrete surface for cavities and chisel all the loose portions to expose the sound core concrete and the rusted reinforcement.

Step 2   Clean the rusted reinforcement and exposed concrete surface by sand blasting or by using mechanical devices.

Step 2(a)   If the rusted reinforcement exhibits a reduction in area of reinforcing bar by more than 10%, weld additional steel on either side of the loss, taking care to see that the weld length (L_d) on either side is adequate based on the diameter of the bar.

Step 3   Apply two coats of mineral-based polymer-modified corrosion-inhibiting primer.

Step 4   Apply two-component polymer-based bond coats with zinc-rich primer.

Step 5   Apply polymer-modified, ready-to-use mortar of low permeability or gunite with guniting aid.

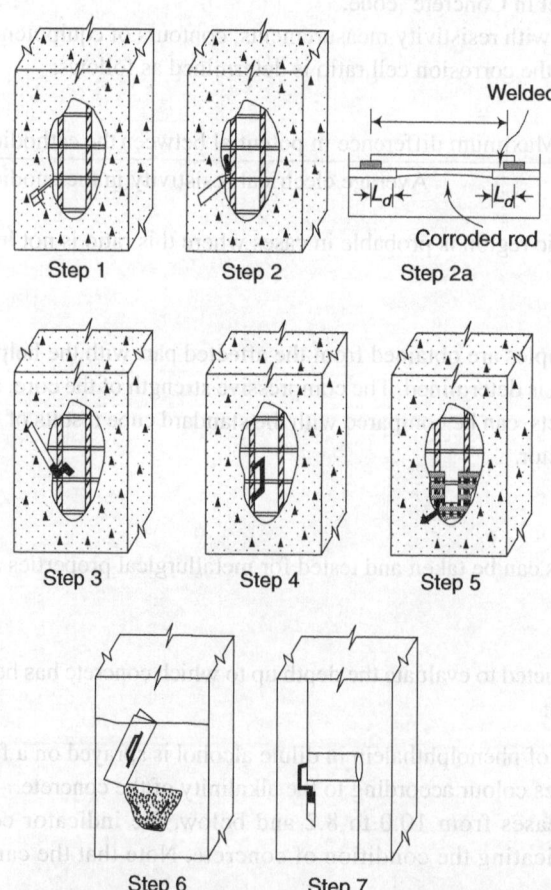

Fig. 11.12 Steps for repairing the damaged portion

Step 6   Apply carbonation-resistant, polymer-modified, ready-to-use fine mortar over the whole surface.

Step 7   Apply a final protection coat to concrete if required.

While dismantling carbonated concrete, the steel should be exposed completely all around the rebar as shown in Fig. 11.13.

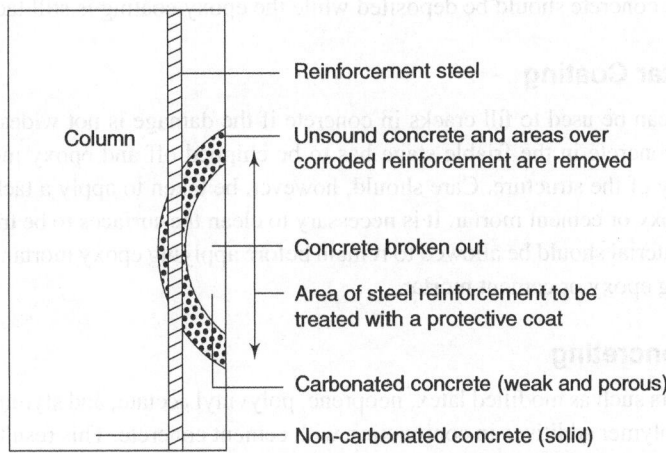

Fig. 11.13 Dismantling—Completely exposing the steel around the rebar

11.6.1 Cement Grouting

High-pressure cement grouting to seal minute cracks and voids has been practised for a long time. Low-pressure gravity grouting can be very effectively used in making permeable concretes impermeable, thereby increasing their resistance to corrosive environments. This is used as a repair technique to treat superficial or deep-rooted cracks in concrete developed due to aggressive environments. For treatment of wide cracks, medium-pressure grouting, with the help of a grease gun or any other suitable air pressure gun, can be used. Cement grout is used in the ratio of 1:2 (1 portion cement: 2 portions water). This is suitable for the early stages of deterioration, when cracks have just started appearing and concrete has not yet been reduced to a friable mass. This technique ensures the reinstatement of the alkaline environment around the steel reinforcement, and helps to restore the passivating film around the steel reinforcement, inhibiting further rusting. Cement grouting imparts the following properties to the restored concrete:

1. A high degree of structural adequacy with respect to the environmental conditions
2. Improved durability
3. Increased functional lifespan in the presence of aggressive environments at low costs

11.6.2 Polymer Grouting

Polymer grouting is a preventive technique used to make concrete more durable. Polymer compounds such as polyester resin, styrene, and polypropylene are used during surface impregnation to make concrete impermeable. The surface to be impregnated has to be completely dry and thoroughly cleaned. Vacuum is then created by sucking air out of the pores and polymer grouting is achieved under pressure. The polymer has to be cured with hot water to achieve polymerization.

11.6.3 Epoxy Grouting

Epoxy grouting is especially used in treating deep-rooted cracks.

11.6.4 Epoxy Coating

Epoxy coating is highly effective against corrosion of concrete, used advantageously as a rehabilitation or repair technique in conjunction with other methods of repair suiting the damaged structure. It can be applied with brushes or specially designed spray guns. Whenever epoxy coating is used in conjunction with guniting (or) new concreting, the epoxy-treated surface should be properly roughened with emery paper to improve the bond of the coat, or the concrete should be deposited while the epoxy coating is still tacky.

11.6.5 Epoxy Mortar Coating

Epoxy mortar coating can be used to fill cracks in concrete if the damage is not widespread. If the damage is extensive, then the concrete in the friable stage has to be chipped off and epoxy mortar coating applied to reinstate the integrity of the structure. Care should, however, be taken to apply a tack coat of epoxy resin before applying the epoxy or cement mortar. It is necessary to clean the surfaces to be treated thoroughly and no loose or unsound material should be allowed to remain before applying epoxy mortar. The tack coat should be tacky while applying epoxy or cement mortar.

11.6.6 Polymer Concreting

In this method, polymers such as modified latex, neoprene, polyvinyl acetate, and styrenebutadiene or acrylic polymers are used as polymer additives in conjunction with cement concrete. This results in very high imper-meability and high acid resistance, thereby making polymer concrete a very suitable material for counteracting corrosive environments. This method can be used as a repair method to replace the affected concrete.

11.6.7 Shotcreting or Guniting

One of the most extensively used, and by far the most successful, methods of repair and rehabilitation of concrete structures is guniting. The repairs are carried out after completely chipping off the damaged concrete from the surface and augmenting the reinforcement wherever necessary. A cement–sand mix, in the ratio 1:2 or 1:3 depending upon the requirement, is applied at high pressure over the surface with the help of specialized equipment. Guniting is extensively used to rehabilitate concrete bridges, dams, spillways, buildings, marine structures, etc. This method is quite costly and cannot be used in isolated locations.

11.6.8 Cement Concrete Jacketing

An affected old reinforced concrete member can be repaired by providing a 100–150 mm thick new concrete layer all around. The new layer may contain reinforcement as per requirement. Such a technique of providing an additional layer of concrete with reinforcements is known as 'jacketing' a member by cement concrete. Two different approaches can be used for cement concrete jacketing.

First approach Epoxy mortar coating is applied to the old concrete with proper tack coats or even epoxy painting resorted to at times before bonding new concrete to it. The old concrete is cleaned very thoroughly with water jets, air pressure, and/or sand blasting. Next tack coat/epoxy paint is applied. New concrete having very low water–cement ratio (not exceeding 0.45) is laid.

Second approach High-grade new concrete of rich proportions [e.g., 1:1:2 (1c:1s:2ca)] and low water–cement ratio is deposited over old concrete with the help of proper formwork. The old surface is cleansed of loose materials with the help of water jets, air pressure, and/or sand blasting. It is then copiously sprinkled with cement slurry of low water–cement ratio. If the new concrete is not very thick, no new cracks are likely to be seen at the junction of the old and new concretes and delamination is not likely to be experienced. Any delamination or presence of cracks can be detected by tapping the surface with a hammer and checking for a hollow sound. This technique is highly cost effective, especially in isolated coastal regions, where pressure guniting may not be feasible.

A coat of epoxy or a water-repelling, silicon-based agent such as sodium methylsiliconate or ethylsiliconate can be applied to keep the structure safe from the effect of aggressive environments. A coat of sodium silico-fluoride, which gives a high degree of impermeability to the rejuvenated structure, can also be used.

11.6.9 Modification of the Environment

The methods available for rendering the environment less corrosive include the removal or elimination of harmful constituents in the pores of the concrete around the steel. Its constituents may be in the form of chemicals such as water and chlorides, gases such as oxygen and hydrogen sulphide, and electrical currents. Water can often be eliminated by facilitating drainage away from rather than through a structure.

Chlorides can be eliminated by a process known as *electrochemical chloride removal*. This process has been used to decontaminate structures such as bridge decks in research studies. Harmful gases such as oxygen and hydrogen sulphide can be eliminated using chemical processes from the electrolyte in the pores making the pore fluid less corrosive. This method is predominantly applied to structures exposed to aqueous solutions.

Deep polymer impregnation of the critically contaminated concrete around the reinforcing steel is also being practised. The method ties up the existing contaminants and prevents the intrusion of additional contaminants. Although it is promising, the practicality and economic feasibility of deep polymerization in concrete structures is restricted to special applications, such as underwater structures.

11.6.10 Cathodic Protection

The electrochemical method of corrosion prevention refers to cathodic and anodic protection. These are applicable to systems where a continuous electrolyte exists. Soil, water, and chemical solutions form conducting paths and constitute a continuous electrolyte during electrochemical protection. Figure 11.14 illustrates cathodic protection. Note that the sacrificial anode gets 'consumed' during protection and hence requires to be replaced periodically.

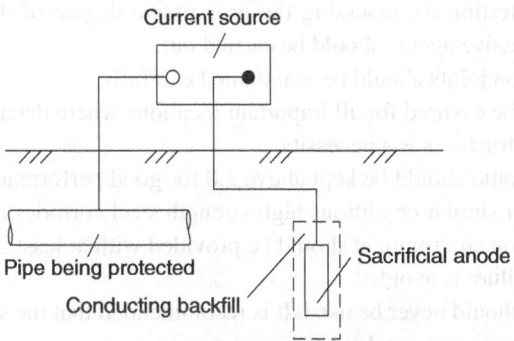

Fig. 11.14 Cathodic protection

11.6.11 Development of an Accelerated Test

Corrosion measurements using the half-cell $Cu/CuSO_4$ were made on buildings of different ages and compared with laboratory-accelerated test results after 90 days. Though an exact correlation could not be made, the laboratory and field readings indicate the same trend when laboratory results at 90 days are compared with 25-year-old buildings.

Alternate wetting and drying

After curing for 28 days, the specimens are put into 3% NaCl solution. They are kept in the salt solution for 12 hours for wetting and then kept outside for the next 12 hours for drying. This cycle is repeated continuously for 3 months.

Galvanostatic method

In this method, a specimen of the reinforcement embedded in concrete is taken as the anode. Another mesh or stainless steel is taken as the cathode. The specimens are then immersed in 3% NaCl solution and a direct current of 1.5 A is passed through the specimen. In a relatively short time, a sufficient quantity of electricity flows due to the current density, so that the effects of corrosion are measurable. This induces accelerated corrosion.

Gravimetric method

At the end of the study period, the embedded steel is dismantled, cleaned, and weighed. The actual weight loss indicates the extent of corrosion.

Pull-out test

The strength required to pull out the rebar from the specimen at the end of the study period indicates the extent of corrosion when compared with the virgin pull-out strength.

Based on a critical review and analysis of the various factors that influence the durability of reinforced cement concrete structures, the following points should be emphasized during the design, construction, and maintenance stages to enhance the life of important RC structures.

- At the planning, designing, and construction stages, the environment should be accounted for while choosing the mix proportions, so that the concrete is able to perform the service it has been designed for.
- The use of dense concrete (above grade M20) and concrete with adequate cover over reinforcement should be ensured.
- The congestion of reinforcement at joints must be avoided by proper detailing.
- The use of coated steel as reinforcement for severe exposure conditions is recommended.
- Periodic non-destructive testing for assessing the progressive degree of damage under the cumulative deleterious effect of aggressive agents should be carried out.
- Construction and expansion joints should be maintained carefully.
- Inspection facilities must be ensured for all important locations where deterioration is likely.
- Periodic maintenance of structures is a necessity.
- Cover (c) to diameter (d) ratio should be kept above 2.0 for good performance.
- It must be noted that under similar conditions high-strength steel corrodes more than mild steel.
- Beams exposed to corrosion environment should be provided with at least 25% excess shear capacity, so that unfavourable shear failure is avoided.
- Water–cement ratio >0.5 should never be used. It is recommended that the water–cement ratio be limited to 0.4 for important structures in coastal areas.

11.7 Laboratory Testing for Assessment of Corrosion

For measurement of the corrosion rate of reinforcing steel in concrete, many electrochemical and non-destructive techniques are available for monitoring corrosion of steel in concrete structures.

11.7.1 Tests on Concrete

Rebar corrosion on existing structures are assessed by the following methods:

- Open circuit potential (OCP) measurements
- Surface potential (SP) measurements
- Concrete resistivity measurements

- Linear polarization resistance (LPR) measurements
- Tafel extrapolation
- Galvanostatic pulse transient method
- Electrochemical impedance spectroscopy (EIS)
- Harmonic analysis
- Noise analysis
- Embeddable corrosion monitoring sensor
- Cover thickness measurements
- Ultrasonic pulse velocity technique
- X-ray, gamma radiography measurement
- Infrared thermograph electrochemical method
- Visual inspection

Many of these tests are described in Chapter 27. However, the Embeddable Corrosion Instrument (ECI) monitoring sensor is of particular interest for monitoring because it provides real-time information on structural condition. Figure 11.15 shows the application of an ECI sensor in an RCC structure by monitoring five key factors: linear polarization resistance, open circuit potential, resistivity, chloride ion concentration, and temperature.

It also communicates these through a digital network. Thus ECI provides comprehensive, real-time information on structural conditions.

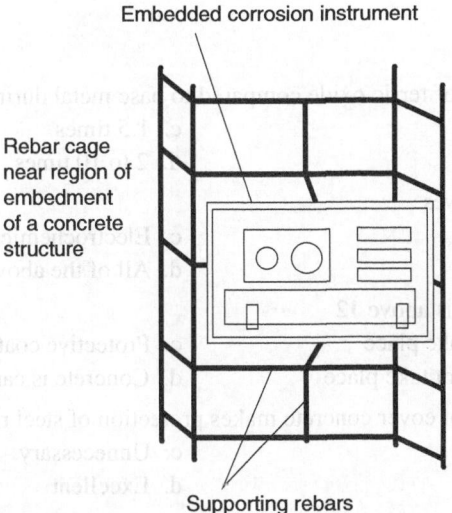

Embedded corrosion instrument

Rebar cage near region of embedment of a concrete structure

Supporting rebars

Fig. 11.15 ECI sensor during operation

11.7.2 Tests on Steel Reinforcement for Weight Loss

In the laboratory gravimetric methods are used to assess the metal loss in rebars. Due to corrosion, the bar weight reduces and also the bond deteriorates. These can be assessed and the influence on service life evaluated. The loss in weight of steel due to corrosion is evaluated by liquid displacement technique shown in Fig. 11.16. In this figure the lost metal embedded in concrete is seen. The white arrows indicate the diffusion of fluid sent in through the pores in the cover concrete. The diffused liquid migrates and occupies the corrosion pit as shown by the elliptical liquid drop. The amount of metal lost is evaluated as a function of liquid displaced which is assessed based on the fluid sent in by pressure.

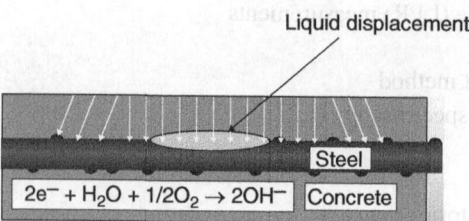

Fig. 11.16 Weight loss measurement by liquid displacement

Exercises

Review Questions

1. Describe the causes for corrosion of steel in concrete.
2. What are the factors which influence corrosion?
3. How is the depth of carbonation tested?
4. What preventive measures will ensure good protection for new structures?
5. What is cathodic protection and when is it applied?
6. Describe the various remedial measures and their relative efficiency against corrosion of steel in RC structures.

Multiple Choice Questions

1. The increase in the volume of ferric oxide compared to base metal during corrosion is
   a. 1.1 times
   b. 1.25 times
   c. 1.5 times
   d. 2 to 10 times

2. Corrosion mechanism involves
   a. Chloride infestation
   b. Chemical action
   c. Electrochemical action
   d. All of the above

3. When pH value of concrete is above 12
   a. Corrosion of rebar will take place
   b. Corrosion of rebar will not take place
   c. Protective coating of steel is required
   d. Concrete is carbonated

4. Larger water–cement ratio of cover concrete makes protection of steel reinforcement against corrosion
   a. Better
   b. Worse
   c. Unnecessary
   d. Excellent

5. A half-cell potential (mV) more negative than (−)350 mV indicates that corrosion of steel is
   a. Uncertain
   b. Certain 90%
   c. Certain 10%
   d. None of the above

6. Carbonation test on concrete is conducted using
   a. Carbon tetrachloride
   b. Phenolphthalein
   c. Sodium chloride
   d. Ammonium nitrate

7. Fusion-bonded epoxy coating to protect steel
   a. Reduces the bond and hence cannot be used
   b. Increases the bond and hence can be used
   c. Does not affect the bond and hence can be used
   d. Can be judiciously used though it decreases the bond

8. Cathodic protection involves
   a. A sacrificial anode
   b. A current source
   c. Continuous monitoring
   d. All of the above

9. The cover used for RCC column is
   a. 15 mm
   b. 20 mm
   c. 25 mm
   d. 40 mm

10. For protecting steel reinforcement against corrosion, the water–cement ratio for concrete should be less than
   a. 0.3
   b. 0.4
   c. 0.5
   d. 0.6

Answers to Multiple Choice Questions

1. d
2. d
3. b

4. b
5. b
6. b

7. d
8. d
9. d

10. b

CHAPTER 12

Durability

This chapter discusses important durability problems that arise during the service life of a concrete structure. These durability problems could be a result of improper design, execution, or specification at the time of tendering for the work. There is no material which is 100% resistant to chemical action and deterioration due to physical actions such as abrasion or impact. Concrete deteriorates with age and use. However, under normal conditions, good quality concrete has a long life. Concrete made with natural pozzolans has been in use for more than 1000 years.

Concrete gets deteriorated due to external agencies such as sulphates and chlorides, aggressive coastal environment, and cold weather freeze–thaw cycles. Internal factors such as impurities and alkali–aggregate reaction also affect concrete. These problems relate to durability issues.

Often concrete is associated with not only reinforcing steel but also other metals such as copper and aluminium. These metals are sometimes used as inserts. Corrosion of embedded metals leads to cracking and disintegration which is also a serious problem. This defect is mainly due to poor quality of cover concrete. Because of this problem, the concrete structure becomes unserviceable and over a period of time may collapse.

This chapter considers the durability problems of concrete under various environmental conditions, as well as the behaviour of concrete under high temperature caused by fire. However, this chapter does not discuss failures or problems associated with overloading or errors in design or construction.

12.1 Durability and Impermeability

Two important aspects to be considered with respect to concrete are durability and impermeability. For example, concrete used for a water tank should be impermeable as well as durable, whereas for building structures durability is more important. It should also be noted that impermeability of concrete leads to improved durability.

Durability may be defined as the property required for fulfilling all the service requirements of a structure during its intended life with the expected periodic maintenance. Note that maintenance is a must for all concrete structures, and it is incorrect to over-specify the requirements during design and construction in order to avoid proper maintenance during the service life of the structures. It will be an over-expectation to assume that a concrete structure will be in a 'new' condition throughout its service life without maintenance. Concrete, in fact, has the inherent ability to sustain all designed loads and environmental effects without damage. However, it is important to protect it against chemical or physical attack.

The low durability that we witness in many concrete structures, which gets manifested in the form of cracking and spalling, is principally due to inferior design, specification, or construction.

Concrete structures are not made with only concrete. In addition to reinforcing steel, structures are also made with a number of metallic and non-metallic constituents, which may need replacements more than once during their service life. Such replacements have to be planned for even at the time of the original design and construction. Certain parts of a concrete structure may also be subjected to physical wear and tear. Parking garages, concrete roads, and breakwater walls are examples of structures subjected to repeated wear.

In cold climates and in mountainous regions, the physical effects of freeze–thaw cycles are inevitable. These call for special designs of concrete mixes using air-entraining agents and vigilant inspection at regular time intervals. Hence, a regular maintenance schedule becomes imperative in order to maintain the integrity of concrete structures.

Concrete possesses a pore structure. It is this pore structure which distinguishes concrete from metals which can be made airtight or watertight. The capillary pore structure of concrete allows the permeation of gases or liquids, especially under pressure. The macrostructure of concrete reveals that it consists of (a) coarse and fine aggregates, (b) hydrated cement paste, and (c) entrapped air voids. The macrostructure also reveals visible cracks in the hydrated cement paste and aggregates, mainly due to the volume changes caused by shrinkage, settlement, and expansion/contraction due to temperature.

A closer look at the microstructure of concrete (Fig. 12.1) reveals that its hydration products consist of C–S–H gel, ettringite crystals, and monosulphate, and its gel pores, capillary pores, and entrapped and entrained air voids contribute to its porosity. In addition to the aforementioned characteristics, the interface between the aggregate and the surrounding cement paste known as the transition zone is weak and porous.

The permeability of concrete depends on its pore structure. Even the best of concrete is not airtight or watertight unless the pores are closed somehow. The capillary pore structure of concrete allows water to penetrate at a slow rate. The pressure in most liquid-retaining structures is about 10 to 12 m head of water

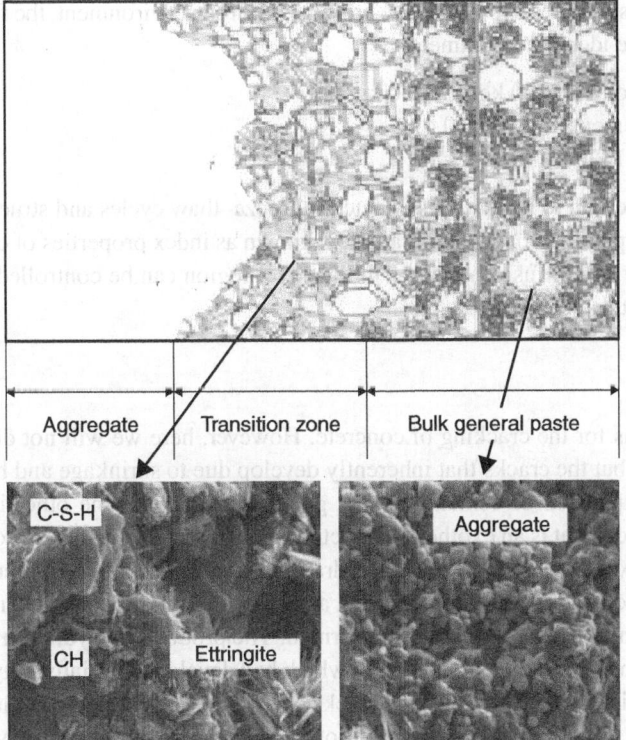

Fig. 12.1 Microstructure of concrete

(about 1.0 to 1.2 atm). This slow permeation does not reduce the alkalinity in well-made concrete. Research by the US Navy on hollow reinforced concrete spheres submerged at a depth of 1000 m in sea showed little seepage into the spheres, as reported by Green and Perkins (1980). The factors governing the permeability of concrete can be summarized as follows:

a. Quality of constituent materials, i.e., cement, sand, and aggregates
b. Quality of the pore structure, which is based on the water–cement ratio, the admixtures used, and the degree of hydration
c. Quality of the interfacial transition zone
d. Degree of compaction
e. Cracking arising due to different causes (structural as well as non-structural)
f. Adequacy of curing

It is incorrect to mention that all concretes produced for various applications should be impermeable. Depending on functional use, different extents of permeability can be permitted in the specification. Again, it is difficult to enforce permeability requirements in the field based on permeability values or on the coefficient of permeability of concrete.

This is equivalent to making a permeability test mandatory, which cannot be carried out in the field. The test carried out in the laboratory may not be 100% relevant to a structure in the field. Thus, other parameters which control permeability are fixed. The following three parameters are included in the specification for concrete work, to ensure the required impermeability of concrete for the job on hand:

a. Minimum cement content
b. Maximum water–cement ratio
c. Minimum cube strength

For general engineering structures exposed to the present-day urban environment, the following limiting values are recommended for the identified parameters:

a. Minimum cement content: 360 kg/m^3 (400 kg/m^3)
b. Maximum water–cement ratio: 0.5 (0.4)
c. Minimum strength: M25 (M30)

The values in brackets pertain to structures subjected to freeze–thaw cycles and structures in coastal regions.

However, for very important works, the parameters known as index properties of concrete which are associated with permeability and diffusion of gases in the cover region can be controlled. We will discuss about these properties later in this chapter.

12.1.1 Cracking

There are several reasons for the cracking of concrete. However, here we will not discuss cracks that occur due to defective design, but the cracks that inherently develop due to shrinkage and heat of hydration.

The coefficient of expansion of cement paste is greater than that of the other ingredients of concrete. Moreover, the setting of cement is an exothermic reaction. The heat of hydration is proportional to the cement content; hence richer mixes develop more heat of hydration and give rise to more extensive cracking. To avoid excessive heat of hydration and consequent cracking, a maximum limit for cement content has been specified. Initially, due to evolution of heat a few cracks are formed. The products of hydration get formed later during curing. These products may heal some of the cracks which had developed initially. A typical pattern of shrinkage cracks is shown in Fig. 12.2. Normally such cracks occur on the surface and are just skin deep. However, in mass concreting work, the effect of heat of hydration is considerable and some preventive measures should be taken to avoid such cracks. These aspects are discussed in Chapter 19.

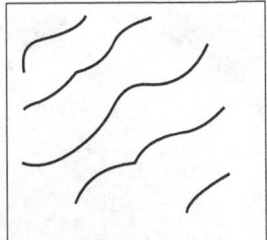

Fig. 12.2 Pattern of shrinkage cracks

Depending on the importance of the structure, the crack control measures listed below can be taken to ensure better long-term durability of the structure:

a. The structure should be designed with minimum or no cracking conditions.
b. Thermal cracking should be avoided by reducing the heat of hydration using one or more of the following:
   - storing aggregates in the shade during execution
   - using blended cement
   - replacing cement partially by pozzolans
   - cooling aggregates by water, ice, etc.
c. Appropriate curing methods should be used to effect reduction in heat of hydration.

12.1.2 Carbonation

Concrete made with Portland cement is highly alkaline due to the presence of calcium hydroxide. This alkalinity present in the pore water of concrete can be reduced by the acidic compounds in the atmosphere, especially carbon dioxide and sulphur dioxide. The effect of reduction of the pH value of concrete by these chemicals is known as *carbonation*. Concrete gets carbonated on the surface, including on the sides of cracks and wherever it is in contact with the atmosphere. This carbonated concrete does not provide the necessary protection to steel reinforcement.

The rate of carbonation is determined by the chemical reaction between CO_2 and $Ca(OH)_2$ in pore water, resulting in the formation of $CaCO_3$ as per the following equation:

$$CO_2 + Ca(OH)_2 \rightarrow CaCO_3 + H_2O$$

Thus, the rate of diffusion of CO_2 inwards through concrete is an important parameter. This diffusion takes place through the pore structure of concrete. The chemical reaction takes place fast and the period of diffusion is the real defence against carbonation. The more the water in the concrete (with larger water–cement ratio), the more rapidly carbonation takes place. Hence, the use of a low water–cement ratio for making concrete ensures a definite defence against carbonation.

A simple test for carbonation can be made on concrete by spraying phenolphthalein on it. The carbonated concrete remains unchanged in colour whereas the 'good' concrete 'blushes' with a bright pink colour.

12.1.3 Alkali–Silica Reaction

The reaction between the alkalis in cement (mainly sodium and potassium hydroxides) and the reactive silica (SiO_2) found in aggregates in the presence of water (hydroxyl ions) can cause expansion and serious cracking in hardened concrete. The reaction can be described as follows:

$$SiO_2 + 2NaOH + H_2O \rightarrow Na_2SiO_2 2H$$

Silica   Alkali   Water   Alkali–silica gel

Fig. 12.3 Alkali–silica gel (*Source*: Nemati 2015. Reproduced with permission from Prof. Kamran M. Nemati)

The alkali–silica gel (Fig. 12.3) that gets formed is expansive in nature, because of which considerable internal bursting force gets generated, which can result in serious cracking. The circumstances required for the alkali–silica reaction (ASR) to take place are

a. the presence of reactive aggregate,
b. high alkali content in the cement used for making the concrete, and
c. concrete in a wet condition.

All these conditions are required for the damage due to ASR to occur. These conditions exist in humid (dams, bridge piers, sea walls, etc.) and exposed (roads and building exteriors) environments.

ASR leads to expansion and cracking of concrete, loss of strength, and pop-outs and exudation of alkali–silica gel. When it is necessary to use aggregates from a previously unknown source, it is important to investigate it for reactive silica. The following methods may be used to prevent ASR damage:

a. Limiting alkali content:
  • Use of low-alkali cement
  • Limiting other sources of salt such as contaminated aggregates, penetration of sea water, and use of chemical-free solution
  • Limiting the maximum cement content in concrete to a low value
b. Limiting reactive aggregate:
  • Size
  • Quantity
  • Reactivity
c. Limiting the presence of moisture:
  • Relative humidity <75%
  • Repairing cracks, leaking joints, etc.
d. Using pozzolanic mineral admixture:
  • Blast-furnace slag
  • Silica fume, volcanic ash, metakaolin, etc.
e. Using air entrainment to allow expansion
f. Structural design—limiting access to water
  • Avoiding de-icing salt
  • Ensuring adequate compaction
  • Obtaining good finished surfaces with proper curing

12.1.4 Chemical Attack

Chemical aggression to concrete results from the following:

a. Chemicals in subsoil and groundwater
b. Chemicals in the atmosphere where the structure is built
c. Chemicals in the liquid stored in tanks
d. Chemicals added during manufacture of concrete, knowingly or unknowingly

Chemicals in subsoil and groundwater

Dry chemicals are less harmful compared to those dissolved in groundwater. Naturally occurring groundwater contains chemicals and industrial effluents which affect concrete. Industrial effluents may contain sulphuric, hydrochloric, nitric, phosphoric, and phenolic acids, ammonium compounds, and sulphates. The range of chemicals is quite vast; here we discuss only the common ones. It is always prudent to test groundwater to assess the type of chemicals present.

Acids in groundwater

A pH value of less than 7 indicates the presence of acids in groundwater. The severity of attack depends on

a. the type and quantity of acid,
b. the continuity of replenishment (as in industrial pollution),
c. the velocity of flow of groundwater (still or flowing), and
d. the cement content and the impermeability of the structure.

Table 12.1 gives the variation of the pH values of some typical acids with concentration. Table 12.2 gives a rough indication of the severity of attack. This table is to be treated only as a rough guide based on experience. Each case of pH < 7.0 should be treated as a separate problem and further investigation should be carried out.

The following points will be useful in the assessment with respect to a pH value less than 7.0.

a. Chemical analysis should be carried out for pH <7.0.
b. When chemical analysis indicates a probable slight attack, it is necessary to use blast-furnace slag cement. The cement content should be above 350 kg/m^3 and water–cement ratio less than 0.5.
c. When analysis shows appreciable attack (pH 5.0–4.5), the cement content should not be less than 360 kg/m^3 and the water–cement ratio should be less than 0.45. Plasticizers are necessary to obtain a good workable mix.
d. When severe attack is indicated by analysis, in addition to the precautions in (c), the surface should be protected by coating it with an inert material such as epoxy resin or polyester resin reinforced with a glass fibre membrane. The coating should not be less than 0.75 mm thick and should be applied without any holidays (pin holes and defects).

Table 12.1 Variation of pH value

Concentration in solution	pH value				
	Hydrochloric acid	Acetic acid	Sulphurous acid	Carbonic acid	Boric acid
1N	0.10	2.4	0.3	2.4	4.5
N/10 = 0.1N	1.07	2.9	1.2	3.8	5.2
N/100 = 1% = 0.01N	2.02	3.4	2.1	4.5	5.9

Table 12.2 Severity of attack

Probable aggressiveness to Portland cement concrete			
Significant attack unlikely	**Slight attack probable**	**Appreciable attack probable**	**Severe attack probable**
pH 7.0–6.5	pH 6.5–5.0	pH 5.0–4.5	pH below 4.5

It should be noted that sulphate-resisting cement is not really acid proof. High-alumina and supersulphated cements are resistant to a range of weak acids; supersulphated cement is more resistant to sulphate attack than ordinary Portland cement. Either of these two cements can be used in foundation concrete.

The dense concrete suggested above can be supplemented by the treatment discussed here: A provision of 1000 gauge polyethylene sheets with bituminous adhesive bonded polyisobutadiene sheeting or bonded chlorinated polyethylene along the bottom and carried up to top ground level can be expected to provide reasonable protection under very severe conditions.

Sulphate attack

Sulphate attack is caused by the chemical reaction between sulphate ions and hydration products, leading to the formation of ettringite and gypsum. Monosulphate, CH, and water combine to form ettringite. There are different sources of sulphate ions, such as sea water, sewage, industrial waste, salts in groundwater, and delayed release of clinker. The following equation gives an example of how monosulphate CH combines with a source of sulphate ion:

$$C_4AH_{10} + 2CH + 12H \rightarrow C_4AS_2H_{32}$$

The expansive forces generate tensile stresses in concrete. This leads to severe damage and cracking. The x-ray diffraction analysis of damaged and undamaged concrete is shown in Fig. 12.4. Figure 12.5(a) shows a typical etched surface of concrete due to chemical attack.

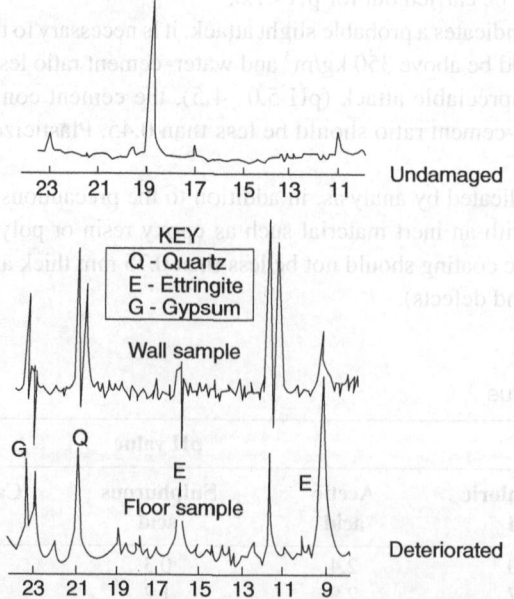

Fig. 12.4 X-ray diffraction analysis of undamaged and deteriorated concrete

(a)   (b)

Fig. 12.5 (a) Etched surface of concrete; (b) Freeze–thaw scaling
of a railroad bridge (*Source*: Nemati 2015. Reproduced
with permission from Prof. Kamran M. Nemati)

Chloride attack

Chloride attack and the consequent corrosion of steel in concrete were discussed in the previous chapter.

12.1.5 Physical Aggression

Physical aggression and the consequent deterioration of concrete structures are caused by

a. freezing and thawing of structures, for example structures exposed to harsh environment in the northern parts of India, near the Himalayas, and
b. abrasion from grit-laden water striking the concrete at high velocity.

Frost damage due to freezing and thawing

Pores and voids get introduced into concrete due to various factors. Some are intentionally introduced such as by air-entraining agents. Others are accidentally formed during manufacture. Following are a few typical cases of pores and voids which influence the freezing and thawing behaviour of concrete:

• Entrapped during mixing; 10 µm to 1 cm; usually devoid of water
• Introduced by admixtures; 0.1–0.2 mm; usually dry
• Capillary porosity: Cavities of excessive, unreacted mixing water; 0.01–5 µm; contain water; freezing point depends on pore solution chemistry, and its temperature ranging from –1°C to –8°C
• Gel pores: Very fine internal C–S–H pores; 1–10 nm; contain chemically bound water; resist freezing due to chemical bonding; typical freezing temperature: –78°C

The frost damage mechanism can be identified to have the following three phases:

Phase I: Hydraulic pressure

1. Prior to freezing, water is at low pressure in both upper and lower pores.
2. The cold front enters the upper pore, which increases the pressure of water. The surrounding concrete is exposed to high-pressure water.
3. The cold front moves to cross the upper pore. The high-pressure area reaches the lower pore, causing fluid to enter the lower pore. Hydraulic pressure and accelerated damage are caused by the movement of fluid through highly restricted channels between capillary pores.

Phase II: Osmotic pressure

4. The freezing increases the osmotic pressure in the pores.

Phase III: Capillary ice growth

5. Slowly the capillaries are filled with ice formation, and this causes the surrounding concrete to crack due to bursting pressure.

Figure 12.5(b) shows the typical freeze–thaw damage at a railroad bridge site. The following methods can be used to prevent the frost damage of the cement paste as well as the aggregates:

a. Paste damage can be reduced by using a low water–cement ratio, ensuring proper curing, and using air-entraining agents.
b. Aggregate damage may be controlled by using frost-resistant aggregates such as granite.

Abrasion from grit-laden water

This occurs in walls and floors opposite inlets in tanks, weir crests, and floors of channels carrying high-velocity water, particularly if the water contains grit. With very high velocity, 'cavitation' is a serious danger. Turbulence is a vital factor in promoting cavitation.

For resisting abrasion in the velocity range 4–6 m/s, concrete should be manufactured using a water–cement ratio of 0.45 and should have a strength not less than 40 MPa. For the velocity range 6–15 m/s, special high-performance concrete having 400 kg/m^3 of cement and a maximum water–cement ratio of 0.4 in which the microstructure is modified by suitable mineral admixtures should be used.

However, where the flow is continuous over long periods and has a velocity exceeding 5 m/s, the special protective measures suggested below are used.

a. A 75-mm-thick sacrificial layer of high-strength concrete or granite integrated with the concrete below.
b. Three coats of 75 mm thick epoxy resins applied evenly.
c. A protective layer of epoxy resin mortar made with abrasion-resistant aggregates.
d. In extreme cases, steel plates fixed on the wall or lined with concrete made with steel fibre reinforcement are used. Steel-fibre-reinforced concrete is especially suitable in tunnels where changes in cross section occur.

12.2 Design for Durability using Performance Specification

Reinforced concrete structures deteriorate because of lack of attention to durability during execution. Such execution is done without any adherence to durability specification. The current codes do not specify performance levels. The specifications are rather prescriptive than performance based. Such an approach has led to poor long-term performance. On the other hand, both performance-based specifications (PBS) and performance-based design (PBD) ensure a specified level of behaviour in terms of definitive quantities such as chloride penetration level. The desired result is achieved by mathematical modelling and predicting service life. Indeed such an approach is considered to give a tangible solution to durability problems.

Permeability and diffusion of fluids or gases through concrete depends on variables which are known as *transport parameters*. The design based on codes at present does not take into account such variables or parameters. The *Durability Index approach* can be used to specify performance limits to ensure durability. This approach is based on three tests, described in detail in the subsequent Section 12.2.3, which can be used to guarantee long-term service performance. The actual performance measured can be linked to mechanism of concrete deterioration.

12.2.1 Performance based Specification

Conventionally concrete strength has been used as a guiding parameter of durability. However, concrete strength does not address properties such as curing and compaction. The quality of outer cover layer is important and controls agents of aggressive chemicals getting into the concrete. We cannot control the environment; therefore, measuring transport properties especially in the cover region will ensure knowledge about durability. Thus, the specification should include performance indicators. With such a specification, a service life model can be postulated.

12.2.2 Durability Index (DI) Method

This method of design considers the following:

i. The quality of outer layer (cover) is important to protect embedded steel.
ii. Measurement of durability parameter is essential.
iii. Quality of cover is related to transport and movement properties such as water absorption and atmospheric/ionic diffusion.
iv. Identification of tests links these durability properties with transport mechanism such as absorption and diffusion.
v. These identified index tests are used as reference against actual performance of the structure. Thus, Durability Index (DI) is quantifiable like diffusion, permeability, or sorptivity.

12.2.3 Durability Index Tests

The following DI tests are used as benchmark tests:

Oxygen Permeability Test (OPT)

Permeability of concrete is due to interconnected capillary pores with sizes of at least 120 nm. Permeation is governed by Darcy's Law.

$$\frac{dq}{dt}\frac{1}{A} = K\frac{dh}{dL}$$

where

$\dfrac{dq}{dt}$ is the rate of fluid flow

$\dfrac{dh}{dL}$ is the pressure gradient

A is the surface area

K is Darcy coefficient of permeability

This test is used to determine carbonation resistance. The specimen for this test consists of oven-dried (50°C for 7 days) concrete samples of size 68 mm diameter and 25 to 30 mm thick. The test set-up is shown in Fig. 12.6. The specimen is placed inside the permeability cell in a secure rubber gasket. It is then subjected to a pressure of 100 kPa. This is done through a pressurized oxygen cell. The pressure decay is monitored after isolation. Using the decay, the Darcy coefficient of permeability K is found. The oxygen permeability index (OPI) is defined as follows:

$$\text{OPI} = (-) \log K$$

OPI ranges from 8 to 11. The larger the index, the less permeable the concrete.

Water Sorptivity Test (WST)

Sorptivity is the rate of movement of a wetting front through a porous concrete material. In the water sorptivity test, the unidirectional absorption of water into one face of a pre-conditioned concrete disc sample is assessed. The diagram in Fig. 12.7 shows this test. The sample is weighed to determine the mass of water absorbed at selected time intervals. The plot of mass of water expressed in terms of equivalent thickness in mm absorbed versus square root of time is made. The lower water sorptivity index indicates better potential for durability of

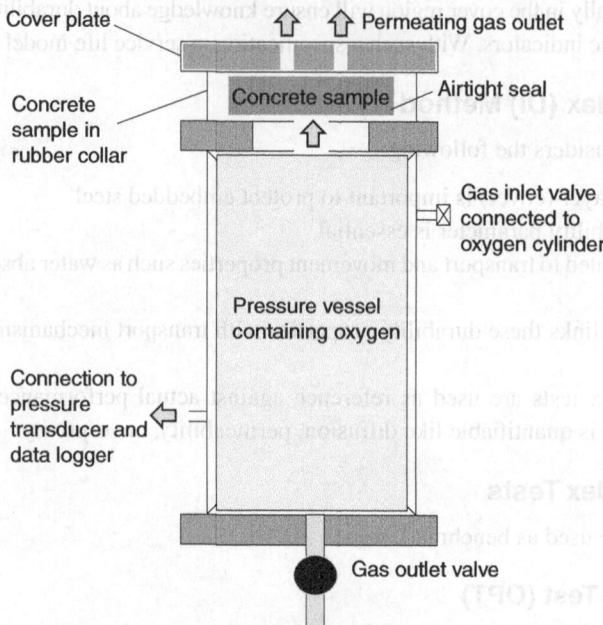

Fig. 12.6 Oxygen permeability test

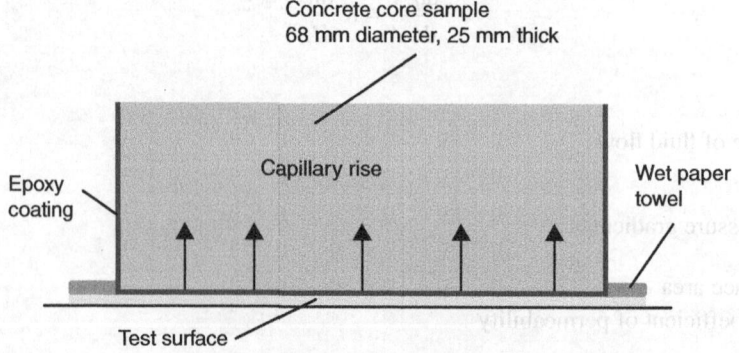

Fig. 12.7 Water sorptivity test

the concrete. Sorptivity value is of the order of approximately 5 mm/√h for well-cured M30–M50 concretes, to 15–20 mm/√h for poorly cured M20 concrete.

Rapid Chloride Penetration Test (RCPT)

The apparatus to conduct this test has a two-cell conduction rig. Each cell consists of a 5M NaCl solution so that there is no concentration gradient across the sample. The concrete disc sample is pre-conditioned using vacuum saturation with a 5M NaCl solution. A 10 V potential applied causes the chloride conductivity. The chloride migration is the result of conduction from the applied potential difference as can be seen in Fig. 12.8.

Einstein's equation based on diffusion and conduction allows the conductivity test to be used as an index of concrete diffusivity. The test is dependent upon the changes in the pore structure and cement chemistry (mainly binder type), which might appear to be insignificant when using the permeation process. Typical chloride conductivity index values range from >3 mS/cm for M20–M30 OPC concretes, to < 0.75 mS/cm for M40–M50 slag or fly ash concretes. The lower index indicates better durability of the concrete.

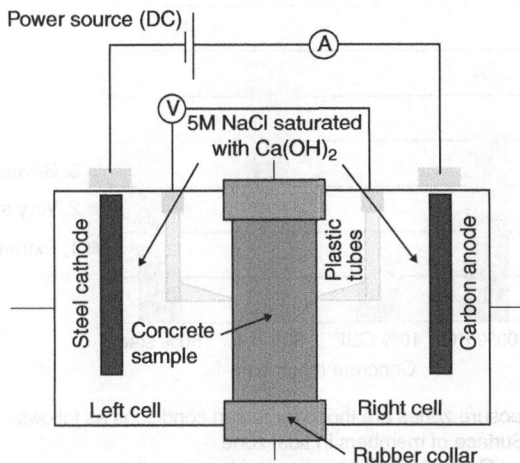

Fig. 12.8 Chloride conductivity test apparatus

Suggested acceptance limits of index test values are shown schematically in Fig. 12.9. The exposure classes shown in Fig. 12.10 are those suitable for Indian marine conditions, and the various binder blends are all in regular use in India. It is immediately obvious that limiting chloride conductivity values depend on both the exposure conditions and the binder type. The values in Fig. 12.10 in the form of a bar chart approximately show equal "protection" against chloride ingress, but a single nominal value is considered an oversimplification.

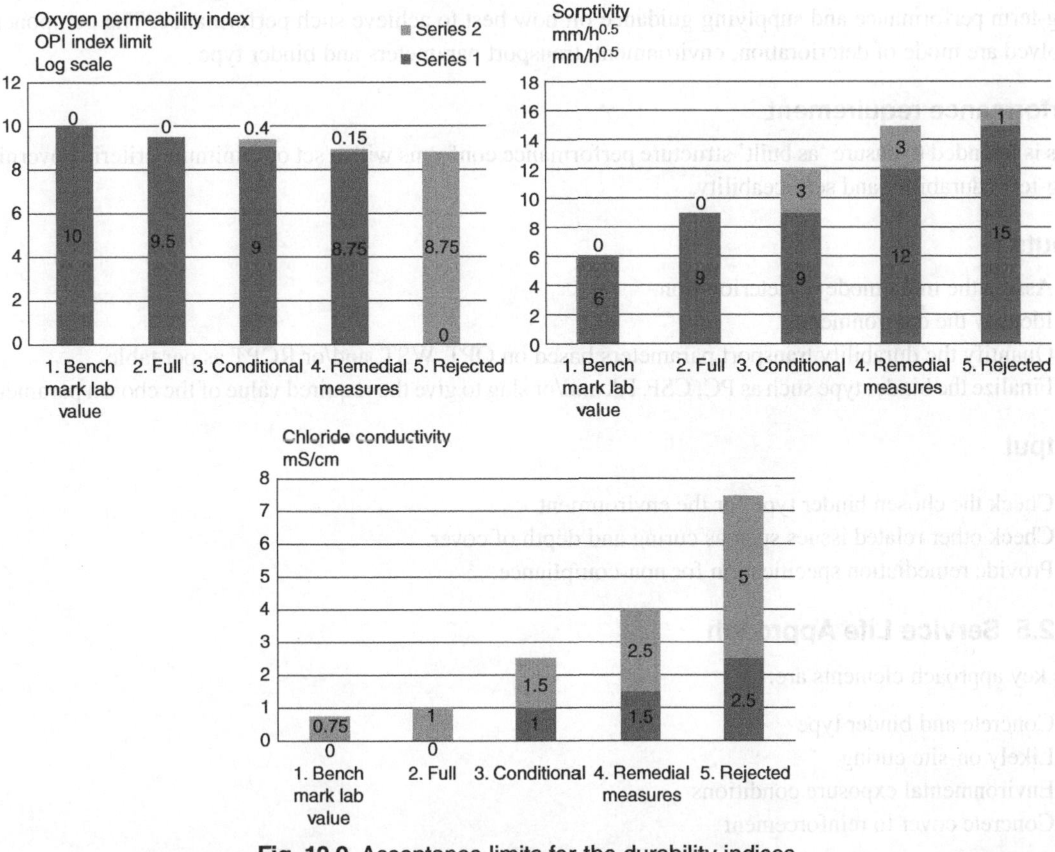

Fig. 12.9 Acceptance limits for the durability indices

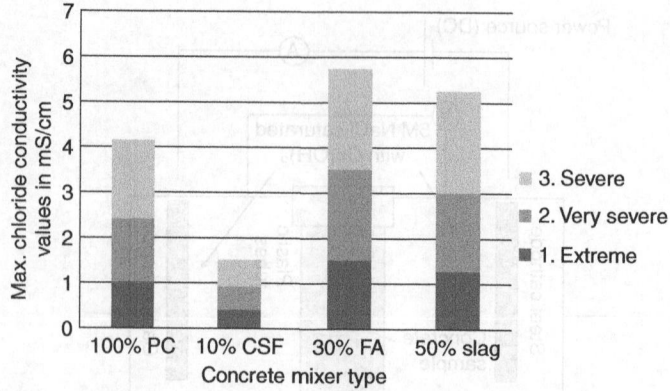

Marine exposure zones are those for Indian conditions as follows:
Extreme: Surface of members in tidal zone
Very severe: Concrete surfaces exposed to sea water spray

Fig. 12.10 Allowable maximum chloride conductivity values (mS/cm) at 28 days (marine exposure) such as in Chennai or Mumbai

12.2.4 Framework for Durability Specification

Framework for a durability specification requires that specifications comprise of both performance and prescriptive elements. This provides a well-balanced and effective specification, giving assurance of adequate long-term performance and supplying guidance on how best to achieve such performance. The components involved are mode of deterioration, environment, transport parameters and binder type.

Performance requirement

This is intended to ensure 'as built' structure performance conforms with a set of minimum criteria governing long-term durability and serviceability.

Inputs

- Assess the main mode of deterioration.
- Identify the environment.
- Quantify the durability transport parameters based on OPT, WST, and/or RCPT as per table.
- Finalize the binder type such as PC, CSF, FA and/or slag to give the required value of the chosen parameter.

Output

- Check the chosen binder type for the environment.
- Check other related issues such as curing and depth of cover.
- Provide remediation specification for non-compliance.

12.2.5 Service Life Approach

The key approach elements are:

- Concrete and binder type
- Likely on-site curing
- Environmental exposure conditions
- Concrete cover to reinforcement

- Notional design life, or "service life", of the structure
- Optimization for economy

An example is given in Fig. 12.11 for marine environments, assuming a 50 year service life (Alexander and Beushausen 2009).

Exposure	Cover (mm)	10% CSF	100% PC	30% FA	50% Slag
Extreme	40	0.25	0.45	0.75	0.85
	60	0.30	0.95	1.35	1.55
	80	0.60	1.30	1.80	2.00
Very severe	40	0.35	0.45	0.90	1.10
	60	0.50	1.15	1.75	2.00
	80	0.85	1.65	2.30	2.60
Severe	40	0.55	1.00	1.85	1.95
	60	1.10	1.85	2.95	3.05
	80	1.55	2.50	3.75	3.85

	Mixes that may be impractical: Concrete grade exceeds 60 MPa
	Mixes requiring nominal grades less than 30 MPa, and/or w/b > 0.55; not recommended
	Mixes that are acceptable and practical. Grades vary from 30 to 60 MPa

Assumptions: (i) Chloride threshold is 0.4% by mass of binder. (ii) Three days wet curing.
Note: For conditions indicated by the light grey shading, the indicated binder types may be used, but w/b should not exceed 0.55 for any marine zone.

Fig. 12.11 Maximum 28-day chloride conductivity values (mS/cm) for 50 year design life in SA marine conditions (for avoidance of corrosion activation at 50 years)

12.3 Performance of Concrete under Fire

During a fire, the temperature may reach up to 1100°C in buildings and even up to 1350°C in industrial structures and tunnels, causing severe damage to the concrete structure. Owing to concrete's fairly low coefficient of thermal conductivity, the movement of heat through concrete is slow. Reinforcing steel, which is sensitive to high temperature, is fairly well protected for a relatively long period of time due to provision of adequate cover. However, with sustained increase in temperature, concrete gets damaged and steel starts buckling outward losing its capacity. Hence it is necessary to study the behaviour of concrete under high temperature.

12.3.1 Behaviour of Concrete under High Temperature

When concrete is heated under conditions of fire, the increase in temperature in the deeper layers of the material is progressive, but because this process is slow, significant temperature gradients are produced between the concrete member's surface and core inducing additional damage to the element.

Table 12.3 lists the changes that take place in concrete containing cement paste and aggregates. When concrete containing siliceous aggregates is heated to temperatures between 300°C and 600°C, it will turn red; between 600°C and 900°C, whitish-grey; and between 900°C and 1000°C, a buff colour is seen. The colour change of heated concrete results principally from the gradual water removal and dehydration of the cement paste, and also transformations occurring within aggregates. While siliceous aggregates turn red when heated, the aggregates containing calcium carbonate get whitish. Due to calcination process, $CaCO_3$ turns to lime and gives pale shades of white and grey. In Fig. 12.12, the colour changes in heated concrete made with the riverbed aggregates are shown.

Table 12.3 List of changes taking place in concrete during heating

Temperature range	Changes
20–300°C	Slow capillary water loss and reduction in cohesive forces as water expands; 80–150°C ettringite dehydration; C–S–H gel dehydration; 150–170°C gypsum decomposition ($CaSO_4$–$2H_2O$); physically bound water loss
300–400°C	Approx. 350°C break up of some siliceous aggregates (flint); 374°C critical temperature of water
400–500°C	460–540°C portlandite decomposition $Ca(OH)_2 \rightarrow CaO + H_2O$
500–600°C	573°C quartz phase change $\beta - \alpha$ in aggregates and sands
600–800°C	Second phase of the C–S–H decomposition, formation of β-C_2S
800–1000°C	840°C dolomite decomposition; 930–960°C calcite decomposition $CaCO_3 \rightarrow CaO + CO_2$, carbon dioxide release; ceramic, binding initiation which replaces hydraulic bonds
1000–1200°C	1050°C basalt melting
1300°C	Total decomposition of concrete, melting

Source: Hager, 2013, table 1. Reproduced with permission from I. Hager.

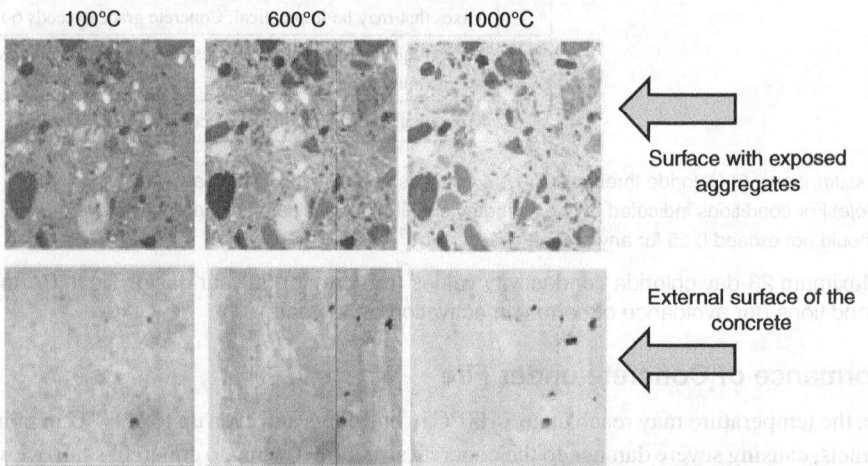

Fig. 12.12 Colour change of heated concrete (*Source*: Hager, 2014. Reproduced with permission from I. Hager). Refer the Oxford University Press India website for the colour image

The factors affecting the material damage level when concrete is heated are as follows:

- Heating rate
- Maximum temperature
- Time of exposure to temperature
- Load applied during heating
- Moisture content of the material

The testing method used to study the influence of high temperature on the properties of concrete has an important influence on the evaluation of the properties of heated concrete. The most common method is to expose the concrete to high temperature, cool it down to room temperature, and then carry out testing, such as compression or tensile tests.

However, this method gives the "post fire" or "post exposure to the high temperature" properties of concrete. Nevertheless, one must consider that the most appropriate procedure to test the mechanical properties at high temperature is to determine the properties of the material at elevated temperature (tested "hot").

The tested "hot" properties are higher than the residual ones (see Fig. 12.13). The testing of material at the "hot" stage is concerned with the determination of a material's properties under fire conditions, while testing after cooling gives the residual values corresponding to the post-fire performance of concrete. The lower values of residual mechanical properties are attributed to supplementary damage due to additional stresses caused by cooling and the development of cracks.

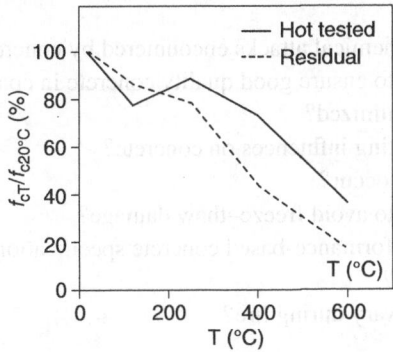

Fig. 12.13 Relative compressive strength as a function of temperature for HPC "hot" and residual behaviour (HPC, 0.9 kg/m^3 fibres, $f_{c20°C}$ = 91 Mpa) (*Source*: Hager, 2013, figure 9. Reproduced with permission from I. Hager)

12.3.2 Codal Requirements of Concrete under Fire

All the properties of concrete deteriorate with increase in fire temperature. Figures 12.14 and 12.15 show the reductions that should be considered for compressive strength and elastic modulus for normal concrete as per CEB recommendations.

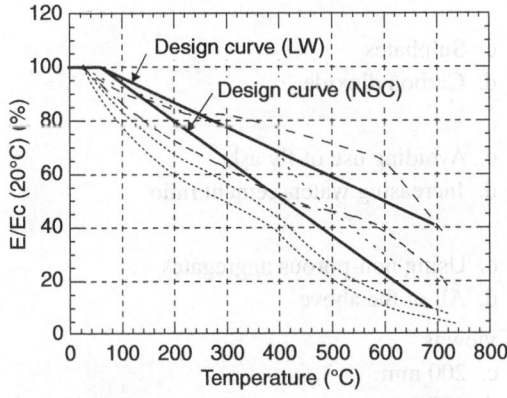

Fig. 12.14 CEB design curve for compressive strength of siliceous normal strength concrete subjected to elevated temperature (*Source*: Phan 1996, figure 4.9)

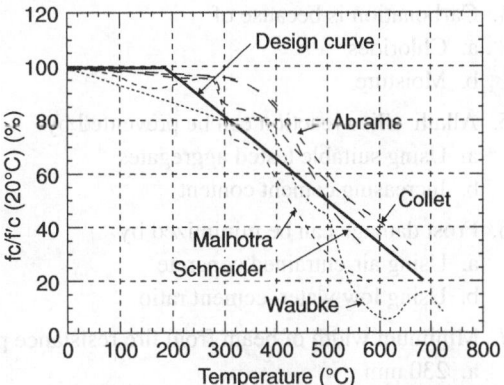

Fig. 12.15 CEB design curve for effect of elevated temperature on modulus of elasticity of lightweight and normal-strength concretes (*Source*: Phan 1996, figure 4.11)

Exercises

Review Questions

1. How does the microstructure of concrete affect the durability of a structure?
2. How impermeable is concrete? What parameters affect the permeability of concrete?
3. What are the reasons for the cracking of concrete and how does it affect durability?
4. What do you understand by carbonation of concrete? How is it tested?
5. Why does alkali–silica reaction disrupt concrete? What precautions can be taken to avoid alkali–silica reaction?
6. What are the various types of chemical attacks encountered by concrete?
7. What precautions can be taken to ensure good quality concrete in coastal structures?
8. How can sulphate attack be minimized?
9. What are the physical deteriorating influences on concrete?
10. How does freeze–thaw damage occur?
11. What precautions can be taken to avoid freeze–thaw damage?
12. What do you understand by performance-based concrete specification?
13. What are Durability Index tests?
14. How does strength of concrete vary during fire?

Multiple Choice Questions

1. Durability problems involve deterioration of concrete due to
   a. Frost action
   b. Chemical attack
   c. Environmental attack
   d. All of the above

2. Reinforced concrete can be made durable by
   a. Increasing water–cement ratio
   b. Decreasing water–cement ratio
   c. Decreasing cement content
   d. Decreasing cover depth

3. Shrinkage cracks occur because of
   a. Heat of hydration
   b. Use of blended cements
   c. Replacing cement by pozzolana
   d. Improper cover to reinforcement

4. Carbonation is because of
   a. Chlorides
   b. Moisture
   c. Sulphates
   d. Carbon dioxide

5. Alkali–silica reaction can be prevented by
   a. Using suitable tested aggregates
   b. Increasing cement content
   c. Avoiding use of fly ash
   d. Increasing water–cement ratio

6. Frost damage can be minimized by
   a. Using air-entrained concrete
   b. Using low water–cement ratio
   c. Using non-porous aggregates
   d. All of the above

7. Minimum width of beam from fire resistance point of view is
   a. 230 mm
   b. 300 mm
   c. 200 mm
   d. 150 mm

8. Chlorides in mixing water for RCC work should be limited to
   a. 2000 mg/L
   b. 1000 mg/L
   c. 1500 mg/L
   d. 500 mg/L

9. Sulphates in mixing water should be limited to
   a. 100 mg/L
   b. 400 mg/L
   c. 500 mg/L
   d. 1000 mg/L

10. pH value for mixing water as per IS: 456-2000 should be
    a. > 7
    b. > 6
    c. < 7
    d. < 6

Answers to Multiple Choice Questions

1. d
2. b
3. a
4. d
5. a
6. d
7. c
8. d
9. b
10. b

CHAPTER 13

Lightweight Concrete

Lightweight concrete is produced by including large quantities of air in the aggregate, in the matrix, or in between the aggregate particles, or by a combination of these processes. According to the method used, the various forms of lightweight concrete can be classified as follows:

Type of lightweight concrete	Method of manufacture
Air entrained in between aggregate particles	No-fines concrete made with dense aggregate
	Partially compacted concrete made with dense aggregate
Air contained in between and within aggregate particles	No-fines concrete made with lightweight aggregate
	Partially compacted concrete made with lightweight aggregate
Air contained within the aggregate particles	Fully compacted concrete made with lightweight aggregate (natural or artificial)
Air introduced by autoclaving or by foaming agents	Foamed concrete

Aggregates that weigh less than about 1000 kg/m^3 are generally considered lightweight and find application in the production of various types of lightweight concrete. The light weight is due to the cellular or highly porous microstructure. However, cellular organic material such as wood chips cannot be used as aggregates because of lack of durability in the moist alkaline environment in Portland cement concrete.

Natural lightweight aggregates are made by processing igneous volcanic rocks such as pumice, scoria, and tuff. Synthetic lightweight aggregates can be manufactured by thermal treatment from a variety of materials such as clay, shale, slate, diatomite, perlite, vermiculite, blast-furnace slag, and fly ash pellets.

Actually, there is a whole spectrum of aggregates weighing 75–1000 kg/m^3. Very porous aggregates, which are at the lighter end of the spectrum, are generally weak and therefore more suitable for making non-structural insulating concretes. At the other end of the spectrum are those aggregates that are relatively strong, less porous, and capable of producing structural concrete. A typical spectrum is illustrated in Fig. 13.1.

Designing concrete mixes involves selecting the most economical mix proportions of cement, water, and fine and coarse aggregates to produce concrete having the required properties. The two basic approaches to the design of concrete mixes are mathematical and empirical; different authorities favour different methods. Both the methods are based on results obtained from experimental work. The mathematical analysis of the result is valuable, but complex equations relating the properties of the materials and the mix proportions to properties of concrete are not in common practice. Unless the research work is very comprehensive, the

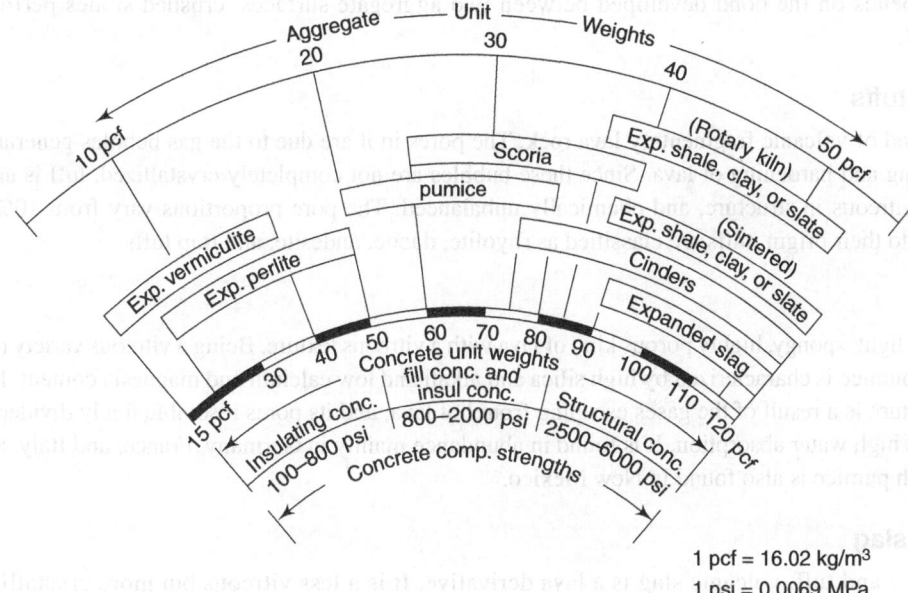

Fig. 13.1 Range of lightweight aggregate concrete

equations derived may only be applicable to a particular material or situation. When different cements and aggregates are used at sites, their properties vary, and this must be allowed for in the mix design. Mathematical expressions also often relate to the properties of the cement or aggregate which cannot conveniently be measured under normal site conditions. It has been shown that the strength of concrete produced on-site may vary by a considerable margin. In addition, it is the view of some workers in this field that the most practical methods of mix design are based on empirical rules derived from experimental research. We agree with this view because the assumptions made in the course of research will be based on the properties of the materials used and the utility of the particular mix in a given environment. This is especially true for lightweight concrete.

Lightweight aggregate can be divided into two categories: natural aggregate and artificial aggregate.

13.1 Natural Aggregates

Natural aggregates are materials that are available to us from a natural source. Some of the most common natural aggregates are silica sand, gravel, crushed lightweight stone, volcanic tuffs, pumice, etc. We will discuss some of them in the following sections.

Silica sand

Any silica sand suitable for ordinary concrete is a good aggregate for lightweight concrete. Sand must not contain clay, silt, or organic impurities. Its silt content should not exceed 6%. All lightweight concretes of the no-fines and graded types are made using mainly silica sand.

Gravel and crushed stone

Gravel and crushed stones conforming to the required standards are suitable for no-fines concrete. Aggregates should be cleaned by flushing them with water before use. Since the strength of no-fines concrete mainly depends on the bond developed between two aggregate surfaces, crushed stones perform better than gravel.

Volcanic tuffs

Tuff is a kind of volcanic fragmentary lava rock. The pores in it are due to the gas bubbles generated by the rapid cooling and hardening of lava. Since these bubbles are not completely crystallized, tuff is amorphous in nature, vitreous in structure, and chemically unbalanced. The pore proportions vary from 10% to 60%. According to their origin, tuffs are classified as rhyolite, dacite, andesite, and trap tuffs.

Pumice

Pumice is a light, spongy, highly porous kind of lava with a vitreous texture. Being a vitreous variety of rhyolite and silica, pumice is characterized by high silica and alkali and low calcium and magnesia content. Its spongy cellular texture is a result of the gases escaping from hot lava, and its pores resemble finely divided bubbles. Pumice has high water absorption. It is found in abundance mainly in Germany, France, and Italy. Relatively low strength pumice is also found in New Mexico.

Volcanic slag

Like pumice and tuff, volcanic slag is a lava derivative. It is a less vitreous but more crystallized slag-like material. Volcanic slag has properties similar to foamed blast-furnace slag. It contains a high proportion of ferrous oxide, lime, and organic oxide. Its alkali content is low. Its sulphur and sulphate content must not exceed 10.5% and 1.0%, respectively. The proportions of the components of volcanic slag are as follows:

$$\text{Silica (SiO}_2\text{): 43–55\%}$$

$$\text{Alumina (Al}_2\text{O}_3\text{): 18–24\%}$$

$$\text{Ferrous oxide (Fe}_2\text{O}_3\text{): 8–20\%}$$

The bulk density of completely dried volcanic slag is as follows:

$$\text{0–7 mm diameter: 0.90 g/cm}^3$$

$$\text{7–15 mm diameter: 0.75 g/cm}^3$$

Crushed volcanic slag powder, when in contact with lime, turns into a valuable chemical binder, which adds strength to lightweight concrete.

13.2 Artificial Aggregates

Artificial aggregates comprise materials such as brick rubble, cinder, sintered cinder, blast-furnace slag, foamed blast-furnace slag, expanded clay, sintered fly ash perlite, and vermiculite. Most of these materials are produced in a rotary kiln. We will discuss these next.

Unscreened material

In the rotary kiln process, raw material is introduced in its natural form and size at the top of a slightly tilted horizontal cylinder with a refractory lining. The cylinder is rotated at slow speed. As the raw material moves slowly down to the burning lower end, the heat level is about 1000°C. This process causes the simultaneous formation of gases and the onset of a pyroplastic condition in the material. The level of viscosity of the softened mass is regulated to enable gases to be entrapped to form internal cellular structures. This material becomes vitrified on cooling, which takes place inside the kiln.

Intermediately removed material

The hot material from the rotary kiln is cooled externally and then crushed. It is then screened to segregate the required size aggregates.

Pre-sized material

The material is pre-sized by crushing and screening before introducing into the rotary kiln.

Crushed material

The pre-moistened mixture is carried by a travelling belt or pan to the drying and ignition hood. The burner is below the hood. The burning starts at the surface and penetrates though the entire depth producing gases. This simultaneously causes expansion and the onset of pyroplasticity. The material becomes viscous, trapping gases and creating a cellular structure. The clinker formed is crushed and sized.

Pelletized material

The raw material (clay/shale) is mixed with moisture and finely ground coal and extruded before burning to produce pellets. These pellets are used as aggregates. The various types of pellets used are described as follows.

Brick rubble   Brick rubble became an important aggregate in many countries after World War II. It requires breaking and grading devices. The bulk density varies as follows:

Size (mm)	Bulk density (kg/m³)
0–1	1120
1–3	900
3–7	800
7–15	750
15–30	700

The sulphate content of brick rubble is normally limited to 1%. The relation between cube strength and crumbling coefficient is as follows:

Brick type	Cube strength (MPa)	Crumbling coefficient
Slurried and twice burnt brick	10–12	0.23–0.30
High-strength brick	8–10	0.50–0.60
Masonry brick	3–8	0.85–1.05

Cinder   Cinder aggregate is a residue of coal burnt in industrial boilers. The residue is melted and sintered to form cinders. Cinders are classified according to the fuel used in the binder as anthracite, coal, and brown coal cinder. The properties of these cinders depend on the mineralogical composition of rocks with coal deposits and the coal-burning method. The major harmful ingredients in these aggregates are unburnt coal, sulphur compounds, and burnt lime content. The poor quality of cinders that contain a high proportion of impurities can be improved by methods such as screening, electromagnetic separation, washing and spreading, and agglomeration by second burning.

Fly ash aggregate Granulated and sintered ashes from pulverized coal burners are important aggregates. Large quantities of this industrial by-product are derived from pulverized-coal-operated boilers of thermal stations. These ashes are extensively used by the building industry. Ash as pozzolana is used for cellular and other types of concrete as a mineral admixture. Sintered fly ash was first tried in the UK and the USSR. The fly ash used for sintering should have good air permeability, adequate mineral composition, and necessary fuel content for the second burning process. It must contain shale, clay, ferrous oxide, and at least 5% unburnt coal. If the percentage of unburnt coal is less than 5%, sintering can be done by introducing fuel or other burning methods. The two sintering methods are the shaft furnace method and the sintering belt method. For ashes containing 5–8% unburnt coal, the shaft furnace method is suitable. If the content of coal is higher due to the increased temperature in the furnace, smelting and agglomeration take place and its recovery from the furnace is almost impossible. This problem can be solved by the sintering belt method. It is found that the necessary temperature for sintering is 1200–1250°C. The sintering method consists in mixing fuel ash with mine culm of 30–40% coal content in a proportion that achieves the best burning capacity. The mixture is further improved by grinding the mine culm in a ball mill or edge runner mill and spraying it with water. The ground-moist mix is then crushed to the required size by a granulating drum. The size of the pellet depends upon the sloping angle of the container: steeper the slope of the drum, smaller the pellet size. These granules are dried in order to keep them from sticking with each other and to increase their strength before feeding them into the furnace chamber, where they are sintered at 1200–1250°C. The temperature and burning rate are maintained by adequate air supply. The sintered granules can easily be recovered from the shaft furnace. Sintered fly ash pellets have a hard, coarse, red shell and a fine porous structure. Pellets are round and come in sizes varying from 7 to 30 mm. Since large particles are not suitable for concrete, the pellets must be crushed to the required size. The bulk density of fly ash aggregates varies with their origin, combustion, and sintering temperature.

Foamed blast-furnace slag In the operation of a blast furnace, silica and alumina constituents combine with lime (included in the furnace charges) to form molten slag. This slag collects at the top of the iron. Its composition is given below:

CaO: 30–50%

SiO_2: 23–38%

Al_2O_3: 8–24%

MgO: 1–18%

Fe_2O_3: 0.5–1%

SO_3: 2–8%

In the furnace, slag in the molten state will be at 1400–1600°C. If allowed to cool slowly, it solidifies into a grey, crystalline, stone-like material known as *air-cooled slag*. This is normally used for laying roads and as aggregate for heavy duty concrete. Cooling of slag with excessive water produces *granulated slag*—a more friable material. Chilling with a controlled amount of water to trap steam in the mass gives a porous product that has a pumice-like character. This product is known as *foamed slag* or *expanded slag*. Upon cooling, this material is used as lightweight aggregate. The bulk density varies as follows:

Size (mm)	Bulk density (kg/m³)
12–20	300
3–12	500
<3 (dust)	700

The chemical content of foamed slag is same as that of the original molten slag. However, the sulphate content in molten slag is lower. The mineralogical composition changes during the transition from the liquid state to the solid state. Standards for foamed blast-furnace slag for concrete aggregate limit the CaO content and sulphate content to 50% and 0.5%, respectively. Granulated slag is not active, but many foamed slags do exhibit hydraulic activity when mixed with lime or Portland cement; they act as mineral admixtures with pozzolanic activity and hence contribute to the strength. In the UK, foamed slag is used as an aggregate in concrete building blocks, insulations of concrete roof screeds, and in reinforced units such as windows and door frames.

Expanded clay aggregates Certain clays and shale when heated to the semi-plastic stage, called the *point of incipient vitrification*, expand or bloat, owing to the formation of gas at the fusion temperature, to seven times their original volume. The cellular structure so formed is retained even after cooling. These materials are used as lightweight aggregates. Expanded clay and shale have been used in the USA and continental Europe for many years. The following are different forms of expanded clay aggregates:

Hydite is an angular fragmental bloated clay aggregate obtained (used first in 1920 in the USA) from shale or clay in rotary kilns fired by pulverized fuel oil or gas.

Rocklite is produced from shale. Each of the size fractions is fired separately in the rotary kiln. This produces discrete, well-foamed, rounded particles, which do not require crushing.

Lytag is sintered, pulverized fly ash (PFA) lightweight aggregate. It is manufactured by pelletizing PFA and heating the pellets until they fuse into expanded lightweight aggregate.

Aglite is a recent and very successful aggregate manufactured in the USA from clay expanded in the sintering hearth.

Leca or light expanded clay aggregate is a Danish product. It is a light, rounded, smooth material produced in a rotary kiln.

Keramzit is a product similar to Leca produced in Eastern Europe.

Fragmented aggregates with sharp edges give greater concrete strength, whereas rounded materials give better workability with a lower water–cement ratio. With careful grading and good compaction, the lighter types of expanded clay aggregate can produce remarkably strong concrete at modest density.

Expanded slates and shale Certain types of slate expand when they become pyroplastic upon being exposed to high temperature. These are similar to plastic clay but are produced differently. Slates are pre-crushed in a jaw-crusher into 50- to 80-mm lumps and refined to about 25 mm grain size. Expansion takes place in a rotary or continuous furnace or in a capsulated environment. The strength of these aggregates depends on the relative expansion coefficient.

Expanded perlite It is a vitreous volcanic product produced from rapidly cooled lava. The water content of perlite varies from 3% to 5%. If exposed to 900–1200°C, the material melts into a pyroplastic state and expands owing to the entrapped steam. Perlites can form strong or weak bonds with water. Perlite must be evaluated based on its temperature, the expansion conditions, and the expansion rate. The size best suited for expansion is 0.6–1.2 mm. Grains measuring 5–10 mm will have a bulk density of 220–280 kg/m^3, and those measuring 10–20 mm will have a bulk density of 160–220 kg/m^3.

Vermiculite Lightweight concretes made with vermiculite are valuable thermal insulators. Vermiculite is a type of mica characterized by high magnesia content (25–30%). Preheated (75–80°C), crushed, and graded vermiculite is exposed to 750–1100°C. The laminar texture of the material consisting of flaky layers loosens under heat and expands in volume up to 15–20 times its original volume.

13.3 Physical and Mechanical Characteristics of Aggregates

The physical and mechanical properties of lightweight aggregates are different from normal-weight aggregates with respect to shape, surface texture, unit weight, and porosity. Because of this, lightweight aggregates generally absorb water and retain it in their pores. This section describes the method of assessing these differences and accounting for them when mix proportioning.

13.3.1 Particle Size, Shape, and Surface Texture

The *granularity* of lightweight aggregate concrete is related to the quality of aggregates. The normal sizes adopted are given in the following table:

Fraction of aggregates	Particle size specifications (mm)			
	Germany	France	Hungary	India
Fine	0–3	0–7	0–7	5
Medium	3–12	7–15	7–15	10
Coarse	12–25	15–25	15–30	20

The size and shape of the particles of a typical lightweight aggregate adopted in India are shown in Fig. 13.2. The shape of the particle can vary depending on the type of the manufacturing process. It can be rounded, angular, or irregular. The surface texture may be smooth with small pores or irregular with large pock-marked surfaces. These are important parameters in mix design because they control the degree of water absorption and affect the workability of fresh concrete.

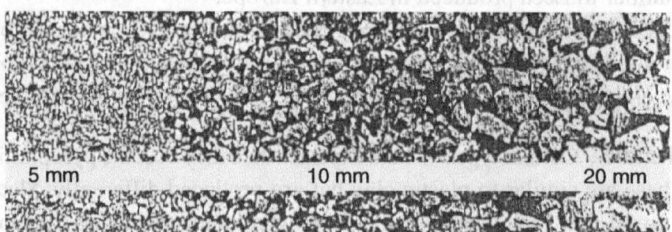

| 5 mm | 10 mm | 20 mm |

Fig. 13.2 Three different sizes of manufactured lightweight aggregate

13.3.2 Specific Weight, Density, and Bulk Density of Aggregates

The specific weight of a finely ground material passing through a 900-mm mesh sieve is determined using the pycnometer as follows:

$$\gamma = \frac{S}{W} \ \text{g/cm}^3$$

where S is the weight in grams and W is the water displaced by the dried material in cm^3.

The density γ_v of aggregates is equivalent to the weight of the unit volume of the porous material (including the pores in the particles):

$$\gamma_v = \frac{S}{V}$$

where V is the volume of the dry material in cm^3. The density of lightweight aggregate is lower than that of normal-weight aggregate. The specific gravity also varies with the particle size—it decreases with increase in the size of aggregates.

Bulk density is equivalent to the weight of unit volume of the aggregate bulk:

$$\gamma_{bv} = \frac{S}{V_a + V_b} \, g/cm^3$$

where S is the weight of the bulk, V_a is the volume of the particles (cm^3/m^3), and V_b is the volume of the interstices between the particles (cm^3). The bulk density is based on the weights of 7- to 15-mm grains and can vary from 1/3 to 2/3 times that of normal-weight aggregate (crushed stone). The bulk density of lightweight aggregate is of the order of 700–800 kg/m^3.

13.3.3 Compactness and Porosity

The compactness of an aggregate is given by

$$t_1 = \frac{\gamma_v}{\gamma}$$

The bulk compactness is

$$t_2 = \frac{\gamma_{bv}}{\gamma}$$

The internal porosity is

$$p_1 = \left[1 - \frac{\gamma_v}{\gamma} \right] \times 100$$

The total porosity is

$$p_2 = \left[1 - \frac{\gamma_{bv}}{\gamma} \right] \times 100$$

The volume of voids between the particles is given by

$$p = p_2 - p_1$$

13.3.4 Moisture Content and Water Absorption

The moisture percentage is given by

$$m = \frac{G_m - G_d}{G_d} \times 100$$

where G_m is the weight of the damp specimen and G_d is the weight of the oven-dried specimen. The water absorption is given by

$$W = \frac{G_w - G_d}{G_d} \times 100$$

where G_w is the weight of the saturated specimen and G_d is the weight of the oven-dried specimen.

Example 13.1

The void ratio of a typical sample of expanded coarse aggregates is calculated as follows:

$$\text{Bulk density}(\gamma_{bv}) = 720 \text{ kg/m}^3$$
$$\text{Specific density}(\gamma) = 2400 \text{ kg/m}^3$$
$$\text{Particle density}(\gamma_v) = 1440 \text{ kg/m}^3$$

These data give the pore content of an individual particle. Find the volume of voids between the particles.
Solution

$$p_1 = \frac{2400 - 1440}{2400} \times 100\% = 40\%$$

and the bulk voids

$$p_2 = \frac{1440 - 720}{1440} \times 100\% = 50\%$$

The volume of voids between particles

$$p = p_2 - p_1 = 50 - 40 = 10\%$$

13.3.5 Strength Properties

There are four methods for testing aggregate strength: (a) compression test of small cubes cut from large units, (b) determination of the impact strength of particles, (c) determination of the compressibility of the particle bulk under pressure, and (d) determination of the crumbling coefficient of particles under pressure.

Test cubes cut from larger units show the strength of individual cubes only and are not related to the strength of the bulk aggregates. The impact strength of the particles is tested by dropping a 10 kg weight 20 times from a height of 20 cm onto 0.5 L of 7- to 15-mm aggregates. The strength of the particles is determined by the percentage of weight of particles passing through a 7-mm mesh sieve. The compressibility of particle bulk under pressure is determined by the specific force needed to compress loose aggregates kept in a 12.7-cm-high metal container of diameter 7.6 cm to within four-fifths of the height of the cylinder.

The crumbling coefficient of particle bulk is determined by gradually loading 5 tonnes onto 0.5 L of 7- to 15-mm aggregates in a standard 5-L container. The maximum loading is reached in one-and-a-half minutes. After loading, the material is removed and passed through 7.3- and 1-mm sieves and the weight of the residue is related to the original quantity. The crumbling coefficient is given by

$$S = F_b - F_a$$

where F_b and F_a are the fineness moduli before and after loading. According to Ujhelyi (1975),

$$C_a = -0.00294 B_d + 3.058$$
$$B_d = -420 C_a + 1125$$

where B_d is the bulk density of the aggregate and C_a is the crumbling coefficient of the aggregate. Table 13.1 shows a comparison between the physical properties of lightweight and natural aggregates.

Table 13.1 Comparison of physical properties of natural stone and low-density (lightweight) aggregates

Physical properties of aggregate	Lightweight aggregate			Natural aggregate		
	Solite coarse 20–5 mm	Solite medium 10–2 mm	Solite fine 5–0 mm	Natural sand 5–0 mm	Natural stone 15–5 mm	Natural stone 20–2 mm
Gradation (% passing by weight)						
25 mm	100					100
20 mm	95				100	98
15 mm	50	100			95	70
10 mm	25	95			55	40
5 mm	5	40	100	100	10	8
2.4 mm		5	85	95	2	4
1.2 mm			55	75		
0.6 mm			40	50		
0.3 mm			25	10		
0.15 mm			12	2		
Bulk dry loose density (kg/m³)	720	800	880	1440	1520	1520
Dry particle density (kg/m³)	1420	1500	1650	2620	2650	2790
Moisture content (wt%) for 1-day soak	9	10	12	2	1	1

13.4 Factors Influencing the Strength and Density of Lightweight Aggregate Concrete

The strength and density of lightweight aggregate concrete is influenced by a number of parameters associated with both the aggregates and the method of making concrete. The issues involved differ from the methods adopted for manufacturing normal-weight concrete. These factors are discussed below.

13.4.1 Grade and Granulometry of Aggregates

The strength of concrete depends on the strength and, to a certain extent, the surface and shape of aggregates. The density and strength of aggregates with high crumbling coefficients decrease if the proportion of fines is reduced. Experiments show an increased strength of lightweight concrete if the proportion of 0–1 mm fines is greater than or equal to 30% of the aggregates. For increased strength and density, and improved weather resistance, the fine grains of lightweight aggregates are substituted with sand. This in turn increases the cement requirement.

13.4.2 Grade and Proportion of Cement

Increase in strength and density is not related to the quantity of cement linearly. Low cement content (150–300 kg/m^3) may result in relatively lower density and higher strength than for high cement content (350–450 kg/m^3) if the concrete is compacted well. Rich concrete if not compacted well may result in the same strength as that of a well-compacted weak concrete.

13.4.3 Mixing Water and Consistency

In order to obtain thorough compaction and flowability around the reinforcement, the concrete must have a good consistency, which can be checked using the slump or penetration test. Due to the high rate of water absorption of aggregates, the consistency of lightweight aggregate concrete is less affected by the quantity of water than that of ordinary concrete. Without adequate water, the required strength cannot be obtained; this is true also if too much water is absorbed by the aggregate. The consistency of concrete can best be determined by trial. Tests carried out by Lewis (1982) to determine the consistency of foamed-slag concretes showed that a plastic consistency can be induced by adding 2–4% (by weight of cement) vinsol resin, an entraining agent, without any loss in the strength of concrete.

13.4.4 Method of Mixing of Concretes

Gravity-type mixers are adequate for concretes having a low content of fine aggregates (10–15%). Positive-type mixers are most effective if the fine aggregate proportion is 30–50%; they lead to a strength increase of 30–100% as compared to gravity-type mixers. The final strength of concrete also depends on the sequence in which the components are fed to the mixer. The best result can be obtained by adding cement to aggregate already saturated with water.

13.4.5 Compaction

Strength and density are closely related to the degree of compactness. Compaction varies with bulk density, proportion of 0–1 mm fines, cement content, and consistency of material. The degree of compactness is the density ratio of compactly placed mix to loose concrete mix. Continuously graded concrete can be compacted by a vibration needle, whereas no-fines concrete can be compacted by ramming (or rodding). The strength varies with the vibrating frequency and amplitude as well as with the type of vibrator used. Compaction efficiency can be improved by applying pressure during vibration.

13.4.6 Curing of Concrete

Fresh concretes are extremely sensitive to intensive sunshine and wind and, hence, must be protected by damp sheet covers and by spraying water. Cold weather is also harmful, causing the hardening to slow down. Precast units if autoclaved or steam-cured should not be exposed to temperatures less than 15°C.

13.5 Properties of Lightweight Concrete

The suitability of a particular type of lightweight aggregate concrete depends on its desired properties: compressive strength, density, elastic properties, thermal properties, durability, and cost. We will look at some of the important properties of lightweight concrete in this section.

13.5.1 Compressive Strength

The compressive strength of lightweight aggregate concrete is generally lower than that of normal-weight aggregate concrete. To achieve higher strength, normal sand can be used; to achieve even higher strength, smaller aggregates can be used. It has also been observed that the rate of strength gain for lightweight aggregate at later ages is more. This is due to the longer period of hydration owing to the moisture trapped in the porous aggregate. This internal curing process can be fully realized if soaked aggregates are used.

By using proper admixtures, both mineral and chemical, the voids in the matrix can be reduced to achieve higher strength and density. Figure 13.3 shows an increase in strength from 30 to 60 MPa in lightweight aggregate when its density is improved from 1360 to 1900 kg/m^3. By proper mix proportioning and using mineral and chemical admixtures, lightweight high-strength concrete can be produced.

Table 13.2 shows the typical properties of lightweight high-strength aggregate concretes. Table 13.3 shows the (typical) mixture proportions for M35 and M45 concrete made with lightweight aggregate (solite) and both mineral and chemical admixtures for enabling pumping of concrete.

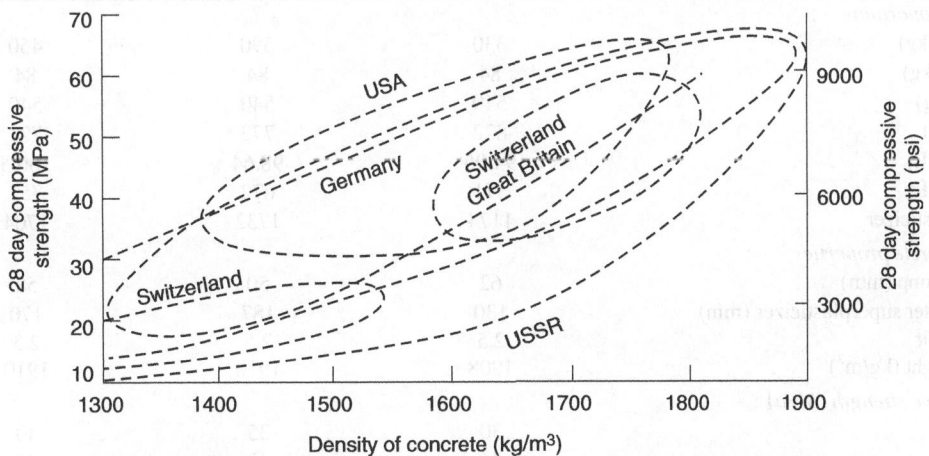

Fig. 13.3 Compressive strength versus density of lightweight aggregate concrete
(*Source*: Nawy 2001)

Table 13.2 Typical properties of lightweight high-strength concrete (HRWRAs plus commercial retarder)

Property	Ordinary Portland cement			Pozzolana cement (fly ash interground)		
	1	2	3	4	5	6
Mixture proportions						
Cement (kg/m³)	560	950	950	500	500	500
Fly ash interground (kg/m³)				60	60	60
Water (L)	210	335	295	180	190	190
Natural sand (kg/m³)	660	1290	1220	660	760	720
Coarse aggregate type	LWA	LWA	Natural stone	LWA	LWA	Natural stone
Size (mm)	10	20	15	10	20	15
Quantity (kg/m³)	650	900	1670	650	530	1670
Concrete						
Air content (%)	2.25	2.25	2.0	2.0	2.0	2.0
Slump (mm)	100	150	100	100	140	130
Density (kg/m³)						
Fresh	2020	1960	2410	2040	1990	2370
Equilibrium	1950	1930	2380	1990	1920	2360
Oven-dry	1920	1900	2360	1960	1880	2330
Strength properties						
Compressive strength at 28 days (MPa)	67.4	52.5	53.3	61.8	46.8	54.5
Compressive modulus of elasticity at 210 days (MPa)	24	21.9	43.8		27.1	43.9
E_{test}/E_{calc}	0.76	0.82	1.21		0.99	0.90

Table 13.3 Mixture proportions for fly ash lightweight aggregate pumped concrete

	Mixture 1	Mixture 2	Mixture 3
Mixture proportions			
Cement (kg)	330	390	450
Fly ash (kg)	84	84	84
Solite (kg)	540	540	540
Sand (kg)	822	772	721
Water (mL)	94.06	98.64	100.5
WRA (mL)	571	672	757
Superplasticizer	1174	1732	1704
Fresh concrete properties			
Initial slump (mm)	62	50	58
Slump after superplasticizer (mm)	130	187	170
Percent air	2.5	2.5	2.3
Unit weight (kg/m³)	1908	1910	1910
Compressive strength (MPa)			
4 days	30	35	40
7 days	34	40	45
28 days	44	48	52
Splitting tensile strength (MPa)	3.64	3.78	4.0

13.5.2 Modulus of Elasticity and Poisson's Ratio

The modulus of elasticity and Poisson's ratio are important elastic properties of concrete. Figure 13.4 shows the relationship between modulus of elasticity and strength of lightweight aggregate concrete. For concrete of density 1500–2500 kg/m^3 and strength up to 35 MPa, the ACI code recommends the following relationship for modulus of elasticity of concrete:

$$E_c = w^{1.5} \times 0.043 \sqrt{f_c'} \text{ MPa}$$

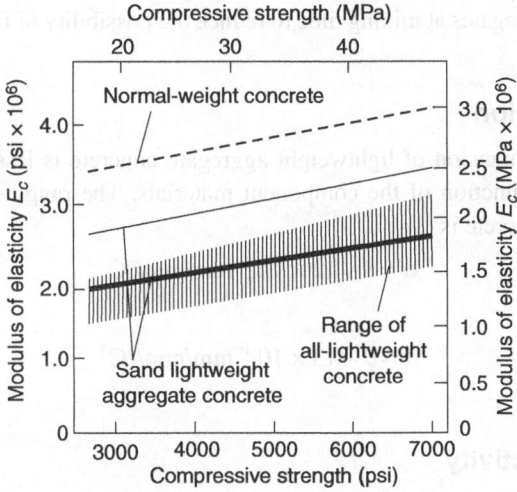

Fig. 13.4 Relationship between modulus of elasticity and strength of concrete
(*Source*: Nawy 2001)

For high-strength lightweight concrete, the following equation for strength in the range 40–50 MPa is applicable:

$$E_c = \left(3.32\sqrt{f_c'} + 6895\right)\left(\frac{w^{1.5}}{2320}\right) \text{ MPa}$$

In the above equation f_c' is the cylinder compressive strength of concrete. Note that $f_c' = 0.7 f_{ck}$ is approximately, where f_{ck} is the characteristic cube compressive strength of concrete.

Poisson's ratio (v) based on tests varies between 0.15 and 0.25 and for practical purposes $v = 0.2$ can be adopted. Both the modulus of elasticity and Poisson's ratio are less for lightweight concrete when compared with normal-weight concrete.

13.5.3 Water Absorption and Moisture Content

Lightweight coarse aggregate has a high degree of absorption because of its cellular structure. It generally absorbs 5–20% water compared to the 2% absorbed by normal-weight aggregate. Hence it is important to account for this absorption for evaluating workability in the mix design.

A slump of 75–100 mm in lightweight aggregate concrete maintains cohesiveness, preventing lighter coarse particles from migrating towards the surface when the concrete is vibrated under wet conditions. This behaviour is different from that of normal-weight concrete, where sand particles can migrate upwards to cause segregation.

13.5.4 Creep and Shrinkage

The values of both creep and shrinkage are greater for lightweight concrete as compared to normal-weight concrete. However, when the strength of concrete is increased, the value of creep for lightweight concrete becomes less. For ultrahigh-strength lightweight concrete, the values of creep for both normal-weight and lightweight concrete become equal.

13.5.5 Durability

In order to make a structure using lightweight aggregate concrete durable, it is necessary to take all the precautions one will take for a structure made of normal-weight concrete. In addition, special care has to be taken to protect saturated aggregates at mixing time to reduce the possibility of frost damage during the initial hardening stages of concrete.

13.5.6 Thermal Expansion

The coefficient of thermal expansion of lightweight aggregate concrete is less than that of normal-weight concrete and is primarily a function of the component materials. The range of the coefficient of thermal expansion for lightweight concrete is

$$\alpha = 8 - 12 \times 10^{-6} \text{ mm/mm/°C}$$

whereas it is

$$\alpha = 9 - 13 \times 10^{-6} \text{ mm/mm/°C}$$

for normal-weight concrete.

13.5.7 Thermal Conductivity

Lightweight aggregate concrete has excellent insulating properties. Thermal conductivity is a measure of the rate at which heat energy passes through a unit area of material of unit thickness per degree temperature gradient:

$$k = 0.0072e^{0.00125w} \text{ (J/m}^2\text{s °C/m)}$$

where $e = 2.71828$ and w is the unit weight of concrete (in kg/m^3). Figure 13.5 gives the average k values for the oven-dry density of concrete in the range 400–2400 kg/m^3.

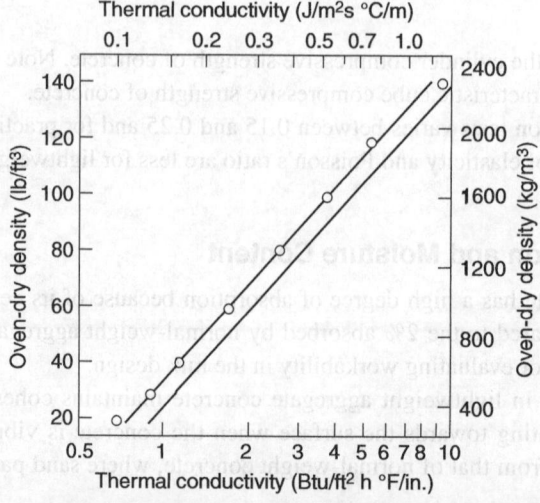

Fig. 13.5 Thermal conductivity versus density of concrete (*Source*: Nawy 2001)

Example 13.2 Proportioning a sand–lightweight aggregate concrete mixture using the weight method

Design a concrete mixture by the weight method using lightweight coarse aggregate and normal-weight fine aggregate (sand) for a structural lightweight concrete slab with 28-day compressive strength f_{ck} = M35. Use the following data in the mixture design:

Coarse aggregate size = 20 mm

SG factor = 1.5

Absorption = 11.0%

Fine aggregate: absorption = 1.0%

Fineness modulus = 2.80

Oven-dry loose weight of coarse aggregate = 780 kg/m³

Solution

Step 1   Unit weight of constituent materials From Table 13.4, the desired slump is 75–100 mm. Use air entrainment particularly if the structure is exposed to severe weathering during construction. From Table 13.5, the water–cement ratio required to produce M35 concrete is 0.40. From Table 13.6, the mixing water requirement is 168 kg/m³. This value falls within the maximum permissible limit shown in Table 13.7.

$$\text{Required cement content} = \frac{168}{0.40} = 420 \text{ kg}$$

Table 13.4 Recommended slumps for various types of construction

Type of construction	Slump (mm)	
	Maximum	**Minimum**
Beams and reinforced walls	100	25
Building columns	100	25
Floor slabs	75	25

Note: Slump may be increased when chemical admixtures are used, provided that the admixture-treated concrete has the same or lower water–cement or water–cementitious materials ratio and does not exhibit segregation potential or excessive bleeding. The slump may be increased by 25 mm for methods of consolidation other than vibration.

Table 13.5 Relationship between water–cement ratio and compressive strength

Compressive strength at 28 days (MPa)	Approximate water–cement ratio	
	Non-air-entrained concrete	**Air-entrained concrete**
42	0.41	0.38
35	0.48	0.40
28	0.57	0.48
21	0.68	0.59
14	0.82	0.74

Note: Values are estimated average strengths for concrete containing not more than 2% air for non-air-entrained concrete and 6% total air content for air-entrained concrete. For a constant water–cement or w/(c+p), the strength of concrete is reduced as the air content is increased. The 28-day strength values may be conservative and may change when various cementitious materials are used. The rate at which the 28-day strength is developed may also change. Strength is based on a 15-cm cube moist cured for 28 days. Practice for making and curing concrete test specimens in the field: These are cubes moist cured at 25 ± 2°C prior to testing. The relationships in this table assume a nominal aggregate size of about 10–20 mm. For a given source of aggregate, strength produced at a given water–cement or w/(c+p) increases as the nominal maximum size of aggregate decreases.

Table 13.6 Approximate mixing water and air content requirements for different slumps and nominal maximum sizes of aggregates

Slump (mm)	Water (kg/m³ of concrete) for nominal maximum sizes (mm) of aggregate		
	10	15	20
Air-entrained concrete			
25–50	183	177	168
75–100	204	195	183
125–150	213	201	189
Recommended average total air content (%)			
Mild exposure	4.5	4.0	4.0
Moderate exposure	6.0	5.5	5.0
Extreme exposure	7.5	7.0	6.0
Non-air-entrained concrete			
25–50	210	201	189
75–100	231	219	204
125–150	240	225	210
Approximate amount of entrapped air in non-air-entrained concrete (%)	3	2.5	2

Notes: The quantities of mixing water given for air-entrained concrete are based on typical total content requirements as shown for 'moderate exposure' in the table. These quantities are for use in computing the cement or cementitious materials content for trial batches at 25–30°C. They are maximum for reasonably well-shaped angular aggregates graded within limits of accepted specifications. The use of water-reducing chemical admixtures may also reduce mixing water content by 5% or more. The volume of the liquid admixtures is included as part of the total volume of the mixing water. The slump values of 175–275 mm are obtained only through the use of water-reducing chemical admixture; they are for concrete containing nominal maximum-size aggregate not larger than 25 mm.

Additional recommendations for air content and necessary tolerances on air content for control in the field are given in a number of ACI documents, including ACI 201, 345, 318, 301, and 302 C 94 for site-mixed and ready-mixed concrete.

These values are based on the criteria that 9% air is needed in the mortar phase of the concrete. If the mortar volume is substantially different from that determined in this recommended practice, it may be desirable to calculate the needed air content by taking 9% of the actual mortar value.

Table 13.7 Maximum permissible water–cement ratios for concrete

Type of structure	Structure wet continuously or frequently and exposed to freezing and thawing	Structure exposed to seawater or sulphates
Thin sections (railings, curbs, sills, ledges, ornamental work) and sections with less than 25 mm cover over steel	0.45	0.40
All other structures	0.50	0.45

Note: Concrete should also be air-entrained. If sulphate-resisting cement is used, the permissible w/c or w/(c + p) ratio may be increased by 0.05.

From Table 13.8, for unit volume, the coarse aggregate content is $1 \times 0.50 = 0.50$ m³.

$$\text{Dry weight of coarse aggregate} = 0.50 \times 780 = 390 \text{ kg}$$

Since the coarse aggregate has an absorption of 11.0%,

$$\text{Saturated weight} = 1.11 \times 390 = 432.9 \approx 433 \text{ kg}$$

Table 13.8 Volume of coarse aggregate per unit volume of concrete

Maximum size of aggregate (mm)	Volume of oven-dry loose coarse aggregates per unit volume of concrete for different fineness modulus of sand			
	2.40	**2.60**	**2.80**	**3.00**
20	0.56	0.54	0.50	0.49
50	0.67	0.65	0.63	0.61
75	0.74	0.72	0.70	0.68

Note: Volumes are based on aggregates in oven-dry loose condition. The volumes are selected from empirical relationships to produce concrete with a degree of workability suitable for usual reinforced construction. For more workable concrete, such as sometimes required when placement is by pumping, they may be reduced up to 10%.

The remaining constituent material would be fine normal-weight aggregate (sand) in this case. From Table 13.9, the weight of 1 m³ of air-entrained concrete made with lightweight aggregate having a specific gravity factor of 1.5 is estimated to be 1788 kg/m³ for 6% air entrainment. Hence,

$$\text{Saturated dry weight of sand (SSD)} = 1788 - (168 + 420 + 433)$$

$$= 767 \text{ kg}$$

$$\text{Oven-dry weight of sand} = \frac{767}{1.01} = 759 \text{ kg}$$

The details of laboratory trial batches for producing 1 m³ of saturated surface-dry aggregate concrete are given in the second column of Table 13.10.

Step 2 *Adjustment of initial mixture* Tests indicate that the total moisture content for lightweight coarse aggregate and fine aggregate is 15% and 6%, respectively. The absorbed water does not become a part of the mixing water and must be excluded from the adjustment of added water. Therefore, the surface water contributed by coarse lightweight aggregate (low-density coarse aggregate) is 15.0 – 11.0 = 4% and that by fine normal-weight aggregate (normal-density fine aggregate–sand) is 6.0 – 1.0 = 5%. Adjust the aggregate for this free moisture:

$$\text{Fine aggregate} = \frac{767}{1.01} \times 1.06 = 804 \text{ kg}$$

$$\text{Coarse aggregate} = \frac{433}{1.11} \times 1.15 = 448.6 \approx 449 \text{ kg}$$

The adjustment for the added water, to account for the moisture added to the aggregate, becomes

$$\text{Water from fine aggregate} = 804 - 767 = 37 \text{ kg}$$

$$\text{Water from coarse aggregate} = 449 - 433 = 16 \text{ kg}$$

Consequently, the amount of water to be added to the batch is equal to 168 – (37 + 16) = 115 kg. The weights to be used for a 1 m³ wet aggregate trial batch are given in the third column of Table 13.10. By trial and adjustment, the quantity of water that can give 75–100 mm of slump was found in the laboratory to be 152. Hence,

$$\text{Additional water} = 152 - 115 = 37 \text{ kg}$$

$$\text{Total weight of adjusted batch in table} = 1788 + 37 = 1825 \text{ kg}$$

Table 13.9 First estimate of the weight of fresh lightweight concrete comprising of low-density coarse aggregate and normal-density fine aggregate (sand)

Specific gravity factor	First estimate of weight of lightweight concrete (kg/m³) for air-entrained concrete (%)		
	4	6	8
1.00	1614	1578	1536
1.20	1698	1662	1626
1.40	1788	1746	1710
1.60	1872	1830	1794
1.80	1956	1920	1878
2.00	2046	2004	1962

Note: Values for concrete of medium richness 330 kg/m³ and medium slump with water requirements based on values for 75–100 mm of slump in table. If desired, the estimate weight may be refined as follows (if necessary information is available): For each 5 kg difference in mixing water from table, correct weight per m³, 3 kg in the opposite direction; for each 40 kg difference in cement content from 330 kg, correct the weight per m³, 3 kg in the same direction.

Table 13.10 Laboratory trial batches

	Content (kg/m³)		
	(a) Saturated surface dry aggregate trial batch	(b) Wet aggregate trial batch	(c) After adjustment
Cement	420	420	435
Fine aggregate (SSD)	767	804	776
Coarse aggregate (SSD)	433	449	403
Water (net mixing)	168	115	174
Total	1788	1788	1788

Step 3 *Adjustment for proper batch yield* It was found further that the concrete mixture, although satisfactory as per finishing and workability properties, had a slump of 50 mm and a unit weight of 1788 kg. To provide for proper yield for a future trial batch, make the following adjustments:

$$\text{Yield of trial batch} = \frac{1825}{1788} = 1.02 \text{ m}^3$$

Mixing water actually used = 115 (added) + 37 (fine aggregate) + 16 (coarse aggregate) = 168. Mixing water required for 1 m³ of concrete with the same 50-mm slump at the trial batch is 168/1.02 = 164 kg. For each 25 mm of the desired increase in slump, the amount of mixing water should be increased at the rate of approximately 6 kg/m³ when the initial slump is 75 mm and at a lower rate when the initial slump is higher. Hence, add about 10 kg/m³ of water to raise the slump from the measured 50 mm to about 75–100 mm, resulting in a net mixing water amount of 164 + 10 = 174 kg. The adjusted cement weight because of the added water while maintaining a w/c value of 0.40 is 174/0.40 = 435 kg. Since the workability was found to be satisfactory, we maintain the quantity of lightweight coarse aggregate per unit volume of concrete used in the trial batch. The amount of coarse aggregate per m³ becomes 449/1.01 = 444.55 (wet), which is equal to 444/1.15 = 387 kg (dry) or 463 (SSD). As calculated before, the estimate for the weight of a unit volume of concrete is 1788 kg. Therefore, the required amount of fine aggregate per m³ is

$$1788 - [174 \text{ (water)} + 435 \text{ (cement)} + 403 \text{ (coarse SSD)}]$$
$$= 776 \text{ (SSD)} \text{ or } 776/1.01 = 768 \text{ kg (dry)}$$

Step 4 *Final proportions* The adjusted batch weights for dry and SSD aggregates are given in Table 13.10.

13.6 Design of Lightweight Concrete

The strength of lightweight concrete depends on various factors, such as the shape of the aggregate surface grains, while the bulk density remains unchanged. Thus, there is no standard method of design, and even the method suggested by G. Rothfuchs does not give exact results for all types of aggregates. The principle of Rothfuchs' method is that the strength of concrete depends on the relation between the cement content C and porosity P of the aggregates. The ratio is an important parameter as defined below.

$$\frac{C}{P} = \frac{\left[\text{cement content (kg/m}^3) \right]}{\left[\text{porosity (dm}^3/\text{m}^3) \right]}$$

Cement porosity and concrete strength are affected by the proportion of fines (0–1 mm), cement content, and compactness. To design concrete of required strength, the weight of aggregate per m³, the density of aggregate, and the quantity of cement must be ascertained. The compacted volume of aggregate can be determined from the weight and density of the aggregate contained in 1 m³ of concrete using the graph shown in Fig. 13.6(a). The concrete composition can be read from the graph in Fig. 13.6(b). A straight line can be started from the point denoting the concrete strength in Fig. 13.6(c) until it meets the curve, following the arrow, the curve of

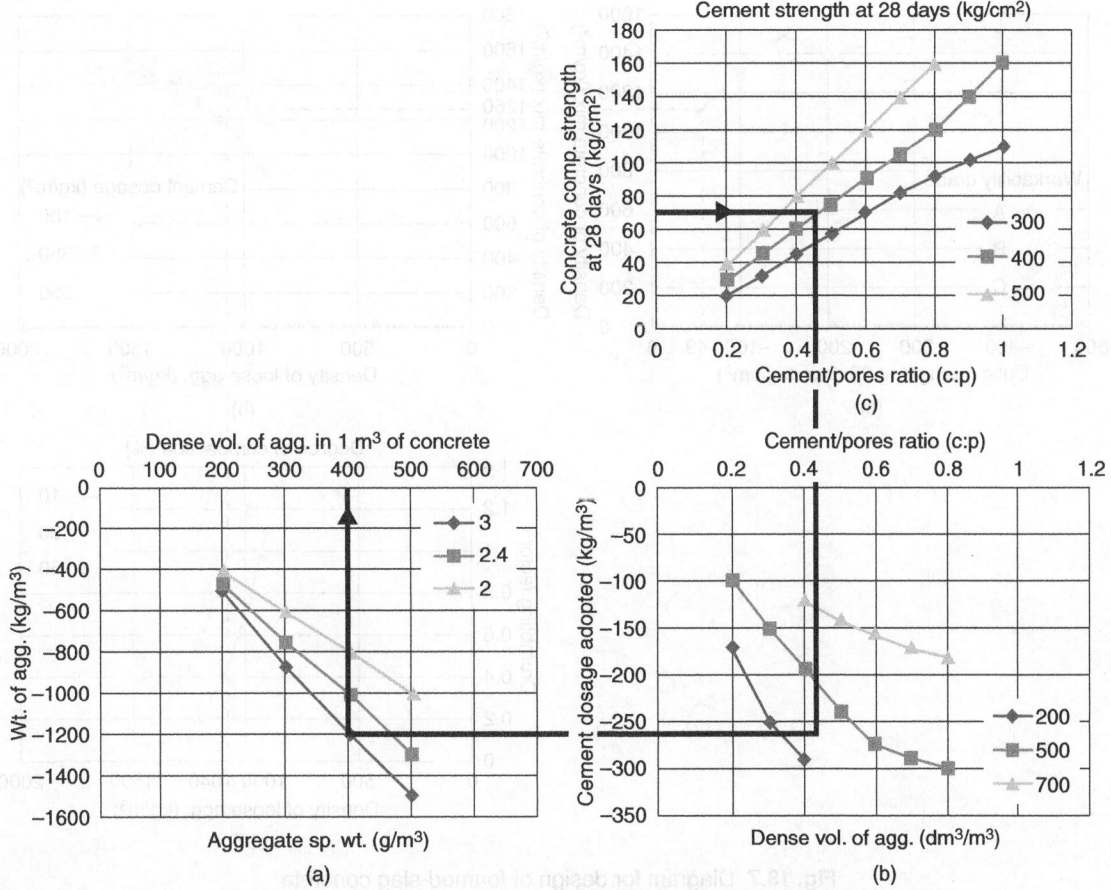

Fig. 13.6 Diagram for design of lightweight concretes

given cement strength 400 kg/m². The point of intersection will be projected to the curve for dense volume and a horizontal line started at this second intersection point indicates the required quantity of cement.

13.6.1 Ujhelyi's Modification

Ujhelyi suggested that the shape and surface effect of aggregates need not be considered; he evaluated concrete strength taking into consideration the granulometry and strength of aggregates, the quantity and grade of cement, and the degree of compactness, while introducing a parameter M, called *workability modulus*

$$M = T_f \times C$$

where T_f is the degree of compactness in per cent and C is the cement content in kg/m³. Estimating the concrete strength by using this parameter requires the following:

• Bulk density of aggregate
• Proportion of fines (0–1 mm)
• Cement content
• Density of fresh concrete (which can be determined by knowing the weights of the material used from their unit weights)

Based on these data, concrete strength can be assessed by using the graphs shown in Fig. 13.7.

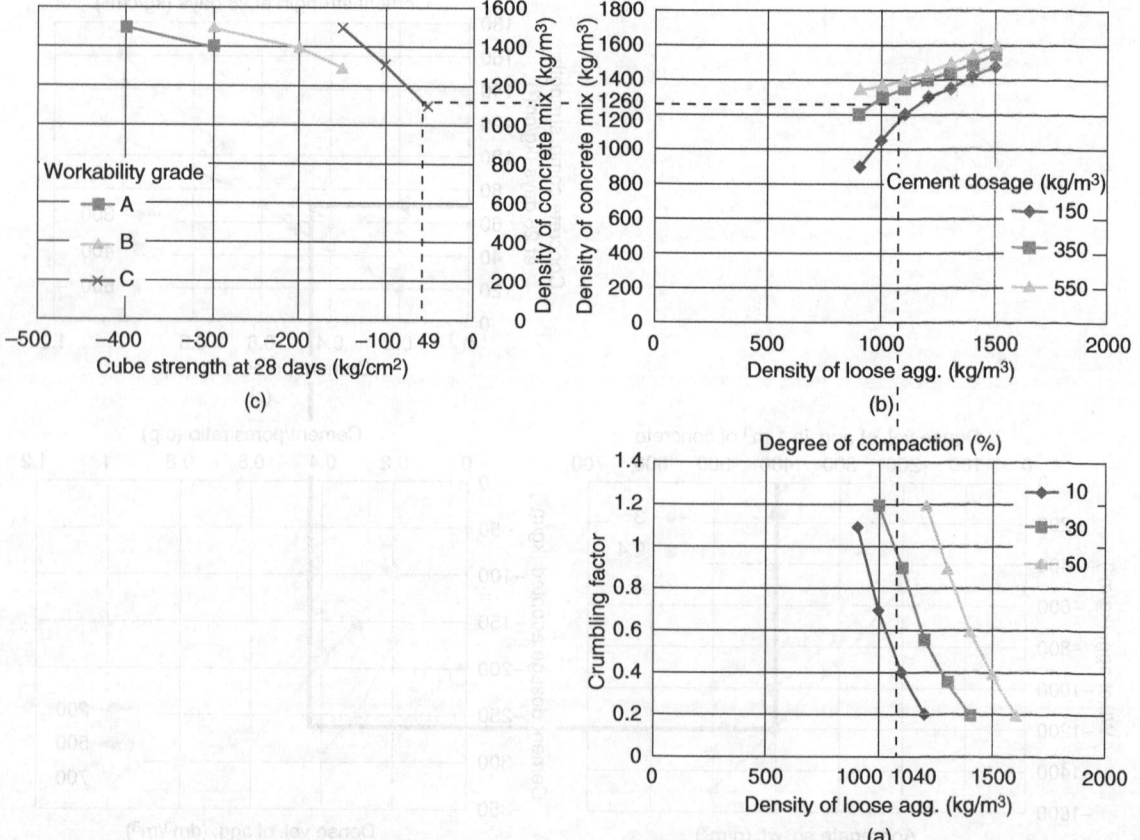

Fig. 13.7 Diagram for design of foamed-slag concrete

Example 13.3

Find the expected concrete strength at 28 days if the bulk density of 7–15 mm aggregate particles is 0.7 g/cm^3, proportion of fines (0–1 mm) is 25%, cement content is 250 kg/m^3, and density of on-site concrete is 1500 kg/m^3.

Solution

From graph (a) in Fig. 13.7 the bulk density of loose aggregate is 1040 kg/m^3, and the value of density for loose concrete mix is 1260 kg/m^3 as given in graph (b). The degree of compactness

$$\frac{1500 - 1260}{1260} \times 100 = 19.1\%$$

Workability modulus

$$M = 19.1 \times 2.50 = 47.750$$

Reading from graph (c) in Fig. 13.7, the required concrete strength is 49 kg/cm^2.

13.6.2 Mix Design Methods used in the United Kingdom

In the UK, *Road Note 4* or its modified version is used for the design of concrete mixes with natural aggregates. A separate *Road Note 4* is needed for lightweight aggregate concrete. *Road Note 4* gives a curve relating the compressive strength to the water–cement ratio at various ages for mix proportions made with ordinary and with rapid-hardening Portland cement. Table 13.11 relates the water–cement ratios to the mix proportions for aggregates of different maximum sizes, shapes, and grading.

J.W. Westley (of Lytag Ltd) has developed a new method of designing lightweight aggregate concrete mixes based on the net or effective water–cement ratio. The details of his mix proportions to give four degrees of workability concrete of average grading are given in Table 13.11. The relationship between the compressive strength of concrete at various ages and the effective water–cement ratio can be seen from Fig. 13.8.

Example 13.4

Design a suitable lightweight concrete mix. Assume that the mean strength required is 400 MPa.

Solution

Referring to Fig. 13.8, for this required strength at 28 days, the effective water–cement ratio will be 0.53.
    Referring to Table 13.11, for medium/high workability the mix proportions are

Cement content = 390 kg/m^3
Aggregate–cement ratio by weight = 2.79 (with 38% fines)

The strength of lightweight aggregate (LWA) is the predominant factor in the strength of LWA concrete (LWAC). So LWAC will fail if the aggregate is weak in strength. In the present case, concrete is considered as a two-component material consisting of mortar and aggregate. The strength of mortar is taken as the unit strength. Figure 13.9 shows the variation of the strength of concrete (σ)/strength of mortar (σ_m) with the variation of the strength of aggregate (σ_a)/strength of mortar (σ_m). A basic model is arrived at using the formula

$$\frac{\sigma}{\sigma_m} = \left(\frac{\sigma_a}{\sigma_m}\right) n$$

where n is the volume fraction. It can be seen that the strength of concrete having a high volume fraction of weaker aggregate is only one-third to one-fifth of the mortar strength. If the aggregate strength is not small compared to the strength of the mortar, the concrete behaves like ordinary concrete. This is probably the case

Table 13.11 Detail of mix proportions to give four degrees of workability of concrete made with Lytag of average grading

Effective water–cement ratio by weight	Low			Low-medium			Medium-high			High		
	Compacting factor											
	0.80			0.85			0.90			0.95		
	Slump (in.)											
	0			0–1/2			(1/2)–3			3–6		
	a/c volume (weight)	% fines volume (weight)	Cement content (kg/m³; lb/yd³)	a/c volume (weight)	% fines volume (weight)	Cement content (kg/m³; lb/yd³)	a/c volume (weight)	% fines volume (weight)	Cement content (kg/m³; lb/yd³)	a/c volume (weight)	% fines volume (weight)	Cement content (kg/m³; lb/yd³)
0.30	2.56(1.60)	28(32)	554(934)	2.39(1.50)	27(31)	597(1006)	2.20(1.38)	27(31)	654(1102)	2.31(1.45)	26(30)	657(1108)
0.35	3.11(1.96)	29(33)	490(826)	2.84(1.78)	28(32)	533(899)	2.55(1.59)	27(31)	588(991)	2.65(1.66)	27(31)	589(994)
0.40	3.72(2.34)	31(35)	429(723)	3.38(2.13)	30(34)	475(796)	3.00(1.88)	28(32)	524(883)	3.03(1.90)	28(32)	532(897)
0.45	4.34(2.75)	34(39)	376(634)	3.93(2.48)	32(36)	420(707)	3.50(2.20)	30(34)	468(788)	3.50(2.20)	30(34)	474(798)
0.50	5.11(3.28)	40(45)	328(553)	4.57(2.90)	35(40)	370(624)	4.04(2.55)	32(36)	418(705)	4.01(2.53)	32(36)	424(715)
0.55	6.15(4.04)	52(57)	276(464)	5.35(5.44)	41(46)	325(548)	4.63(2.95)	35(40)	372(626)	4.55(2.89)	34(39)	383(645)
0.60	6.76(4.55)	63(68)	245(413)	6.05(3.96)	49(54)	288(486)	5.35(3.43)	40(45)	332(559)	5.23(3.35)	38(43)	341(575)
0.65	7.64(5.30)	79(82)	208(351)	6.91(4.63)	61(66)	250(421)	6.05(3.96)	49(54)	293(494)	5.99(3.88)	44(49)	306(515)
0.70				8.02(5.52)	73(78)	219(370)	6.83(4.55)	58(63)	259(437)	6.75(4.55)	53(58)	272(499)
0.75							8.02(5.47)	69(74)	229(386)	7.55(5.10)	64(69)	243(410)
0.80										8.55(5.90)	74(79)	224(378)
0.85												

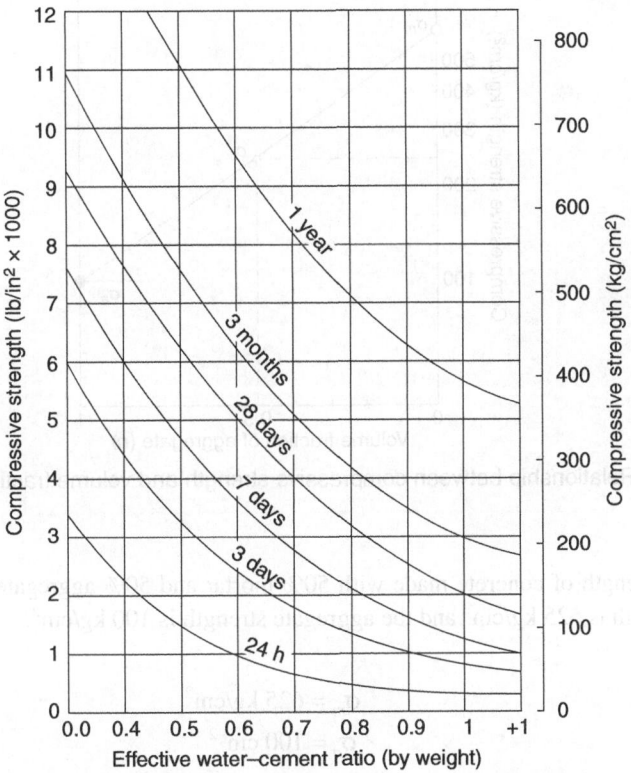

Fig. 13.8 Relationship between compressive strength and effective water–cement ratio by weight

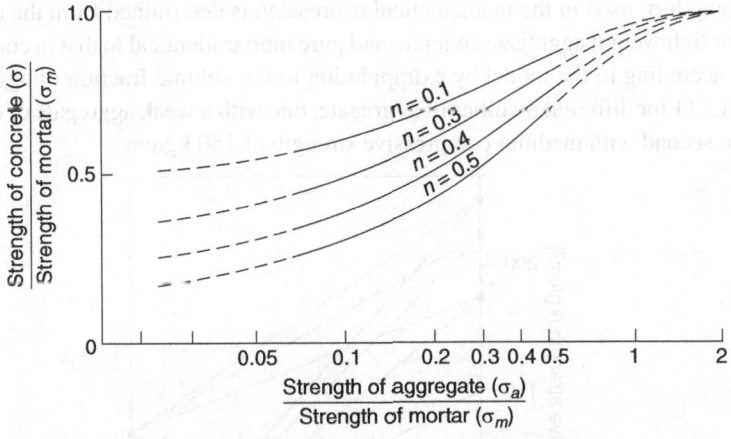

Fig. 13.9 Lightweight aggregate concrete as a two-component material

with concrete made with some strong lightweight aggregate in the UK and the USA. In India, the use of lightweight aggregate is still in its infancy.

In Fig. 13.10, the compressive strength is shown in log scale against the volume fraction of aggregate in linear scale. The relation between the compressive strength and the volume fraction of aggregate is given by

$$\sigma \cong \sigma_a^n \times \sigma_m^{1-n}$$

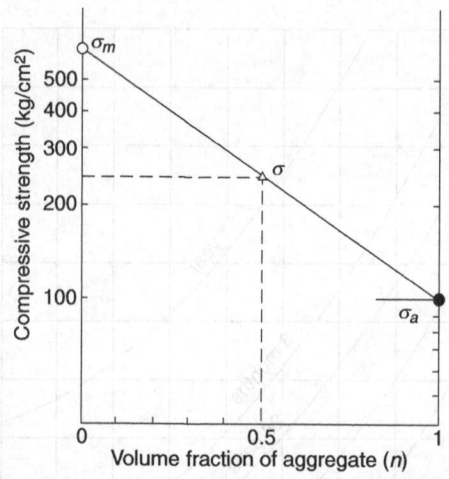

Fig. 13.10 Relationship between compressive strength and volume fraction of aggregate

Example 13.5

Calculate the strength of concrete made with 50% mortar and 50% aggregate by volume. Assume that the mortar strength is 625 kg/cm² and the aggregate strength is 100 kg/cm².

Solution

$$\sigma_m = 625 \text{ kg/cm}^2$$
$$\sigma_a = 100 \text{ cm}^{-2}$$

Therefore, the strength of concrete

$$\sigma = 100^{0.5} \times 625^{(1-0.5)} = 250 \text{ kg/cm}^2$$

The aggregate strength σ_a used in the mathematical expression is determined from the compressive strength values obtained for lightweight aggregate concrete and pure mortar identical to that in concrete. The aggregate strength is found according to the model by extrapolation to the volume fraction of aggregate $n = 1$. This is explained in Fig. 13.11 for different qualities of aggregate, one with a weak aggregate of compressive strength 50 kg/cm² and the second with medium compressive strength of 150 kg/cm².

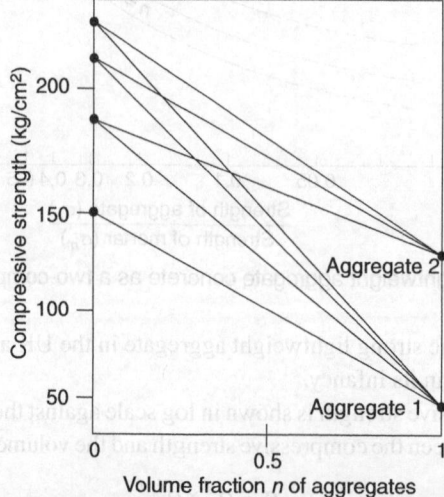

Fig. 13.11 Influence of weak and strong aggregates

Determining aggregate strength has the following two advantages:

1. The strength of aggregates may be related directly to the strength of the concrete using the two-component model.
2. The aggregate strength is independent of the mortar used. This is normally not the case when one tries to evaluate aggregate strength from tests on the concrete.

With the knowledge of the strength–weight relationship of the aggregate, the theoretical model makes it possible to predict the aggregate quality. This is not suitable for given concrete qualities, as is illustrated in Fig. 13.12.

The best aggregate for very high-strength concrete of 850 kg/cm^2 is the aggregate that is strong and heavy (shown by the outer line in Fig. 13.12). Such heavy-strength concrete demands a mortar of strength 1000–1500 kg/cm^2. At the more practical strength level of 400–500 kg/cm^2, we cannot get the full benefit of the strong aggregate. So the lighter medium strength is the best here (Fig. 13.13). If a low unit weight of concrete is required, lightweight aggregate is used. Medium-strength, lightweight aggregate, using sandstone mortar, can be recommended for concrete with a unit weight above 1600 kg/m^3.

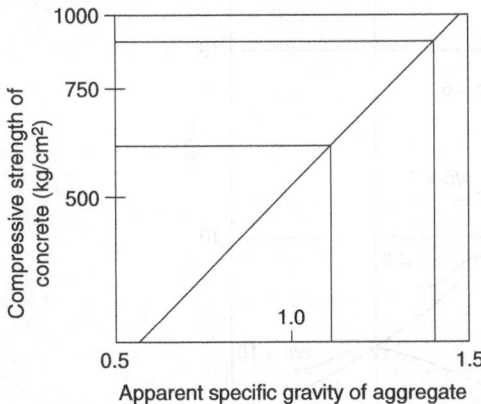

Fig. 13.12 Strength–weight relationship of aggregates

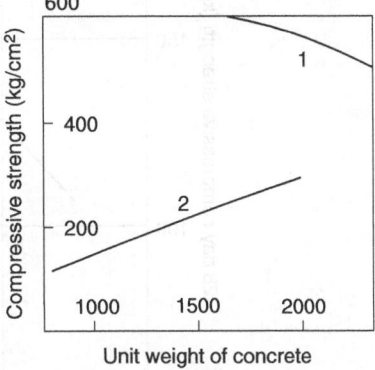

Fig. 13.13 Compressive strength versus unit weight of concrete relationship

13.6.3 Design of No-fines Concrete Mixes

No-fines concrete is normal concrete in which finer fractions of aggregates are missing. It can be considered a form of lightweight concrete. The absence of finer fractions creates voids uniformly distributed throughout the mass of the concrete. This reduces the density of the concrete. The advantages of no-fines concrete are the high degree of thermal insulation, speedy construction, low density, and less shrinkage. Since the segregation in no-fines concrete is practically nil due to the absence of the fine aggregate, it can be safely dropped from height while casting.

The absence of fine aggregates results in a lesser surface area, which helps reduce the cement requirement. The cement quantity required will be of the order of 70–130 kg/m^3.

The density of no-fines concrete depends on the type of grading of the coarse aggregate used. Generally, it will be in the range of 60–75% of that of normal concrete. For lightweight aggregate, it can be as low as 1800 kg/m^3. All types of aggregate can be used to produce no-fines concrete.

The compressive strength of no-fines aggregate depends on its density and varies between 70 kg/cm^2 for a density of 1900 kg/m^3 to 140 kg/cm^2 for a density of 2100 kg/m^3. The strength of no-fines concrete increases with age, as is the case with normal concrete. The water–cement ratio is not the only controlling factor that determines the strength. However, for a given aggregate–concrete ratio, there is an optimum water–cement ratio that yields the highest strength. A high water–cement ratio results in segregation, while a low water–cement ratio adversely affects workability.

13.6.4 Mix Design

The design of no-fines concrete is based on the strength required at a particular age. The relations between the water–cement ratio and strength and between the aggregate–cement ratio and strength are given in Fig. 13.14. The proportioning is based upon the cellular structure required and the surface area of particles to be coated with cement paste. Trial mixes are made using the estimated proportions. If workability is too low, the cement content may be increased. If workability is high and bleeding takes place, the water–cement ratio or cement content may be reduced. Gravel or crushed stone do not absorb water, whereas lightweight aggregate absorbs water as much as 10–20% by weight in 24 h. Necessary allowances for this absorption should be made in the water content of these mixes.

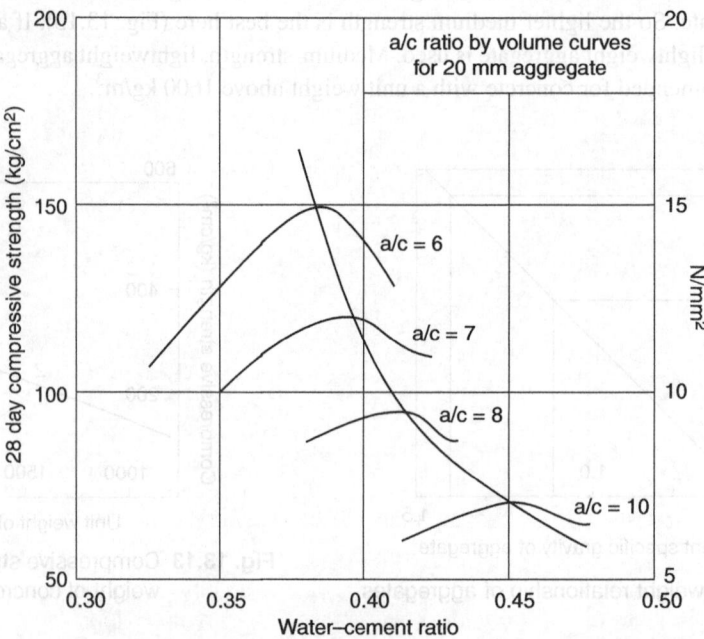

Fig. 13.14 Compressive strength, water–cement ratio, and aggregate–cement ratio for no-fines concrete

Example 13.6

Design a no-fines concrete mix for a cube strength of 90 kg/cm^2 using the following data:

Control	= Good (75%)
Maximum size of coarse aggregate	= 20 mm
Type of coarse aggregate	= Irregular gravel
Cement	= Portland cement
Bulk density of cement	= 1472 kg/m^3
Bulk density of coarse aggregate	= 1520 kg/m^3

Solution

Design

Average cube strength at 28 days = 90/0.75 = 120 kg/cm^2

From the chart in Fig. 13.14,

Optimum w/c = 0.39 (by weight)

a/c ratio by volume = 7.00, corresponding density of concrete
= 2050 kg/m³

$$\text{a/c ratio by weight} = \frac{7 \times 1520}{1472} = 7.23$$

Proportions

Cement	:	Coarse aggregate	:	Water
1		7.23		0.39

Quantities per cubic metre of concrete

$$\text{Cement} = \frac{1}{8.62} \times 2050 = 238 \, \text{kg}$$

$$\text{Coarse aggregate} = \frac{7.23}{8.62} \times 2050 = 1719 \, \text{kg}$$

$$\text{Water} = \frac{0.39}{8.62} \times 2050 = 93 \, \text{kg}$$

Example 13.7

Design a no-fines concrete mix given the following:

Average compressive strength at 28 days = 75 kg/cm²
Coarse aggregate   = Irregular gravel, 20 mm size
Cement   = Portland cement
Bulk density of cement   = 1472 kg/m³
Bulk density of gravel   = 1600 kg/m³

The aggregate contains 1.5% moisture content by weight. Estimate quantities required for field mix.

Solution

Design

Average compressive strength = 75 kg/cm²

From the chart in Fig. 13.14,

w/c ratio required = 0.45
a/c ratio by volume = 9.5
Corresponding density of concrete = 1955 kg/m³

$$\text{a/c ratio by weight} = \frac{9.5 \times 1600}{1472} = 10.3 \, \text{kg}$$

Proportions of materials

Cement	:	Coarse aggregate	:	Water
1		10.3		0.45

Note 1 + 10.3 + .45 = 11.75

Weight of materials

$$\text{Cement} = \frac{1}{11.75} \times 1955 = 166 \, \text{kg}$$

$$\text{Water} = \frac{0.43}{11.75} \times 1955 = 72 \, \text{kg}$$

$$\text{Coarse aggregate} = \frac{10.4}{11.75} \times 1955 = 1730 \, \text{kg}$$

$$\text{Moisture content in coarse aggregate} = \frac{1.5}{100} \times 1730 = 26 \, \text{kg}$$

Field mix by weight after adjusting the moisture content in coarse aggregate:

$$\text{Cement} = 166 \, \text{kg}$$

$$\text{Water} = 72 - 26 = 46 \, \text{kg}$$

$$\text{Coarse aggregate} = 1730 + 26 = 1756 \, \text{kg}$$

13.7 Proportioning of Lightweight Aggregate Concrete

Lightweight concrete is designed based on water/powder [w/(c + p)] ratio. The c + p gives the cementitious material in the mixture. The trial mixture may be made based on the volume of cement, cementitious material, coarse aggregate and sand for ordinary strength levels. For high-strength lightweight concrete, in addition to the above, addition of chemical admixtures and adjustment of water to get desired slump is necessary. Typical water-reducing admixtures and/or superplasticizers are used. Small amounts of entrainment of air help in improving workability.

Well-graded aggregates ensure proper particle packing. Table 13.12 gives a typical comparison of particle size distribution for lightweight aggregate concrete by weight and volume. If the resulting mixture is satisfactory from workability (Table 13.4) and finishability considerations, the mix is further assessed from yield point of view. Table 13.6 gives mixing water and air required for various maximum sizes of aggregates for the desired slump. Table 13.5 gives the relationship between the water–cement ratio and compressive strength. Table 13.7 shows the water–cement ratio limitations based on the severity of environment. Table 13.8 gives the volume of coarse aggregate as a function of its size. Table 13.9 is a guide to make a first estimate of the weight of fresh lightweight concrete made of low-density coarse aggregate and sand. Figures 13.15(a) and (b) show the relationship between concrete compressive strength and cement content for all-lightweight and sand-lightweight aggregate concrete, respectively. Yield is the total batch weight divided by actual unit weights of materials per unit volume (kg/m³).

Table 13.12 Comparison of fineness modulus by weight and by volume for typical lightweight concrete

Sieve size (no.)	Opening (mm)	Per cent retained by weight	Cumulative per cent retained by weight	Bulk specific gravity (SSD basis)	Per cent retained by volume	Cumulative per cent retained by volume
4	0.474	0	0	0	0	0
8	2.36	22	22	1.55	26	26
16	1.18	24	46	1.78	25	51
30	0.59	19	65	1.90	19	70
50	0.36	14	79	2.01	13	83
100	0.18	12	91	2.16	10	93
Pan	0.00	9	100	2.40	7	100

Note: Fineness modulus (by weight) 3.03; fineness modulus (by volume) 3.23.

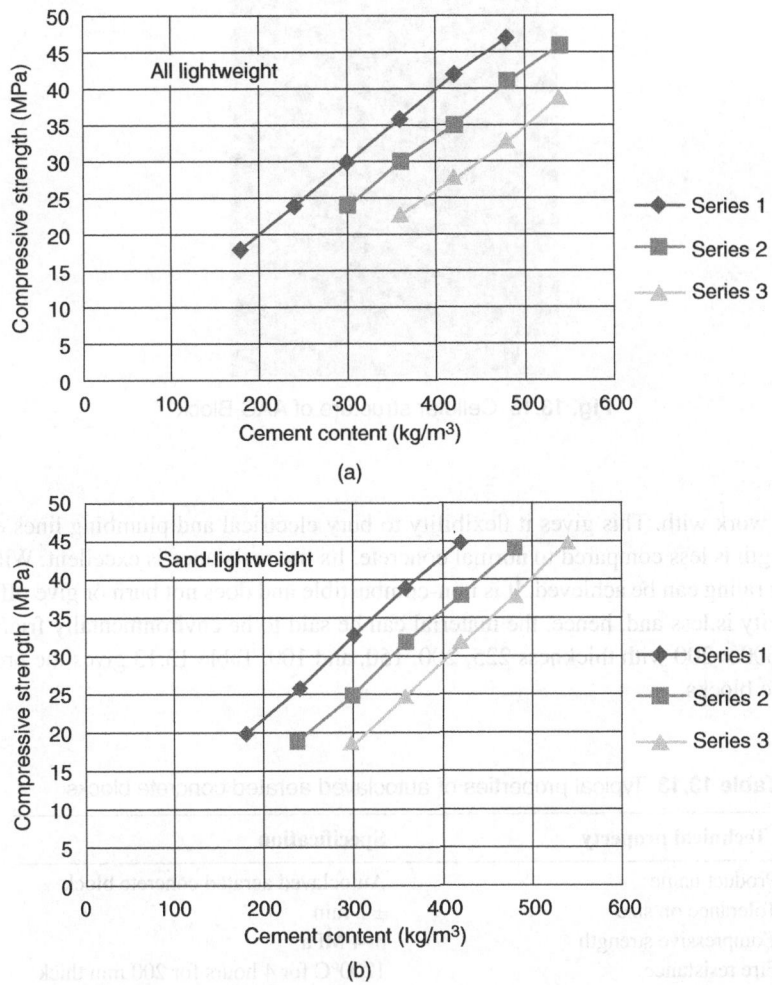

Fig. 13.15 Compressive strength versus cement content of lightweight aggregate concrete mixture

13.8 Cellular Concrete

Cellular concrete is mainly composed of cement, water, and air pores with fillers (such as fly ash and sand) without any coarse aggregates. The expansion agent that is added causes the fresh mixture to rise like bread dough when autoclaved. Cellular concrete blocks involve the use of aerated concrete which is made by introducing air or other gas into a slurry composed of cement or lime and a siliceous filler so that when the mixture sets hard after autoclaving, a uniform cellular structure is obtained. There are several ways in which air cells or other voids may be formed in the slurry as to result in a cellular structure after autoclaving. The typical size of air bubbles is around 0.3–0.4 mm in diameter. In the factory where it is made, the material is moulded and cut into precisely dimensioned units. The blocks conform to IS standard IS 2185 Part 3:1984. Cellular concrete is characterized by its low density when compared with normal concrete. The density of cellular concrete is around 400–1600 kg/m³. It is referred to as autoclaved aerated concrete (AAC).

A well-made autoclaved aerated concrete combines insulation and structural capability in one material for walls, floors, and roofs. Its light weight makes it easy to transport, and cellular structure (Fig. 13.16) makes

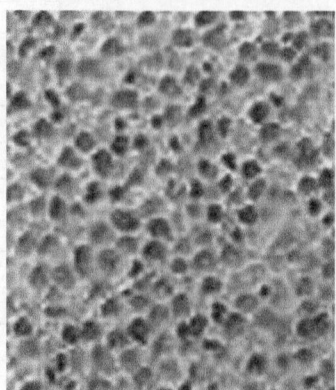

Fig. 13.16 Cellular structure of AAC Block

it easy to cut and work with. This gives it flexibility to bury electrical and plumbing lines easily. However, its structural strength is less compared to normal concrete. Its fire resistance is excellent. With 200 mm thick block, a four-hour rating can be achieved. It is non-combustible and does not burn or give off toxic fumes. Its thermal conductivity is less and, hence, the material can be said to be environmentally friendly. It comes in convenient sizes 600×200 with thickness 225, 200, 150, and 100. Table 13.13 gives the properties of commercially available blocks.

Table 13.13 Typical properties of autoclaved aerated concrete blocks

Technical property	Specification
Product name	Autoclaved aerated concrete block
Tolerance on size	± 2 mm
Compressive strength	3–4 MPa
Fire resistance	1200°C for 4 hours for 200 mm thick
Density	600–700 kg/m^3
Sound reduction	42 dB for 200 mm thick
Thermal conductivity	0.16 KW/m°C
Colour	Light grey
Applicability	For load and non-load bearing walls

13.9 Foam Concrete

Foam concrete is made by adding a special foam to the cement mortar slurry. This foam is made from foaming agent, a highly concentrated, highly efficient, low dosage liquid which is fed with water into a foam generator. Inside the generator, the concrete foaming agent is diluted with water to make a pre-foaming solution which is then forced at high pressure through the foaming nozzle. This produces uniform, stable foam which has a volume of about 20 to 25 times that of the pre-foaming solution.

Foamed concrete is also very useful for void filling. As it is very fluid, it will pour into even the most inaccessible places. It can be used for planned work, but also in emergencies to provide stability and support for earth pressure very quickly. Because of its liquid nature, the application of foam concrete is simple. It will spread to fill any cracks or gaps and will also level itself (see Fig 13.17).

Fig. 13.17 Application of foam concrete (*Source*: Reproduced with permission from Foam Concrete Ltd)

Exercises

Review Questions

1. What are the typical ranges of aggregate unit weight for making structural lightweight concrete?
2. What are the various methods of producing lightweight aggregate artificially?
3. What is the significant difference between mixture proportioning of normal-weight concrete and light-weight concrete?
4. Why is lightweight concrete preferred for constructing partitions in multi-storey buildings? Explain with respect to their physical characteristics of lightweight aggregate concrete.
5. Compare the thermal conductivities of normal-weight and lightweight concretes.
6. Discuss the importance and effects of water absorption and moisture content of lightweight aggregate concrete.
7. Discuss the environmental impact of normal-weight concrete and lightweight aggregate concrete.
8. Compare the elastic properties of normal-weight concrete and lightweight concrete.
9. Design a concrete mix by weight method using lightweight aggregate, coarse aggregate, and sand for a structural lightweight slab having a 28-day compressive strength of M30. Use the following data: Coarse aggregate size is 20 mm; specific gravity factor is 1.5; absorption is 10%; fine aggregate water absorption is 1%; fineness modulus is 2.78; oven-dry, loose weight of coarse aggregate is 800 kg/m^3.
10. What is the difference between cellular concrete and foam concrete?

Multiple Choice Questions

1. Lightweight concrete can be made with
   a. Natural lightweight aggregates
   b. Artificial lightweight aggregates
   c. Entraining air in the mortar
   d. All of the above

2. Lightweight aggregates weigh about
   a. 1000 kg/m^3
   b. 2000 kg/m^3
   c. 2500 kg/m^3
   d. 4000 kg/m^3

3. Moisture held in the pores of lightweight aggregate is
   a. Less than in normal aggregates
   b. More than in normal aggregates
   c. Equal as in normal aggregates
   d. Uncertain

4. M15 concrete using lightweight aggregates weighs approximately in kg/m^3
   a. 2000
   b. 1500
   c. 500
   d. 2500

5. Foam concrete can float on water
   a. True
   b. False
   c. Depends on surface area to weight ratio
   d. Depends on the cellular structure

6. Foam concrete is used
   a. As a precast block
   b. For filling trenches
   c. For reinforced concrete works
   d. For high-strength applications

7. Pumice is
   a. A dense stone
   b. A porous stone
   c. Used for high-density concrete
   d. A sedimentary rock broken and used as aggregate

8. Porous concrete is
   a. Never used
   b. Factory made
   c. Used for RCC works
   d. Used in lightweight concrete paving for rain harvesting

9. Fire resistance of autoclaved aerated concrete
   a. Is better than normal concrete
   b. Is worse than normal concrete
   c. Depends on exposure conditions
   d. Depends on embedded steel

10. Sound insulation of lightweight concrete
    a. Is better than normal concrete
    b. Is worse than normal concrete
    c. Depends on dB level
    d. Depends on source distance

Answers to Multiple Choice Questions

1. d	4. b	7. b	10. a
2. a	5. a	8. d	
3. b	6. b	9. a	

CHAPTER 14

High-strength Concrete

During the last few decades, developments in admixtures and mixing and placing methods have made it possible to produce concretes with much higher strengths (70–100 MPa). The aim of this chapter is to summarize the developments in materials, production methods, and mechanical properties of high-strength concrete and their uses.

High-strength concrete has compressive strength of up to 100 MPa as against conventional concrete which has compressive strength of less than 50 MPa. Concrete having compressive strength greater than 200 MPa is classified as ultrahigh-strength concrete.

Table 14.1 summarizes the materials used in various types of concrete. The ingredients of high-strength concrete are the same as those used in conventional concrete with the addition of one or two admixtures, both chemical and mineral. However, there are two crucial aspects to be considered while deciding upon the ingredients to be used. The first relates to the use of an extremely low water–cement ratio and the second to the use of a proper mix in order to produce concrete with minimum or no voids. A proper mix can be obtained using a proper particle packing method. The table shows that it is essential to use a plasticizing chemical admixture to produce high-strength concrete. Generally water-reducing admixtures (WRAs) are used. The mix requires high paste volume, which often leads to shrinkage and high evolution of heat of hydration, besides increasing the cost. The substitution of cement by supplementary cementitious materials such as mineral admixtures partially introduces favourable behaviour with respect to the above-mentioned defects and incidentally reduces the cost. The materials that are commonly used are fly ash, ground granulated blast-furnace slag, silica fume, rice husk ash, and metakaolin. The use of such materials not only improves the properties of fresh concrete but also enhances the long-term durability characteristics.

High-strength concrete essentially has a low water–binder ratio. A value of 0.3 is suggested as the boundary between normal and high-strength concrete. For ultrahigh-strength concrete, this value is further reduced by the use of high-range water-reducing admixtures (HRWRAs) or viscosity modifying agents (VMAs).

Although the compressive strength of concrete has been steadily improving in recent times, the potential to increase it further has become evident with its use in columns of high-rise buildings and long-span bridge girders all over the world. Table 14.1 indicates that a fourfold increase in compressive strength is possible by using conventional cement along with a waste product such as silica fume and a relatively small amount of chemical admixture in the form of an HRWRA or VMA.

14.1 Classification of High-strength Concrete

Table 14.1 gives a general classification of high-strength concrete with the typically required water–binder ratios and other materials used. This table should be used as a general guideline; it is possible to get high and very high strengths in several other ways also. It has been possible to produce micro-defect-free cement (MDFC) with a compressive strength of nearly 600 MPa. MDFC is not really suitable for structural concrete and hence is not considered here.

Table 14.1 Classification of high-strength concrete

Parameter	Conventional	High strength	Very high strength	Ultrahigh strength
Compressive strength (MPa)	<50	50–100	100–200	>200
Water–binder ratio (typical)	0.45–0.55	0.30–0.45	0.24–0.30	<0.24
Chemical admixture	Not necessary	WRA/HRWRA necessary	HRWRA &VMA	HRWRA &VMA essential
Mineral binder addition	Not necessary	Fly ash/slag/metakaolin/ rice husk ash	Silica fume	Silica fume and fly ash
Aggregate type	Gravel/crushed stone/ lightweight aggregate	Crushed stone	Crushed stone/ artificial aggregate	Artificial aggregate
Maximum size of aggregate (mm)	Any size	15	10	5
Fibres	Optional	Beneficial	Beneficial	Essential
Air entrainment	Necessary	Necessary	Necessary	Not necessary
Processing	Conventional	Conventional	Conventional	Heat and pressure required
Permeability coefficient (cm^2/s)	10^{-10}	10^{-11}	10^{-12}	$<10^{-14}$
Chloride diffusivity (cm^2/s)	10^{-10}	10^{-9}	10^{-8}	10^{-7}

It is a known fact that conventional concrete greatly underutilizes the potential of cement. By careful selection of the amount of free water, additives, aggregate, and minerals, it is possible to obtain higher strength at a very little additional material cost.

14.2 Composition of High-strength/Ultrahigh-strength Concrete

The composition of high-strength concrete aims at using all the constituents of concrete fully. In ordinary concrete, not all the cement that is added gets hydrated. Hence, some form of mineral admixture is added to obtain high-strength concrete. This acts as a cementitious material and hence reduces the water–binder ratio.

Two examples of the composition of 200 MPa concrete, taken from Sauzeat et al. (1996) and Aitcin and Richard (1996), are given in Table 14.2. The key ingredients of this type of concrete are the following:

a. Low water–binder ratio
b. Large quantity of silica fume and/or fine mineral powder
c. Aggregates containing fine sand with good particle packing characteristics
d. High dosage of superplasticizers

Table 14.2 Composition* of concrete with compressive strength of 200 MPa

	Sauzeat et al.	Aitcin and Richard
Cement	1.0	1.000
Water	0.28	0.150
Superplasticizer	0.06	0.044
Silica fume	0.33	0.250
Fine sand	1.43	1.100
Quartz flour	0.30	

*Composition given in mass ratio.

14.3 Microstructure of High-strength Concrete

A number of microstructure studies have revealed dense distribution of particles in high-strength concrete (Fig. 14.1) as compared to those in conventional (ordinary) concrete (Fig. 14.2). Based on these observations, the following conclusions can be made:

a. The material is more homogeneous on a millimetre scale.
b. No pronounced transition zone between the sand and the paste is seen.
c. Low porosity (1–3%).
d. Proportionately lower percentage of capillary porosity (10% of total porosity).

These features lead to low water absorption, gas permeability, and chloride diffusivity.

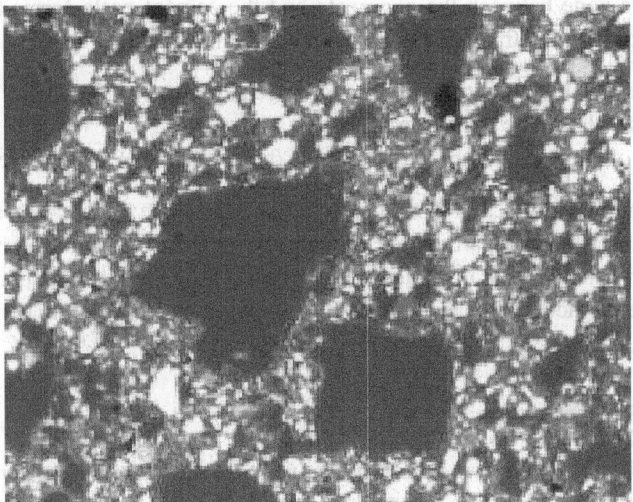

Fig. 14.1 Microstructure of high-strength concrete

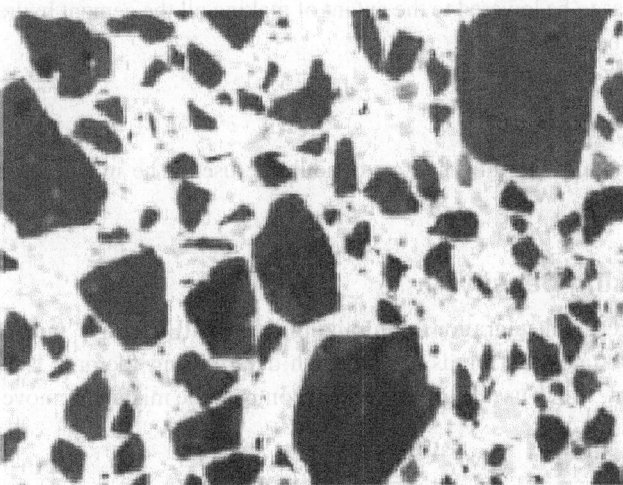

Fig. 14.2 Microstructure of ordinary concrete

14.4 Manufacturing Considerations

It is not possible to obtain high-strength concrete by manipulating a particular parameter alone. Strength can be thought of as a chain; all weak links should be eliminated in order to get the desired strength. There are two approaches for developing high-strength concrete.

1. The macro-defects can be eliminated by introducing polymers and then subjecting the mix to an extrusion process. It produces micro-defect-free concrete. This approach is not suitable for *in situ* construction.
2. Using dense packing theory, high-strength concrete can be developed with available equipment and carefully chosen available material. The voids can be eliminated by the careful process of densification with smaller and smaller particles.

Choosing the second alternative, one should eliminate all the weak zones in conventional concrete. It is well known that the transition zone between the aggregate and the cement paste forms the weakest link in the strength chain. The reasons for the transition zone being weak are the following:

a. Accumulation of bleed water below the larger aggregate particles
b. Presence of weak Ca(OH) in the transition zone
c. Less dense packing of particles in the transition zone due to the wall effect of the larger aggregates

These defects in the transition zone can be overcome by the use of pozzolanic material, e.g., fly ash. However, silica fume gives a much better particle packing ability, and with the addition of superplasticizers, we can use really low water–binder ratios to achieve nearly defect-free high-strength practical concrete which can be readily used in field practice. The techniques for achieving higher strength concretes are summarized in Fig. 14.3.

14.4.1 Free Water Content

Using currently available hyperplasticizers and/or viscosity modifying agents, the water content can be reduced to 140 L/m^3 while maintaining a slump of about 75 mm.

14.4.2 Water–Binder Ratio

The water–binder ratio has to be lowered to the extent of making all the cement hydrate. However, it may not be advisable to reduce it below 0.25.

14.4.3 Binder Content

The amount of binder that can be fully hydrated should be used. The maximum amount known is about 540 kg/m^3.

14.4.4 Particle Packing Density

Voids are sources of stress concentration; minimizing them will improve the particle packing density. Griffith's (1921) theory states that voids are strength-determining factors. Theoretically, one can use macro-defect-free cement, which uses high pressure to compact the mix and remove all the pores. However, this is not practical.

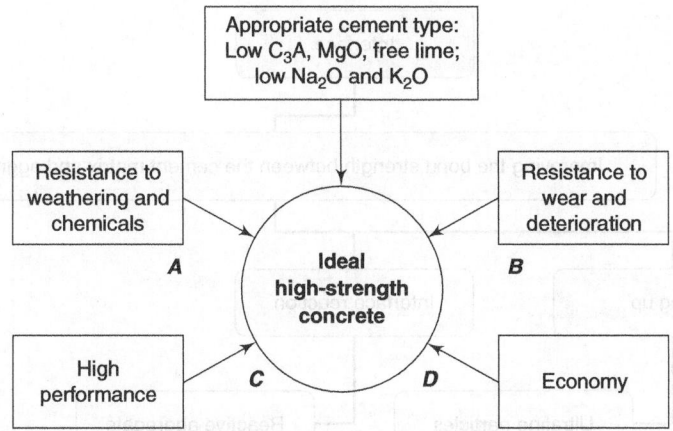

A Controlled proportions
Appropriate cement type, low water–cement ratio, proper curing, alkali-resistant aggregates, suitable admixture, use of super-plasticizers, fly ash polymer and/or silica fume as mineral admixtures, air entrainment

B Controlled material quality
Low water–cement ratio, proper curing, dense homogeneous concrete, high-strength, water-resisting aggregate with good surface texture

C Controlled placing and curing
Good quality paste, low water–cement ratio, optimal cement content and cementitious material, sound aggregate grading and vibration, low air content, high strength

D Controlled handling
Large maximum size of aggregate, efficient grading, minimum slump, minimum cement content, optimum automated plant operation, admixtures and entrained air, quality assurance and control

(a)

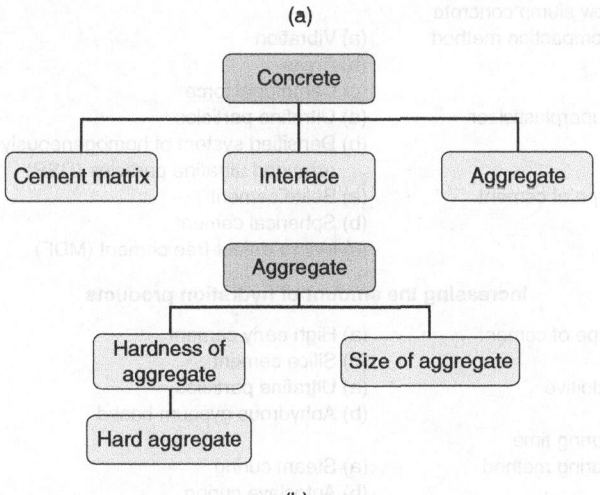

(b)

Fig. 14.3 Principal parameters for the manufacture of high-strength concrete

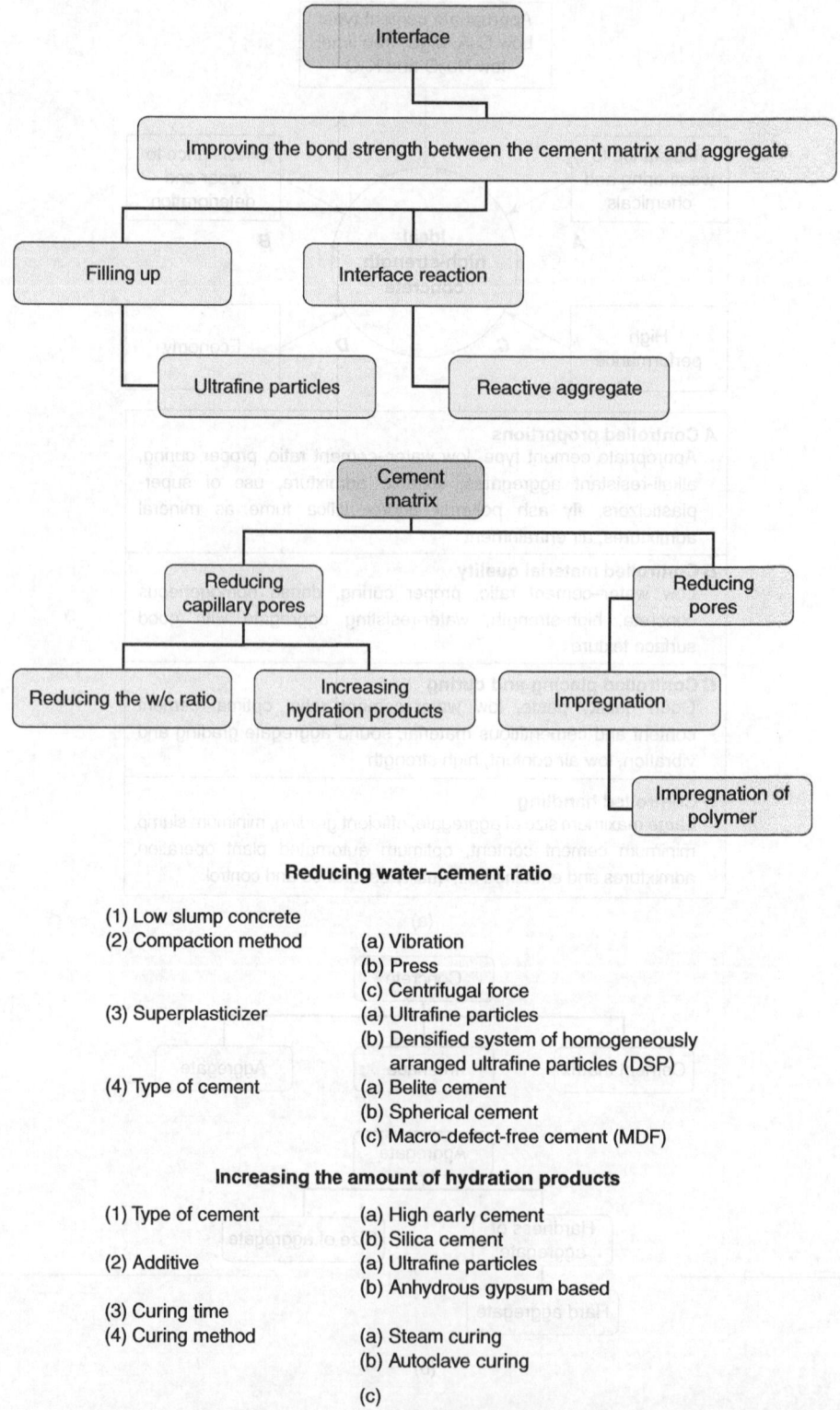

Reducing water–cement ratio

(1) Low slump concrete
(2) Compaction method   (a) Vibration
                             (b) Press
                             (c) Centrifugal force
(3) Superplasticizer   (a) Ultrafine particles
                             (b) Densified system of homogeneously
                                 arranged ultrafine particles (DSP)
(4) Type of cement   (a) Belite cement
                             (b) Spherical cement
                             (c) Macro-defect-free cement (MDF)

Increasing the amount of hydration products

(1) Type of cement   (a) High early cement
                             (b) Silica cement
(2) Additive   (a) Ultrafine particles
                             (b) Anhydrous gypsum based
(3) Curing time
(4) Curing method   (a) Steam curing
                             (b) Autoclave curing

                             (c)

Fig. 14.3 Principal parameters for the manufacture of high-strength concrete

Practically, one can select a combination of sizes such that the free space between the particles is reduced when the material is mixed. This concept is explained in Fig. 14.4. When single-sized (mono-sized) particles are used, the percentage of voids can be as high as 30%. Also, the result is an open (weak) microstructure. When the radii of the micro-filler spheres are reduced by one-eighth, we get the most closed microstructure with only about 3% voids.

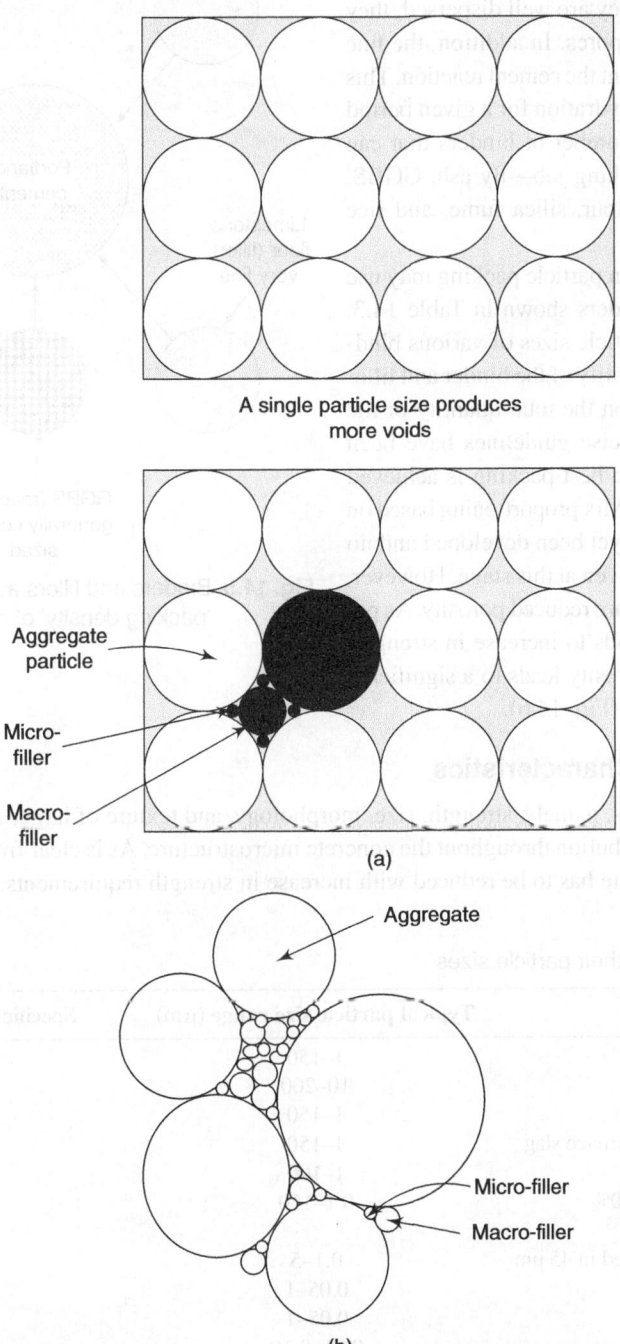

A single particle size produces more voids

(a)

Aggregate

Micro-filler

Macro-filler

(b)

Fig. 14.4 (a) Concept of particle packing; (b) Effect of fillers between aggregate particles

Cubic particles are capable of being packed without any voids, but this is not possible for wet concrete in a colloidal suspension. Generally, the key particle sizes affecting packing density are those smaller than 125 µm. These particles are smaller than the capillary pores. When they are well dispersed, they will block the capillary pores. In addition, the fine powder effect will augment the cement reaction. This increases the degree of hydration for a given period of curing. There are a number of binders that can perform the particle packing job—fly ash, GGBS, metakaolin, limestone flour, silica fume, and rice husk ash (Fig. 14.5).

A mix design based on particle packing may use the combinations of binders shown in Table 14.3. This table shows the particle sizes of various binders/fillers. The exact quantity of the binder and filler to be used will depend on the total quantity of the binder required. No precise guidelines have been developed as yet and the best packing is achieved mostly by trial and error. Mix proportioning based on particle packing has not yet been developed and no reliable method can be given at this stage. However, our attempt here will ensure reduced porosity. As per Griffith's theory, this leads to increase in strength. Note that increase in porosity leads to a significant increase in permeability (Fig. 14.6).

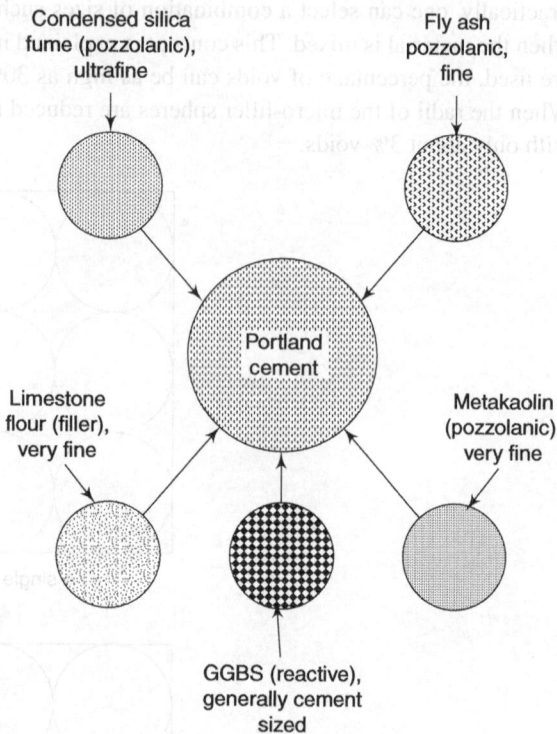

Fig. 14.5 Binders and fillers available to enhance 'packing density' of cement paste

14.4.5 Aggregate Characteristics

Aggregate characteristics, namely, strength, size, morphology, and texture of larger aggregates, significantly influence the stress distribution throughout the concrete microstructure. As is clear from Table 14.1, the maximum size of the aggregate has to be reduced with increase in strength requirements. Up to 100 MPa normal

Table 14.3 Binders and their particle sizes

Binder	Typical particle size range (µm)	Specific surface area (m²/kg)
Rice husk ash	1–150	350
Portland cement (grade 33)	10–200	280
Portland cement (grade 43)	1–150	350
Ground, granulated blast-furnace slag	1–150	350
Portland cement (grade 53)	1–100	450
Pulverized fly ash (PFA), 10% retained in 45 µm	0.1–150	500
Ultrafine fly ash, 1% retained in 45 µm	0.1–5	5,000
Ultrafine limestone flour	0.05–1	10,000
Metakaolin	0.05–1	15,000
Condensed silica fume	0.01–0.10	20,000

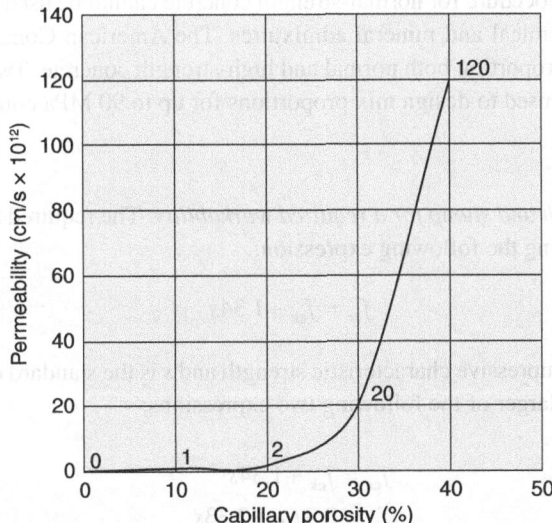

Fig. 14.6 Effect of porosity on permeability of concrete

good quality aggregates can be used, including lightweight aggregates. Beyond this strength, special qualities are required for aggregates. To obtain a strength of 150 MPa, broken granite aggregates can be used with the maximum size limited to 10 mm. Care should be taken to ensure that there is no internal micro-cracking in the aggregates during crushing.

Very high-strength artificial aggregates can be obtained using calcined bauxite having a maximum size of 3–5 mm. These are very strong aggregates having strength of 300 MPa. Their external texture is very rough and hence they produce excellent interfacial bonds with the matrix. For fine aggregates (FAs), well-graded sand is suitable. As with cement paste, continuous grading is crucial with increase in strength.

14.5 Selection of Mix Proportions

The selection of cementitious materials and the optimization of the proportions of the ingredients are more of art than pure science. The traditional methods of mix proportioning discussed previously need to be modified based on strength requirements. High strength can be produced by reducing the water–cement ratio below than that normally used for ordinary concrete. This can be made possible by using admixtures. A low water–binder ratio can be achieved in one of the following ways:

- Increasing the binder content
- Decreasing the water content
- Increasing the binder content as well as decreasing the water content

While using these methods, it is essential to add a superplasticizer to obtain the required slump. In a high-strength concrete mix, for a given water–binder ratio, the reduction in water content is associated with reduced cement content, but the requirement of the plasticizer increases. Hence, the cost saving achieved by the decrease in cement content is lost to the added admixture cost. Thus, a careful balance must be arrived at. It is also essential to ensure that the mix obtained is cohesive, to satisfy the placement requirements.

The mix-proportioning procedure for normal-strength concrete cannot be used to proportion high-strength concrete involving both chemical and mineral admixtures. The American Concrete Institute method (ACI 211.4R-93) can be used to proportion both normal and high-strength concrete. Two methods are discussed in the following. These can be used to design mix proportions for up to 90 MPa concrete.

14.5.1 Method 1

Step 1   *Selection of strength and slump for a required workability*: The required strength f_{cr} to be used for a trial mix can be obtained using the following expression:

$$f_{cr} = f_{ck} + 1.34s$$

where f_{ck} is the specified compressive characteristic strength and s is the standard deviation of the test results. f_{cr} should be chosen as the larger of the following two expressions:

$$f_{cr} = f_{ck} + 1.34s$$
$$f_{cr} = 0.9f_{ck} + 2.33s$$

The laboratory trial should give a strength of

$$f_{cr} = \frac{f_{ck} + 9}{0.9} \text{ MPa}$$

The slump is chosen based on field experience and laboratory batching trials. The recommended values of slump are given in Table 14.4.

Table 14.4 Recommended values of slump

Type of concrete	Slump (mm)
Made without water reducer	25–60
Made with high-range water reducer	50–100

Step 2   *Selection of the maximum size of aggregate*: The maximum size of aggregate (MSA) is selected based on the physical requirements and the need for making good concrete while allowing for efficient particle packing. The following physical requirements must be satisfied: The maximum size of the aggregates should not exceed (a) one-fifth the narrowest dimension of the sides of forms, (b) one-third the depth of the member (slab), and (c) three-fourth the clear distance between the individual reinforcing bars, the bundles of bars, and the prestressing strands. For packing requirements, the suggested maximum sizes of aggregates are given in Table 14.5. When using high-range water reducers and selected coarse aggregates (CAs), compressive strengths in the range 62–85 MPa can be attained using aggregates of sizes larger than the suggested nominal maximum size of up to 25 mm.

Table 14.5 Suggested maximum size of coarse aggregates

Required concrete strength (MPa)	Maximum size of aggregate
< 62	20–25
≥ 62	10–12.5

Step 3 *Selection of CA content*: The CA content required to obtain optimum packing with sand of fineness modulus in the range 2.5–3.2 is given in Table 14.6 (ACI 211.4R). We can get the weight by multiplying the value given in the table with the unit weight. From this value, the oven-dry weight of CA per unit volume of concrete can be calculated.

Table 14.6 Volume of CA per unit volume of concrete

Maximum size of aggregate (mm)	Fractional volume of oven-dry-rodded CA
10.0	0.65
12.5	0.68
20.0	0.72
25.0	0.75

Step 4 *Water and air content*: Table 14.7 shows the mixing water requirements for FAs with 35% voids. For other void percentages, some adjustment in the water content is necessary.

Table 14.7 Estimates of mixing water and air content for fresh concrete based on the use of sand with 35% voids

Slump (mm)	Mixing water (L)			
	MSA = 8.47	MSA = 12.7	MSA = 19.05	MSA = 25.4
25.4–50.8	182.9	174.05	168.15	165.2
50.8–76.2	188.8	182.9	174.05	171.1
76.2–101.6	194.7	188.8	179.9	177.0
Entrapped-air content	3	2.5	2	1.5
Mixes with HRWRA	2.5	2	1.5	1

Step 5 *Water–cementitious powder ratio*: The water–cementitious powder ratio, w/(c + p), is calculated by dividing the mixing water content by the combined weight of cement and fly ash or other mineral admixtures. Tables 14.8 and 14.9 give the recommended maximum w/(c + p) ratios for concrete as functions of the maximum size of aggregate for achieving different strengths, respectively, without and with the use of HRWRAs at 28 and 56 days.

Table 14.8 Maximum w/(c + p) ratios for concrete without HRWRA

Field strength (MPa)		w/(c + p)			
		MSA = 8.47	MSA = 12.7	MSA = 19.05	MSA = 25.4
47.6	28 day	0.42	0.41	0.40	0.39
	56 day	0.46	0.45	0.44	0.43
54.4	28 day	0.35	0.34	0.33	0.33
	56 day	0.38	0.37	0.36	0.35
61.2	28 day	0.30	0.29	0.29	0.28
	56 day	0.33	0.32	0.31	0.30
68	28 day	0.26	0.26	0.25	0.25
	56 day	0.29	0.28	0.27	0.26

Table 14.9 Maximum w/(c + p) ratios for concrete with HRWRA

Field strength (MPa)		w/(c + p)			
		MSA = 8.47	MSA = 12.7	MSA = 19.05	MSA = 25.4
47.6	28 day	0.50	0.48	0.45	0.43
	56 day	0.55	0.52	0.48	0.46
54.4	28 day	0.44	0.42	0.40	0.38
	56 day	0.48	0.45	0.42	0.40
61.2	28 day	0.38	0.36	0.35	0.34
	56 day	0.42	0.39	0.37	0.36
68	28 day	0.33	0.32	0.31	0.30
	56 day	0.37	0.35	0.33	0.32
74.8	28 day	0.30	0.29	0.27	0.27
	56 day	0.33	0.31	0.29	0.29
81.6	28 day	0.27	0.26	0.25	0.25
	56 day	0.30	0.28	0.27	0.26

Step 6 *Content of cementitious material*: The weight of cementitious material per cubic metre of concrete is calculated by dividing the amount of mixing water per cubic metre of concrete by w/(c + p). However, if the specification includes a minimum limit on the amount of cementitious material permissible per cubic metre of concrete, this must also be satisfied.

Step 7 *Basic mixture with no mineral admixture*: The first trial mixture is made using ordinary Portland cement as the only cementitious material. The weight of the cement is the same as the weight of the binder arrived at in the previous step. The sand content is calculated by subtracting the quantities of all other materials required to produce 1 m³ of concrete.

Step 8 A companion mixture is proportioned using a mineral admixture. The new sand content, as explained in the previous step, is determined.

Step 9 Trial mixes are produced and the workability and strength characteristics are determined. The weights of sand, CA, and water are adjusted to correct for the moisture condition of the aggregates.

Step 10 The mixture proportion is further adjusted to achieve the required slump by changing the ratio of the contents and adjusting the percentage of the HRWRA for several trial mixtures.

Step 11 Once the trial mixture proportions have been adjusted to produce the desired workability and strength properties, specimens are cast from the trial batches. The results of the strength tests are compared with the requirements of the client to enable the selection of acceptable proportions for a job based on strength and cost considerations.

14.5.2 Method 2

Step 1 The concrete mix has to be designed for a somewhat higher target average strength f'_{ck}, so that not more than the specified proportion of test results fall below the characteristic strength. The margin over the characteristic strength depends upon the degree of quality control in the field (expressed in terms of the standard deviation) and the accepted proportion of strength test results below the characteristic strength f_{ck}, given by the relation

$$f'_{ck} = f_{ck} + ts$$

where f'_{ck} is the target average compressive strength at 28 days, f_{ck} is the specified characteristic compressive strength at 28 days, t is a statistical value depending upon the accepted proportion of low results and the number of tests (according to IS: 456-2000, $t = 1.65$), and s is the standard deviation depending on the grade of concrete and degree of control obtained using Table 14.10.

Table 14.10 Target mean compressive strength when there is insufficient data to establish the standard deviation

Specified characteristic compressive strength f_{ck} (MPa)	Target average compressive strength f'_{ck} (MPa)
<20.5	$f_{ck} + 6.9$
20.5–34.5	$f_{ck} + 8.3$
>34.5	$f_{ck} + 9.7$

Step 2 *Selection of maximum size of aggregates*: The maximum size of aggregate is chosen using Table 14.5 as in Method 1.

Step 3 *Estimation of free water content*: The water content required to obtain the desired workability/slump depends on the amount of superplasticizer used and its characteristics. The superplasticizer dosage is optimized based on its saturation point. The efficiency of a superplasticizer can be assessed based on a flow test through a flow funnel. A typical behaviour curve is shown in Fig. 14.7. Note that increasing the dosage beyond the saturation point does not increase the flow significantly. Hence it is advisable to limit the dosage to that at the saturation point. Table 14.11 gives the water dosage at the saturation point.

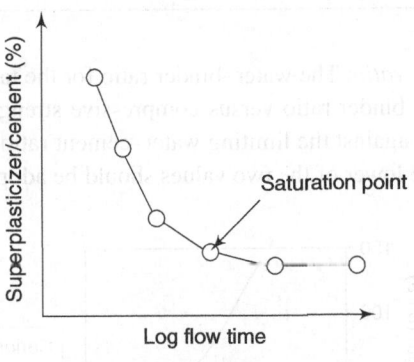

Fig. 14.7 Saturation point

Table 14.11 Minimum water dosage

Saturation point (%)	Water dosage (L/m³)
0.6	120–125
0.8	125–135
1	135–145
1.2	145–155
1.4	155–165

Step 4 The superplasticizer dosage is obtained from the dosage at the saturation point. If the saturation point is not known, then a dosage of 1% can be adopted as first trial.

Step 5 *Air content*: The value of air content is taken from Table 14.12. The initial value of entrapped air can be taken as 1% and refined in subsequent trials.

Table 14.12 Approximate entrapped-air content

Maximum size of aggregate (mm)	Entrapped air (% volume of concrete)
10	2.5
12.5	2
20	1.5
25	1

Step 6 *The CA content*: The CA content is obtained from Table 14.13 as a function of the typical particle shape. The CA selected should satisfy the requirements of grading as well as preferable shape according to IS: 383-1970.

Table 14.13 Coarse aggregate content

Particle shape	Dosage (kg/m³)
Elongated or flat	950–1000
Average	1000–1050
Cubic	1050–1100
Rounded	1100–1150

Step 7 *Selection of water–binder ratio*: The water–binder ratio for the target mean strength is chosen from Fig. 14.8, which shows the water–binder ratio versus compressive strength relationship. The water–binder ratio so chosen should be checked against the limiting water–cement ratio for the requirements of durability according to IS: 456-2000, and the lower of the two values should be adopted.

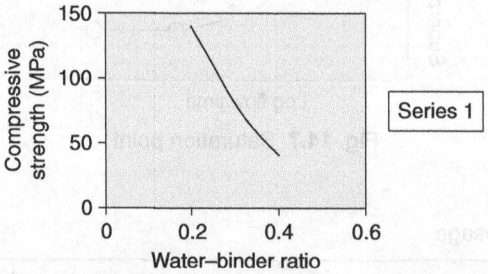

Fig. 14.8 Compressive strength versus water–binder ratio

Step 8 *Binder content*: The total binder content is calculated by dividing the water content by the water–binder ratio. This is checked against the minimum cement content specified after taking durability considerations into account according to IS: 456-2000. The larger value of cement content is adopted.

Step 9 *Superplasticizer dosage*: The mass of solids in a superplasticizer, M_{sol}, is calculated as

$$M_{sol} = \frac{\text{Binder content} \times \text{Superplasticizer dosage}}{100}$$

The volumes of the liquid superplasticizer (V_{liq}), the water in the liquid superplasticizer V_w, and the solids in the liquid superplasticizer (V_{sol}) are evaluated using the following relationships:

$$V_{liq} = \left(\frac{M_{sol} \times 100}{s} \right) SG$$

$$V_w = V_{liq} \times SG \times \left(\frac{100 - s}{100} \right)$$

$$V_{sol} = V_{liq} - V_w$$

where s is the percentage of the total solid content in the superplasticizer and SG is the specific gravity of the liquid superplasticizer.

Step 10   Estimation of FA content: By knowing the volumes of cement, mineral admixture, water, superplasticizer, air, and CA, the volume of FA can be calculated by deducting all these from the total volume of concrete.

Step 11   Moisture adjustment: The actual quantities of CAs, FAs, and water are calculated after allowing for the necessary corrections for water absorption and free moisture as in Method 1. The volume of moisture included in the liquid plasticizer is calculated and subtracted from the mixing water content.

Step 12   Unit mass of concrete: The mass of concrete per unit volume is calculated by summing up the masses of concrete's ingredients.

Step 13   The mixture proportion: Using the masses of the ingredients for $1\,m^3$ of concrete, the trial mix proportion by weight is derived.

Step 14   Trial batch adjustments: Owing to the many assumptions underlying the foregoing theoretical calculations, trial mixes must be checked. If necessary, the mix proportions can be modified to meet the workability and strength criteria. This can be done by adjusting (a) the percentage replacement of cement by silica fume, (b) the dosage of the superplasticizer, (c) the solid content of binder, and (d) the air content and unit weight by laboratory trial batches to optimize the mixture proportioning. Fresh concrete should be tested for slump, freedom from segregation, unit weight, and air content. Hardened concrete cured under standard conditions should be tested for strength at the specified age.

14.6 Properties of High-strength Concrete

The properties of high-strength concrete are significantly different from those of normal-strength concrete. These properties are examined in this section when the concrete is setting and hardening as well as in the hardened state. These properties should be taken into account while designing structures using high-strength concrete.

14.6.1 Setting and Hardening

When the concrete mixture is in the liquid phase, there are isolated solid grains in a connected structure. Hydration starts from the surface of the grain. As the outer crust grows thicker, it retards the hydration process. The formation of hydrates around each grain changes the liquid into a continuous solid.

In ordinary concrete, the hydration continues: the anhydrous core remains at the centre of the grain for a long time. The post-setting hydration process leads to the internal growth of a skeleton structure (hardening) and reduction of the water content in the pores (self-desiccation). The reduction of pore water has the same effect as drying. For high-strength concrete with a low water–cement ratio, the shrinkage caused by the reduction in pore water causes internal compression, which is developed due to the surface tension of the pore water at the liquid–vapour interface. The mobility of the fluid state decreases as the water content decreases due to the reaction. Thus, self-desiccation leads to the reduction of the liquid phase mobility. The mobility becomes much less as the concrete hardens. Hence, high-strength concrete with a low water–cement ratio is more sensitive to early drying. Sometimes cracks may appear even before setting. However, these cracks disappear if the concrete is cured properly. The appearance of cracks can be attributed to desiccation. Since high-strength concrete exhibits high sensitivity to early drying, it is absolutely necessary to have an efficient curing system.

14.6.2 Heat Development

The heat of hydration of concrete is affected by the cement content, the water–binder ratio, and especially the silica fume content. Increasing the cement content leads to significant heat of hydration. Experiments done by Smeplass and Maage (Paillere et al. 1987) have shown that at a water–binder ratio of 0.27, silica fume does not affect hydration significantly. However, above the water–binder ratio of 0.3, the effect of silica fume in increasing the heat evolution is as strong as that of increase in cement content.

14.6.3 Shrinkage

High-strength concrete exhibits sensitivity to early drying and faster autogenous shrinkage. The high autogenous (self-desiccation) shrinkage of silica fume in high-strength concrete was noticed by Paillere, Buil, and Serrano (1989). This effect is due to the high paste volume used in high-strength concrete. A high autogenous shrinkage leads to early age cracking. This occurs in concrete with a low water–cement ratio as also in concrete in which silica fume is used. However, this disadvantage can be overcome by proper mix proportioning and early and efficient curing. Dhir (1997) has shown that shrinkage strain is not really a problem and its value is less than that of conventional ordinary strength concrete (see Table 14.14 and Fig. 14.9).

Table 14.14 Comparative values of indicative shrinkage strains (mm/m)

	Conventional concrete	High-strength concrete
Total shrinkage strain	470	320
Extrapolation to the ultimate	650	340
Autogenous shrinkage strain	120	200
Extrapolation to the ultimate	120	220
Desiccation shrinkage strain	350	120
Extrapolation to the ultimate	530	120

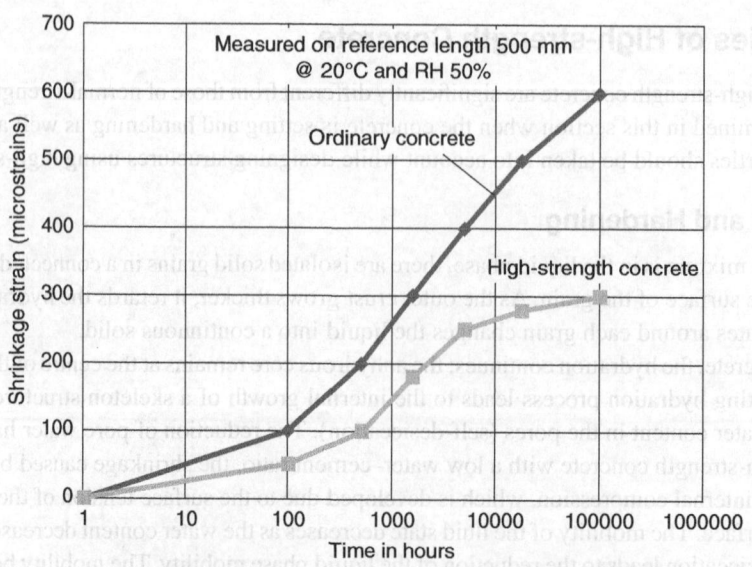

Fig. 14.9 Shrinkage of high-strength concrete

14.6.4 Creep

Information on the creep of concrete of higher strength is limited. The creep coefficient for high-strength concrete is less than that for normal concrete (see Table 14.15). However, it has been observed that high-strength concrete is likely to creep faster than normal-strength concrete.

Table 14.15 Comparative creep and shrinkage strains for normal and high-strength concrete (mm/m)

	Normal-strength concrete	High-strength concrete
Creep plus shrinkage	700	360
Basic creep	132	93
Additional drying creep	320	102
Creep–elastic strain ratio	0.65	0.49

14.6.5 Elastic Deformation and Crack Growth Characteristics

High-strength concrete is considerably more brittle than normal concrete. Figure 14.10 shows the uniaxial compressive stress–strain curve for ultrahigh-strength concrete compared with normal concrete. To obtain the 'falling branch' of the stress–strain curve of high-strength concrete, it is necessary to test the specimens in a closed-loop digitally strain controlled testing machine (Shah and Ahamed 1994). The post-peak response of high-strength concrete is characterized by a steep descent compared to the gradual decrease witnessed in normal concrete.

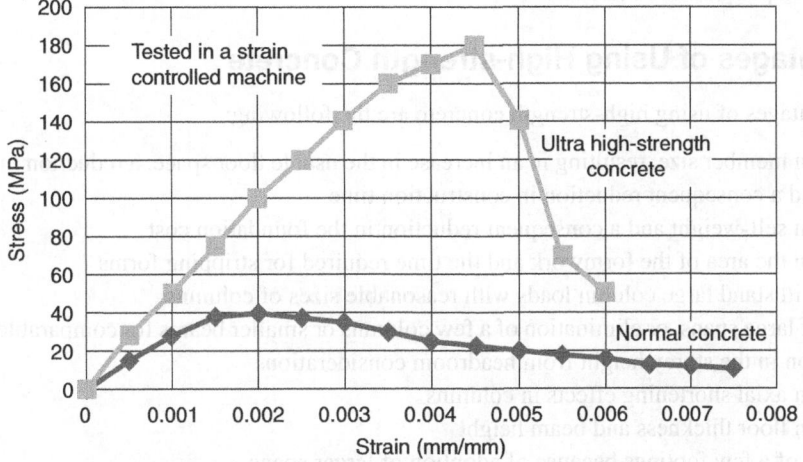

Fig. 14.10 Stress–strain curve for ultrahigh-strength concrete

The values of fracture energy at cracking can be used to compare the various types of concrete with a view to assessing their post-peak performance. For a classically brittle material like cast iron, once the critical value of fracture energy is reached, the cracks propagate catastrophically. However, for concrete, continued crack propagation requires additional energy to overcome the friction between the grain boundaries sliding along each other. Crack propagation is related to the effectiveness of aggregate interlock, grain boundary sliding, crack branching, and such other microstructural toughening mechanisms. The fracture resistance curve (Fig. 14.11), known as the *R*-curve, depicts the stable range of post-peak performance. Behavioural stability related to crack propagation in various concretes can be observed in this figure. The *R*-curve for conventional concrete shows stability over a much larger range of strain than that for the more brittle micro-defect-free (MDF) material responsible for very high strength.

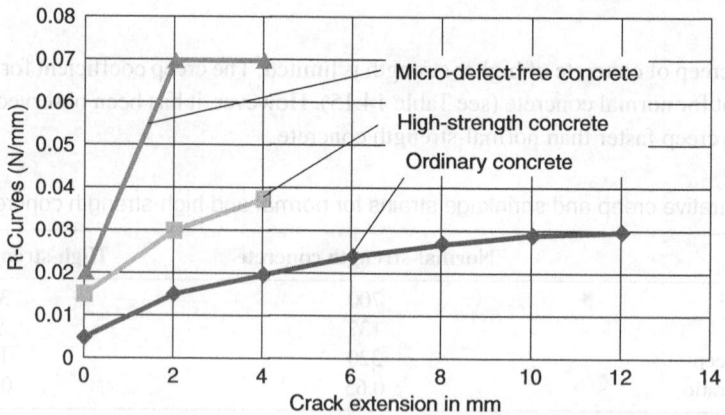

Fig. 14.11 Fracture resistance of concretes

14.6.6 Performance During Fire

It is considered that the low vapour permeability of dense concrete could lead to the build-up of internal steam pressure, aiding sudden failure and collapse. However, since the free water content in these concretes is less, this contingency may not arise. Moreover, the inclusion of a small percentage of polypropylene fibres helps in reducing this risk. The fibres also enhance the ductility of high-strength concrete. These aspects are discussed in detail in the chapter on fibre concrete.

14.7 Advantages of Using High-strength Concrete

The main advantages of using high-strength concrete are the following:

- Reduction in member size, resulting in an increase in the usable floor space, a reduction in the quantity of concrete, and a consequent reduction in construction time
- Reduction in self-weight and a consequent reduction in the foundation cost
- Reduction in the area of the formwork and the time required for stripping forms
- Ability to withstand large column loads with reasonable sizes of columns
- Provision of large spans, or elimination of a few columns or smaller beams for comparable spans, leading to a reduction in the storey height from headroom considerations
- Reduction in axial shortening effects in columns
- Reduction in floor thickness and beam height
- Elimination of a few footings because of adoption of larger spans
- Superior durability and long-term performance
- Lower creep and shrinkage
- Larger stiffness as a result of a larger value of Young's modulus of concrete
- Higher resistance to crack propagation, chemical attack, etc.
- Reduced maintenance costs

14.8 Applications of High-strength Concrete

In the context of tall buildings and large-span bridges, it is clear that use of high-strength concrete leads to reduced column sizes and beam depths. In addition, it results in improved performance in terms of creep, shrinkage, and other elastic properties that yield a more favourable deformation pattern for tall buildings. The improved elastic properties also limit the sway and reduce elastic shortening and other secondary effects.

Hence, high-strength concrete has found application in various structures used by various industries. Table 14.16 summarizes the applications of high-strength concrete in different industries around the world.

Table 14.16 Applications of high-strength concrete in different industries

Industry	Design strength (MPa)	Technique used for achieving high strength
Railway bridges	50–80	Superplasticizer + high early cement
Highway bridges	60–70	Superplasticizer + high early cement
Diaphragm walls	50–55	Superplasticizer + low heat cement
Oil drilling rigs	60–65	Superplasticizer + silica fume
Abrasion-resistant concrete	40–80	Superplasticizer + silica fume or anhydrous gypsum additive
High-rise reinforced concrete (RC) buildings	40–80	Superplasticizer or VMA + silica fume + ultrafine particles
RC piles	50	Superplasticizer + silica fume
Prestressed concrete piles	80	Superplasticizer or VMA + silica fume
Spun piles	85	Superplasticizer + silica fume + autoclave curing
Steel concrete composite piles	80	Superplasticizer + silica fume OR anhydrous gypsum additive + expansive additives
Centrifuged RC piles	50–80	Superplasticizer + silica fume OR anhydrous gypsum additive + low-slump concrete
Railway sleepers	40–50	Superplasticizer
Machine foundations	80–120	Superplasticizer

High-strength concrete has been used in bridges, oil drilling rigs, diaphragm walls, and tall buildings as also in prestressed concrete piles. The mix proportions used in some of these applications are summarized in Table 14.17.

Table 14.17 Mix proportions used in different applications

(a) Mix proportions used in different bridges [Slump: 12 + 2.5 cm G(max):20 mm]*

Name	Type (span) (m)	Design strength (MPa)	Type of cement	w/c ratio (%)	s/a ratio† (%)	Dosage of SP‡ (%)	Average strength (MPa)
Da-ni Ayaragigawa (Japan 1973)	49	59	Normal	30	40	0.75	65
Iwahana (Japan 1973)	45	79	High early	23	38.5	1.5	83
Akkagawa (Japan 1976)	46	79	Normal	30	39.5	1.5	93¶

*G denotes the maximum size of aggregate.

†s/a denotes sand–aggregate.

‡SP denotes superplasticizer.

¶Autoclave curing.

(b) Mix proportions used in high-rise buildings

Design strength	Target strength	G(max)	Slump (mm)	w/c ratio (%)	s/a ratio (%)	Test strength
59	71	20	25	29	40.1	64

(c) Mix proportions for concrete products

Product	Design strength	Target strength	Cement type	G(max) (mm)	Slump (mm)	w/c ratio (%)	s/a ratio (%)	Unit weight		SP dosage
								Water	Cement	
Centrifugal										
RC pipes	45	54	Normal	20	7	37	43	168	460	0.95
PSC sleepers	49	64	High early	20	4	32	40	140	440	0.85
PC beams	49	64	High early	20	3	33	35	147	450	0.85
Segments	49	61	Normal	20	5	33	40	146	450	1.05

14.9 Cost Implications

Cost comparisons are site-specific but a few general observations reveal the advantages of using high-strength concrete. The large load-carrying capacity of high-strength concrete is ideal for the heavily loaded columns of high-rise structures and the girders of long-span bridges. However, while comparing the advantages, one should take into account the requirement of greater quality control and the operational capability of handling and curing high-strength concrete.

Table 14.18 compares the relative cost of different column options. It is possible to achieve a good composite action by skin friction between tube concrete interfaces. This reduces or eliminates the need for the provision of shear studs, and thus reduces the cost of skin-reinforced systems. It has been observed that high-strength concrete filled steel tubular columns are the most advantageous for high-rise buildings.

Table 14.18 Comparison of the relative cost of different column options

Column type	Configuration	Reinforcements	Relative cost factor
Square	A	60-MPa RCC Size 840 × 840 8 Y 24 main R 10 ties	1.0
Square	A	120-MPa RCC Size 600 × 600 8 Y 32 main R 10 ties	0.80
Square	A	120-MPa RCC Size 660 × 660 8 Y 32 main R 10 ties	0.77
Circular	B	120-MPa concrete in grade 250 steel tube 740 × 8 CHS	0.98

(contd)

(contd)

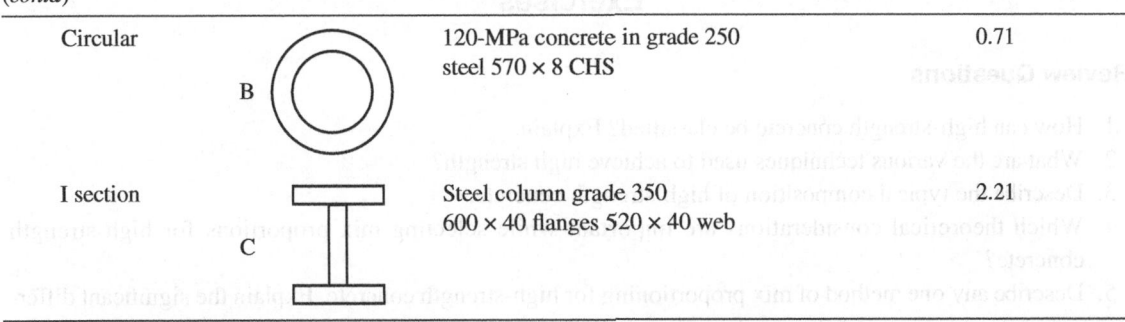

Circular B	120-MPa concrete in grade 250 steel 570 × 8 CHS	0.71
I section C	Steel column grade 350 600 × 40 flanges 520 × 40 web	2.21

Note: CHS indicates circular hollow steel section.

Table 14.19 compares the use of conventional concrete and high-strength concrete in high-rise construction. In the table, the cost per cubic metre is apportioned to different systems such as formwork, reinforcement, columns, floor slabs, and equipment recycling. Column concrete cost increases with the strength of concrete. However, the increase in concrete cost is more than offset by a reduction in the reinforcement cost and the saving in equipment recycling enabled by the use of high-strength concrete. A comparison of the material required for a 275-m chimney is made in Table 14.20. The quantity of concrete required in the optimized method of using high-strength concrete in lower levels and ordinary concrete in upper levels is compared with the conventional method of using ordinary concrete for the entire height. It is seen that the material requirement can be reduced by more than 30% in the optimized method. It is therefore evident that the use of high-strength concrete is advantageous.

Table 14.19 Cost apportioned to different items for a five-storey building

Column size 600 × 600	Breakdown of costs in terms of total cost	Cost ratio
30-MPa concrete		
Formwork	0.25	1.0
Reinforcement	0.77	
Column concrete	0.08	
Floor slab concrete	0	
Shoring equipment	0	
Total	*1.10*	
80-MPa concrete		
Formwork	0.43	0.52
Reinforcement	0.29	
Column concrete	0.34	
Floor slab concrete	0.08	
Shoring equipment	– 0.04	
Total	*1.10*	

Table 14.20 Comparison of concrete quantities for a 275-m chimney

Height (m)	M25 conventional concrete (m³)	Optimized high-strength concrete of varying strength (m³)			
		M70	M50	M40	M25
0–130	7130	3680	—	—	—
130–190	1665	—	1255	—	—
190–240	950	—	—	895	—
240–275	525	—	—	—	515
Total	*10,270*		*6345*		

Exercises

Review Questions

1. How can high-strength concrete be classified? Explain.
2. What are the various techniques used to achieve high strength?
3. Describe the typical composition of high-strength concrete.
4. Which theoretical considerations are important while selecting mix proportions for high-strength concrete?
5. Describe any one method of mix proportioning for high-strength concrete. Explain the significant difference between mix proportioning for conventional concrete and that for high-strength concrete.
6. Describe the important fresh state properties of high-strength concrete.
7. Distinguish between the elastic properties of high-strength and conventional concrete.
8. What are the important long-term properties of high-strength concrete? Compare them with those of conventional concrete.
9. What are the advantages of using high-strength concrete?
10. Describe the various applications of high-strength concrete in India.
11. What are the cost implications of using high-strength concrete for tall buildings in India?

Multiple Choice Questions

1. The strength range of high-strength concrete is
   - a. <50 MPa
   - b. 50–100 MPa
   - c. 100–200 MPa
   - d. >200 MPa

2. To produce high-strength concrete, we should
   - a. Control proportions
   - b. Control material quality
   - c. Control placing, handling, and curing
   - d. All of the above

3. The water–binder ratio for high-strength concrete is
   - a. <0.3
   - b. 0.3 to 0.4
   - c. 0.4 to 0.45
   - d. 0.45 to 0.55

4. Specific surface area of silica fume in m^2/kg is
   - a. 450
   - b. 5000
   - c. 10,000
   - d. 20,000

5. It is advantageous to use high-strength concrete for
   - a. Reducing the beam depth for long span bridges
   - b. Reducing the column dimensions in high-rise structures
   - c. Releasing the formwork early to gain cycle time in high-rise construction
   - d. All of the above

6. Densification of high-strength concrete mix is achieved by
   - a. Using finer particle mineral admixture
   - b. Grinding the cement finer
   - c. Chemical means
   - d. Particle packing technique

7. Saturation point of superplasticizer dosage represents
   - a. Increase in dosage is advantageous
   - b. Increase in dosage is not advantageous
   - c. Increase in dosage is not permitted
   - d. Increase in dosage is permitted

8. Compared to ordinary concrete, high-strength concrete has
   a. Self-desiccation
   b. More heat of hydration
   c. More early drying shrinkage
   d. All of the above

9. Compared to ordinary concrete, failure of high-strength concrete is
   a. Brittle
   b. Ductile
   c. Gradual
   d. Explosive

10. Compared to ordinary concrete, performance of high-strength concrete under fire is
    a. Favourable
    b. Unfavourable
    c. Ductile
    d. Explosive

Answers to Multiple Choice Questions

1. b
2. d
3. a
4. d
5. d
6. d
7. b
8. d
9. a
10. b

CHAPTER 15

High-performance Concrete

It is well known that conventional concrete designed on the basis of compressive strength does not adequately meet many functional requirements such as impermeability, resistance to frost, and thermal cracking. While high-strength concrete, discussed in Chapter 14, aims at enhancing strength and consequent advantages owing to improved strength, the term high-performance concrete (HPC) is used to refer to concrete of required performance for the majority of construction applications. The American Concrete Committee on HPC (ACI 1993) includes the following six criteria for material selection, mixing, placing, and curing procedures for concrete:

1. Ease of placement
2. Long-term mechanical properties
3. Early-age strength
4. Toughness
5. Life in severe environments
6. Volume stability

The above-mentioned performance requirements can be grouped into the following three general categories:

a. Attributes that benefit the construction process
b. Attributes that lead to enhanced mechanical properties
c. Attributes that enhance durability and long-term performance

The performance requirements of concrete cannot be the same for different applications. Hence the specific definition of HPC required for each industrial application is likely to vary. The Federal Highway Administration (FHWA), Washington DC, has defined HPC for highway application on the following strength, durability, and water–cement ratio criteria.

1. It should satisfy one of the following strength criteria:
   a. 4 h strength $\geq$ 17.5 MPa (2500 psi)
   b. 24 h strength $\geq$ 35.0 MPa (5000 psi)
   c. 28 days strength $\geq$ 70.0 MPa (10,000 psi)
2. It should have a durability factor greater than 80% after 300 cycles of freezing and thawing.
3. It should have a water–cement ratio of 0.35 or less.

Though the freezing and thawing resistance is indicated as a measure of high performance by the FHWA, the durability with respect to chloride ingress has been a very important parameter for marine-exposed structures especially in India. At present, there is virtually no structure in coastal regions that has achieved its designed lifespan without repairs. Moreover, national standards based on codes of practice are not definite in respect of durability requirements. Therefore, defining HPC for all applications precisely is difficult and has a bearing on the purpose for which the concrete is used. Table 15.1 gives examples of structures constructed using HPC in various countries.

From the experiences of adoption of HPC, it is evident that performance definition becomes important. The following eight parameters (the first four related to strength and the next four related to durability) are used by the FHWA to grade HPC.

1. Compressive strength
2. Modulus of elasticity
3. Creep
4. Shrinkage
5. Freeze–thaw resistance
6. Abrasion
7. Chloride permeability
8. Scaling resistance

Table 15.2 summarizes the required grade for the various environmental exposure conditions, and Table 15.3 shows grades of performance requirement based on the FHWA specifications. The test procedure is discussed later in the chapter.

Table 15.1 Examples of HPC structures

Structure	Performance Requirement	Admixtre used during construction	Remarks
Thames Barrier, UK	Resistance to marine attack and abrasion	PFA up to 50%, admixtures	A movable flood barrier in the River Thames, east of Central London
Arctic oil drilling rig, Japan	Freeze–thaw resistance	CSF admixtures	Drill barge that was used for oil exploration in the Arctic waters
Gullfaks offshore platform, Norway	Strength 65–70 MPa pump to 180 m	CSF admixtures	Sits 217 metres below the waterline. The height of the total structure is 380 m, making it taller than the Eiffel Tower
National Bank Building, USA	Pump to 268 m Strength 50 MPa	PFA admixtures	Tall building where concrete was pumped to 268 m
Channel Tunnel, UK	Resistance to severe exposure, long-term strength gain	PFA, GGBS admixtures	A rail tunnel linking London and Paris beneath the English Channel at the Strait of Dover
Baynunah Tower, UAE	High resistance to chloride attack, hot weather concreting	CSF admixtures	One of the tallest buildings in UAE
Stoerbælt Tunnel, Denmark	Resistance to freeze–thaw, chlorides, sulphates	Low alkali cement PFA, CSF	Great Belt Fixed Link runs between the Danish islands of Zealand and Funen
T Sing Ma Bridge, Hong Kong	To control heat of hydration	GGBS PFA, CSF admixtures	World's ninth-longest span suspension bridge
Shanghai Yangpu Bridge, China	To effect high durability M50 concrete	CSF admixtures	It is among the world's longest bridges, with a total length of 8354 m. Its longest span of 602 m makes it one of the longest cable-stayed bridges in the world
Ganga Bridge, India	Under-water concreting	Admixture and placing method	Bridge across river Ganga, connecting Digha Ghat in Patna and Pahleja Ghat in Sonpur

Table 15.2 Recommended HPC grades for various exposure conditions

Exposure condition	Recommended HPC grade			
	N/A	Grade 1	Grade 2	Grade 3
Freeze–thaw (F/T) durability (x = F/T cycles/year)	$x < 3$	$3 \leq x \leq 50$	$50 \leq x \leq 100$	$x > 100$
Scaling resistance applied salt (x = tonnes/lane/mile/year)	$x < 5$	$5 \leq x$	$10 \leq x \leq 12$	$x \geq 12$
Abrasion resistance (x = average daily traffic)	—	$x \leq 50,000$	$50,000 \leq x \leq 1,00,000$	$1,00,000 \leq x$
Chloride permeability applied salt (x = tonnes/plane/mile/year)	$x < 1.0$	$1.0 \leq x \leq 3.0$	$3.0 \leq x \leq 6.0$	$6.0 \leq x$

Table 15.3 Grades of performance characteristics for high-performance concrete

Performance characteristic	Performance grade			
	1	2	3	4
Freeze–thaw durability (x = relative dynamic modulus of elasticity after 300 cycles)	$60\% < x < 80\%$	$80\% < x < 90\%$	$90\% < x < 95\%$	$x > 95\%$
Scaling resistance (x = visual rating of interface after 50 cycles)	$x = 5 - 6$	$x = 4 - 3$	$x = 2 - 1$	$x = 0$
Abrasion resistance (x = avg. depth of wear in mm)	$2.0 > x > 1.0$	$1.0 > x \geq 0.5$	$0.5 > x > 0.25$	$x < 0.25$
Chloride permeability (x = coulombs)	$3000 \geq x \geq 2000$	$2000 > x > 800$	$800 \geq x \geq 400$	$x < 400$
Strength (MPa) (x = compressive strength)	$41 \leq x < 55$	$55 \leq x < 69$	$69 \leq x < 97$	$x \geq 97$
Elasticity (GPa) (x = modulus of elasticity)	$28 \leq x \leq 40$	$40 \leq x \leq 50$	$50 \leq x \leq 60$	$x > 60$
Shrinkage (x = microstrain)	$800 > x \geq 600$	$600 > x \geq 400$	$400 > x \geq 200$	$x < 200$
Creep (x = microstrain/pressure unit) (MPa)	$75 \geq x \geq 60$	$60 \geq x > 45$	$45 > x \geq 30$	$30 \geq x$

15.1 Methods for Achieving High Performance

In general, better durability performance has been achieved by using high-strength, low water–cement ratio concrete. Though in this approach the design is based on strength and the result is better durability, it is desirable that high performance, namely durability, is addressed directly by optimizing critical parameters such as the particle size of the required materials. Two approaches to achieve durability are as follows:

1. Reducing the capillary pore system such that no fluid movement can occur is the first approach. This is very difficult to realize and all concrete will have some interconnected pores.
2. Creating chemically active binding sites which prevent transport of aggressive ions such as chlorides is the second and more effective method.

These two approaches are shown in Fig. 15.1.

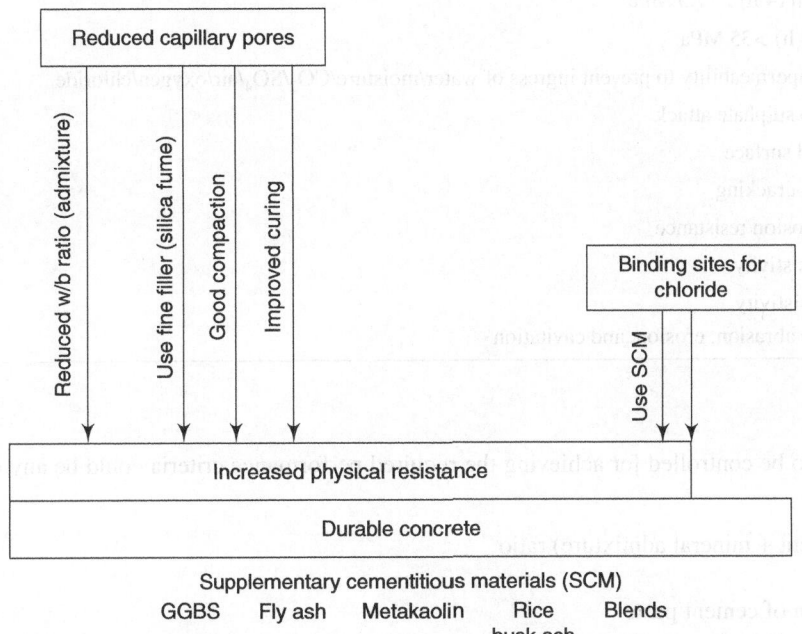

Fig. 15.1 Techniques of production of HPC

15.2 Requirements for High-performance Characteristics

Permeation is a major factor that causes premature deterioration of concrete structures. The provision of high-performance concrete must centre on minimizing permeation through proportioning methods and suitable construction procedures (curing) to ensure that the exposure conditions do not cause ingress of moisture and other agents responsible for deterioration.

Permeation can be divided into three distinct but connected stages of transportation of moisture, vapour, air, gases, or dissolved ions. These stages are schematically shown in Fig. 15.2.

It is important to identify the dominant transport phenomenon and design the mix proportion with the aim of reducing that transport mechanism which is dominant to a predefined acceptable performance limit based on permeability.

Like the requirement of permeation characteristics, there can be other performance characteristics which may become the specific need for which HPC is used. Table 15.4 gives a list of such desired characteristics for which HPC has been used.

Concrete takes in water by capillary suction. The rate at which water enters is called *sorptivity*.	The ease with which fluid passes through concrete usually under a pressure differential is referred to as *permeation*.	Vapour or gas ions are sucked through concrete under the action of ion concentration differential known as *diffusion*.

Fig. 15.2 Three stages of transportation of fluids and gases

Table 15.4 Salient high-performance requirements

Compressive strength >70 MPa

Very early strength (4 h) >17.5 MPa

Early strength (24 h) >35 MPa

High degree of impermeability to prevent ingress of water/moisture/CO_2/SO_4/air/oxygen/chloride

High resistance to sulphate attack

Smooth structured surface

Absence of micro-cracking

High level of corrosion resistance

High electrical resistivity

High chemical resistivity

High resistance to abrasion, erosion, and cavitation

The parameter to be controlled for achieving the required performance criteria could be any one or more of the following:

- Water/(cement + mineral admixture) ratio
- Strength
- Densification of cement paste
- Elimination of bleeding
- Homogeneity of the mix
- Particle size distribution
- Dispersion of cement in the fresh mix
- Stronger transition zone
- Low free lime content
- Very little free water in hardened concrete

Concrete properties in the fresh state have been extensively researched towards optimum energy consumption during mixing and filling ability under various concreting conditions by Tangtermsirikul (1994). Mixing intensity is linearly proportional to the solid volume fraction of cement, and filling ability is relatively dependent on flowability, shape of mould, and flow direction. Note that with an increase in the free water content, though the deformability of fresh concrete increases, the filling ability decreases since too much water produces voids. Similarly, an increase in the free water content decreases segregation resistance. Optimum ranges for HPC are with good deformability and segregation resistance as shown in Fig. 15.3. This figure can be used to adjust the mix proportion for the required flow performance characteristics.

Concrete in the hardened state in compression can be tested based on the standard 15 cm cylinder. The stress–strain curves for HPC show fairly good agreement with those recommended by Collins et al. (1993) (Fig. 15.4).

15.3 Factors Controlling Performance

There are many factors that play a role in the performance of concrete. Of these, material selection and mechanisms that affect performance are important. In this section, we discuss the aspects of material selection. The mechanisms affecting performance are discussed in the next section.

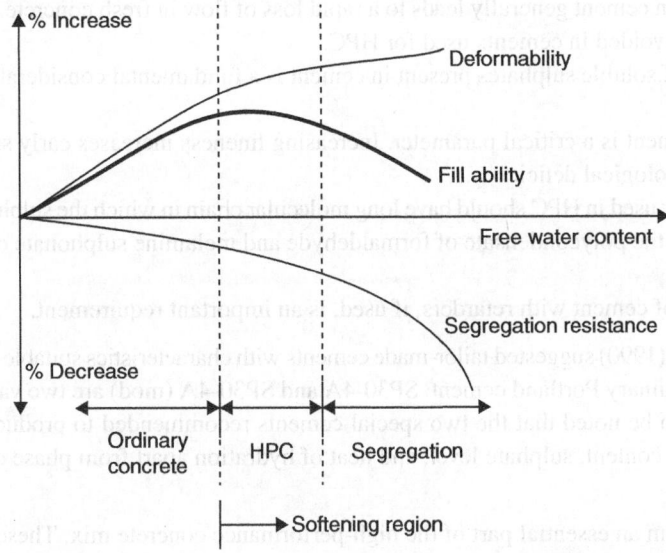

Fig. 15.3 Deformability, segregation resistance, and flowability of concrete

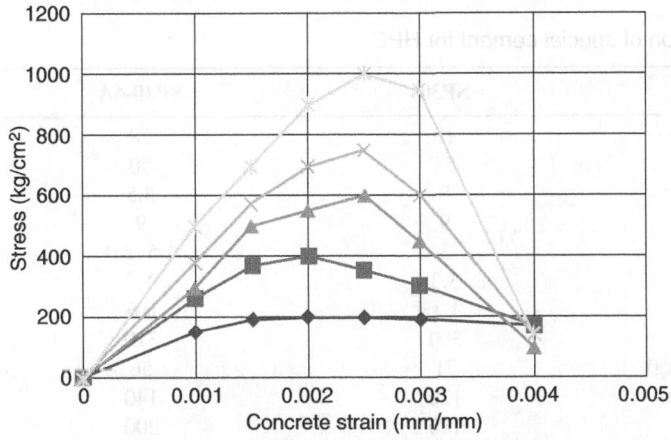

Fig. 15.4 Stress–strain relationship of HPC

15.3.1 Material Selection

The main ingredients of HPC are almost the same as that of conventional concrete. These are

- Cement
- Fine aggregate
- Coarse aggregate
- Water
- Mineral admixtures (fine filler and/or pozzolanic supplementary cementitious materials)
- Chemical admixtures (plasticizers, superplasticizers, retarders, air-entraining agents)

Cement

There are two important requirements for any cement: (a) strength development with time and (b) facilitating appropriate rheological characteristics when fresh. Studies done by Perenchio (1973) and Hanna et al. (1989) have led to the following observations:

- High C_3A content in cement generally leads to a rapid loss of flow in fresh concrete. Therefore, high C_3A content should be avoided in cements used for HPC.
- The total amount of soluble sulphates present in cement is a fundamental consideration for the suitability of cement for HPC.
- The fineness of cement is a critical parameter. Increasing fineness increases early strength development, but may lead to rheological deficiency.
- The superplasticizer used in HPC should have long molecular chain in which the sulphonate group occupies the beta position in the polycondensate of formaldehyde and melamine sulphonate or that of naphthalene sulphonate.
- The compatibility of cement with retarders, if used, is an important requirement.

Ronneburg and Sandrik (1990) suggested tailor-made cements with characteristics suitable for HPC (Table 15.5). Note that SP30 is an ordinary Portland cement. SP30-4A and SP30-4A (mod) are two varieties of tailor-made special cements. It is to be noted that the two special cements recommended to produce very high-strength concrete have low C_3A content, sulphate level, and heat of hydration apart from phase composition.

Mineral admixtures
Mineral admixtures form an essential part of the high-performance concrete mix. These are used for various purposes, depending upon their properties. More than the chemical composition, mineralogical and granulometric characteristics determine the influence of the mineral admixture's role in enhancing properties of concrete.

Table 15.5 Composition of special cement for HPC

	SP30	**SP30-4A**	**SP30-4A (mod)**
C_2S (%)	18	28	28
C_3S (%)	55	50	50
C_3A (%)	8	5.5	5.5
C_4AP (%)	9	9	9
MgO (%)	3	1.5–2.0	1.5–2.0
SO_3 (%)	3.3	2–3	2–3
Na_2O equivalent (%)	1.1	0.6	0.6
Blaine fineness (m^2/kg)	300	310	400
Heat of hydration (kcal/kg)	71	56	70
Setting time critical (min)	120	140	120
Final	180	200	170

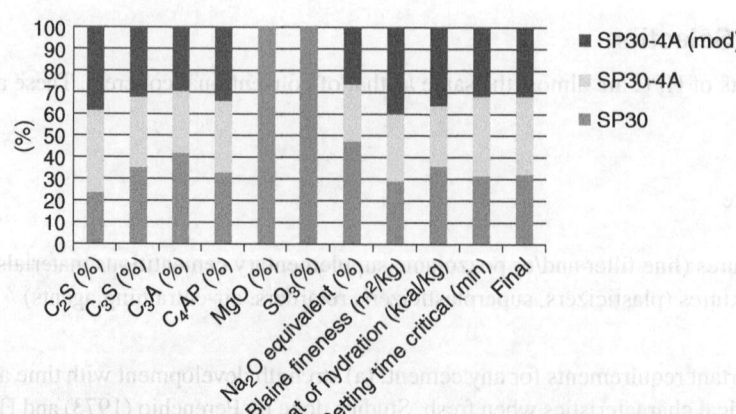

Chemical composition   The chemical compositions of mineral admixtures such as natural pozzolans, fly ash, silica fume, rice husk, ash, and metakaolin are presented in Fig. 15.5.

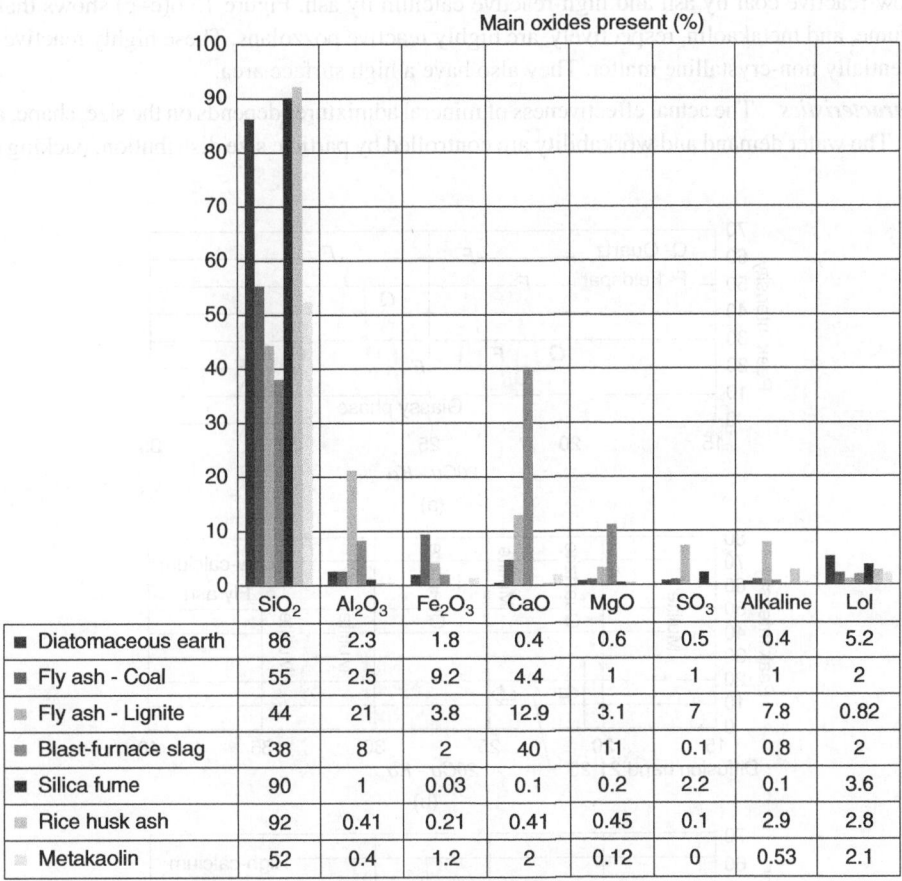

Main oxides present (%)

	SiO₂	Al₂O₃	Fe₂O₃	CaO	MgO	SO₃	Alkaline	Lol
■ Diatomaceous earth	86	2.3	1.8	0.4	0.6	0.5	0.4	5.2
■ Fly ash - Coal	55	2.5	9.2	4.4	1	1	1	2
▦ Fly ash - Lignite	44	21	3.8	12.9	3.1	7	7.8	0.82
■ Blast-furnace slag	38	8	2	40	11	0.1	0.8	2
■ Silica fume	90	1	0.03	0.1	0.2	2.2	0.1	3.6
▦ Rice husk ash	92	0.41	0.21	0.41	0.45	0.1	2.9	2.8
▦ Metakaolin	52	0.4	1.2	2	0.12	0	0.53	2.1

Fig. 15.5 Typical oxide analysis of mineral admixtures

A look at this figure reveals that silica and alumina content vary widely. However, these chemical differences do not significantly influence the properties of concrete. Fly ashes generally contain less silica and more alumina compared to natural pozzolans. Both fly ash and blast-furnace slag have high calcium and magnesium oxide content. Highly active admixtures such as silica fume and rice husk ash contain high content of silica. Metakaolin contains roughly equal proportions of silica and alumina.

Mineralogical composition   The mineralogical composition is best examined with the help of x-ray diffraction (XRD) analysis based on reflection angle and peak intensity. The XRD patterns of various mineral admixtures are scaled and plotted in Fig. 15.6. The XRD patterns reveal the differences between crystalline and mono-crystalline mineral admixtures.

Figure 15.6(a) shows a typical XRD analysis of a natural pozzolana—santorin earth. 90% contains glassy phase contributed by aluminosilicate glass. The non-crystalline admixtures are composed of feldspar and quartz. The low-calcium fly ash from coal [Fig. 15.6(b)] has quartz and mullite. These do not have pozzolanic activity. On the other hand, in high-calcium (such as the one from Neyveli) fly ash [Fig. 15.6(c)] the crystalline minerals are quartz tricalcium aluminate, free calcium oxide, and calcium alumino sulphate. Except quartz, other minerals react with water even at ordinary temperature and hence have good pozzolanic activity. Figure 15.6(d) shows the typical XRD of granulated blast-furnace slag. Its chemical composition is mullite glass and its reactivity depends on other reactive minerals such as dicalcium aluminosilicate (2CaO, Al_2O_3, SiO_2). In addition, we have non-reactive minerals such as gehlenite and diopside. Thus, we could classify these pozzolans as low-reactive coal fly ash and high-reactive calcium fly ash. Figure 15.6(e–g) shows that rich husk ash, silica fume, and metakaolin, respectively, are highly reactive pozzolans. These highly reactive pozzolans contain essentially non-crystalline matter. They also have a high surface area.

Particle characteristics The actual effectiveness of mineral admixtures depends on the size, shape, and texture of particles. The water demand and workability are controlled by particle size distribution, packing effect, and

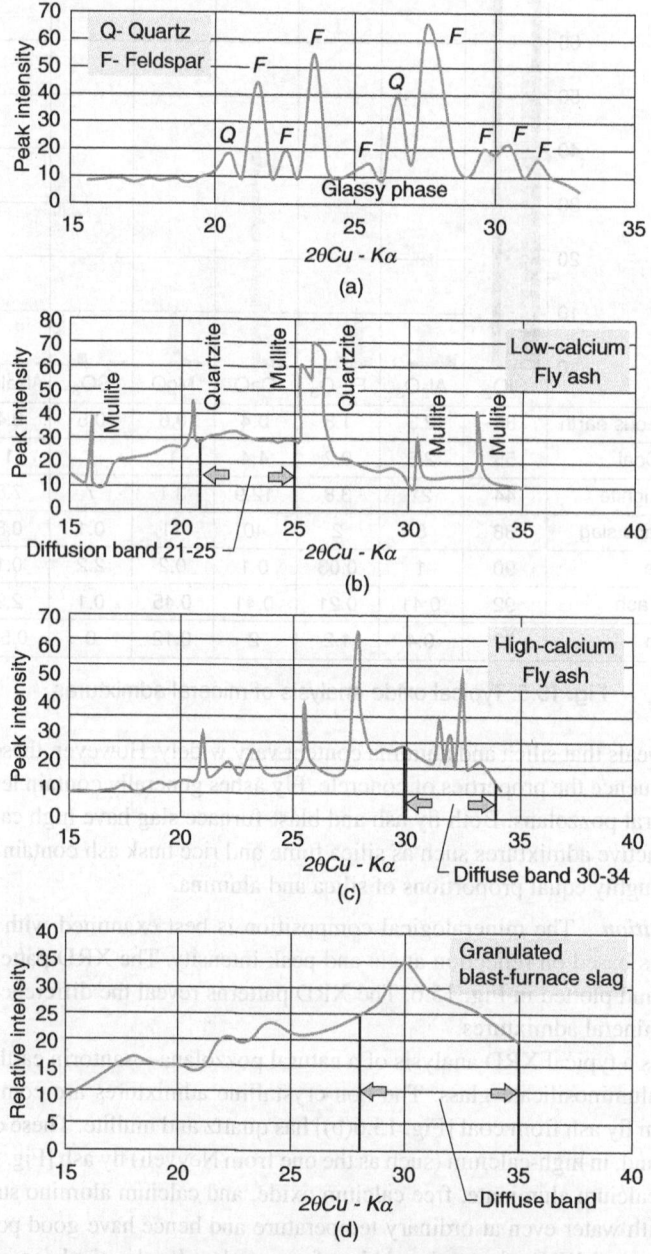

Fig. 15.6 (*contd on next page*)

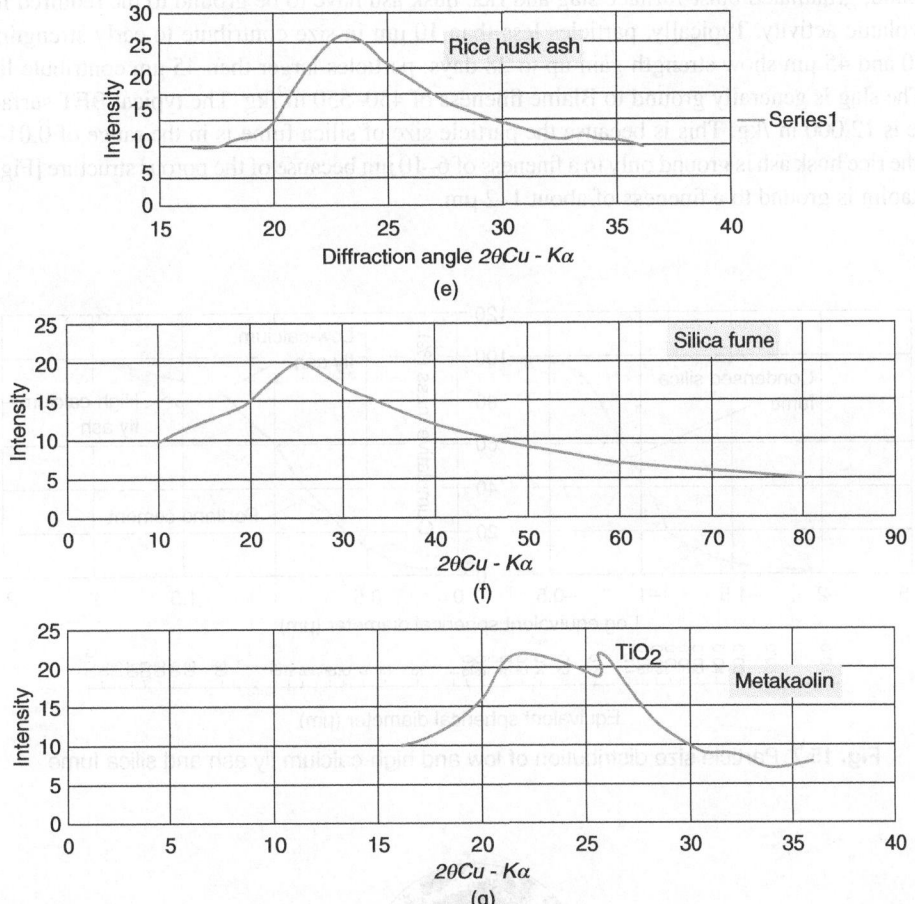

Fig. 15.6 XRD analysis of (a) natural pozzolana, (b) coal-based fly ash, (c) high-calcium fly ash, (d) GGBS, (e) rice husk ash, (f) silica fume, and (g) metakaolin

surface texture. In fact the control of particle size distribution is important for ensuring pozzolanic/cementitious nature of the admixture. The size of particles larger than 45 μm or those retained on sieve number 325 should be limited to about 33%. Particles having sizes larger than 45 μm do not have reactivity under normal conditions. It has been observed that pozzolanic activity is proportionate to the percentage of particles less than 10 μm in these highly reactive pozzolans.

For determining the surface area of particles of cement, Blaine air permeability apparatus is used. This cannot be successfully used for pozzolans because they are ultra-fine. For these materials, the BET (Brunaer, Emmet, and Taylor) nitrogen absorption method is useful. Figure 15.7 shows the particle size distribution of low and high-calcium fly ash and silica fume. The particle size of fly ashes ranges from 2 to 100 μm, whereas that for silica fume ranges from 0.01 to 0.45 μm.

The texture of the particles can be well assessed by examining the scanning electron microscope photographs [Fig. 15.8(a–d)] of different mineral admixtures. These micrographs clearly show the typical porous texture of volcanic ash and rice husk ash [Fig. 15.8(a–b)]. Fly ash and silica fume consist of solid spherical particles [Fig. 15.8(c–d)]. The difference in surface textures of various mineral admixtures can be clearly seen in these figures.

The ground, granulated blast-furnace slag and rice husk ash have to be ground to the required fineness to assist pozzolanic activity. Typically, particles less than 10 μm in size contribute to early strength; particles between 10 and 45 μm show strength gain up to 28 days; particles larger than 45 μm contribute little to the strength. The slag is generally ground to Blaine fineness of 450–550 m²/kg. The typical BET surface area of silica fume is 12,000 m²/kg. This is because the particle size of silica fume is in the range of 0.01–0.45 μm. However, the rice husk ash is ground only to a fineness of 6–10 μm because of the porous structure [Fig. 15.8(b)]. The metakaolin is ground to a fineness of about 1–2 μm.

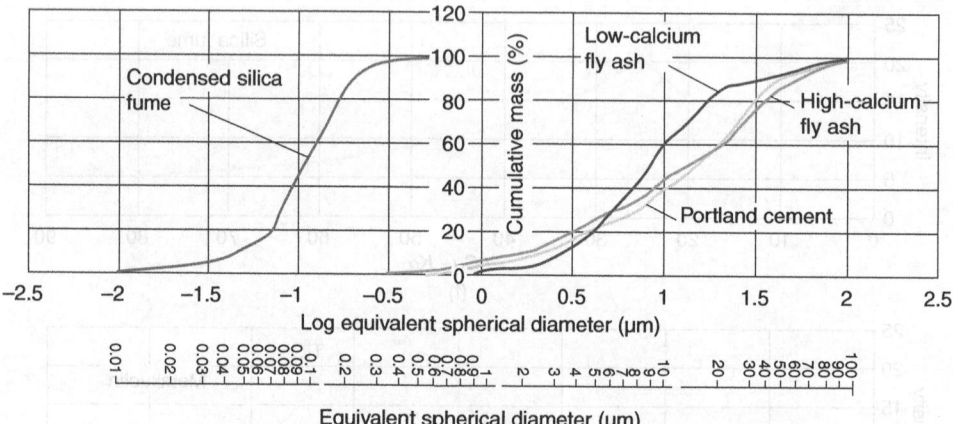

Fig. 15.7 Particle size distribution of low and high-calcium fly ash and silica fume

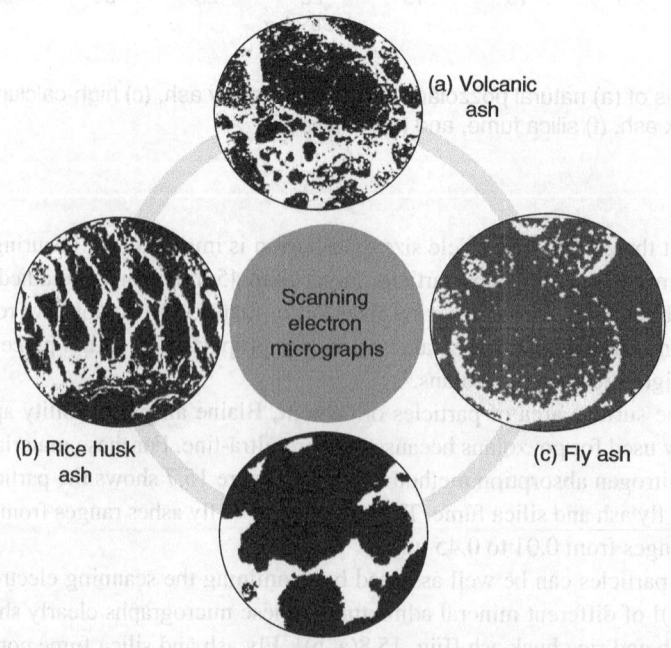

Fig. 15.8 Scanning electron micrographs of (a) volcanic ash, (b) rice husk ash, (c) fly ash, and (d) silica fume

The pozzolanic material with large surface area shows excellent reactivity. It imparts stability and cohesiveness to the mixture and prevents bleeding as well as segregation.

A summary of the characteristics of different mineral admixtures is given in Table 15.6.

Table 15.6 Characteristics of mineral admixtures

Type	Classification	Chemical composition	Particle characteristics
Ground, granulated blast-furnace slag (GGBS)	Cementitious and pozzolanic	Silicate glass containing calcium magnesium silicate	Unprocessed material are grains like sand. These are ground to size <45 μm (500 m²/kg Blaine) particles and have a rough texture
Calcium-rich fly ash	Cementitious and pozzolanic	Silicate glass containing mainly calcium, magnesium, and aluminium alkides. Also contains C_3A, CaO, C_3S, C_4A_3S traces, unburnt carbon 1–2%	Powder consists of particles <45 μm. However, 10–15% are more than 45 μm. Particles are solid spheres of 20 μm. Surface is generally smooth
Condensed silica fume	Highly active pozzolana	Pure silica of non-crystalline form	Extremely fine powder consisting of solid spheres of 0.1 μm average diameter, about 20 m²/g surface areas estimated by nitrogen absorption method
Rice husk ash	Highly active pozzolana	Consists essentially of pure silica in non-crystalline form	Particles are <45 μm but have cellular and porous structure
Low-calcium fly ash	Cementitious and pozzolanic	Mostly silicate glass containing aluminium and iron and alkides, small quantities of quartz, haematite, etc.	Powder having particles of 15–30% > 45 μm. Most particles are solid sphere. Cenospheres and plerospheres may be present
Natural material	Natural pozzolana	Contains aluminosilicate glass, natural pozzolans containing quartz, feldspar, and mica	Particles are <45 μm and have rough texture. Consists of crystalline silicate material
Slowly cooled blast-furnace slag, bottom ash, field burnt rice husk ash	Weak pozzolana	Consists of crystalline silicate material	Pulverized to fine powder, and ground materials have rough texture

15.4 Mechanisms Affecting Performance

In this section we will discuss some of the mechanisms that are used to enhance the performance characteristics of concrete.

Permeation

Partial replacement of cement by mineral admixtures reduces water requirement to obtain a particular consistency. The water reduction and increase in mobility is caused by both spherical particle shape and adsorption–dispersion mechanism that is similar to water-reducing chemical admixtures. Fine particles of mineral admixtures get adsorbed on oppositely charged surfaces of cement particles and prevent them from flocculation. In addition, particle packing is also responsible for water reduction and, hence, impermeability. The fine particle size reduces the void space, and hence less water is required for plasticizing. Figure 15.9 shows the operation of this mechanism in the cement–silica fume system. Fine particles of cement cannot behave like silica fume as they dissolve in water and hence cannot take the role played by filler. Note the relative sizes of cement grain and silica fume.

Figure 15.10 shows the absorption levels of concrete made with different percentages of PFA content. The results show that excellent permeation properties can be achieved with different strengths depending upon the percentage of PFA used. Clearly, a particular permeation level can be achieved with a lower strength, provided

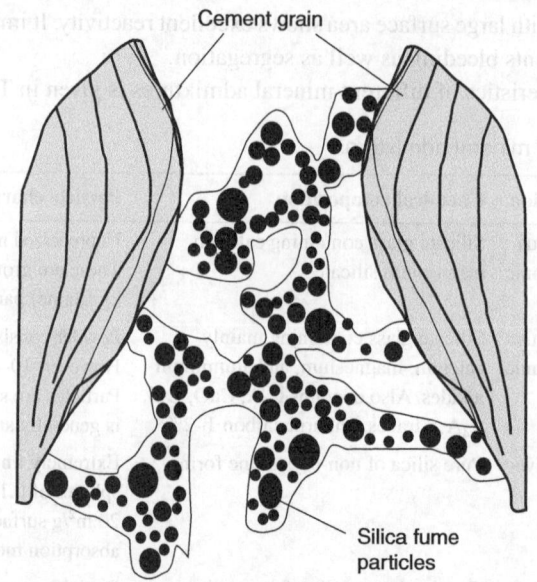

Fig. 15.9 Mechanics of bleeding reduction by silica fume

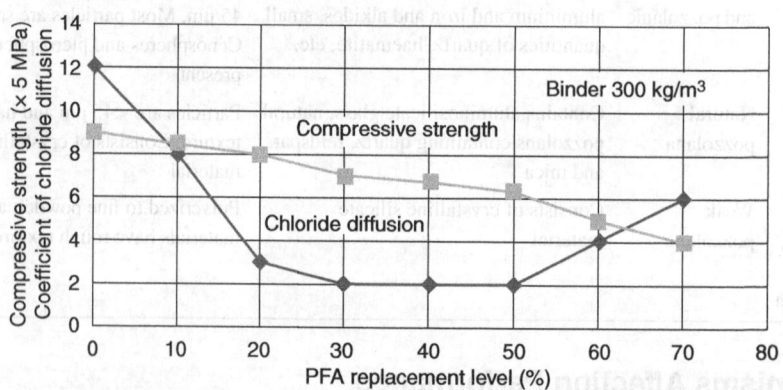

Fig. 15.10 Absorption levels of concrete made with different percentages of PFA replacement

the percentage of PFA is more. Thus, reduction in permeability can be achieved by proper choice of binder material rather than by increase in strength.

Chloride ingress

Chloride ingress causes damage in structures located in marine environments. It can be minimized by the following methods:

i. Use of large percentages of fly ash (PFA) and GGBS. These physically bind chloride ions and thus block their passage through the concrete cover.

ii. Particle packing designed to minimize the volume of voids physically prevents permeation.

Very low levels of chloride diffusion can be achieved by the above two methods. Low PFA/GGBS in high-strength concrete and high PFA/GGBS contents in ordinary strength concrete provides the required resistance against chloride ingress.

Carbonation

Mineral admixtures cannot control carbonation. Results of a study by Jones et al. (2000) are shown in Fig. 15.11. Concrete made with Portland cement has lowest carbonation depth. It is evident that higher strength can control carbonation depth.

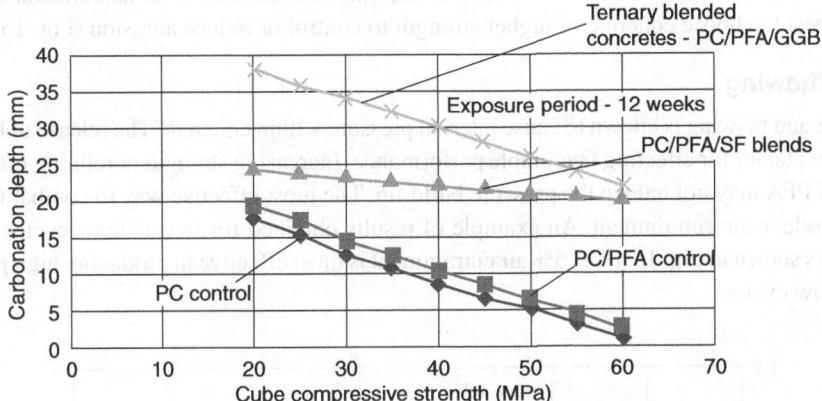

Fig. 15.11 Importance of concrete strength in controlling carbonation depth

Corrosion of steel

The most important factor that influences corrosion rate is the environmental condition (more so, humidity). The material and type of concrete have minor effect compared to humidity. To reduce corrosion, it is advisable to control the agent/process that causes corrosion. These aspects have been dealt in Chapter 11.

Sulphate attack

Damage to concrete in underground structures resulting from sulphate attack is caused mainly by diffusion of water from the surroundings into the concrete. The sulphates react with aluminate phases of hydrated cement paste and Ca(OH), leading to the formation of ettringite and gypsum. The accompanying reaction increases the cement paste volume and hence generates internal bursting stresses, which causes the concrete to disintegrate. To control this, both impermeability and binder chemistry have to be improved.

For controlling sulphate ingress, increase in the level of replacement of GGBS has shown (Fig. 15.12) to reduce the percentage relative linear expansion; though there is associated reduction in strength. The success

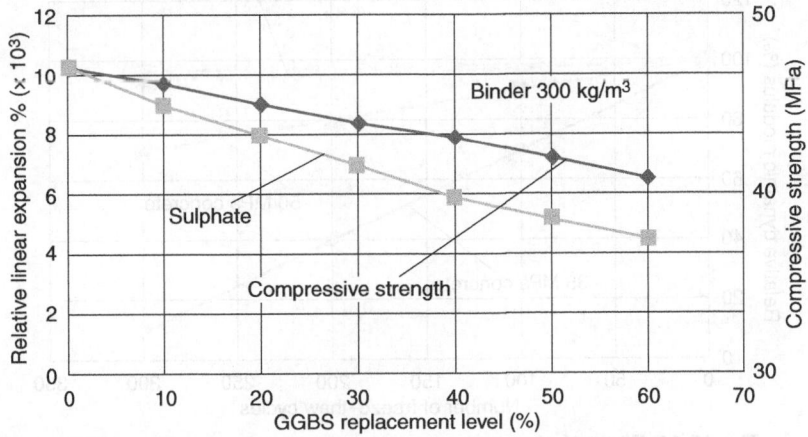

Fig. 15.12 Controlling sulphate ingress

of combating sulphate attack depends on the reduction in available aluminates (cement-based material) and efficient mix design to control permeation properties (Dewar 1994).

Abrasion resistance

The effect of binder addition is less critical than the strength of concrete as far as abrasion resistance is considered. It is best to choose concrete of higher strength to control or reduce abrasion (Fig. 15.13).

Freezing and thawing

Repeated freezing and thawing is known to cause internal pressure within concrete. The release of this pressure build-up is a critical factor for effecting favourable performance. Increasing strength or refining microstructure by the addition of PFA may not reduce the pressure build-up. The most effective way to combat freezing and thawing is to introduce air entrainment. An example of results obtained for freeze–thaw cycles of different types of concrete is shown in Fig. 15.14. A 5% air entrainment is most effective in producing high performance against freeze–thaw cycles.

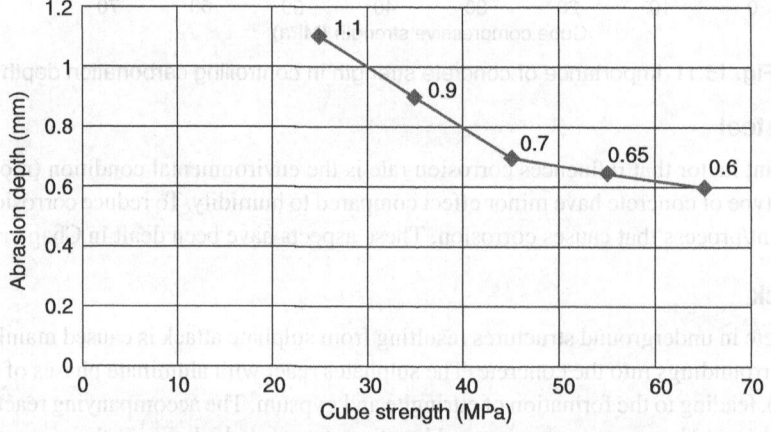

Fig. 15.13 Importance of strength in reducing abrasion

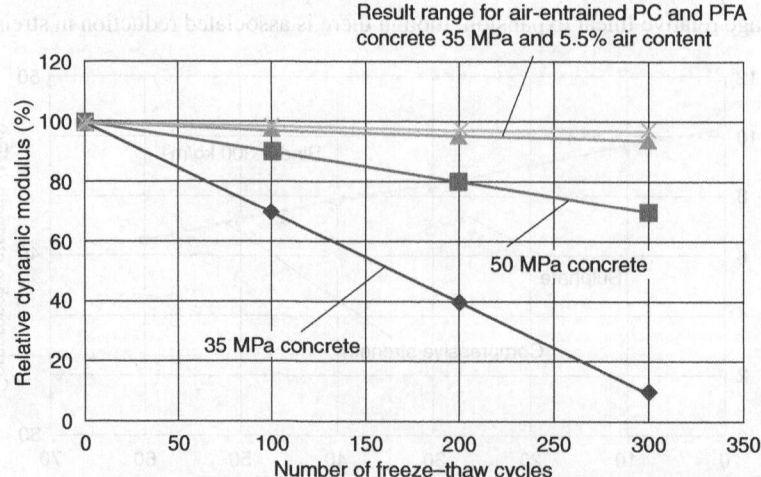

Fig. 15.14 Effect of air entrainment in controlling freeze–thaw effects

Complex exposures

The exposure conditions may be complex in certain situations. In marine atmosphere both chlorides and sulphates may be present. In addition, the concrete surface may also be subjected to severe abrasion. Chloride binding is a complex phenomenon. In highly durable concrete, high levels of pozzolanic binders are used. These poly-blends provide both chloride binding and reduced capillary pore size. Figure 15.15 shows the efficiency of poly-blends in reducing the diffusion index. The blend PC/GBS/SF has a significantly lower particle density index compared to other blends for the environment considered. Figure 15.16 shows that good performance in a chloride-only environment fails to perform well in complex environmental conditions. For complex environment situations, it is necessary to rank the materials and then choose them based on real environmental conditions.

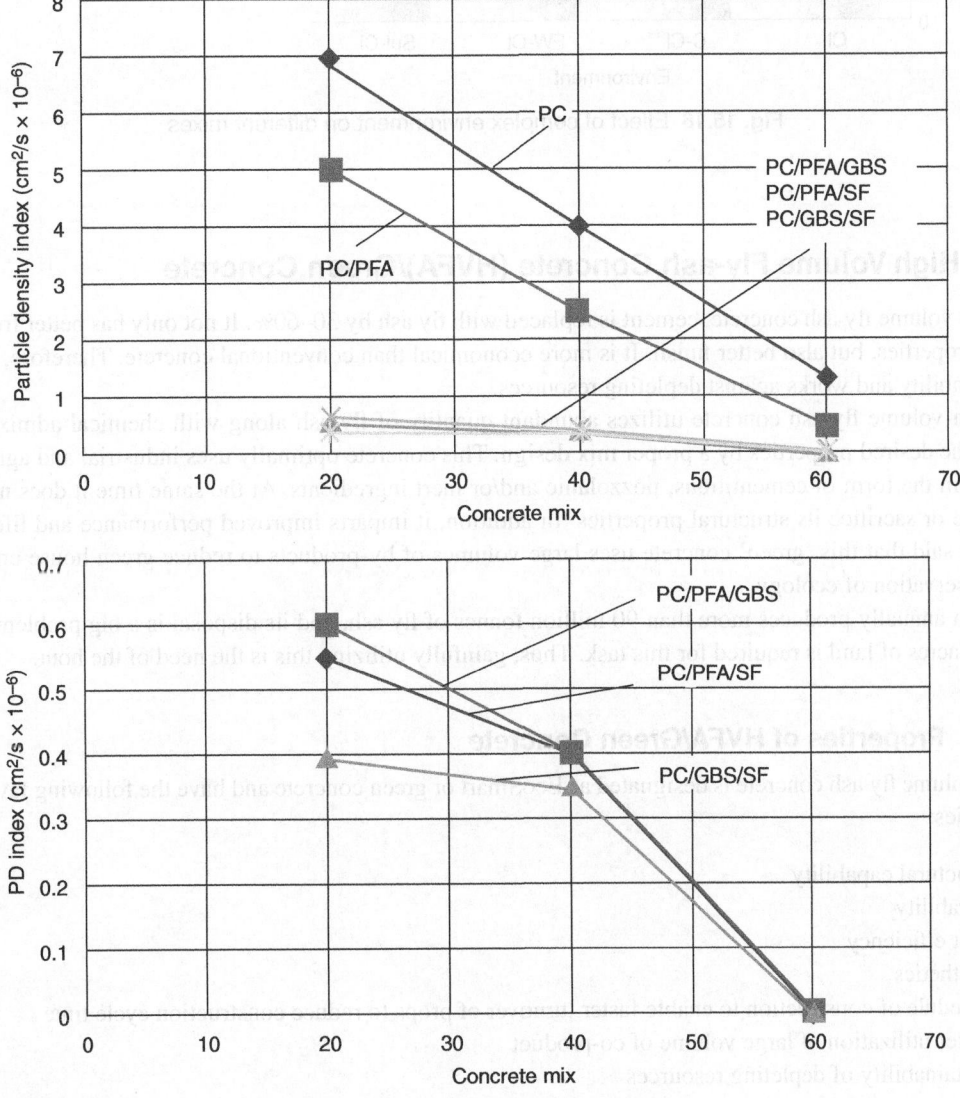

Fig. 15.15 Efficiency of poly-blends

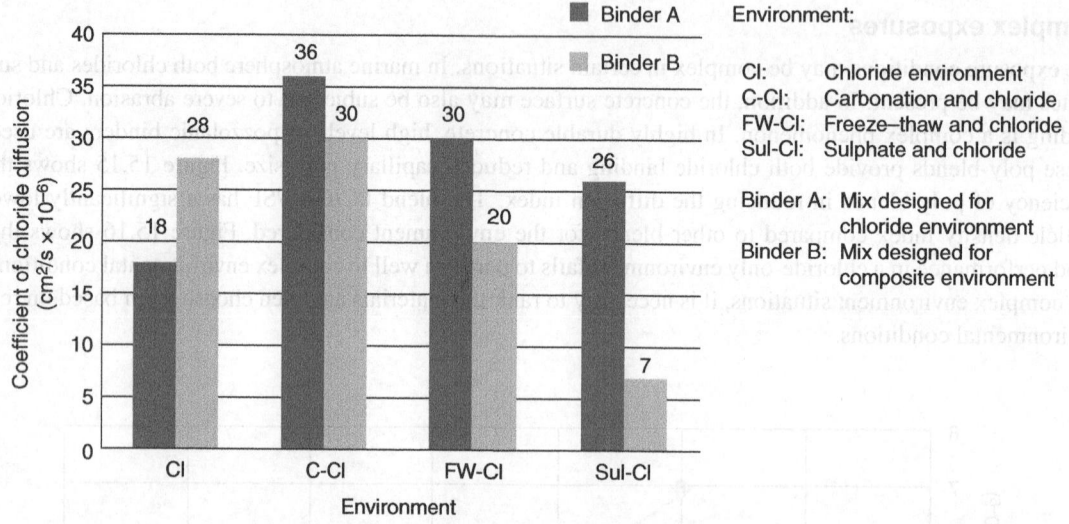

Fig. 15.16 Effect of complex environment on different mixes

15.5 High Volume Fly-ash Concrete (HVFA)/Green Concrete

In high-volume fly ash concrete, cement is replaced with fly ash by 30–60%. It not only has better fresh concrete properties, but also better finish. It is more economical than conventional concrete. Therefore, it offers sustainability and works against depleting resources.

High-volume fly ash concrete utilizes abundant quantity of fly ash along with chemical admixtures to attain the desired properties by a proper mix design. This concrete optimally uses industrial and agriculture wastes in the form of cementitious, pozzolanic and/or inert ingredients. At the same time it does not compromise or sacrifice its structural properties. In addition, it imparts improved performance and life. It can thus be said that this 'green' concrete uses large volumes of by-products to reduce green house emissions for preservation of ecology.

India annually produces more than 90 million tonnes of fly ash, and its disposal is a big problem. About 65,000 acres of land is required for this task. Thus, gainfully utilizing this is the need of the hour.

15.5.1 Properties of HVFA/Green Concrete

High-volume fly ash concrete is designated as EcoSmart or green concrete and have the following favourable properties:

- Structural capability
- Durability
- Cost efficiency
- Aesthetics
- Schedule of construction to enable faster turnover of props to reduce construction cycle time
- Better utilization of large volume of co-product
- Sustainability of depleting resources

15.5.2 Mixture Proportions

Typical range of component materials for different levels of strength in high-performance HVFA concrete is shown in Fig. 15.17. The control of water content is most essential because the amount of water is varied within a narrow range between 100 and 130 kg/m^3 by using a combination of one or more tools such as a superplasticizing admixture, a high-quality fly ash, or a well-graded aggregate. As the water content between the different strength levels does not vary much, it is necessary to increase the cementitious materials substantially to achieve higher strength. When very high strength is needed at an early age, it can be obtained by adopting one or more of the following methods: a higher ratio between Portland cement and fly ash, substitution of a high-early strength Portland cement with ordinary Portland cement, and replacement of a portion of the fly ash with a more reactive pozzolana such as silica fume or rice husk ash.

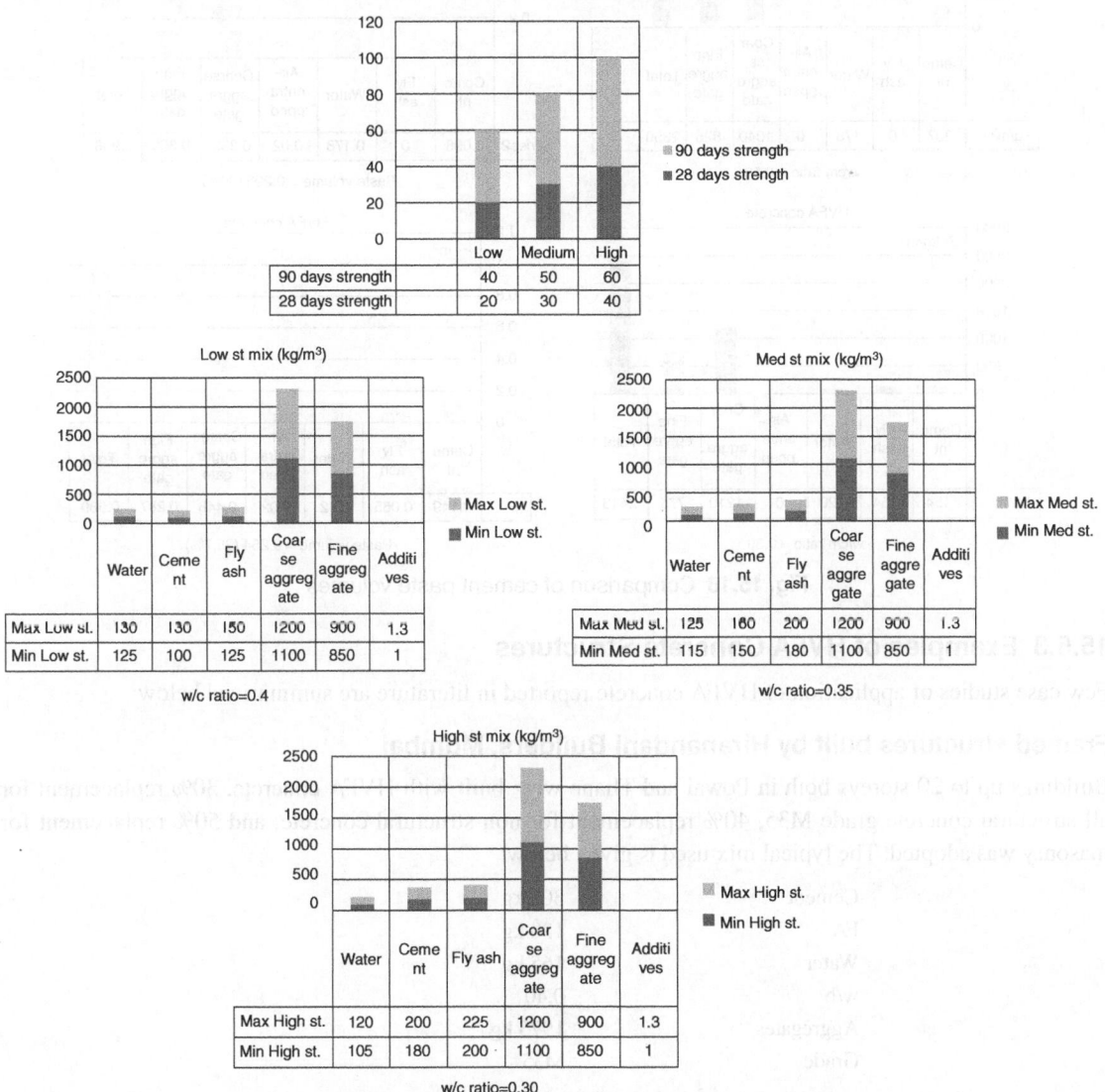

	Low	Medium	High
90 days strength	40	50	60
28 days strength	20	30	40

Low st mix (kg/m^3)

	Water	Cement	Fly ash	Coarse aggregate	Fine aggregate	Additives
Max Low st.	130	130	150	1200	900	1.3
Min Low st.	125	100	125	1100	850	1

w/c ratio=0.4

Med st mix (kg/m^3)

	Water	Cement	Fly ash	Coarse aggregate	Fine aggregate	Additives
Max Med st.	125	160	200	1200	900	1.3
Min Med st.	115	150	180	1100	850	1

w/c ratio=0.35

High st mix (kg/m^3)

	Water	Cement	Fly ash	Coarse aggregate	Fine aggregate	Additives
Max High st.	120	200	225	1200	900	1.3
Min High st.	105	180	200	1100	850	1

w/c ratio=0.30

Fig. 15.17 Typical mix proportions for different strength levels

Figure 15.18 shows mix proportions of conventional 25 MPa concrete compared with HVFA concrete with similar strength but with higher slump. Due to reduction in water requirement, the total volume of cement paste in HVFA concrete is only 25% compared to 29.6% for conventional Portland cement concrete. This represents a 30% reduction in cement paste to aggregate volume ratio.

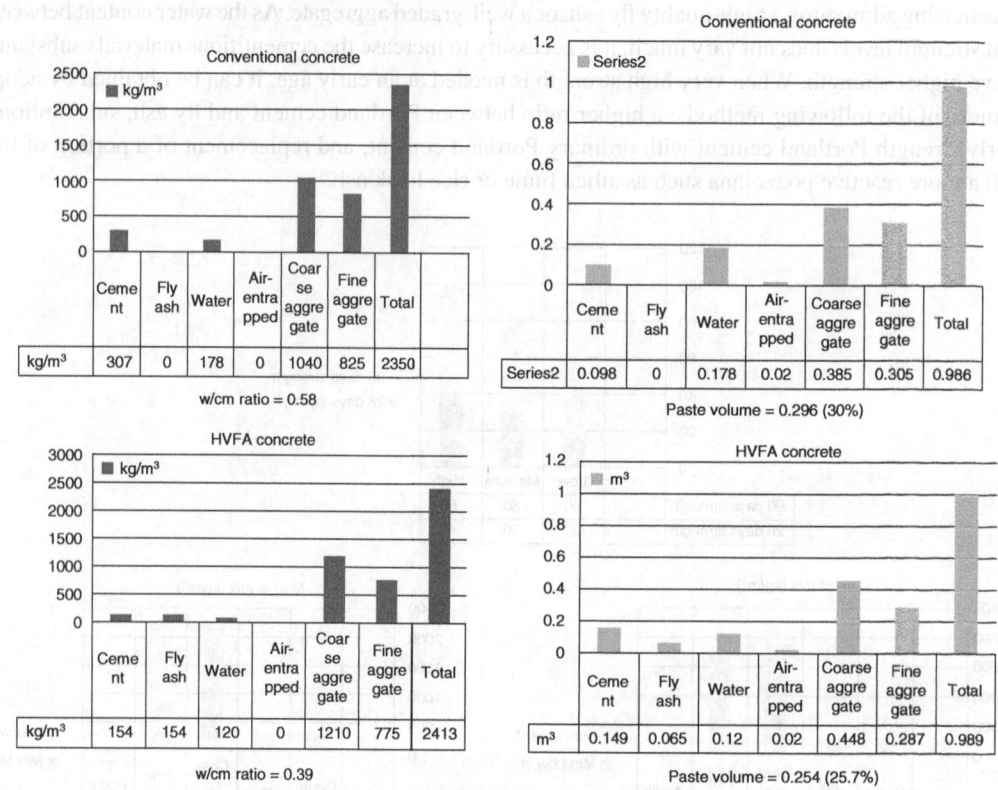

Fig. 15.18 Comparison of cement paste volumes

15.5.3 Examples of HVFA Concrete Structures

Few case studies of application of HVFA concrete reported in literature are summarized below:

Framed structures built by Hiranandani Builders, Mumbai

Buildings up to 20 storeys both in Powai and Thane were built with HVFA concrete. 30% replacement for all structural concrete grade M35, 40% replacement for non-structural concrete, and 50% replacement for masonry was adopted. The typical mix used is given below:

Cement	300 kg
FA	130 kg
Water	165 kg
w/b	0.40
Aggregates	1975 kg
Grade	M35

Delhi metro rail

Delhi metro rail has about 100 km of underground section, which is subjected to aggressive environment. 30% fly ash in all structural concrete and 70% slag in underground sections were used.

Concrete road in Ropar, Punjab

700 m long, 7 m wide demonstration project undertaken with the details shown in Fig. 15.19.

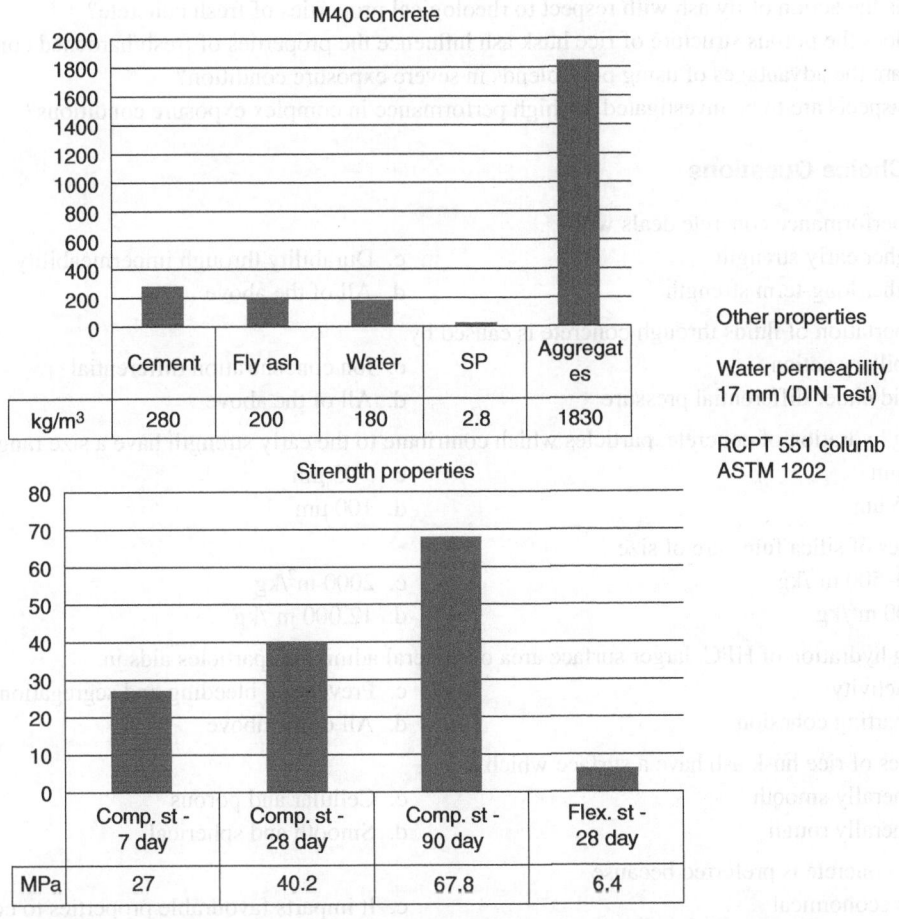

M40 concrete

kg/m³	Cement	Fly ash	Water	SP	Aggregates
	280	200	180	2.8	1830

Other properties

Water permeability
17 mm (DIN Test)

RCPT 551 columb
ASTM 1202

Strength properties

MPa	Comp. st - 7 day	Comp. st - 28 day	Comp. st - 90 day	Flex. st - 28 day
	27	40.2	67.8	6.4

Fig. 15.19 Typical details of HVFA concrete used for a concrete road in Ropar

It is possible to produce HVFA concrete and use it in construction. It is required from the point of view of sustainability of the concrete industry. The ideal conditions for applying it are warm weather, large pores, and no form stripping constraints. The mix should be properly designed.

Exercises

Review Questions

1. What are the important aspects of performance of concrete which the ACI Committee has identified?
2. How has the FHWA defined HPC?
3. List the aspects of HPC that are related to strength and durability separately.
4. What are the important approaches for achieving durable concrete?
5. List the requirements of concrete performance generally demanded by clients.
6. What are the methods of transportation of fluids and gases which aid permeation in concrete?

7. What are the factors which control the performance of HPC?
8. List different mineral admixtures and their characteristics.
9. How does silica fume help in reducing bleeding?
10. What is the action of fly ash with respect to rheological properties of fresh concrete?
11. How does the porous structure of rice husk ash influence the properties of fresh/hardened concrete?
12. What are the advantages of using poly-blends in severe exposure condition?
13. What aspects are to be investigated for high performance in complex exposure conditions?

Multiple Choice Questions

1. High-performance concrete deals with
   a. Higher early strength
   b. Higher long-term strength
   c. Durability through impermeability
   d. All of the above

2. Transportation of fluids through concrete is caused by
   a. Capillary action
   b. Fluid under differential pressure
   c. Ion concentration differential
   d. All of the above

3. During hydration of concrete, particles which contribute to the early strength have a size range of
   a. 10 μm
   b. <45 μm
   c. >45 μm
   d. 100 μm

4. Particles of silica fume are of size
   a. 450–500 m²/kg
   b. 1000 m²/kg
   c. 2000 m²/kg
   d. 12,000 m²/kg

5. During hydration of HPC, larger surface area of mineral admixture particles aids in
   a. Reactivity
   b. Imparting cohesion
   c. Preventing bleeding and segregation
   d. All of the above

6. Particles of rice husk ash have a surface which is
   a. Generally smooth
   b. Generally rough
   c. Cellular and porous
   d. Smooth and spherical

7. HVFA concrete is preferred because
   a. It is economical
   b. It helps in sustainability of the concrete industry
   c. It imparts favourable properties to concrete
   d. All of the above

8. Blaine air permeability apparatus cannot be used to determine the particle size of
   a. Silica fume
   b. Low-calcium fly ash
   c. High-calcium fly ash
   d. Cement

9. The surface area of silica fume particles is determined by
   a. Blaine permeability apparatus
   b. BET nitrogen absorption method
   c. Sieve analysis
   d. Hydrometer analysis

10. XRD analysis of mineral admixtures is used for determining
    a. Mineralogical composition
    b. Particle size
    c. Permeability coefficient
    d. Relative reactivity

Answers to Multiple Choice Questions

1. d	4. d	7. d	10. a
2. d	5. d	8. a	
3. a	6. c	9. b	

CHAPTER 16

Polymers in Concrete

Polymers or epoxies are used for imparting certain special properties to concrete. They can be used for the following reasons:

- To improve strength and durability of hardened concrete
- To improve chemical resistance and impermeability of hardened concrete
- To modify the flow characteristics of fresh concrete
- To improve the bond characteristics between old and new concrete for repair work

The price of various polymers varies considerably, but as a rough guide it can be assumed that polymers cost approximately 20 times that of ordinary cement. However, direct cost should not be the sole basis of an economic assessment of their use because the treatment could reduce the fabrication cost and substantially improve performance and durability.

Some popularly used polymers are listed below:

Urethanes: These are polymers and copolymers produced by the reaction of isocyanates with polyols.

Acrylics: These are polymers and copolymers of the esters of acrylic and methacrylic acids.

Styrene butadiene resins (SBR): These resins are basically synthetic rubber in solution.

Vinyl: This is a general term for substituted ethylenes and their copolymers such as polyethylene and polystyrenes (basically copolymers than homopolymers).

Epoxies: Synthetic polymers which are condensates of epichlorohydrin and a suitable polyhydroxyl material; most commonly used polyhydroxyl material is bisphenol–A.

Polymers are used in the following ways with concrete:

- Polymer-impregnated concrete
- Polymer concrete
- Polymer-modified concrete/mortar
- Polymer as protective coating
- Polymer as bonding agent
- Other applications

A comparison of major properties of some polymers is shown in Fig. 16.1. Figure 16.2 illustrates compressive stress as a function of strain in concrete at different levels of polymer content [polymer used is methyl methacrylate (MMA)].

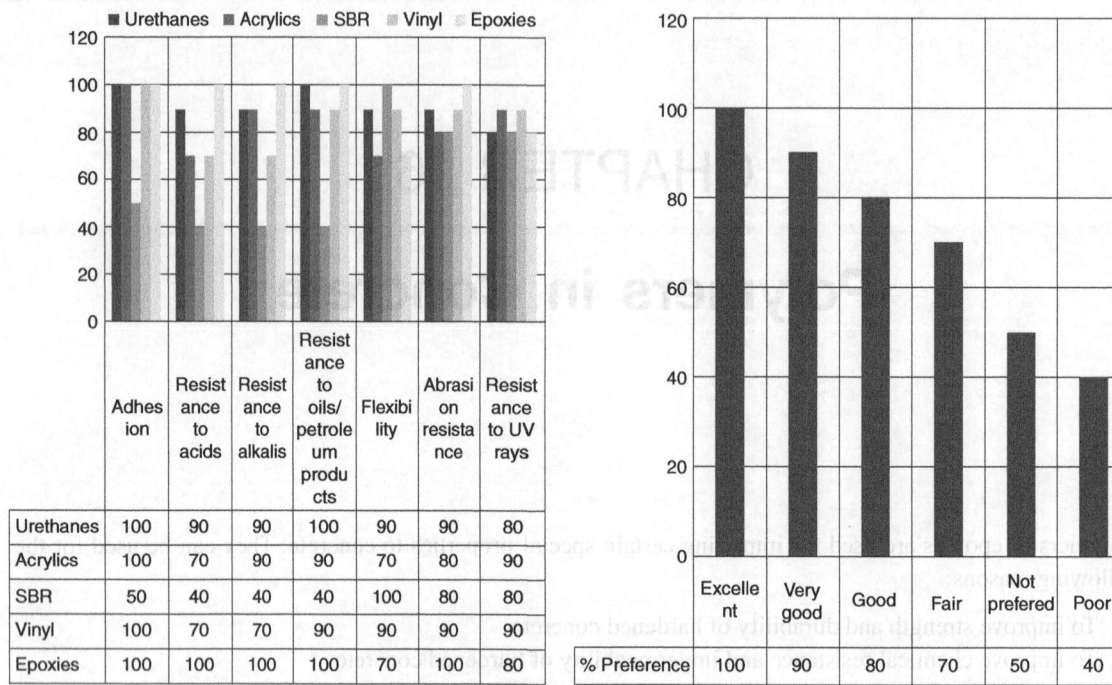

	Adhesion	Resistance to acids	Resistance to alkalis	Resistance to oils/petroleum products	Flexibility	Abrasion resistance	Resistance to UV rays
Urethanes	100	90	90	100	90	90	80
Acrylics	100	70	90	90	70	80	90
SBR	50	40	40	40	100	80	80
Vinyl	100	70	70	90	90	90	90
Epoxies	100	100	100	100	70	100	80

	Excellent	Very good	Good	Fair	Not prefered	Poor
% Preference	100	90	80	70	50	40

Fig. 16.1 Comparison of properties of different polymers and their preferences

The price of various polymers varies considerably, but as a rough guide it can be assumed that polymers cost approximately 20 times that of ordinary cement. However, direct cost should not be the sole basis of an economic assessment of their use because the treatment could reduce the permeation content and substantially improve performance and durability.

Some popularly used polymers are listed below.

Urethanes: These are polymers and copolymers involving urethane, which is also known as carbamate.

Acrylics: These are polymers based on acrylic or methacrylic acid that can be made into resins.

Styrene butadiene resins (SBR): These resins are based on synthetic rubber from petroleum.

Vinyl: This is a general term for unsaturated emulsions and their copolymers such as polyvinyl acetate and polyvinyl chloride, as well as copolymers polyvinyls.

Epoxies: Synthetic polymers which are cured using epoxides. Epoxies are a class of hydroxyl materials, most commonly used polyhydroxyl material is bisphenol-A.

Polymers are used in the following ways with concrete:

- Polymer-impregnated concrete
- Polymer concrete
- Polymer-modified concrete/mortar
- Polymer as protective coating
- Polymer as bonding agent
- Other applications

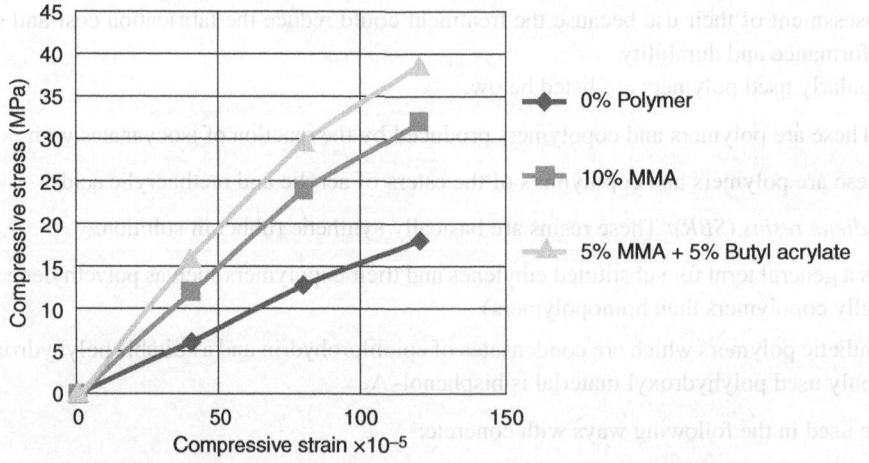

Fig. 16.2 Compressive strain bearing capacity of polymer concretes at different polymer loadings

16.1 Polymer-impregnated Concrete

The ways in which the polymer is introduced into the hardened concrete vary widely and depend upon the commercial objective. These include the following operations:

- The concrete is thoroughly dried, usually by heating.
- The dry concrete is evacuated.
- The concrete is immersed in the chosen monomer (or the monomer is applied to the surface of the concrete).
- Pressure is applied.
- The impregnated concrete is sealed to avoid loss of monomer.
- The monomer is converted into polymer either by gamma radiation or by thermal-catalytic method.
- The concrete is cooled.

The full sequence of operations can only be followed in a precast concrete factory. In the site work it is normally possible to dry the concrete only partially, apply the monomer to the surface, and use heat to control the polymer conversion.

A wide range of monomers are being used in concrete. These include acrylonitrile, ethyl acrylate, MMA, polyester styrene, styrene, and vinyl chloride. A mixture of 70% MMA and 30% of trimethylolpropane trimethacrylate has also been used for high-temperature applications in de-salination plants.

Improvements in compressive strength and tensile strength result from the introduction of the polymer. However, the quantity of polymer that can be introduced depends upon the porosity of concrete, and hence the potential improvement of a particular concrete is substantial if the original concrete is weak (i.e., has a high porosity) but is relatively small if the basic concrete is of high strength and low porosity. In fact, by careful mix design, it is not difficult to make workable plain concretes from ordinary Portland cement and strong natural aggregates with a 28-day cube crushing strength of $100 \, \text{N/mm}^2$, whereas most polymer-impregnated concretes, irrespective of the strength of the basic unmodified concrete, have cube strengths in the range of $120–150 \, \text{N/mm}^2$. These high strengths are stable and do not increase further with age. But the strengths of plain concretes continue to rise and at one time the advantage shown by the polymer-impregnated concretes would have largely disappeared. On the other hand, for very high early-age strengths (at 7 days, say), the polymer impregnation technique could be used. However, even if the polymer–cement ratio is only 1:20, the material costs are about doubled and, with the addition of extra handling, curing, and polymerization costs, the resulting product is necessarily expensive.

16.1.1 Properties

Polymer impregnated concretes normally have cube crushing strength in excess of $100 \, \text{N/mm}^2$ irrespective of the strength of the original untreated concrete. The weaker concretes absorb a higher proportion of the monomer and hence have higher material costs. The flexural strength is usually about $15 \, \text{N/mm}^2$, which is slightly higher than that for the highest strength plain concrete that can be made from normal constituents. The elastic modulus lies in the range of $30–60 \, \text{kN/mm}^2$, which is similar to those for high-strength plain concrete (about $45 \, \text{kN/mm}^2$).

As the strength and elastic modulus of high-strength plain and polymer-impregnated concretes are not very different, the failure and cracking strains are unlikely to differ significantly. High-strength concrete of both types tends to be brittle, and cracks, once initiated, propagate rapidly and frequently run through the aggregate. This can mean that the total energy expended in fracturing high-strength material is less than that demanded by more conventional (medium-strength) concretes, in which the aggregate delays the propagation of cracks and failure is relatively ductile and not explosive or catastrophic.

During the manufacturing cycle, polymer-impregnated concretes are often heated to 150°C. Because of the reduced porosity and permeability, these concretes have low shrinkage and creep characteristics. If the temperature is above the ambient temperature, it is possible that the creep will be larger.

The polymer-impregnated concretes tested have improved resistance to sulphate, chloride, and acid attacks compared with the plain concretes from which they are made. While significant improvements have been observed in the resistance to cycles of freezing and thawing, it is important to note that similar improvements are shown by good (high-strength) air-entrained concretes devoid of any polymer.

16.1.2 Applications

It is likely that the greatest commercial potential for polymer-impregnated concrete will depend upon the enhanced resistance to damage from aggressive environments. It is impossible to make a general recommendation for the use of polymers in any particular application. Each must be judged separately and alternative solutions compared. With regard to the use of polymer impregnation to improve the durability of concrete in sub-zero temperatures, it is important to emphasize the adequate performance of correctly designed plain concrete and to dispel the view that polymer impregnation is essential in these conditions. The fact that dense concretes, such as prestressed kerbs, are also resistant to de-icing salts is important to note in this context.

Polymer impregnation can be used to repair damaged concrete, but it is impossible to make a general recommendation regarding the viability of the technique. Every potential application is different and must be considered separately. Sufficient information is already available for a rational decision to be reached for most applications.

There is some evidence that polymer impregnation can improve the resistance of the concrete surface to abrasive wear. So in factories where heavy equipment is likely to damage the floor, or in dense traffic areas in cities where frequent repair work would seriously impede the flow of traffic, these polymer-based special techniques can be advantageously employed.

16.2 Polymer-modified Cement Concrete

Concretes with polymers added during mixing to modify the properties of the hardened concrete are classified as polymer-modified cement (PMC) concrete. Polymers are added to concrete mixes either as an aqueous emulsion or in a dispersed form in an attempt to improve one or more of the following properties:

- tensile strength and extensibility of concrete
- impact resistance
- abrasion resistance
- durability and resistance to aggressive fluids
- bond between old and new concrete

16.2.1 Properties

One of the earliest polymers used was polyvinyl acetate (PVA), but the range of polymers that are currently being used is now extremely wide and includes PVA copolymers, acrylics, vinyls, natural rubber, and styrene butadiene rubber. The proportions of polymer incorporated also vary considerably and range from under 1% to over 30% of the solid volume of cement.

The addition of these polymers has certain common effects. Concrete mixes become more workable and thus the water content can be reduced. Despite the higher workability, more air is entrained in polymer cement concretes. PMC concretes usually contain at least 3% more entrained air than plain concretes of similar workability. Reduction in the water content increases the crushing strength, but the extra voids have the reverse effect and, in consequence, the polymer admixtures generally have only a small effect upon the crushing strength. However, there is normally a significant increase in the flexural strength of the concrete, which may be attributed to an improved bond between the aggregate and the matrix.

As the elastic moduli of polymers are generally lower than those of cement pastes and concretes, the moduli of polymer cement concretes are lower than those of the equivalent plain cement concrete. The additional entrained air reduces the moduli further.

The durability of concretes with polymer admixtures depends primarily upon the properties of the polymer used and the mix proportion. Hydrated Portland cement is alkaline (having pH value greater than 12.5) and some of the polymers hydrolyse in moist cement environments. PVA is particularly sensitive to such an environment and its use is recommended only in dry conditions. PVA copolymers have been used to resist hydrolysis in alkaline solutions, but these are more expensive.

At least 5% polymer by weight of cement is required to obtain substantial changes in the properties of the hardened concrete. The addition of polymer usually increases the setting time of the cement significantly.

Polymer concretes have a greater resistance to abrasion than plain concretes, wear rates being reduced by as much as 75%. For this reason, and because of improved durability, polymer cement concretes have been used for factory floors, where the abrasion resistance of conventional concretes may be inadequate. Polymer concrete floors provide a neat and clean floor usually needed by electronics industry.

16.2.2 Applications

Polymer cement concretes are several times more expensive than plain concretes. Therefore, they are used only for special applications. The principal advantages shown by these concretes are (a) greater failure strain, (b) good bond with old concrete, (c) improved resistance to abrasion, and (d) improved durability and resistance to chemical attack. Some typical applications in which these properties have been worthwhile are the following:

- For factory floors, particularly where chemicals or oils are liable to be spilt
- For repair of old or damaged concrete
- For surfacing steel bridges or ship decks
- For flooring in frozen-food factories
- For loading ramps, where the abrasive wear of concrete is high
- For cementing ceramic tiles to concrete (the extra bond and flexibility are advantageous here)
- For concretes subjected to large doses of de-icing salts

16.3 Polymer Cement Concrete

Just as cement is used as a binder in cement concrete, monomer or resin is added to bind preheated aggregates consisting of coarse, fine, ultrafine, and other particle sizes. The commonly used binders are styrene, MMA, polyesters, and epoxies.

In the prepack method graded dry aggregates are packed in moulds and polymer is poured into the voids and, if necessary, impregnated by vacuum process. In the premix method polymer and aggregates are mixed in conventional mixers and the mix is transferred to moulds. The mix is vibrated for compaction.

16.3.1 Properties

They are highly resistant to chemical attack and freeze–thaw cycles. Permeability and absorption are almost zero. The typical properties of polymer concrete and plain concrete are shown in Fig. 16.3 for comparison. Figure 16.4 compares some of the key properties achieved with cement concrete and polymer concrete.

16.3.2 Applications

Even though the initial cost of polymer cement concrete is high, the material cost efficiency is estimated to be 400% compared to ordinary cement concrete. Hence polymer cement concrete is used to manufacture pipes for carrying chemicals in industries.

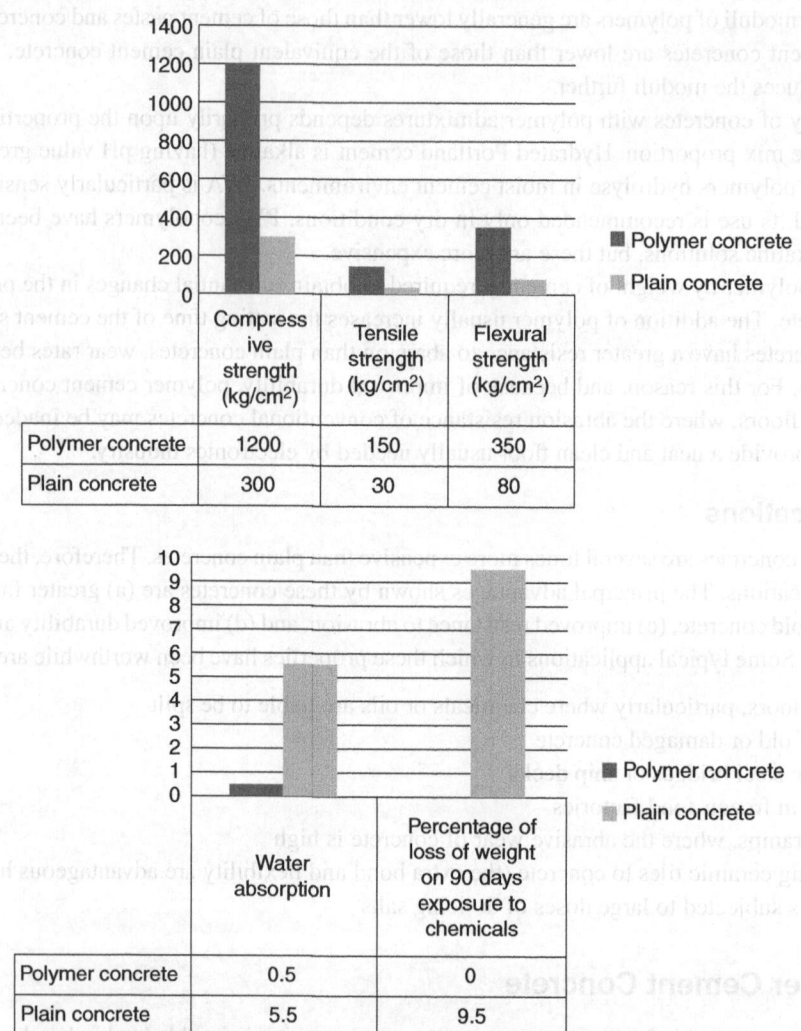

	Compressive strength (kg/cm²)	Tensile strength (kg/cm²)	Flexural strength (kg/cm²)
Polymer concrete	1200	150	350
Plain concrete	300	30	80

	Water absorption	Percentage of loss of weight on 90 days exposure to chemicals
Polymer concrete	0.5	0
Plain concrete	5.5	9.5

Fig. 16.3 Comparison of properties of polymer concrete and plain concrete

16.4 Polymer Composites

These concretes are produced using polymers with cement, sand, and aggregates. The addition of polymers to concrete has been shown to improve its (a) compressive strength, (b) resistance to wear and tear, (c) fatigue resistance, (d) impact resistance, (e) impermeability, (f) durability, and (g) chemical resistance. Because of these properties, they have found application in the following areas:

1. Precast products such as kerb stone
2. Bridge decks
3. Chequered plates for industrial structures
4. Manhole covers
5. Sewers
6. Tunnel linings
7. Pipes carrying chemicals

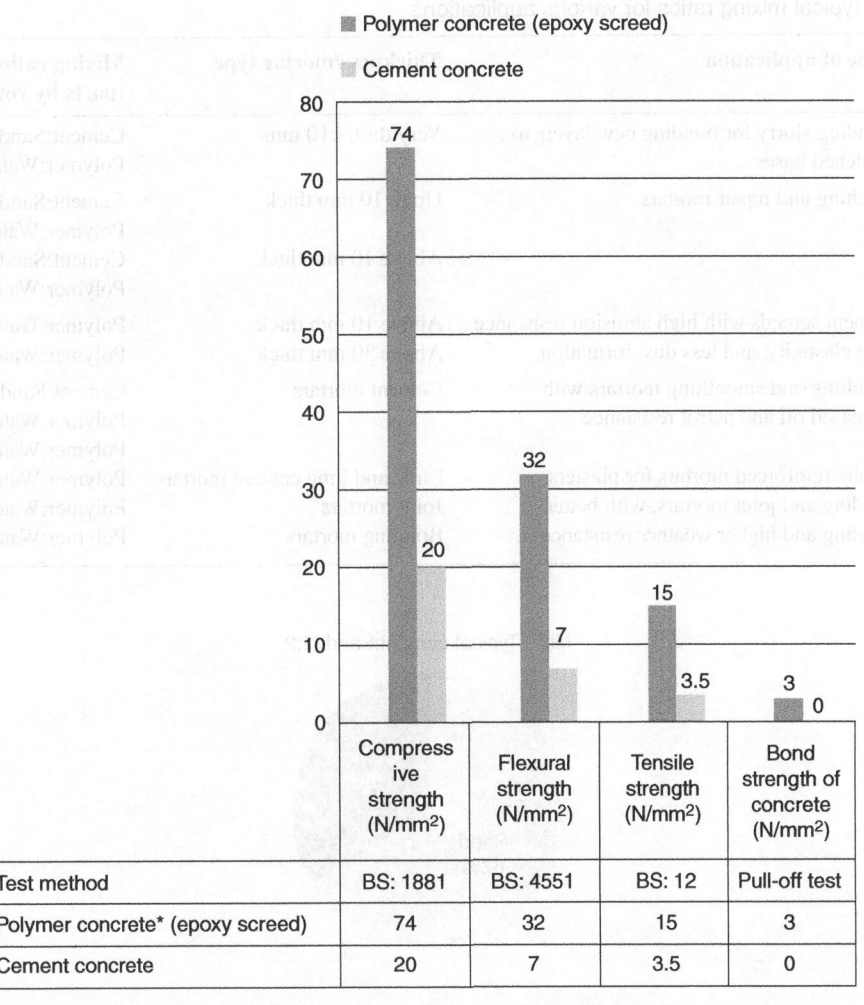

	Compressive strength (N/mm²)	Flexural strength (N/mm²)	Tensile strength (N/mm²)	Bond strength of concrete (N/mm²)
Test method	BS: 1881	BS: 4551	BS: 12	Pull-off test
Polymer concrete* (epoxy screed)	74	32	15	3
Cement concrete	20	7	3.5	0

Fig. 16.4 Comparison of key strength properties

16.5 Proportioning of Polymer Concrete

While using polymers, it should be noted that polymer dispersion is water based and the mixing ratio varies from manufacturer to manufacturer, depending upon solid contents. Ultimately, it is the proportion of solid content of polymer and cement content that reflects on the quality and cost of polymer concrete. Typical mixing ratios for various applications are shown in Table 16.1.

Economy is one of the most decisive guiding factors in determining the amount of polymer, in a particular polymer concrete. More the content of polymer, the more is the enhancement in the properties of polymer concrete and the durability of the repaired structure.

Depending on the type of the polymer and its contents, the degree of elasticity also varies.

16.6 Tests on Polymer Concrete

The best way to ensure the advantages of polymers is to conduct a series of tests on the polymer-modified mortars and to compare the results with mortars without polymers. In most of the practical cases, it suffices

Table 16.1 Typical mixing ratios for various applications

S. No.	Type of application	Thickness/mortar type	Mixing ratios (parts by volume)
1.	Bonding slurry for bonding new layers to hardened bases	Very thin, <10 mm	Cement:Sand = 1:1 Polymer:Water = 1:1
2.	Patching and repair mortars	Up to 10 mm thick	Cement:Sand = 1:2 Polymer:Water = 1:2
		Above 10 mm thick	Cement:Sand = 1:3 Polymer:Water =1.3
3.	Cement screeds with high abrasion resistance, high elasticity, and less dust formation	Above 10 mm thick Above 30 mm thick	Polymer:Water = 1:4 Polymer:Water = 1:6
4.	Levelling and smoothing mortars with increased oil and petrol resistance	Cement mortars	Cement:Sand = 1:2 Polymer:Water = 1:2 Polymer:Water = 1:5
5.	Plastic reinforced mortars for plasters, bonding and joint mortars, with better bonding and higher weather resistance	Lime and lime cement mortars Joint mortars Bonding mortars	Polymer:Water = 1:10 Polymer:Water = 1:2 Polymer:Water = 1:2

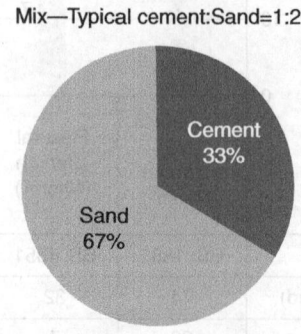

Mix—Typical cement:Sand=1:2

only to conduct the tests on polymer-modified concrete if the base concrete values are known. It is normally sufficient that the repair mortar has 10% higher strength than the base concrete. The following tests can be conducted to determine the suitability of the formulation:

- Compressive strength test
- Flexural strength test
- Bond strength test
- Air entrainment test
- Alkali resistance test
- Chloride content test

In the case of specialized repairs, the following tests may also be conducted:

- Water permeability test
- Vapour permeability test
- Carbonation resistance test
- Wear resistance test
- Impact resistance test
- Chloride ion penetration test

- Shrinkage characteristics test
- Bond and shrinkage tests in typical repair case
- UV resistance test
- Modulus of elasticity test
- Dynamic modulus of elasticity test
- Coefficient of thermal expansion test

Only those tests which have a direct bearing on the given repair situation should be conducted.

Figures 16.5 and 16.6 show some typical test results for polymer-added mortars. The polymer used in these cases is an acrylic dispersion with solid content of 33%. The mix used is as follows:

Gauging solution—1 part of polymer dispersions:2 PBW of water

Mixing ratio—100 PBW of mortar:12.5 PBW of gauging solution

Mortar ratio—1 PBW cement:3 PBW of well-graded sand (PBW = parts by weight)

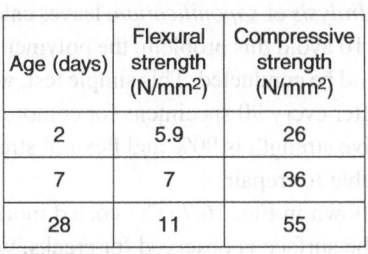

Age (days)	Flexural strength (N/mm^2)	Compressive strength (N/mm^2)
2	5.9	26
7	7	36
28	11	55

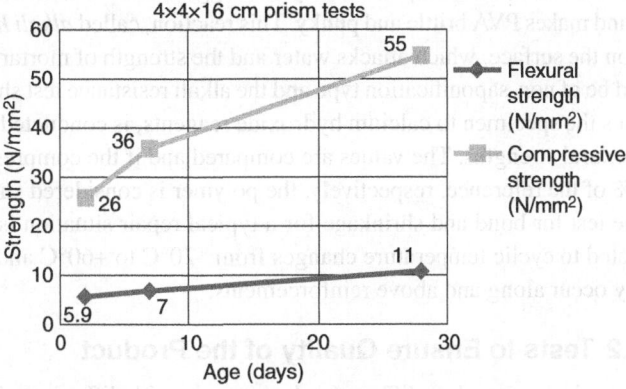

Fig. 16.5 Flexural and compressive strength of the polymer cement concrete

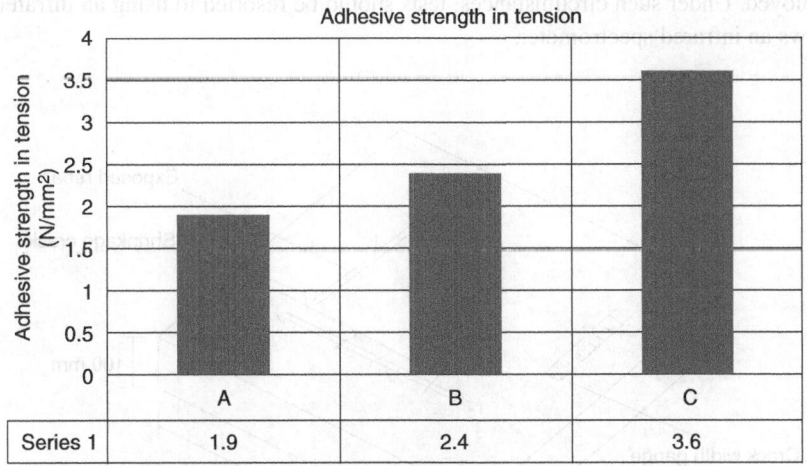

A: after 7 days, B: after 28 days, C: after 90 days

Fig. 16.6 Typical bonding strengths under standard condition

Notes: Stored at constant conditions = 2.1 N/mm^2

Temperature cycling = 2.2 N/mm^2

Temperature cycling + humidity cycling = 1.9 N/mm^2

Fractures occurred mainly in concrete

16.6.1 Precautions While Testing

Since the setting behaviour and mixing proportions are peculiar to polymer mortars, it is advisable to observe a few precautions while testing polymer mortars. It is preferable to conduct an air-entrainment test because some formulations tend to entrap air during the mixing process. The air entrapped is at times 20%. Therefore, if there is marked loss in the strength of polymer mortars as compared to cement sand mixture, it can be due to the air entrainment. The air entrained by polymer addition should not be 1–1.5% more than that in the control mix.

Secondly, since the polymer-modified mortars are used as thin overlays, it is preferable to test the compressive and flexural strengths on thinner sections. Normally, as per Deutsches Institute für Normung (DIN) [German Institute for Standardization] specification, the tests are conducted on prisms of 40 × 40 × 160 mm. The prisms are tested for flexural strengths and the broken halves of the prisms are tested for compressive strengths, with special attachments in which the load is transferred to 40 × 40 mm area. If prisms are not available, 50 × 50 × 50 mm cubes can be utilized for compression strength tests.

One of the most important properties of the polymers to be used in concrete should be the resistance to saponification. If, for example, the polymer suggested is PVA, the alkali reacts with the ester molecular group of PVA and makes PVA brittle and punky. This reaction, called *alkali hydrolysis* or *saponification*, leaves calcium soap on the surface, which attacks water and the strength of mortars. To avoid this problem, the polymer used should be of non-saponification type and the alkali resistance test should be conducted. This simple test, which exposes the specimen to calcium hydroxide reagents, is conducted after every 90 specimens for compressive and flexural strengths. The values are compared and if the compressive strength is 90% and flexural strength is 75% of the reference, respectively, the polymer is considered suitable for repair.

The test for bond and shrinkage for a typical repair situation is shown in Fig. 16.7. The coated mortar is subjected to cyclic temperature changes from –20°C to +60°C and the surface is observed for cracks, which mostly occur along and above reinforcements.

16.6.2 Tests to Ensure Quality of the Product

There are instances when different polymer types with different solid content are used. Since polymer dispersion is based on water, it can have different concentrations. One should consider the solid content of the type of polymer employed. Under such circumstances, tests should be resorted to using an infrared spectrometer. Figure 16.8 shows an infrared spectrometer.

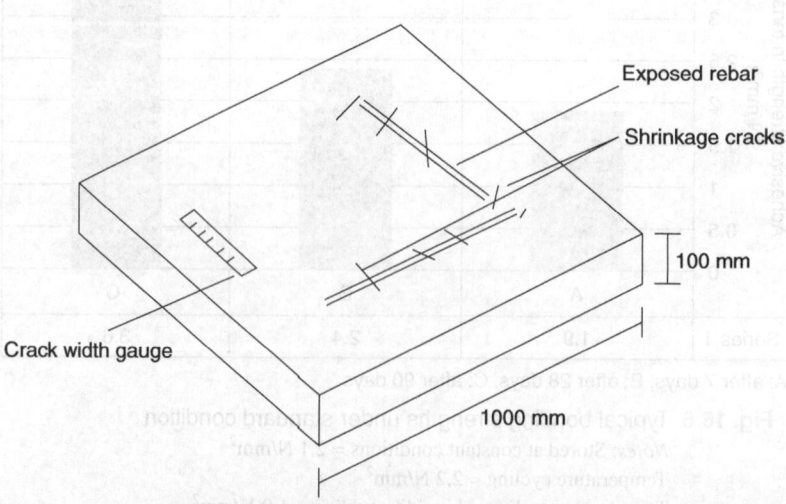

Fig. 16.7 Typical specimen for bond or shrinkage used in repairs

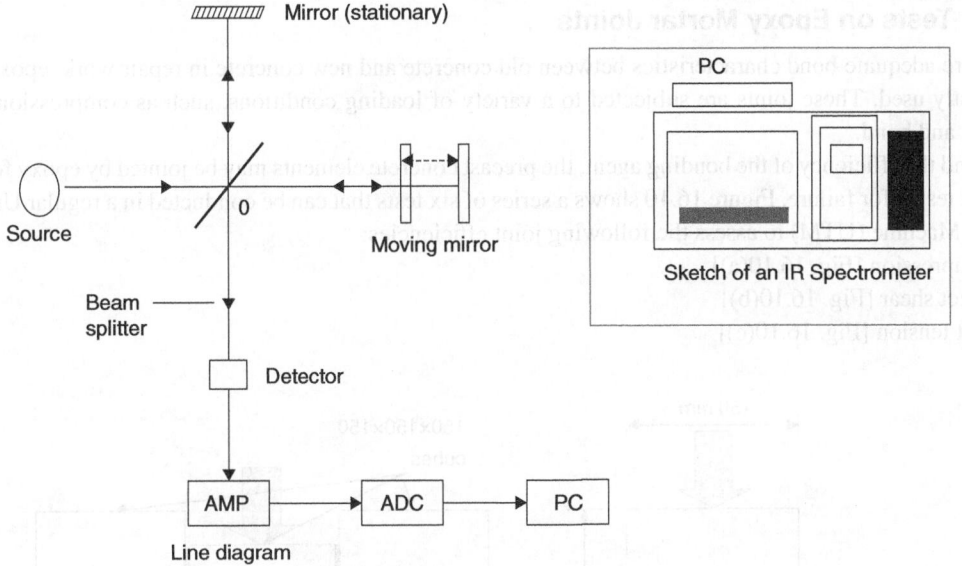

Fig. 16.8 Infrared spectrometer

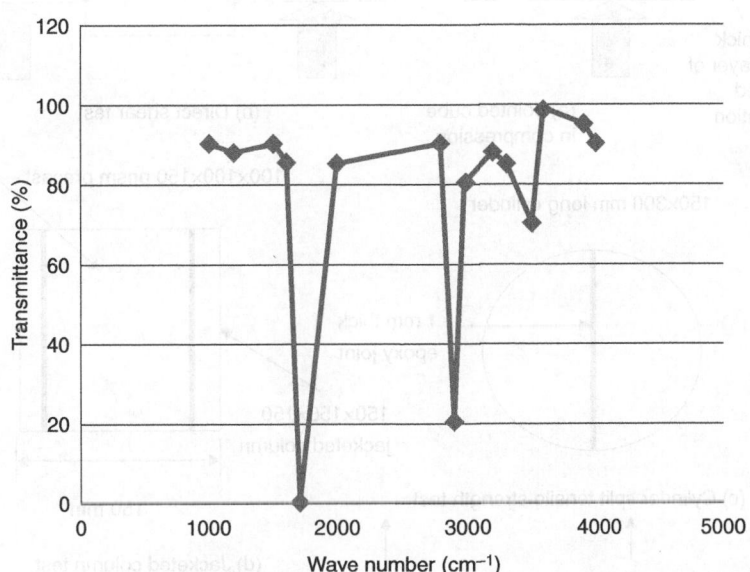

Fig. 16.9 A typical infrared spectrum

The compositions of different samples can be qualitatively compared by *infrared spectrometry*. A typical spectrum for acrylic dispersion is shown in Fig. 16.9. The spectrum of the product intended for use should be compared with a chosen reference spectrum to ensure quality of the product used for repair.

Every batch of the product for repair should be tested to determine the solid content by the simple method of oven-drying. These tests ensure the quality of the product and give a comparative evaluation procedure.

16.6.3 Tests on Epoxy Mortar Joints

To ensure adequate bond characteristics between old concrete and new concrete in repair work, epoxy joints are mostly used. These joints are subjected to a variety of loading conditions, such as compression, shear, tension, and bond.

To find the efficiency of the bonding agent, the precast concrete elements may be jointed by epoxy formulation and tested for failure. Figure 16.10 shows a series of six tests that can be conducted in a regular Universal Testing Machine (UTM) to assess the following joint efficiencies:

- Compression [Fig. 16.10(a)]
- Direct shear [Fig. 16.10(b)]
- Split tension [Fig. 16.10(c)]

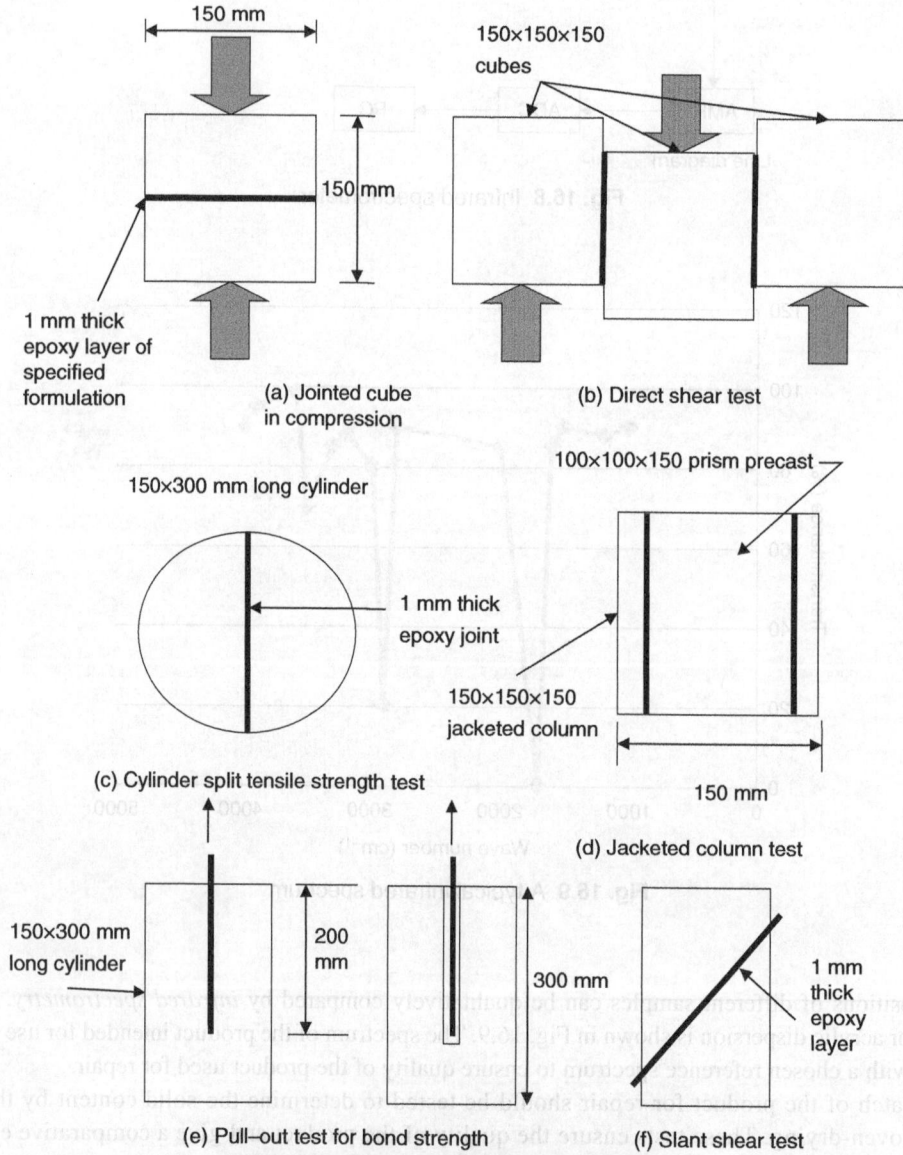

Fig. 16.10 Determination of joint efficiencies

- Jacketing efficiency [Fig. 16.10(d)]
- Pull-out efficiency [Fig. 16.10(e)]
- Slant shear [Fig. 16.10(f)]

Exercises

Review Questions

1. What conditions will justify the costly use of polymers in concrete?
2. What are the different types of polymers used in concrete?
3. What are the key property improvements one can realize by use of polymers in concrete?
4. List the differences between polymer-impregnated concrete, polymer-modified concrete, and polymer concrete.
5. What are the uses of polymer concrete?
6. How is the mix proportioning of polymer concrete decided?
7. What are the various quality control tests done to ensure good performance of polymer concrete?
8. What physical tests can be done to confirm the efficiency of an epoxy joint?
9. What is saponification? How is it detected and assessed?
10. Compare the important properties of normal concrete with those of polymer concrete.

Multiple Choice Questions

1. Polymers are used in concrete to
   a. Improve strength
   b. Improve chemical resistance
   c. Modify flow characteristics in fresh concrete
   d. All of the above

2. Polymer which exhibits poor performance against acids and alkalis is
   a. Urethanes
   b. Epoxies
   c. Acrylics
   d. SBR

3. The compressive strength of polymer-impregnated concrete is about
   a. 20 MPa
   b. 40 MPa
   c 100 MPa
   d. 60 MPa

4. Polymer-modified concrete relates to
   a. Impregnating porous concrete with polymer under pressure
   b. Impregnating porous concrete with polymer under high temperature
   c. Adding polymer during mixing concrete
   d. Pasting new and old concrete

5. Compared to conventional concrete, polymer concrete has
   a. Improved strength
   b. Reduced water absorption
   c. Better resistance to chemicals
   d. All of the above

6. The shear resistance in terms of bond characteristics of old and new concrete is assessed by
   a. Slant shear test
   b. Split tension test
   c. Pull-out test
   d. Compression test

7. To attach old concrete to a layer of new concrete, it is necessary to use
   a. Polymer concrete
   b. Polymer-impregnated concrete
   c. Polymer bonding agent
   d. Polymer-modified concrete

8. Compared to conventional concrete, polymer concrete has
   a. Better fire resistance
   b. Almost same fire resistance
   c. Lower fire resistance
   d. Excellent fire resistance

9. Compared to conventional concrete, the ductility of polymer concrete is
   a. More
   b. Same
   c. Less
   d. Extremely low

10. Addition of polymer makes the setting time of cement
    a. More
    b. Less
    c. Extremely low
    d. Same as without addition of polymer

Answers to Multiple Choice Questions

1. d	4. c	7. c	10. a
2. d	5. d	8. c	
3. c	6. a	9. c	

CHAPTER 17

Steel-fibre-reinforced Concrete

Advanced cement-based materials and improved concrete construction techniques make it possible to design structures which can resist severe loads resulting from earthquakes, impact, fatigue, and blast environments. Conventional concrete cracks easily, but when concrete is reinforced with random dispersed fibres, we get favourable behaviour for repeated loads. Fibres prevent microcracks from widening. Addition of fibres also makes components ductile and tough. This chapter systematically describes the basic properties and the theoretical background for application of steel-fibre-reinforced concrete (SFRC) for structural components subjected to dynamic loads. It also gives experimental evidence which shows the potential of the extensive use of this material for special applications in earthquake-prone and blast-resistant structures.

Research carried out in various parts of the world has established that addition of fibres improves the static flexural strength, fatigue, ductility, and fracture toughness of the material. Recent investigations have also given rise to highly reinforced SFRC containing up to 20% volume of steel fibres. The recent developments are due to the introduction of a new generation of additives such as superplasticizers and microsilicas, which allow the use of high volume of steel fibres and high-strength concrete. Figure 17.1 is a four-phase representation which defines fibre-reinforced mortar and concrete. Table 17.1 gives typical ranges of the application of fibre-reinforced material.

17.1 Basic Properties

The durability of concrete when reinforced with conventional rebars is a major concern in aggressive environments. To address this problem, there have been efforts, in recent years, to develop alternatives to conventional rebars.

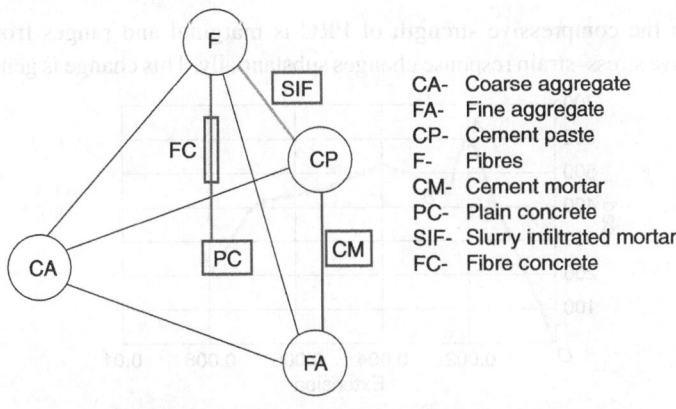

CA- Coarse aggregate
FA- Fine aggregate
CP- Cement paste
F- Fibres
CM- Cement mortar
PC- Plain concrete
SIF- Slurry infiltrated mortar
FC- Fibre concrete

Fig. 17.1 Four-phase representation of FRC

Table 17.1 Ranges of application of fibre-reinforced concrete

S. no.	Volume fraction (V_f,%)	Matrix	Application
1	0.5	Concrete	Pile caps, tripods
2	0.5–3	Concrete	Machine foundation, joints
3	3–8	Mortar	Sheets repair works and joints
4	8–20	Slurry	Thin sheets

Fibre-reinforced plastics and fibre-reinforced concrete (using different types of fibres) have shown better behaviour because of their inherent ability to stop or delay crack propagation. Reinforcing fibres stretch more than concrete under loading. Materials used in fibre reinforcing include acrylic fibres, asbestos, cotton, glass, nylon, polyester, polyethylene, polypropylene rayon, rock wool, and steel. Acid-resistive glass and steel are common. Plastic fibres are used because of their ability to resist corrosion. Natural fibres have little value because they are prone to decay.

The percentage of fibres in concrete mix is based on volume and is expressed as a per cent (fraction) of the mix. 1–2% of fibres are common.

The main properties of FRC in tension, compression, and shear are influenced by the type of fibre, volume fraction of fibres, aspect ratio (the length of the fibre divided by its diameter), and the orientation of the fibre in the matrix.

In this section, we consider the properties of steel fibre concrete with random orientation, which is the most common one for civil engineering applications.

17.1.1 Behaviour in Tension

The most significant effect of incorporating fibres is to delay and control the tensile cracking of the composite material. The fibres provide a ductile member in a brittle matrix and the resulting composite has ductile properties. The fibre and the matrix share the tensile load until the matrix cracks and then almost the full force gets transferred to the fibre. This is a predominant feature of FRC. This mechanism gives rise to favourable dynamic properties such as energy absorption and fracture toughness that distinguish FRC from conventional concrete.

Tensile strength of fibrous composites have been studied by Mangat (1976). The principal effect of fibres in a cementitious material is to cause relief of tensile stress at the crack tip and prevent unstable crack propagation. Kelly (1970) investigated the mechanism of fibre pull-out. The typical load-extension curve presented by him is shown in Fig. 17.2. Debonding of fibre characterizes the straight line portion of the curve *OA*. In the case of short fibres, the debonding occurs at maximum load. The debonding energy per unit area is obtained by dividing the area *OAB* under the stress–strain curve by the surface area of the fibre. The additional energy dissipation of fibre concrete results from the pull-out energy as well.

17.1.2 Behaviour in Compression

Though the increase in the compressive strength of FRC is marginal and ranges from 0% to 20%, the post-cracking compressive stress–strain response changes substantially. This change is generally characterized

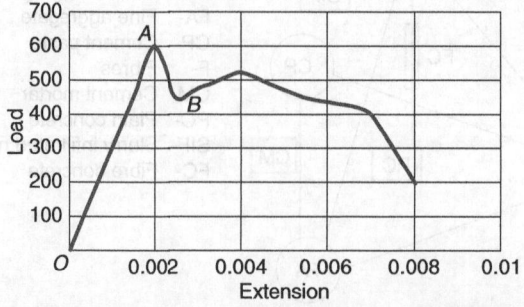

Fig. 17.2 Load–extension curve for fibre pull-out test

by a noticeable increase in strain at peak load and a significant increase in ductility beyond ultimate load, resulting in substantially higher toughness. This increased toughness is advantageous in preventing sudden and catastrophic failures especially under earthquake and blast type of loading. The typical increase in the toughness index varies between 200% and 300% (Ramakrishnan 1987). There is no appreciable change in the linear part of the stress–strain curve in compression for randomly oriented fibres as can be seen in Fig. 17.3.

The improvements in ductility and energy absorption capacity resulting from the increase in fibre volume fraction are comparable to those improvements obtained by confining steel of conventional concrete by transverse steel as shown in Fig. 17.4. Since confinement by transverse steel produces improvements of the same nature

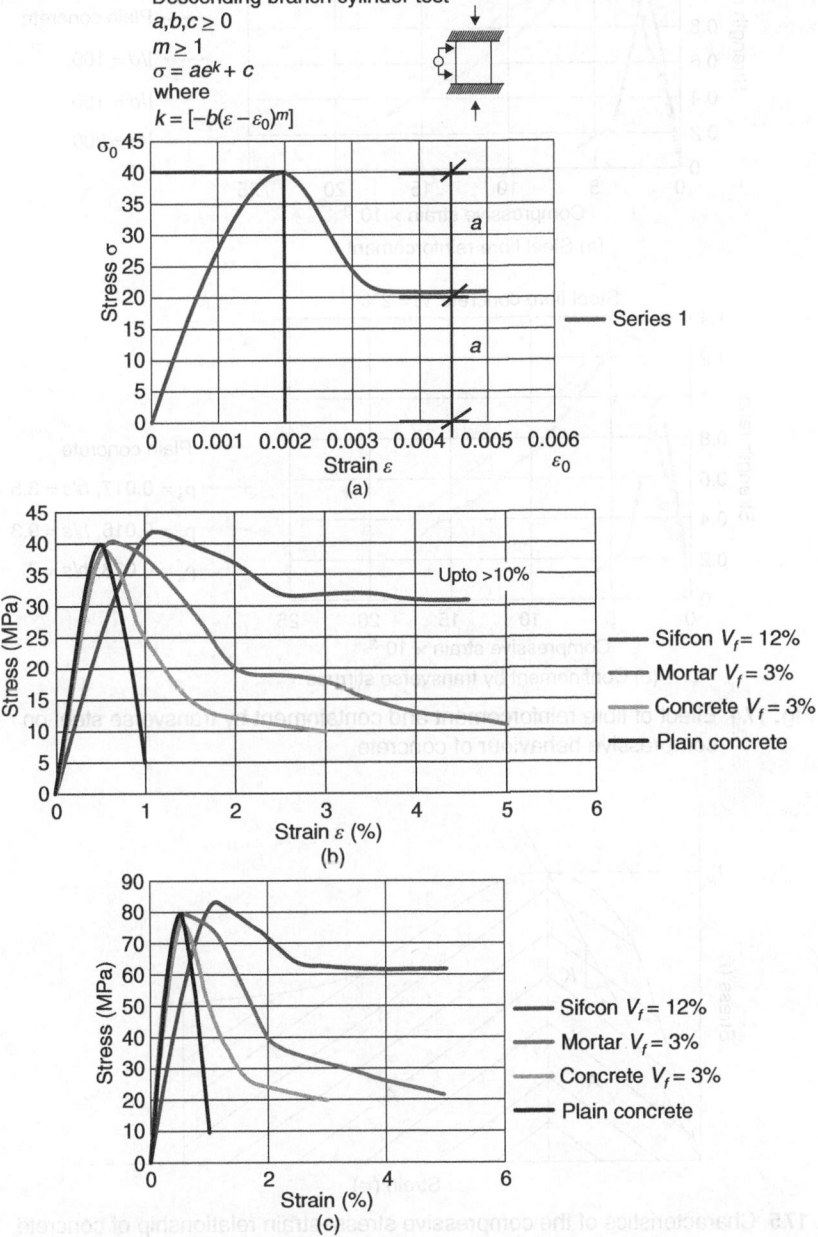

Fig. 17.3 Typical stress–strain curves for fibre-reinforced concrete in compression

as fibre reinforcement in the compressive behaviour of concrete for a certain reinforcement index of each fibre type, there exists a confinement condition which results in comparable compressive stress–strain relationship of FRC. Thus, the characteristic model for FRC consists of two curvilinear ascending and descending branches similar to confined concrete as shown in Fig. 17.5. It is evident that the improvement in material toughness

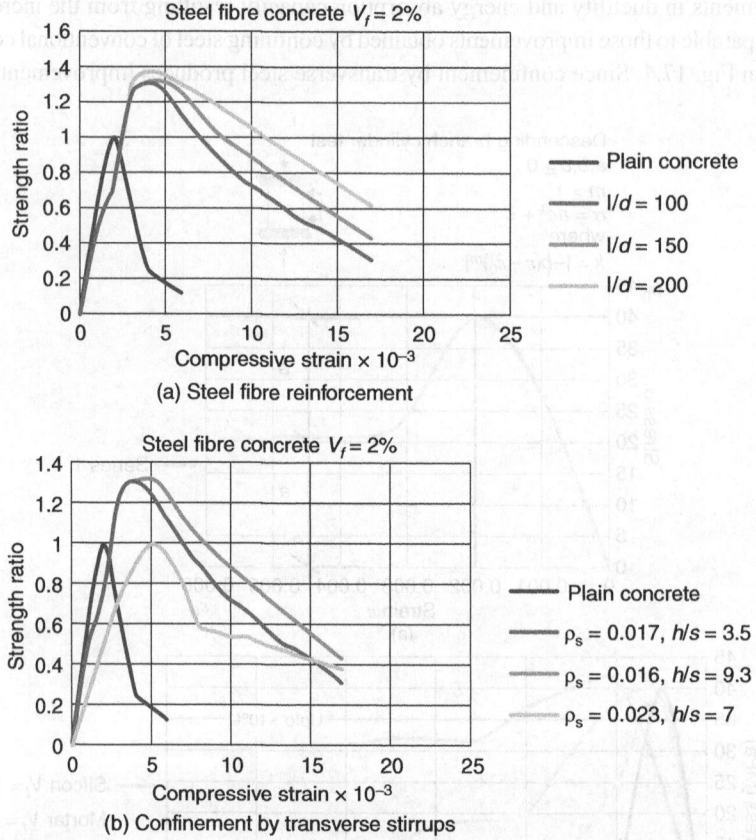

(a) Steel fibre reinforcement

(b) Confinement by transverse stirrups

Fig. 17.4 Effect of fibre reinforcement and containment by transverse steel on compressive behaviour of concrete

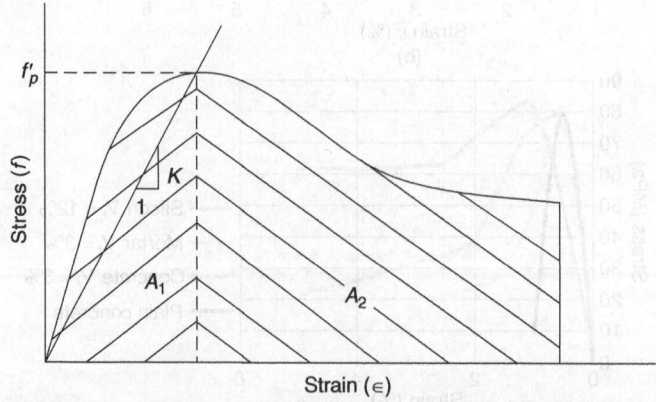

Fig. 17.5 Characteristics of the compressive stress–strain relationship of concrete and fibre-reinforced concrete

can be assessed as the ratio of total area $(A_1 + A_2)$ under the stress–strain curve up to a strain limit beyond the peak stress to the area up to its peak (A_1). Thus, the benefits of confining steel in the conventional earthquake resistance design can also be attained by suitable inclusion of fibres, i.e., by using fibre-reinforced concrete.

17.1.3 Behaviour of Reinforced FRC

When fibres are used in members which are also reinforced with conventional steel bars, it is necessary to examine their behaviour in tension more closely (Liquin and Guofan 1987). The typical tensile stress–strain curve of steel-reinforced FRC is shown in Fig. 17.6. Though more ductile, as seen from Fig. 17.6, the contribution of fibres to strength is not significant (Fig. 17.7).

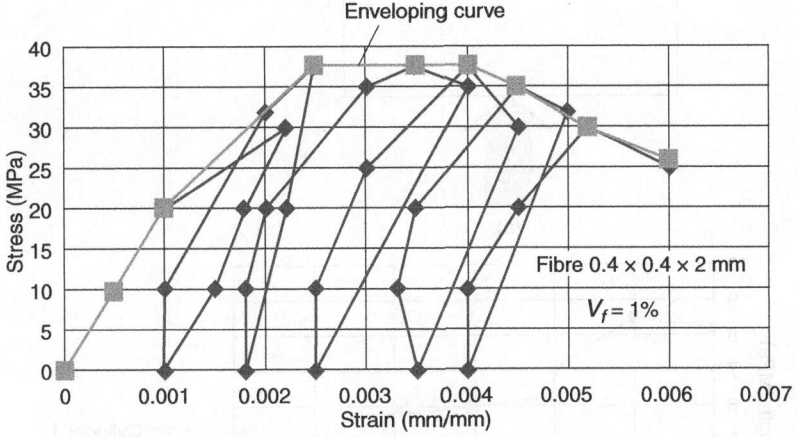

Fig. 17.6 Typical stress–strain curve for steel-reinforced FRC

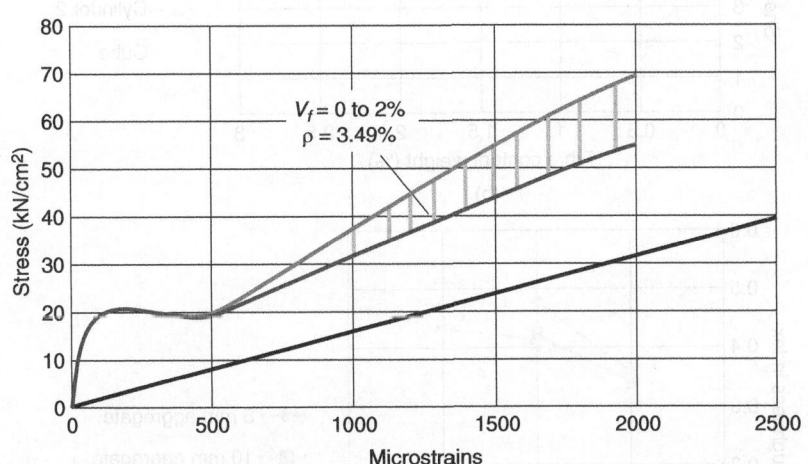

Fig. 17.7 Stress–strain relationship for steel-reinforced FRC

17.1.4 Behaviour of FRC Under Shear

The shear strength and toughness index of compact cube specimens (Barr 1987) are shown in Fig. 17.8. An examination of the graph shown in Fig. 17.8 reveals that the shear strength is not affected by fibre volume. However, the post-cracking toughness increases uniformly with increase in fibre content. This again shows a favourable FRC behaviour in earthquake- and blast-prone areas.

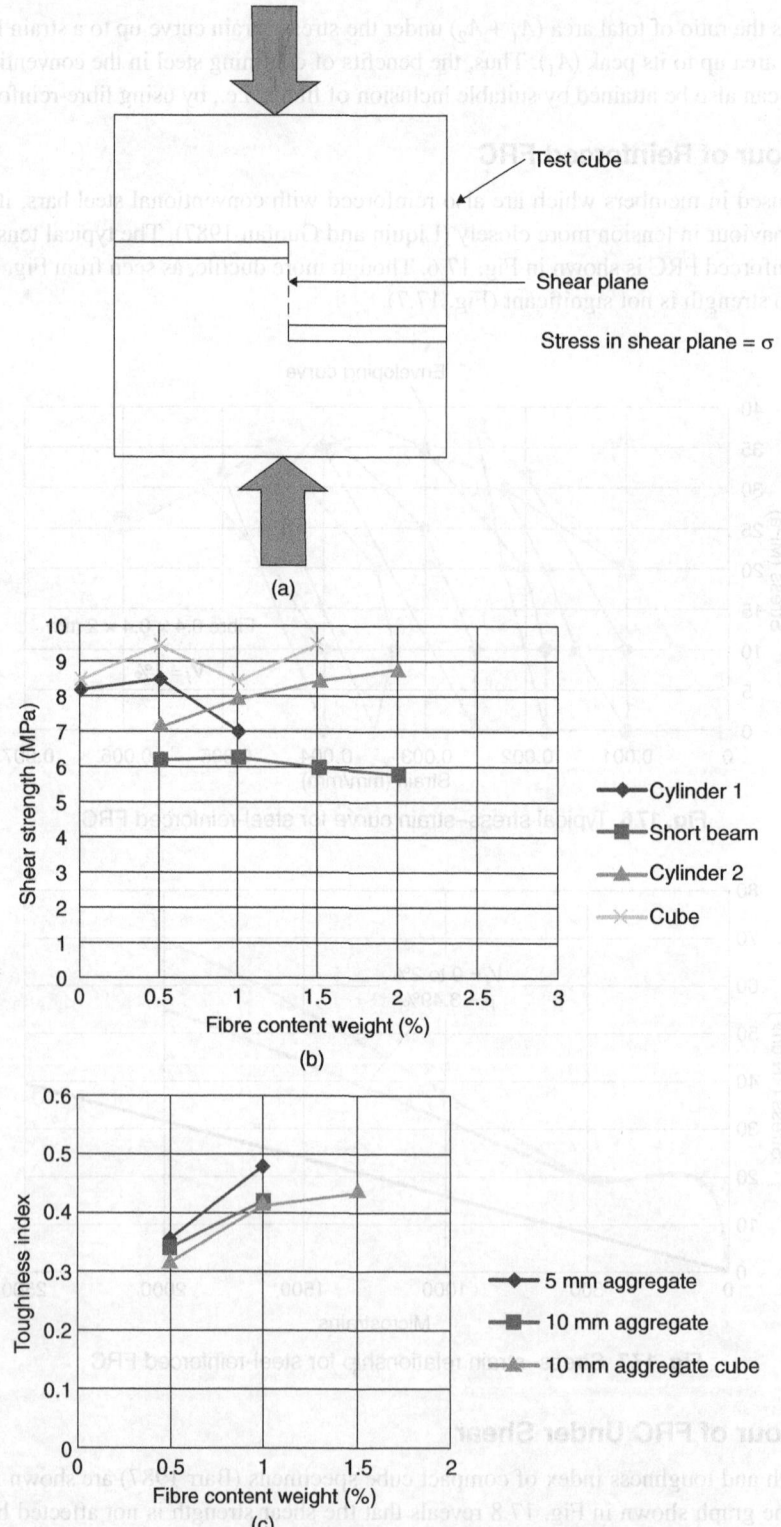

Fig. 17.8 Shear strength and toughness indices of SFRC

17.1.5 Behaviour of FRC Under Flexure and Cyclic Loading

Fibre-reinforced concrete by virtue of improved toughness functions better under cyclic loading. Figure 17.9 shows the load–deflection behaviour of slurry infiltrated fibre concrete (SIFCON) beams. Note the excellent behaviour of beams under cyclic and reversed cyclic loads.

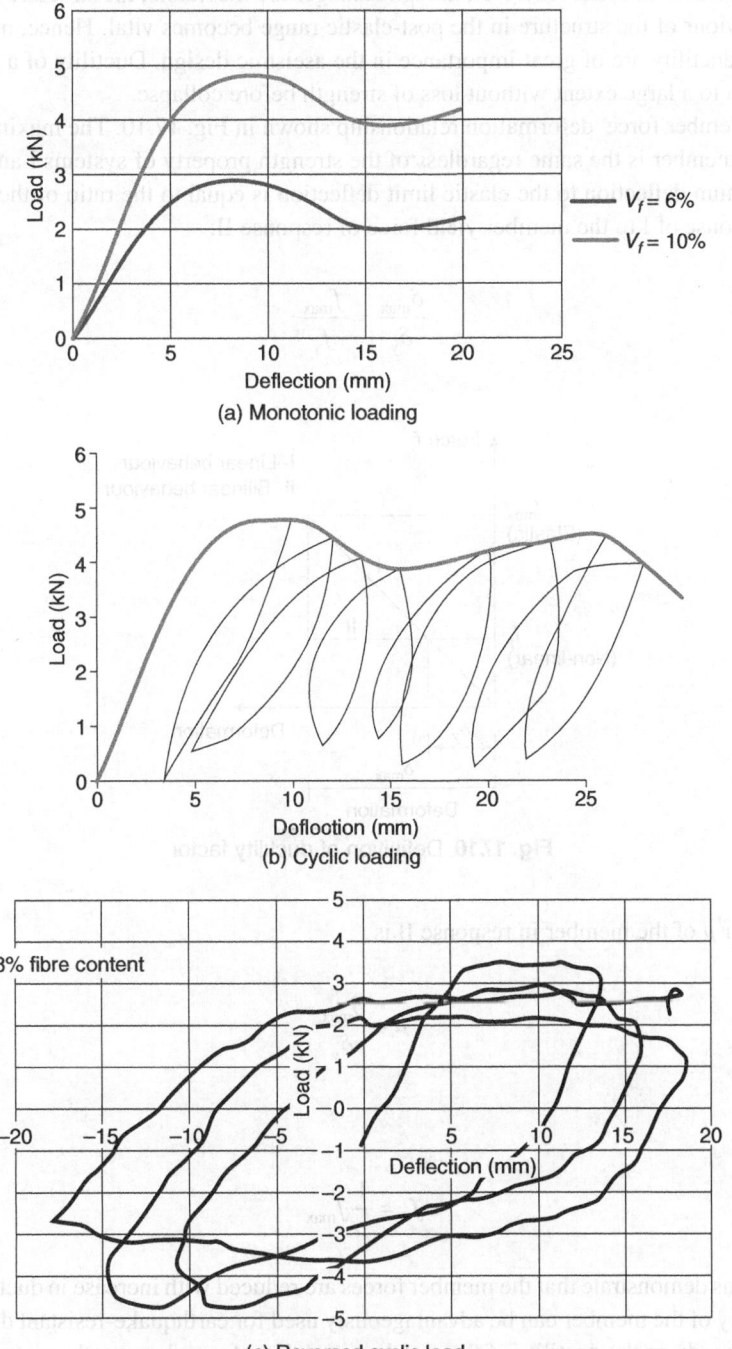

(a) Monotonic loading

(b) Cyclic loading

(c) Reversed cyclic load

Fig. 17.9 Flexural behaviour of SIFCON beams

17.2 Design Requirements

Earthquake-resistant design involves the determination of expected seismic force and designing the structural members to resist this force. Aseismic codes do not intend to ensure that the structure shall suffer no damage during strong earthquakes. This is because a structure that can withstand strong earthquakes without any damage will be too expensive to build. However, though damages are inevitable, the structure should not collapse. For this, the behaviour of the structure in the post-elastic range becomes vital. Hence, non-linear behaviour and toughness or ductility are of great importance in the aseismic design. Ductility of a structure means the capacity to deform to a large extent without loss of strength before collapse.

Consider the member force–deformation relationship shown in Fig. 17.10. The maximum deflection δ_{max} developed by the member is the same regardless of the strength property of systems I and II compared. The ratio of the maximum deflection to the elastic limit deflection is equal to the ratio of the force developed in purely elastic response of I to the member yield force of response II:

$$\frac{\delta_{max}}{\delta_y} = \frac{f_{max}}{f_y}$$

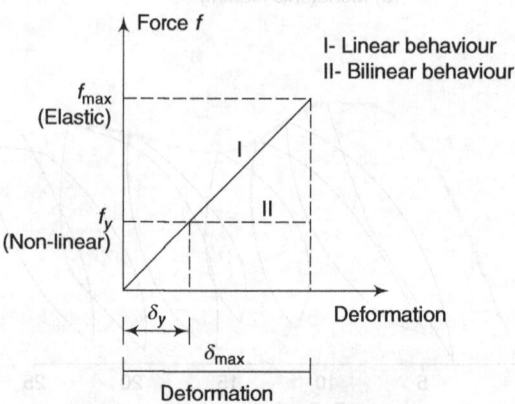

Fig. 17.10 Definition of ductility factor

The ductility factor μ of the member in response II is

$$\mu = \frac{\delta_{max}}{\delta_y}$$

i.e.,

$$f_y = \frac{1}{\mu} f_{max}$$

The above equations demonstrate that the member forces are reduced with increase in ductility factor, indicating that the ductility of the member can be advantageously used for earthquake-resistant design. The ductility of the structure depends on the ductility of the members and that of members on the material, though a direct correlation cannot be established.

In fact, concrete as we all know is not a ductile material, but, if reinforced with dispersed steel fibres properly, can be made to behave in a ductile manner. This can be achieved by careful design of reinforcements. The toughness possessed by FRC can be advantageously used for imparting ductility to concrete members.

17.3 Theoretical Considerations

Beyond the cracking limit, the following two limit states can be identified for both RC and FRC beams.

1. The limit state of the incipient yield of the section indicated by the yielding in steel and concrete or fibrous concrete reaching a maximum strain of 0.002.
2. The limit state of collapse indicated by concrete and fibrous concrete reaching a strain limit of 0.0035 and 0.0095, respectively. While 0.0035 is an accepted limit for concrete, 0.0095 for fibrous concrete is chosen based on tests (Fig. 17.3).

With the above assumptions, the following basic equations have been worked out:
For concrete

1. At yield, the yield curvature

$$\phi_y = \frac{0.002}{kd}$$

where the depth of the neutral axis is

$$kd = \frac{A_s f_y - A_s' f_y}{0.85 f_s' b(2/3)}$$

2. At ultimate, the ultimate curvature

$$\phi_u = \frac{0.0035}{c}$$

where the depth of the neutral axis is

$$c = \frac{A_s f_y - A_s' f_s}{0.85 \times 0.85 f_c' b}$$

For FRC

1. At yield, the depth of the neutral axis is

$$kd = \frac{A_s f_y + 0.5\sigma_t - A_s' f_s'}{0.85 f_c' b(2/3) + 0.5\sigma_t b}$$

where σ_t is the tensile strength of FRC and the yield curvature is

$$\phi_y = \frac{0.002}{kd}$$

2. At ultimate, the depth of the neutral axis is

$$c = \frac{A_s f_y + 0.5(D-c)b\sigma_t - A'_s f'_s}{0.85 \times 0.85 f'_c b}$$

and the ultimate curvature is

$$\phi_u = \frac{0.0095}{c}$$

In ductility predictions it is necessary to determine the deformation that has occurred when the ultimate moment is reached. The actual curvature distribution at the ultimate moment can be idealized into elastic and inelastic regions of a typical beam of span $2l$ as shown in Fig. 17.11.

It may be noted that the curvature ductility is

$$\mu_\phi = \frac{\phi_u}{\phi_y}$$

whereas the member ductility is

$$\mu_\Delta = \frac{\Delta_u}{\Delta_y}$$

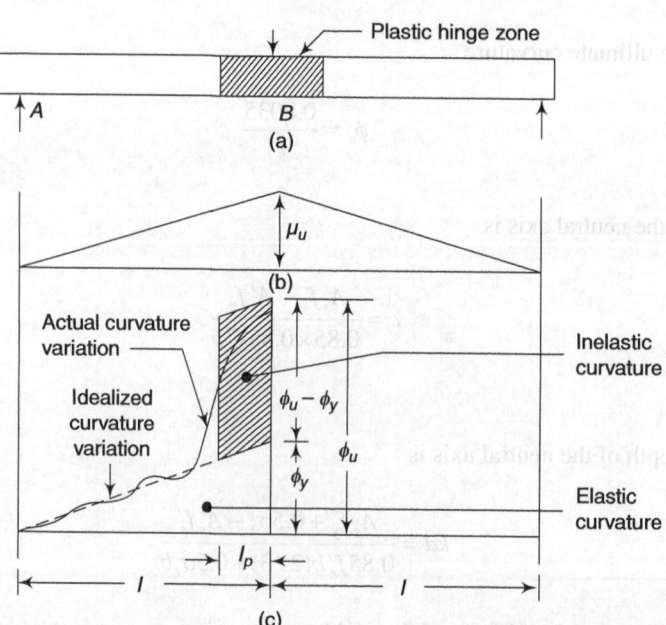

Fig. 17.11 Curvature distribution along the length of the beam: (a) beam, (b) bending moment diagram, and (c) curvature diagram (*Source*: Lakshmipathy 1983)

Wherein the deflection at yield

$$\Delta_y = (\phi_y l / 2) \times (2/3) l$$

and the deflection at ultimate

$$\Delta_u = (\phi_y l / 3) \times (2/3) l + (\phi_u - \phi_y) l_p (l - l_p / 2)$$

l_p is the length of plastic hinge and may be obtained from the well-known expression by Mattock (1965):

$$l_p = 0.5d + 0.05z$$

where d is the effective depth and z is the distance of critical section to the point of contraflexure. l_p can be approximately taken as equal to $2d$.

17.4 Experimental Evidence

Repeated loading tests have been performed on relatively small, conventionally reinforced cantilever beams containing steel fibres. The specimens were subjected to displacements into the inelastic range to attain a maximum ductility of the order of 4. A significantly greater ductile response was observed in FRC beams than in RC beams.

Henager (1977) has reported an experimental comparison between two exterior beam–column joints of a typical multistorey frame. One joint was conventionally reinforced while in the other, fibrous concrete was used in addition to stirrups at farther spacing. Superior performance of fibre concrete joint was evident from the results of his investigation.

Lakshmipathy (1983) has studied the replacement of stirrups with fibre concrete in beam–column junctions with various amounts of replacements as shown in Table 17.2.

Typical ductilities attained in RC and FRC joints are shown in Fig. 17.12. The increase in ductility is approximately threefold for fibrous joints.

Lakshmipathy (1983) has also reported an elaborate experimental investigation of two-quarter full-size seven-storey frame, one made of RC and the other of FRC joints. The frames were subjected to earthquake-type lateral load. Figure 17.13 shows a comparison of cumulative energy absorption capacities for the FRC and RC frames. It is evident that the FRC frame exhibits more than twice the cumulative energy absorption capacity in comparison to the RC frame.

17.5 Applications of Steel Fibre Reinforced Concrete

Generally, SFRC is very ductile and particularly well suited for structures which are required to exhibit:

- High fatigue strength, resistance to impact, blast and shock loads
- Shrinkage control of concrete
- High tensile strength due to flexure or shear
- Erosion and abrasion resistance and resistance to splitting
- Temperature resistance and high thermal gradient
- Earthquake resistance

Table 17.2 Loading tests on RC and FRC joints

Joint cycle	RC J	FR J-1	FR J-2	FRJ-3	FR J-4	Remarks
1	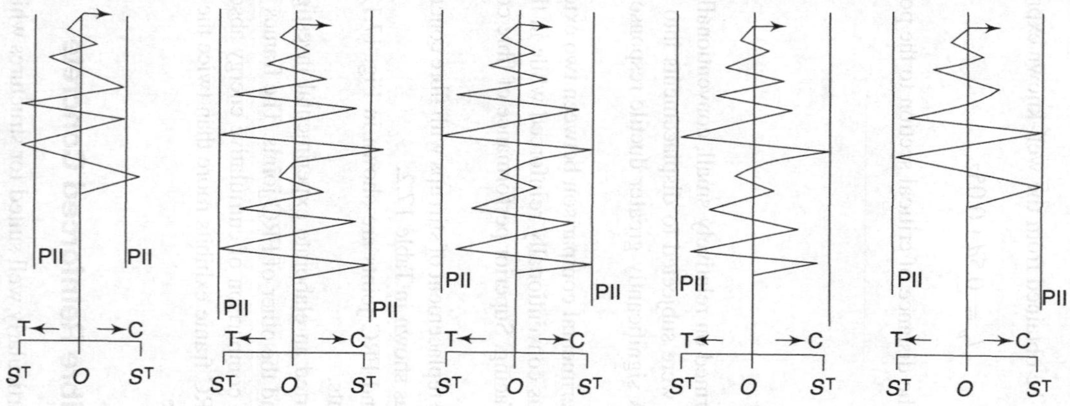					Precracking elastic yield
2						
3						
4						Post yield and inelastic deformation
5						
6						
7						d is the depth of beam

Each cycle column shows a load–displacement hysteresis plot with axes labelled S^T, O, S^T, with C (Compression, up arrow) and T (Tension, down arrow) directions and points marked PII.

Displacement amplitudes: RC J = $d/4$; FR J-1 = $d/3$; FR J-2 = $d/2$; FRJ-3 = $d/3$; FR J-4 = $d/4$.

T—Tension   C—Compression

Spacing of stirrups up to 50 cm from the column face

Source: Lakshmipathy 1983

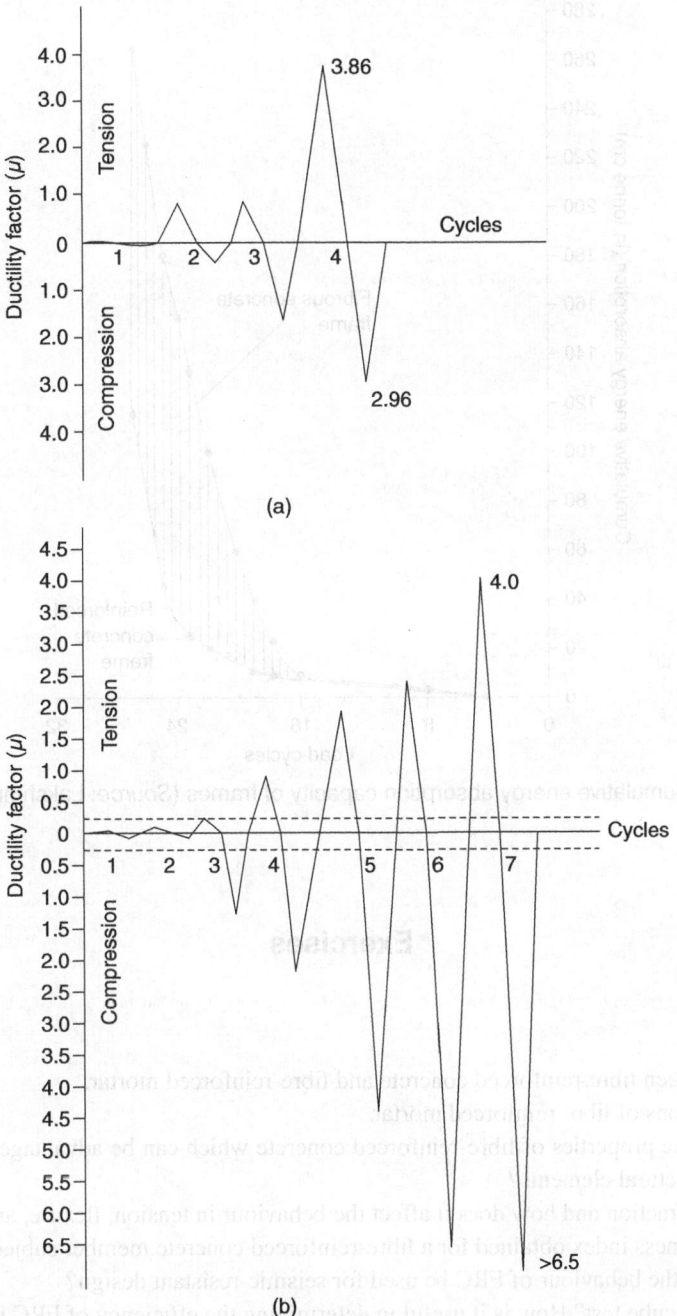

Fig. 17.12 Ductility of (a) RC and (b) FRC joints (*Source*: Lakshmipathy 1983)

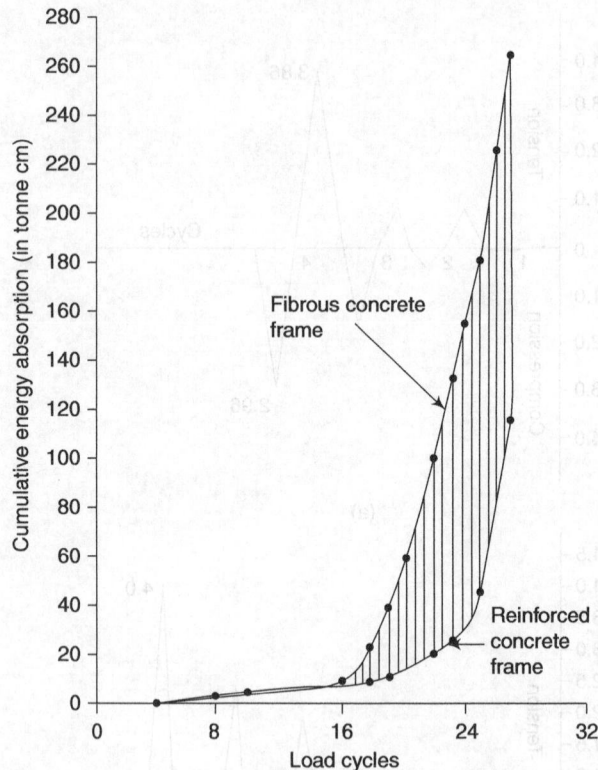

Fig. 17.13 Cumulative energy absorption capacity of frames (*Source*: Lakshmipathy 1983)

Exercises

Review Questions

1. Distinguish between fibre-reinforced concrete and fibre-reinforced mortar.
2. List the applications of fibre-reinforced mortar.
3. What are the basic properties of fibre-reinforced concrete which can be advantageously made use of in the design of structural elements?
4. What is volume fraction and how does it affect the behaviour in tension, flexure, and shear?
5. How is the toughness index obtained for a fibre-reinforced concrete member subjected to compression?
6. In what way can the behaviour of FRC be used for seismic-resistant design?
7. What is compact cube test? How is it useful in determining the efficiency of FRC in shear?
8. Are confinement of concrete and fibre reinforcement similar? If so how to use fibre concrete advantageously in resisting shear?
9. What are the application areas of FRC?
10. Why should FRC be used only with regular reinforcement?

Multiple Choice Questions

1. Steel-fibre-reinforced concrete is normally made with
   - a. 1–2% fibres
   - b. 5% fibres
   - c. 10% fibres
   - d. 20% fibres
2. Steel fibre concrete exhibits
   - a. More compressive strength
   - b. More toughness
   - c. More impermeability
   - d. Less ductility
3. Steel-fibre-reinforced concrete has better
   - a. Crack resistance
   - b. Ductility
   - c. Toughness
   - d. All of the above
4. The aspect ratio (*l*/*d*) of the fibres used is about
   - a. 100
   - b. 500
   - c. 1000
   - d. 10
5. SFRC exhibits ductile performance similar to
   - a. Steel
   - b. Confined concrete
   - c. Reinforced concrete
   - d. Masonry block
6. Shear strength of SFRC is determined by
   - a. Beam test
   - b. Split cylinder test
   - c. Cube test
   - d. Compact cube test
7. It is advantageous to use SFRC for
   - a. Earthquake resistance
   - b. Dynamic load applications
   - c. Fatigue load applications
   - d. All of the above
8. The fibres used in concrete are
   - a. Steel fibres
   - b. Glass fibres
   - c. Polypropylene fibres
   - d. All of the above
9. Fibres are used in concrete for
   - a. Increasing durability
   - b. Increasing ductility
   - c. Increasing strength
   - d. Increasing impermeability
10. In fibre concrete the following failure mechanism is considered to be premature
    - a. Fibre snapping
    - b. Fibre pull-out
    - c. Fibre yielding
    - d. Fibre buckling

Answers to Multiple Choice Questions

1. a	4. a	7. d	10. b
2. b	5. b	8. d	
3. d	6. d	9. b	

CHAPTER 18

Ready Mixed Concrete

Ready mixed concrete (RMC) is a specialized material in which cement, aggregate, and other ingredients are weigh batched at a plant in a central or truck mixer before delivery to the construction site in a condition ready for placing by the customer. RMC is manufactured at a place away from the construction site, the two locations being linked by a transport operation. As per IS: 4926-2003, ready mixed concrete is defined as 'concrete mixed in a stationary mixer in a central batching and mixing plant or in a truck mixer and supplied in a fresh condition to the purchaser either at site or into purchaser's vehicle'.

The 'short life' of fresh concrete, with only 2–3 h before it must be placed, results in ready mixed concrete being a very much local delivery service, with rarely more than 60–120 min journey to the construction site. The need for supply of RMC to fit in with the customer's construction programme means that RMC has to be both a product and a delivery service. This means that the ready mixed supplier is involved in two separate businesses—firstly, processing materials and secondly, transporting product with a very short life.

When we refer to the customer, we are speaking in effect of two customers. As far as the product is concerned, concrete must satisfy not only the person who is using it, i.e., the builder or contractor, but also the authority responsible for defining the properties. However, the ready mix supplier has only one contract and that is with the builder or contractor and relies on the latter to define exactly the requirements of the specifying authority (engineer).

The basic product in ready mixed concrete is fresh concrete, which is placed on site by the customer. It is distinct from hardened, precast concrete units. The introduction of RMC has gradually replaced the operation in which contractors made their own concrete on site. When RMC was first introduced, engineers and contractors with considerable expertise in concrete production and quality control were doubtful of the quality of this new product, whose manufacture was no longer under their control. RMC suppliers need to have stringent quality control for the product and its delivery so that customers' apprehensions regarding the quality of concrete supplied by them are taken care of. It will take a while before customers place their confidence and trust in the product and services offered by the supplier.

Experience shows that the specifying authority or engineer will be satisfied with the ready mixed concrete if

- The supply complies with the specification for fresh and hardened concrete.
- They are assured of continuity of supplies from experienced and reliable RMC companies.

In turn, the contractor or builder will be satisfied if

- The deliveries are always on time and concrete is supplied at the required rate.
- The workability is correct and appropriate for the placing method used.
- The quantities are correct.

- On those occasions when concrete proves to be defective, the supplier bears their fair share of the cost of removal and replacement of the defective material.
- The total cost of concrete, including supply, handling, and placing, is economic. From this, it is seen that the specifying authority (engineer) is concerned primarily with the quality of the product, whereas the user, i.e., builder or contractor, is mainly concerned with the service and its cost, i.e., value for their money.

18.1 Historical Growth of Ready Mixed Concrete

Ready mixed concrete was first patented in Germany in 1903, but means of transporting the RMC was not sufficiently developed by then to enable the concept to be utilized commercially. The first commercial delivery of RMC was made in Baltimore, USA, in 1913 and the first revolving-drum-type transit mixer, of a much smaller capacity than those available today, was used in 1926. In the 1920s and 1930s, RMC was introduced in some European countries.

Some early plants were of very small capacities. In 1931, an RMC plant set up at what is now Heathrow airport, London, had 1.52 m^3 capacity central mixer, supplying six 1.33 m^3 capacity agitators with an output of 30.58 m^3/h. Aggregates were stored in four compartments, each of 76.45 m^3 capacity. Cement was handled manually in bags. Till the beginning of World War II, there were only six firms producing RMC in the UK. After the war, there was a boost to the RMC industry in whole of Europe. In the mid-1990s, there were as many as 1100 RMC plants in the UK, consuming about 45% of cement produced in the country.

European Ready Mixed Concrete Organization (EMRO) was formed in 1967. In 1997, some 5850 companies having an output more than 100 m^3/h were represented by it. Cement consumption in RMC plants ranged from 33% to 62% of the total cement sales.

In the USA, till 1933, only 5% of cement produced was utilized through RMC. ASTM published first specification for ready mixed concrete in 1934. The RMC industry in the USA progressed steadily. During 1950–1975, RMC industry consumption of total OPC in the USA increased from 33% to 66% and by 1990 to 72.4%. There were 5000 RMC companies in that country by 1978.

In Japan, the first RMC plant was set up in 1949. Initially, dump trucks were used to haul concrete of low consistency for road construction. In the early 1950s mixing-type trucks were introduced. Since then there has been a phenomenal growth of the industry in that country. By the end of the 1970s, there were 4462 RMC plants in Japan, and by 1992, Japan was the largest producer of RMC, producing 181.96 million tonnes of concrete. In many countries, including some developing countries such as Taiwan, Malaysia, and Indonesia, as well as certain countries in the Gulf region, RMC industry is well developed today.

Ready mixed concrete plants arrived in India in the early 1950s, but their use was restricted to only major construction projects such as dams. Later RMC was used for other projects such as construction of long-span bridges and industrial complexes. These were, however, captive plants which formed an integral part of the construction projects. It was during the 1970s when the Indian construction industry spread overseas, particularly in the Gulf region, that an awareness of RMC was created among Indian engineers, contractors, and builders. Indian contractors in their works abroad started using RMC plants with a capacity 15–60 m^3/h. Some of these plants were later brought to India in the 1980s. Currently there are many ready mix plants operating in different parts of India, especially in metropolitan cities and towns.

18.2 Advantages of Ready Mixed Concrete

Advantages of RMC are well recognized. Some of these are given below:

- *Uniform and assured quality of concrete*: Since RMC is factory produced, the raw material and production process quality is better than conventional site mixed concrete.

- *Durability of concrete*: RMC ensures that correct water–cement ratio is maintained. Hence the durability of RMC is consistent and better.
- *Faster construction speed*: In site mixed concrete, the contractor needs to mobilize labour for mixing as well as placing. In RMC, fresh concrete is supplied in a placeable condition and can directly be placed by pumping. Hence, a faster construction speed can be achieved.
- *Elimination of storage needs at the construction site*: In the case of site mixed concrete, all raw materials such as aggregates, sand, and cement have to be stored at the site. In urban situations and when the work is progressing close to the highways, there is a problem of storage of raw materials affecting smooth flow of traffic. In the case of RMC, this problem is completely avoided as the storage of materials takes place at the central plant.
- *Easier admixture addition*: In RMC, admixtures can be added in a controlled manner because of the use of sophisticated computer-controlled methods of releasing exact quantities needed. This is not possible in normal concreting.
- *Documentation of mix designs*: The contractor purchases fresh concrete from the supplier of RMC, who is responsible not only for documentation but also for maintaining the records.
- *Reduction in wastage of material*: In RMC, materials are stored in bulk and used in bulk. Hence, wastage that occurs in loose handling of cement is avoided.
- *RMC is eco-friendly*: The production of RMC is done in an environmentally assessed and licensed central plant. Hence, dust and noise pollution which is inevitable in site mixed concrete is avoided.

18.3 Components of an RMC Plant

A plant producing RMC consists of (a) an RMC plant with auxiliary/supporting equipment, (b) transit mixers, and (c) site equipment for handling concrete.

18.3.1 Ready Mixed Concrete Plant

Generally, an RMC plant with a capacity of 55–60 m³/h would meet the requirements of a small town. Initially the plant could work on an 8–10 h shift; with a growing market, 16–18 h operation is feasible. Operations can be expanded by an additional output of 55–60 m³/h at the same site. The typical layout of an RMC plant is shown in Fig. 18.1. Figure 18.2(a) shows the schematic diagram of a central ready mix plant, and Fig. 18.2(b) shows the various components of a typical batching plant.

The auxiliary/supporting equipment for an RMC plant would include the following:

1. Arrangement for unloading 50 kg bags/jumbo bags/bulk cement at site. Jumbo bags are flexible bags made of synthetic fibre. They are of a square or drum shape. They have openings at top and bottom with covers and chutes for discharging material. These bags carry 500–2000 kg of granular powder material. Figure 18.3(a) shows a jumbo bag, and Fig. 18.3(b) shows a filling machine. Figures 18.4(a) to 18.4(e) show different methods of bulk handling of cement in an RMC plant.
2. Fork lift truck
3. Front-end loader
4. Weigh bridge
5. Telecommunication system
6. Captive diesel set
7. Storage cement silos for 2–3 days capacity
8. Admixture tank
9. Site laboratory

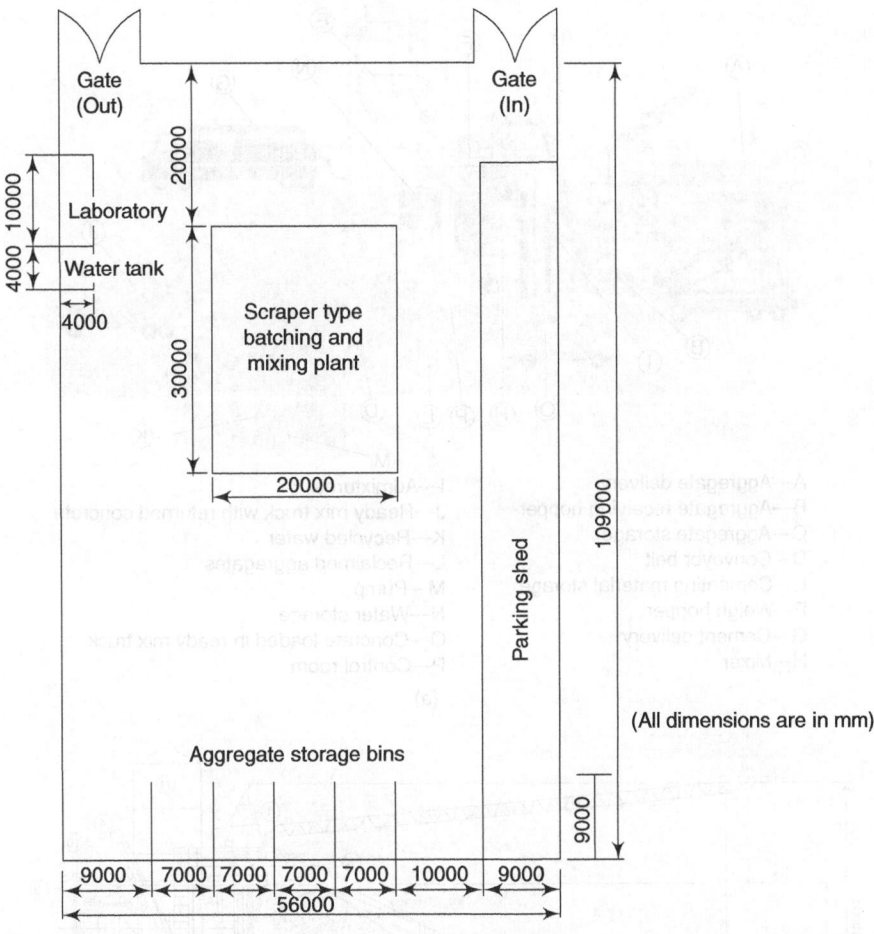

Fig. 18.1 Typical layout of an RMC plant

The list of equipment in a site laboratory is given in Table 18.1.

Table 18.1 List of control laboratory equipment at site

Name of equipment	Specification
Vicat apparatus	IS: 5513
Blaine's apparatus for fineness of cement	IS: 5516
Mortar cube vibrator	IS: 4031
50 cm^2 cube moulds	IS: 4031
150 mm cube mould	IS: 516
$100 \times 100 \times 400$ beam mould	IS: 516
Slump test apparatus	IS: 7320
Needle vibrator	IS: 2505
Compacting factor apparatus	IS: 5515
A set of IS sieves	IS: 460
Curing tank	IS: 9013
Compression testing machine	IS: 516
Platform balances	

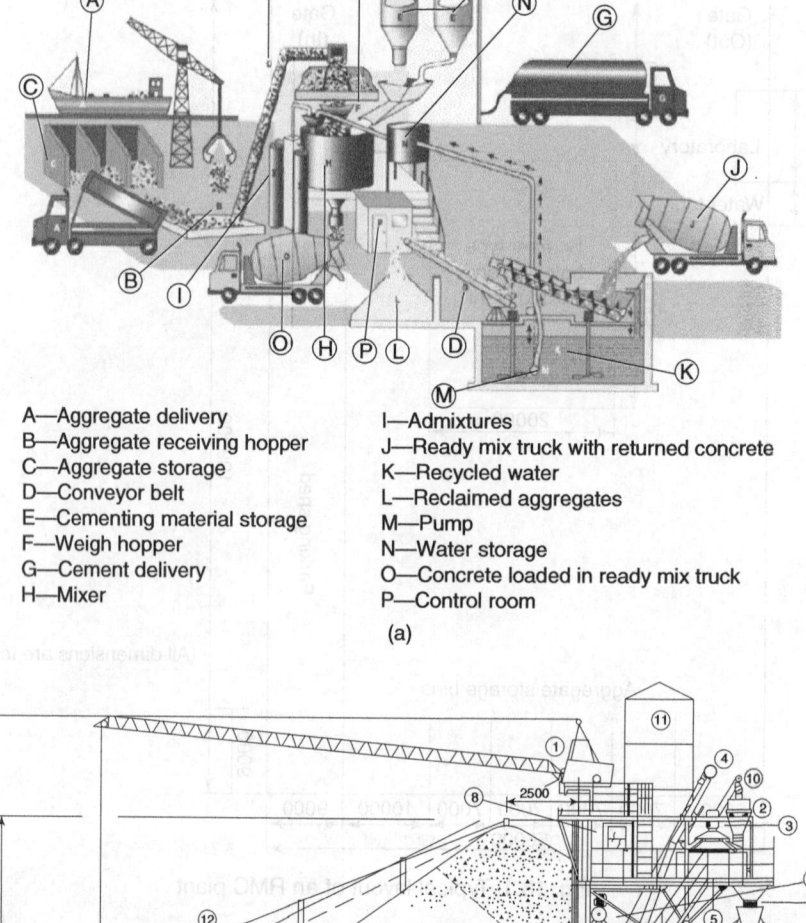

A—Aggregate delivery
B—Aggregate receiving hopper
C—Aggregate storage
D—Conveyor belt
E—Cementing material storage
F—Weigh hopper
G—Cement delivery
H—Mixer

I—Admixtures
J—Ready mix truck with returned concrete
K—Recycled water
L—Reclaimed aggregates
M—Pump
N—Water storage
O—Concrete loaded in ready mix truck
P—Control room

(a)

1. Overhead crane
2. Cement weight hopper
3. Pan mixer
4. Skip hoist winch

5. Service walkway
6. Weighing cage
7. Skip
8. Control cabin

9. Electrocompressor
10. Loading screw feed
11. Cement silo
12. Aggregates partition walls

(b)

Fig. 18.2 (a) Schematic of a central ready mix plant; (b) Components of a batching plant

18.3.2 Transit Mixers

A mixer is generally mounted on a self-propelled chassis, capable of mixing the ingredients of concrete and agitating the mixed concrete during transportation. A typical transit mixer is shown in Fig. 18.5. Generally a 4–6 m³ capacity transit mixer would meet the requirement of most construction sites. Using a 6 m³ capacity

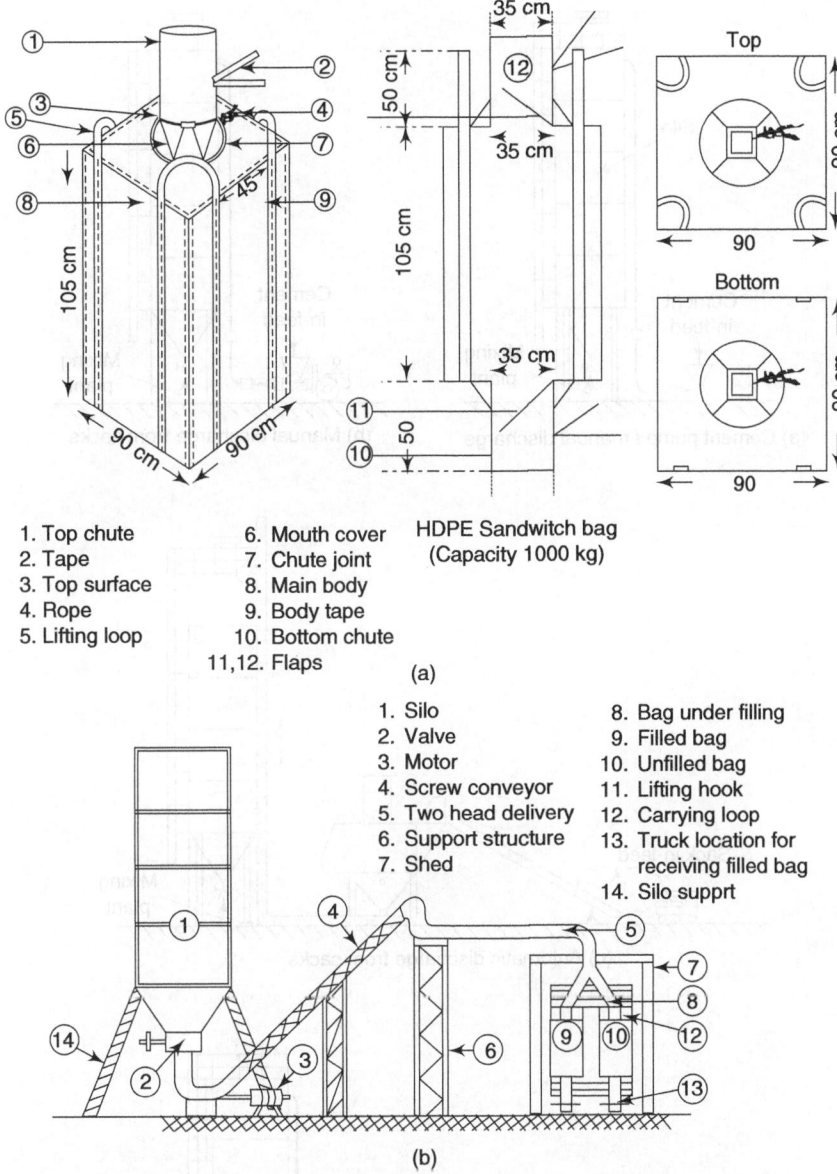

1. Top chute
2. Tape
3. Top surface
4. Rope
5. Lifting loop
6. Mouth cover
7. Chute joint
8. Main body
9. Body tape
10. Bottom chute
11,12. Flaps

HDPE Sandwitch bag
(Capacity 1000 kg)

(a)

1. Silo
2. Valve
3. Motor
4. Screw conveyor
5. Two head delivery
6. Support structure
7. Shed
8. Bag under filling
9. Filled bag
10. Unfilled bag
11. Lifting hook
12. Carrying loop
13. Truck location for receiving filled bag
14. Silo supprt

(b)

Fig. 18.3 (a) A typical jumbo bag and (b) jumbo bag filling plant

transit mixer would substantially reduce the number of trips. However, for manoeuverability in small streets, 4 m^3 would be ideal. Judicious combination of 6 and 4 m^3 transit mixers is necessary, depending on the anticipated demand, road condition, handling of concrete at the site of work, etc. Minimum overall fleet average of six trips per shift could be achieved. In big cities, 6 m^3 capacity transit mixers will be suitable and in small towns 4 m^3 capacity transit mixers can also be used.

18.3.3 Equipment for Site Handling of Concrete

Normally the responsibility of the RMC supplier ceases once concrete is unloaded at site from the transit mixer. The handling and placing of concrete is done by construction contractors. However, it may be desirable for

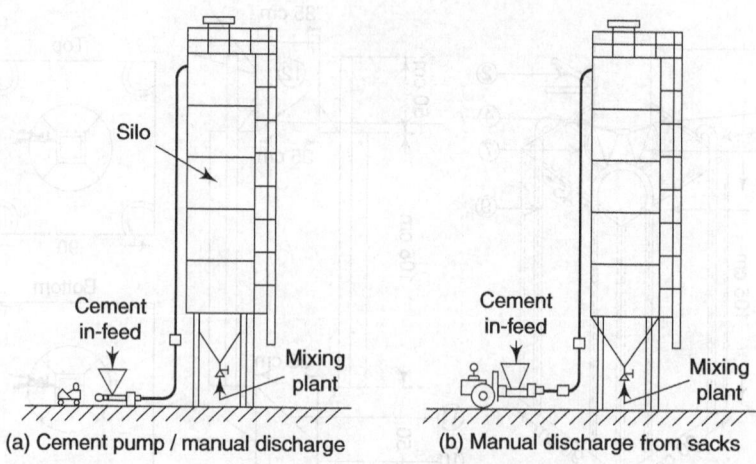

(a) Cement pump / manual discharge

(b) Manual discharge from sacks

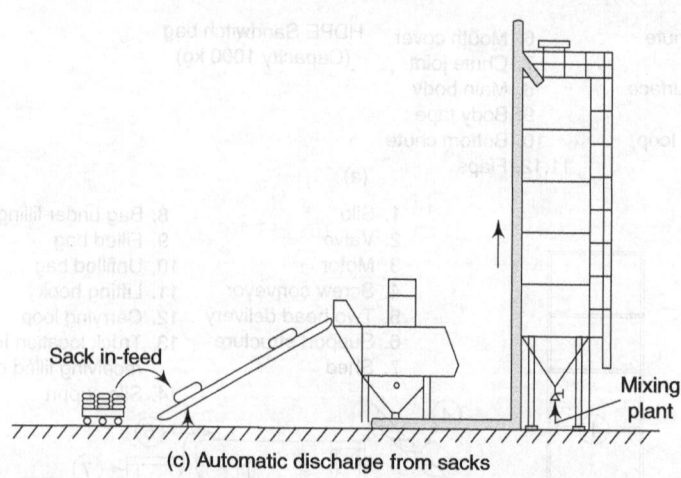

(c) Automatic discharge from sacks

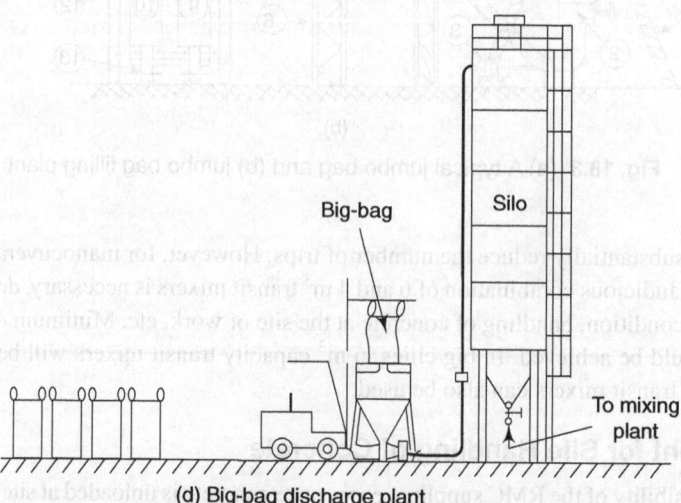

(d) Big-bag discharge plant

Fig. 18.4 *(contd on next page)*

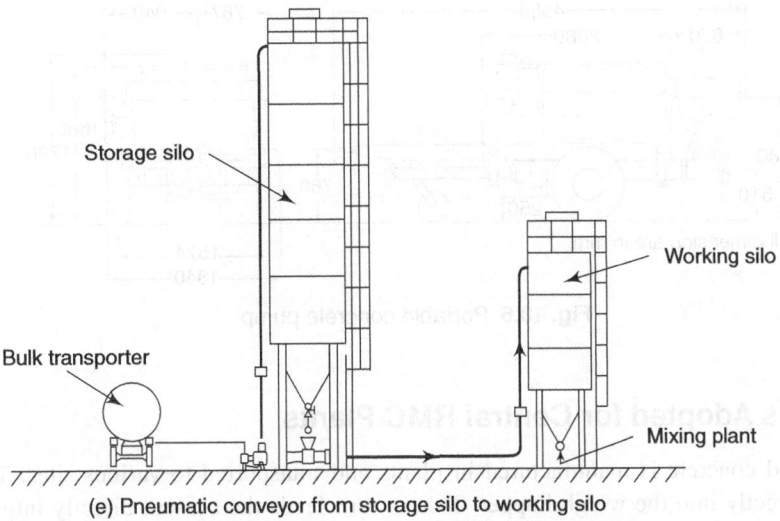

(e) Pneumatic conveyor from storage silo to working silo

Fig. 18.4 Methods of bulk handling of cement

Fig. 18.5 Transit mixer (truck mixer)

RMC operators to have some equipment for handling and placing concrete at site. The concrete shall be placed as soon as possible, after delivery, as close as is practical to its final position to avoid re-handling or moving the concrete horizontally by vibration. Depending on the equipment for handling and placing, the unloading and placing time for a 6 m³ capacity transit mixer would be approximately as follows:

Truck-mounted concrete pump:	6–8 min
Trailer-mounted pump:	15–20 min
Tower crane and bucket:	25–30 min
Skip and hoist:	35–45 min

However, the full load of concrete should be discharged within 30 min of arrival at site. The typical portable pump used for the operations at site is shown in Fig. 18.6.

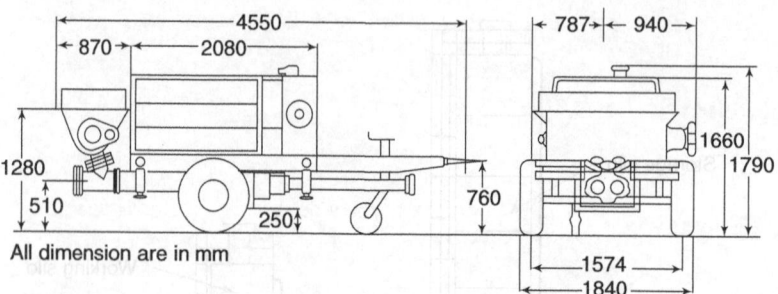

Fig. 18.6 Portable concrete pump

18.4 Process Adopted for Central RMC Plants

Centrally batched concrete is manufactured in plants and transported to various sites. The cement silo feeds cement directly into the weigh hopper. This in turn feeds the cement directly into the transit mix truck. Generally, the plant is capable of manufacturing all types of concretes used by today's construction industry. The aggregates are accurately weighed by the front end loader into the weigh hopper and then transferred directly into the truck by means of a conveyor system. Bulk admixture drums are generally kept adjacent to the conveyor. Another method is to use 'live' overhead storage. Storage can range from 100 to 1000 tonnes of four to six different materials. This is used when a higher production rate of ready mixed concrete is required.

With the load-cell-type weigh batcher, the digital scale display can easily be read by the front-end loader operator. This is to ensure that a high degree of accuracy of the order of ±10 kg is achieved during batching. A similar type of weigh batcher using the traditional knife-edge lever scale system and an analogue scale read-out is also effective, and can easily be maintained. This type of plant can be erected and operated within 1–2 h of it arriving on the site.

The cement weigh hopper valves and installed aerators ensure that an accurate, even flow of cement is weighed before discharging it into the mixing bowl of the truck. There are various admixtures that can enhance the performance of concrete. These admixtures must be carefully measured for each batch. This is achieved by using precision measuring devices. A full range of air entrainers, superplasticizers, retarders, water reducers, and accelerators can easily be handled with such devices attached to the equipment. The controls for these activities are located in the batching control room. The amount of instrumentation will depend upon the size, desired performance, production rate, and the degree of automation of the plant.

Whether the plants are situated at temporary sites for large civil construction projects such as dams, airport runways, and bridges, or located in permanent city sites, great care should be taken to ensure that dust does not pollute the atmosphere. This is important from the environmental protection consideration. The plant with cement capacity of 225 tonnes and live aggregate storage of 300 tonnes and a production rate of 60–70 m³/h using an automatic batching system is used normally in a city site set-up. Trucks are washed and cleaned prior to loading each time. The yard is generally designed to conserve all water by directing spillover to ponds so that water can be reused. Ready mix trucks coming back after delivery may have small quantities of concrete remaining in their bowls. This is washed out and stored in specially designed containment ponds. These ponds allow all solids to settle while water circulates through the ponding system. Solid material can be removed from the site and reused as a base coarse material or as a fill. The water is used as mixing water for concrete production through the plant after thorough testing. The water storage plant can have a capacity ranging from 40 to 80 m³, depending on plant size. The depth of ponds ranges from 1.5 to 2.5 m. Therefore, care is taken to ensure that these ponds are fully fenced for safety.

Compressive strength cubes are made after checking workability using slump test. This should be done both at the batch plant and at the customer's site. Normally six cubes are taken and stored in water baths at a controlled temperature (21°C) until testing. Three cubes are normally tested on the 7th day and the other three are tested on the 28th day. Special customer requirements may warrant testing to allow for early and safe 'deproping' of beams or floors. If slow setting cement is used, testing may be done on the 42nd and 63rd day. For special works, cylinder specimens are also cast and tested. Capping of the cylinder ends is required to ensure proper quality results. The capping material is the same as that used in the manufacture of false teeth and dentures. For normal works, modified Plaster of Paris can be used for capping purpose. The tests are conducted in a compression machine. One dial measures the hydraulic pressure required to break the cylinder and another indicates the preset rate at which the pressure is applied. The compression machine should be checked and calibrated regularly.

To ensure quality of all materials used in the production of ready mixed concrete, they are tested for size, shape, cleanliness, density, and moisture content. Sieve analysis on all aggregates and sands is performed regularly to ensure particle size distribution. A special accelerated 24 h cube or cylinder test can be conducted to predict the 28-day compressive strength. In this test, concrete cylinders are placed in a special curing water tank, immediately upon compaction (heated to 35°C) for 24 h. They are then tested and the results are compared with earlier information relating to tests conducted on concrete of the same strength and slump. This testing procedure is used as an indicator of the 28-day strength of concrete.

A wide range of mix design can easily be supplied from the central batch plant. Fly ash can be used to partially replace cement and produce a product of desired strength and performance. If necessary one can also make self-levelling concrete. For sites with a number of different buried services, the flowable fill using self-levelling concrete can be coloured with suitable oxide to identify the particular service in the trench. Another specialist product of concrete is used to fill hollow masonry blocks. This is pumped through a 50 mm pipe, at a high slump, so as to flow into the cavities. For this purpose M25 grade concrete is used. Cement and silica fume can also be used with conventional aggregate to make precast products. The concretes in such cases are designed to be light coloured and to be very dense. The central batching plant can be designed as simple, robust, low-cost structure, or as a sophisticated high-speed plant with several cement options with an overhead 'live' storage aggregate bin. The plant can utilize both wet and dry admixtures.

Any of the increasing number of special, sophisticated concretes demanded by the construction industry can be produced from centralized batching plants with appropriate quality control to ensure a consistent product of requisite standard.

18.5 Concrete Specification

Since ready mixed concrete is merely concrete manufactured away from the site, with the exception of requirements for delivery, there should be no difference between the specification for site mixed concrete and the specification for ready mixed concrete. This is recognized in several countries, including in Germany and the UK, where there is one specification for concrete, which has to be complied with both by site mixed and ready mixed concrete.

In 1990, the European standards organization, CEN, considered the introduction of the CEN standard for ready mixed concrete. It has now been agreed that this should be amplified to be a comprehensive specification for concrete and not just for ready mixed concrete.

The basic aim of specification for structural concrete is to control the water–cement ratio at different levels to produce concretes of different qualities. However, there is still no quick direct method of measuring the water–cement ratio of any concrete mix. As a result, two independent methods of specification have been established. With the mix proportion specifications, water–cement ratio is controlled through the mix proportions

Table 18.2 Concrete mix information

Particulars	Information required for each mix
RMC supplier	
Contractor	
Site address	
Mix code	
Grade (MPa)	
Chemical additives	
Mineral additives	
Water–cement ratio	
Maximum aggregate size	
Cement type and grade	
Target workability	
Temperature of concrete	
Class of sulphate resistance	
Exposure condition	
Class of finish	
Mix application	
Other requirements (stripping time, etc.)	
Concrete testing frequency	
Material testing	
Alternatives offered	
Method of curing	
Quantity ordered	

and the workability of fresh concrete. With the performance specification, the end product requirements of strength and durability are controlled through the strength of concrete in the hardened state. Table 18.2 shows the concrete mix information to be furnished by the contractor to the RMC supplier.

The responsibility for the quality of the final concrete in the structure in the case of ready mixed concrete is shared between the purchaser (contractor) and the supplier. The following are the three alternative methods of order and supply of ready mixed concrete:

Performance batch: The purchaser specifies all the performance requirements, such as strength and work-ability, and the supplier takes the full responsibility for the mix proportions and various ingredients that go into a batch.

Recipe batch: The purchaser specifies the mix proportion indicating cement content, water–cement ratio, percentage of air and admixture required, and amount and type of coarse and fine aggregates. The purchaser takes the full responsibility for the resulting strength, durability, and workability properties. The supplier provides the stipulated amount of ingredients at a stipulated time in a stipulated condition as specified.

Part performance and part recipe batch: The purchaser specifies the cement content, admixture, and strength requirements. This allows the supplier to proportion and supply the concrete mixture within the constraints imposed by the purchaser.

In India, most purchasers use the third approach. It ensures durability while still allowing flexibility to the supplier to use optimal economical mix proportions.

18.6 Distribution and Transport

The ready mixed plant comprises materials storage and weighing and batching equipment, which feed the materials into a static mixer. The mixed concrete is then discharged into the transport vehicle. Due to its short life some users of ready mixed concrete consider agitation of the mixed concrete during journey is essential. This requires special agitating vehicles, which either slowly mix the concrete during the journey or remix the concrete on arrival at site. The special vehicle comprises a mixing drum on the lorry or truck chassis. One widely used approach is to dispense with the central static mixer and use truck mixers for mixing the concrete immediately after loading at the factory.

In the past 30–40 years there have been several attempts to introduce new and revolutionary truck mixer designs. These attempts have given rise to three basic methods of production and delivery as follows:

1. The central mixer discharges into the truck mixer or agitating vehicle.
2. The central mixer discharges into a special non-agitating vehicle as a bowl or dump truck.
3. Dry materials are weigh batched into a truck mixer drum and mixing is completed at the factory or finished on site.

18.7 Handling and Placing

Efficient use of ready mixed concrete depends on a rapid turnaround of truck mixers and proper facilities for rapid discharge and placing of concrete. With proper access, the modern truck mixer can position itself and discharge its full load in only 10–15 min. This represents a potential delivery rate of at least 25–30 m³/h to a particular location. Whether the contractor can deal with this rate of delivery depends on the type of equipment and the number of personnel available. Often the contractor handles concrete with too few workers, using a crane and a bucket or with manual labour only. This has led to the introduction of continuous handling methods such as the small-diameter mobile pump and the conveyor system.

Despite the obvious advantage of these mobile pumps, statistics show that only about 20% of ready mixed concrete is placed by advanced pumping methods. There are no simple reasons for this, but contributory factors include the following:

1. Some RMC companies which provide their own pumps as part of a 'supply and pump' service find that contractors are unwilling to pay the cost of pumping.
2. Where pumping is operated by a separate specialist company, difficulties occur between the supplier, the pumper, and the contractor when there are blockages or delay in supply.
3. Most pumping contractors under ideal conditions can place less than 50 m³ concrete in 3 h or less. In practice, the average placing rates are only about 10 m³/h.

18.8 Code Recommendations

The specification for making ready mixed concrete is given in IS: 4926-2003 and ASTM C94. The most important parameter is the time that elapses from the instant of adding water to the placement of concrete. ASTM C94 permits a maximum of 90 min, or before the drum has made 300 revolutions, whichever is earlier. However, IS: 4926-2003 permits concrete to be discharged from the truck mixer within 2 h from the time of loading. It also permits longer periods if suitable retarding admixtures are used or in cold, humid weather or when chilled concrete is produced.

Transit mixers of varying capacities ranging from 6 to 10 m³ are available. The most popular one in India is the 6 m³ mixer. The mixers are capable of mixing concrete thoroughly in about 100 revolutions and operate at a speed of 8–10 rpm.

Mixing during transit usually stiffens the mixture. ASTM C94 allows addition of water at site to restore the workability by remixing. ACI 304 recommends that some water out of the total water required as per prescribed water–cement ratio be withheld at the batching plant and added finally at the site so that the water–cement ratio is not altered. It should be ensured that after the final addition of water at site, an additional 30 revolutions of the drum are made for thorough mixing. To offset stiffening, addition of water at site may be permitted, provided the design water–cement ratio is not violated. Such re-tempering of concrete is required especially in hot weather conditions in tropical countries such as India.

The workability and compacting factor should be within the following limits of the specified value by the purchaser.

Slump	±25 mm or ±0.33 of the specified value, whichever is less
Compacting factor	±0.03 where the specified value is 0.9 or larger
	±0.04 where the specified value is less than 0.9 but more than 0.08
	±0.05 where the specified value is 0.08 or less

Ready mixed concrete should be checked for uniformity by taking samples at different locations of a project and examining variations in important performance parameters. The parameters to be investigated for uniformity are unit weight, air content, slump, coarse aggregate content, and average compressive strength. The uniformity requirements of ready mixed concrete with respect to these parameters are shown in Table 18.3.

BS: 1881 Parts 101 and 102 describe sampling methods for ready mixed concrete. Figure 18.7 shows a standard scoop, which collects about 50 N weight of concrete. BS: 1881 Part 101 specifies the standard method of sampling. Figure 18.8 shows the cross-section of the transit mixer drum, which contains different parts of the load. The sample should be taken using the scoop from different parts of the load. As shown in the figure, the first and last 1/6th portion of the load should be regarded as unrepresentative. The sample taken is then thoroughly remixed on a non-absorbent surface before carrying out individual tests.

Table 18.3 Uniformity requirements of ready mixed concrete

Parameter tested	Requirement expressed as maximum permissible difference in the results of samples tested from at least two locations in a batch
Average compressive strength at 7 days of each sample based on average of all comparative test specimens	7.5%
Weight of air-free mortar based on average for all comparative samples tested	1.6%
Coarse aggregate content (by weight retained on standard sieve of the aggregate size used)	6%
Slump	
If average slump is 100 mm or less	25 mm
If average slump is 100–150 mm	37 mm
Air content	1%
Self-weight of concrete (air-free basis)	15 kg/m³

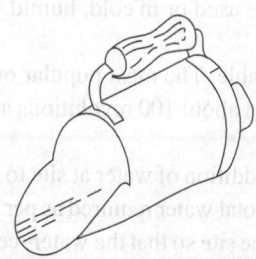

Fig. 18.7 A standard scoop of 50 N capacity

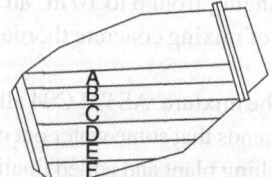

A—Let very first concrete go
B—Take scoop full from part 1
C—Take scoop full from part 2
D—Take scoop full from part 3
E—Take scoop full from part 4
F—Let last concrete go

Fig. 18.8 Standard method of collecting samples

BS: 1881 Part 102 gives an alternate method of collecting samples (Fig. 18.9). When this method is used, an initial discharge of at least 0.3 m³ is made before a sample of six full scoops is collected from a moving stream. The collected sample is then remixed and split into two parts. Slump test is performed on each of these parts. This method of sampling is applicable only for workability tests. These tests will ensure assessment of correct properties of the ready mixed concrete.

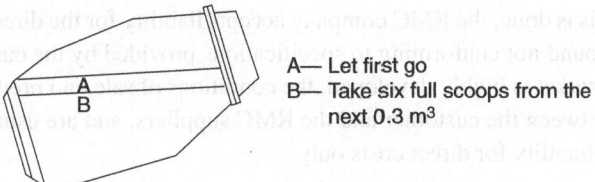

A— Let first go
B— Take six full scoops from the next 0.3 m³

Fig. 18.9 An alternative method of collecting samples

IS: 4926-2003 stipulates one sample for every 50 m³ of production or every 50 batches, whichever has a larger frequency. In order to get a relatively quicker idea of the type of concrete produced, optional tests on beams for the modulus of rupture at 72 + 2 h or 7 days or compressive strength test at 7 days may be carried out in addition to 28-day compressive strength tests on cubes.

The supplier of ready mixed concrete should retain the following records for at least one year after placing the concrete at the contractor's site.

a. Production and delivery
  • Batching instructions
  • Batching records
  • Delivery tickets
  • Equipment calibration
  • Plant maintenance
b. Material production and control
  • Concrete production, material purchase, usage, and stocks
  • Certificate of test results of materials
c. Production quality control
  • Control test results
  • Standard deviation of test results

18.9 Workforce

Before the introduction of ready mixed concrete, the contractors had to carry out batching, mixing, and handling of concrete. On any job, concreting is an intermittent process and, to avoid having trained workforce idle for much of the job, the contractor tends to use those who are employed on less skilled works on the site. The introduction of ready mixed concrete has enabled a properly trained and experienced workforce to be established. Training courses for batching operators and truck-mixer drivers are vital. There are different courses on technical and production subjects for personnel in the industry. Such training programmes are necessary to ensure training of workforce engaged in this industry.

18.10 Conditions of Sale and Product Liability

Ready mixed concrete is a peculiar material which is supplied to the customer in a partially finished form. The customer then has to handle it to place, compact, and cure in order that the potential of concrete can be fully realized. The hardened property on which the characteristic of concrete is based is normally the 28-day compressive strength. Difficulties occur when, on the basis of 28-day tests, concrete already placed in the structures is proved to be not in compliance with the specification. In the early days of ready mixed concrete, this often led to difficulties in the interpretation of the specification and of the contract. Many RMC companies still limit their liability to free replacement of the material involved. However, more responsible companies contract to supply concrete in accordance with the specification as confirmed on the quotation. For strength mixes this is on the basis that the 28-day strength will be made on a specimen properly sampled, compacted, cured, and tested. Provided this is done, the RMC company accepts liability for the direct cost of removal and replacement of the concrete found not conforming to specifications provided by the customer.

In countries where RMC market is highly developed, the conditions of sale and product liability are properly understood and agreed between the customer and the RMC suppliers, and are usually based on detailed quotations and acceptance of liability for direct costs only.

Exercises

Review Questions

1. What are the advantages of using ready mixed concrete instead of site mixed concrete?
2. Give a list of components of a typical ready mixed concrete plant.
3. How is ready mixed concrete specified to satisfy the requirement in the fresh and hardened states?
4. In your opinion what are the reasons for delay in the use of ready mixed concrete in the Indian construction industry?
5. Give a typical layout of ready mixed concrete plant.
6. What are the special precautions to be adopted on the site for efficiently using ready mixed concrete?
7. Give a list of laboratory equipment required to ensure the quality of ready mixed concrete supplied to the contractor.
8. Comment on the conditions for sale for ready mixed concrete with respect to product liability.
9. What are the special features of transportation of ready mixed concrete from the plant to the site?
10. What special features are to be considered while handling and placing ready mixed concrete?

Multiple Choice Questions

1. Ready mixed concrete ensures
   a. Supply complies with fresh and hardened properties
   b. Supply of assured quantity
   c. Supply at specified time and location
   d. All of the above

2. In urban situations RMC is inevitable because of
   a. Traffic problems
   b. Problems related to strength of concrete
   c. Problems related to site storage of materials
   d. Problems related to availability of materials

3. One of the major problems faced by RMC industry is
   a. Lab facilities
   b. Traffic causing time delays in supply
   c. Ensuring the correct workability
   d. Ensuring correct strength

4. Normal time required for truck mounted concrete pump for unloading concrete at site is
   a. 6–8 min
   b. 15–30 min
   c. 25–30 min
   d. 35–45 min

5. The purchaser specifies performance batch to ensure
   a. Proper strength
   b. Proper workability
   c. Adequate durability
   d. All of the above

6. Recipe batch means
   a. Strength is specified
   b. Workability is specified
   c. Mix proportion with w/c ratio is specified
   d. Durability requirement is specified

7. The workability of concrete supplied measured at site should be
   a. Exactly as specified
   b. ±25 mm
   c. ±50 mm
   d. ±10 mm

8. For quality control in RMC concrete collected for testing should be selected from
   a. Top layer in the drum
   b. Bottom layer in the drum
   c. Both top and bottom layers
   d. Moving stream avoiding top and bottom layers

9. Variation permitted in self-weight of concrete is
   a. 10 kg/m^3
   b. 15 kg/m^3
   c. 50 kg/m^3
   d. 100 kg/m^3

10. Quality control of supplied concrete is assessed based on
   a. Strength
   b. Slump values
   c. Time of supply
   d. Uniformity of all parameters of the supplied concrete

Answers to Multiple Choice Questions

1. d
2. c
3. b
4. a
5. d
6. c
7. b
8. d
9. b
10. d

CHAPTER 19

Mass Concrete

Mass concrete can be defined as 'any volume of concrete with dimensions large enough to require measures to be taken to deal with the generation of heat from hydration of the cement and attendant volume change to minimize cracking'. Mass concreting technique is used for massive structures such as dams and large bridge piers and foundations. Mass concrete is basically concrete with a higher proportion of coarse aggregate and a lesser proportion of cement. Large size aggregates are preferred for mass concrete. Strength, economy, uniformity, and all other factors considered for normal concrete should also be taken into account for mass concrete. Mass concrete should be properly designed, placed, and cured to obtain a durable structure with more economy. Temperature shrinkage is a major problem in mass concrete work.

Concrete intended for large structures must have excellent control of early setting in order to prevent shrinkage, cracking, and the problems associated with placement and workability of concrete. The following factors are important in mass concrete work:

- Engineering design of mass concrete work
- Selection of material and proportioning of mixture
- Control of concrete during batching and placement

19.1 Materials for Mass Concrete

The following materials are preferred for mass concrete work:

Portland cement with low heat of hydration   Minimum amount of Portland cement should be used to achieve the desired heat of hydration and economy.

Pozzolans   The addition of pozzolans such as fly ash assists in reducing the heat of hydration, workability, and delayed strength gain.

Aggregates   Both fine and coarse aggregates should meet the IS requirement of grading. In the case of structural concrete for large beams and columns, 20–30 mm aggregate is suitable. For dams 40 mm aggregate can be used. The fine aggregate to total aggregate ratio should be low (nearly 25%).

Mixing water   Mixing water should be very low to give low slump (0–5 cm).

Admixture   Water-reducing and retarding admixtures are preferable. Mass concrete to be pumped should be provided with high-range water-reducing and retarding admixtures to provide a slump of 12–18 cm after the addition of superplasticizers.

Coolants   The temperature of the mass concrete dam should be controlled between 5°C and 20°C especially in hot weather conditions. Addition of finely chipped ice instead of water will help. Aggregates should be kept damp and under shade. Steel forms should be sprayed with cold water. Cast concrete should be cured with cold water.

19.2 Properties of Mass Concrete

The properties of mass concrete are the same as that of normal concrete. Some of these properties are discussed next.

19.2.1 Workability

Workability is the property of concrete which determines the amount of useful internal work necessary to produce full compaction. Workability should be optimum for mass concrete. Uniformity of workability is essential in mass concrete works because economy generally requires low cement content. Lack of uniformity leads to difficulty in mixing, placing, and compaction. Slump of mass concrete should be as low as possible and depends upon the dimensions of the member, spacing of reinforcement, method of placement, and consolidation. Even zero-slump concrete can be used.

19.2.2 Durability

The ability of concrete to resist weathering action, chemical attack, and abrasion is known as durability. It depends on mix design, workmanship, placing, and curing. Concrete with low water–cement ratio which is correctly consolidated and properly cured provides durable concrete. Durability of concrete is dependent on the properties of the constituent materials. So materials used for concrete should be selected carefully. Chemical resistance of concrete is improved by careful selection of materials. Weathering durability can be improved further by entrainment of minute air bubbles into concrete. Air-entrained concrete is more resistant to freezing–thawing and wetting–drying actions, which are common to most massive hydraulic structures. Table 19.1 shows the recommended air content for concrete with severe exposure conditions.

In special structures, such as dams, cement content is increased in the exterior face where weathering action, erosion, and chemical attacks are frequent. Rich mix concrete increases temperature control problems. Cavitation damages are difficult to control. The exact solution to prevent cavitation damage is to adopt a proper hydraulic design. In places where cavitation damages are likely, concrete should be treated and made highly resistant to erosion. This can be accomplished with fibre reinforcement, by coating the concrete with proper epoxy mortar, or by using high-strength concrete.

Mix design should also consider the durability of concrete. The quality of concrete is related to cement and water content and decides both strength and durability of concrete. Any likely chemical action in concrete should be considered. Reaction of alkali with aggregates will cause abnormal expansion, which will affect the

Table 19.1 Recommended air content for concrete subjected to severe environment conditions

Maximum size of coarse aggregate (mm)	Air content (% by volume)
150	4
75	4.5
40	5.5
20	6
10	7.5

strength of concrete and water-tightness. To avoid this, low-alkali cement or non-reactive aggregate should be used. Alkali–aggregate reaction can be controlled by using pozzolans as a mineral admixture.

In severe conditions sulphate attack is possible. This can be avoided by using sulphate-resisting cement. Any concrete exposed to sulphates should be air-entrained. It should not contain calcium chloride which significantly reduces the resistance of concrete to sulphate attack.

19.2.3 Water-tightness

Water-tightness is an important property of concrete in hydraulic structures such as dams. Concrete should be impermeable and this can be achieved by using good quality aggregate, low water–cement ratio, good consolidation during placing, and proper curing. In the case of lean mixes used in mass concrete for dams, pozzolans can be added to the concrete mix to improve water-tightness.

19.2.4 Strength

Strength of concrete is one of the most important factors. Concrete is used as a structural element, and all structural uses are associated with its compressive strength. Strength of concrete is defined as the resistance that concrete provides against load so as to avoid failure. It is dependent on the water–cement ratio, quality of aggregates, compaction, and curing. The primary factor that affects the strength of concrete is the quality of cement paste, which, in turn, depends on the quantity of water and cement used. Figure 19.1 shows the relation between the water–cement ratio and the 28-day compressive strength of air-entrained concrete made with Portland cement. Table 19.2 gives the recommended water–cement ratio for various exposure conditions for mass concrete.

Sometimes it is economical to add pozzolana or use Portland pozzolana cement instead of ordinary cement concrete. Pozzolans are materials that have little cementing value but react with calcium hydroxide to form compounds that are cementitious. This reaction contributes to the ultimate strength and water-tightness of concrete. Pozzolans also increase the plasticity and workability of concrete. But excessive addition of pozzolans affects durability. So it should be used along with cement as a partial replacement or in small percentages.

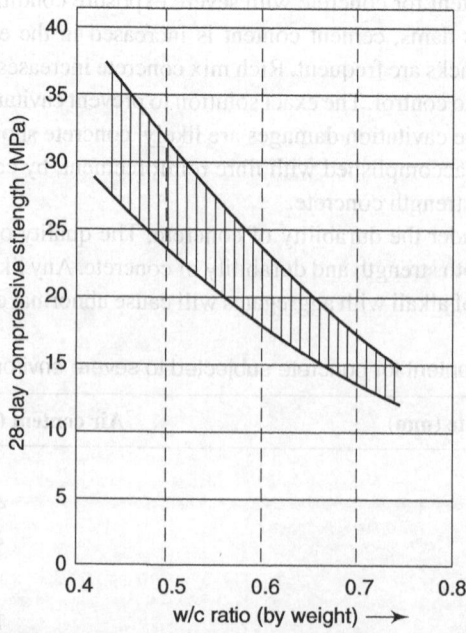

Fig. 19.1   Relationship between 28-day compressive strength and water–cement ratio (by weight) for air-entrained mass concrete

Table 19.2 Recommended maximum water–cement ratio for mass concrete

Type of structure and location of concrete	Maximum water–cement ratio by weight	
	Severe climate (freezing and thawing)	Mild climate (temperature rarely below freezing)
Concrete for interior portions of dams or other unexposed portions of massive units	Water–cement ratio is governed by required strength, workability, thermal property, creep, and volume change (ideally should not exceed 0.8)	
Ordinary exposed structures or concrete at or near waterline in hydraulic or waterfront structures where intermittent saturation is possible	0.5	0.55
Concrete deposited underwater	0.45	0.45
Concrete continuously submerged in water	0.58	0.58
Concrete subjected to high velocity of flow (>12 m/s) or exposed to corrosive liquid, salt, or sea water	0.45	0.45

19.2.5 Temperature Effects

Concrete generates heat during hardening as a result of internal chemical reactions. This heat generated is referred to as heat of hydration, and the amount of heat generated depends on various factors such as ambient temperature, water–cement ratio, characteristics of cement, use of chemical admixtures, size of structural element, and surrounding environment. Dissipation of this heat depends on the type of form, amount of the exposed surface, and ambient temperature at various surfaces of concrete. In small structures, the temperature increase due to hydration is dissipated rapidly, but in massive structures, it leads to thermal expansion and cracking.

Temperature rise within mass concrete causes an outward expansion during the early stage of setting. Due to hydration, the temperature at the internal or central region of concrete will be higher as compared to outer areas. Thus, as the outer (exterior) surface cools and tends to shrink, compressive stresses develop in the interior and tensile stresses develop in the cooling outer crust areas. When this induced tensile stress crosses the safe strength limit of concrete, cracking occurs. This is further complicated by shrinkage stresses due to the loss of moisture from the outer surface of concrete. Generally, most of the heat is generated in the first 7 days. During this time, concrete gains strength.

The maximum temperature reached in the interior of mass is related to the initial placement temperature of the concrete and the subsequent temperature rise during hydration. The increase in temperature is dependent on heat of hydration and dissipation of heat. Heat of hydration is a function of concrete mix design. Heat dissipation depends on the size of mass being concreted and the mix design. The following equation gives an estimate of the temperature rise:

$$T = CH/S$$

where T is the temperature rise in °F of the concrete due to heat generation of cement under adiabatic condition, C is the proportion of cement in concrete (by weight), H is the heat generated due to hydration of cement, and S is the specific heat of concrete.

At the early stage, heat is generated in concrete at a higher rate than that can be transmitted to the exposed surface. This results in an internal heat build-up. At surface this build-up is less as compared to the interior mass. Interior concrete will ultimately cool to a final temperature, while the temperatures at the surface will influence the exterior concrete. The final stable temperature of the interior concrete depends on the ambient condition over a period.

19.2.6 Temperature Control

Temperature control in mass concrete can be achieved in two ways: (a) control of the maximum temperature of the concrete and (b) artificial cooling of concrete to its final stable temperature. Both these techniques are intended to avoid build-up of thermal stresses.

In general, serious volume change stresses can be avoided by controlling the temperature drop from maximum to final. Restraint is also a factor in limiting stresses. Ordinarily, massive concrete structures are placed in blocks that exercise some restraint on the volume change caused by temperature. Field experiences that have proven to be acceptable are reflected in Table 19.3, which can be used as a guide for design.

Table 19.3 Permissible temperature drop based on block length (*L*) and height above foundation (*H*)

Block length (m)	Temperature drop from maximum concrete temperature to grouting temperature (°C)		
	H = 0.2*L*	*H* = 0.2–0.5*L*	*H* = 0.5*L*
45.75–61	13.87	19.425	22.22
36.6–45.75	16.65	22.22	24.95
27.45–36.6	19.425	24.95	No restriction
18.3–27.45	22.22	No restriction	No restriction
Up to 18.3	24.95	No restriction	No restriction

The time duration for initial and final cooling and the rate of temperature drop are largely dependent upon the diffusivity of concrete. Diffusivity provides a method for computing the temperature rise within a block by taking into consideration the losses of heat to the sides and the exposed top of a lift. A measure of the diffusivity of concrete is necessary for use in post-cooling operations and rapid successive lift constructions. In placing a bridge construction block or pier, good use can be made of the diffusivity of concrete by scheduling construction. Since most of the heat is generated within the first 7 days after placement, these 7 days become the critical time for cooling operation. Foundation construction can be scheduled to allow the water curing method to dissipate the heat generated instead of resorting to other expensive cooling methods.

Temperature rise in concrete mass can be controlled by pre-cooling the concrete materials, limiting the amount of heat liberated, and immediate cooling after placement. Pre-cooling the concrete material can lower the maximum temperature but will not substantially change the amount of temperature rise associated with cement hydration.

Cooling the coarse and fine aggregates is of greatest benefit since these materials compose the bulk of concrete. For maximum benefit the aggregates should be cooled through and not just on their surface. Shading the stock of aggregate will considerably lower the temperature. Immersion of sand in water is not helpful because of the subsequent moisture control problem while batching.

In some cases, crushed ice has been substituted for part of the mix water to lower the batch temperature. Ice is batched with sand to avoid its adherence to the sides of the mixer.

19.3 Mix Design

All concepts and principles used in the mix design of normal concrete are adopted here. In massive structures, savings in small amounts in a batch can result in appreciable total saving for the complete structure. Therefore, mix design is important from both quality and economy considerations.

Three general strength levels are available for the selection of an efficient mix for mass concrete.

1. A high-strength zone where small size aggregate is efficient and the selection of the optimum maximum size aggregate is critical. The use of slightly smaller or larger size aggregate will increase the cement requirement significantly.

2. A medium-strength zone, where there is a wide range of maximum size aggregates, requires essentially the same cement content for a given strength. The use of smaller size aggregates would require more cement.

3. A low-strength zone where large size aggregates are essential for the most economical use of the cement.

For massive structures, mixes of low or medium strengths are used. Therefore, the selection of aggregates is important for obtaining the most efficient concrete. The basic considerations for preparing a concrete mixture particularly for massive structures include the following:

1. Need to provide for variability in strength and exposure zones
2. Water–cement ratios for strength and durability for all those zones
3. Need to add pozzolana or admixture
4. Maximum size of aggregate that can be used

Natural gravels with a rounded shape are preferable because these help reduce the sand content needed. Sand content of 20–35% for concrete in dams is common. A lower sand content reduces cement and water requirement and permits better aggregate interlocking in the consolidated mass.

Several methods are available for mix design. The most direct approach is to select proportions based on past experience and reliable test data that have been established. Where previous data are not available, trail mixes can be made and tested to establish a final mix design.

19.4 Massive Structural Reinforced Concrete

Large reinforced concrete structures are also built with mass concrete. A special technique has been developed for this purpose. This technique is mainly used in the construction of large mat foundations, and is different compared to the traditional method. It requires low cement content, especially low-heat cement, limit on the height of pour and pour size, and frequent construction joints. This technique relies mainly on three considerations:

1. Internal strains must be controlled.
2. External applied restraint must be avoided.
3. The entire reinforced concrete section should be cast in one continuous pour.

19.4.1 Internal Thermal Strain

Control of internal strains can be accomplished by ensuring that no part of hydrating concrete mass becomes cooler than the hottest part by 20°C. Lean cement mass concrete can be kept cool by traditional methods. Structural concrete is usually of richer mix due to its high-strength requirement and cannot be kept cool by the same methods. So the exterior is insulated to satisfy this requirement. By insulating the exterior surface for a number of days, surface cracking is significantly reduced. Internal cracking perpendicular to reinforcements can also be minimized.

The problem of estimating concrete temperature changes resulting from hydration is difficult because of the large number of variables involved. This problem can be easily solved by using the following general rule. The rule states that the maximum temperature possible by hydration of concrete is equal to the temperature of concrete at the time of placing plus 12°C for each 100 kg/m^3 of concrete.

It is necessary to record the temperature at the hottest and coolest places in the concrete mass. Sometimes, temperature may also be recorded at more than two places. To measure temperature, thermocouples are placed at a distance of 25 mm in exposed and interior (sensitive) parts. By recording the temperature, its variation can be studied and insulation or cooling methods can be adopted as per the requirement to prevent cracking (Fig. 19.2). Note that the removal of form on the third day makes $\Delta t > 20°$ and hence the surface of concrete cracks.

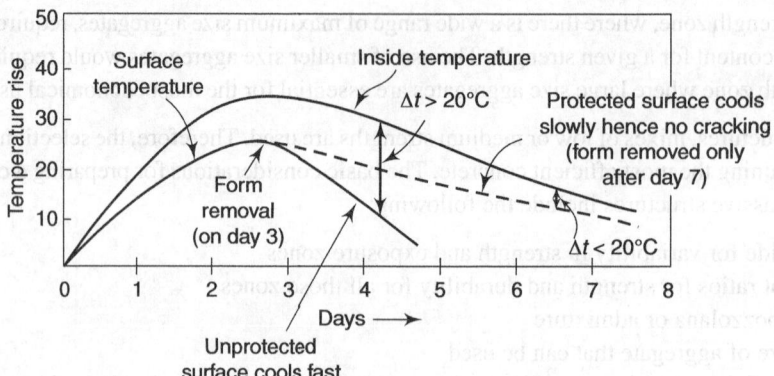

Fig. 19.2 Effect of form removal and temperature differential Δt on surface cracking

19.4.2 External Restraint Cracks

A major source of external strain is concrete formwork or enclosure when placing concrete in between hardened concrete or a substrate which does not allow thermal expansion. The thermal expansion of fresh concrete and inevitable cooling with its accompanying contraction causes tension cracks. So, placing of fresh concrete in between hardened concrete must be avoided.

19.4.3 Continuous Pour

High-volume ready-mix supply equipment is needed for placing large volumes of concrete in one pour. Retarders are usually essential in this kind of work. Preconstruction discussions with everyone involved are essential to ensure proper understanding of the work and for proper coordination. Construction joints may not always be possible to avoid, so the concrete placing sequence must be considered carefully to avoid any trapped section between hardened concrete. In this case both thermal consideration and overall expansion must be taken into account. The large volume pouring technique and associated equipment help builders and engineers to construct massive, crack-free concrete structures economically.

19.4.4 Quality Control

Selection of quality materials and proper batching and placing of concrete are necessary for quality mass concrete work. Equipment which allows rapid placement of large concrete quantities should be used to reduce possibilities of cold joints. The typical low-water, low-slump concrete must be properly consolidated to eliminate cold joints and honeycombing. Segregation of aggregates must be avoided. With judicious handling, there are less chances of segregation. Large-volume buckets with wide discharge gates are effective for delivering large quantities of concrete with minimum segregation. Once placed, the layer of concrete should be properly vibrated to integrate layers together to make them a joint-free matrix. Transportation of concrete using vibrators should be avoided because of the likelihood of segregation and bleeding.

It is worthwhile to have spare units for all equipment such as concrete pump and truck to ensure uninterrupted delivery of mass concrete to the point of placement.

19.5 Roller-compacted Concrete

Mass concrete is consolidated by roller compaction to get roller-compacted concrete (RCC). RCC is widely used for the construction of dams, foundation for nuclear power plants, etc. Very less water is used for RCC; even zero-slump concrete can be used. Long-term properties of this concrete are very similar to mass concrete.

Aggregates processed for use in conventional concrete may be used for making RCC. Generally a vibratory steel wheel roller is used for consolidation. To attain maximum density, strength, and other desired properties, mix should have adequate fine aggregate to fill the voids.

The placement and compaction of RCC is achieved usually laying many layers having thicknesses ranging from 200 to 250 mm. Any vibrating roller that has compacted rock fills will compact no-slump concrete. Self-driven rollers with power-driven drums are more suitable than vibratory rollers which must be handled manually. Depending upon the characteristics (ability) of the roller, the number of layers should be decided and concrete placed. RCC can be advantageously used for construction of gravity dams and concrete pavements.

19.5.1 Concrete Gravity Dams

Gravity dams built using RCC afford economy over conventional concrete dams through rapid placement techniques. Construction procedures associated with RCC require that particular attention be given in the layout and design to water-tightness and seepage control. The designer should take advantage of the latitude afforded by RCC and use judgement to balance cost reductions and technical requirements related to safety, durability, and long-term performance. Figure 19.3 show the cross section of an RCC dam. RCC placement spreading operation is illustrated in Fig. 19.4.

Economic considerations

The factors which make RCC more economical for dam construction are given below.

Material savings   RCC uses less cement content compared to conventional concrete for the same strength because the water content in RCC is low. Hence there is saving in materials. The desired strength is realized by compaction with the help of rollers. The unit cost of concrete for RCC and conventional concrete depends on the volume of construction material. As the RCC volume increases, the unit cost decreases. RCC dams have considerably less volume of construction material than embankments of the same height. Thus, the higher the structure, the more likely the RCC dam will be less costly than the embankment alternative.

Rapid construction   The rapid construction techniques and reduced concrete volume account for the major cost savings in RCC dams. When compared with embankment dams, construction time is reduced by 1–2 years. Other benefits from rapid construction include reduced construction administration costs, earlier project benefits, and possible selection of sites with limited construction seasons. Basically, RCC construction offers economic advantages in all aspects of dam construction that are related to time.

Spillways and appurtenant structures   The location and layout alternatives for spillways, outlet and hydropower works, and other appurtenant structures in RCC dams provide additional economic advantages compared to embankment dams. The arrangement of these structures is similar to conventional concrete dams, but with certain modifications to minimize costly interference to the continuous RCC placement operation. Gate structures and intakes should be located outside the dam mass. Spillways for RCC dams can be directly incorporated into the structure. Generally, the embankment dam spillway is more costly. For projects that require a multiple-level intake for water quality control or for reservoir sedimentation, the intake structure can be readily anchored to the upstream face of the dam. The saving for an RCC dam intake is considerable, especially in high-seismic areas. The shorter base dimension of an RCC dam, as compared to an embankment dam, reduces the size and length of the conduit and penstock for outlet and hydropower works.

Diversion and cofferdam   RCC dams provide cost advantages in river diversion during construction, which reduces damages and risks associated with cofferdam over-topping. The diversion conduit will be shorter when compared with embankment dams. With a shorter construction period, the size of the diversion conduit and cofferdam height can be reduced. These structures may need to be designed only for a seasonal peak flow instead of annual peak flows. With the high erosion resistance of RCC, if over-topping of the cofferdam does occur, the potential for a major failure will be minimal and the resulting damage will be less.

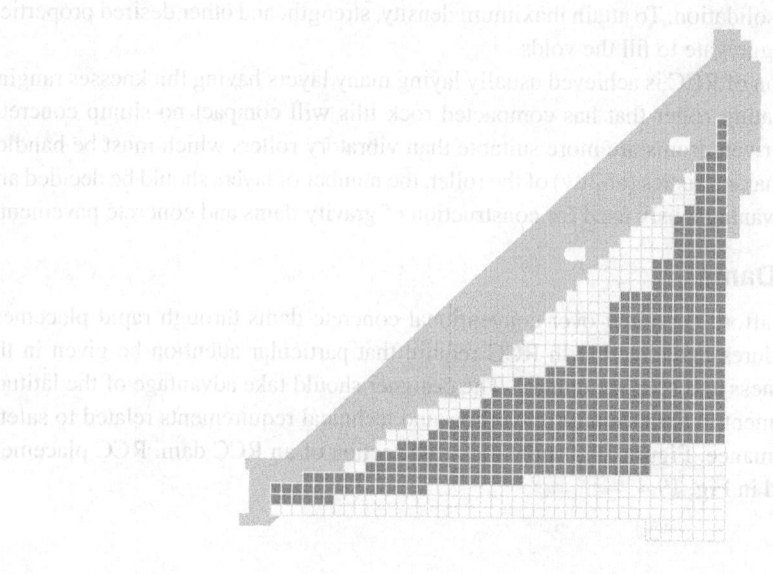

Fig. 19.3 Cross section of a composite RCC lime surki dam

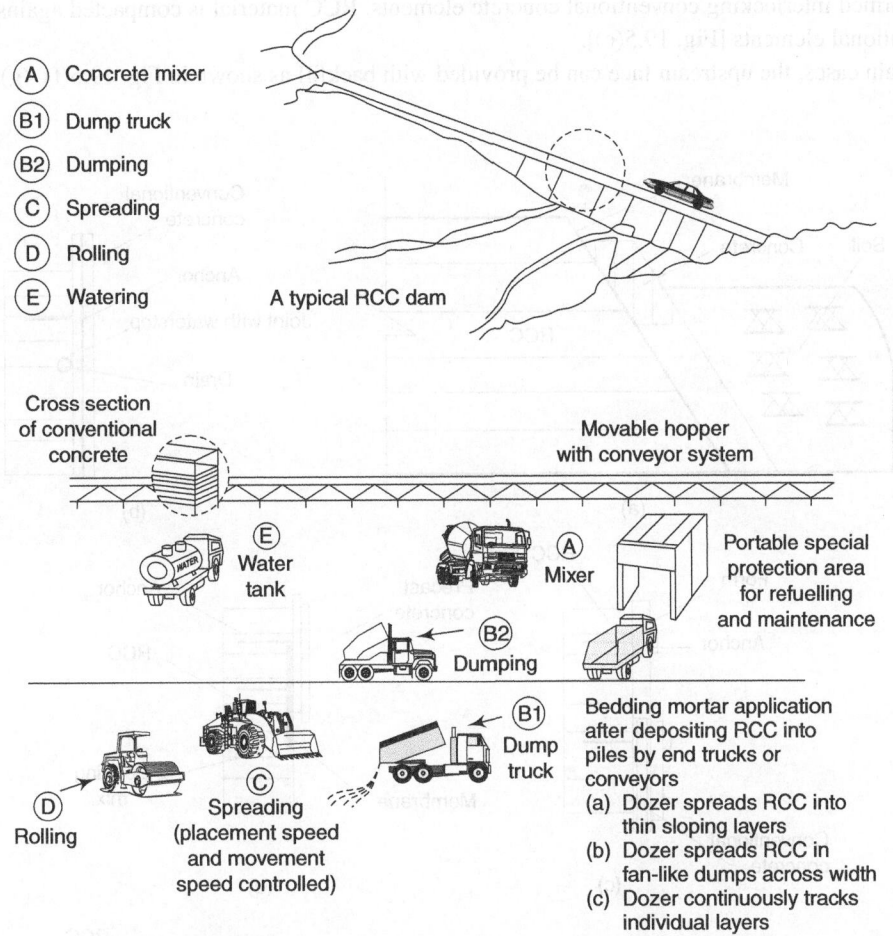

Fig. 19.4 Placement and spreading operation for an RCC dam

Design and construction considerations

Water-tightness and seepage control Excessive seepage is undesirable from the point of view of structural stability and possible long-term adverse effect on durability. Additionally, the loss of water through seepage may at times result in economic loss. RCC that has been properly proportioned, mixed, placed, and compacted should be as impermeable as conventional concrete. Seepage can be controlled by incorporating special design and construction procedures that include contraction joints, with waterstops making the upstream face watertight, sealing the interface between RCC layers, and draining and collecting the seepage water.

Upstream facing RCC cannot be compacted effectively against upstream forms. The operation generally leads to forming of surface voids [Fig. 19.5(a)]. An upstream facing [Fig. 19.5(b)] is required to produce a surface with good appearance and durability. Many facings incorporate a watertight barrier. Facings with barriers include the following:

1. Conventional formwork with a zone of conventional concrete placed between the forms and RCC material [Fig. 19.5(c)].
2. Precast concrete tieback panels with a flexible waterproof membrane placed between the RCC and the panels [Fig. 19.5(d)].

3. Slip-formed interlocking conventional concrete elements. RCC material is compacted against the cured conventional elements [Fig. 19.5(e)].
4. In certain cases, the upstream face can be provided with backfill as shown in Fig. 19.5(f)–(g).

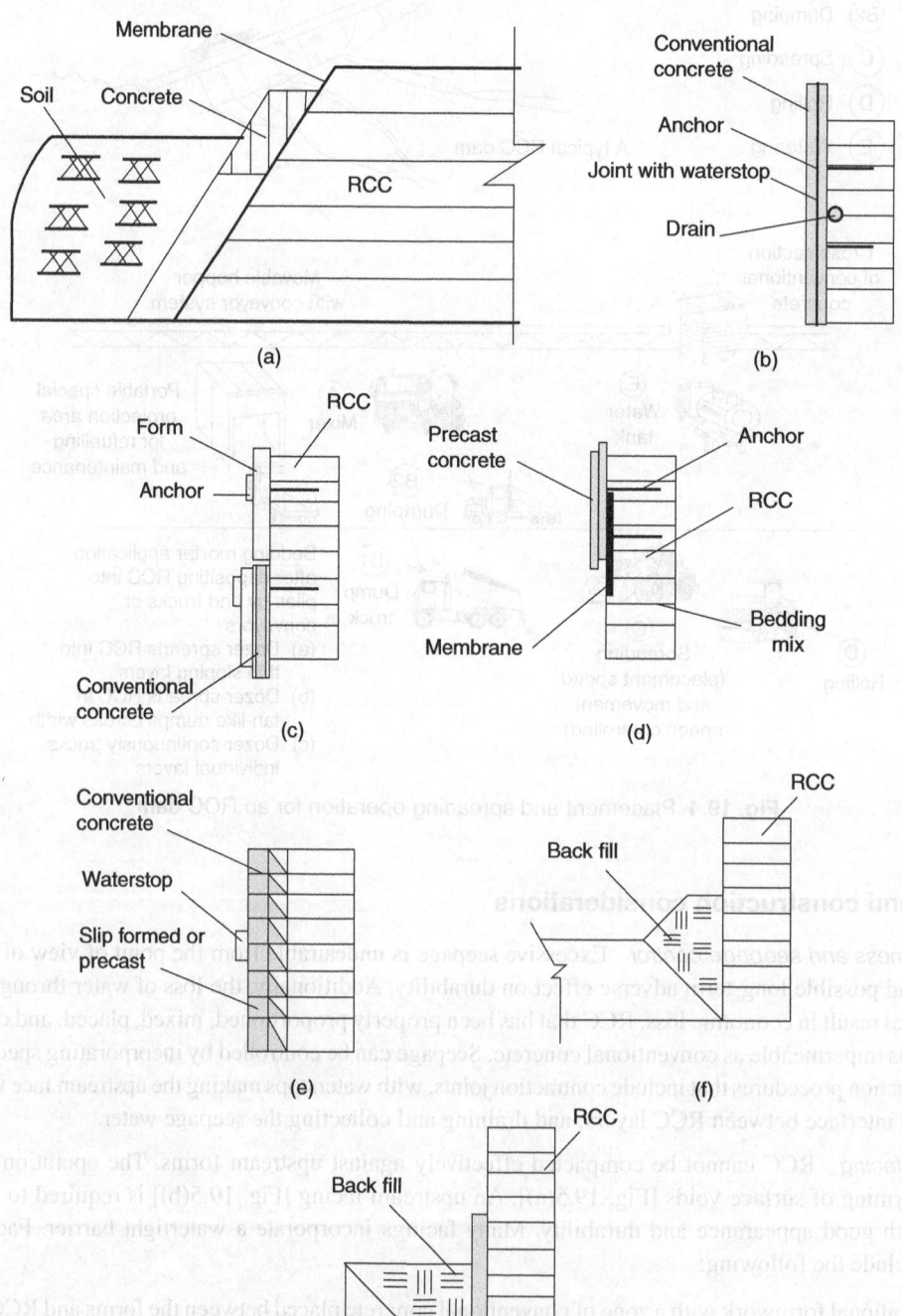

Fig. 19.5 Upstream face options

Horizontal joint treatment Bond strength and permeability are major concerns at the horizontal lift joints in RCC. Good sealing and bonding are accomplished by improving the compactibility of the RCC mixture, cleaning the joint surface, and placing a bedding mortar between lifts. Compactibility is improved by increasing the amount of mortar and fines in the RCC mixture. The lift surfaces should be properly moist cured and protected. Cleaning of the lift surfaces prior to RCC placement is not required as long as the surfaces are kept clean and free of excessive water. Addition of the bedding mortar serves to fill any voids or depressions left in the surface of the previous lift and squeezes into the voids at the bottom of the new RCC lift as it is compacted.

Seepage collection A collection and drainage system is a means for stopping unsightly seepage water from reaching the downstream face and for preventing excessive hydrostatic pressures. It also reduces uplift pressures within the dam and increases stability. Collection systems include vertical drains with waterstops at the upstream face and vertical drain holes drilled from within the gallery near the upstream or downstream face.

Downstream facing Downstream facing systems for non-overflow sections are generally required for aesthetic reasons, maintaining slopes steeper than the natural repose of RCC, and freeze–thaw protection in severe climate locations. Facing is necessary when the slope is steeper than 0.8*H* to 1.0*V*. Thicker lifts require a flatter slope. Figure 19.6 shows the downstream facing options. Downstream facing systems include the conventional vertical slip forming the placement and the horizontal slip forming similar to that used on the upstream face. When this type of slope is used, the structural cross section should include a slight overbuild to account for deterioration and unravelling of material loosened due to severe weather exposure of such structures over the years of service life.

Transverse contraction joints Transverse contraction joints are required in most RCC dams. The potential for cracking may be slightly lower in RCC because of the reduction in mixing water and reduced temperature rise resulting from the rapid placement rate and lower lift heights. In addition, the RCC characteristic of point-to-point aggregate contact decreases the volume shrinkage. Thermal cracking may, however, create a leakage path to the downstream face, which is aesthetically undesirable. Thermal studies should be performed to assess the need for contraction joints. Contraction joints may also be required to control cracking if the site configuration and foundation conditions may potentially restrain the dam. If properly designed and installed, contraction joints will not interfere or complicate the continuous placement operation of RCC.

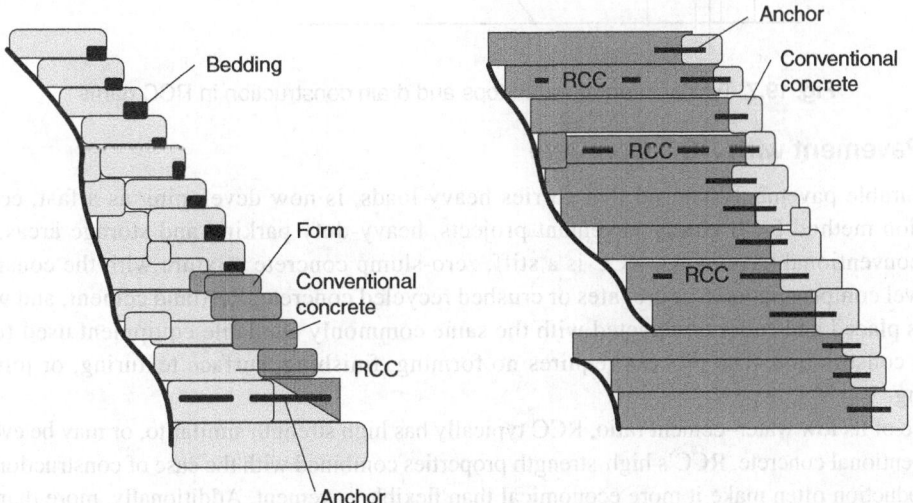

Fig. 19.6 Downstream face options

Waterstops   Standard waterstops may be installed in the internal zone of conventional concrete placed around the joint near the upstream face. Waterstops and joint drains are installed in the same manner as for conventional concrete dams. Typical internal waterstops and joint drain construction in RCC dams are shown in Fig. 19.7.

Curing

Curing should be started immediately following the completion of concreting. The purpose of curing is to retain moisture in concrete, which is needed for continued hydration of cement. To prevent loss of moisture, surfaces must be covered with waterproof paper, wet sand, or wet gunny bags. To obtain better curing, ponding or flooding of water can also be done; this is the most effective method of curing. Curing ideally should be done continuously for 28 days.

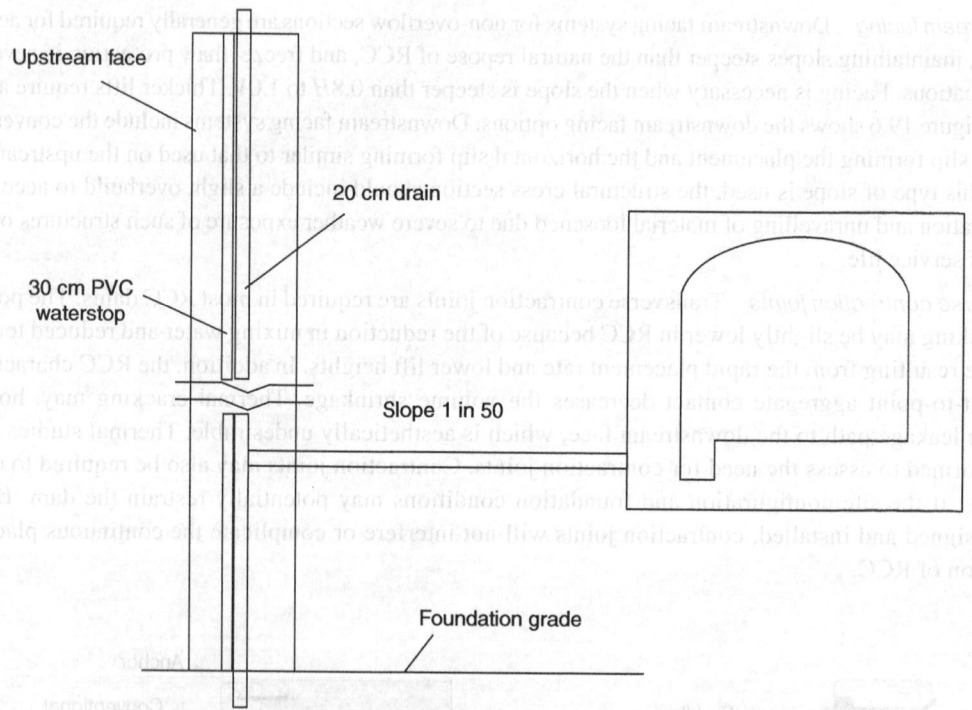

Fig. 19.7 Typical internal waterstops and drain construction in RCC dams

19.5.2 Pavement with RCC Concrete

RCC, a durable pavement material that carries heavy loads, is now developing as a fast, economical construction method for highway pavement projects, heavy-duty parking and storage areas, and as a base for conventional pavements. RCC is a stiff, zero-slump concrete mixture with the consistency of damp gravel comprising local aggregates or crushed recycled concrete, Portland cement, and water. The mixture is placed and roller compacted with the same commonly available equipment used for asphalt pavement construction. The process requires no forming, finishing, surface texturing, or joint sawing and sealing.

Because of its low water–cement ratio, RCC typically has high strength similar to, or may be even greater than, conventional concrete. RCC's high-strength properties combined with the ease of construction and high rate of production often make it more economical than flexible pavement. Additionally, more than 20 years of exposure as logging roads in cold climates in Canada has demonstrated that RCC has adequate resistance to freezing and thawing. But unlike conventional concrete, it has a drier mix stiff enough to be compacted by a vibrator roller. Typically, RCC is constructed without joints. It needs neither forms nor finishing, nor does it contain dowels or steel reinforcement. These qualities make RCC suitable for both specialized applications and mainstream pavement. Today, RCC is used for any type of industrial or heavy-duty pavement. The reason is that RCC has the strength and performance of conventional concrete with the economy and simplicity of asphalt. Coupled with long service life and minimal maintenance, RCC's low initial cost adds up to economy and value. The high strength of RCC pavements eliminates common and costly problems traditionally associated with asphalt pavements.

Advantages of RCC pavements

Constructing pavements using RCC has the following advantages:

- RCC pavements are rust-proof.
- They effectively span soft localized sub-grades.
- They do not deform under heavy, concentrated loads.
- They do not deteriorate from spills of fuel and hydraulic fluids.
- They do not soften under high temperatures.

Method of construction

RCC owes much of its economy to high-volume, high-speed construction methods. Large capacity mixers set the pace. Normally, RCC is blended in continuous mixing pug mills at or near the construction site. These high-output pug mills have the mixing efficiency needed to evenly disperse the small amount of water used (Fig. 19.8).

Dump trucks transport RCC and discharge it into an asphalt paver, which places the material in layers up to 250 mm thick and 6–12 m wide. Compaction is the next and important stage of construction. It provides density, strength, smoothness, and surface texture. Compaction is done immediately after placement and continuously until the pavement meets density requirement. Curing ensures a strong and durable pavement. As with any type of concrete, curing makes moisture available for hydration—the chemical reaction that causes concrete to harden and gain strength. Water cure sprays irrigate the pavement to keep it moist. A spray on membrane can also be used to seal moisture inside. Once cured, the pavement is ready for use. An asphalt surface is sometimes applied for greater smoothness or as a riding surface for high-speed traffic (Fig. 19.9).

When appearance is important, joints can be saw cut into RCC to control crack location. If economy outweighs appearance, RCC is allowed to crack naturally.

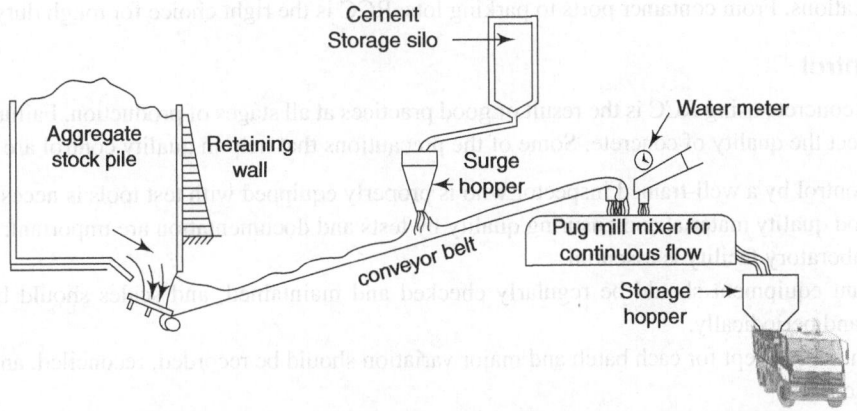

Fig. 19.8 RCC plant layout for pug mill operation (*Source*: PCA 1987)

(a)

(b)

Fig. 19.9 RCC pavement: (a) construction (b) finished surface
(*Source*: PCA 2000)

Economy and performance

For RCC, economy has been the mother of invention. The need for low-cost, high-volume material for industrial pavement led to its development. Low cost continues to draw engineers, owners, and construction managers to RCC. But present-day RCC owes much of its appeal to performance. It has the strength to withstand heavy and specialized loads, the durability to resist freeze–thaw damages, and the versatility to take on a wide variety of paving applications. From container ports to parking lots, RCC is the right choice for tough duty pavements.

Quality control

A good mass concrete using RCC is the result of good practices at all stages of production. Failure of any one stage will affect the quality of concrete. Some of the precautions that help in quality control are given next.

- Quality control by a well-trained inspector who is properly equipped with test tools is necessary.
- Using good quality material and ensuring quality by tests and documentation are important.
- On-site laboratory facility is essential.
- Batch plant equipment should be regularly checked and maintained, and scales should be calibrated properly and periodically.
- Record should be kept for each batch and major variation should be recorded, reconciled, and avoided in future batches.
- Proper placing, compaction, and curing should be done.

Exercises

Review Questions

1. What are the properties of materials used for mass concrete?
2. Explain the importance of temperature effect in mass concrete work.
3. Explain the role of air entrainment for mass concrete with respect to strength and durability of concrete.
4. How is temperature rise controlled in mass concrete?
5. What is roller-compacted concrete?
6. Discuss the economic advantages of roller-compacted concrete.
7. Explain the design considerations of roller-compacted concrete.
8. Explain the importance of curing.
9. What should be done to ensure good quality of mass concrete?
10. What are the advantages of RCC pavements?

Multiple Choice Questions

1. Concrete is defined as mass concrete if
   a. Chloride problem is addressed
   b. Sulphate problem is addressed
   c. Cracking problem is addressed
   d. Heat of hydration problem is addressed

2. The appropriate water–cement ratio used for mass concrete is in the range of
   a. 0.5 to 0.55
   b. 0.3 to 0.35
   c. 0.40 to 0.45
   d. 0.25 to 0.30

3. The temperature rise in mass concrete normally permitted within a block is
   a. 40°C
   b. 30°C
   c. 35°C
   d. 20°C

4. Using roller-compacted concrete for dam construction leads to
   a. Less cement content
   b. Rapid construction
   c. Reduced cost
   d. All of the above

5. The use of roller-compacted concrete for pavements is preferred because
   a. Concrete does not deform under heavy concentrated loads
   b. Concrete does not deteriorate under oil spillage
   c. Concrete does not soften under increase in temperature
   d. All of the above

6. Mass concrete is used for
   a. Raft slabs
   b. Large bridge piers
   c. Pavements and dams
   d. All of the above

7. Mass concrete consists of
   a. Higher proportion of coarse aggregates with less cement content
   b. Higher proportion of coarse aggregates with higher cement content
   c. Less proportion of coarse aggregates with less cement content
   d. Less proportion of coarse aggregate with higher cement content

8. The heat of hydration causes
   a. Volume change in concrete
   b. Rise in temperature
   c. Thermal cracking
   d. All of the above

9. The method of controlling heat of hydration in mass concrete is
   a. To organize concreting work at night when the temperature is lower
   b. To store the aggregates under shade or cool them by adding moisture
   c. To add ice or chilled water instead of water at room temperature
   d. All of the above

10. It is preferable to use the following type of cement in mass concrete works
    a. Ordinary Portland cement
    b. Rapid hardening cement
    c. Slag cement
    d. Low heat cement

Answers to Multiple Choice Questions

1. d	4. d	7. a	10. d
2. a	5. d	8. d	
3. d	6. d	9. d	

CHAPTER 20

Formwork

Formwork comprises self-supporting structures that give shape and geometrical dimensions to the shapeless fresh concrete. It takes the load of wet concrete as well as other loads caused by construction activities. Formwork plays an important role in safety, quality, and cost of any reinforced concrete construction.

Concrete when placed in a mould entraps air. In stiff or less workable concrete, the entrapped air could be 20%; in easily workable concrete, it could be 5%. It is essential to remove the entrapped air because of the following reasons:

a. Voids reduce the strength of concrete (even 1% of entrapped air reduces the concrete strength by 6%).
b. Voids increase permeability, which in turn reduces durability.
c. Voids reduce the bond between concrete and reinforcement.

Therefore, concrete should be properly compacted. To facilitate proper compaction, formwork should be planned and designed properly. Formwork should serve the following purposes:

a. Compacting exerts pressure, so formwork must be strong and stable. (Generally vibrators are used for compaction, which exerts significant pressure on formwork.)
b. The cement slurry tries to come out during compaction, so formwork should be leak-proof to retain the entire cement paste.

Good quality concrete can be produced by proper mix design, placing, and adequate compaction. We can achieve good compaction only if formwork is properly designed and installed. Thus formwork is the first step in ensuring good construction practice.

The rate of pour of ready-mix concrete is very high compared to site-mixed concrete. Consequently the load on formwork will be high. Therefore, when ready-mix concrete is used, it is very essential to have engineered formwork. When concrete is to be compacted by vibrators, the effect of vibration on formwork should be considered. Engineered formwork is essential for safety at site and for ensuring quality construction.

Generally, formwork construction is left to the carpenters. Carpenters acquire their skill from practice, experience, and by trial and error. Therefore, training for carpenters and engineering supervision in the process of formwork erection are essential.

Formwork should be designed in such a way that it does not cause any difficulty for bar benders. Many accidents take place in RCC construction because of faulty formwork. Formwork with adequate factor of safety will avoid such accidents. Because of its temporary nature, formwork does not get due importance either in design, making, installation, or removal. Neither the engineer nor the builder pays the attention it deserves. Defective and leaky formwork has been identified as one of the important causes for premature deterioration of many concrete structures.

20.1 Materials

Timber and steel are common materials used for formwork. The quality of concrete produced with steel formwork is better than that produced with timber forms. Mild steel pipes, struts, braces, and steel sheets are used for making steel formwork. These are commercially available in the market and generally satisfy the codal requirements. Therefore, quality of steel formwork for concrete construction can be easily maintained and their use is preferred at site.

Timber formwork is widely used in India. Timber used for formwork should satisfy IS: 3629-1966 requirements. Some of the requirements that the structural timber formwork should have are listed below:

a. The modulus of elasticity of a clear specimen of timber tested in bending should not be less than 5600 MPa.
b. Timber should be seasoned to have moisture content ranging from 12% to 20%.
c. Timber should not have defects such as knots and cracks.
d. Timber should be suitable in respect of durability and treatability.
e. The size of the timber member should be designed based on prescribed permissible stresses.

In order to obtain a smooth concrete surface, boards are made as panels with tongue-and-groove joint systems. There are possibilities for gaps when these boards are placed side by side and fixed. To avoid these gaps, skilled personnel should be employed in making these boards. If these gaps are very small, these can be repaired with thin reapers or with oil-based putty.

The thickness of timber sheeting should be such as to ensure a rigid, plain surface without bending and denting. For timber, the following minimum thicknesses are recommended even if design calculations indicate adequacy of a smaller value:

For walls and vertical sides of beams	25 mm
For bottom floors supporting normal loads	30 mm
For floor slabs for heavy construction loads	37 mm

Figure 20.1 shows the basic wood member arrangements for a floor system. Various components of formwork arrangement, such as shores, stringers, joists, scabs, and decking cleats, are shown in the figure.

Plywood can also be used for making formwork. A minimum thickness of 12 mm is recommended for plywood. Plywood is generally preferred for curved or non-planar formwork. Plywood of lesser thickness can also be adopted for accommodating the curvature.

Posts used to support formwork should have a minimum thickness that prevents them from buckling under load. If the vertical spacing between horizontal braces is 1.8 m, a minimum thickness of 75 mm will ensure an l/d ratio of 24. Since the end conditions of posts or struts are such that they are not restrained against rotation and lateral movement, the effective length of the posts should be twice the distance between the braces. A maximum slenderness ratio for long timber columns is 50 as per IS: 3629-1966. This means the l/d ratio of the post cannot exceed 25 under any circumstance. Even for light loads a minimum size of 100 mm is recommended for posts. As a corollary, it is to be noted that intermediate bracing is very essential to prevent the post from buckling for even a floor height of 3 m.

Plywood and timber used for sheeting should have optimum moisture content. If the moisture content is very low, water from concrete could be absorbed by plywood or timber and the layer of concrete near the shuttering surface will have insufficient water for hydration and for attaining the required strength. Sometimes shuttering is laid in position and left exposed to hot weather conditions and concreting is delayed unduly. Moistening the surface of formwork under such conditions is essential. Use of mould oil on the surface not only helps in easy stripping of forms but also prevents undue drying of the surface of forms. It also ensures preservation of moisture content in the wet concrete.

The material used for sheeting should not deteriorate on moisture absorption. Generally, plywood and timber boards crack, spall, or become brittle due to repeated cycles of wetting and drying and the wood fibres break and lose their strength. Therefore, shuttering should be stored with proper care for subsequent reuse.

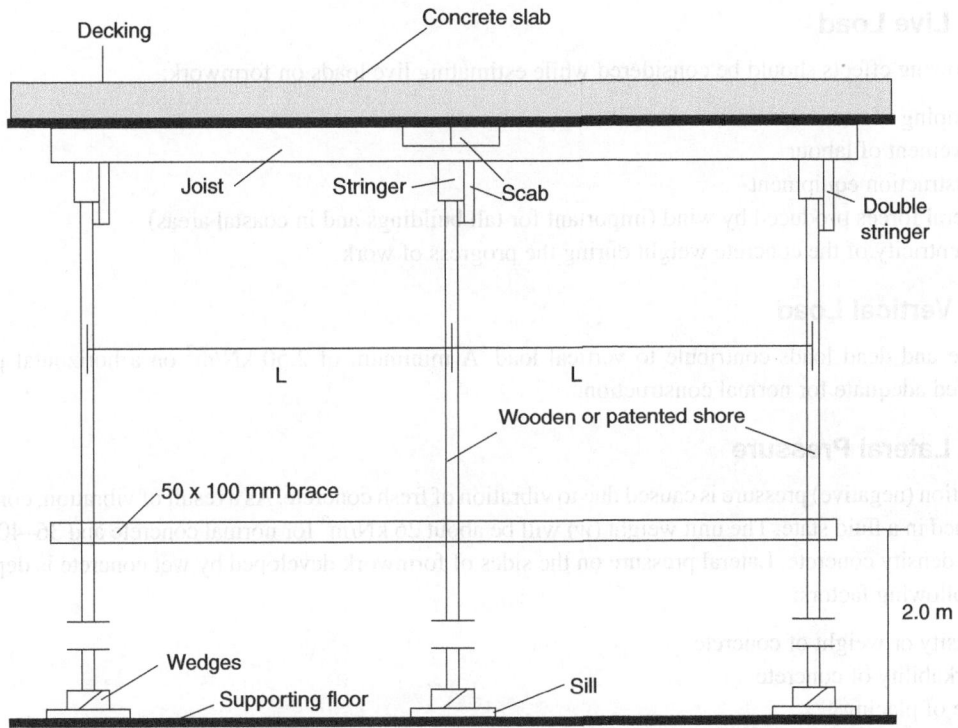

Forms for a flat slab concrete floor

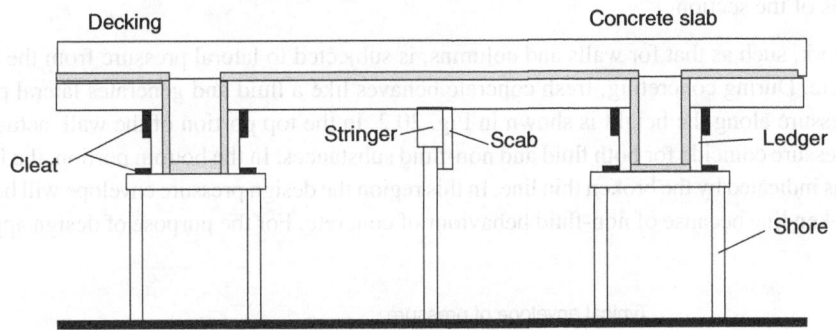

Forms for concrete beams and slab with intermediate stringers

Fig. 20.1 Formwork elements for a typical concrete floor

20.2 Forces on Formwork

Shuttering or formwork plays an essential role in concrete work. Many accidents take place because of faulty formwork. A major cause for failure of structures during construction is insufficient load-bearing capacity of the formwork. Formwork for concreting should support all vertical and horizontal loads. We will discuss these loads next.

20.2.1 Dead Load

The dead load includes the weight of fresh concrete and reinforcing steel. The unit weight of fresh concrete including steel is 26 kN/m^3. The self-weight of form is a small fraction. Generally 8–10% of concrete is considered as the weight of formwork.

20.2.2 Live Load

The following effects should be considered while estimating live loads on formwork:

a. Dumping of concrete
b. Movement of labour
c. Construction equipment
d. Lateral forces produced by wind (important for tall buildings and in coastal areas)
e. Eccentricity of the concrete weight during the progress of work

20.2.3 Vertical Load

Both live and dead loads contribute to vertical load. A minimum of 2.50 kN/m^2 on a horizontal plane is considered adequate for normal construction.

20.2.4 Lateral Pressure

An agitation (negative) pressure is caused due to vibration of fresh concrete. As a result of vibration, concrete is maintained in a fluid state. The unit weight (w) will be about 26 kN/m^3 for normal concrete and 36–40 kN/m^3 for high-density concrete. Lateral pressure on the sides of formwork developed by wet concrete is dependent on the following factors:

a. Density or weight of concrete
b. Workability of concrete
c. Rate of placing
d. Temperature of concrete
e. Method of consolidating the concrete
f. Dimensions of the section

Vertical formwork, such as that for walls and columns, is subjected to lateral pressure from the accumulated depth of concrete. During concreting, fresh concrete behaves like a fluid and generates lateral pressure. The variation of pressure along the height is shown in Fig. 20.2. In the top portion of the wall, actual and design envelopes of pressure coincide for both fluid and non-fluid substances. In the bottom portion, the fluid pressure varies linearly as indicated by the broken thin line. In this region the design pressure envelope will be as indicated by the thick broken line because of non-fluid behaviour of concrete. For the purpose of design approximation,

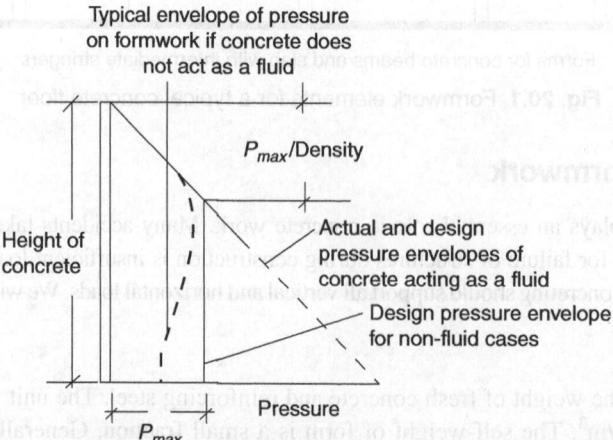

Fig. 20.2 Typical and assumed distributions of concrete lateral pressure

the variation in pressure can be taken as linear up to P_{max}/density, and below that level a constant. Thus, the maximum pressure is limited to P_{max} as shown.

With the above approximation, ACI committee 347 has developed formulae for calculation of lateral pressure on formwork for prescribed conditions of temperature, rate of placement, vibration, unit weight of concrete, and slump.

Column

$$P = C_w C_c \{7.2 + [785(R/T) + 17.8]\}$$

With a maximum of 150 $C_w C_c$ kN/m^2, a minimum of 30 C_w kN/m^2, but in no case greater than *wh*. Note that *wh* is the maximum liner pressure at the base such as the one developed by hydrostatic pressure.

Wall

$$P = C_w C_c [7.2 + (1156/T)17.8 + 244(R/T) + 17.8]$$

With a maximum of 100 $C_w C_c$ kN/m^2, a minimum of 30 C_w kN/m^2, but in no case greater than *wh*. Here

P = lateral pressure
h = depth of plastic concrete from the top of placement to the point under consideration (in m)
w = unit weight (in kN/m^3)
R = rate of displacement (in m/h)
T = temperature of concrete during placement
C_w = unit weight coefficient (as per Table 20.1)
C_c = chemistry coefficient (as per Table 20.1)

Construction Industry Research and Information Association (CIRIA) has developed a formula to find pressure exerted on formwork by wet concrete. This formula takes into account the dimensions of the cross section of members cast in addition to other parameters:

$$P_{max} = D \left[C_1 \sqrt{R} - C_2 K \sqrt{(H - C_1 \sqrt{R})} \right] \text{ or } Dh \text{ kN/m}^2, \text{ whichever is smaller.}$$

Here

P_{max} = maximum concrete pressure on formwork (kN/m^2)
C_1 = coefficient dependent on the size and shape of formwork
C_2 = coefficient dependent on the constituent material of concrete
D = weight density of concrete (kN/m^3)
H = vertical form height (m)
h = vertical pour height (m)

Table 20.1 C_c and C_w coefficients to be used in pressure equation

Type of blends used	Chemistry coefficient (C_c)	Unit weight coefficient (C_w)
Blends without retarders containing less than 70% fly ash	1.2	Concrete weighing less than 22.50 kN/mm^3 $C_w = (1 - w/23.2)$ but not less than 0.8
Blends with retarders containing less than 70% slag or 40% fly ash	1.4	Concrete weighing 22.50–24 kN/mm^3 [$C_w = 1$]
Blends containing more than 70% slag or 40% fly ash	1.4	Concrete weighing more than 24 kN/mm^3 [$C_w = w/23.2$]

K = temperature coefficient taken as $K = [36/(T + 16)]^2$

R = rate at which the concrete rises vertically up the form (m/h)

T = concrete temperature at placing (°C)

When $C_1\sqrt{R} > H$, fluid pressure Dh should be taken as the design pressure. Term $C_1\sqrt{R}$ incorporates the effect of vibration and workability, because these factors are largely dependent on size, shape, and rate of rise. Term $C_2K\sqrt{(H - C_1\sqrt{R})}$ incorporates the effects of the height of discharge, cement type, admixture, and concrete temperature.

The value of coefficient C_1 is taken as 1 for walls and 1.5 for columns. The value of the coefficient C_2 is given in Table 20.2. In some special cases, formwork can be inclined. In such cases the following formulae can be used.

When only lateral pressure acts,

$$P = wh \sin^2 \theta$$

When the weight of concrete also acts,

$$P = 2500t \cos\theta + wh \sin^2 \theta$$

where

P = lateral pressure

θ = angle of formwork to horizontal

h = vertical depth

w = pressure acting normal to the slope

t = thickness of slab

When designing formwork for retaining walls, earth pressure should be considered. Rise in water table can also increase pressure on formwork.

20.2.5 Lateral Load

Wind load falls into this category. In general, horizontal load (W) acting in any direction on any floor should not be less than 150 kg/linear metre or 2% of the total dead load (Fig. 20.3). While designing bracings for vertical structures

Table 20.2 Value of coefficient C_2

Concrete with	C_2
OPC, RHPC, or SRPC without admixture	0.3
OPC, RHPC, or SRPC with admixture, except retarder	0.3
OPC, RHPC, or SRPC with retarder	0.45
LHPBC, PBFC, PPFAC, or blends containing less than 70% GGBFS or 40% PFA without admixture	0.45
LHPBC, PBFC, PPFAC, or blends containing less than 70% GGBFS or 40% PFA with any admixture, except a retarder	0.45
LHPBC, PBFC, PPFAC, or blends containing less than 70% GGBFS or 40% PFA with a retarder	0.6
Blends containing more than 70% GGBFS or 40% PFA	0.6

Notes

OPC: Ordinary Portland cement

LHPBC: Low-heat Portland blast-furnace cement

PBFC: Portland blast-furnace cement

RHPC: Rapid hardening Portland cement

PPFAC: Portland pulverized fuel ash cement

SRPC: Sulphate-resisting Portland cement

GGBFS: Ground-granulated blast-furnace slag cement

PFA: Pulverized fuel ash

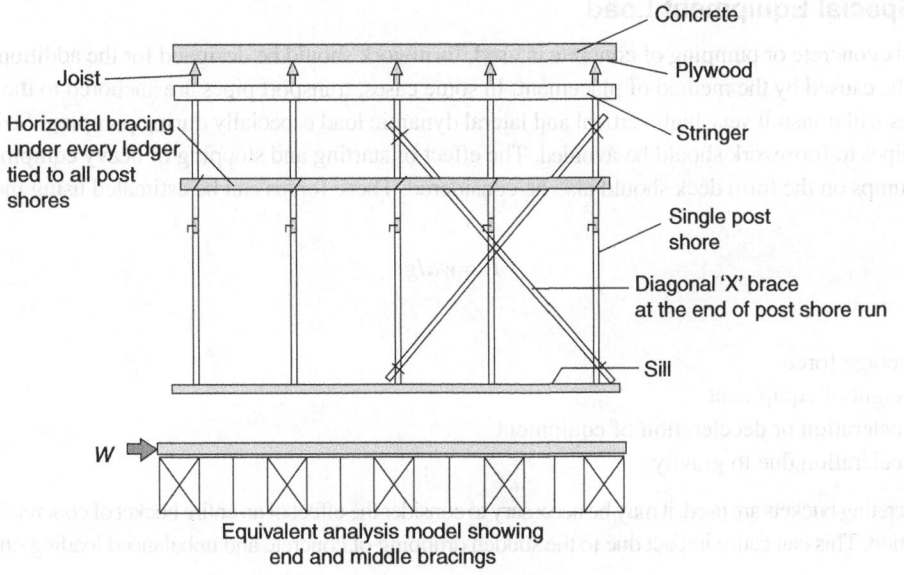

Concrete
Plywood
Joist
Stringer
Horizontal bracing under every ledger tied to all post shores
Single post shore
Diagonal 'X' brace at the end of post shore run
Sill

W

Equivalent analysis model showing end and middle bracings

Fig. 20.3 Schematic of bracings for slab formwork

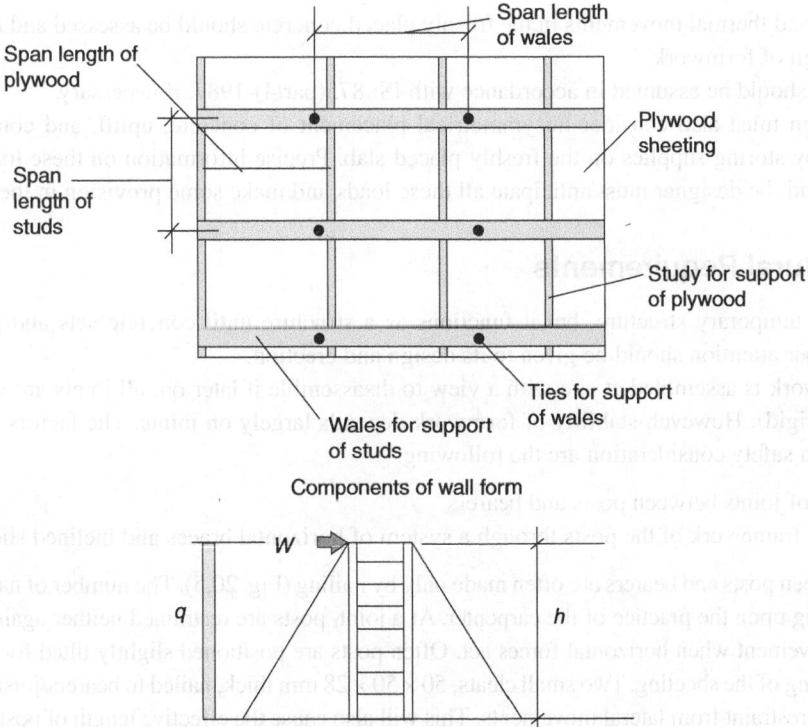

Span length of wales
Span length of plywood
Plywood sheeting
Span length of studs
Study for support of plywood
Ties for support of wales
Wales for support of studs

Components of wall form

W
q
h

Fig. 20.4 Schematic of wall formwork and its bracing

such as wall, wind load requirement should also be considered. Bracing for wall forms should be designed for a horizontal load (*W*) of at least 150 kg/linear metre of the wall at the floor edge or 2% of vertical load distributed uniformly at the slab edge or $qh/2$ (where q is the actual wind pressure) (Fig. 20.4), whichever is larger.

20.2.6 Special Equipment Load

If ready-mix concrete or pumping of concrete is used, formwork should be designed for the additional pressure that could be caused by the method of placement. In some cases, transport pipes are anchored to the formwork. These pipes will transmit very high vertical and lateral dynamic load especially during pumping. So connecting transport pipes to formwork should be avoided. The effect of starting and stopping of heavy equipment such as concrete pumps on the form deck should also be considered. These forces can be estimated using the following formula:

$$F = wa/g$$

where

F = average force
w = weight of equipment
a = acceleration or deceleration of equipment
g = acceleration due to gravity

If large concreting buckets are used, it may be necessary to consider the effect of an entire bucket of concrete in one place for distribution. This can cause impact due to the sudden dropping of concrete and unbalanced loading on formwork.

20.2.7 Other Loads

In addition to the already discussed loads, the formwork must also be able to withstand the following loads.

1. Shrinkage and thermal movements in the freshly placed concrete should be assessed and accommodated in the design of formwork.
2. Snow load should be assumed in accordance with IS: 875(part4)-1987, if necessary.
3. Form design must also consider unsymmetrical placement of concrete, uplift, and concentrated load produced by storing supplies on the freshly placed slab. Precise information on these loads will not be available and the designer must anticipate all these loads and make some provision in the design.

20.3 Structural Requirements

Formwork is a temporary structure, but it functions as a structure until concrete sets and gains strength. Therefore, proper attention should be given to its design and erection.

Since formwork is assembled at site with a view to disassemble it later on, all joints are very temporary (joints are not rigid). However, stability of formwork depends largely on joints. The factors to be specially considered from safety consideration are the following:

a. Flexibility of joints between posts and bearers
b. Stability of framework of the posts through a system of horizontal braces and inclined shores

The joints between posts and bearers are often made only by nailing (Fig. 20.5). The number of nails ranges from 2 to 4, depending upon the practice of the carpenter. At a joint, posts are restrained neither against rotation nor from lateral movement when horizontal forces act. Often posts are positioned slightly tilted for the purpose of adjusting levelling of the sheeting. Two small cleats, 50 × 50 × 28 mm thick, nailed to bearers/joists on either side of a post ensure restraint from lateral movements. This will also cause the effective length of posts to be smaller, thus contributing to structural stability. Gross movement is prevented when accidental knocking of post occurs.

Joints between the beam bottom and posts need special attention (Fig. 20.6). In this case, the beam bottom does not rest directly on the post, but rests on a cross member which is first nailed to the post. The cross bearer is nailed on either side of the post so as to reduce the effect of constructional eccentricity. Figure 20.7 shows the spacing of shores for a typical beam and slab floor.

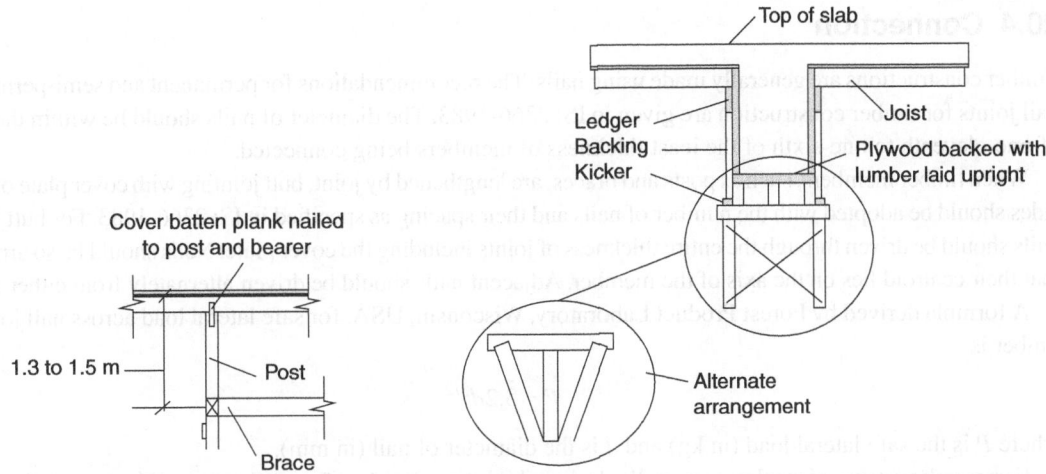

Fig. 20.5 Location of horizontal brace in a floor   **Fig. 20.6** Alternate arrangements for beam bottom support

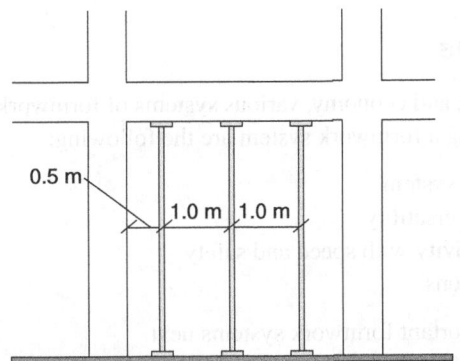

Fig. 20.7 Spacing of shores for a concrete frame

It is absolutely necessary that posts are connected to the adjacent ones by a system of horizontal braces. Figures 20.1 and 20.5 show the position of the brace. The braces are fixed at 1.3–1.5 m below the bearer of sheeting which is slightly above the mid-height of posts. The braces also serve as a formwork on which a working platform can be made for enabling the form-stripping operation. Bracing in both directions is essential to connect all vertical posts. This will prevent buckling of the post in X and Y directions in the horizontal plane. Figure 20.1 shows the location of brace just above the floor.

Diagonal braces, inclined both ways, also need to be provided at least in one bay, along a row of posts constituting several bays. The appropriate location will be between the top of the post and the mid-height or bottom in an end bay (Fig. 20.3). These inclined braces will be the most crucial members to ensure stability during unforeseen severe wind gusts or when excessive differential settlement takes place under the posts.

Inclined struts should be used along with high wall and tall column formwork in order to ensure stability and verticality. At least two struts on either side of the formwork should be provided so as to ensure a stable configuration (Fig. 20.4). Struts will usually be solid timber sections connected to the bed plate so as to distribute their pressure evenly to the soil on which they are embedded. For column formwork, inclined struts on all four sides are installed to ensure stability.

20.4 Connection

Timber constructions are generally made using nails. The recommendations for permanent and semi-permanent nail joints for timber construction are given in IS: 2366-1983. The diameter of nails should be within the limit of one-eleventh to one-sixth of the least thickness of members being connected.

When timber members, such as posts and braces, are lengthened by joint, butt jointing with cover plate on both sides should be adopted with the number of nails and their spacing as specified in IS: 2366-1983. For butt joints, nails should be driven through the entire thickness of joints including the cover plate. Nails should be so arranged that their centroid lies on the axis of the member. Adjacent nails should be driven alternately from either face.

A formula derived by Forest Product Laboratory, Wisconsin, USA, for safe lateral load across nail jointing timber is

$$P = 3.2d^{3/2}$$

where P is the safe lateral load (in kg) and d is the diameter of nail (in mm).

Formwork systems of steel use specially designed jointing couplers for connecting to the braces, for transferring loads to the ground, for floor lengthening, for inserting at an intermediate brace, etc. Instructions given by the manufacturer of such systems should be strictly followed to ensure proper connections.

20.5 Formwork Systems

In order to ensure safety, quality, and economy, various systems of formwork have been developed. Some of the considerations while selecting a formwork system are the following:

- Simplicity in assembling the system
- Flexibility, ease of use, and versatility
- Labour and material productivity with speed and safety
- Suitability for most applications

We will discuss some of the important formwork systems next.

20.5.1 Flex System

This system is adopted for slabs and beams up to a floor height of 4.4 m. The components used are very light and easy to erect and dismantle. This system is flexible and can be used for various heights and spacings. This system enables re-propping and frequent reuse of materials (Fig. 20.8).

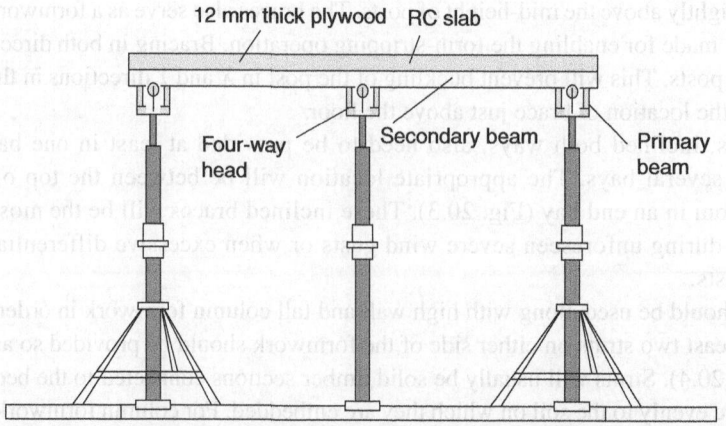

Fig. 20.8 Flex system

20.5.2 Beam-forming System

A major advantage of beam-forming system is that it eliminates the use of timber for beam sides and beam bottom. This system is suitable for beams with depth of 300 mm onwards. It is easy to align, fix, and remove. Such formwork systems have adjustable depth in steps of 10 mm increment and ensure right angle for side formwork with respect to beam bottom (Fig. 20.9).

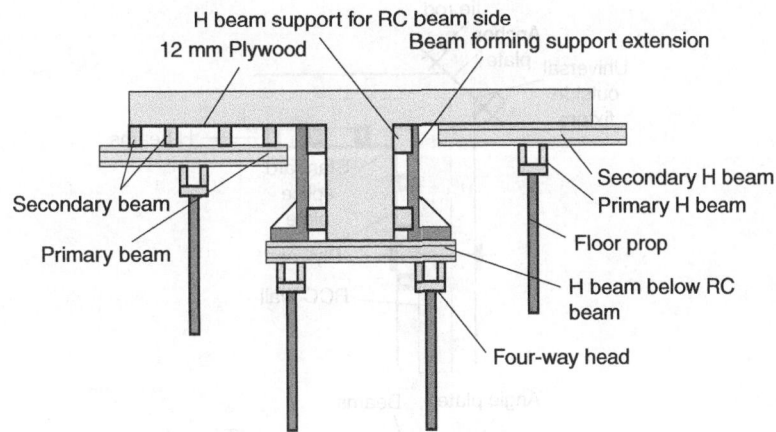

Fig. 20.9 Beam-forming system

20.5.3 Heavy Duty Tower System

This system, shown in Fig. 20.10, can be used up to a height of 100 m with a load-carrying capacity of 25 tonnes. It is rigid and stable with minimum bracings. This system is easy to handle, fix, and remove with unskilled labour. A tower as a whole can be shifted from one place to another using crane or transport devices. This system has good flexibility in height adjustment and can also be converted into stair tower with additional standard components. This system is ideal for heavy floors involving considerable heights.

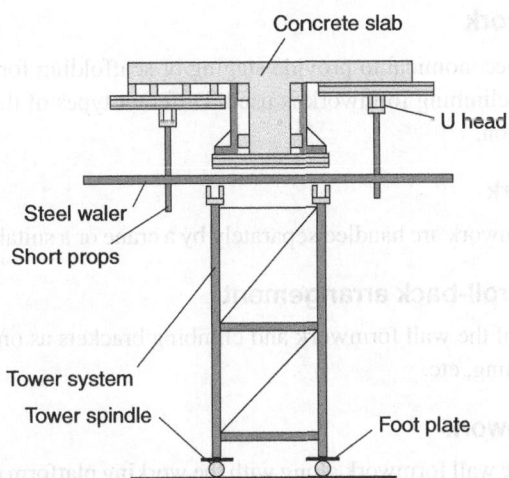

Fig. 20.10 Heavy-duty tower system

20.5.4 Wall/Column Formwork System

The advantage of this formwork is that it minimizes the number of sheeting and ensures excellent concrete finish. Large area panels are used as single units. This is an ideal system for speedy construction (50 uses in 6 months). This is more suitable when high-strength concrete is used. This system is designed to take pressure up to 35 kN for walls and up to 90 kN for columns (Fig. 20.11).

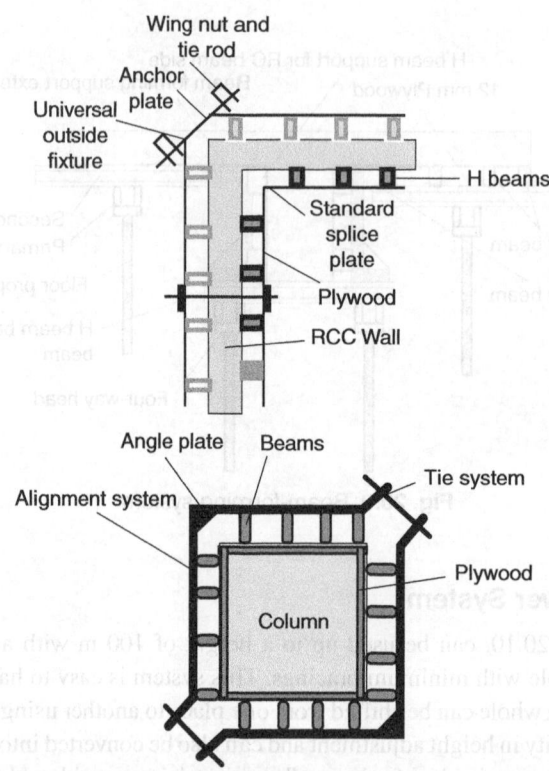

Fig. 20.11 Wall/column formwork system

20.5.5 Climbing Formwork

For tall structures it is very uneconomical to provide staging or scaffolding for supporting the external wall formwork. In such a case, the climbing formwork is used. Different types of the climbing formwork system used are discussed in this section.

Simple climbing formwork

Here the brackets and wall formwork are handled separately by a crane or a suitable lifting device (Fig. 20.12).

Climbing formwork with roll-back arrangement

This system facilitates lifting of the wall formwork and climbing brackets as one unit and allows rollback of shutters for deshuttering, cleaning, etc.

Automatic climbing formwork

As the name suggests, this is the wall formwork along with the working platform climbing along the wall using an electric motor/hydraulic pump, thus eliminating the need for a crane to handle the wall panels and brackets (Fig. 20.13). This system is very often used for very tall structures such as chimneys and cooling towers.

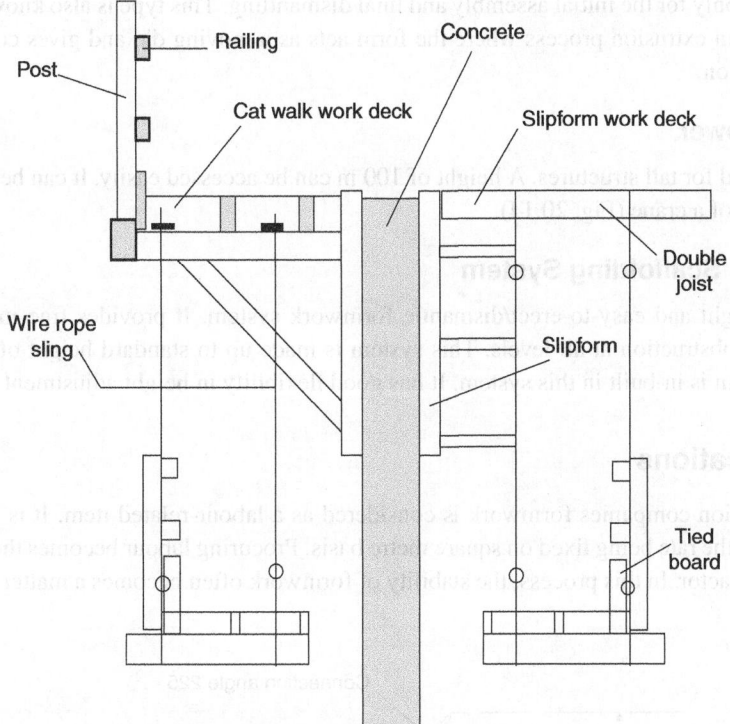

Fig. 20.12 Simple climbing form

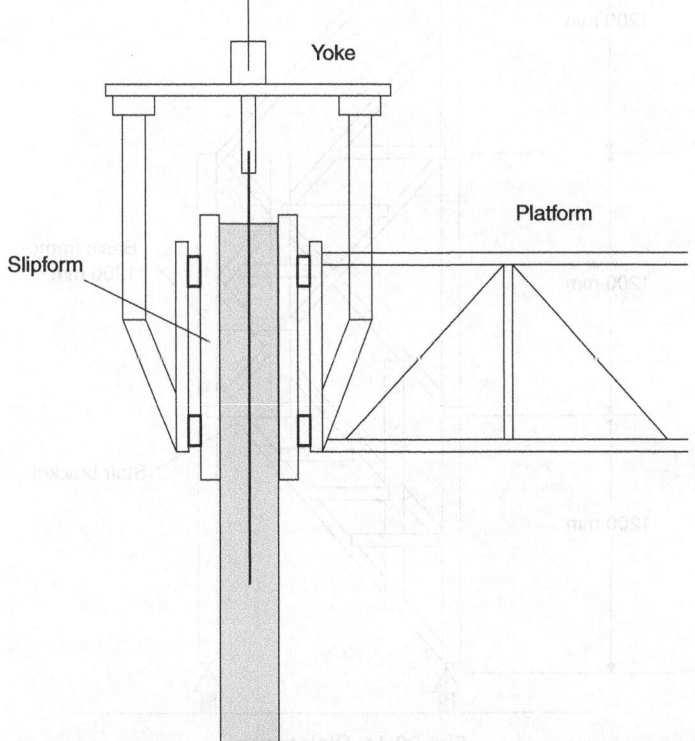

Fig. 20.13 Automatic climbing slipform

Crane is required only for the initial assembly and final dismantling. This type is also known as *slipform*. This formwork is like an extrusion process where the form acts as a moving die and gives concrete the required shape and dimension.

20.5.6 Stair Tower

This system is used for tall structures. A height of 100 m can be accessed easily. It can be handled as a single unit with the help of a crane (Fig. 20.14).

20.5.7 Access Scaffolding System

This is a lightweight and easy-to-erect/dismantle formwork system. It provides free movement for workmen without any obstruction at all levels. This system is made up to standard height of 0.9, 1.2, or 1.8 m. A working platform is in-built in this system. It has good flexibility in height adjustment (Fig. 20.15).

20.6 Specifications

In many construction companies formwork is considered as a labour-related item. It is sub-contracted to a labour contractor, the rate being fixed on square metre basis. Procuring labour becomes the dominant concern of the labour contractor. In this process, the stability of formwork often becomes a matter of good luck rather

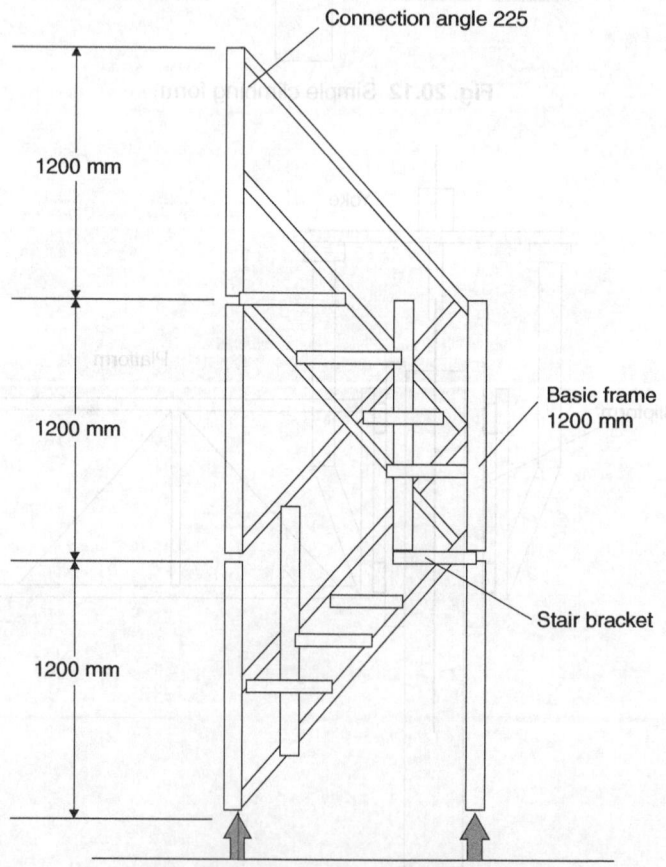

Fig. 20.14 Stair tower

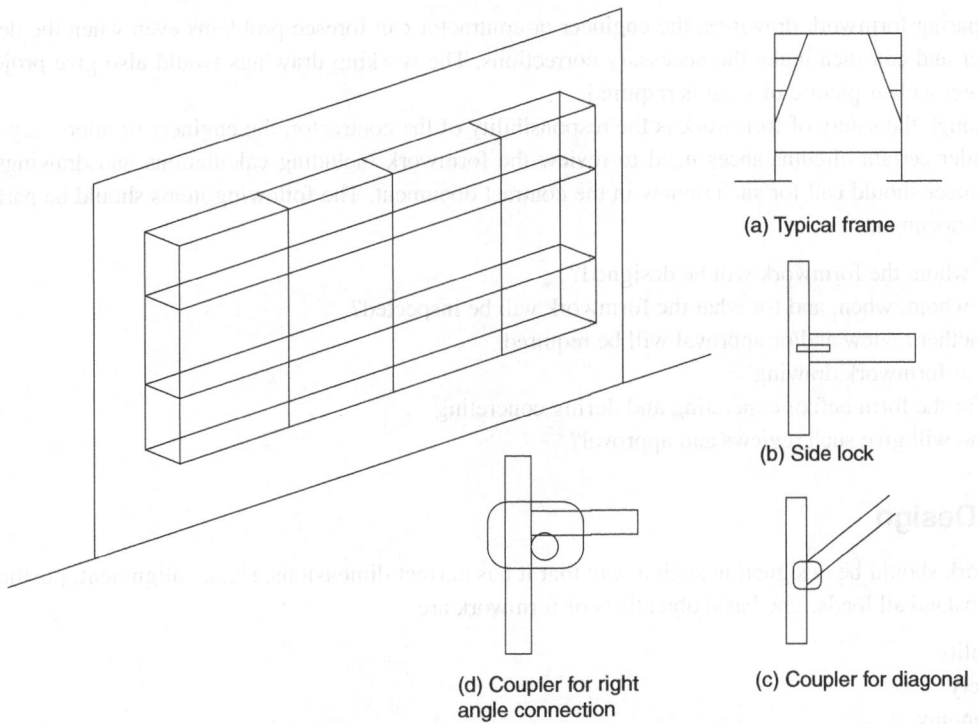

(a) Typical frame

(b) Side lock

(c) Coupler for diagonal

(d) Coupler for right angle connection

Fig. 20.15 Scaffolding system with connections

than a result of good engineering practice. It should not be argued that formwork needs engineer's attention only for a special or high-value structure. Formwork stability even for a normal building is a matter of great importance. It should be ensured that the formwork contractor uses the appropriate material optimally. A proper specification should be prepared before entrusting the formwork to a qualified contractor.

Specifications should not impose on the contractor to use specific material unless a particular design requires such material. The contractor should normally be free to choose the material. But the material chosen by him should meet the formwork requirements. When the contractor chooses any standard material, he should comply with the manufacturer's stipulations regarding size, spacing, etc. Tolerance specified for the dimensions of structural elements and deviations from lines and levels should be met by the formwork construction. There should be no additional tolerance other than that given for the dimensions of structural elements and their deviations from line and levels.

Safety and construction of formwork is the responsibility of the builder or contractor. It is necessary that formwork design and drawings are reviewed and approved. Contract documents should include all information necessary for the contractor to design the formwork and prepare its drawings. These would include the following:

a. Live load of wet concrete and superimposed load for which the formwork is designed.
b. Sequence of concreting in special structures such as shells, domes, folded plates, etc.
c. Location and detail of construction and expansion joints.
d. The necessity of reshoring for structures such as shells, domes, folded plates, and long-span structures.
e. Provision essential for formwork meant for special construction methods.
f. For post-tensioned concrete members, the effect of load transfer during tensioning and special precautions to be taken for formwork design and erection.
g. Requirement of embedment, inserts, built-in frames for opening and holes through concrete necessary for effective planning of formwork and its erection.

By preparing formwork drawings, the engineer or contractor can foresee problems even when the design is on paper and can then make the necessary corrections. The working drawings would also give project site employees a clear picture of what is required.

Although the safety of formwork is the responsibility of the contractor, the engineer or approving agency may under certain circumstances need to review the formwork including calculations and drawings. If so the engineer should call for such review in the contract document. The following items should be part of the contract document:

a. By whom the formwork will be designed?
b. By whom, when, and for what the formwork will be inspected?
c. Whether review and/or approval will be required
   - for formwork drawing
   - for the form before concreting and during concreting
d. Who will give such reviews and approval?

20.7 Design

Formwork should be designed in such a way that it has correct dimensions, shape, alignment, position, and can withstand all loads. The basic objectives of formwork are:

- Quality
- Safety
- Economy

To achieve these objectives, the following aspects should be considered in the formwork design:

a. Correct assessment of load on formwork
b. Selection of proper material considering its strength, stability, and cost
c. Selection of proper supporting system
d. Effect of vibration on formwork

Formwork for concrete must support all vertical and lateral loads that may be applied until such time as these loads can be carried by the hardened concrete structure itself. Various loads and forces that can act on formwork have been discussed already. The detailed design of formwork should take into account the following:

a. Position of construction joints
b. Pour size of concrete
c. Reuse of form material
d. Clear span (or spans) with multiple supports in between
e. Ease of stripping and re-propping of formwork
f. Need for back propping and re-propping
g. Supporting material

20.7.1 Assumptions for Formwork Design

In order to make the design of formwork simple, some basic simplifying assumptions are made. These are as follows:

a. All loads are assumed to be uniformly distributed.
b. Beam supports over three or more spans are regarded as continuous.
c. Strength of nails is neglected in determining the size of the main formwork. This does not apply when considering splices, braces, and brackets.

20.7.2 Permissible Stresses

Working stress of material should be based on applicable codes. Some of the common materials used for formwork and the Bureau of Indian Standard codes to be followed for the same are given in Table 20.3.

Table 20.3 BIS codes to be followed for various materials

Material	Code
Timber	IS: 399-1963, IS: 833-1994
Plywood	IS: 4990-1989
Steel	IS: 800-2007, IS: 2750-1964
Tubular section	IS: 806-1968
Brickwork/stone	IS: 1905-1987, IS: 1597(part1)-1967, IS: 2212-1981
Concrete	IS: 456-2000

20.7.3 Deflection

Formwork must be so designed that the various parts do not deflect beyond the permissible limit. The permissible deflection depends on the desired finish as well as location. The acceptable and frequently used values for permissible deflections are given below.

For sheeting:	1.6 mm
For members spanning (up to 1.50 m):	3.0 mm

(Note that the span is not for the final structure but for the formwork.)

20.7.4 Analysis

For proper design of formwork, analysis should be done first. It should incorporate the following aspects:

a. Design load of the slab or structural member including live, partition, and other loads. Wherever a reduced live load has been taken into account for the design of a certain member and allowance has been made for construction loads as well, these should be shown on the structural drawing and taken into consideration in performing the calculations

b. Dead weight of concrete and formwork

c. Specified design strength of concrete

d. Strength of concrete at the time it is required to support shoring loads from above

e. Cycle time between the placements of each floor

f. Span of the slab or structural member between permanent supports

g. Type of formwork system, i.e., span of horizontal formwork components and load distribution among props

h. Age at which required maturity is attained. Stresses in the material should be limited to the following values

The maturity depends on bending and shear stresses in the material.

Bending stress f_b in beams:

$$f_b = M/s$$

where M is the bending moment and s is the section modulus.

Shear stress f_v in solid rectangular beams:

$$f_v = 1.5V/(bd)$$

where V is the shear force.

Shear stress f_v in thin sheets such as plywood:

$$f_v = VQ/(Ib) = V/(Ib/Q)$$

where Ib/Q is the rolling shear constant.

Bearing f_{cl} or axial compression f_c:

$$f_{cl} \text{ or } f_c = P/A$$

When timber or lumber is used, the allowable stresses are adjusted for the following variables as shown in Table 20.4. The various adjustment factors for stresses are computed using the following:

$$\{k_{adj}\}_{bending} = C_D \times C_M \times C_L \times C_F \times C_{fu} \times C_T \times C_t \times C_v \times C_c \times C_f$$

$$\{k_{adj}\}_{shear} = C_D \times C_M \times C_H \times C_T$$

$$\{k_{adj}\}_{bearing} = C_M \times C_b \times C_t$$

$$\{k_{adj}\}_{compression} = C_D \times C_M \times C_F \times C_P \times C_t$$

$$\{k_{adj}\}_{elastic\ modulus} = C_M \times C_T \times C_t$$

Table 20.4 Adjustment factors for formwork design

Item	Distribution factor	Factor for which allowance is made
C_D	Load factor	Allows for distribution of load based on stiffness
C_T	Time-dependent creep factor	Allows creep effect
C_H	Shape factor	Allows for end cracking
C_F	Size factor	Allows size effect
C_{fu}	Flat use factor	Allows loading along wide face
C_L	Beam stability factor	Based on d/b
C_P	Column stability factor	Based on column buckling susceptibility
C_b	Bearing area factor	Bearing perpendicular to the grain
C_M	Moisture factor	Based on moisture content
C_t	Temperature factor	Based on ambient temperature
C_v	Volume factor	Based on the volume of member
C_c	Curvature factor	Allowance for bent members
C_f	Form factor	Allows for different shapes

20.7.5 Design of Beam Forms

Design of beam form (Fig. 20.16) depends on the type of load it is to be subjected to and is generally dependent on bending moment, shear force, and deflection. There are various load conditions and some of them are listed in Table 20.5.

20.7.6 Design of Column Forms

When designing a column form (Fig. 20.17), care must be taken in calculating stresses and allowable load because many failures have resulted due to inadequate shoring and bracing of column formwork. The maximum load that a vertical shore can safely support depends on the following factors:

a. Slenderness ratio
b. Allowable stress in compression parallel to grain
c. Net area of shore cross section

The safe allowable compression load on the column can be arrived at based on the following steps:

Step 1 Calculation of compression stress

$$F_c = F_p \left\{ \left[1 + (F_{ce}/F_p) \right]/2 - \sqrt{\left\{ \left[1 + (F_{ce}/F_p) \right]/2 \right\}^2 - (F_{ce}/F_p)/c} \right\}$$

where

F_c = allowable compression stress in the column
F_p = allowable compression stress parallel to grain
F_{ce} = impact of Euler buckling, $0.3E/(l/d)^2$
E = Young's modulus
l/d = slenderness ratio
c = a factor for the type of lumber—0.8 for sawn lumber

Step 2 Calculation of allowable compression load

$$P_c = F_c A$$

where

P_c = allowable compression load
F_c = allowable compression stress
A = net cross-sectional area

Note that in addition to the above, appropriate adjustment factors are used.

20.7.7 Design of Slab Forms

The design of slab form (Fig. 20.18) includes design of decking, joists, and stringer. The design involves the following steps:

Step 1 Calculation of loads

Loads that will be applied on the slab form should be calculated. Various loads that act on the form have been discussed earlier.

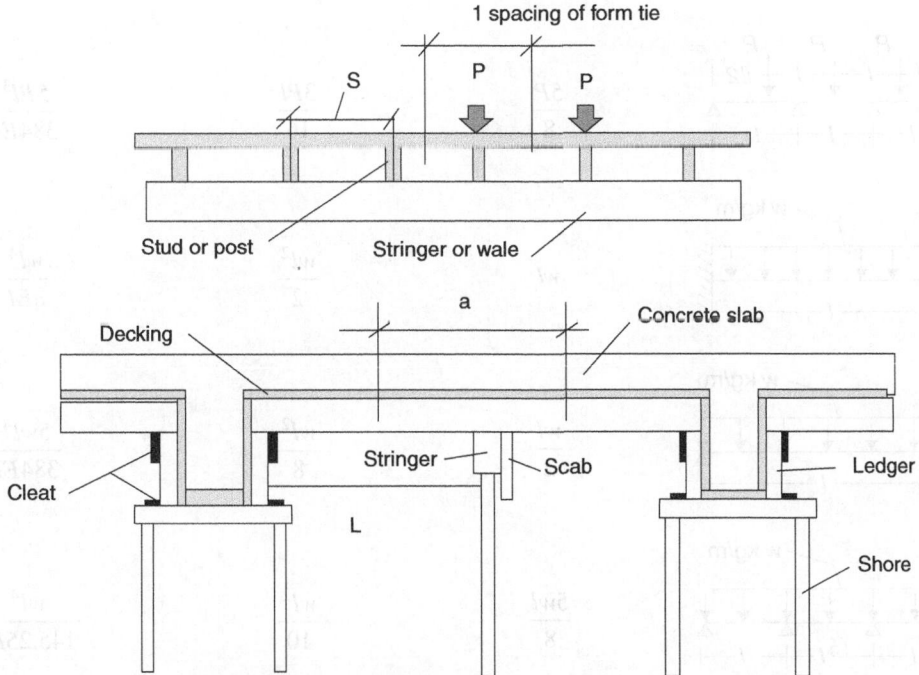

Fig. 20.16 Beam forms and its components and load modelling

Table 20.5 Maximum bending moment, shear, and deflection for beams

No.	Load diagram P (kg)	Shear force V (kg)	Bending moment M (kg m)	Deflection Δ (m)
1.		P	Pl	$\dfrac{Pl^3}{3EI}$
2.		$\dfrac{P}{2}$	$\dfrac{Pl}{4}$	$\dfrac{Pl^3}{48EI}$
3.		$\dfrac{Pa}{l}$	$\dfrac{Pab}{l}$	$\dfrac{Pa^2b^2}{3EIl}$
4.		P	Pa	$\dfrac{Pa}{24EI}\left[3l^2 - 4a^2\right]$
5.		$\dfrac{3P}{2}$	$P\left[\dfrac{l}{4}+a\right]$	$\dfrac{P}{48EI}\left[l^3 + 6al^2 - 8a^3\right]$
6.		$\dfrac{5P}{8}$	$\dfrac{3Pl}{16}$	$\dfrac{5Pl^3}{384EI}$
7.		wl	$\dfrac{wl^2}{2}$	$\dfrac{wl^4}{8EI}$
8.		$\dfrac{wl}{2}$	$\dfrac{wl^2}{8}$	$\dfrac{5wl^4}{384EI}$
9.		$\dfrac{5wl}{8}$	$\dfrac{wl^2}{10}$	$\dfrac{wl^4}{145.25EI}$

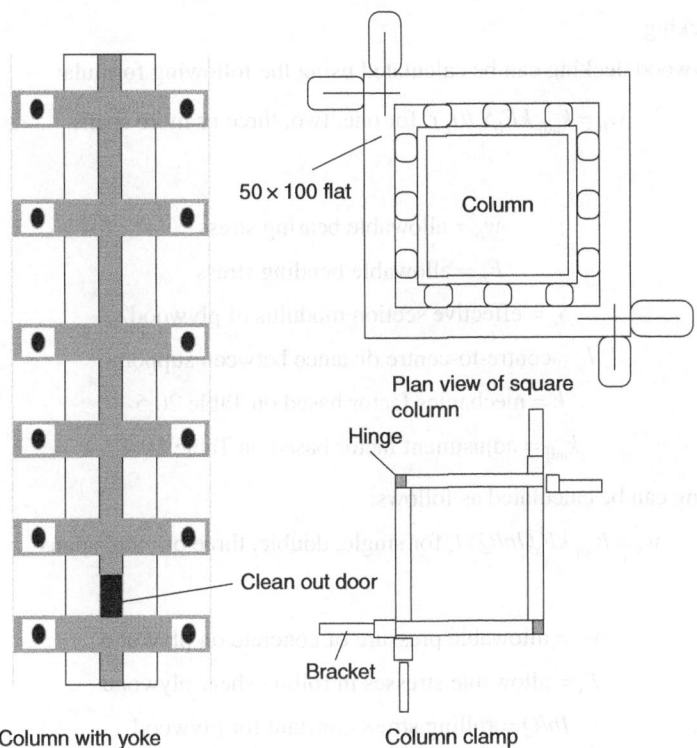

50 × 100 flat

Column

Plan view of square column

Hinge

Clean out door

Bracket

Column clamp

Column with yoke

Forms for a square column

Fig. 20.17 Column forms and its components

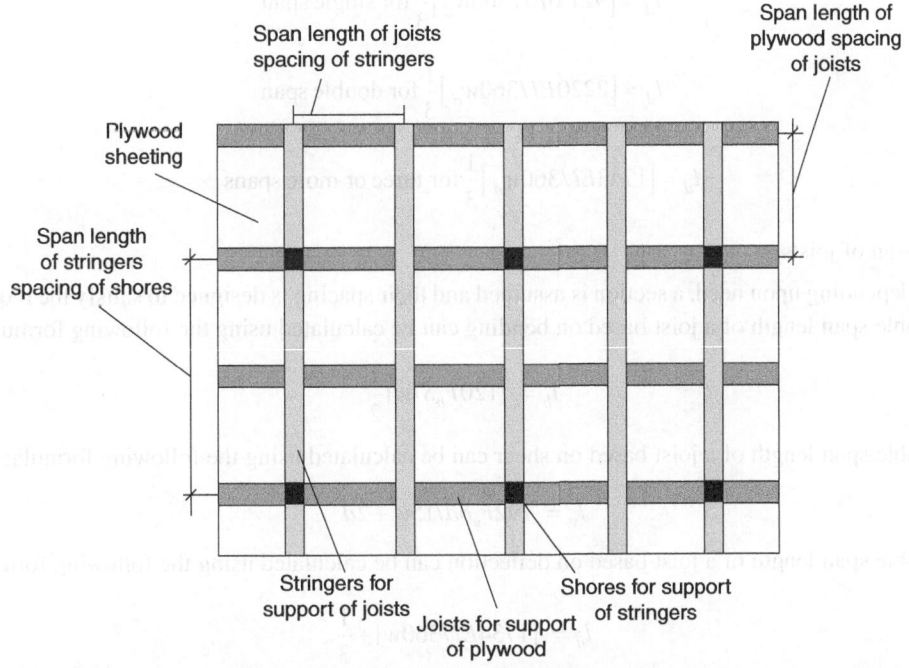

Span length of joists spacing of stringers

Span length of plywood spacing of joists

Plywood sheeting

Span length of stringers spacing of shores

Stringers for support of joists

Shores for support of stringers

Joists for support of plywood

Fig. 20.18 Slab form details

Step 2 Design of decking

Bending stress in plywood decking can be calculated using the following formula:

$$w_b = k_{adj}\, kF_b S_c/(l_b)^2 \text{ for one, two, three or more spans}$$

where

$$w_b = \text{allowable bearing stress}$$
$$F_b = \text{allowable bending stress}$$
$$S_c = \text{effective section modulus of plywood}$$
$$l_b = \text{centre-to-centre distance between supports}$$
$$k = \text{mechanics factor based on Table 20.5}$$
$$k_{adj} = \text{adjustment factor based on Table 20.4}$$

Shear stress in decking can be calculated as follows:

$$w_s = k_{adj}\, kF_s(Ib/Q)/l_s \text{ for single, double, three or more spans}$$

where

$$w_s = \text{allowable pressure of concrete on plywood}$$
$$F_s = \text{allowable stresses in rolling shear plywood}$$
$$Ib/Q = \text{rolling stress constant for plywood}$$
$$l_s = \text{clear distance between supports}$$

Bending deflection in decking can be calculated as follows: For a permissible deflection of $\Delta_d = l/360$,

$$l_d = \left[921.6EI/360w_d\right]\frac{1}{3} \text{ for single span}$$

$$l_d = \left[2220EI/360w_d\right]\frac{1}{3} \text{ for double span}$$

$$l_d = \left[1743EI/360w_d\right]\frac{1}{3} \text{ for three or more spans}$$

Step 3 Design of joists

For joists, depending upon need, a section is assumed and their spacing is designed to satisfy the requirement. The allowable span length of a joist based on bending can be calculated using the following formula:

$$l_b = [120F_bS/w]\frac{1}{2}$$

The allowable span length of a joist based on shear can be calculated using the following formula:

$$l_v = 192F_vbd/15w + 2d$$

The allowable span length of a joist based on deflection can be calculated using the following formula:

$$l_d = [1734EI/360w] + \frac{1}{3}$$

Minimum of these three spacings should be adopted for use.

Step 4 Design of stringer

Stringers can be designed in the same manner as joists. A stringer section is selected first and their spacing should be found using the same formula as given for joists.

20.7.8 Formwork Drawing and Calculation

All major design values and loading conditions should be shown on the formwork drawing. These include the assumed values of the live load, the compressive strength of concrete for formwork removal and for application of construction loads, the rate of placement, temperature, height and drop of concrete, weight of moving equipment which may be operated on formwork, design stresses, camber diagram, and other pertinent information.

In addition to specifying the type of material, size, lengths, and connection details, formwork drawings should be provided for applicable details such as the following:

- Procedure, sequence, and criteria for removal of forms, shores, and reshores
- Design allowance for construction loads on new slabs should be shown when such allowances affect shoring and reshoring schemes
- Anchors, form ties, shores, lateral bracing, and horizontal plan bracing
- Field adjustment of forms
- Waterstops, keyways, and inserts
- Working scaffolds and runways
- Weep holes or vibrator holes where required
- Screeds and grade strips
- Location of external vibrator mountings
- Crush plates or wrecking plates where stripping may damage concrete
- Removal of spreaders or temporary blocking
- Cleanout holes and inspection openings
- Construction joints, contraction joints, and expansion joints conforming to design drawings
- Sequence of concrete placement and minimum permitted 'elapsed time' between adjacent placements
- Chamfer strips or grade strips for exposed corners and construction joints
- Camber
- Mudsills or other foundation provision for formwork
- Special provision such as safety, fire, drainage, and protection from ice and debris at water crossings
- Formwork coatings

20.7.9 Common Design Deficiencies

Some of the most common formwork design deficiencies are as follows:

- Lack of allowance in design for such loading as caused by placing equipment and temporary material storage
- Inadequate provision to prevent dislocation of beam forms where the slabs frame into them on only one side
- Inadequate anchorage against uplift in inclined (battered) form faces
- Insufficient allowance for eccentric loading caused by the placement sequence
- Failure to investigate safe bearing stresses in members which are in contact with shores, struts, or props
- Failure to provide lateral bracing, or lack of the shoring/props
- Failure to investigate the adequacy of the slenderness ratio of compression members
- Inadequate provision to tie the corners of intersecting cantilevered forms together

- Failure to account for loads imposed on anchorage during gap closure in the process of aligning formwork forcibly
- Failure to take into account the conditions of materials to be actually used for the formwork, environment, and site consideration
- Inadequate attention to the detailing of connections. It should be ensured that local failure, if any, does not lead to a progressive collapse involving the entire formwork and the freshly placed structure. The formwork scheme should preferably be so designed that the vertical members are subjected to compressive force only under the action of combined horizontal (lateral) and vertical loads

20.8 Shores

Shores are used in formwork which supports beams, floor slabs, bridge decks, etc., until these members gain sufficient strength and become self-supporting. Shores are installed as single-member units, which may be tied together at one or more intermediate points with horizontal and diagonal braces to give them greater stiffness and to increase their load-carrying capacity. In order to prevent movement while they are in use, shores should be fastened at the top and bottom. There are many types of shores in use. Some of them are explained here.

20.8.1 Wood Post Shores

This type of shore [Fig. 20.19(a)] is commonly used. Joints can be easily made with nails. The shore holder shown in Fig. 20.19(b) keeps the shore in place. The purlin splicer is provided at the top to enable good support for the purlin [Fig. 20.19(c)]. The screw jack at the bottom enables height adjustment with ease [Fig. 20.19(d)]. Figure 20.19(e) shows the reshoring spring. While preparing wooden shores, proper inspection is needed because carpenters are employed in making wooden shores and there are possibilities for error. Some of the advantages and disadvantages of this type of shore are listed in this section.

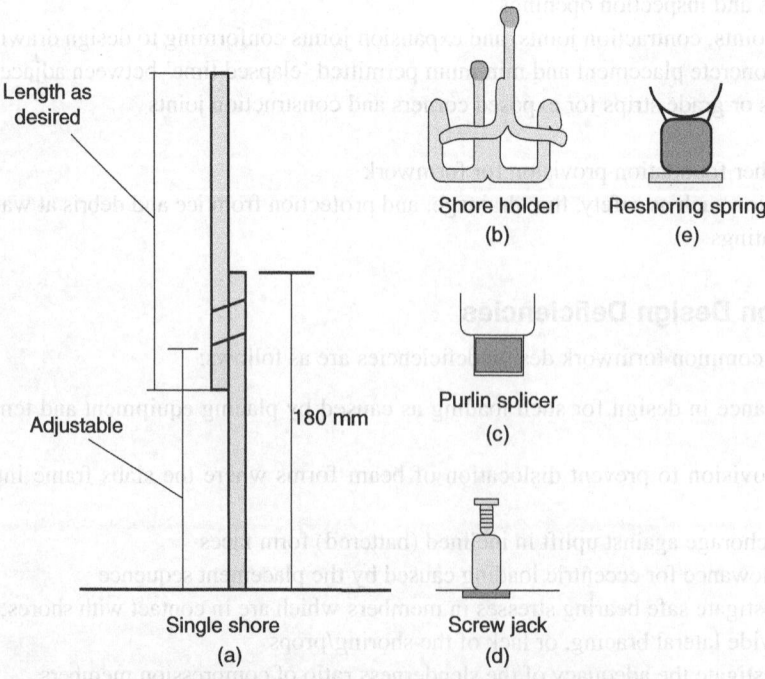

Length as desired

Adjustable

180 mm

Single shore
(a)

Shore holder
(b)

Reshoring spring
(e)

Purlin splicer
(c)

Screw jack
(d)

Fig. 20.19 Wooden post shores and accessories

Advantages

1. The initial cost is low.
2. It is easy to attach and remove braces.

Disadvantages

1. The cost of labour for installing wooden shores may be higher than for installing patented shores.
2. Unless they are stored carefully, they may develop permanent bows, which will reduce the load-carrying capacity.
3. If wooden shores are too long, they should be sawed to the required length, which will result in additional labour and wastage. If wooden shores are short, splicing will have to be adopted.
4. When compared to unspliced shores, the strength of spliced shores is less and splicing will also add to the labour cost.

20.8.2 Adjustable Shores

As the name indicates, these shores can be adjusted as per requirement. These shores are generally made of steel (Fig. 20.20), with provision for adjustment. This simplifies the problem of fine adjustment. Adjustable shores have various fittings which can be adjusted and interchanged at the top to extend height. Generally used fittings are flat-bearing plate, U-head, and T-head. For support of stringers or other horizontal timber in slab forming, the U-head is preferable because it permits nailing laterally into the stringer. These adjustable shores are generally patented by their developers. Patented shores are generally made of steel and can be considered as a substitute to wooden shores or as advancement of wooden shores. Some of their advantages and disadvantages as compared with wooden shores are given in this section.

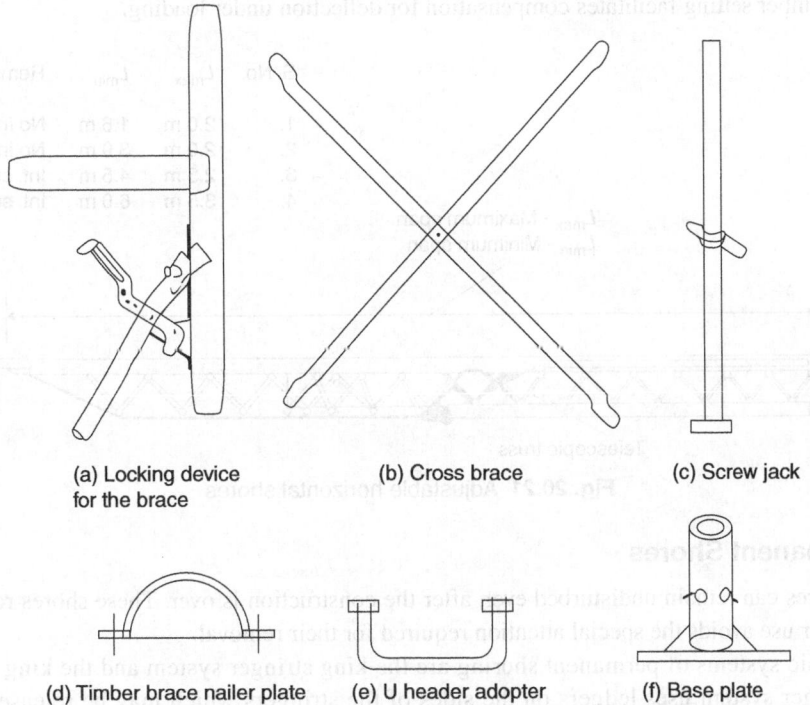

(a) Locking device for the brace   (b) Cross brace   (c) Screw jack

(d) Timber brace nailer plate   (e) U header adopter   (f) Base plate

Fig. 20.20 Adjustable steel shores and accessories

Advantages

1. These are available in various lengths.
2. These can be adjusted for various ranges of length.
3. In general, these are rugged, which assures long life.
4. Shore area is usually large enough to give large bearing area between the shores and the stringers that rest on them.

Disadvantages

1. Their initial cost is very high as compared to wooden shores.
2. Attaching bracings is difficult as compared to wooden shores.
3. Because of their slenderness, some of them are less resistant to buckling than wooden shores.

20.8.3 Horizontal Shores

Horizontal shores (Fig. 20.21) are steel members which support slabs, beams, bridge decking, etc. These can be used as joists directly under the sheeting. These units are manufactured with an adjustable built-in camber setting device to compensate deflection under loading.

Advantages

1. These are lightweight shores.
2. Their length is adjustable, which permits a wide range of use.
3. These can be installed and removed with great speed.
4. The number of vertical shores needed is reduced and this leads to more construction work space below the deck in the lower floor.
5. Built-in camber setting facilitates compensation for deflection under loading.

S. No.	L_{max}	L_{min}	Remarks
1.	2.0 m	1.6 m	No int. sup.
2.	2.0 m	3.0 m	No int. sup.
3.	2.5 m	4.5 m	Int. sup. reqd.
4.	3.5 m	6.0 m	Int. sup. reqd.

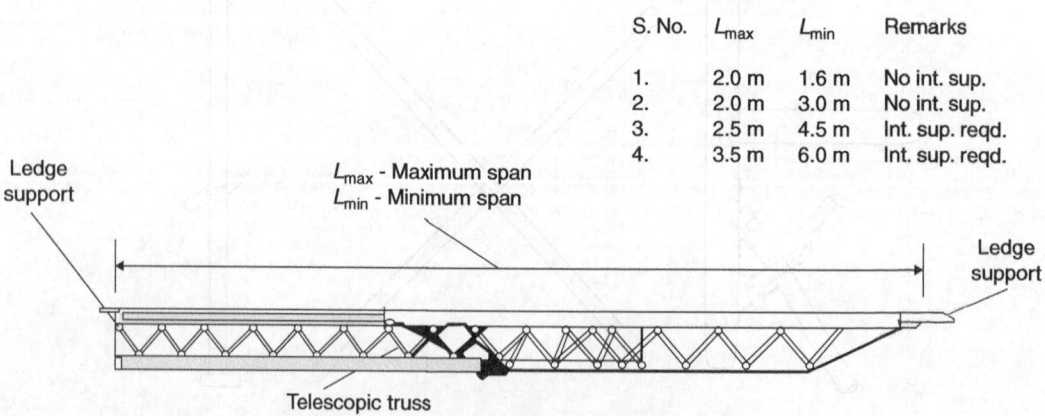

L_{max} - Maximum span
L_{min} - Minimum span

Ledge support

Ledge support

Telescopic truss

Fig. 20.21 Adjustable horizontal shores

20.8.4 Permanent Shores

Permanent shores can remain undisturbed even after the construction is over. These shores remain with the structures. Their use avoids the special attention required for their removal.

The two basic systems of permanent shoring are the king stringer system and the king shore system. The king stringer system uses ledgers on the sides of the stringers which may be released on the side, permitting the removal of the joist between the stringers. In the king shore system, the stringer may be attached to the side of the shores so that the stringer may be removed, permitting the release of the joists and the decking from contact surfaces. The shores and a trapped strip of contact decking are all that remains in place.

20.8.5 Backshores

Backshores are shores placed under a concrete slab or a structural member after the original forms and shores have been removed from a small area, without allowing the slab or member to deflect or support its own self-weight or additional construction loads from above.

20.9 Removal of Forms and Shores

Removal of formwork is also as important as erecting it. Even though formwork is the general responsibility of the contractor, the time of removal should be specified by the engineer. Generally, early removal of form-work is proposed for reusing forms. In warm weather conditions, early removal is possible. The advantage of early removal is that the surface repair or treatment can be done when concrete is still green. In cold weather conditions, early removal is not possible.

A major factor to be considered in the removal of formwork is the strength of concrete at the time of removal. Formwork should not be removed until the concrete has achieved enough strength—at least twice the stress to which it is to be subjected on formwork removal. The strength gain depends on several factors, such as the type of cement used, temperature and weather conditions, and curing regime. The use of blended cements may delay the strength gain. Stripping is the last operation, but it is the first and most important aspect that needs to be taken into account even during the design stage.

When forms are stripped, there must be no excessive deflection or distortion and no evidence of crack-ing or damage to concrete. In order to avoid these, often stripping can be partial, leaving some formwork in place.

Another major factor that influences the removal of form is the sequence of its removal. For example, consider a cantilever structure, while removing the formwork and shores, forms should be removed from the free end (overhanging end) towards the fixed end only. If the forms are removed from the opposite end, the cantilever member will behave like a simply supported beam—a condition normally not envisaged by the designer. This will make the cantilever fail in a brittle manner and may even lead to failure of the whole structure.

20.10 Reshoring

Reshores are shores firmly placed under a stripped concrete structural member where the original formwork has been removed. These are required till the new structural member supports its own weight. Reshoring is done to maximize the reuse of forms. Premature reshoring and inadequate size and spacing of reshores have been responsible for a number of construction failures. Reshoring is a highly critical operation, and it is essential that the procedure be planned in advance.

When removal of formwork and reshoring is being done, the live load for which the completed structure is designed, as well as the actual strength of the partially cured concrete, must be considered. Allowance must be made for additional live and dead load. Location, spacing, and type of reshores should also be considered. Reshoring will also change the distribution of the bending moment along the length of a member.

All reshores must be straight without any kink, bend, or warp. When placing reshores, care should be taken for the load to be applied by reshores on the floor below. Reshores should never be located where these will significantly alter the pattern of stress determined in the structural analysis (i.e., these should not induce ten-sion in places where tension steel is not provided).

20.11 Construction Loads

Structural formwork and its supports need to be given consideration in respect of (a) loads that are applied to formwork and its supports and (b) loads that the formwork and its supports apply to the structure.

It is convenient to express construction loads as a factor times the sum of the self-weight of the floor and the dead load of the formwork and shores. The term floor-loading ratio is used for this factor. Progressive slab deflection and cracking in slabs cause problems in some flats during construction. This is because a rational method of analysis of the construction sequence and loading has not been adopted.

Figure 20.22 shows a typical construction cycle using three levels of shores. Figure 20.23 represents the construction of a high-rise building using three levels of shores and forms. A 7-day construction cycle and 5-day stripping are used. The loads carried by the slabs and shores, in terms of the loading ratio, are shown in the figure adjacent to the element concerned. Floors 1, 2, and 3 supported by shores and stiff ground cannot deflect and hence carry no load. All the loads are carried by the shores to the foundation. At 26 days the lowest level shores are removed, allowing the three slabs to deflect and carry their self-weight; the removed shores are placed on the

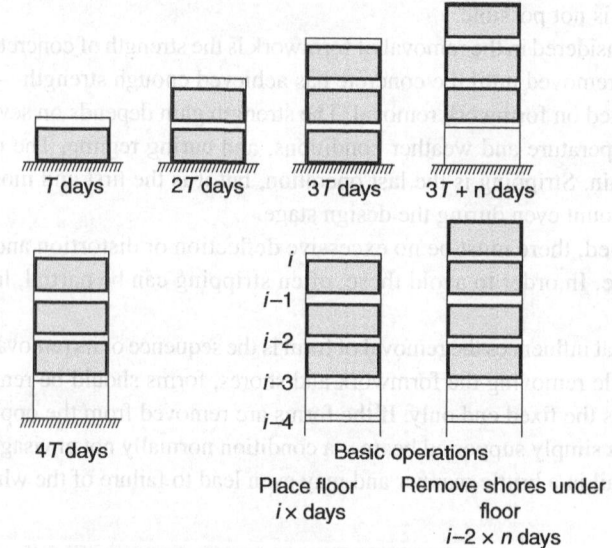

Fig. 20.22 Construction sequence using three levels of shores

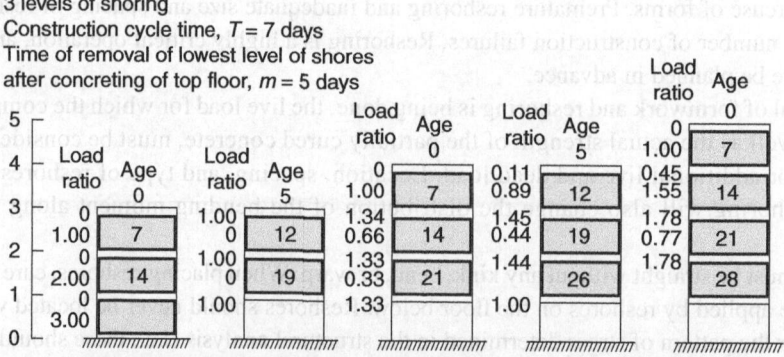

Fig. 20.23 Load ratio versus time for three levels of shores

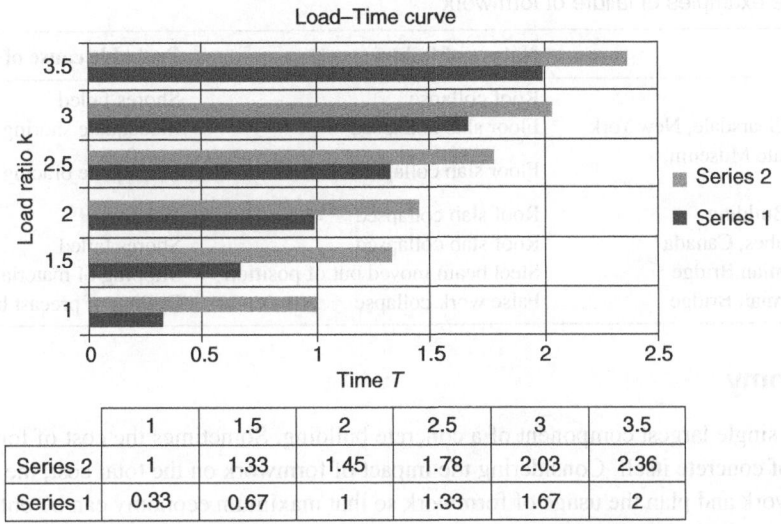

Fig. 20.24 Load–time history for three levels of shores

	1	1.5	2	2.5	3	3.5
Series 2	1	1.33	1.45	1.78	2.03	2.36
Series 1	0.33	0.67	1	1.33	1.67	2

third floor slab and the fourth floor is poured. As all the three supporting slabs have an equal stiffness, the weight of the newly poured slab is carried equally by the three lower slabs. The absolute maximum load ratio occurs when the shores connecting the supporting assembly with the ground level are removed. However, the load level converges for upper floors. The converged value for the upper floor level is approximately 2.00 (Fig. 20.23). The most heavily loaded slab is always the last slab near the base. The load–time history of the third floor slab and a typical slab are shown in Fig. 20.24. The load ratio for a slab assuming three-level shores can be as high as 2.36.

20.12 Failure of Formwork

The failure of formwork is always embarrassing and expensive for everyone involved in the project. Accidents affect not only workers but also their families. So everyone involved in the project must be alert to unsafe conditions, and all work must be performed with safety regulations and requirements specified in the design.

Failure may be collapse of the entire formwork or a part of it. Sometimes there may also be movement or small distortion of formwork, which may require removal and replacement. If formwork is properly designed and constructed, failure can be avoided. Some of the reasons for the failure of formwork are given below:

a. Improper or inadequate shoring
b. Inadequate bracing of members
c. Lack of control of placement of concrete
d. Improper connections
e. Premature stripping of formwork
f. Improper design
g. Failure to follow codes and standards
h. Negligence of workers or supervisors

In order to prevent failure of formwork, the following precautions should be taken:

a. The formwork should be designed properly.
b. Erection and stripping should be done only under engineering supervision.

The sequence of removal should be pre-designed and correctly executed. Some of the major projects in which formwork failure has taken place and the probable cause of failure are given in the Table 20.6.

Table 20.6 Some examples of failure of formwork

Project	Nature of failure	Probable cause of failure
ESSO	Roof collapse	Shores failed
Concrete building Searsdale, New York	Floor slab collapsed	Inadequate shoring
Smithsonian Institute Museum, Washington	Floor slab collapsed	Inadequate bracing of shores
Californian Bank Building	Roof slab collapsed	Unknown
Tornado-subway tubes, Canada	Roof slab collapsed	Shores failed
Oceanside, Californian Bridge	Steel beam moved out of position	Slipping of material during removal
San Bruno, Californian Bridge	False work collapse	Rolling of precast beam

20.13 Economy

Formwork is the single largest component of a concrete building. Sometimes the cost of formwork can also exceed the cost of concrete itself. Considering the impact of formwork on the total cost, the engineer should design the formwork and plan the usage of formwork so that maximum economy can be obtained.

Economy of formwork begins with the design development of the structure itself. Following points should be considered while designing formwork for a building structure.

a. While designing the structure, consider the material and tools that will be required to make, erect, and remove the formwork.
b. Design the structure with standard dimensions that will be unit multiple of forms and centring sheets.
c. Use the same size of columns from the foundation to the roof; this will permit the use of column forms without alteration.
d. Use beams of the same depth and spacing in every floor; this will permit the reuse of beam forms without alteration.
e. Specify the same width for columns and column-support girders in order to reduce or eliminate the cutting and fitting of girder forms into column forms.

Any type of formwork is more economical if used in a repetitive manner. Repetition involves removal and reuse of formwork. Forms should be removed as soon as possible to provide the greatest number of reuses, but removal should be after sufficient strength has been attained by fresh concrete. Therefore a careful planning of the construction operation is required to balance safety and economy.

Some of the important points to achieve economy in formwork expenditure are as follows:

a. While designing formwork, maximum usage of material should be achieved.
b. High-quality finish on concrete surface is not required for sides that will not be exposed.
c. When planning forms, consider the sequence and method of stripping.
d. Use prefabricated panels wherever possible.
e. Strip forms as soon as it is safe in order to facilitate maximum reuse of forms.
f. Create cost awareness among carpenters and other workers involved in formwork construction.
g. Use long-length timber or plywood without cutting, where their extending beyond the working area is not objectionable.
h. After removal, clean the panels and store them at a safe place so that they can be reused.

Exercises

Review Questions

1. Explain the role of formwork in the quality of concrete construction.
2. What are the functions of formwork?

3. What are the forces that act on the formwork?
4. How is lateral pressure on wall formwork assessed? Explain.
5. What are the basic objectives of formwork design?
6. What are the various systems of formwork?
7. Explain the role of a contractor in formwork erection.
8. What are the basic assumptions made in the design of formwork?
9. List some of the common deficiencies in formwork which lead to the failure of structures.
10. What are the reasons for the failure of formwork?
11. Explain different types of shoring and their relative merits.

Multiple Choice Questions

1. Defective formwork leads to
   a. Poor concrete quality
   b. Cracking when forms are relaxed
   c. Accidents at site
   d. All of the above

2. As per IS: 456-2000 formwork should be
   a. Designed
   b. Not designed but executed based on practice
   c. Dependent on the structure
   d. Made with available material at site

3. The maximum deviation in cross section of formwork permitted by IS: 456-2000 is
   a. 12 mm
   b. 50 mm
   c. 25 mm
   d. 6 mm

4. Vertical forms should not be released before
   a. 16–24 hours
   b. 7 days
   c. 3 days
   d. 10 days

5. Formwork for beams up to 6 m can be stripped after
   a. 7 days
   b. 14 days
   c. 21 days
   d. 28 days

6. Soffit forms for slabs can be released (if props are re-fixed immediately after removal of formwork)
   a. 3 days
   b. 7 days
   c. 10 days
   d. 24 hours

7. For beams with spans larger than 6 m, formwork can be released only after
   a. 3 days
   b. 7 days
   c. 14 days
   d. 21 days

8. While contemplating early removal of formwork, one should consider
   a. Type of cement used
   b. Type of curing
   c. Temperature influencing hydration
   d. All of the above

9. If PPC cement is used, formwork removal should be
   a. Delayed
   b. Hastened
   c. As specified in IS: 456-2000
   d. None of the above

10. While stripping formwork for a beam, the stripping sequence should be
    a. From support towards the mid-span
    b. From mid-span to the support
    c. From quarter span in either direction
    d. Independent of the way it is stripped

Answers to Multiple Choice Questions

1. d   4. a   7. d   10. b
2. a   5. b   8. d
3. a   6. a   9. a

CHAPTER 21

Structural Concrete Block Masonry

Masonry is one of the oldest building materials, but one which is least understood. Misconceptions regarding its behaviour have led, over the years, to a serious misuse of the material through inadequate or even non-existent design procedures. Added to this, the poor construction practices have worsened the situation. However, perhaps because of considerable amount of information and data available today, both as to its properties and structural performance, sound design techniques and improved manufacturing processes, vastly improved construction practices have evolved in the recent years. From the traditional masonry blocks, over the years, new forms of hollow structural masonry blocks have evolved. A wall does more than just enclose a building in an attractive form. It must have strength to support floors and roofs and should resist wind and seismic loads. It must give adequate protection against noise, heat, cold, and fire damage. This chapter traces the history and development of structural concrete block masonry.

21.1 Historical Development

Brick is the oldest man-made building material invented almost 10,000 years ago. Some of the oldest bricks in the world taken from archaeological digs at the site of ancient Jericho (Palestinian territory) resemble long loaves of bread, sometimes with bold patterns of Neolithic thumb print impression in their rounded tops (see Fig. 21.1). Excavations of ruins of Babylonia exposed a masonry arch believed to have been built around 1400 BC. Arch construction reached a high level of refinement under the Romans, and the later developments were primarily in the adoption of different profiles. By the early twentieth century the demand was for high-rise construction but the technology of stone and masonry was not so developed to meet the requirements of high-rise structural skeletal systems. So masonry was relegated to secondary use as facings, infills, and for fire proofing.

The famous Monadnock building in Chicago is one of the tallest buildings built with traditional brick masonry. This 16-storey building has 6-m-thick walls to provide stability against wind loads. The height reached by this building is considered as the limit for unreinforced brick masonry construction techniques. During the same period, Spanish Architect Antoni Gaudí came out with a number of masonry designs based on efficiency of forms which included innovative brick arch and shell structures. It is to be noted that these structures utilized the brick masonry in compression. However, the turn of the twentieth century saw lightweight high-rise multi-storey buildings replace the ancient and traditional brick masonry architecture.

The first experimentally reinforced concrete building, the Eddystone Lighthouse (1774), was actually constructed of concrete and stone. However, the practice of reinforcing with steel was entirely limited to concrete with a few exceptions.

0 2 4 6 in

0 5 10 15 cm

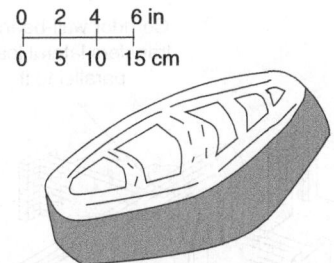

Fig. 21.1 Sun-dried brick, circa 8000 BC

As buildings increased in height, the lateral force imposed due to wind or seismic effect placed a formidable limitation on height. This challenge turned out to be so severe (as demonstrated by the 1933 Long Beach, California earthquake) that the use of plain masonry was prohibited for the Pacific coast region in the USA.

As a consequence, the demand for the continued use of masonry led to the evolution of plain masonry into 'composite systems', which we now term 'reinforced masonry'. In 1923, A. Brebner, serving as undersecretary in the Public Works Department of the Government of India, published his report (2 years of experimental work). His report is credited to mark the true beginning of the modern development of reinforced brick masonry. It is highly significant to note that his report, which included a rational design theory, brought about a change in the use of this material form in buildings of earthquake-prone countries such as India and Japan. Engineers in these countries found that reinforced masonry, if properly constructed, offered excellent resistance to seismic forces. Brebner indicated that nearly 3 million square feet of reinforced brick masonry was constructed in the three years prior to 1922.

S. Kanamori, Civil Engineer, Department of Home Affairs, Imperial Japanese Government, reported in July 1930: 'There is no question that reinforced brick work should be used instead of unreinforced brick work when any tensile stress would be incurred in the structure. We can make them more safe and strong, saving much cost. Further, I have found that reinforced brick work is more convenient and economical in building than reinforced concrete and what is still more important, there is always very appreciable saving in time.' Those early years of application included a variety of structures: public/private buildings, retaining and sea walls, chimneys, bridges, culverts, and storage bins.

21.2 Modern Trends

The present concept of reinforced masonry utilizes the floors and roofs as diaphragms acting as a horizontal flanged girder to distribute lateral loads to walls, which in turn provides horizontal shear resistance needed in addition to carrying the normal vertical live and dead loads. This type of structure may be defined as a box system or shear wall system (Fig. 21.2). These walls if constructed of plain masonry would be incapable of resisting the magnitude of horizontal shear and bending forces imposed on them. For this reason, modern reinforced masonry contains reinforcing steel to resist the shear and tensile stresses so developed. When these walls are subjected to lateral forces acting normal to them, they behave as flexural members spanning vertically between floors and horizontally between pilasters. Therefore reinforcement must also be provided to develop resistance on the tension face.

Had these techniques and construction methods been available in the olden days, the designers of Monadnock might have used only 30-cm(1 ft)-thick or less instead of 2-m(6 ft)-thick masonry walls. Compare that massive structure with the modern Hanalei Hotel in San Diego, California, with only 20-cm(8 in)-thick walls at the bottom storey (Fig. 21.3).

21.3 Modern Masonry

Modern masonry may take any of several forms. Structurally, it may be divided into load-bearing, non-load-bearing, or veneer construction. These forms may be solid masonry, solid walls of hollow units, or cavity walls. Finally these may be reinforced, partially reinforced, or plain. These forms may be designed empirically or analytically.

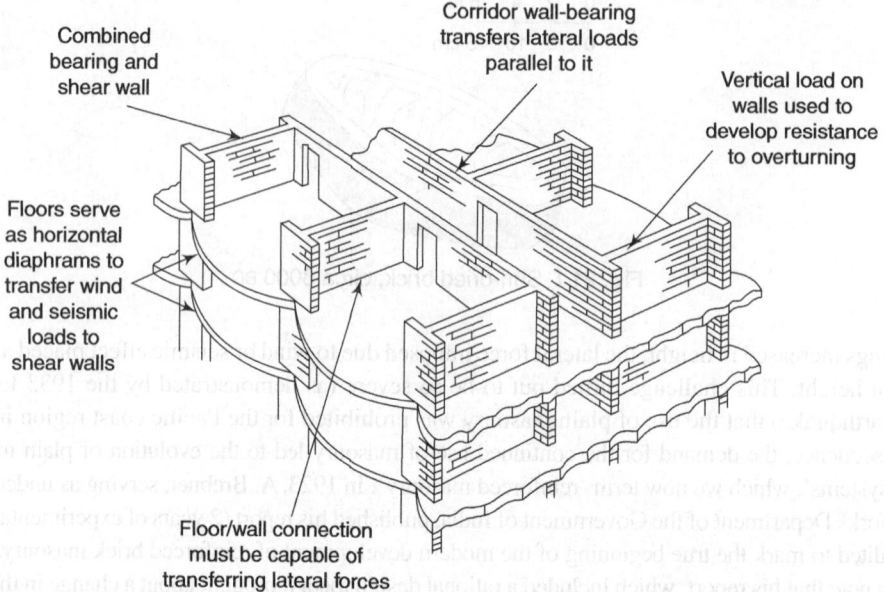

Combined bearing and shear wall

Corridor wall-bearing transfers lateral loads parallel to it

Vertical load on walls used to develop resistance to overturning

Floors serve as horizontal diaphrams to transfer wind and seismic loads to shear walls

Floor/wall connection must be capable of transferring lateral forces

Fig. 21.2 Concept of shear wall system (*Source*: PCA 1976)

Fig. 21.3 Hanalei Hotel, San Diego (*Source*: PCA 1976)

Load-bearing masonry supports its own weight as well as the dead and live loads of the structure and all lateral wind and seismic forces. Non-load-bearing masonry (including veneers) also resists lateral loads and may support its own weight for the full height of the structure or be wholly supported by the structure at each floor level. Solid masonry is built of solid units or fully grouted hollow units in multiple wythes with the collar joint between wythes filled with mortar or grout. Solid walls of hollow units have open cores in units, but have grouted collar joints. Cavity walls have two or more wythes of solid or hollow units separated by an open collar joint or cavity (Fig. 21.4). These are not common in India.

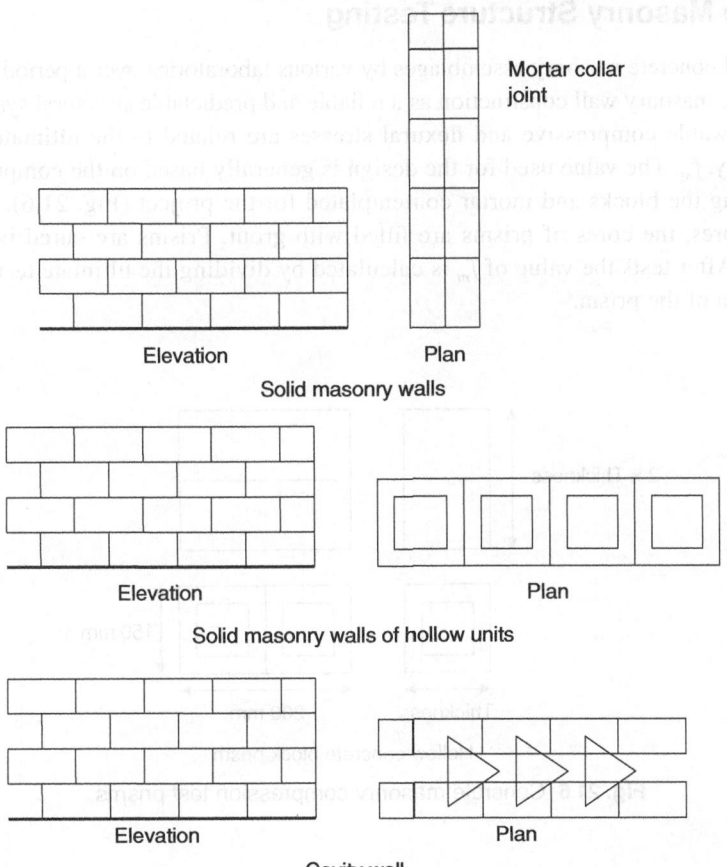

Mortar collar joint

Elevation   Plan

Solid masonry walls

Elevation   Plan

Solid masonry walls of hollow units

Elevation   Plan

Cavity wall

Fig. 21.4 Types of masonry construction

The concrete masonry manufacturing process consists of six standard stages of production:

1. Receiving and storing raw materials
2. Batching and mixing
3. Moulding unit shapes
4. Curing
5. Cubing and storage
6. Delivery of finished units (Fig. 21.5)

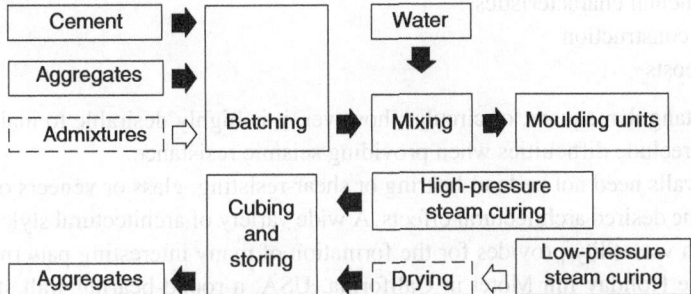

Fig. 21.5 Concrete masonry manufacturing process

21.4 Concrete Masonry Structure Testing

Extensive testing of concrete masonry assemblages by various laboratories over a period of many years has established concrete masonry wall construction as a reliable and predictable structural system.

In designs, allowable compressive and flexural stresses are related to the ultimate net compressive strength of masonry, f_m. The value used for the design is generally based on the comprehensive strength test on prisms using the blocks and mortar contemplated for the project (Fig. 21.6). If the structure is to have grouted cores, the cores of prisms are filled with grout. Prisms are cured in accordance with the field practice. After tests the value of f_m is calculated by dividing the ultimate test load by the solid cross-sectional area of the prism.

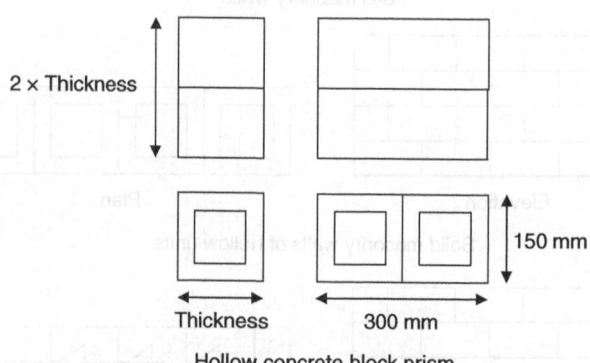

Fig. 21.6 Concrete masonry compression test prisms

21.5 High-rise Masonry Construction

Reinforced masonry bearing walls are ideally suited for high-rise construction. The development of high-strength masonry block along with improvements in grouting and reinforcing techniques has made masonry bearing walls practical for multi-storey construction especially when combined with precast floor or roof slabs. The basic concept involves designing every floor to act as a horizontal diaphragm in transferring wind or seismic loads to transverse shear walls, which in turn carry these forces to the foundation as shown in Fig. 21.7. The shear walls are reinforced to develop the moment and shear forces generated due to the lateral loads. High-rise masonry construction has several desirable features. The important of these are

- simplicity of design
- excellent environmental characteristics
- speed and ease of construction
- reduced building costs

Buildings may be rectangular, square, or circular; however, it is highly desirable to maintain as much symmetry as possible to preclude difficulties when providing seismic resistance.

Since all exterior walls need not be load-bearing or shear-resisting, glass or veneers of various types may be used in achieving the desired architectural effects. A wide variety of architectural styles and layouts can be readily achieved. Such versatility provides for the formation of many interesting patterns in which masonry units may be laid. The Holiday Inn Motel in California, USA, a round-bearing wall structure (Fig. 21.8), provides an excellent example.

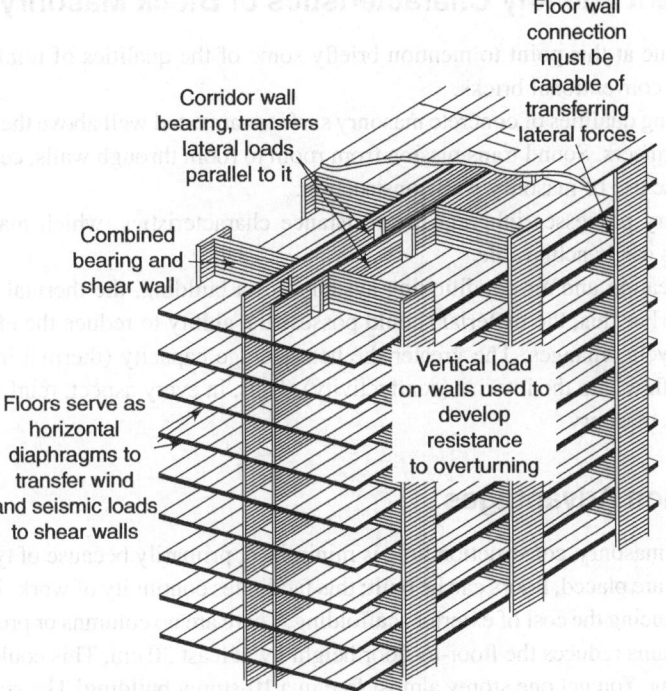

Floor wall connection must be capable of transferring lateral forces

Corridor wall bearing, transfers lateral loads parallel to it

Combined bearing and shear wall

Vertical load on walls used to develop resistance to overturning

Floors serve as horizontal diaphragms to transfer wind and seismic loads to shear walls

Fig. 21.7 High-rise concept in block masonry (*Source*: PCA 1976)

Fig. 21.8 Holiday Inn Motel, a round bearing-wall multi-storey structure (*Source*: PCA 1976)

21.6 Environment Friendly Characteristics of Block Masonry

It would be appropriate at this point to mention briefly some of the qualities of reinforced masonry blocks which are superior to conventional bricks.

The sound-absorbing qualities of concrete masonry surfaces are rated well above the effective levels recommended by sound engineers. Sound transmission from room to room through walls, ceiling, and floors can be minimized through the use of masonry wall construction.

Reinforced masonry possesses inherent fire-resistance characteristics, which make the fire ratings for multi-storey buildings easily achievable.

In the design of heating and air-conditioning systems for a building, the thermal inertia of the building material is important. For this, the material should possess the ability to reduce the effect of maximum heat gain or loss during cyclic changes. The greater the heat storage capacity (thermal inertia), the smaller the instantaneous rate of flow into the interior (conductivity). Thus, in every aspect, reinforced block masonry is better than solid brick masonry.

21.7 Construction Advantages

With the use of block masonry, construction time is minimized, primarily because of typical repetitive nature of layout. When walls are placed, floors can be built; this facilitates continuity of work. The floors next provide work area, thereby reducing the cost of exterior scaffolding. There are no columns or projecting beams to form. The elimination of beams reduces the floor-to-floor height by at least 30 cm. This could add up to 3 m height for a 10-storey building. You get one storey almost free in a 10-storey building! The structural masonry walls have a surface that can be painted, stained, or left natural. No plastering is really required. Last but not least, occupancy of lower stories is possible even before upper stories are completed.

21.8 Applications of Concrete Block Masonry

The most common application of concrete masonry is for built-in-place walls for buildings of all kinds. However, there are a number of other common applications as described here.

One-storey commercial buildings

Structural masonry blocks are very much suitable for use in one-storey industrial buildings. The typical plan of an industrial building completed with structural masonry blocks is shown in Fig. 21.9.

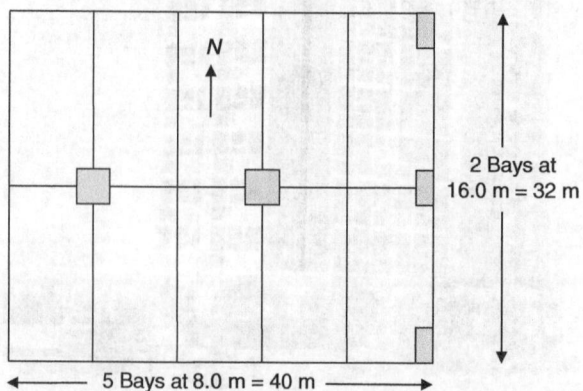

Fig. 21.9 Plan of a typical single-storey commercial building

High-rise buildings

High-rise buildings need to have shear walls to resist wind and earthquake loads. The bearing walls also function as shear walls. Figure 21.10 shows some examples of two-directional bearing/shear wall layouts. Figure 21.11 shows multi-directional shear/bearing walls used for typical non-rectangular, high-rise structures.

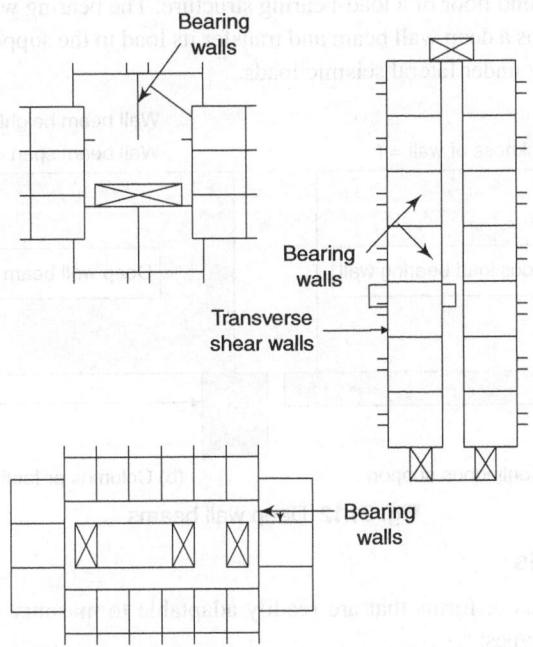

Fig. 21.10 Examples of two-directional bearing/shear wall layouts (*Source*: PCA 1976)

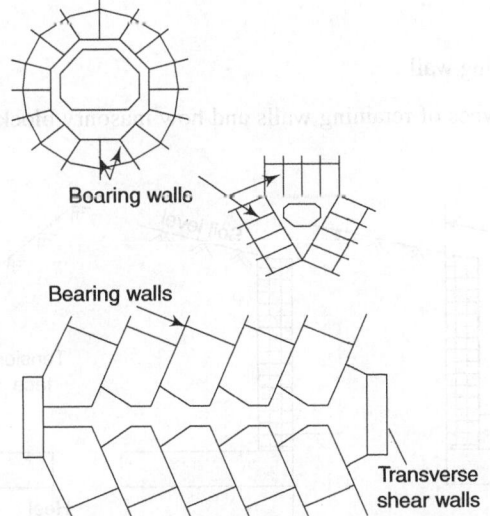

Fig. 21.11 Examples of multi-directional bearing/shear wall layouts (*Source*: PCA 1976)

Medium-rise buildings (up to 4 storeys)

In medium-rise buildings, the concept of deep wall beams is used extensively. The concept is based on wall spanning between columns or footings instead of having a continuous line support (Fig. 21.12). If soil-bearing capacities permit this type of concentrated load, the wall may be designed as a flexural member. Bearing, non-load-bearing, and shear walls may employ this principle to advantage. Deep wall beams may also be used to advantage to open up the ground floor of a load-bearing structure. The bearing wall on the floor above can be supported on columns to act as a deep wall beam and transfer its load to the supports. However, their stability should be examined carefully under lateral seismic loads.

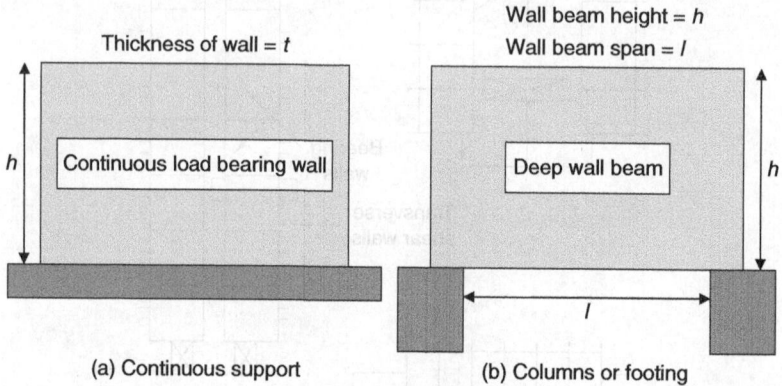

Fig. 21.12 Deep wall beams

Masonry retaining walls

There are several basic shapes or forms that are readily adaptable to masonry retaining wall construction. These include the following types:

- gravity
- cantilever
- counterfort
- buttressed wall
- supported retaining wall
- intersecting effect of retaining wall

Figure 21.13 shows different types of retaining walls and how masonry blocks can be used advantageously in these walls.

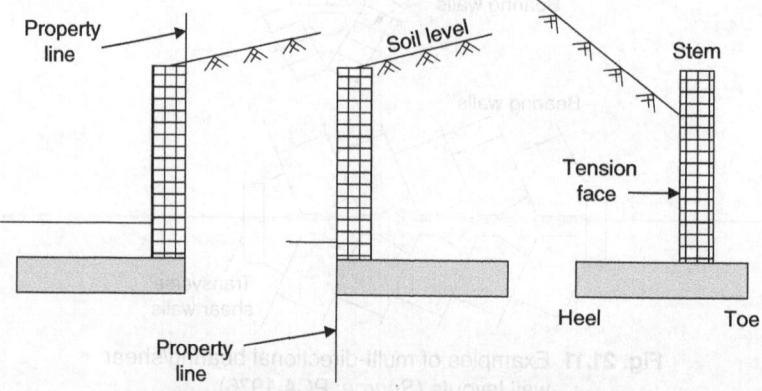

Fig. 21.13 Different types of retaining wall

Swimming pools

Masonry has been used in the construction of swimming pools for many years and has provided satisfactory, low-maintenance installations for both public and private facilities. Figure 21.14 shows a typical brick pool design completed during 1936 and giving satisfactory performance even today.

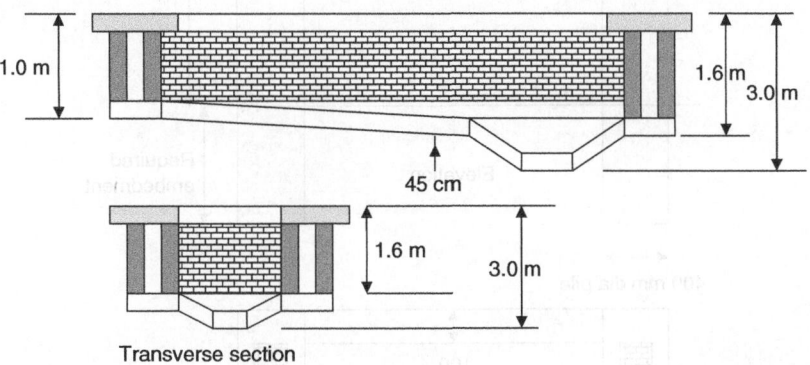

Fig. 21.14 Brick pool design

Chimneys

Even commercial chimneys of nearly 25 m base width and 30 m height have been successfully built with concrete block masonry. Figure 21.15 shows the plan diagram of a reinforced masonry industrial chimney completed and functioning satisfactorily.

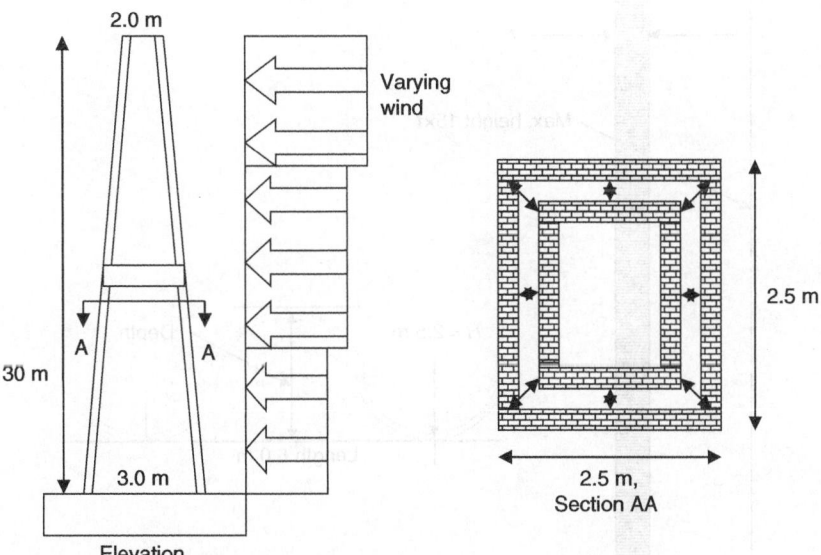

Fig. 21.15 Reinforced masonry industrial chimney

Compound walls

Brick garden walls may take different forms. A straight wall with pilasters must be designed with reinforcement. Figure 21.16 shows brick pier and panel garden wall. Serpentine and folded plate walls provide more lateral stability. A typical serpentine wall is shown in Fig. 21.17.

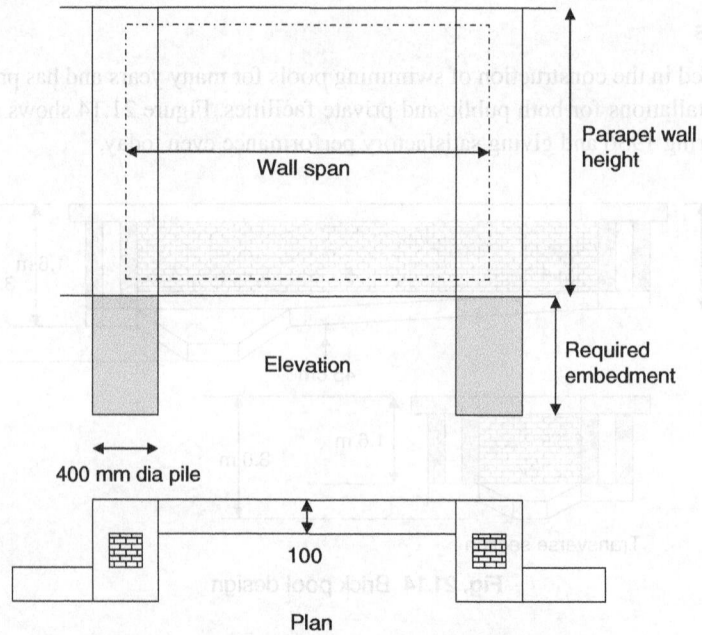

Fig. 21.16 Brick pier and panel garden wall

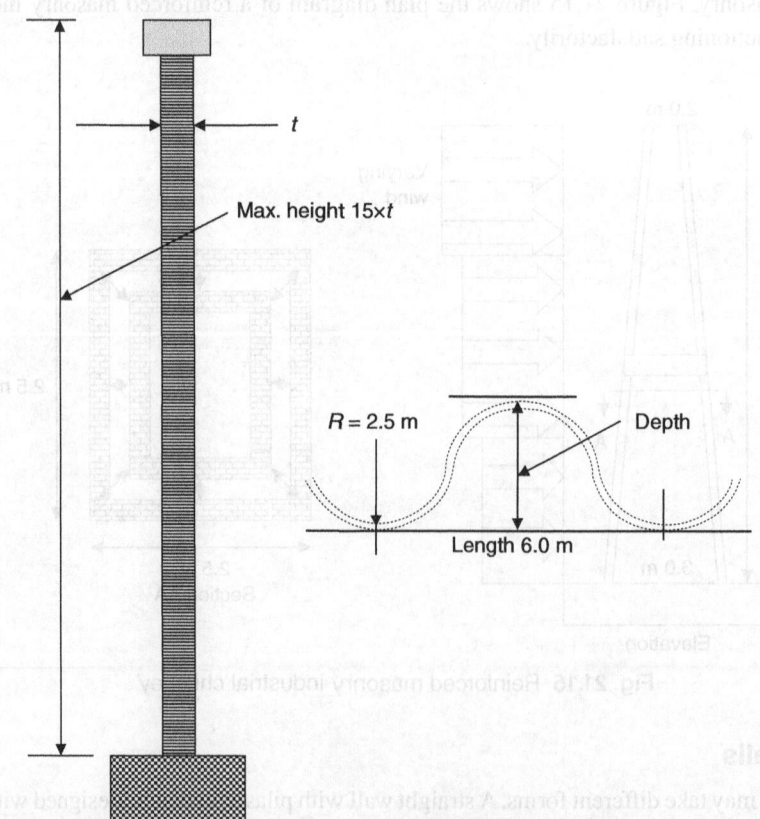

Fig. 21.17 Serpentine wall

Exercises

Review Questions

1. What are the functions of a wall?
2. Trace the historical development of reinforced brick work in India and abroad.
3. What are the various types of masonry wall construction?
4. What are the advantages of high-rise masonry construction?
5. What are the various applications of concrete block masonry?
6. Distinguish between load-bearing wall, non-load bearing wall, and shear wall?
7. What do you understand by a wall beam? Explain.
8. What are the various types of retaining walls that could be built with reinforced masonry?
9. What are the different ways in which a compound wall can be made strong against wind load?

Multiple Choice Questions

1. The number of storeys in the tallest building built with brick masonry is
   a. 10
   b. 4
   c. 16
   d. 20

2. The first masonry block was made approximately _____ number of years ago.
   a. 1000
   b. 10000
   c. 5000
   d. 8000

3. Hollow block masonry is efficient in providing
   a. Thermal insulation
   b. Sound insulation
   c. Savings in foundation due to reduced self-weight
   d. All of the above

4. Compressive strength of block masonry is determined based on
   a. Block strength
   b. Mortar strength
   c. Strength of prism made with block and mortar
   d. Code prescribed value

5. Block masonry construction uses the concept of _____ to resist lateral loads
   a. Frame action
   b. Diaphragm action
   c. Truss action
   d. Shear wall action

6. In masonry construction walls can be supported by
   a. Slabs
   b. Walls
   c. Frames
   d. Deep wall beams

7. In compound wall construction, walls are provided in a serpentine plan to effect
   a. Better appearance
   b. Cost savings
   c. Better lateral load resistance
   d. Ease of construction

8. It is possible to have high-rise masonry construction because
   a. Masonry is cheap
   b. Masonry can resist tension due to lateral loads
   c. Masonry can be selectively reinforced to carry required tension
   d. Masonry can be grouted and strengthened

9. Bearing walls can resist lateral loads because
   a. They can resist tension
   b. They are light
   c. They are located in plan such that tension does not occur
   d. They are designed such that compression due to gravity loads is more than tension developed by lateral loads

10. Reinforced hollow block masonry is efficient in resisting earthquake loads because
    a. They are light
    b. They are reinforced selectively as per requirement
    c. They can mobilize shear wall action
    d. All of the above

Answers to Multiple Choice Questions

1. c	4. c	7. c	10. d
2. b	5. d	8. c	
3. d	6. d	9. d	

CHAPTER 22

Quality Control of Concrete Construction

The quality control of concrete construction involves two aspects. First, it should ensure that concrete as a material has desired characteristics such as durability, serviceability, safety, and aesthetics. Second, the structure constructed using concrete should have good performance. The needs of structural performance vary based on the use to which a structure is put to and the environment in which it has to perform. Hence, uniform common standards for all structures are not practical. Therefore, concrete as a material should satisfy production quality requirements, which are generally detailed for a job in the specification for field control. In most instances, compressive strength of concrete is taken as the sole representative or measure of performance, which has time and again proved to be inadequate, especially when dealing with issues related to durability. Therefore, each step in production (of material) and construction (of structure) has to be specified. Each of the attributes of the material and the structure is variable. Hence, probability-based specification is essential to account for tolerances of these variables. Thus, quality control ensures compliance of specifications. It also ensures that the structure produced performs as per specification.

For the manufacture of concrete, the quality control process involves material, personnel, equipment, and process (workmanship) in all stages of concreting. The stages include procurement of material, batching, mixing, transporting, placing, compacting, curing, stripping forms, inspection, and testing.

For the construction of the structure as a whole, quality will be ensured only if all the uncertainties are scientifically managed. This will have to start at the planning stage itself and should continue through the stages of design, detail engineering, procurement, construction, and life-long maintenance. Such an elaborate compliance plan for managing uncertainties requires a quality management system which will have quality control as a subsystem.

22.1 Statistical Parameters and Variability

Statistical parameters involved in concreting, such as cube compressive strength, are not constant, and specifications permit these to vary within certain limits. The concreting process itself makes this variability inevitable. However, the acceptance criteria of the finished product ensure that this variability does not affect performance requirements.

22.1.1 Sampling

It is neither practical nor feasible to test all concrete produced. Hence, any parameter (e.g., compressive strength) is determined based on a few samples as representative of all concrete. Since samples should represent the whole concrete produced, these should be chosen at random. Good or bad samples should not be chosen purposely. IS: 456-2000 specification warrants that a minimum number of samples based on the total quantity of work are necessary so that the cube compressive strength obtained out of them represents the compressive strength of the whole concrete produced and used in the structure. This is further discussed later in the chapter.

22.1.2 Measure of Variability

In Fig. 22.1, concrete compression cube test results are plotted in a histogram (compressive strength $\bar{\sigma}_{cu}$ on the x axis and the number of results, N, on the y axis). Table 22.1 shows typical values of the cube compressive strength. It can be seen that the results vary from a minimum to a maximum. The following parameters are used to quantify variability.

Mean The average $\bar{\sigma}_{cu}$ for n samples, having $\sigma_{cu1}, \sigma_{cu2}, ..., \sigma_{cu3}$, is expressed as

$$\bar{\sigma}_{cu} = \sum_{i=1}^{n} \sigma_{cui}/n \tag{22.1}$$

As the sample size increases, $\bar{\sigma}_{cu}$ approaches the mean of the whole concrete.

Range It is the difference between the minimum and maximum strengths of all cubes tested.

Deviation Generally each cube result will have strength different from the mean of the whole concrete represented by $(\sigma_{cui} - \bar{\sigma}_{cu})$ as shown in Table 22.1. This 'deviation' does not give an idea of the quality of the whole sample but deals with only the quality of sample 'i' considered.

Standard deviation It is the root mean square (rms) of the deviation of the whole concrete having $\bar{\sigma}_{cu}$ as the mean. That is,

$$S = \left[\sum_{i=1}^{n} (\sigma_{cui} - \bar{\sigma}_{cu})^2 /(n-1) \right]^{1/2} \tag{22.2}$$

The square of S is called *variance*. Standard deviation increases with increasing variability. When all the results are in a narrow band, S will be minimum, which indicates good quality control. On the other hand, a large S indicates spread of results and, hence, poor control. The mean of two batches of concrete may be the same but the spreads of results (larger S) indicate poor control (Fig. 22.2). Table 22.1 shows a gradual slackening of control from Batch 1 to Batch 5. Thus, the spread of strength indicates that the standard deviation is more. The overall standard deviation of all the five batches is 2.51. The standard deviation of Batch 5 is 2.99. It indicates that Batch 5 has poorer control since its spread of results is more.

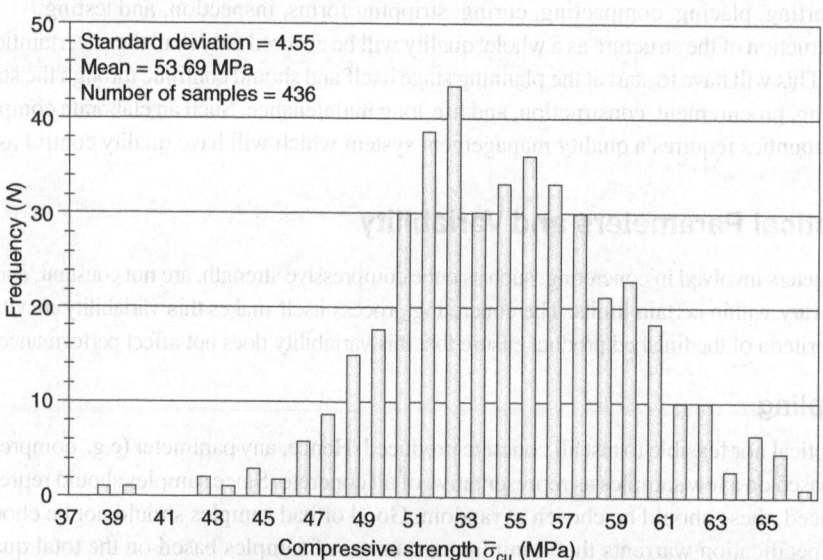

Standard deviation = 4.55
Mean = 53.69 MPa
Number of samples = 436

Fig. 22.1 Test results plotted in a histogram

Table 22.1 Standard deviation calculation

Sample no. i	Concrete strength (MPa) σ_{cui}	Deviation $\sigma_{cui} - \bar{\sigma}_{cu}$	Square deviation $(\sigma_{cui} - \bar{\sigma}_{cu})^2$
1	38.3	0.27	0.072
2	38.1	0.07	0.004
3	37.6	−0.43	0.184
4	36.7	−1.33	0.768
5	39.2	1.17	1.368
6	37.4	−0.63	0.396
7	36.1	−1.93	3.724
8	41.2	3.17	10.048
9	40	1.97	3.880
10	35.7	−2.33	5.428
Total	380.3		26.881
Mean	$\sigma_{cu} = 38.3$		$S^2 = 2.986$
	SD		$S = 1.728$
	Cov.		$v = 4.544$

Calculation of SD for various batches

	Mean	Standard deviation	Cov.	Sum. sq dev.
Batch 1	38.03	1.73	4.54	26.93
Batch 2	38.24	2.61	6.82	61.30
Batch 3	39.55	2.76	6.97	68.55
Batch 4	37.45	2.83	7.55	72.08
Batch 5	39.06	2.99	7.65	80.46
Sum	192.33			309.34
Total	1923.3			6.31
Overall	38.46	2.51		6.53

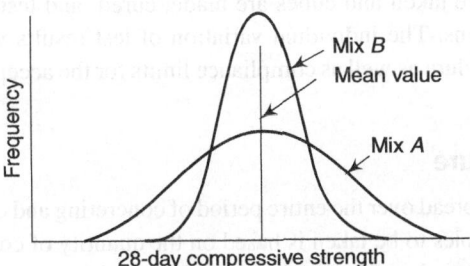

Fig. 22.2 Spreads of results indicate poor control

An important limit of strength, say $(\bar{\sigma}_{cu} \pm kS)$, can be set to prevent majority of samples from having very low or very high strength, which is not within the acceptable limit. In the above expression limit k is the probability factor. For different values of k, the percentage of results falling above or below a particular value is shown in Fig. 22.3 in relation to the area bounded by the normal probability curve. The value of k is given in Table 22.2 for various percentages of defective results.

The variation in results can be expressed as a percentage of standard deviation with respect to mean. This is defined as the coefficient of variation,

$$v = (S/\bar{\sigma}_{cu})100 \qquad (22.3)$$

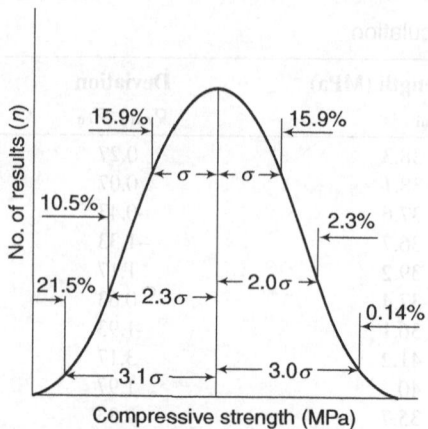

Fig. 22.3 Percentage of results falling above or below
a particular value

Table 22.2 Probability factor *k* for various percentages of defective results

Percentage of defective results	50	16	10	5	2.5	1	0.5	0
Number of defective results of the total	1 in 2	1 in 6	1 in 10	1 in 20	1 in 40	1 in 100	1 in 200	0
Probability factor *k*	0	1.0	1.28	1.65	1.96	2.33	2.58	Infinity

With a constant coefficient of variation, standard deviation *S* increases with strength. The compressive strength test results are found to follow a bell-shaped curve (shown in Fig. 22.2), called normal distribution curve.

22.2 Recommendations of IS: 456-2000

Samples from fresh concrete are taken and cubes are made, cured, and tested. The test result of a sample is the average of three specimens. The individual variation of test results within a sample should be less than ±15%. The sampling procedure as well as compliance limits for the acceptance of concrete are stipulated in IS: 456-2000.

22.2.1 Sampling Procedure

A random sampling procedure spread over the entire period of concreting and covering all mixing units should be adopted. The number of samples to be taken is based on the quantity of concrete, as given in Table 22.3.

At least one sample should be taken from each shift. Three specimens should be taken for each sample for getting the 28-day strength. Additional specimens are generally required to determine the 7-day strength, strength at the time of stripping the formwork, or to check the effect of the duration of curing.

22.2.2 Target Mean Strength

Since the strength test results follow a normal distribution variation, some results will fall below the specified strength. The design of the structure is based on characteristic strength. The term 'characteristic strength' indicates the strength of the material below which not more than 5% of the test results are expected to fall. In the design of concrete mix, therefore, the target strength aimed (which is the average strength) should be higher than the characteristic strength. The target mean strength is obtained as

$$f_t = f_{ck} + kS \tag{22.4}$$

where

f_t = target mean strength
f_{ck} = characteristic strength
k = probability factor
S = standard deviation

The probability factor k depends on the number of specimen results which are likely to fall below the characteristic strength. Table 22.2 and Fig. 22.3 show the relationship between k and the number of acceptable defective results (5% or 1 in 20 in the case of characteristic strength as per IS: 456-2000).

Table 22.3 Frequency of sampling

Quantity of concrete work (m³)	No. of samples
1–5	1
6–15	2
16–30	3
31–50	4
51 and above	4 + one additional sample for each additional 50 m³ or part thereof

22.2.3 Acceptance Criteria

IS: 456-2000 stipulates the following conditions for cube compression strength compliances:

a. The mean strength determined from any group of four consecutive test results complies with the appropriate limit in the second column of Table 22.4.
b. Any individual test result complies with the appropriate limit in the third column of Table 22.4.

Both the following conditions have to be met for the compliance of flexural strength of concrete:

i. The mean strength of any group of four consecutive test results should exceed the characteristic flexural strength by at least 0.3 N/mm², that is, mean strength $\geq f_{flexural} + 0.3$ N/mm².
ii. Any individual flexural strength test result should not be less than the characteristic flexural strength by 0.3 N/mm², that is, individual flexural strength $\geq f_{flexural} - 0.3$ N/mm².

Note that $f_{flexural}$ indicates the characteristic flexural strength of concrete (5% or 1 in 20 as per Table 22.2).

The quantity of concrete represented by a group of four consecutive test results will include the batches from which the first and last samples were taken together with the in-between batches. For individual test results, only the particular batch from which the sample was taken is at risk. The acceptance limits mentioned above are shown in Fig. 22.4.

Determination of standard deviation   Standard deviation of each grade should be assessed based on the test strength results of samples.

Number of test results of samples   The total number of test strength samples required to constitute an acceptable record for calculation of standard deviation should not be less than 30. Where sufficient test results for a particular grade of concrete are not available, the value of standard deviation given in Table 22.5 may be

Table 22.4 Characteristic compressive strength compliance

Specified grade	Mean of the group of four non-overlapping test results (N/mm²)	Individual test result (N/mm²)
M15	$\geq f_{ck} + 0.825\,S$ (rounded to nearest 0.5) or $f_{ck} + 3$, whichever is larger	$\geq f_{ck} - 3$
M20 or above	$\geq f_{ck} + 0.825\,S$ (rounded to nearest 0.5) or $f_{ck} + 4$, whichever is larger	$\geq f_{ck} - 4$

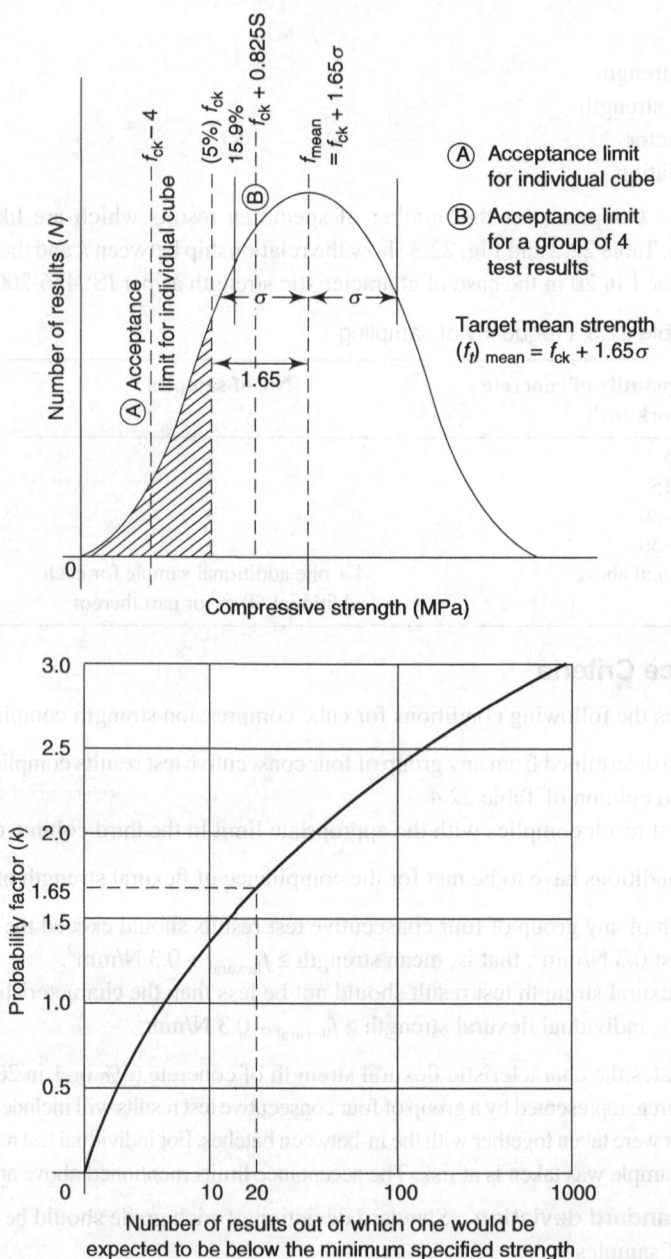

Fig. 22.4 Various important limits with respect to normal distribution curve

Table 22.5 Assumed standard deviation

Concrete grade	Assumed standard deviation (N/mm²)
M10–M15	3.5
M20–M25	4.0
M30–M50	5.0

assumed in the first instance. As soon as the results of test samples are available, actual calculated standard deviation can be used for the design of mix. The values given in Table 22.5 correspond to site control having proper storage of cement, weigh batching of all materials, controlled addition of water, regular checking of all materials, aggregate grading, appropriate moisture content, and periodic checking of workability and strength. Where the quality control is slackened, the values given in Table 22.5 should be increased by 1 N/mm^2.

22.3 Errors in Concrete Construction

In concrete construction, very often errors are committed by people involved in the job, not fully understanding the consequences of their errors/neglect. In this section we will discuss the various errors resulting from lack or neglect of quality control over processes, materials, and workmanship in the production of concrete.

22.3.1 Cement

Ordinary Portland cement (OPC) and Portland Pozzolana cement (PPC) are available in the market and are permitted to be used in concrete, reinforced concrete, and prestressed concrete. What is often forgotten is that at early ages the PPC gains strength slowly and, therefore, the removal of formwork should be delayed, say, up to 21 days instead of 14 days. Those who follow cube tests may not have to take this precaution. Yet another error is in the old (even current) practice of volume batching on the basis of a 50-kg bag of cement without any consideration for weight loss caused by multiple handling of the bag.

Adulterated cement is hard to identify beforehand. Once used in concrete, it will be too late to correct the problem after setting. Distinction between OPC and PPC can be established through a check on the pozzolana content. The method used is codified in IS: 3501. Table 22.6 lists some field tests to identify adulterated cement.

22.3.2 Aggregates

No serious checks are made on the dirt and silt contents in the aggregates. Often stored coarse aggregates allow silt to filter to the bottom, and the bottom layer aggregate is used in concrete assuming it to be of the same quality as that of the top layer. Concrete is to be mixed/designed considering the fineness and size of sand. However, very little of sieving is done at site to maintain uniformity of the sand size. Moisture content in sand is also not measured. Therefore, the water–cement ratio varies from batch to batch; without accounting for the varying moisture content in sand, a constant quantity of water is added in the mixer for every batch. Bulkage due to moisture in sand is also often ignored or wrongly assessed.

22.3.3 Water–Cement Ratio

Although the water–cement ratio is considered and generally understood as the most critical element for the strength and durability of concrete, scant attention is paid to maintain a predetermined water–cement ratio on

Table 22.6 Field tests on cement

S. No.	Test	Procedure
1.	Test of cement adulterated with stone dust	Take a sample of the doubtful cement and heat it on a clean steel plate with the help of a stove for about 20 min. If the cement changes colour, it is indicative of adulteration with stone dust. Genuine cement does not undergo any colour change when heated.
2.	Test of cement adulterated with fly ash	Take a small quantity of the doubtful sample in a glass tumbler and add water till the tumbler is half full. After vigorous agitation of the contents, allow the mixture to settle. If the cement is adulterated with fly ash, it could be seen that the ash particles either float or remain in suspension with the cement particles settling down.

the production line. The mixture operator, the loader, the mason, and the finisher who finishes the concrete all add water at free will to maintain what they call 'workability'.

22.3.4 Mixture Type and Mixing Time

A mixing time of 60s in the drum is generally considered adequate for mixing concrete. However, it is not fully understood that this mixing time depends on the type of mixer and the position of the axis of the mixer. Pan mixers take only 30s, whereas the inclined-axis mixers (tilting drum) may need as much as 120s. It is also to be understood that the sequence of loading of the constituent materials into the mixer also has significant bearing on the quality of concrete. The right sequence will be to load part of mixing water into the drum in the first instance and to charge the loader with part of the coarse aggregate, sand, cement, and the rest of the aggregate in that order. A change in the sequence of loading is a uniform error that is committed at all sites of construction under the mistaken belief that what gets out of the mixer 'looks good' and, therefore, concrete is good.

22.3.5 Placing by Mortar Pans

Usually a series of mortar pans are dumped on the reinforcement to effect easy discharge of the concrete. This results in dispersed depositing of coarse aggregates and mortar remaining in isolated locations. The mortar pan is to be faced inwards from the shutter and slowly discharged to get an even distribution of the concrete matrix.

22.3.6 Reinforcement

With the advent of different types of reinforcement, the site staff often mistakes all deformed bars as high-strength bars. The error is not in identifying the type of bar (whether it is hot-rolled or cold-worked)—whether the cold-worked bars are made to uniform pitch, whether the bars have their deformations uniformly, or are corroded away. The presence of heavy rust on the surface of bars can endanger the life of the concrete structure. Providing cover blocks and maintaining a uniform distance from the formwork and between layers of reinforcement through chairs remain as statements made in textbooks only. The site staff seldom appreciate that adequate and uniform cover is essential to prevent corrosion of steel in concrete. Yet another mistake often made is with bending and re-bending of high-strength deformed (HSD) bars, treating them at par with mild steel bars. Cold-worked HSD bars have a lower ductility and, therefore, such bars will not be able to withstand repeated bending and re-bending for long.

22.3.7 Formwork

Maintaining lines and levels for concrete formwork is as essential as the mould for casting. Neglect of this aspect is often based on the presumption that a subsequent plaster layer can cover up all defects. It is not uncommon to see lines and levels, including plumb lines, being off by more than 2–3 cm. Joints of formwork are seldom packed effectively. The scanty sealing of the joints through coir, mud, etc., will last only till the first pouring of water is made to wet the form surface. All the cement slurry escapes through these joints. Coating of the formwork with suitable materials as form release agents is essential. The release agents differ in their absorption capacity and are different for timber formwork, plastic formwork, and steel formwork. Heights of columns and walls to be connected in one go are often points of dispute at site. While old specifications restrict the height to around 1 m, new practices tend to allow this height to be around 1.8 m. A larger drop is alright if the concrete has been designed specifically for preventing segregation. Dropping of concrete from the top of the formwork in heights more than this can lead to segregation. Hydrostatic pressure of vibrated concrete also demands heavy ties for formwork at the bottom of the pour. This often leads to gaping forms during construction. If greater heights are to be concreted in one go, the use of a chute or pumped delivery to the bottom of the formwork becomes necessary. For the same reason, concrete should only be poured in layers to ensure proper vibration. Stripping of formwork is often done without proper consideration of transfer of loads to the supporting structure. One mistake committed often is the removal of the props for cantilevers from the support outwards. Often one attempts to remove the props from below at an early date, believing that a re-propping can re-establish the supported conditions. It is not realized that while removing the props the self-weight of the elements would have already acted and there are no means by which this loading due to self-weight can be reversed.

22.3.8 Vibration

One assumes that if there is a vibrating machine available at site, all compaction problems can be solved. Spacing of needle points far apart does not ensure compaction. The size of the aggregate and the congestion of reinforcement will decide the capacity of the vibrator to effect compaction. For the fear of disturbing poorly done formwork scant vibrations are done. Some supervisors and builders believe that slab concrete is not to be vibrated. This is not correct. Over-vibration and quick withdrawal of needle vibrators are yet another set of errors committed at the site. Such operations end up creating pockets of mortar or cavities. Improper compaction around steel bars can leave cavities on the underside of these bars, thus preventing proper bonding between concrete and steel. A displacement of steel from its designed position through indiscrete walking, stamping, and placing of heavy loads during concreting can seriously alter the resisting moment capacity of a given concrete section. Thus, steel at the top of a cantilever often ends up being at the bottom by the time concreting is completed.

22.3.9 Finishing

Concreting done with indiscrete addition of water will require a lot of reworking to finish the surface. Such concrete with a lot of slurry on the top leads to shrinkage, cracking, and crazing on top. The top layer of concrete in such cases will be very weak.

22.3.10 Curing

Wet concrete needs to remain in a humid atmosphere to retain the water for it to complete the hydration reactions. Such reactions generate considerable heat in the concrete. It is for this reason that curing by proper means is to be insisted upon. At majority of the construction sites, one sees very little attention paid to curing. The concrete surface remains dry for most of the time, and it is this inadequate curing that leads to cracking of concrete and spalling of the surface layer.

22.3.11 Expansion and Construction Joints

During construction, concreting may have to be stopped for purposes of rest, change of crew, etc. A partially set concrete surface is to be raked up and treated with cement slurry and mortar, and then overlain by fresh concrete. The new layer of concrete should be well vibrated to integrate it with the old concrete layer. Such an exercise often is not followed strictly. Neglecting this could give rise to points where leakage of water and development of temperature cracks occur. The need to concrete a large surface area in squares not extending 3 m in length is not well understood by the site staff. This is a measure to reduce shrinkage effects. Expansion joints are to be provided as free joints, through the structure, to allow for differential and temperature movements. Such joints are to be at 30 m spacing in the case of masonry structures and at 45 m spacing for reinforced concrete structures. Though often such a provision is shown in the design drawings, the site staff seldom make sure that these joints are provided for, and if provided for, they are kept completely free to allow movement. The errors are by way of allowing mortar to fall through these gaps and set, as also by allowing debris to get deposited in these joints. This restricts movement and results in cracking of the building at or near the joint.

Cracks in buildings founded on weak soils often exert forces across RCC elements also. Horizontal continuous bands in RCC are provided in masonry structures under such conditions to take care of differential settlements.

22.3.12 Precast Concrete

The general errors normally made in small precasting works are highlighted here. Moulds with re-entrant angles can make the release of concrete product difficult. Closed precise elements should have sides with adequate batter. Surface projections/depressions on masonry moulds are often not attended to this way, resulting in locking of products. Release agents are not applied uniformly. Cover on thin elements is to be carefully attended to. Stripping pieces are to be inserted to facilitate easy stripping of vertical and horizontal form surfaces. One often neglects the action of self-weight in handling an RCC or prestressed element. Removing the mould supporting elements at the centre and at diagonal corners before getting adequate strength, as well as giving improper lines of support while stacking, are common errors. Often precast elements are kept upside down, assuming these products to behave like steel—with more or less equal strength in tension and compression. During handling and erecting, the horizontal load distribution directly from the slings, rather than through a distributor beam, is a common mistake, which results in accidents and loss of strength due to cracking. Table 22.7 shows a checklist to be verified before commencing any concreting work, and Table 22.8 shows common defects and possible remedies.

Table 22.7 Checklist before commencing concreting

1.	Final levels and lines of centring checked	Yes/No
2.	Reinforcement and centring approval by the engineer-in-charge	Yes/No
3.	Cover blocks for reinforcement provided	Yes/No
4.	Necessary tightening of supports and bracing completed	Yes/No
5.	Required quantities of cement, metal, and sand (approved quality) for the day's work brought to site	Yes/No
6.	If construction joint is planned, necessary stop board prepared and brought to site	Yes/No
7.	Concrete mixer and vibrator with operators and mechanic available	Yes/No
8.	Necessary wooden benches for walking over slab reinforcement available	Yes/No
9.	Water–cement ratio fixed and measuring can for water available	Yes/No
10.	Supervisor at mixing point tasked with sufficient briefing to ensure production of quality concrete	Yes/No
11.	Cube moulds for making test cubes and slump cone kept ready	Yes/No
12.	Bulkage test is conducted and the percentage of bulkage decided for sand	Yes/No
13.	All inserts to be embedded in concrete checked and placed in position including electrical conduits	Yes/No

Table 22.8 Common defects/problems in concrete and their remedies

S. No.	Defect	Cause	Suggested solution
1.	Segregation and bleeding	Mix is lean Over-vibration	Use richer mix Avoid over-vibration Reduce water–cement ratio
2.	Permeability and shrinkage	High water content in mix Lack of compaction Improper grading of aggregate Improper mix design	Resort to sufficient compaction Use proper graded aggregate Use correct water–cement ratio
3.	Blow holes on exposed faces	Inadequate cover between reinforcement and mould face which restrains local flow of concrete between them Lack of sufficient vibration	Use air-entraining to improve workability if the water–cement ratio is to be kept low Use adequate cover for reinforcement Use larger and cubical coarse aggregate and sufficient vibration Resort to adequate mechanical vibration using a vibrator head of larger circumference (it is to be noted that the vibrator needle has no effect on the concrete below the tip of the needle) Resort to knifing at the contact face of mould Resort to vigorous tamping with a wooden mallet on all sides of formwork as the concreting proceeds

(contd)

(contd)

S. No	Defect	Cause	Suggested solution
4.	Plastic cracks (small, near horizontal cracks at faces)	During compaction higher particles tend to settle down and water rises up and collects below certain points of concrete, remaining higher due to arching or interlocking. This causes cracks during drying below such points	Use cohesive mix Place and compact concrete in layers, avoiding any local points of arching
5.	Crazing (map of fine cracks on the surface up to 3 cm long)	Shrinkage of surface due to surface carbonation (caused due to the reaction of free lime released during hydration with CO_2, which reduces the resistance of the surface to drying shrinkage)	Reduce water–cement ratio Use pozzolana cement Resort to membrane curing and keep the surface covered and moist
6.	Efflorescence (white patches)	High water–cement ratio Large quantity of free lime released during hydration	Reduce water–cement ratio Use pozzolana cement
7.	Spalling	Use concrete of low strength Form lining is not absorbent Cement used is slow setting	Use concrete of sufficient strength Use absorbent form lining Use cement of appropriate setting time or use accelerators
8.	Cement not hardening	Cement used is adulterated	In case of suspected adulteration, carry out necessary field tests to ascertain whether the cement is adulterated or not, prior to using it in the work
9.	Loss in workability and mix becoming harsh	Evaporation of moisture in concrete, which is required for hydration: due to severe temperature conditions coupled with wind Improper grading or insufficient cement sand mortar in the mix	In case the initial setting time has not lapsed, the concrete may be tempered with by adding a little quantity of water having regard to the water–cement ratio and then remixing before use

22.4 Tools for Quality Management

A quality management plan documents how an organization can plan, implement, and assess the effectiveness of its quality assurance and quality control operations. With respect to the concrete construction industry, the management tools discussed here will be useful to give leverage over common concreting problems.

22.4.1 Control Chart

Variation of quality is inherent as in everything we do. A control chart, having superimposed alert indicators, is a graph for monitoring the variation in a process activity. The alert indicators are not guessed but calculated using statistical theories of probability. As soon as performance (in this case concrete strength) approaches or crosses the action lines, steps should be taken to investigate this variation in order to correct it and remove the source of variation. Over a period of time, these reactive actions, if investigated and standardized, will improve the capability of the process (concreting). This will help to meet the requirement (required strength) and reduce the likelihood of failure.

Control charts were originally developed for monitoring the manufacturing process. However, they have been found to be useful for monitoring quality by all concerned (i.e., the contractor, the engineer, the builder, and the owner). Figure 22.5 shows a typical control chart in which the upper and lower control limits (UCL and LCL) have been shown. The alert indicators, also called 'action lines', clearly define when 'action' should be taken. They are set within the control limits so that corrective action is taken before the output (control strength) becomes unacceptable. The control limits are set at the required number of standard deviations either side of the mean. These clearly define the expected deviation in the behaviour process and hence are helpful in preventing unnecessary intervention. Control charts enable process monitoring between prescribed limits so that action can be taken when the data (Fig. 22.6) shows that action lines or control limits are breached.

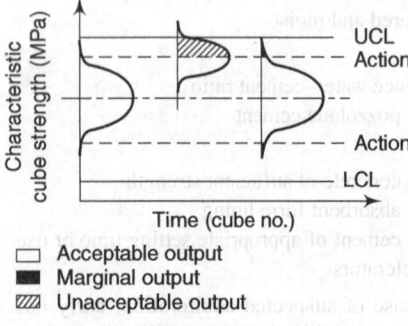

Fig. 22.5 A typical control chart

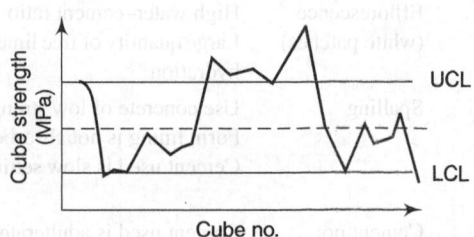

Fig. 22.6 Data variation with respect to control limits

22.4.2 CUSUM chart and its application

CUSUM control systems (short for CUmulative SUM) have found widespread use in the concrete industry. In CUSUM charts, the central line does not represent a constant mean value but is a zero line for the assessment of the trend in the results. For concrete production the following CUSUMs are used:

- CUSUM M, for the control of mean strength
- CUSUM R (range), for the control of standard deviation
- CUSUM C, for the control of correlation

The CUSUM method involves subtracting the test result from a target value. Thus, one can produce an ongoing running sum (the CUSUM) of the differences. If the process is in control, the points on the CUSUM plot are distributed randomly (positive and negative differences cancelling each other out), to give a cumulative sum that is close to zero. However, if the process slips out of control, CUSUM will move towards the upper control limit (UCL) or lower control limit (LCL).

Consider the data for compressive strength in Table 22.9 for a plant operating on a standard deviation of 3 N/mm^2 and with a target mean strength of 30 N/mm^2. Column 2 of the table gives the strength of concrete cube sequentially for 18 samples. Column 3 gives the deviation with respect to target strength of 30 N/mm^2. The calculated values of CUSUM are shown in column 4. A plot of CUSUM chart is shown in Fig. 22.7. In the same chart the upper and lower control limits (UCL and LCL) based on standard deviation are also shown. The standard deviation assumed is 3.0. It is seen that CUSUM approaches LCL at point 11. CUSUM plot gives a clear visual picture of the trend. It shows that a change in the process has occurred at point 11. CUSUM chart can be used to control mean strength, standard deviation, or for that matter correlation.

Table 22.9 CUSUM data

| Sample sequence | Target strength 30 N/mm² | | |
	28-day strength (N/mm²)	Deviation	CUSUM (N/mm²)
1	27	−3	−3
2	32	2	−1
3	26	−4	−5
4	25	−5	−10
5	32	2	−8
6	28	−2	−10
7	29.5	−0.5	−10.5
8	30	0	−10.5
9	25	−5	−15.5
10	30	0	−15.5
11	24	−6	−21.5
12	34	4	−17.5
13	36.5	6.5	−11
14	32	2	−9
15	34.5	4.5	−4.5
16	35	5	0.5
17	34	4	4.5
18	38	8	12.5

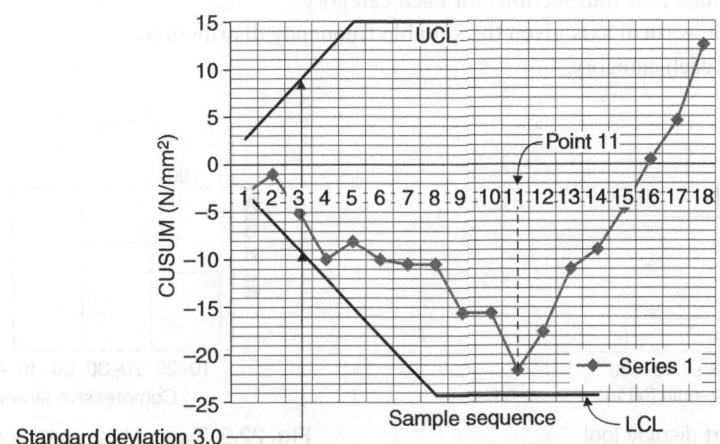

Fig. 22.7 Example for CUSUM plot

22.4.3 Data Display

Data display methods show information that is easy to use. These are important tools in quality management and highlight patterns, trends, and relationship. The main types of display charts are

    a. Bar chart
    b. Histogram
    c. Pie chart
    d. Scatter diagram
    e. Trend (run) graph
    f. Spider diagram

The patterns and trends of different quality parameters, such as cube strength and slump, can be displayed, which make data display methods user-friendly for a wider use by all concerned.

Bar chart   This display tool enables comparison of several discrete items with one another. For example, Fig. 22.8 shows the average strengths of different types of concrete produced over a two-month period.

To produce a bar chart, the following steps are taken:

1. Identify the parameter to be measured (strength).
2. Design a method of collecting data (sampling).
3. Review collected data and select a horizontal scale.
4. Plot items (strengths of different types of concrete, steel, brick, etc., along the vertical scale).
5. Draw bars (in different colours or shades) to show different performance parameters (strength) for each item (material).

Histogram   A histogram shows the range of data which have been collected on a particular process characteristic (e.g., concrete strength). It shows the frequency distribution in bar forms. The data used in a histogram are also the starting point for developing control charts as these are linked to variations.

Figure 22.9 shows a typical histogram of concrete strength. To develop a histogram, the following procedure is followed:

1. Identify the variable to be measured (i.e., 28-day cube strength, water–cement ratio, slump, etc.).
2. Set up a method to collect data (sampling, item frame, checksheet, etc.).
3. Total number of times (frequency) that each category of variable (strength) has occurred.
4. Break the horizontal axis into sections for each category.
5. Choose a suitable vertical axis given the suitable frequency distribution.
6. Plot the bars for each category.

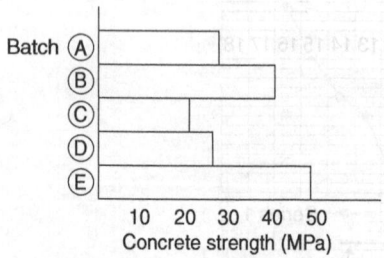

Fig. 22.8 Bar chart display tool

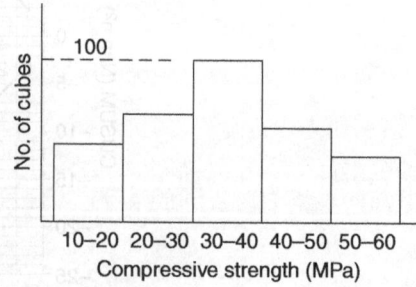

Fig. 22.9 Typical histogram for concrete strength

Pie chart   A pie chart displays properties in relation to the whole item and allows easy visual comparison among the parts (Fig. 22.10).

To develop a pie chart, the following procedure is followed:

1. Identify the characteristic to be subdivided into components.
2. Collect data on all attributes (weight, volume number, etc.).
3. Identify how this is broken into subdivisions already identified.
4. Convert subdivisions into percentages of the total.
5. Break up the chart into proportions of the percentages and label accordingly.

Scatter diagram   This kind of display identifies the relationship between two variables by plotting changes. As an example, a scatter diagram can be drawn between slump and water–cement ratio (Fig. 22.11).

To develop a scatter diagram, the following steps are followed:

1. Identify the two variables.
2. Collect performance data from site/lab.
3. Plot a suitable scale on each axis and express the points showing the relationship between causes and effects of the chosen variable on the vertical axis.
4. Plot each pair of measures.
5. The more closely the dots are aligned, the stronger is the correlation between the two. A random pattern of dots indicates poor relationship.

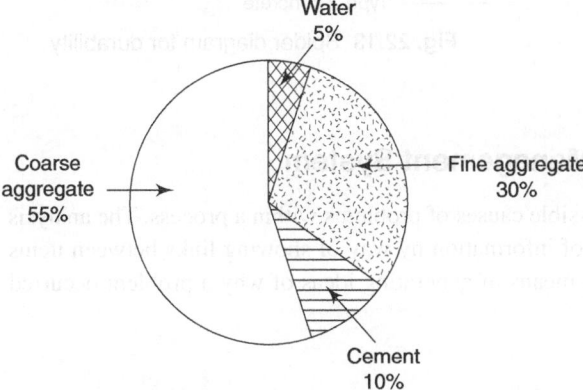

Fig. 22.10 Pie chart showing materials in concrete

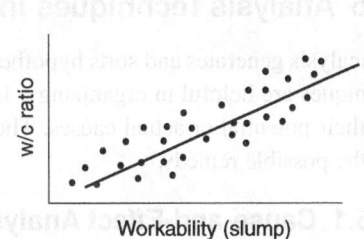

Fig. 22.11 Scatter diagram for slump with respect to w/c ratio

Trend chart (run/line graph)   This display method is used to display the trend of one of the variables against time. For example, it can be strength gain of concrete with days (Fig. 22.12).

To develop the chart, the following sequence is adopted:

1. Identify the variable whose trend needs to be ascertained.
2. Collect data (based on tests) on the variable.
3. Draw suitable axes, time on *x* and variable on *y*.
4. Plot each measure either retrospectively or allow the graph to build over time as data become available.

Spider diagram   A spider diagram is a way of showing performance on a range of dimensions, each of which is a component of the overall issue (Fig. 22.13). An example is the durability of concrete.

To develop a spider diagram, the steps given below need to be followed:

1. Identify the parameter to be measured and the components affecting it.
2. Collect data, which allow information to be plotted on a linear scale.
3. Collate the data and summarize the performance (for instance the ultrasonic pulse velocity [UPV] value as a function of durability).
4. Draw a spider diagram with the appropriate number of legs using the relevant scale for each.
5. Plot the performance of each component and label the same.
6. Add one or more performances as a comparison.

Data display methods help to display relationships, patterns, and trends in the data collected. The methods described above are visual, interesting, and easier to use than tables or raw data. They help to turn raw data into useful information.

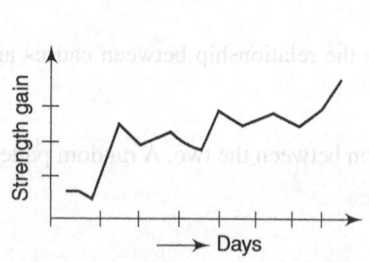

Fig. 22.12 Trend chart showing concrete strength gain with age

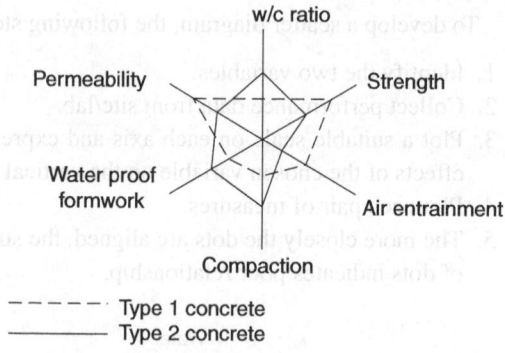

Fig. 22.13 Spider diagram for durability

22.5 Analysis Techniques in Quality Management System

An analysis generates and sorts hypotheses about possible causes of problems within a process. The analysis techniques are helpful in organizing a large amount of information by way of showing links between items and their potential or actual causes. These provide a means of generating ideas of why a problem occurred and the possible remedy.

22.5.1 Cause-and-Effect Analysis

The cause-and-effect analysis is a technique for identifying all the possible causes (inputs) associated with a particular problem (output). This technique helps one to narrow down to identify the exact characteristic relationship that is responsible for an effect.

A cause-and-effect diagram, also known as the fishbone diagram or Ishikawa diagram, graphically illustrates the results of the analysis and is constructed in steps. A cause-and-effect analysis is carried out by groups who have the relevant experience and knowledge of the cause being analysed.

The analysis starts with the selection of a problem, such as a honeycomb present in concrete. The problem is defined in definite terms. The next step is to brainstorm and arrive at the most probable causes that lead to honeycombing. Next a fishbone diagram, as shown in Fig. 22.14, is drawn. Then categories of causes are established by reviewing the output of the brainstorming sessions to determine the major cause categories (Fig. 22.15), which can be

- workmen
- equipment
- material
- environment
- method or process

There is no single set of categories. One must adapt the categories to suit the problem being analysed.

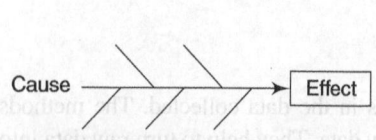

Fig. 22.14 Fishbone diagram

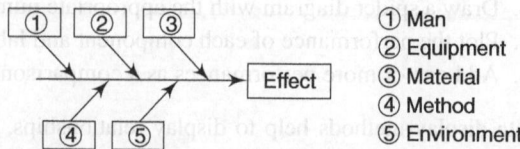

① Man
② Equipment
③ Material
④ Method
⑤ Environment

Fig. 22.15 Major causes for honeycombs in concrete

Potential causes from the brainstorming session are then transferred to the fishbone diagram, placing them in the appropriate category. If causes fit in more than one category, it is important to duplicate them. Related causes are plotted as twigs on the branches. Figure 22.16 shows a typical fishbone diagram used for analysing the problem of honeycombing due to leaky formwork in concrete construction.

The cause-and-effect analysis is a variable tool for

- focusing on causes and not symptoms
- capturing the collective knowledge of a group
- providing a picture of why a defect is happening
- establishing a sound basis for providing future action to avoid the defect

A fishbone diagram may be developed for all activities within the process that is generating the output, so that causes are linked to particular steps in the process as shown in Fig. 22.17.

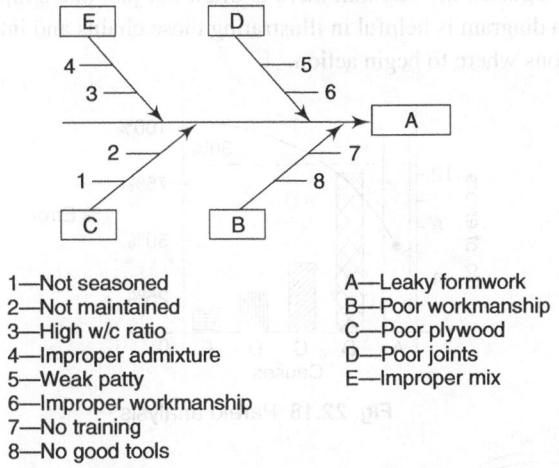

1—Not seasoned
2—Not maintained
3—High w/c ratio
4—Improper admixture
5—Weak patty
6—Improper workmanship
7—No training
8—No good tools

A—Leaky formwork
B—Poor workmanship
C—Poor plywood
D—Poor joints
E—Improper mix

Fig. 22.16 Mirror twigs and branches of fishbone diagram

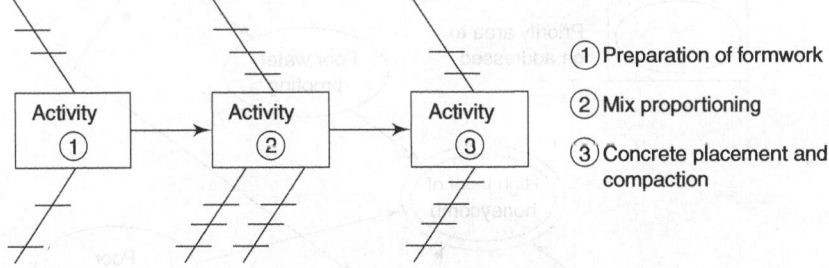

①Preparation of formwork
②Mix proportioning
③Concrete placement and compaction

Fig. 22.17 Fishbone diagram for multiple activities

22.5.2 Pareto Analysis

A Pareto analysis often reveals that a small number of failures are responsible for the bulk of quality costs, a phenomenon known as 'Pareto principle'. The pattern is also called '80–20 rule' and shows itself in many ways. For example, 80% of weaknesses are the result of 20% of causes. 80% of quality costs are caused by 20% of problems. 20% of defects in materials cause 80% of problems.

A Pareto diagram allows data to be displayed as a bar chart and enables the main contribution to the problem to be identified and highlighted (Fig. 22.18). The first step to build a Pareto diagram is to gather data. Next draw errors as a bar chart due to different causes. It may be helpful to also depict cumulative errors. This helps to identify the categories contributing up to 80% of the quality cost. The most frequent error is not always the most important. The Pareto analysis identifies the most important error or mistake, such as a leaky formwork.

22.5.3 Relational Diagram

A relational diagram is a technique to relate complex causes and effects. It allows one to investigate the multiple chains of effects and causes which often exist. This is illustrated graphically in Fig. 22.19.

At first one should define the effect to be analysed. This effect is to be placed in the middle of the diagram. Identify the key factor which causes the effect. Write these across. Cross-links are added. Based on these causes are identified.

A relational diagram recognizes the fact that there is often not just one simple causal relationship at the root of the problem. Such a diagram is helpful in illustrating these chains and interconnections and therefore useful in identifying locations where to begin action.

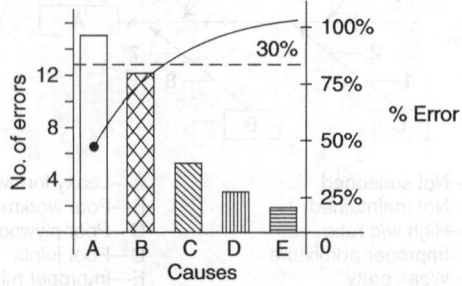

Fig. 22.18 Pareto analysis

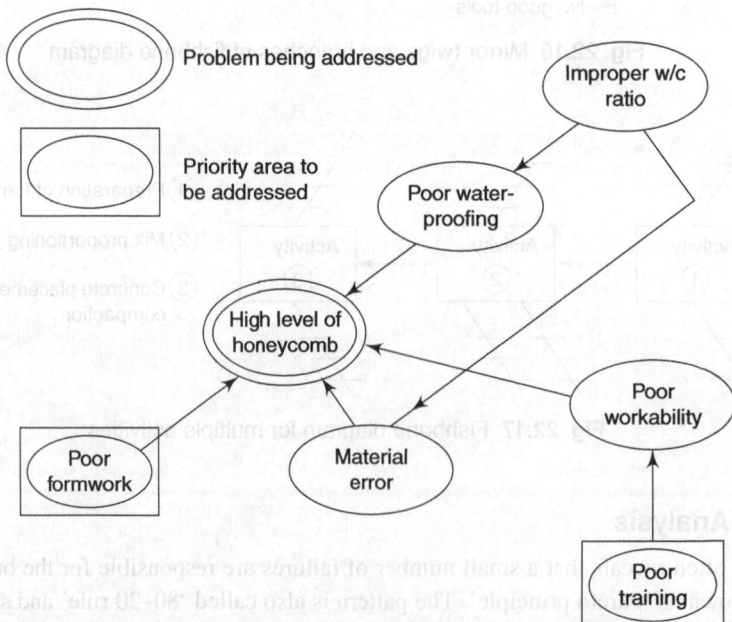

Fig. 22.19 Typical relational diagram

22.5.4 SWOT Analysis

A SWOT analysis is a graphical way of summarizing a particular process, product, department, or organization in terms of its strengths, weaknesses, opportunities, and threats. A central item that is to be analysed is chosen. Then the following four areas are analysed and discussed in detail among the group involved (Fig. 22.20).

Strengths   The behaviour and effects of performance which are strong.

Weaknesses   The behaviour, effects and aspects of performance which are weak.

Opportunities   Events and external changes that create an improvement in condition.

Threats   Events and changes external to the body being analysed which could adversely affect performance.

A SWOT analysis is useful for summarizing the various forces at play in a situation as a starting point for identifying the areas for action.

Fig. 22.20 SWOT analysis for concrete construction

22.6 Quality Management System for Construction

Quality management ensures that every component of the structure keeps performing throughout its lifespan. In fact, quality is a measure of the degree of excellence and is indeed related to fulfilment enjoyed by the user. In concrete construction, even if rigid quality control is not followed, the material performs for a short while without loss of strength. On account of this property of concrete, many in the construction industry have been operating under the misconception that rigid quality management, which is essential for mechanical industries, is not so important for concrete manufacture. This is not correct. The quality management in the current day context is based on the fact that the probability of failure of structures must be as low as possible and definitely lower than a prefixed accepted limit. Hence, quality management in essence is the management of uncertainties inherent in the construction industry.

The quality management system in a true sense should have the following three components (Fig. 22.21):

1. Quality assurance plan (QAP)
2. Quality control process (QC)
3. Quality audit (QA)

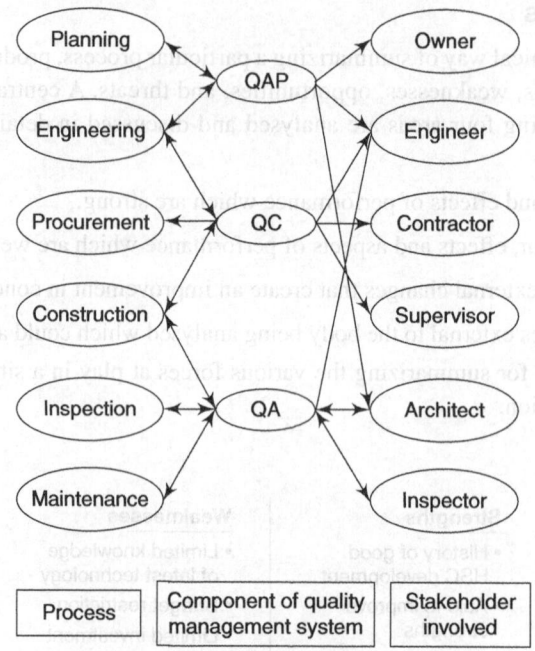

Fig. 22.21 Quality management system

22.6.1 Quality Assurance Plan (QAP)

The following aspects should be addressed by any QAP:

a. organizational set-up
b. responsibilities of personnel
c. coordinating personnel
d. quality control measure
e. control norms and limit
f. acceptance/rejection criteria
g. inspection programme
h. sampling, testing, and documentation
i. material specification and qualification
j. corrective measure for non-compliances
k. resolution of disputes/difficulties
l. preparation of maintenance record

The quality assurance plan starts right from the planning and design stage itself, and it can be defined as a procedure for selecting a level of quality required for a project.

22.6.2 Quality Control Process

It is a system of procedures and standards by which the contractor, the product manufacturer, and the engineer monitor the properties of the product. Generally the contracting agency is responsible for the QC process. A contractor responsible for quality control incurs a cost for it, which is less than the uncontrolled cost for correcting the defective workmanship or replacing the defective material. Hence, it is prudent to introduce effective quality control.

22.6.3 Quality Audit

This is the system of tracing and documentation of quality assurance and quality control programme. It is the responsibility of the process owner. Both design and construction processes come under this process. The concept of QA encompasses the project as a whole. Each element of the project comes under the preview of quality audit.

Exercises

Review Questions

1. Explain the significance of quality control.
2. What are the recommendations given in IS: 456-2000 to produce good quality concrete?
3. List the various errors in concrete construction.
4. What are the common defects in making concrete and what are the remedies for these defects?
5. Distinguish between quality management system and quality control.
6. What are the three steps involved in quality management?
7. List the various data display methods used in quality assessment.
8. What are the various analysis techniques used in quality management system?
9. What are the statistical parameters of cube strength? Explain.
10. How do you assess the quality of two batches of concrete having the same average strength?
11. What is the significance of characteristic strength with respect to probability factor k? Explain.
12. How is the target strength for mix design fixed based on quality of concrete? Explain.

Multiple Choice Questions

1. The characteristic strength indicates the strength of concrete below which not more than ___ % of test results can occur.
   a. 10   c. 0
   b. 5   d. 5

2. The mean strength of concrete is the strength below which ___ % of test results can occur.
   a. 10   c. 0
   b. 5   d. 50

3. The target strength for concrete of grade M30 with a standard deviation 3 is
   a. 25 MPa   c. 30 MPa
   b. 35 MPa   d. 40 MPa

4. Number of test samples required to constitute an acceptable record for calculation of standard deviation should not be less than
   a. 6   c. 12
   b. 3   d. 30

5. Assumed standard deviation as per IS: 456-2000 increases with
   a. Increase in strength   c. Decrease in strength
   b. Decrease in quality control   d. Increase in w/c ratio

6. Control limits in charts for quality control are intended to
   a. Keep the variation in quality within limit   c. Check whether corrective measures are working
   b. Alert changes in the process adopted   d. All of the above

7. IS: 456-2000 stipulates that the quantity of cement and aggregates should be determined by
   - a. Volume
   - b. Weight by any weighing machine
   - c. Weight or volume
   - d. Weight by a weighing machine with 2% accuracy

8. CUSUM charts are used to check
   - a. Mean strength
   - b. Standard deviation
   - c. Range of strength
   - d. All of the above

9. The negative tolerance for cover permitted by IS: 456-2000 is
   - a. 0 mm
   - b. −10 mm
   - c. −5 mm
   - d. −1 mm

10. The mixing time in a mixer machine for concrete should be at least
    - a. 2 min
    - b. 5 min
    - c. 1 min
    - d. 10 min

Answers to Multiple Choice Questions

1. b	4. d	7. d	10. a
2. d	5. a	8. d	
3. b	6. d	9. a	

CHAPTER 23

Repair Materials

The recent developments in material science and technology has enabled us to repair and rehabilitate old structures. These developments allow us to restore the old and weak structures to their pristine state in a cost-effective manner. Concreting a new structure and rehabilitating an old and existing one differ substantialy from each other in respect of the process involved. The technology to be adopted in a repair work has to be advanced as we have to consider the aspects which primarily led to the failure. The behaviour of a composite old–new system is totally different compared to the original old system. Rehabilitation if not handled properly will end up with repair failures.

Structures are subjected to variation in temperature and humidity because of changes in environmental conditions. The severity of the atmosphere varies depending on wind direction and location of industries in an urban location. If these considerations are overlooked at the design stage, the structures are at a greater risk of premature failures.

23.1 Repair Methodology

There is a variety of materials available that can be used to repair and rehabilitate dilapidated structures. The materials for rehabilitation include the following:

Polymer concrete   Polymer impregnation in concrete by compounds such as polymers, monomers, styrene, polyesters, methyl methacrylate, and similar compounds is a useful method to repair structures.

Epoxy grouts, mortars, and coatings   These are used

- for bonding plastic concrete to hardened concrete
- for bonding rigid materials to each other
- for patchwork
- for coating over concrete to give a preferred colour and to resist the action of chemicals, water penetration, abrasion, etc. Epoxy coatings are available for marine, underwater, and moisture-resistant conditions

Latex-modified concrete   Such concrete is used to improve

- bond strength
- shear strength
- durability of the structure

Miscellaneous materials and techniques   In addition, there are some other materials and techniques which can be used for repairing structures under special circumstances.

- *Coating:* Bituminous compounds such as linseed oil, fluorosilicate compounds, paints, styrene–butadiene rubber (for protection from rain), etc.

- *Jacketing:* Materials used for jacketing purposes include rubber, metals, plastic, concrete, fibre-reinforced concrete, fibre-reinforced plastics, ferrocement, polypropylene, etc.
- *Sand blasting:* This is used to remove foreign materials and stains.
- *Bentonite:* Bentonite and kaolinite are basically clay products. They prevent penetration of water through masonry soil or concrete structures.
- *Shrinkcomp grouting:* This is used for preventing problems of differential shrinkage between hardened and fresh concrete.

Proper and timely repair and rehabilitation ensure safety and serviceability of structures.

The selection of material is the most important step in the repair and rehabilitation programme. The rehabilitation engineer has to choose from the different materials available in the market. The choice of the material has a chemical angle. The manufacturer's literature normally highlights the composition of materials rather than its performance characteristics. Since repair schemes are not the same everywhere, different materials have to be employed at different sites for different conditions. The selection of material is guided by the type of structure, nature and extent of deterioration, and economic consideration. Commonly, material requirement for repair are corrosion inhibiting coat, bond coat, polymer patch mortars (coarse and fine), and flexible crack bridging surface protection coatings.

It is preferable to have all the materials based on one generic polymer so as to make the materials more compatible. The patching materials commonly used are cement-based materials or epoxy mortars. Due to similar coefficients of thermal expansion, cementitious materials are preferred over epoxy materials. The major problem in cement-based materials is the shrinkage characteristic. Therefore, formulations of patch mortar incorporate in the cement matrix several specialty chemicals to mitigate the shrinkage effects. The drying shrinkage should be reduced by using low binder content and low water–cement ratio.

23.2 Issues Related to Material Technology

The damage that occurs to a structural system stems from the deterioration of material. It may manifest in the form of cracking and/or disintegration of a member of a structural system. The cause needs to be examined at a microstructural level.

Concrete repair involves the use of a wide variety of materials with different physical and chemical properties and application techniques. Its compatibility with original construction material, availability, ease of use, and toxicity are some of the important attributes to be examined while choosing a repair material.

Compatibility is a measure of matching of physical, chemical, electrochemical, and dimensional properties of the repair materials with those of the original construction material. The original construction material is normally designated as the substrate. Figure 23.1 shows the effects of the mismatch between

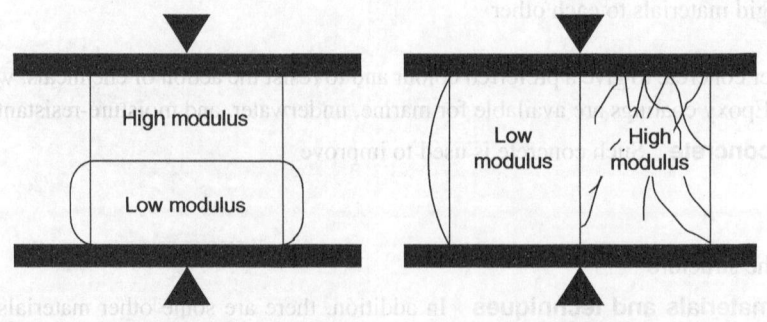

(a) Load perpendicular to the interface (b) Load parallel to the interface

Fig. 23.1 Effects of mismatched elastic moduli

the original low-modulus material (e.g., concrete) and the new repair material (e.g., patch mortar made of epoxy). It is necessary to choose the correct repair material for achieving the desired service life of the repaired structure. Note that the low-modulus material is likely to peel off in case the member load is parallel to patchwork. However, any loading perpendicular to the surface of patchwork will not affect the load transfer.

Table 23.1 shows the general requirements of a patch repair material with respect to the properties of the substrate.

The performance of the repair material should be much superior to the original material. It should be capable of being easily placed at a desired location. Thus, it should be possible to engineer the material for a specific repair performance. Repair materials are special material whose microstructural properties are doctored so that a desired macro-behaviour results. These materials are costly. However, their higher cost is justified by their superior long-term lifecycle behaviour with minimum or no maintenance.

Table 23.1 General compatibility requirements of patch repair materials

Property	Relationship of repair material (R) with concrete substrate (C)
Shrinkage strain	R < C
Creep coefficient (for repairs in compression)	R < C
Creep coefficient (for repairs in tension)	R > C
Thermal expansion coefficient	R = C
Modulus of elasticity	R = C
Poisson's ratio	R = C
Tensile strength	R > C
Fatigue performance	R > C
Adhesion	R > C
Porosity and resistivity	R = C
Chemical reactivity	R < C

23.3 Desired Properties of Repair Materials

In order to be used as repair materials, the materials need to have some desired properties. Some of the desired properties of these special materials are listed below:

1. Engineered materials with high performance, high durability, but low maintenance. The examples are
   - composites
   - block copolymer
   - high-performance concrete
2. Materials that are easy to use, have increased productivity, and reduced construction cycle time. The examples are
   - high-flow self-levelling concrete/mortar
   - set controlling materials
   - materials with reduced sensitivity for size, storage, and substrate condition such as temperature, moisture, and place of application
3. Safe materials which are environment-friendly: those which do not release harmful fumes during application and during service.
4. Materials that do not add to the dead weight of the repaired component or structure.

23.4 Materials for Repair

The various materials for repair are tabulated in Table 23.2.

Portland cement based materials Cement, sand mortar and cement, and sand and aggregate concrete have been used for repair. The application of these materials for repair jobs poses several problems such as shrinkage, cracking, and eventual failure of the repair work.

Resin-based products Epoxy resin formulations are used predominantly for repair work. The resin used for repair is typically a light-amber-coloured liquid having the consistency of motor oil. It must be combined with an amine or polyamide which is a hardener or curing agent. Once combined, a molecular crust linkage takes place. The following types of epoxies are in use:

a. Normal grade epoxy is suitable for bonding.
b. Low-viscosity epoxy is used if the crack width is less than 0.1 mm or less.
c. Normal epoxy grade mixed with sand (to increase the volume) is useful for grouting and patchwork. This gives fast development of strength. Especially when injection is involved, core samples should be taken to check the efficiency of execution.

Table 23.3 shows the properties needed for epoxy resin to bond fresh concrete to old hardened concrete.

Table 23.2 Materials used for repairs

S. No.	Material (generic type)
1.	Portland cement mortar
2.	Portland cement concrete
3.	Polymer-modified cement
4.	Polymer-modified mortar
5.	Polymer-modified concrete
6.	Polymer and fibre-modified mortar
7.	Polymer mortar
8.	Polymer concrete
9.	Resin–fibre composites
10.	Resin–polymer mixture (for injection grouting)

Table 23.3 Properties needed for epoxy resin to bond fresh concrete to hardened concrete

S. No.	Property	Type I	Type IV
1.	Viscosity, centipoises (N s/m^2), maximum	2000 (2)	2000 (2)
2.	Consistency (mm), maximum	6.4	6.4
3.	Gel time (min)	30	30
4.	Bond strength, minimum, psi (kg/cm^2)		
	2-day, moist cure	1000 (70.37)	1000 (70.37)
	14-day, moist cure	1500 (105.55)	1500 (105.55)
5.	Absorption, 24 h maximum (%)	1	1
6.	Heat deflection temperature, 7-day minimum °F (°C)	—	120 (49)
7.	Linear coefficient of shrinkage on cure, maximum	0.005	0.005
8.	Compressive yield strength 7-day minimum, psi (kg/cm^2)	8000 (562.96)	10000 (703.7)
9.	Compressive modulus, minimum, psi (kg/cm^2)	150000 (10555)	200000 (14047)
10.	Tensile strength, 7-day minimum, psi (kg/cm^2)	5000 (351.85)	7000 (492.59)

Polymer-modified cement products   The classification of polymers is shown in Fig. 23.2. Acrylic polymer products are widely used because of their better durability under long-term exposure to UV radiation. In cement mortar or concrete, the polymer can be incorporated as a second binder into the mix. These polymer mortars are two-phase systems which form matrix with cement. In the cementitious water phase, fine polymer particles of size 0.1–0.2 mm are dispersed.

In the cement–polymer system, the polymer particles join and form chain link reinforcement (Fig. 23.3) increasing tensile and flexural strengths. This helps achieve greater plasticity and reduce the shrinkage stress. Hence the addition of polymers vastly improves the property of plain cement mortar. Table 23.4 gives important properties of polymer mortars.

Micro-concrete   Based on hydraulic binders, readymade formulations of concrete are tailored to give concrete which is flowable and shrinkage-free. Only addition of water is required at site. Such formulations can be applied in complicated locations, and also enable achievement of high early strength. They are available either as type A (normal strength) or as type B (high strength). Table 23.5 gives the important properties of these types of micro-concrete.

Fibre-reinforced cement composites   Fibre-reinforced concretes have improved tensile strength and toughness compared to conventional concrete. They also have improved energy absorption capacity. Advanced composites offer high tensile strength, durability, ductility, and preferred energy absorption capacity. These aspects will be discussed further in Section 23.5.

Table 23.6 summarizes various products useful for repair and their properties.

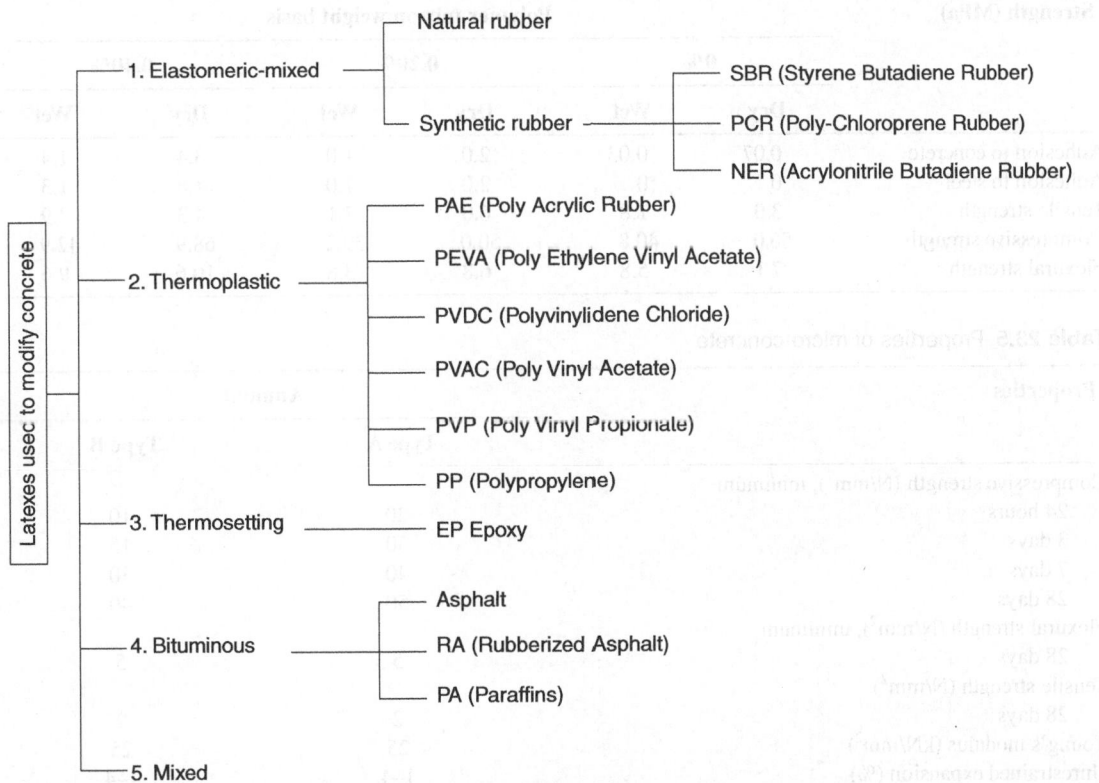

Fig. 23.2 Classification of polymers

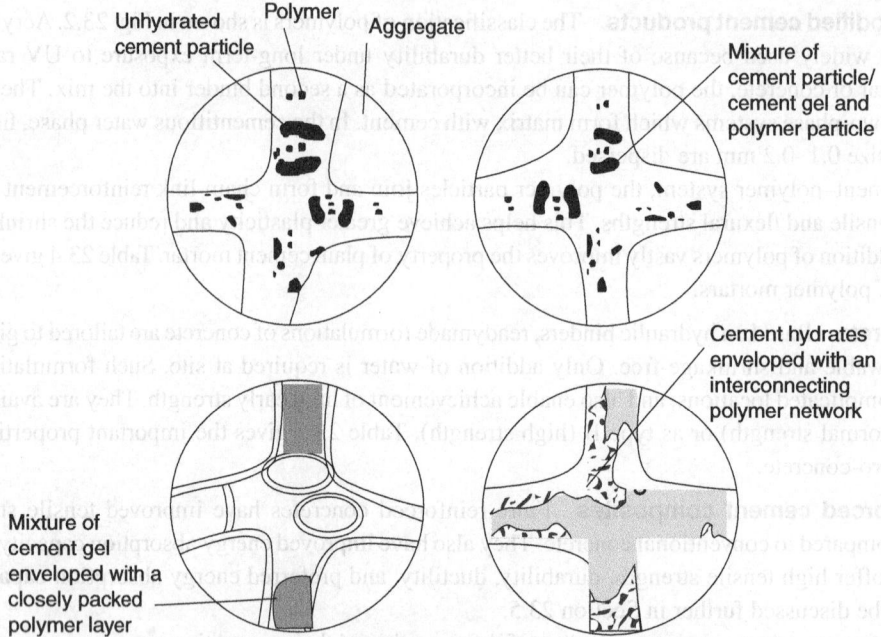

Fig. 23.3 Formation of polymer–cement co-matrix

Table 23.4 Properties of polymer–cement mortar

Strength (MPa)	Polymer mix on weight basis					
	0%		0.20%		0.40%	
	Dry	Wet	Dry	Wet	Dry	Wet
Adhesion to concrete	0.07	0.03	2.0	1.0	3.4	1.4
Adhesion to steel	0	0	2.0	1.0	1.6	1.3
Tensile strength	3.0	1.8	6.0	3.1	4.3	3.9
Compressive strength	56.0	40.8	50.0	39.2	68.9	42.9
Flexural strength	7.1	5.8	6.8	3.8	10.6	9.6

Table 23.5 Properties of micro-concrete

Properties	Amount	
	Type A	Type B
Compressive strength (N/mm^2), minimum		
24 hours	20	10
3 days	30	15
7 days	40	30
28 days	50	40
Flexural strength (N/mm^2), minimum		
28 days	5	5
Tensile strength (N/mm^2)		
28 days	2	2
Young's modulus (kN/mm^2)	25	25
Unrestrained expansion (%)	1–4	1–4
Fresh concrete density (kg/m^3)	2100–2200	2100–2200

Table 23.6 Materials guide

Product	Compressive strength (MPa)		Flexural strength (MPa)	Adhesion/bond to old structure	Performance				Durability		Advantages	Disadvantages
	1 day	7 days			Rate of gain of strength	Volume change	Chemical resistance	Deformation				
OPC-based products	5–12	25–50	2–7	Medium to poor	Slow	High	Poor	Low			Low cost	Poor bond and adhesion; high volume change
Epoxy-based products	30–35	90–100	40–50	Very good	Fast to rapid	Low	Very good	High			Very good performance; possible to grout thin cracks	Application requires skill; costly
Polymer-modified cement products	10–20	20–50	6–15	Very good	Normal to fast	Medium to low	Good	Medium			Good all round performance; affordable cost	Choice of correct formulations essential
Micro-concrete	20	50	5	Medium	Fast	Medium to low	Medium				Easy application	Requires formwork

23.5 New Repair Systems/Products

In this section we will discuss some recently developed repair products and systems.

Composites New and improved composite materials which are stronger, lighter, and durable are constantly being developed for repairing structures. These improved composites are costly. But considering the long service life they provide, their high strength-to-weight ratio, and the ease with which they can be applied in situ without suspending function, their use is well justified.

FRP composite bars FRP composite bars as well as carbon aramid meshes are an effective replacement for defective steel or corroded reinforcement. They are most effective as repair material in atmospheres where corrosion is a major problem. They can provide the required strength and toughness and hence are being increasingly considered for use in structures subjected to severe weather or de-icing salts. Figure 23.4 shows the repair methodology that uses carbon FRP composites to repair a column.

High-performance concretes This type of concrete is finding applications in high-rise building columns, off-shore platforms, and heavy-duty floors with congested reinforcements because of its high workability, strength, toughness, and dimensional stability. Such concretes have been used in places where superior non-porous concrete is needed. Their microstructure properties have been effectively modified by addition of microsilica (silica fume) in order to strengthen the weak transition zone between the aggregate and paste phases of concrete. The properties of HPC have been discussed in detail in Chapter 15.

Special admixtures Many special admixtures are in use in normal concrete. For repair work, following types of admixtures are in use.

  i. *Air-entraining agents* These agents have shown better compatibility with newer plasticizers. Their use has been extensively discussed in Chapter 3.
 ii. *Superplasticizers* New plasticizing agents work on delayed release mechanism and hence the behaviour of these is independent of the time of addition. This property can be advantageously used for repair works.

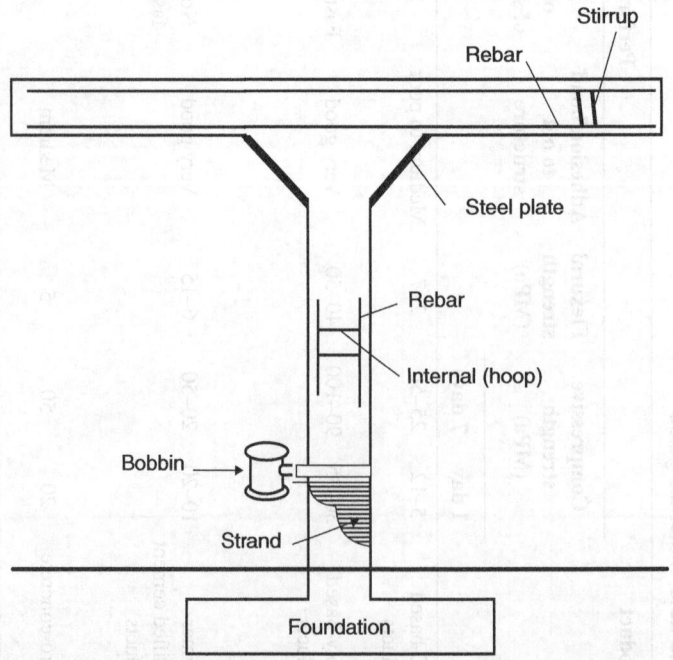

Fig. 23.4 Schematic of column repair with FRP composites

iii. *Shrinkage-reducing agents* The most frustrating problem in repair work is shrinkage, which leads to cracks in repair material. Current method of combating this problem includes the use of very low water–cement ratio and using shrinkage-reducing admixtures (SRA). These agents reduce shrinkage by reducing the surface tension of water in the pores between 2.5 and 50 nm in diameter. Even after the concrete hardens, admixtures remain in the pores and continue to reduce the surface tension, thus reducing shrinkage.

iv. *Viscosity-modifying agents* These admixtures provide pseudoplastic (viscosity decreasing) flow behaviour for concrete or slurries which are pumped or sprayed. Such self-levelling (SL) or self-compacting (SC) concretes are particularly needed for repair works. Highly flowable concrete can be effectively used in place without undergoing significant shrinkage or separation.

v. *Retarders* Retarders that control the hydration process for extended periods enable ready-mixed concrete to be transported over long distances. These also enable repair material to be plastic till the repair is over.

vi. *Corrosion-inhibiting admixture* Corrosion inhibitors provide required protection against reinforcement corrosion of repaired materials. The current practice is to use multifunctional admixtures which provide more than one property modification. For example, corrosion inhibition can be achieved along with retardation or shrinkage reduction.

vii. *High performance cementitious system* High performance cementitious systems are chemically bonded ceramics (CBC) that have properties similar to fired ceramics. These have good potential as they are very strong, dense, are micro-defect free (MDF), and densified systems containing ultra-fine particles (DSUP). These are known as belite cement or sulfo-aluminate cements. These are technology-shaping parameters of repair materials.

Injection materials Cracks may be an indication of a damaged or distressed structure. However, all types of cracks are not to be treated alike. Basically, the cracks have to be repaired for two reasons, namely, for structural purposes and for durability purposes. One of the most prevalent techniques of repairing cracks is injection of different types of materials (depending upon the nature of the defect). The selection of material for injection requires a thorough understanding of the properties of the material and functions that such a repair has to perform. In all the cases, it is imperative that the cause of cracking is properly determined, so that the selection of material is appropriate. Basically, material injection can be used for three purposes. Firstly, injection materials that are used to restore the structural stability of the structures. Secondly, injection materials that are used to protect the reinforcement against moisture and air entering the concrete, thereby lowering the rate of corrosion. Thirdly, injection materials which are used to stop the water from entering the structure.

The selection of material for injecting into cracks largely depends on the following factors:

- pattern of cracks
- width of cracks
- movements in the crack faces
  - due to temperature variations
  - due to dynamic loading
- moisture in cracks
- dirt in cracks

The pattern of cracking helps in ascertaining the reason for cracking, which in turn helps in the selection of base material. The width of crack has a direct bearing on the viscosity of the material required: it depends on the movements in the crack, which reflects on the type of material required, i.e., whether it should act as structural injection or just an elastic seal. In case the injection is used for structural purpose, it should be able to transfer the stresses from one crack face to the other and should have adequate compressive and flexural strengths, at least 10–15% higher than the neighbouring concrete. The moisture in the crack calls for a water-compatible system of injection. Presence of dirt in the crack will affect the choice of the crack preparation system. Table 23.7 shows the criteria for selection of materials.

Table 23.7 Selection of materials

Type of crack	Width (mm)	Movement	Water	Type of material	Mode of application and/or principle
Shrinkage cracks in concrete	0.2	No	No	Two-component epoxy injection	Surface treatment which works through capillary action
Shrinkage cracks in plastic state	0.2	No	No	One-component flexible paint on acrylic base	Coat with roller or brush
Shrinkage cracks in concrete, brickwork	0.2–1	No	No	Two-component epoxy, low viscosity	Low-pressure injection, smaller cracks with high-pressure injection
Shrinkage cracks in concrete, brickwork	1–2	No	No	Two-component epoxy injection and solvent-free epoxy	Low-pressure injection
Shrinkage cracks in concrete, brickwork	2–5	No	No	Solvent-free epoxy thixotropic	Low-pressure injection, with hand pump
Shrinkage cracks in concrete, brickwork	5	No	Dry/wet	Polymer-modified cement-based grout	Grout with injection grout by gravity or hand pump
Shrinkage cracks in concrete, brickwork	15	No	Dry/wet	Non-shrink grout	Cut and fill non-shrink mortar
Moving cracks in concrete, brickwork	0.2–1	Due to temperature changes	Dry/wet	Two-component polyurethane injection and flexible paints when wet joints, primary injection with polymer gel forming	Higher pressure injection or low-pressure injection, then coat with roller/brush
Joints in prestressed concrete (coupling joints)	0.2–2	Vibration	Dry/wet	Two-component polyurethane injection and joint sealant, when wet joint primary injection with polyurethane gel forming	Higher pressure injection or low-pressure injection, then coat with roller/brush

Exercises

Review Questions

1. What are the general requirements of patch repair materials?
2. How are repair materials classified?
3. What are the desired properties for repair materials?
4. What are the advantages of using latex-modified systems?
5. Discuss the properties of micro-concrete.
6. What are new repair systems?
7. What repair materials are used in highly corrosive environment?
8. Write a note on various injection materials and their use for repair work.
9. What is 'compatibility' of repair material? Discuss how its requirements will change with loading?

Multiple Choice Questions

1. Cement–sand mortar exhibits following problem when used as a repair material
   a. Increases dead weight
   b. Exhibits shrinkage cracking
   c. Requires time to gain strength
   d. De-bonds early

2. Epoxy-based materials are preferred as repair material because
   a. They bond well with concrete
   b. They gain strength in a short time
   c. They are stronger than substrate
   d. All of the above

3. Low-viscosity polymers are used as repair material to
   a. Repair cracks when crack width is more
   b. Fill hairline cracks with injection technique
   c. Repair high-strength concrete
   d. Repair cracks under water

4. Acrylic polymers are preferred as repair material for
   a. Repairs to have UV stability
   b. Resisting abrasion
   c. Resisting temperature
   d. Resisting impact

5. Polymer-modified concrete means
   a. Using polymer to bond old and new concrete
   b. Using polymer as a second binder in the mix
   c. Using polymer as the only binder in the mix
   d. Using fibres with polymer in the mix

6. Micro-concrete is used as a repair material because
   a. Regular concrete has less slump
   b. Regular concrete has large size aggregates
   c. Regular concrete requires more effort to consolidate
   d. All of the above

7. Fibre-reinforced cement composites are preferred for repair work because
   a. They have improved energy absorption capacity
   b. They are tougher
   c. They have better crack resistance
   d. All of the above

8. Carbon fibre-reinforced composites are used as surface-bonded reinforcement because
   a. They are lighter
   b. They are stronger
   c. They are durable
   d. All of the above

9. Compatibility of a repair material means
   a. Strength of the repair material matches with the substrate
   b. Properties of the repair material are compatible with the substrate
   c. Deformation capacity of the repair material is same as the substrate
   d. Ability of the repair material to take load in a short time

10. Low-viscosity treatment of shrinkage cracks in concrete involves
    a. Low-pressure injection
    b. High-pressure injection
    c. Gravity injection
    d. Capillary action

Answers to Multiple Choice Questions

1. b
2. d
3. b
4. a
5. b
6. d
7. d
8. d
9. b
10. d

CHAPTER 24

Repair and Rehabilitation

Distresses in buildings are a common occurrence. A building component develops cracks whenever the stress on it exceeds its cracking strength. Stresses in a building component can be caused by externally applied forces, such as dead, live, wind, or seismic loads, or by deformations induced by foundation settlement. Deformations can be caused internally due to thermal movements, moisture changes, chemical action, etc.

Distresses can be broadly classified as

- structural
- non-structural

Structural distresses are generally caused by faulty design, faulty construction, and/or overloading. Such distresses can endanger the safety of a building. Extensive cracking of a reinforced concrete beam due to overloading is an example of structural distress. Non-structural distresses may not endanger the safety of a building, but may look unsightly and create an impression of faulty work or a feeling of instability. Non-structural distresses are caused by internally induced stresses in building components. These generally do not directly result in structural weakening. Non-structural cracks may however (because of penetration of moisture through them or by weathering action) result in the corrosion of reinforcement and thus render the structure unsafe and lead to its collapse.

Cracks are likely to damage the internal finish of a building due to percolation of moisture, thus increasing the cost of maintenance. It is therefore necessary to undertake treatment measures for stitching and sealing of these cracks.

Internally induced stresses in the components of a building lead to dimensional changes. When there is restraint to the movement of induced stresses, cracking and spalling occur.

Building components under stress tend to move away from stiff portions of buildings, which act as fixed points, due to dimensional changes caused by moisture or heat. In symmetrical structures, the centre of the structure acts as a fixed point and movement takes place away from the centre on either side. A building as a whole can move easily in the vertical direction, but its movement in the horizontal direction is restricted by the action of the sub-structure and its foundation. Thus, vertical cracks occur in walls more frequently due to horizontal movement of building components. Volume changes due to chemical action within a component result in either expansion or contraction, thus leading to the occurrence of cracks.

Stresses in building components may be compressive, tensile, or shear. Most of the building materials that are subject to cracking, namely, masonry, concrete, mortar, are weak in tension and shear. Thus, even forces of a small magnitude, capable of causing tension or shear, will result in cracking.

Classification of cracks is done on the basis of their widths. The three broad classes are

- thin—less than 1 mm in width
- medium—1 to 2 mm in width
- wide—more than 2 mm in width

Cracks may be of uniform width throughout or may be narrow at one end, gradually widening at the other. These may be straight, toothed, stepped, map patterned, or random and may be vertical, horizontal, or diagonal. Cracks may be only at the surface or may extend to the full depth of the member. The occurrence of closely spaced fine cracks at the surface of a material is called *crazing*.

Cracks due to different causes have varying characteristics. It is only by careful observation of these characteristics that one can diagnose the cause of cracking and adopt an appropriate remedial treatment. Thin cracks, even when closely spaced, are less damaging to the structure compared to a fewer number of wide cracks.

24.1 Distresses in Concrete Structures

The three basic symptoms of distress in concrete structures are

- cracking
- spalling
- disintegration

Common defects in liquid-retaining structures are

- porous areas of concrete
- loss of bond at joints
- shrinkage crack between joints

Overloading of structures leads to the following:

- excessive flexural cracking
- shear and diagonal tension cracking
- shear bond failure leading to splitting along the reinforcement
- crashing of concrete due to compression
- column cover surface spalling due to excessive compression
- unacceptable large deflection in beams and slabs

Common types of damage A structure is said to have failed not only if it collapses but even if it does not perform the designed or desired functions. The structure may be said to have failed due to

- a total or partial collapse
- discolouration
- spalling leading to reduction in sizes
- deformations which could be in the form of deflection, buckling, twisting, or distortion

24.2 Deterioration of Structures: Causes and Prevention

Structures are subject to various types of stresses, both during their construction and during their service life. Generally all possible precautions are taken to avoid any kind of adverse reactions caused by foreseen and unforeseen distress causing agents. But it is not possible to ensure failsafe structures at all times. We will now discuss some of the possible causes that can lead to the failure of a structure, as also some of the preventive measures aimed at avoiding such failures.

24.2.1 Occurrences Incidental to the Pre-construction/Design Stage

Poor design details at re-entrant corners, changes in the cross section, rigid joints in precast elements, and deflections are some of the practices at the design stage that can lead to the failure of the structures. These design problems can lead to

- leakage through joints
- inadequate drainage
- inefficient drainage slopes
- unanticipated shear stresses in piers, columns, abutments, etc.
- incompatibility of materials at the critical section

Errors in the design of the structure can also prove crucial and may lead to its failure. So the designer needs to be *absolutely* careful while designing a structure.

24.2.2 Occurrences Incidental to the Construction Stage

A majority of structural failures owe their origin to the causes that arise at the construction stage.

Local settlement of subgrade   Consequent upon pouring of fresh concrete, the subgrade beneath it might compress or settle a bit. This leads to uneven stress on the concrete layer over it. This stress gives rise to cracks in the concrete layer (Fig. 24.1). Cracks of this nature can be cured while applying the final finish to the concrete.

Swelling of formwork   Formwork absorbs moisture from the wet concrete or from the atmosphere. This leads to expansion of formwork, thereby leading to a release of lateral pressure and cracks in concrete [Fig. 24.2(a)].

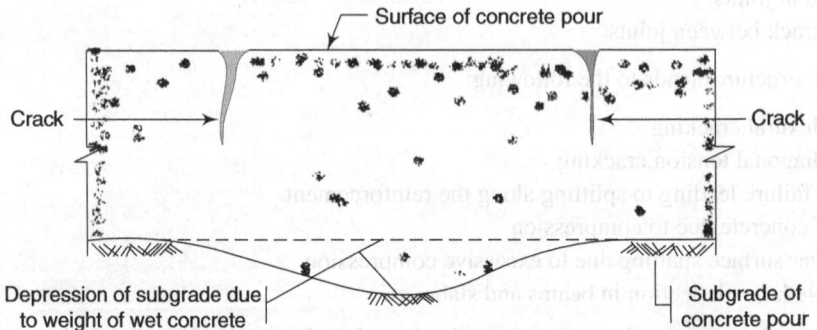

Fig. 24.1 Cracking due to settlement of subgrade during construction

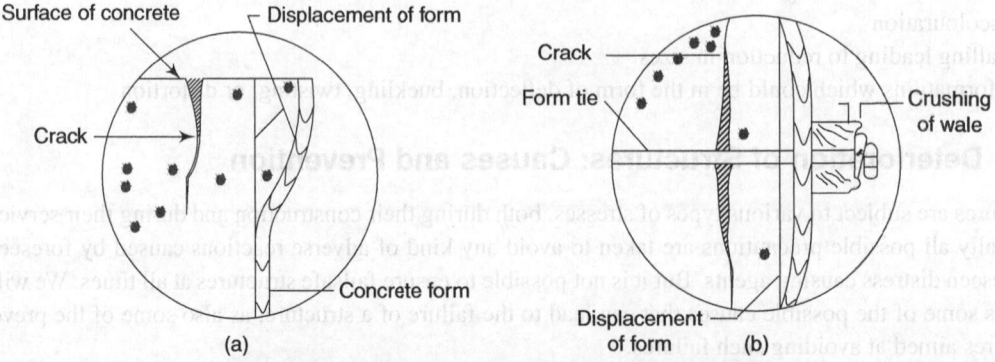

Fig. 24.2 Cracking due to movements of the forms

This can be prevented by providing coatings on the formwork to prevent it from absorbing moisture. This type of cracking can also occur due to crushing of wale [Fig. 24.2(b)]. Unyielding lateral ties with good end anchorage will avoid this problem.

Internal settlement of concrete   Surface cracks appear due to differential settlement between the surface and the interior volume of suspension. The surface cracks (as shown in Fig. 24.3) can be closed by delayed finishing. Curing of concrete has to be started as soon as possible after its placement because this will delay the setting of the surface concrete, thus reducing the amount of differential settlement. Proper compaction of concrete is also necessary.

Setting shrinkage   The cracks caused by shrinkage have the characteristic appearance of alligator scales. To avoid these cracks, the concrete needs to be cured properly and timely.

Premature removal of shores   Removal of shores from newly poured concrete can lead to re-distribution of stress/load on the formwork, which leads to avoidable cracks. Thus, it is necessary to ensure that shores are removed only after the concrete has gained sufficient strength so that there is no possibility of its cracking.

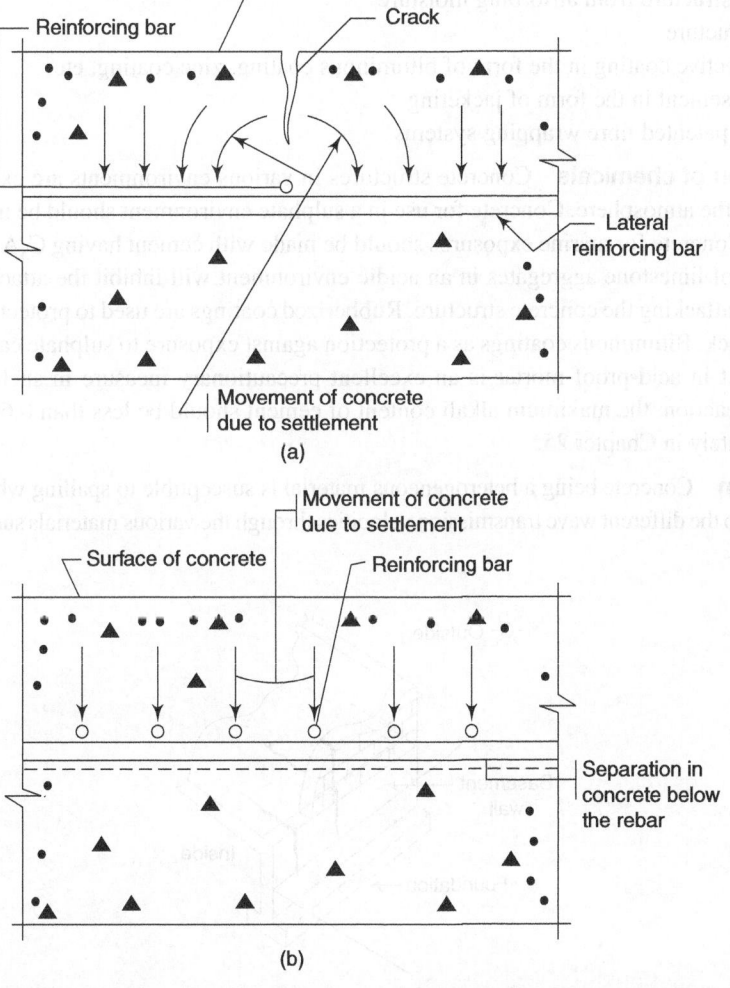

Fig. 24.3 Cracking of concrete due to settlement of concrete suspension

Vibrations At times vibrations might be caused by the worker walking indiscreetly and dumping the pan of concrete on the formwork with force. These vibrations also lead to cracks in concrete. Hence, the labourers on the project should be careful to avoid vibrations caused by dropping of pans, walking over reinforcement, etc., as far as possible.

24.2.3 Occurrences Incidental to the Post-construction Stage

Structures are exposed to deteriorating agents even in the post-construction stage. There is a wide spectrum of causes that affect a structure after it is constructed. We will discuss some of them here.

Temperature stresses Cracks in concrete are caused due to temperature stress, which might be due to (i) difference between temperature on the inside of the building and its outside environment and (ii) variation in the internal temperature of the building or structure (Fig. 24.4).

Corrosion of reinforcement This could be caused by (i) entry of moisture through cracks and (ii) electrochemical action. So it is necessary to take measures to seal cracks to prevent moisture from reaching the reinforcement. The reinforcement needs to be protected against corrosive chemical action. Corrosion of reinforcement can be prevented by

- keeping the structure clean
- preventing the structure from absorbing moisture
- painting the structure
- providing protective coating in the form of bituminous coating, zinc coating, etc.
- providing encasement in the form of jacketing
- wrapping with patented fibre wrapping systems

Aggressive action of chemicals Concrete structures in various environments are exposed to the action of chemicals from the atmosphere. Concrete for use in a sulphate environment should be made with sulphate-resistant cement. Concrete for marine exposures should be made with cement having C_3A content not greater than 8%. The use of limestone aggregates in an acidic environment will inhibit the attack by neutralizing a portion of the acid attacking the concrete structure. Rubberized coatings are used to protect the structure better against an acid attack. Bituminous coatings as a protection against exposure to sulphate can be used. A facing of ceramic tiles set in acid-proof mortar is an excellent precautionary measure in such cases. To prevent alkali–aggregate reaction, the maximum alkali content of cement should be less than 0.6%. We will discuss these more elaborately in Chapter 25.

Weathering action Concrete being a heterogeneous material is susceptible to spalling when subject to shock waves. This is due to the different wave transmission velocities through the various materials such as the aggregates,

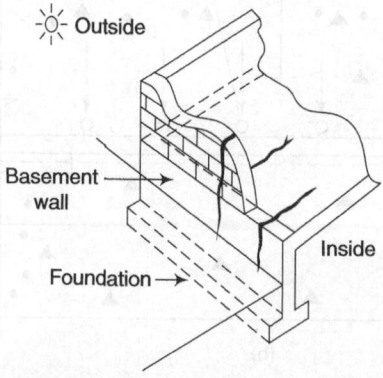

Fig. 24.4 Temperature cracking in a building

matrix, and the reinforcement comprising the structure mass. Sufficiently reinforced sections will provide a concrete structure with excellent resistance to shock waves. Erosion is also an agent that can cause weathering of structures. This can be prevented by the use of high-strength concrete, proper curing of concrete, and by proper finishing methods.

24.2.4 Other Causes

In addition to the aforementioned causes, structures can fail owing to many other causes also. We will list some of them here.

1. Errors in earlier repairs
2. Overloading
3. External influences, which include

   - Earthquake
   - Wind
   - Fire
   - Tornado
   - Flash floods

The following actions are attributable to careless construction practices:

a. Leaching action
b. Internal sulphate attack
c. Chemical attack
d. Permeability of concrete structures

24.3 Crack Repair Techniques

In this section we will discuss some techniques that are used to repair various types of cracks.

24.3.1 Sealing with Epoxies

Cracks in concrete can be sealed by injecting epoxy bonding compounds under pressure into the cracks. The usual practice is to drill into cracks from the face of concrete at several locations. Water or some solvent is injected to flush out the dirt. The surface is then allowed to dry. The epoxy is injected into the drilled holes until it flows out through other holes. The work should proceed from bottom to top.

24.3.2 Routing and Sealing

This method involves enlarging the cracks along their exposed surfaces, filling, and finally sealing them with a suitable material (Fig. 24.5). This is the simplest and most common technique for sealing cracks and is used for sealing both fine pattern cracks and larger isolated defects. The cracks should be dormant unless they are opened up enough to put in a substantial patch, in which case the repair may be more properly termed 'blanketing'.

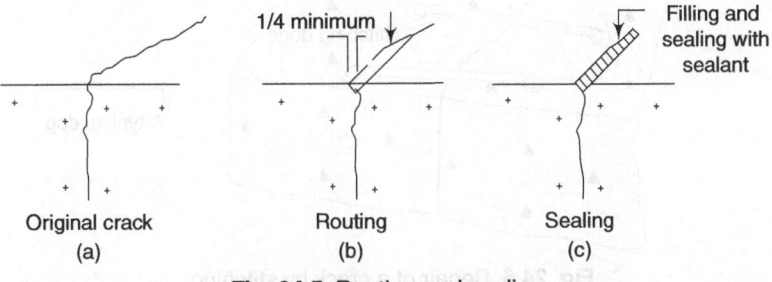

Fig. 24.5 Routing and sealing

On road pavements it is common to see cracks that have been sealed by pouring hot tar over them. This is a simple, inexpensive, and expedient technique. In this technique, watertightness of the joint is not required and appearance is not important.

Routing and sealing of leaking cracks should be done on the pressure face so that the water-aggressive agents cannot penetrate the interior of the concrete and cause side effects such as swelling, chemical attack, and corrosion of the rebars.

24.3.3 Stitching

The tensile strength of a cracked concrete section can be restored by stitching in a manner similar to sewing cloth (Fig. 24.6). The following precautions should be taken while adopting stitching as a treatment measure:

i. Any degree of strengthening can be accomplished, but it must be noted that strengthening also tends to stiffen the structure locally.

ii. Stitching the crack tends to cause its migration elsewhere in the structure. For this reason, strengthening of the adjacent areas of the crack is necessary to take care of additional stresses. Moreover, the stitching dogs should be of variable length, orientation, and so located that the tension transmitted across the crack does not devolve on a single plane of the section but is spread out well over an area. Strengthening of the adjacent sections of concrete may consist in providing external reinforcement embedded in a suitable overlay material.

iii. In places where water ingress is likely, the crack should be sealed as well as stitched so that the stitches do not get corroded. A suitable overlay should be applied to achieve this.

iv. Stress concentrations occur at the ends of cracks, hence the spacing of the stitching dogs should be reduced at such locations. Stress concentrations at each end of the crack can be relieved by drilling suitable holes or making the ends rounded.

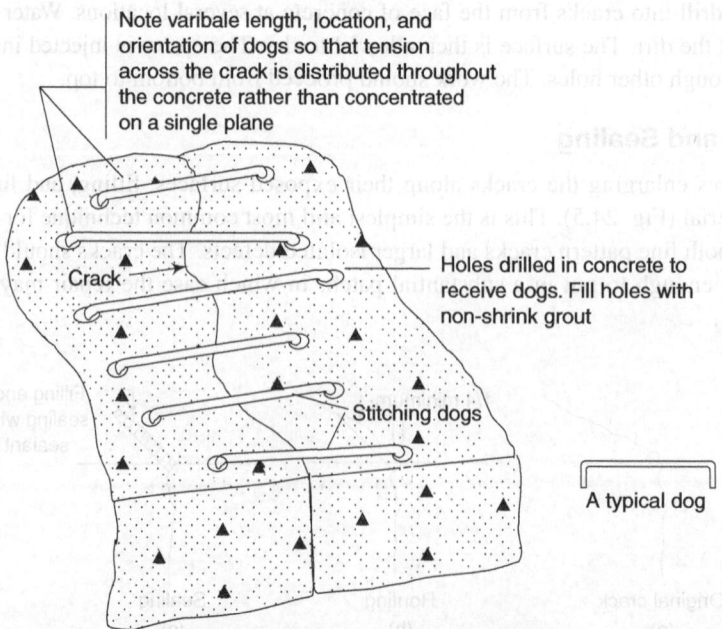

Fig. 24.6 Repair of a crack by stitching

v. Wherever possible both sides of cracks should be stitched to prevent bending action on the dogs due to the movement of the structure. In bending members it is possible to stitch one side of the crack only, but this should be the tension side of the section where movement is originating. If the member is in a state of axial tension, then a symmetrical placement of the dogs is necessary.

vi. If stitching is to supplement the strength of the existing section, the deformation should be compatible. The dogs must be grouted with a non-shrink or expandable mortar so that they have a tight fit. Thus, the movement of the crack will cause simultaneous stressing of both old and new concrete. The holes drilled for accommodating the legs of dogs should be filled with grout.

vii. The dogs are thin and long and so cannot take much of compressive force. These must be stiffened and strengthened by encasing them in an overlay.

24.3.4 External Stress

The development of cracking in concrete is due to tensile stress and can be arrested by suppressing this stress. Further, the cracks can be closed by inducing a compression force sufficient to overcome the tension and to provide a residual compression. Figure 24.7 shows the application of this technique to a Tee beam.

The compressive force is applied by using prestressing wires or rods. The principle is similar to stitching except that the wires are pre-tensioned. This calls for additional anchorages to be provided for prestressing wires (Fig. 24.8). Compressive force also may be applied by wedging, i.e., by opening the crack and filling it with an expanding mortar, by jacking and grouting, or by actually driving wedges (Fig. 24.9). Note that the final deflection due to both loading and expansive forces nearly coincides with the original centerline of the arch.

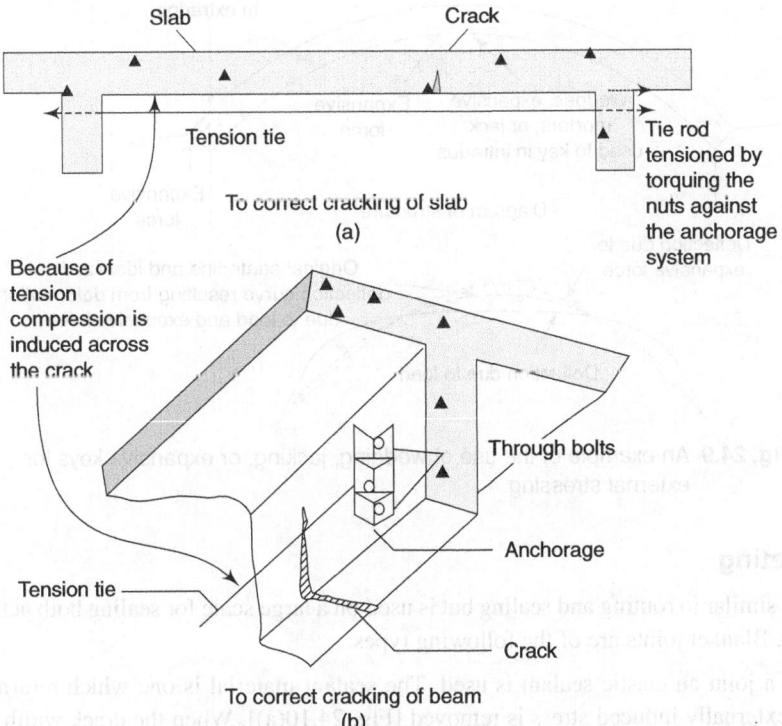

Fig. 24.7 Examples of external stressing

24.3.4 External Stress

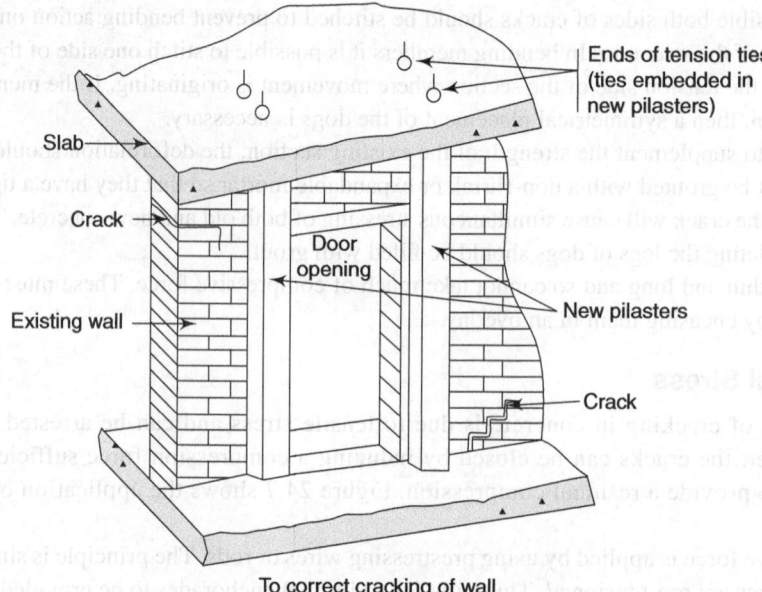

Fig. 24.8 Example of external stressing

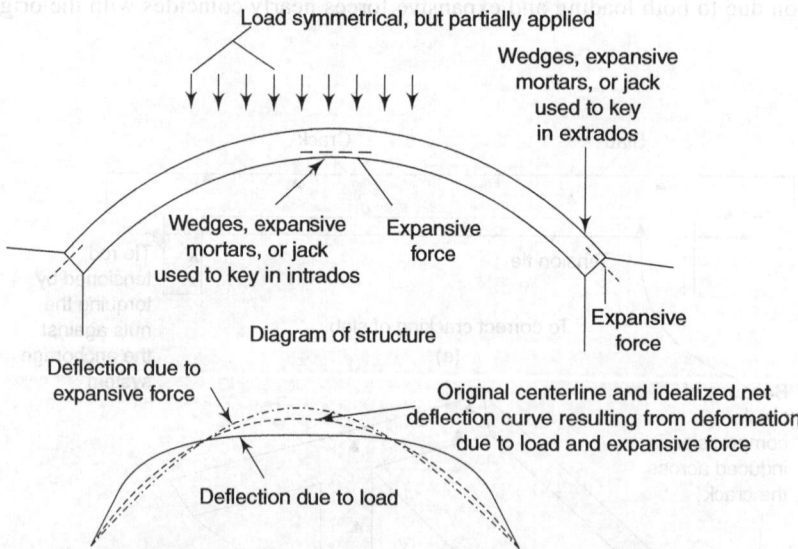

Fig. 24.9 An example of the use of wedging, jacking, or expansive keys for
external stressing

24.3.5 Blanketing

This technique is similar to routing and sealing but is used on a large scale for sealing both active and dormant
cracks and joints. Blanket joints are of the following types:

Type I In such a joint an elastic sealant is used. The sealant material is one which returns to its original
shape when the externally induced stress is removed [Fig. 24.10(a)]. When the crack width is small, a strip

sealant, as shown in Fig. 24.10(b), will be sufficient. The bond breaker ensures separation of the sealant and the bottom, preventing the possibility of cracking of the sealant [Fig. 24.10(c)].

Type II   It is a mastic-filled joint and is similar to the application of an elastic sealant except that the bond breaker is omitted [Fig. 24.10(d)]. The sealant is bonded to the bottom as well as to the sides of the chased dispersion. It is a mastic rather than a compound with elastic properties. This type of joint is used when the anticipated movements are small. In this type of joint there is a risk of tearing of the sealant at the bottom.

Type III   It is a mortar-plugged joint (Fig. 24.11). A recess in the form of a trapezoid to accommodate the mortar plug is made. This recess is filled with mortar. Figure 24.11(a) shows the forces that get generated when the plug is subjected to pressure load externally. Figure 24.11(b) shows the stresses when the pressure is from within. As shown in Fig. 24.11(c), generally a relieving grove is made in the mortar plug. In addition, a second cut at the mouth of the crack tip is made and plugged with quick-set mortar. If the pressure from within is due to fluid, a weep pipe [as shown in Fig. 24.11(d)] helps in draining of the fluid in addition to relieving the pressure.

Type IV   A crimped water bar is shown in Fig. 24.12. Figure 24.12(a) shows the installation details of a crimped water bar. It is generally installed at locations where the bar is not subject to direct loading such as traffic on a pavement. Such detailing is suitable on vertical faces. In a concrete pavement, such a joint is subjected to traffic. In this case, the crimped bar is set inside with a mortar cover protecting the water bar. This is shown in Fig. 24.12(b). It should be noted that in both cases the ends of the water bar should be adequately anchored.

24.3.6 Overlays

Overlays are used to seal cracks. They are useful and desirable when a large number of cracks are present and treatment of each individual crack would be too expensive and laborious.

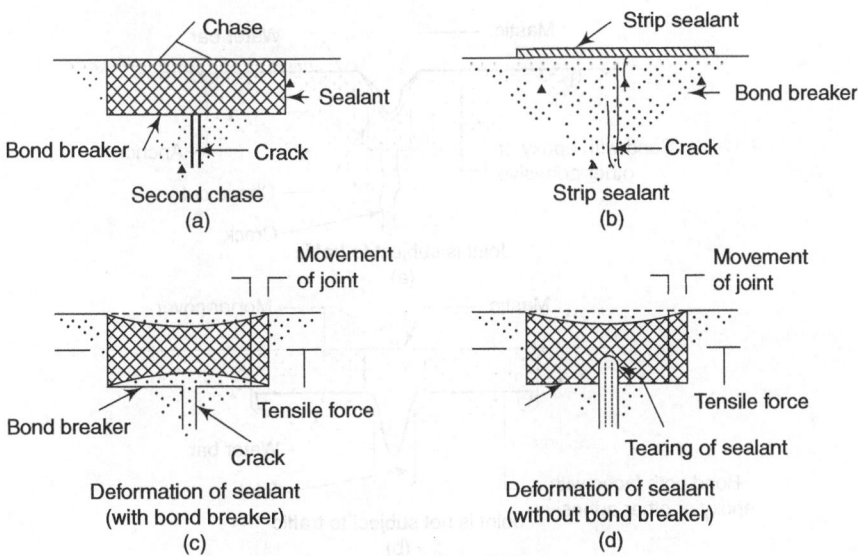

Fig. 24.10 Joint with an elastic sealant

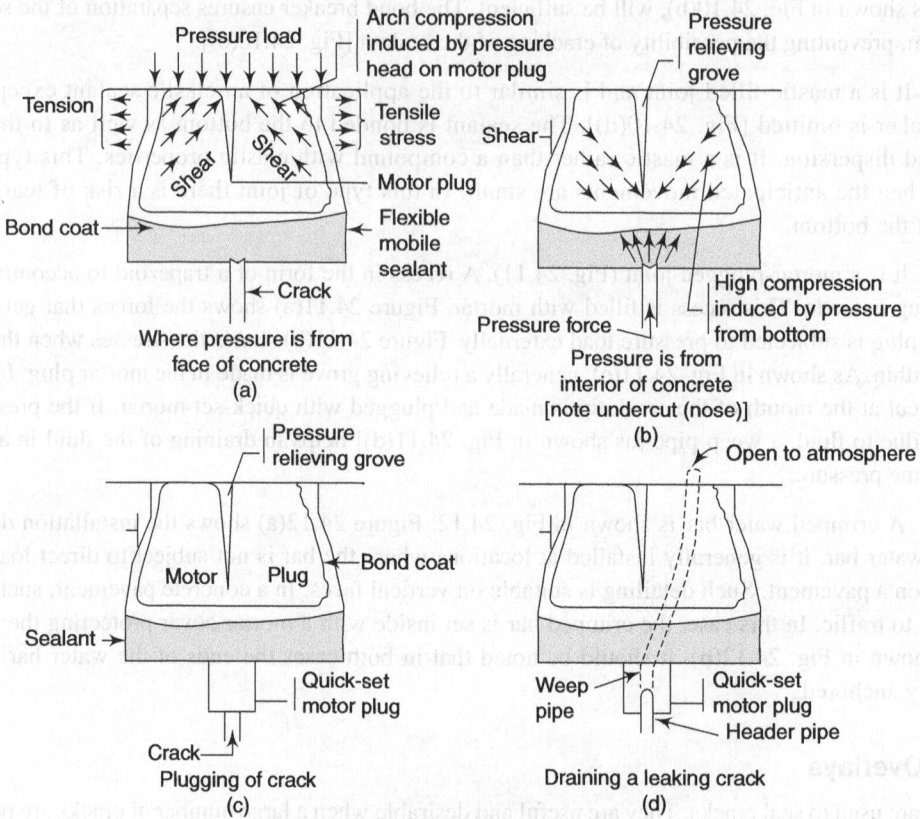

Fig. 24.11 Mortar-plugged joints

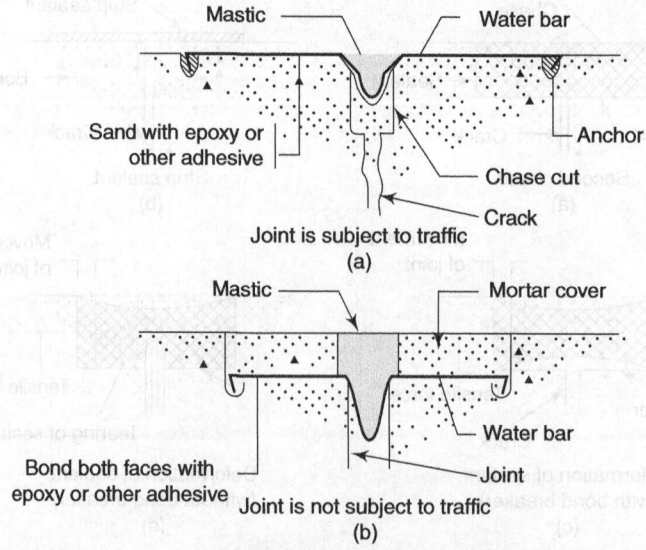

Fig. 24.12 Crimped water bar

Active cracks Sealing of active cracks by overlays should be done with a material which is extensible but not flexible. A two or three-ply polymeric membrane with a top coat of tar or with tar between the plies covered with a protective course of gravel, concrete, or brick, functions very well for this purpose. Gravel is used for roofs.

Dormant cracks If cracks are dormant almost any type of overlay may be used.

24.3.7 Grouting

Grouting is performed in a similar manner as injection of an epoxy. However, the use of an epoxy is a better solution except where considerations of fire resistance or cold weather prevent such use. In these cases, grouting is an effective alternative.

An alternative and better method is to drill down the length of the crack and grout it so as to form a key, as shown in Fig. 24.13. This is applicable only when the crack runs approximately in a straight line and is accessible from one end. The grout key prevents relative transverse movement of the sections of concrete adjacent to the crack. It also prevents leakage through the crack.

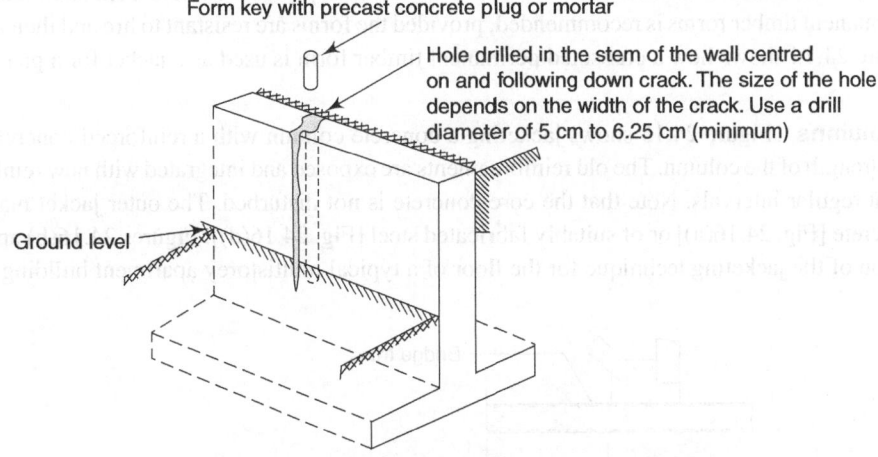

Form key with precast concrete plug or mortar

Hole drilled in the stem of the wall centred on and following down crack. The size of the hole depends on the width of the crack. Use a drill diameter of 5 cm to 6.25 cm (minimum)

Ground level

Fig. 24.13 Sealing a crack by drilling and plugging

24.3.8 Autogenous Healing

The inherent ability of concrete to heal cracks within itself is termed 'autogenous healing'. This is used for sealing dormant cracks such as in the repair of precast units cracked during handling, rectifying cracks developed during the driving of precast piling, sealing of cracks in water tanks, and sealing of cracks which are the result of temporary loading conditions. This property also provides some increase in the strength of concrete damaged by vibration during setting and concrete disrupted due to freezing and thawing.

The mechanism by which autogenous healing occurs is the carbonation of CaO and $Ca(OH)_2$ in the cement paste by CO_2 in the air and water. The resulting $CaCO_3$ and $Ca(OH)$ crystals precipitate, accumulate, and grow through and out from the cracks. The crystals interlace and twine, thus producing a mechanical bonding effect which is supplemented by chemical bonding between adjacent crystals, and between crystals and the surfaces of the paste and aggregates. As a result, some of the tensile strength is restored across the cracked section and the crack is sealed. Dormant cracks such as those caused by shrinkage or fault in construction such as premature removal of forms or settlement of sills supporting shores are self-sealing. This is because cracks get clogged by dirt and debris and the result is that these are plugged and the problems of leakage, especially if it is intermittent, will disappear without any repair. Such self-healing can also be attributed to the autogenous healing property of concrete.

24.4 Repair Techniques/Materials for Structures

Some techniques and materials used for rehabilitating structures are discussed in this section.

24.4.1 Jacketing

Jacketing consists in restoring or increasing the section of an existing member by encasing it in new concrete. The original member need not always be concrete. This method is useful for protection of a section against further deterioration by providing additional strength to the member. This method can be applied to deteriorated columns, piers, and piles and is useful when the entire section or a portion of it needs to be repaired. Concrete can also be jacketed by steel, timber, etc., depending on their relative cost and efficiency.

Types of jacketing

Timber forms For marine environments or where it is desired to protect concrete from chemical reactions, the use of permanent timber forms is recommended, provided the forms are resistant to fire and their appearance is good. Figure 24.14 shows how a creosoted permanent timber form is used as a jacket for a pier of a swing bridge.

Concrete columns Figure 24.15 shows jacketing a concrete column with a reinforced concrete jacket to increase the strength of the column. The old reinforcements are exposed and integrated with new reinforcements by welding at regular intervals. Note that the core concrete is not disturbed. The outer jacket may be made either of concrete [Fig. 24.16(a)] or of suitably fabricated steel [Fig. 24.16(d)]. Figures 24.16(b) and (c) show the application of the jacketing technique for the floor of a typical multistorey apartment building.

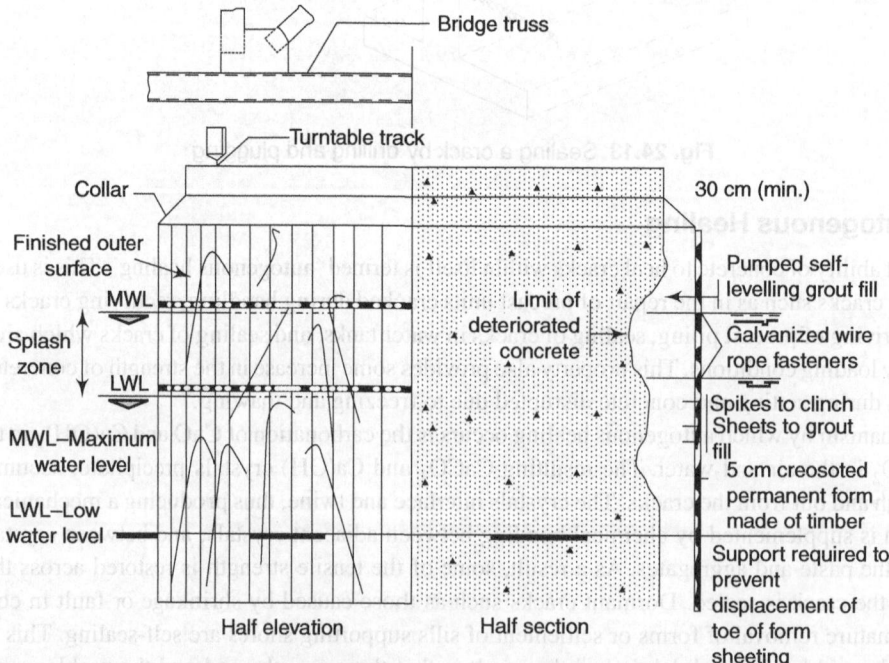

Fig. 24.14 Example of jacketing for repair of a pier of a typical swing bridge

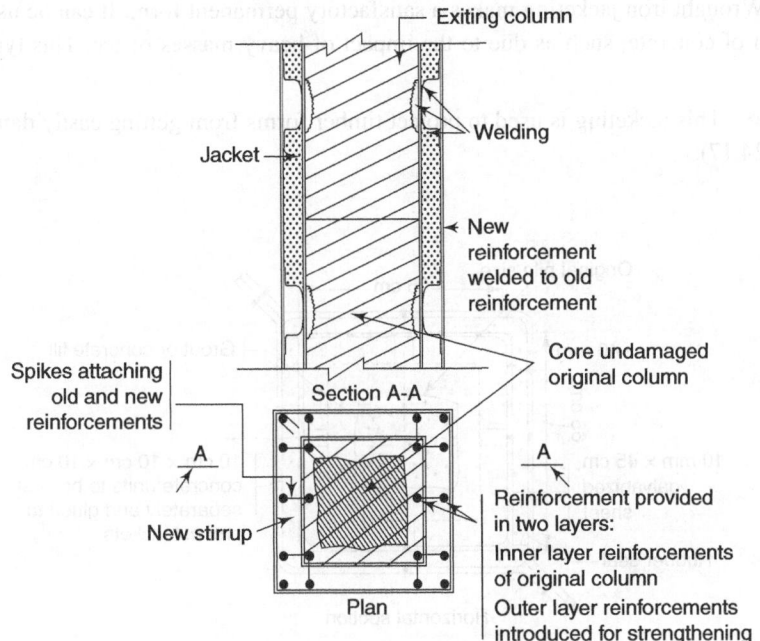

Fig. 24.15 Jacketing of a column

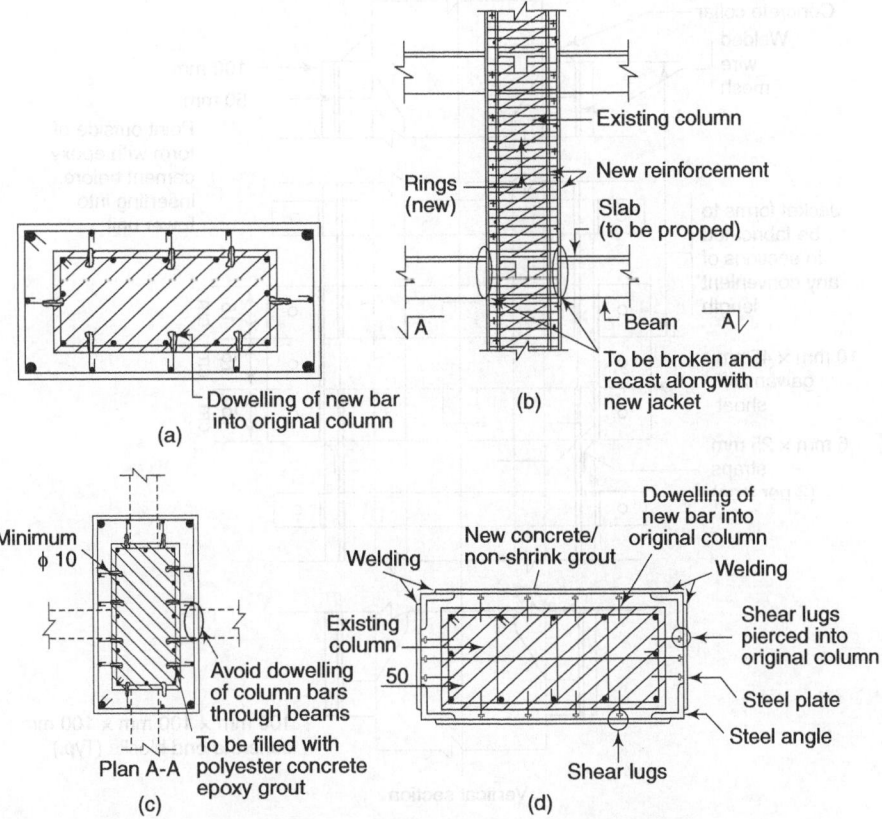

Fig. 24.16 Jackets placed on new concrete members

Wrought iron Wrought iron jacketing makes a satisfactory permanent form. It can be used only in places of severe abrasion of concrete, such as due to the impact of heavy masses of ice. This type of jacketing is generally costly.

Precast concrete This jacketing is used to protect timber forms from getting easily damaged due to fire, borers, etc. (Fig. 24.17).

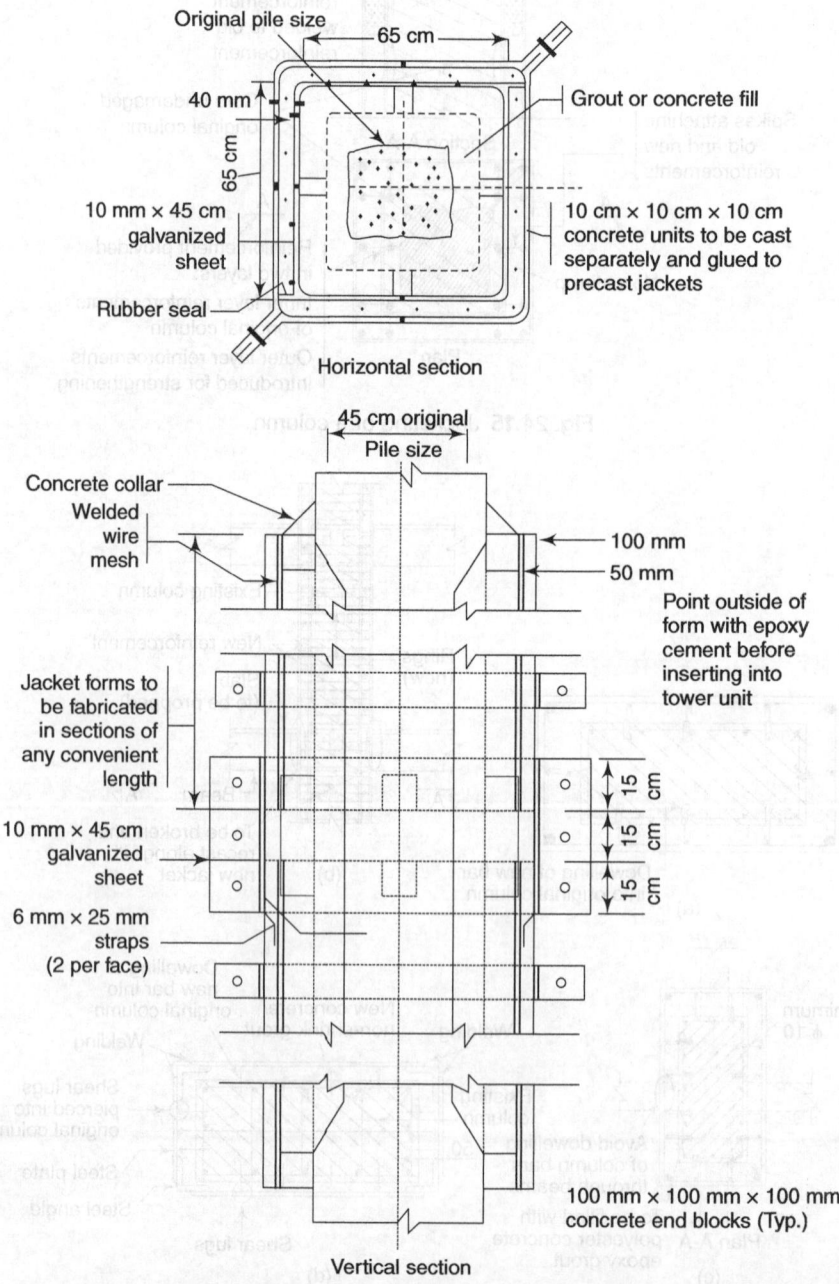

Fig. 24.17 Jacketing—Precast concrete form

Gauge (sheet) metal   Gauge metal and other temporary (Fig. 24.18) forms are used when

  i. the repair is in dry condition and the anticipated corrosion attack is mild,
 ii. it is desired to strip the forms for the sake of appearance,
iii. the repair is intended for a limited life of structure (temporary repair),
 iv. the exposure conditions are so mild that the degree of protection provided by timber or concrete forms is not necessary, or
  v. timber forms will not last if left alone.

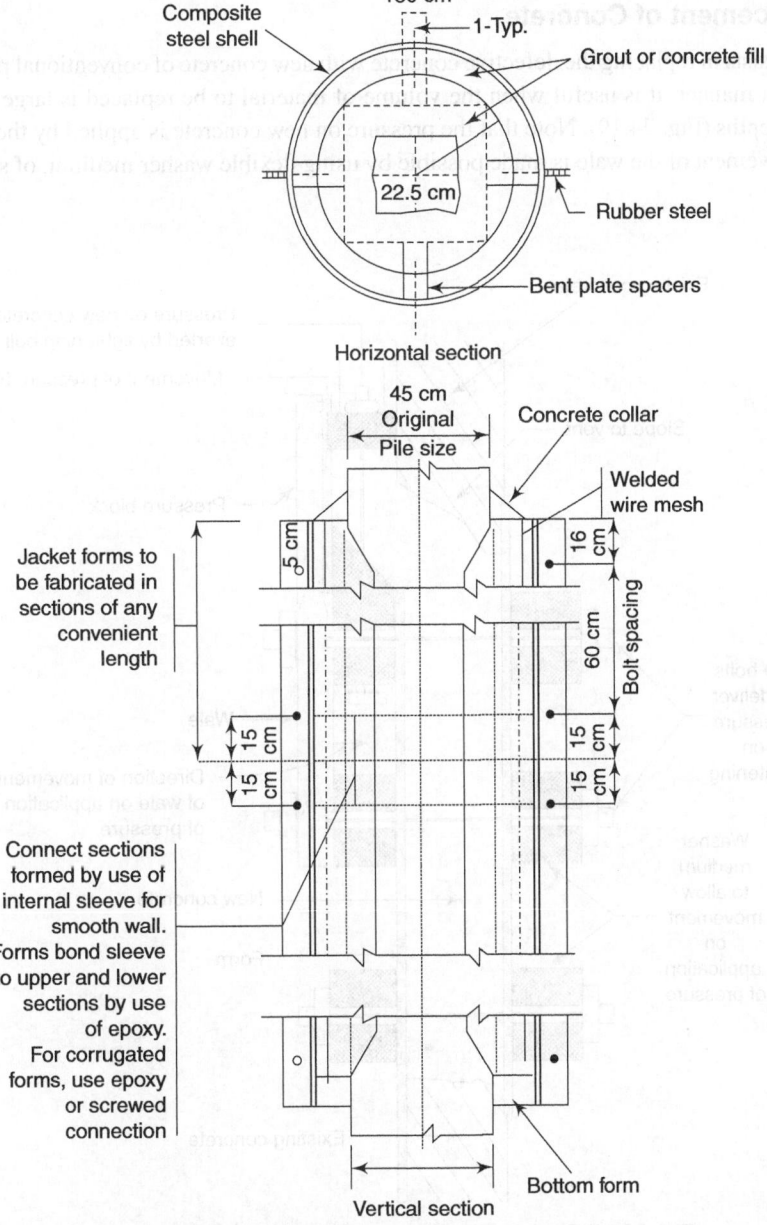

Fig. 24.18 Jacketing—Sheet metal form

24.4.2 Pneumatically Applied Mortar

Pneumatically applied mortar is used for restoration of concrete surfaces when the location of deterioration is at relatively shallow depths. This method of repair can be used for both vertical and horizontal surfaces and is particularly useful for restoring surfaces spalled due to the corrosion of reinforcement.

24.4.3 Prepacked Concrete

This repair technique is well adapted for underwater works and other locations where accessibility is a problem.

24.4.4 Replacement of Concrete

This method consists in replacing the defective concrete with new concrete of conventional proportions placed in a conventional manner. It is useful when the volume of material to be replaced is large and where repair occurs at large depths (Fig. 24.19). Note that the pressure on new concrete is applied by the tightening of the bolts and the movement of the wale is made possible by using flexible washer medium, of specified stiffness.

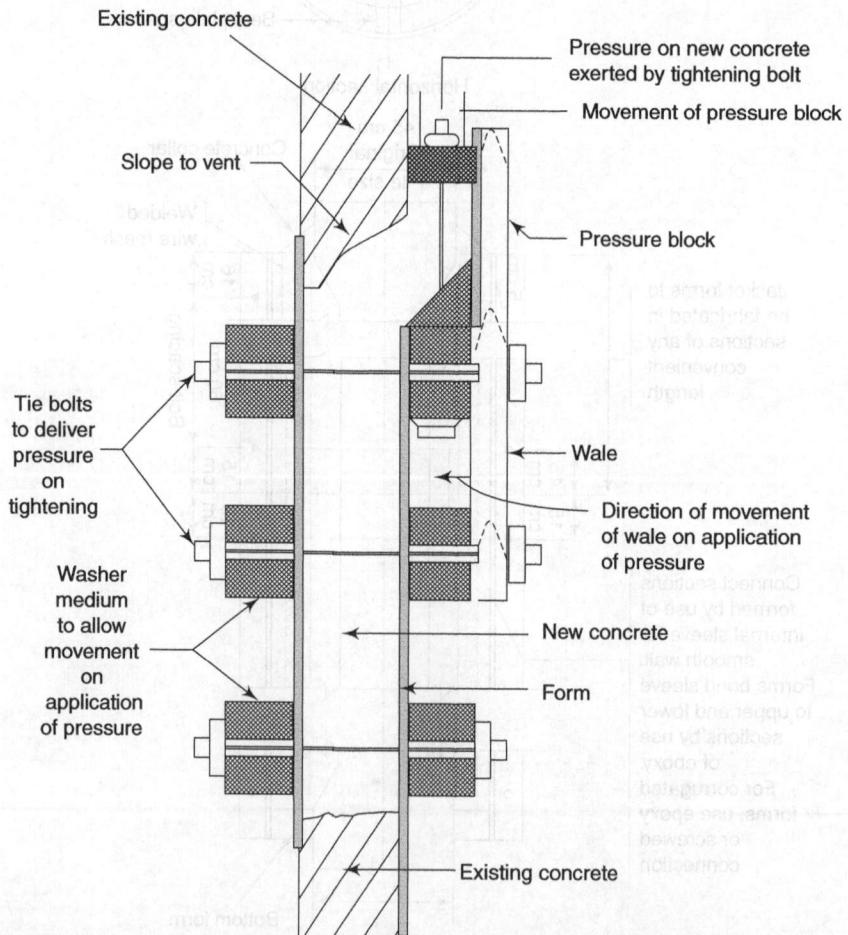

Fig. 24.19 Replacement concrete—Use of pressurized form

Pier walls, hydraulic structures, and similar heavy structures above subgrade and water level are the usual areas of application of this technique.

24.4.5 Dry Pack

Dry packing is the hand placement of very dry mortar and its subsequent tamping or ramming into place, thus producing an intimate contact between the old and new concrete works.

24.4.6 Overlays

Overlays may be used to restore a spalled or disintegrated surface or to protect the existing concrete from the attack of aggressive agents. Overlays used for this purpose include concrete mortar, bituminous compounds, etc. Epoxies should be used to bond overlays to the existing concrete surface in order to have monolithic action between the overlay and the old concrete. The following materials are generally used for this purpose:

- Pneumatically applied mortar
- Sand–cement slurry
- Sand–cement plaster
- Epoxy resins
- Bituminous coating

24.4.7 Epoxy Resins

These are organic compounds which when activated with suitable hardening agents form a strong, chemically resistant layer that has excellent adhesive properties. These are used as binders or adhesives to bond new concrete patches to existing surfaces or to bond together cracked portions. Once hardened, this compound does not melt, flow, or bleed.

24.4.8 Protective Surface Treatments

Durability of concrete can be substantially improved by preventive maintenance in the form of weather-proofing surface treatments. These treatments are used to seal the concrete surface and to inhibit the intrusion of moisture and/or chemicals. Materials used for this purpose are

- oils such as, linseed oils, and petroleum oils
- silicons used to seal concrete and masonry structures against moisture
- epoxies

24.5 Damage Assessment Procedure

While assessing the damage to a structure, the following general considerations should be examined:

- Cause of damage
- Type, shape, and function of the structure
- Construction materials
- Type and extent of damage
- Builder's expertise and capabilities for repair
- Availability of repair materials

The flowchart of activities given in Fig. 24.20 is useful for assessing the work involved. The assessment procedure spans two distinct stages: pre-repair evaluation and post-repair evaluation.

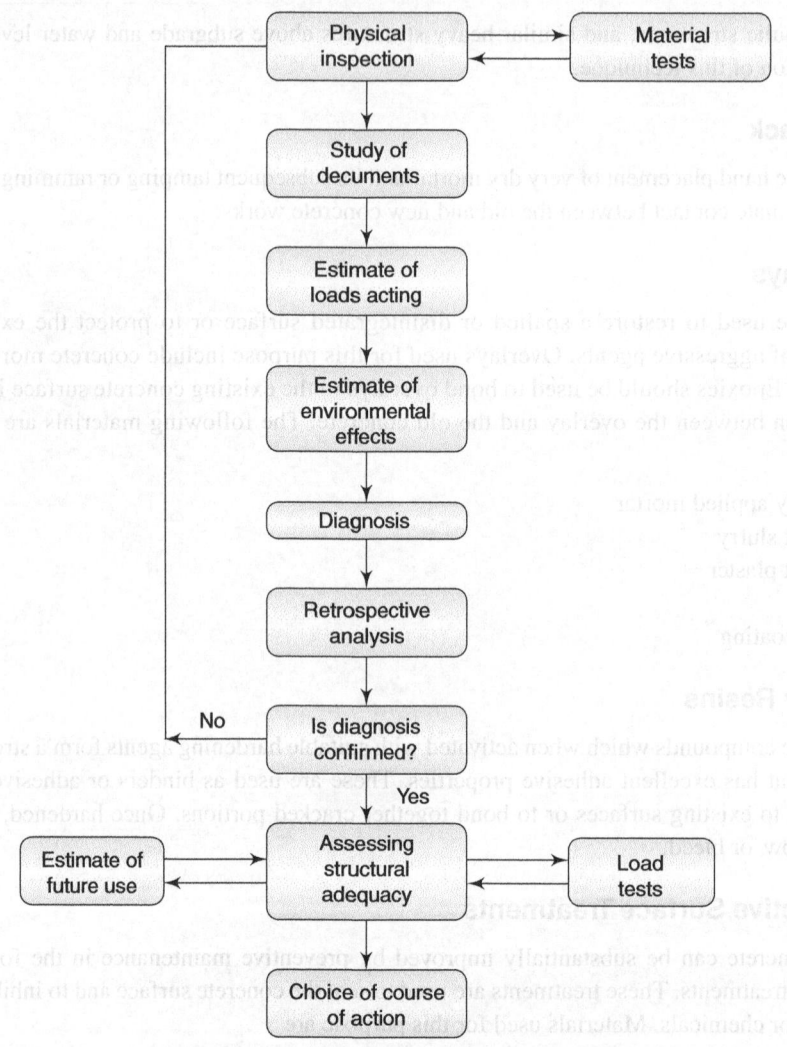

Fig. 24.20 Assessment procedure for damage

24.5.1 Pre-repair Evaluation

Each situation requiring repair and rehabilitation exhibits some unique features which demand specific attention. As such the available documented experience regarding the success or otherwise of procedures used may not fully meet the performance requirements of the structure being repaired. To avoid such a failure, it is necessary to examine the distressed structures before deciding on the type and procedure of repair technique to be adopted.

Reconnaissance A preliminary examination of the distressed area, its location in the structure or system, and the type of exposure involved gives valuable information. This information can be of much help to broadly classify the damage suffered and can help in identifying the likely causes for the deteriorated condition.

Field exploration Once the preliminary examination narrows down the scope of the investigation, a more detailed field study should be undertaken to elaborate and quantify the initial short-listed findings.

Test dismantling   The damaged portion can be carefully chipped off in several locations to expose sound concrete. Based on this, the depth and extent of damage, cracking, delamination, carbonation, and other failures can be evaluated.

Test cores   Wherever possible, 100-mm-diameter diamond drilled cores can be extracted and used to confirm the details as obtained from open dismantling. This will also provide information about the type of workmanship and materials used in the structure.

Non-destructive tests   Ultrasonic pulse velocity tests provide a quick and very convenient method for a general assessment of the concrete condition, extent of deterioration, existence and depth of cracks, etc. Rebound hammer tests and geophysical methods such as seismic and resistivity sounding can also help in evaluating the structure's condition.

Laboratory tests   Where preliminary examination and field examination warrant detailed laboratory tests on the materials, specific tests should be commissioned. Measurement of strength, elastic properties, abrasion, density, absorption, and permeability provide general condition of the structure and can be obtained by suitable laboratory assessment. Petrography and chemical tests on the aggregate matrix and products of distress give useful clues to the possible internal and external causes and mechanism for the observed distress.

24.5.2 Post-repair Evaluation

A post-repair evaluation programme constitutes the only way to make sure that the repair work has been effective and durable. Most of the techniques used for field and laboratory investigation of distress can also be adopted for evaluation of the performance of the finally executed remedial and rehabilitation measures. One sure method of evaluation is to conduct a field test on the rehabilitated structure and document the improvements obtained with respect to stiffness.

24.6 Repair of Structural Components

Various repair techniques have been discussed in the previous sections. In this section we will explain their applicability to various structural components.

24.6.1 Strengthening and Stiffening of Members

Strengthening or stiffening of members such as columns and girders is usually achieved either by replacing the poor quality or defective material or by providing additional better quality material to the member. In either case, the new material is usually reinforcing steel, better quality concrete, thin steel plates, and straps, or various combinations of these materials. The main difficulty in this type of operation is to achieve continuity of the structural action between the original material and the new material. Various techniques of bolting, gluing, dowelling, and keying have been developed to provide positive force transfer and composite action between old and new portions of the structural element.

24.6.2 Column Strengthening

One of the simplest and most effective methods for strengthening a column in an existing building is to partially unload the column by jacking between floors and then inserting two or more props to carry a portion of the axial load. The props are usually rolled steel sections, which may subsequently be encased in concrete to improve fire protection and appearance.

The disadvantage of this method is that considerable floor space is lost. Generally column carries both load and moment. However, the props may not be effective in transferring moment unless positive connection details are introduced at both ends, for example, in the form of end plates bolted through holes drilled in the floor.

24.6.3 Strengthening and Stiffening of Beams and Girders

Commonly used methods for strengthening flexural members include:

- provision of additional concrete on the compression face,
- addition of tensile reinforcement by welding new rebars to the existing rebars with a cast-in-place or gunited cover, and
- bolting steel plates or straps on to the surface of the member to bridge a weakness.

The main problem of strengthening, stiffening, and repair operation is to ensure good composite action between old and new material. Bolting of new steel to the existing member is effective but is time-consuming and expensive. For this reason, gluing procedures have been developed, which provide strong and durable connections between concrete and the surface of steel straps and plates.

24.6.4 Strengthening and Stiffening of Slabs

The simplest and cheapest procedure for strengthening and stiffening slabs is to provide additional props in the vicinity of the existing columns and walls. However, this solution may often be unacceptable for architectural reasons, even though props placed quite close to the existing columns may be sufficient to bring the structure to an acceptable limit of serviceability. The other methods include increasing the depth and/or width by bonding plates of either steel or FRP in order to increase the strength of the supports.

24.6.5 Member Replacement

Replacing a member is time-consuming and expensive. It can cause extensive disruption in the use of the building because of the need for temporary propping. However, the procedure can be effective provided proper care is taken to achieve the continuity of the existing structure with the replaced element.

24.6.6 Strengthening Flexural Members

External prestressing tendons are frequently used to repair and stiffen sagging flexural members such as roof girders in buildings. If the tendons are suitably linked, they can be used to balance a portion of the applied or self-weight load.

Although external prestressing operations are effective in repairing and stiffening flextural members, they are not always so effective in enhancing the ultimate strength. Much of the expenditure on repairs can be saved if care is exercised at the design stage itself by considering the environmental effects, proper selection of materials, and strict and standard inspection programmes throughout the progress of work during construction.

24.7 Research and Development

Considerable amount of research has been reported on the effectiveness of the various treatment measures discussed in this chapter. The following gives some details of the research findings:

a. Liew, S.C., and Cheong, H.K. 1991. 'Flexural behaviour of jacketed RC beams', *Concrete International*, Vol. 13(12), pp. 43-47.

   This paper discusses the effect of jacketing arrangement on RC beams. The cross-sectional area of the member is increased by providing additional reinforcement enclosed within concrete encasement. Experiments conducted show that the treated member is two to three times stronger than the original member.

b. Limaye, R.G., Kamat, M.K., and Sopori, A.K. 1991. 'Experimental investigation on strengthening of reinforced concrete beams with epoxy grouting', *Indian Concrete Institute Bulletin*, No. 37, pp. 25–28.

   Studies on the effectiveness of grouting with new class of epoxy resins on the strength and stability of RC beams have been dealt with. Their findings show that there is a remarkable increase in the strength and stability of RC beams treated with epoxy resins.

c. Neelamegan, M., Dattatreya, J.K., and Parameswaran, V.S. 1992. 'Studies on bonding materials and laminates for strengthening of concrete flexural members', *Civil Engineering Construction Review*, pp. 29–37.

This investigation examines the various types of indigenously available structural adhesives for bonding of external plates/laminates to existing concrete surfaces. It makes a comparative study of various strength and stiffness characteristics of some composite laminates.

d. Lal, A.K. 1991. 'Sulphate attack in buildings', *Indian Concrete Institute Bulletin*, No. 36, pp. 33–34.

This paper examines the factors influencing severity of sulphate attack on buildings and suggests measures to avoid sulphate attack on foundations.

e. George, Z. 1991. 'Failure of concrete and masonry structures: Some case studies', *Indian Concrete Institute Bulletin*, No. 35, pp. 17–25.

This paper examines a few cases of failures and analyses the causes and effects of various parameters on the reported failures.

f. Kukreja, C.B., and Singh, P. 1988. 'Environmental effects on concrete', *Material Testing Laboratory Manual*, Standard Publishers and Distributors, Delhi, pp. 7–10.

Thermal properties of concrete and the effect of cold weather concreting have been discussed in detail in this paper.

g. Pakvor, A. 1992. *Rehabilitation and Reconstruction of Concrete Structures*, ACI Publication SP 128–87, Detroit, MI, pp. 1407–1422.

This study examines when, what, and how the existing concrete structures should be inspected. It also discusses how structures should be repaired, rehabilitated, and reconstructed to ensure safety and serviceability.

h. Raju, N.K., and Nadgir, N.S. 1991. 'Limit state behaviour of reinforced concrete beams strengthened by epoxy bonded steel plates', *Indian Concrete Journal*, Vol. 65(3), pp. 124–129.

The flexural behaviour of RC beams with epoxy-bonded steel plates on the tension side has been investigated experimentally in this paper. The study indicates that the strength and stiffness characteristics and also the ultimate moment-carrying capacity of the flexural member are increased by about 30% because of such bonding.

The repair and rehabilitation procedures described in this chapter will be used for retrofitting the structure, especially for seismic strengthening, which is covered in Chapter 28.

Exercises

Review Questions

1. What are the three main distress symptoms in a structure?
2. List the causes for deterioration and distress in structures. Describe each, explaining what type of distress is witnessed due to each cause.
3. What are the various crack repair techniques? Explain.
4. What is meant by jacketing? Describe different types of jacketing.
5. What is meant by autogenous healing?
6. Describe in detail the damage assessment procedure.
7. How will you assess the effectiveness of repair or rehabilitation executed? Explain.
8. What are the methods for repairing structural components at the element level?
9. What do you understand by the term overlay? Describe different types of overlays used in repair work.
10. What is meant by blanketing? Explain different methods of blanketing.

Multiple Choice Questions

1. A crack whose width is 1 mm to 2 mm is classified as
   - a. Thin
   - b. Thick
   - c. Medium
   - d. Wide

2. Corrosion crack is identified as a crack
   - a. Transverse to the reinforcement
   - b. Along the reinforcement
   - c. At 45 degrees to the reinforcement
   - d. At an arbitrary angle to the reinforcement

3. Autogenous healing means
   - a. Repairing crack with poly mortar
   - b. Repairing crack with steel reinforcement
   - c. Using inherent ability of concrete to self-heal
   - d. Repairing crack with cement mortar

4. Retrofitting a structure means
   - a. Sealing the cracks
   - b. Upgrading the structure to the requirement of the current codes
   - c. Upgrading the structure to its original strength level
   - d. Making the structure serviceable

5. Rehabilitation of a structure involves
   - a. Inspecting the structure
   - b. Designing components for adequate strength
   - c. Execution as per design and validation by testing
   - d. All of the above

6. Jacketing is resorted to
   - a. Enhance existing strength of a member
   - b. Seal the cracks
   - c. Make the element serviceable
   - d. Make an architectural finish

7. An RCC column is plastered for
   - a. Increasing its strength
   - b. Aiding serviceability
   - c. Improving appearance
   - d. Increasing the size

8. The name of the NDT equipment used to check the strength of concrete is
   - a. Half cell
   - b. UPV
   - c. Rebound hammer
   - d. Crack width gauge

9. UPV test is conducted on concrete for determining
   - a. Strength of concrete
   - b. Crack width
   - c. Quality of concrete
   - d. Length and propagation of crack

10. A sure method of qualifying a retrofitted structure is
   - a. Visual inspection
   - b. NDT tests
   - c. Full-scale load test
   - d. Assessing the behavior of finishes like plastering

Answers to Multiple Choice Questions

1. c	4. b	7. c	10. c
2. b	5. d	8. c	
3. c	6. a	9. c	

CHAPTER 25

Concrete Structures in Special Environments

The complex nature of the environment around a concrete structure affects its performance and causes it to deteriorate. To enable the structure to perform throughout its service life, it is not only necessary to improve the characteristics of the building material, but also use good architectural and structural techniques in addition to adopting standard execution, inspection, and maintenance procedures. Serviceability can be ensured only if standard procedures are followed. The concept of concrete durability and its performance is closely related to the action of the physical, chemical, and biological agents of deterioration and the resistance mechanism provided by the structure. The resistance mechanism does not develop automatically, but should be consciously put in place by proper structural design, material selection, execution, and regular maintenance.

Figure 25.1 depicts the degradation of performance of a structure with age, which is inevitable. It also illustrates how by maintenance and/or repair the level of performance can be raised periodically, so that the intended service life is achieved, especially in structures in special environments. Note that the level of performance is raised either by periodic maintenance or by repair and rehabilitation of the structure. The approach should include adopting measures which resist the agents promoting deterioration (shown in Fig. 25.2). This ensures that the performance criteria listed in the figure are satisfied.

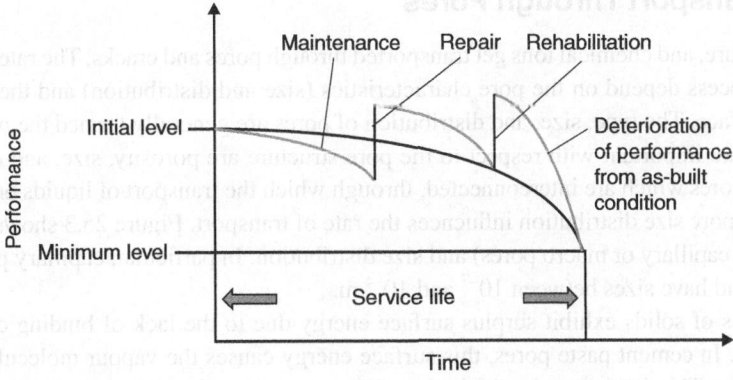

Fig. 25.1 Relationship between performance and service life as affected by maintenance, repair, and rehabilitation

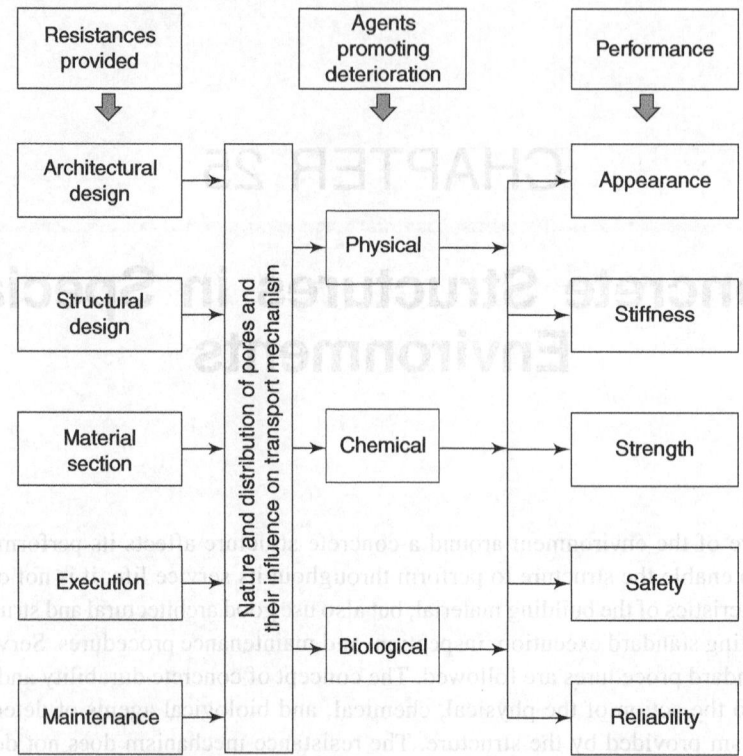

Fig. 25.2 Actions and resistance to achieve performance

It is evident that the combined transportation of heat, moisture from climate, and chemicals (e.g., aggressive chemical ions) within concrete and from the surroundings is the most important factor leading to the deterioration process. The transport of water/moisture within concrete is affected by pore type, size, and distribution in addition to the existing cracks (both micro and macro). Thus, controlling both micro and macro-cracks and pores is essential.

25.1 Water Transport Through Pores

Water, gases, moisture, and chemical ions get transported through pores and cracks. The rate, extent, and effect of the transport process depend on the pore characteristics (size and distribution) and the cracks especially on the concrete surface. The type, size, and distribution of pores are generally termed the *pore structure*. The parameters which are important with respect to the pore structure are porosity, size, and distribution. Open porosity indicates pores which are interconnected, through which the transport of liquids or gases is possible. On the other hand, pore size distribution influences the rate of transport. Figure 25.3 shows pore characteristics (such as micro, capillary or macro pores) and size distribution. In particular, capillary pores considerably reduce durability and have sizes between 10^{-7} and 10^{-4} m.

The free surfaces of solids exhibit surplus surface energy due to the lack of binding components to the adjacent molecules. In cement paste pores, this surface energy causes the vapour molecules to be absorbed into the pore surface. This increases the thickness of the water film depending on the humidity within the pore. Figure 25.4(a) shows the simplified model of a pore, illustrating water absorption. At a particular pore size, as shown in Fig. 25.4(b), capillary condensation takes place which actually depends on the humidity in the air surrounding the concrete.

Increasing the humidity of the surrounding air causes pores of smaller sizes to fill with water and the concrete to saturate. In such water-saturated concrete, there is practically no diffusion of gases (such as CO_2 or O_2). Diffusion is induced by the differences in concentration of chemical ions between the surroundings and either side of concrete as shown in Fig. 25.5. Chemical reaction between CO_2 and concrete at the pore wall causes the concentration of chemicals disolved in the fluid contained in the pores to decrease. This enables CO_2 to diffuse into concrete. Just like CO_2, other chemicals such as chloride and oxygen ions can also diffuse into concrete. The driving force for this diffusion is the differential ion concentration on either side of pore wall.

Owing to capillary suction in the splash zone (Fig. 25.6), saturation is quickly achieved. The extent of capillary rise is determined by an equilibrium condition between the binding forces and the weight of the

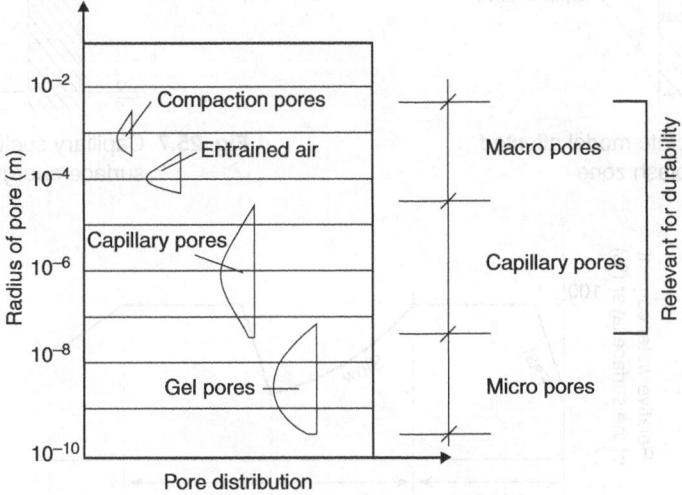

Fig. 25.3 Pore size distribution

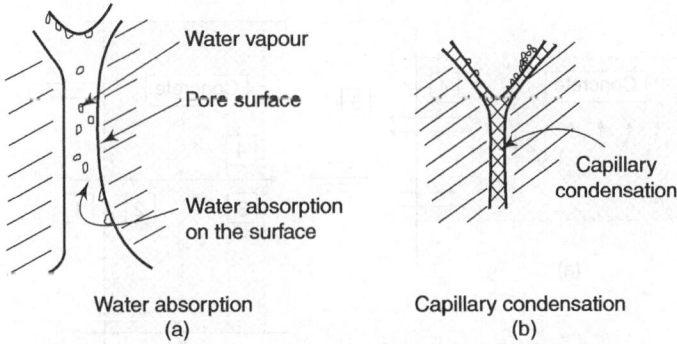

Water absorption
(a)

Capillary condensation
(b)

Fig. 25.4 Capillary pore showing the binding phenomenom

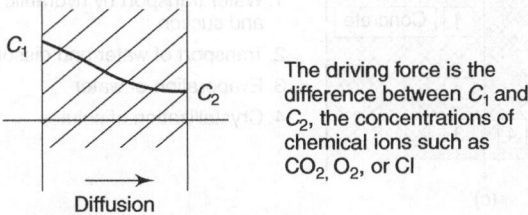

The driving force is the difference between C_1 and C_2, the concentrations of chemical ions such as CO_2, O_2, or Cl

Fig. 25.5 Diffusion through porous concrete

water column (Fig. 25.7) in the capillary. Water loss due to absorption caused by suction is more rapid than the loss due to evaporation (drying) as shown in Fig. 25.8. In the case of a continuously immersed structure, water is a major means of transport. Continuous transport develops only when water evaporates at the surfaces. Different cases are indicated in Fig. 25.9. The water transport depends on evaporation, capillary suction, and hydraulic gradient.

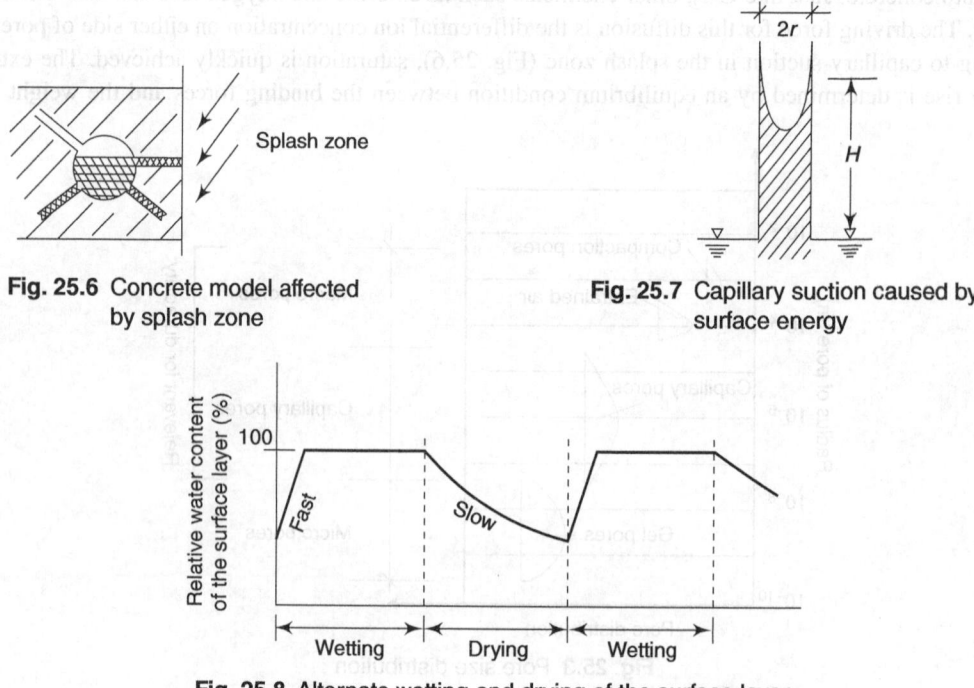

Fig. 25.6 Concrete model affected by splash zone

Fig. 25.7 Capillary suction caused by surface energy

Fig. 25.8 Alternate wetting and drying of the surface layer

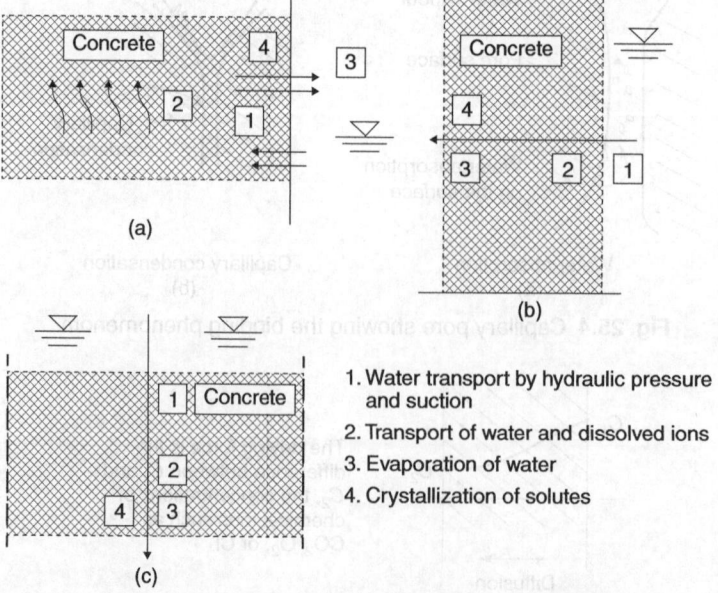

1. Water transport by hydraulic pressure and suction
2. Transport of water and dissolved ions
3. Evaporation of water
4. Crystallization of solutes

Fig. 25.9 Conditions faced by marine structures

Along with water, dissolved ions of carbonates, chlorides, and sulphates are transported. They are left behind in the evaporation zone, and hence a high concentration of ions develops there. These dissolved agents recrystallize and lead to the efflorescence phenomenon. The structure can have different levels of submergence as shown in Fig. 25.9. Figure 25.9(a) shows a marine structure having a significant transport of water and ions. Figure 25.9(b) shows transportation through the width of a slender marine structure, the driving force being the hydraulic gradients. Figure 25.9(c) shows the absence of significant transportation in a fully submerged structure. Note that there is no pressure head driving the transport mechanism in a fully submerged structure.

25.2 Physical Deterioration Agents

Cracking occurs when tensile strain increases beyond the strain capacity of concrete. Tensile strains are generated in concrete because of one of the following reasons:

a. Movement generated within concrete—these may be due to drying shrinkage, temperature change, and plastic settlement or shrinkage
b. Expansion of the material inside concrete—this can be due to corrosion of steel or alkali–aggregate reaction
c. Externally imposed conditions—deformations imposed because of settlement of foundations come under this category

Figure 25.10 lists the various causes of cracks. Figure 25.11 shows the ages at which various types of cracks occur. Figure 25.12 shows examples of various types of cracks which generally occur due to movement and volume changes in concrete. These cracks, known as intrinsic cracks, are classified and explained in Table 25.1.

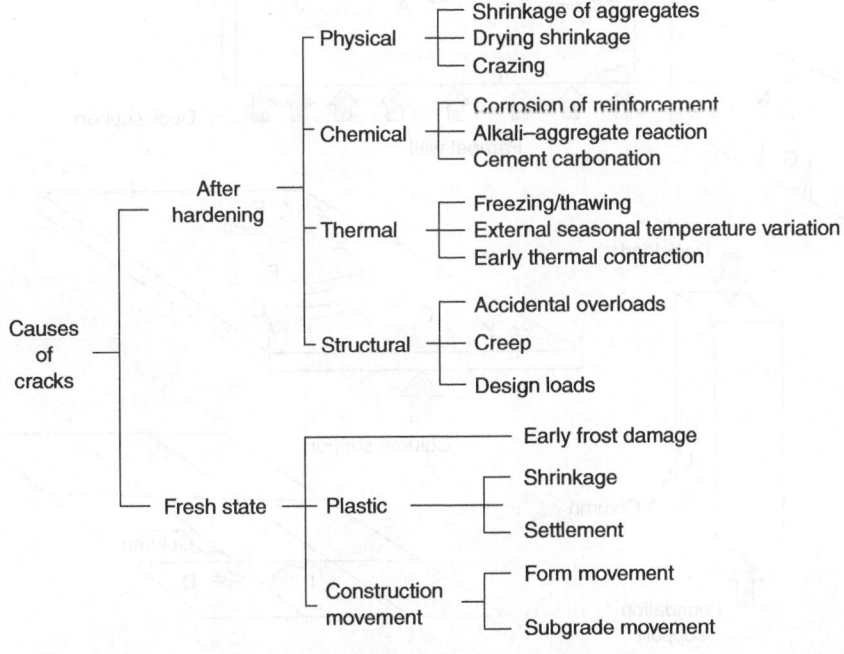

Fig. 25.10 Causes of cracking

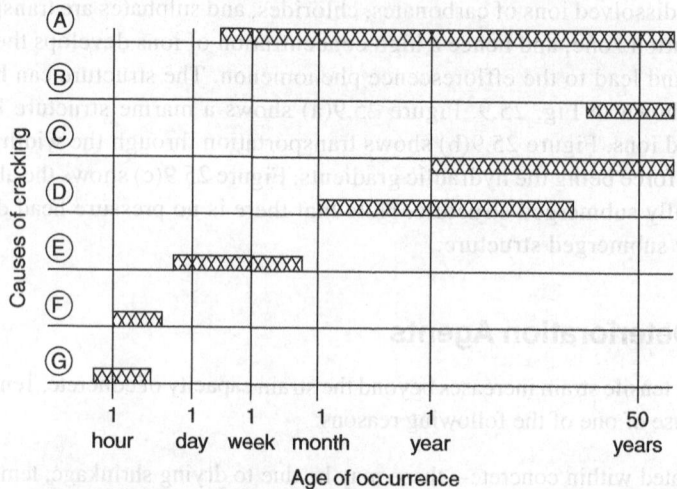

Fig. 25.11 Age of appearance for different types of cracks

Timeline of appearance of cracks from the time of placing concrete

A. Loading and service condition
B. Alkali–aggregate reaction
C. Corrosion
D. Drying shrinkage
E. Early thermal concrete
F. Plastic shrinkage
G. Plastic settlement

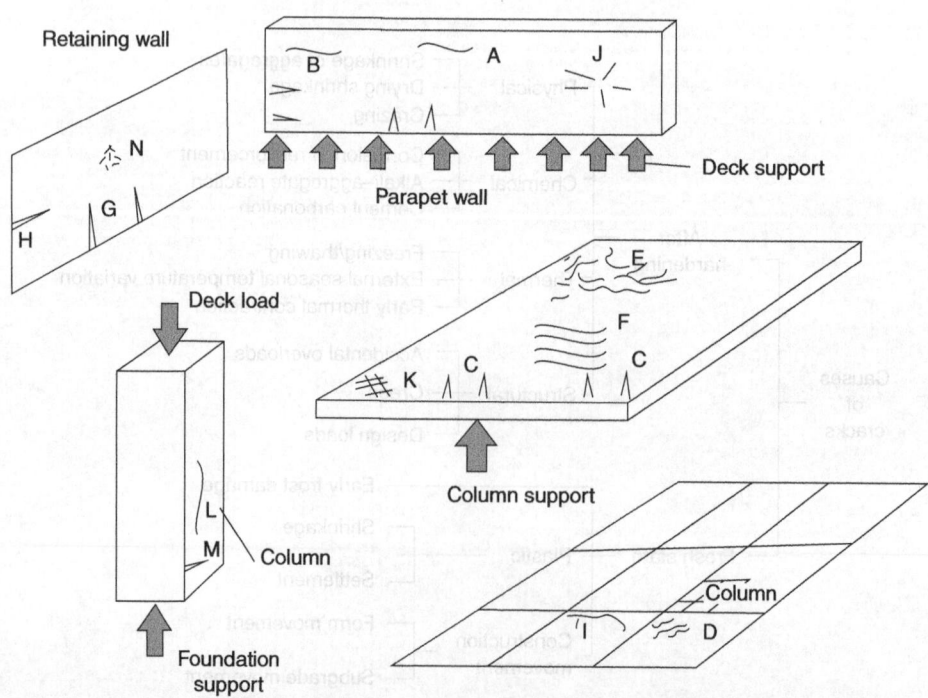

Fig. 25.12 Examples of intrinsic cracks

Table 25.1 Classification and explanation of intrinsic cracks

Type of cracking	Position in Fig. 25.12	Subdivision	Location of prevalence	Primary cause (excluding restraint)	Secondary causes/factors	Remedy (assuming basic redesign is impossible) in all cases	Time of appearance
Plastic settlement	A	Over reinforcement	Deep sections	Excess bleeding	Rapid early drying conditions	Reduce by air entrainment or revibrate	10 min to 3 h
	B	Arching	Top of columns				
	C	Change of depth	Trough and waffle slabs				
Plastic shrinkage	D	Diagonal	Roads and slabs	Rapid early drying	Low rate of bleeding	Improve early curing	30 min to 6 h
	E	Random	Reinforced concrete slabs				
	F	Over reinforcement	Reinforced concrete slabs	Rapid early drying, steel near surface		Curing and proper cover	30 min to 6 h
Early thermal contraction	G	External restraint	Thick walls	Excess heat generation	Rapid cooling	Reduce heat and/or insulate	1 day to 2-3 h
	H	Internal restraint	Thick slabs	Excess temperature gradients			
Long-term drying shrinkage	I		Thin slabs (and walls)	Inefficient joints	Excess shrinkage, inefficient curing	Reduce water-cement ratio and improve curing	Several weeks or more
Crazing	J	Against formwork	'Fair-faced' concrete	Impermeable formwork	Rich mass	Improve curing and finishing	1-7 days, sometime much more
Corrosion of reinforcement	K	Floated concrete	Slabs	Over-trowelling	Poor curing		
	L	Natural	Columns and beams	Lack of cover	Poor quality concrete	Eliminate causes listed	More than 2 years
	M	Calcium chloride	Precast concrete	Excess calcium chloride			
Alkali-aggregate reaction	N		Damp locations	Reactive aggregate plus high-alkali cement	Poor quality porous concrete	Use low w/c ratio	5 years

Figure 25.13 shows the variation of restraint stresses and strength with age. It is seen that the vulnerable period is a band between 2 and 4 hours, beyond which the strength is larger than the stresses imposed. Figure 25.14 shows the ultimate tensile strain as a function of age. The vulnerable period is again seen to be 2 to 4 h after casting.

Plastic shrinkage occurs in green concrete. It is caused by capillary tension in pore water. It occurs 2 to 4 hours after mixing. Shortly after the disappearance of wetness [Fig. 25.15(a) and (b)], if the losses by vapourization exceed the supply of bleed water, the capillary forces shown in Fig. 25.15(c) in the pore water

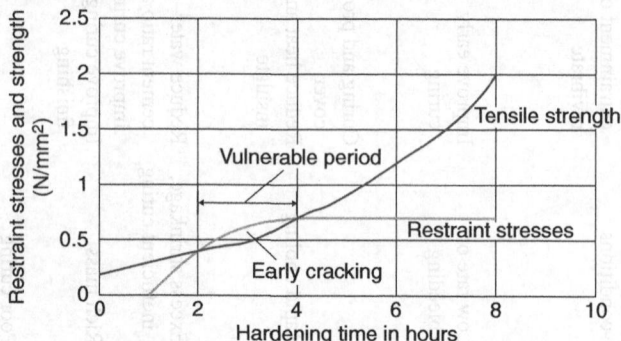

Fig. 25.13 Behaviour of 'young' concrete

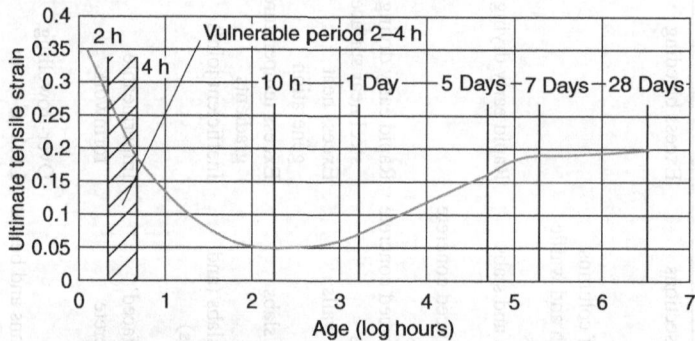

Fig. 25.14 Ultimate tensile strain as a function of age

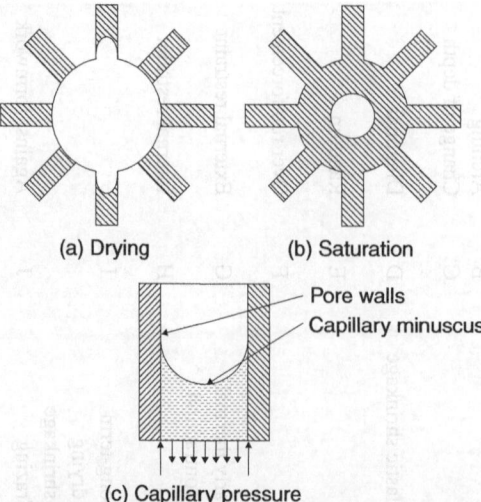

(a) Drying   (b) Saturation

(c) Capillary pressure

Fig. 25.15 Behaviour of water in pores

are activated. Concrete slabs are prone to plastic shrinkage cracks. Parallel cracks in slabs at an angle of about 45° to the slab corner are typical. The spacing of these cracks is generally in the 0.2 to 1 m range. Figure 25.16 shows typical 'map cracking' due to plastic shrinkage. Typical crack widths are of the order of 2 to 3 mm at the surface.

Concrete bleeds during settlement. As a result of gravitational forces, the concrete particles settle and the displaced mixing water surfaces. If the settlement of particles is hampered by reinforcement [Fig. 25.17(a)], cracking occurs. Such cracks are longitudinal and run parallel to the reinforcement as shown in Fig. 25.17(b)–(d).

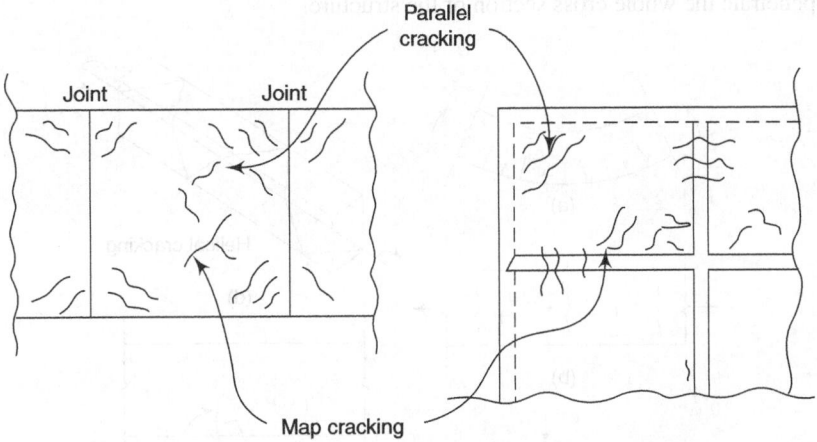

Fig. 25.16 Plastic shrinkage cracks on the surface of pavement and continuous floor slab

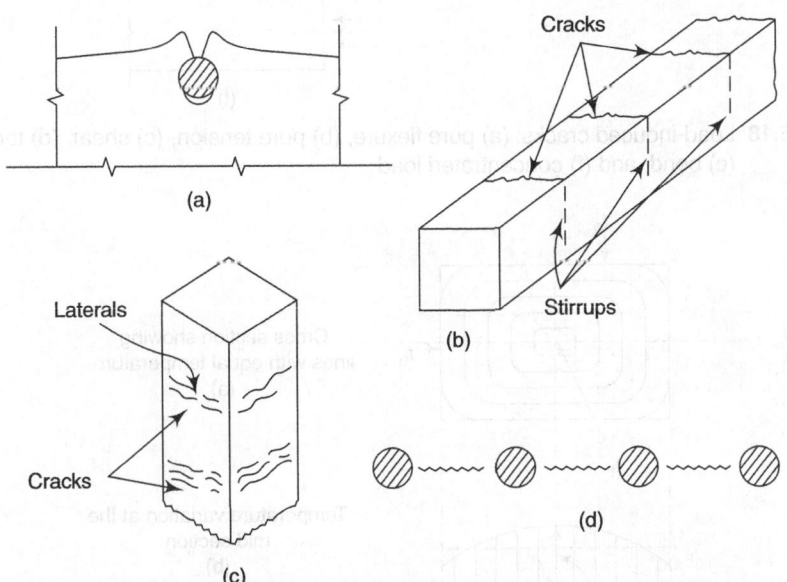

Fig. 25.17 Cracks due to plastic settlement: (a) settlement hampered by reinforcements, (b) cracks parallel to reinforcements on stirrups, (c) cracks parallel to reinforcements, (d) cracks across longitudinal reinforcements

Cracking caused by load may be due to pure flexure, pure tension, shear, torsion, bond, or concentrated load effects. Figure 25.18 summarizes the various forms of cracking due to load effects.

Cracks occur due to the dissipation of the heat generated during the hydration of cement. The hydration heat of cement, which evolves during the hardening process, is dissipated to the surrounding air. This becomes a problem, especially in massive sections. A temperature gradient develops, from a maximum at the core to a minimum at the air surface (Fig. 25.19). Figure 25.19(a) shows the variation of temperature contours indicating t_0 as the temperature at the surface and t_1 as the temperature at the core. Figure 25.19(b) shows the temperature gradient at mid section. This leads to the condition of self-equilibrating internal stresses as shown in Fig. 25.20. The cracks that appear at the surface are generally map cracking.

If a structural element is stressed, especially by axial tension, partition cracks as shown in Fig. 25.21 are formed, which penetrate the whole cross section of the structure.

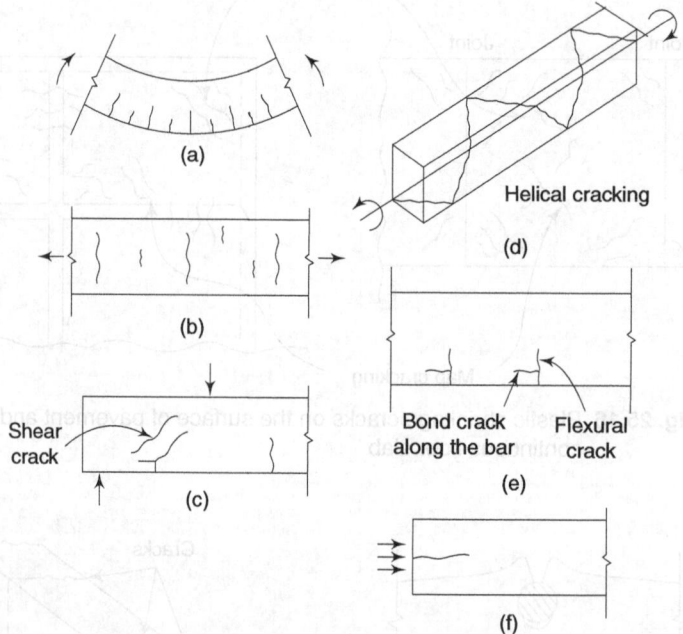

Fig. 25.18 Load-induced cracks: (a) pure flexure, (b) pure tension, (c) shear, (d) torsion, (e) bond, and (f) concentrated load

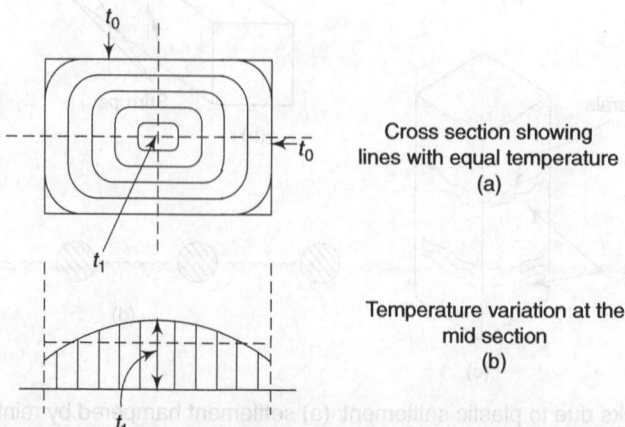

Fig. 25.19 Distribution of temperature due to the heat of hydration

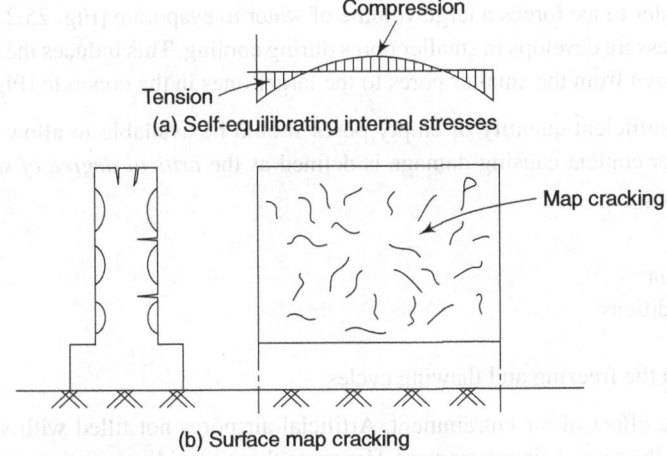

(a) Self-equilibrating internal stresses

Map cracking

(b) Surface map cracking

Fig. 25.20 Self-equilibrating stresses due to temperature gradient causing surface cracking

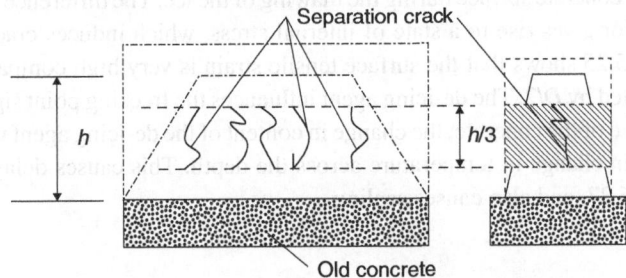

Separation crack

h

$h/3$

Old concrete

Fig. 25.21 Cracking due to early thermal movement

25.3 Frost Attack

Water in the pores freezes when the temperature goes below zero. In such a case the following physical process determines resistance against damage.

a. Transition from water to ice involves a volume increase of about 9%. This causes the concrete to split if the pores are completely filled.
b. The surface energy results in a reduction of the potential energy of the pore water, which results in the depression of the freezing point; see Fig. 25.22(a). Even at (–) 60° only two-thirds of the water freezes. A thin film of water still remains and lubricates the pore walls as shown in Fig. 25.22(b).

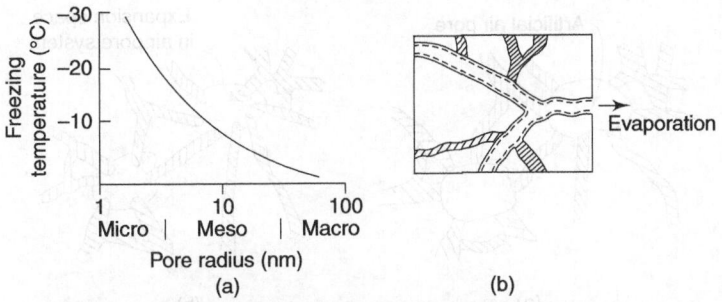

Fig. 25.22 (a) Depression of freezing point due to surface energy;
(b) Evaporation during cooling

c. Transition from water to ice forces a large volume of water to evaporate [Fig. 25.22(b)].
d. A hydraulic low pressure develops in smaller pores during cooling. This induces the diffusion of the water that has not yet frozen from the smaller pores to the larger ones in the concrete (Fig. 25.23).

To prevent damage, a sufficient quantity of empty pores should be available to allow the water to expand. The limit value of water content causing damage is defined as the *critical degree of saturation*. This value depends on the following:

- Age of concrete
- Pore size distribution
- Environmental conditions
- Rate of cooling
- Drying out between the freezing and thawing cycles

Figure 25.24 shows the effect of air entrainment. Artificial air pores not filled with water even under full saturation are present in the case of air entrainment. However, these provide expansion space for freezing water.

The application of chloride as a de-icing agent to a concrete surface covered with ice causes a substantial drop in temperature at the concrete surface during the thawing of the ice. The difference in temperature between the surface and the interior gives rise to a state of internal stress, which induces cracking at the outer layer of the concrete. Figure 25.25 shows that the surface tensile strain is very high compared to the compressive stress at mid depth indicated by $D/2$. The de-icing agent influences the freezing point significantly (Fig. 25.26). Depending on the pore size of the concrete, the change in content of the de-icing agent with increasing distance from the surface results in change in temperature across the depth. This causes delayed freezing across the depth as shown in Fig. 25.27, and also causes scaling.

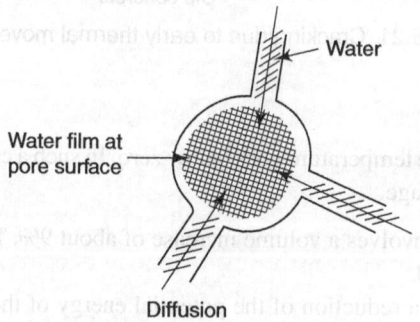

Fig. 25.23 Diffusion during cooling

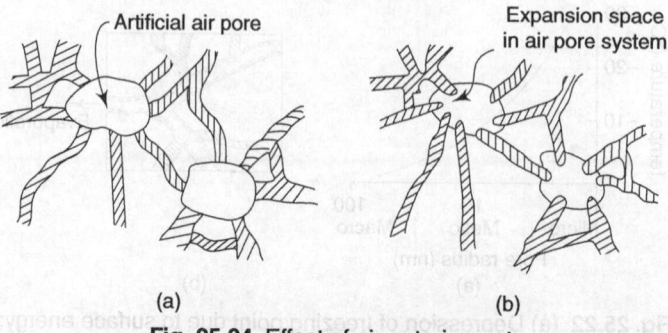

Fig. 25.24 Effect of air entrainment

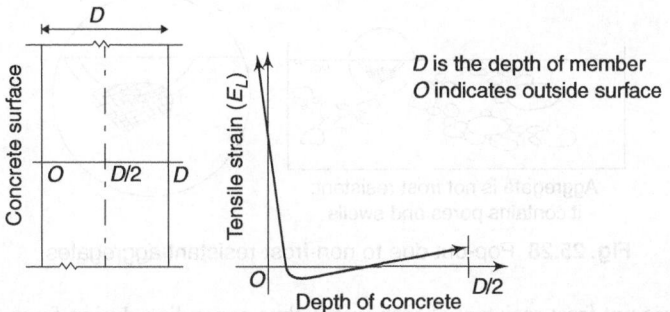

Fig. 25.25 Distribution of tensile strain in concrete experiencing thermal
shock at the surface due to the effects of chlorides
used as de-icing agents

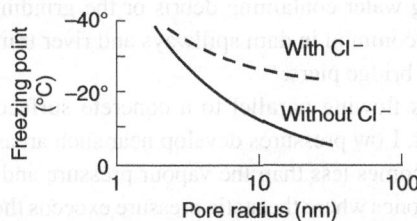

Fig. 25.26 Effect of chlorides (de-icing agents) on the freezing properties of pore water

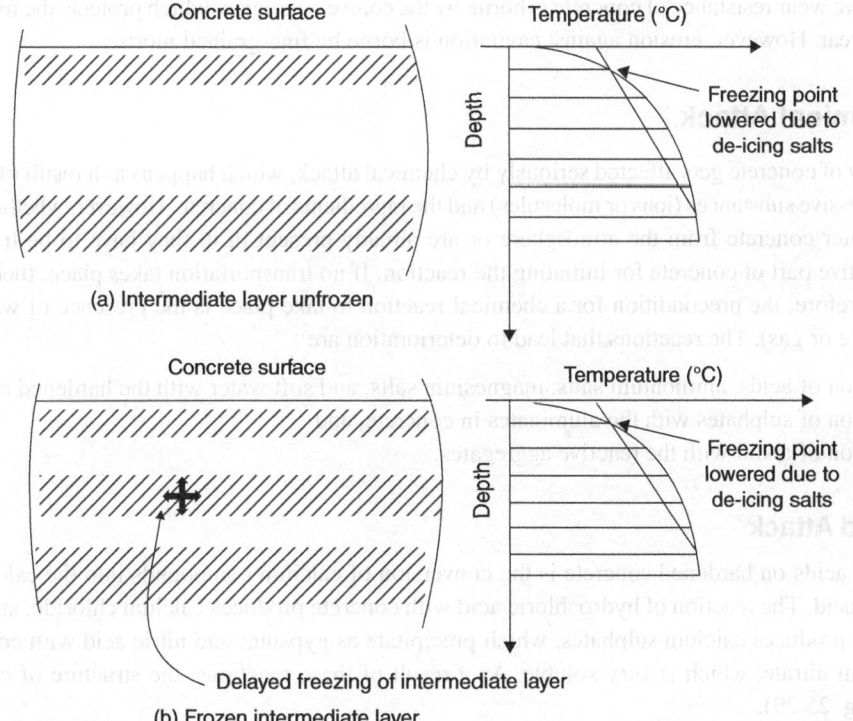

(a) Intermediate layer unfrozen

Delayed freezing of intermediate layer

(b) Frozen intermediate layer

Fig. 25.27 Scaling due to variation in the time taken for the intermediate layers to freeze

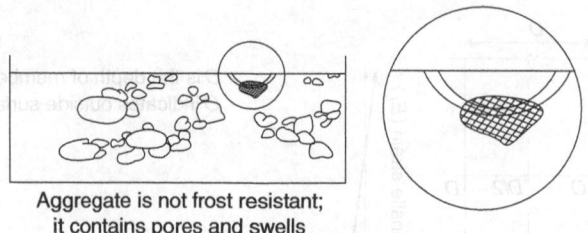

Aggregate is not frost resistant;
it contains pores and swells

Fig. 25.28 Pop-out due to non-frost-resistant aggregates

Aggregates which are not frost resistant absorb water, thus expanding during freezing and destroying the cement paste. This leads to local spalling and results in pop-outs (Fig. 25.28).

25.4 Erosion by Abrasion

Abrasion can be caused by flowing water containing debris or the grinding action of traffic and especially pedestrians on floors. Such wear is common in dam spillways and river training works and in structures protecting embankments, sea walls, or bridge piers.

If water devoid of any solids is flowing parallel to a concrete surface rapidly, any change of path or obstruction causes flow detachment. Low pressures develop near such areas. Under these circumstances, the pressure of the streaming water becomes less than the vapour pressure and vapour-filled bubbles develop in this zone. These bubbles stream to zones where the static pressure exceeds the vapour pressure of the water, the vapour in the bubbles condenses and then they collapse suddenly. This implosion causes an impact resulting in pressure waves. This process is generally known as cavitation and results in damage—pitting and excavation on the concrete surface.

The abrasive wear resistance of concrete is borne by the coarse aggregate, which protects the mortar against mechanical wear. However, erosion against cavitation is borne by fine-grained mortar.

25.5 Chemical Attack

The durability of concrete gets affected seriously by chemical attack, which happens as a result of the reaction between aggressive substances (ions or molecules) and the ingredients of concrete. However, whether aggressive substances enter concrete from the atmosphere or are already present in it, they have to be transported to meet the reactive part of concrete for initiating the reaction. If no transportation takes place, there will be no reaction. Therefore, the precondition for a chemical reaction to take place is the presence of water in some form (moisture or gas). The reactions that lead to deterioration are

a. the reaction of acids, ammonium salts, magnesium salts, and soft water with the hardened cement,
b. the reaction of sulphates with the aluminates in concrete, and
c. the reaction of alkali with the reactive aggregates.

25.5.1 Acid Attack

The action of acids on hardened concrete is the conversion of calcium compounds into the calcium salts of the attacking acid. The reaction of hydrochloric acid with concrete produces calcium chloride; sulphuric acid with concrete produces calcium sulphates, which precipitate as gypsum; and nitric acid with concrete gives rise to calcium nitrate, which is very soluble. As a result of these reactions, the structure of concrete gets destroyed (Fig. 25.29).

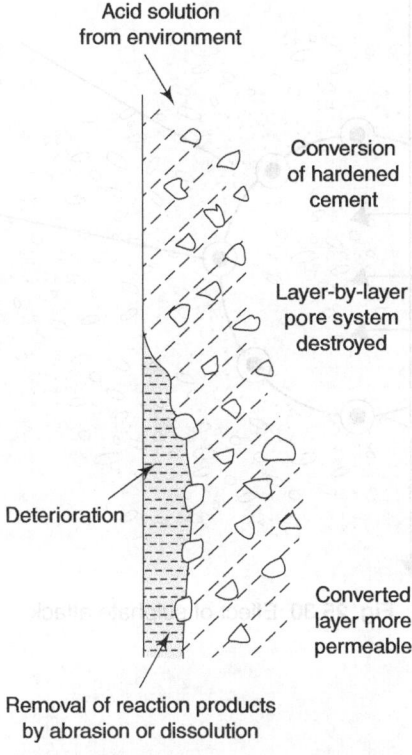

Acid solution
from environment

Conversion
of hardened
cement

Layer-by-layer
pore system
destroyed

Deterioration

Converted
layer more
permeable

Removal of reaction products
by abrasion or dissolution

Fig. 25.29 Effect of acid attack

The rate of the reaction depends on the solubility of the calcium salts that get formed. The less soluble the salts are, the more passive is the reaction. If the salt is soluble, the rate of reaction depends on the rate of dissolution of the salts.

Acid attacks completely convert the hardened cement paste, destroying the pore system. Therefore, in the case of acid attack, the permeability of concrete is less important compared to the reactions that take place. However, in the case of the attack by sulphates and alkalis, permeability is of great importance as described in the following sections.

25.5.2 Sulphate Attack

Sulphates attack only certain compounds in cement. They react with hydrated calcium aluminate to form ettringite and with free calcium ions to form gypsum. These compounds have greater volumes than the reactants, which causes the concrete to expand. This expansion leads to cracking with an irregular pattern (Fig. 25.30). This gives easier access to sulphates for further penetration. The process continues till there is complete disintegration and destruction of concrete. The cracking and disintegration depend on

a. the exposure condition, i.e., the amount of aggressive substance,
b. the permeability of concrete,
c. the susceptibility of concrete (type of cement), and
d. the amount of water available.

Concrete can be protected either by the use of sulphate-resisting cement or by making it highly impermeable. Limiting aluminates in cement ensures protection against sulphate attack.

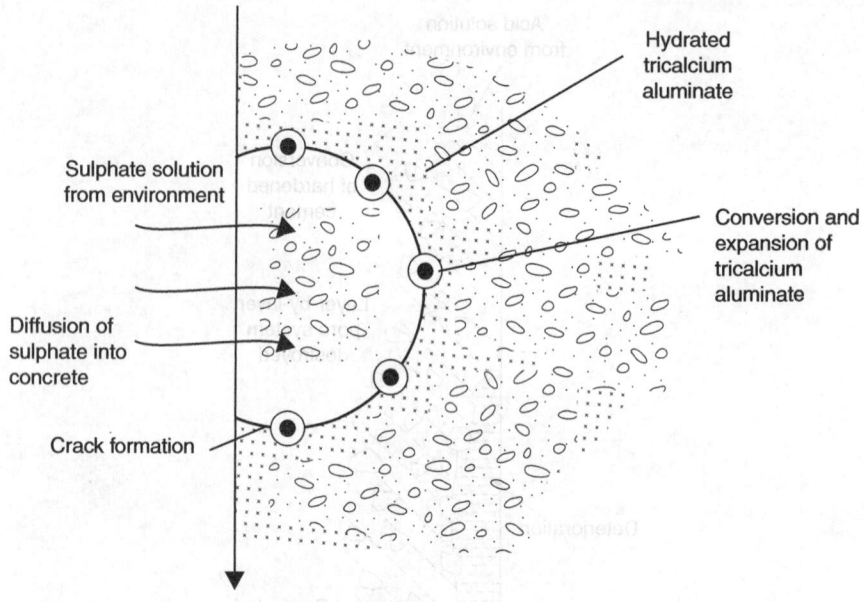

Fig. 25.30 Effect of sulphate attack

25.5.3 Alkali Attack

Alkalies only attack the reactive aggregates in concrete. The reactive substance in this case is not cement (as in sulphate attack) but aggregates which contain silica. The alkaline solution in the pores is lime-saturated and contains potassium and sodium ions. Silica-containing aggregates react with this alkaline solution leading to disruptive expansion (Fig. 25.31). Visible map cracking, pop-outs, and weeping of glassy pearls are the various manifestations of this problem. The reaction depends on

  a. the reactivity of aggregates based on the pressure of reactive amorphous silica,
  b. the grain size of aggregates,
  c. the alkali calcium concentration in the pore water,
  d. the type of cement,
  e. the exposure condition,
  f. the amount of water available, and
  g. the rate of transport.

The use of blended or slag cement limits the reaction by limiting the presence of alkaline solution in the pore water. Good curing helps in reducing the permeability (Fig. 25.32) and thereby controls the alkali–silica reaction.

25.6 Aggressive Environment

The general atmospheric condition near a concrete structure has less importance. The local climate within centimetres (micro-climate) and the conditions around the foundation, piles, and submerged parts have a greater influence on the durability of structure.

As explained earlier, all processes which involve the degradation of the pore structure need water. Under varying climatic conditions, the internal average humidity is greater than the average ambient humidity. This

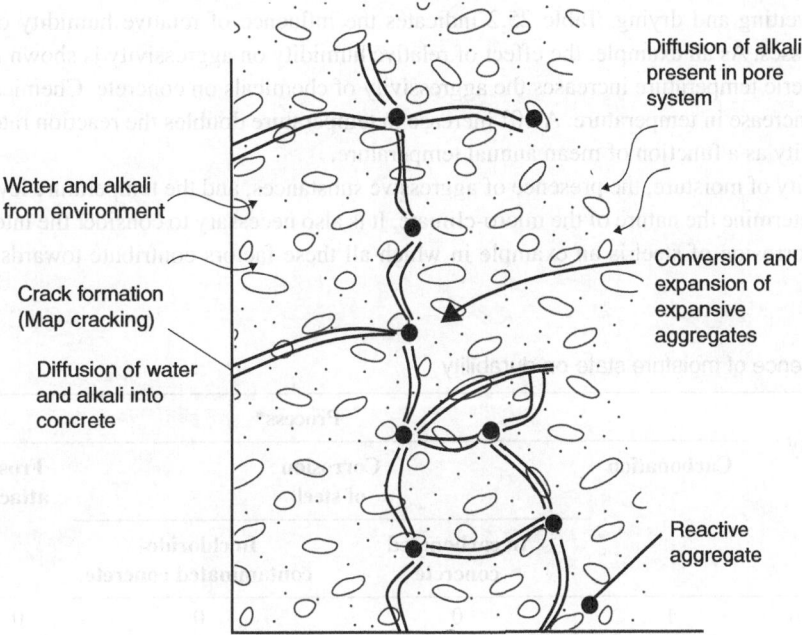

Fig. 25.31 Effect of alkali–silica reaction

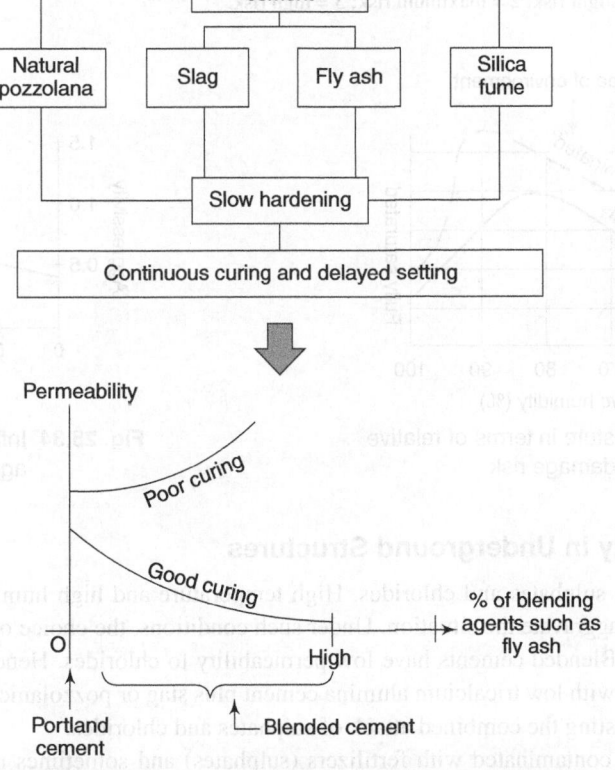

Fig. 25.32 Influence of blending agents on permeability

is true during wetting and drying. Table 25.2 indicates the influence of relative humidity on the different degrading processes. As an example, the effect of relative humidity on aggressivity is shown in Fig. 25.33.

The atmospheric temperature increases the aggressivity of chemicals on concrete. Chemical reactions are accelerated by increase in temperature. A 10° increase in temperature doubles the reaction rate. Figure 25.34 shows aggressivity as a function of mean annual temperature.

The availability of moisture, the presence of aggressive substances, and the temperature level are the main factors which determine the nature of the micro-climate. It is also necessary to consider the interaction among these factors. Corrosion of steel is an example in which all these factors contribute towards increasing the aggressivity.

Table 25.2 Influence of moisture state on durability

Effective relative humidity	Process*				
	Carbonation	Corrosion of steel		Frost attack	Chemical attack
		In carbonated concrete	In chloride-contaminated concrete		
Very low (< 45%)	1	0	0	0	0
Low (45–65%)	3	1	1	0	0
Medium (65–85%)	2	3	3	0	0
High (85–98%)	1	2	3	2	1
Saturated (> 98%)	0	1	1	3	3

*0 = insignificant risk; 1 = slight risk; 2 = maximum risk; 3 = high risk.

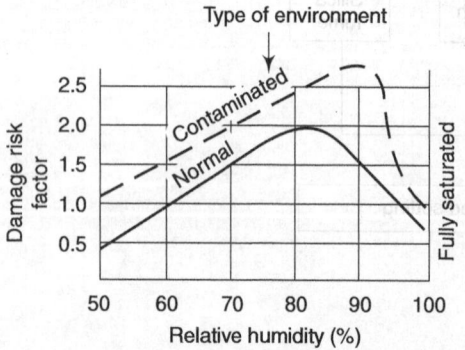

Fig. 25.33 Effect of moisture in terms of relative humidity on damage risk

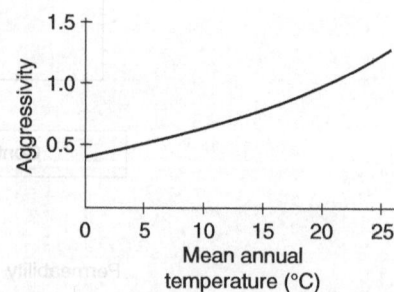

Fig. 25.34 Influence of temperature on aggressivity of environment

25.6.1 Aggressivity in Underground Structures

Soils may contain both sulphates and chlorides. High temperature and high humidity, a regular feature of Indian climate, further aggravate the situation. Under such conditions, the choice of tricalcium alumina content becomes difficult. Blended cements have low permeability to chlorides. Hence, one can specify dense homogeneous concrete with low tricalcium alumina cement plus slag or pozzolanic material or an equivalent blended cement for resisting the combined attack of sulphates and chlorides.

Soil frequently gets contaminated with fertilizers (sulphates) and sometimes mineral oils. Mineral oils containing acidic components should be considered aggressive.

25.6.2 Marine Environment

Seawater contains many dissolved salts. The concentration of these salts varies from place to place. The total salt content is approximately 35 g/L. Figure 25.35 gives the typical relative concentrations of various ions in the Atlantic ocean. In addition, seawater contains dissolved oxygen and CO_2. The micro-climate of a marine environment is shown in Fig. 25.36. The exposure experienced by a structure in a marine environment can be broadly grouped as discussed next.

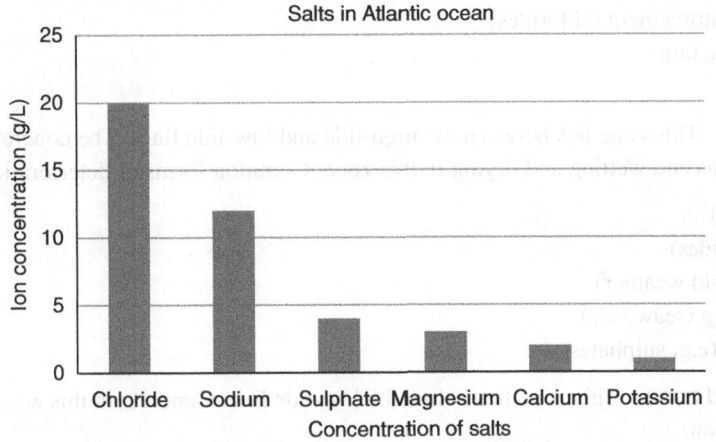

Fig. 25.35 Typical ionic concentration of salts in the Atlantic ocean

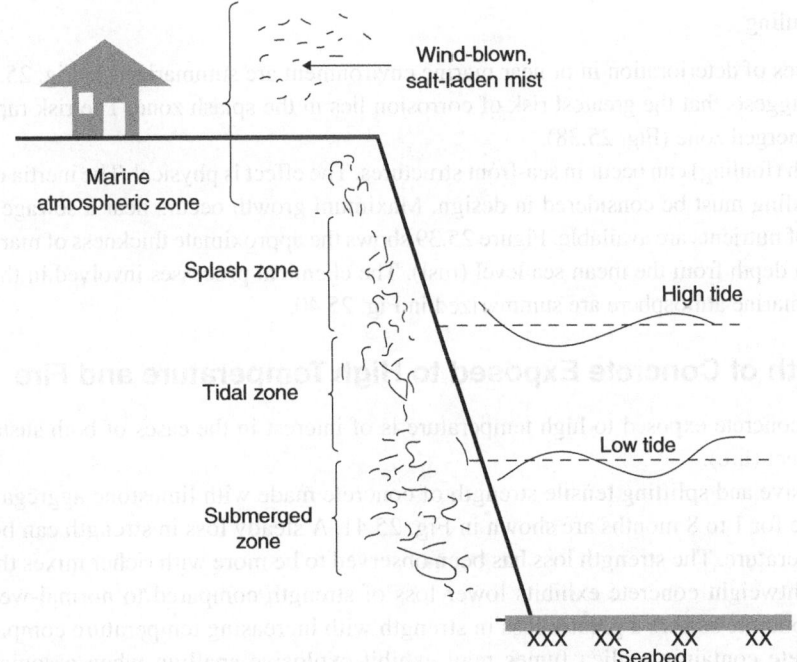

Fig. 25.36 Types of marine environment

Marine atmospheric zone   Concrete is never directly in contact with water but receives wind-blown, salt-laden mist. Based on the prevailing wind, this zone may extend many kilometres into the land. IS: 15498-2004 indicates that cyclonic storms, and hence weather, can extend up to 60 km from the coast line. The most common forms of deterioration in this zone are

- corrosion of reinforcement (chlorides)
- frost damage (cold climate)

Marine splash zone   This zone exists above the high-tide line but is subject to direct wetting by seawater from waves and sprays and may become dry. Common forms of deterioration:

- corrosion of reinforcement (chlorides)
- abrasion (wave action)
- frost damage

Marine tidal zone   This zone lies between the high-tide and low-tide lines. The concrete experiences both submerged conditions and wetting and drying in this zone. Common forms of deterioration:

- abrasion (waves)
- corrosion (chlorides)
- frost damage (cold weather)
- biological fouling (seaweeds)
- chemical attack (e.g. sulphates)

Marine submerged zone   This zone lies below the low-tide line. Concrete in this zone is continuously in a submerged condition.

Marine seabed   In this zone, the concrete is in contact with sand or soil and is also continuously submerged. The common forms of deterioration in the submerged zone and the seabed are

- chemical attack
- biological fouling

The different types of deterioration in or near marine environment are summarized in Fig. 25.37.

Experience suggests that the greatest risk of corrosion lies in the splash zone. The risk rapidly decreases towards the submerged zone (Fig. 25.38).

Marine growth (fouling) can occur in sea-front structures. The effect is physical. The inertia of the members covered with fouling must be considered in design. Maximum growth occurs near a sewage outfall, where large quantities of nutrients are available. Figure 25.39 shows the approximate thickness of marine growth and its variation with depth from the mean sea level (msl). The chemical processes involved in the deterioration of concrete in a marine atmosphere are summarized in Fig. 25.40.

25.7 Strength of Concrete Exposed to High Temperature and Fire

The strength of concrete exposed to high temperature is of interest in the cases of both sustained effect as well as shock effect (fire).

The compressive and splitting tensile strength of concrete made with limestone aggregates exposed to high temperature for 1 to 8 months are shown in Fig. 25.41. A steady loss in strength can be noticed with increase in temperature. The strength loss has been observed to be more with richer mixes than with linear mixes. Also, lightweight concrete exhibits lower loss of strength compared to normal-weight concrete. High-strength concrete suffers a greater loss in strength with increasing temperature compared to normal concrete. Concrete containing silica fumes may exhibit explosive spalling when associated with high temperature.

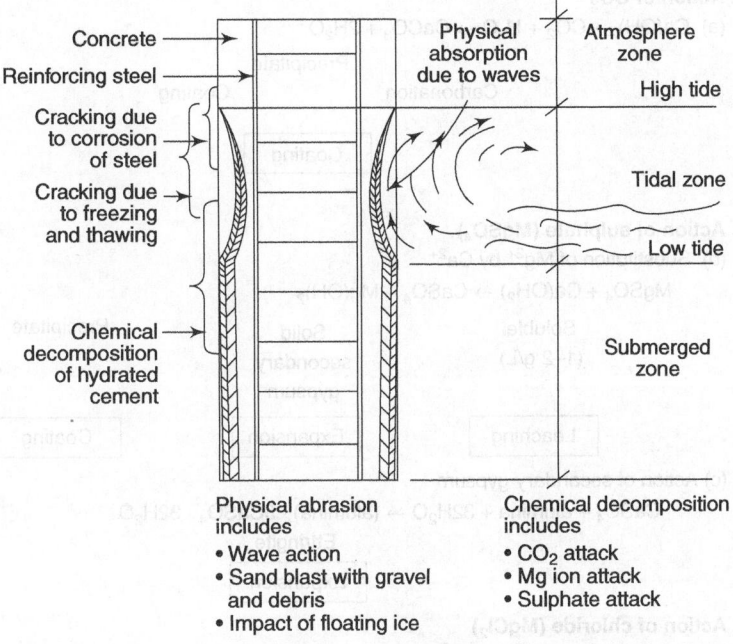

Physical abrasion includes
• Wave action
• Sand blast with gravel and debris
• Impact of floating ice

Chemical decomposition includes
• CO_2 attack
• Mg ion attack
• Sulphate attack

Fig. 25.37 Deterioration of concrete structures near sea environment

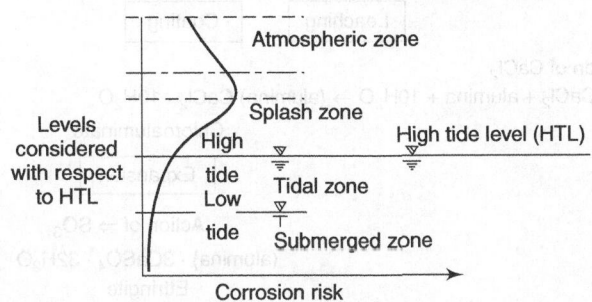

Fig. 25.38 Corrosion risk in marine environment at various levels

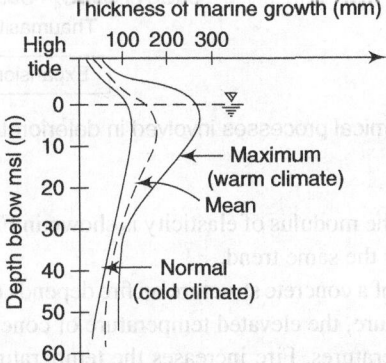

Fig. 25.39 Variation of thickness of marine growth with depth from msl

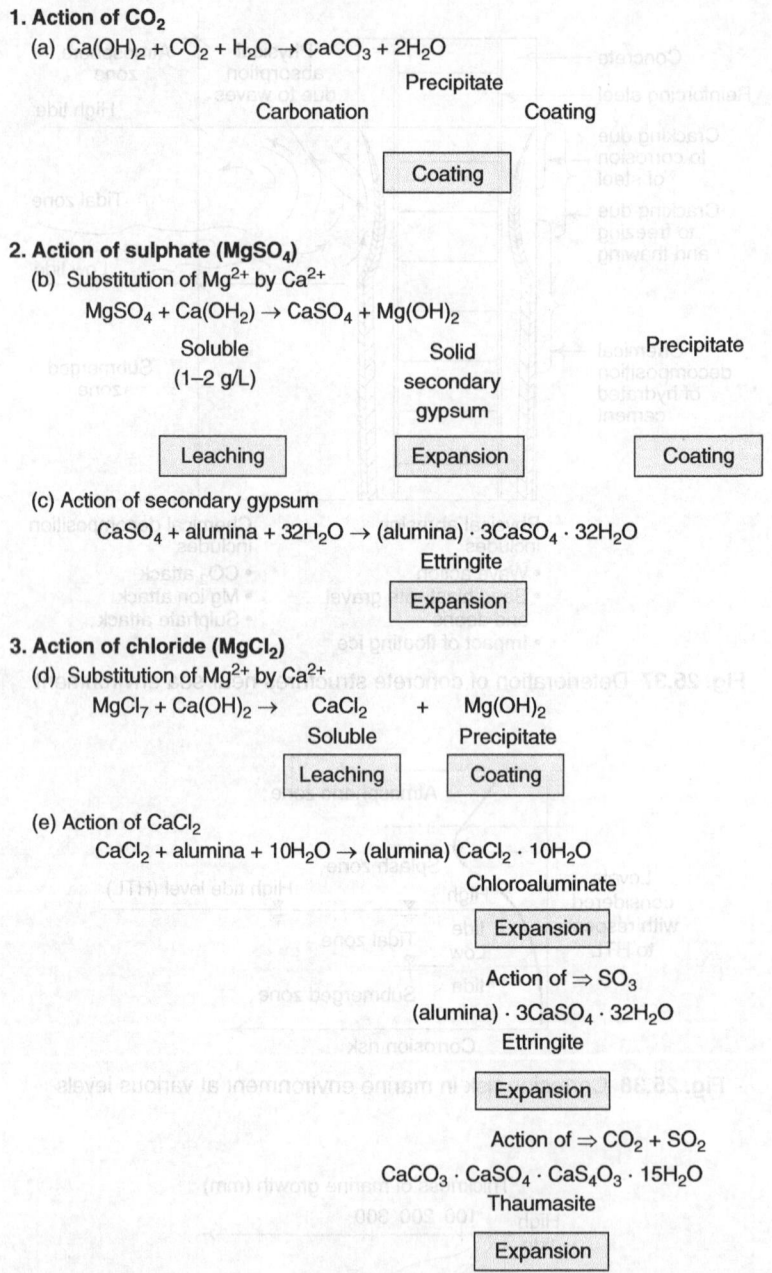

1. Action of CO$_2$

(a) $Ca(OH)_2 + CO_2 + H_2O \rightarrow CaCO_3 + 2H_2O$

Precipitate

Carbonation   Coating

Coating

2. Action of sulphate (MgSO$_4$)

(b) Substitution of Mg^{2+} by Ca^{2+}

$MgSO_4 + Ca(OH_2) \rightarrow CaSO_4 + Mg(OH)_2$

Soluble   Solid   Precipitate

(1–2 g/L)   secondary

gypsum

Leaching   Expansion   Coating

(c) Action of secondary gypsum

$CaSO_4 + alumina + 32H_2O \rightarrow (alumina) \cdot 3CaSO_4 \cdot 32H_2O$

Ettringite

Expansion

3. Action of chloride (MgCl$_2$)

(d) Substitution of Mg^{2+} by Ca^{2+}

$MgCl_7 + Ca(OH)_2 \rightarrow$ $CaCl_2$ + $Mg(OH)_2$

Soluble   Precipitate

Leaching   Coating

(e) Action of $CaCl_2$

$CaCl_2 + alumina + 10H_2O \rightarrow (alumina) CaCl_2 \cdot 10H_2O$

Chloroaluminate

Expansion

Action of $\Rightarrow SO_3$

$(alumina) \cdot 3CaSO_4 \cdot 32H_2O$

Ettringite

Expansion

Action of $\Rightarrow CO_2 + SO_2$

$CaCO_3 \cdot CaSO_4 \cdot CaS_4O_3 \cdot 15H_2O$

Thaumasite

Expansion

Fig. 25.40 Chemical processes involved in deterioration of concrete

The influence of temperature on the modulus of elasticity is shown in Fig. 25.42. The variations of strength and modulus with temperature show the same trend.

The assessment of the behaviour of a concrete structure on fire depends on many factors, the most important being the applied loads on the structure, the elevated temperature of concrete, and the mechanical properties of steel and concrete at those temperatures. Fire increases the temperature of both steel and concrete. This leads to increased deformation and possible failure depending on the applied loads and support conditions.

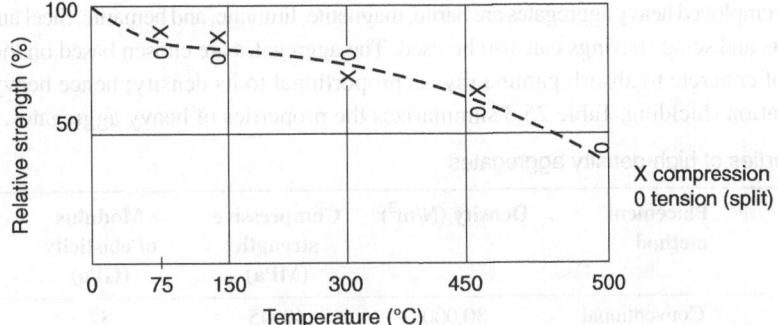

Fig. 25.41 Influence of temperature on compressive and tensile strength of concrete

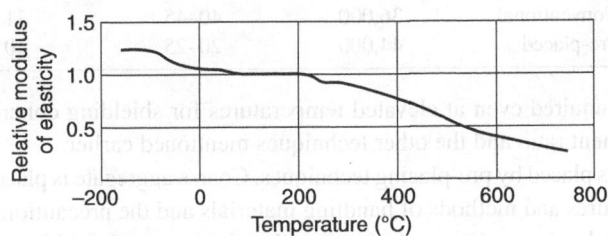

Fig. 25.42 Influence of temperature on modulus of elasticity

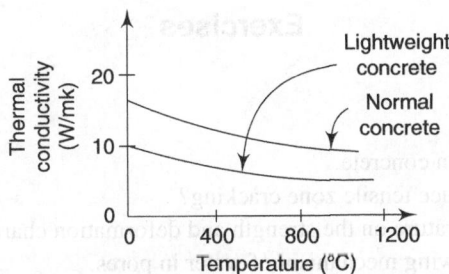

Fig. 25.43 Thermal conductivity of concrete

Initially, when heated to 100°C, the density of most concrete decreases by 100 kg/m^3 due to the evaporation of free water. This has a minor effect on thermal response. The thermal conductivity of concrete is temperature dependent. It decreases with increase in temperature as shown in Fig. 25.43.

The specific heat of concrete also varies with moisture content. The peak occurs between 100°C and 200°C. An average value of 100 J/kg K is appropriate for dry concrete.

25.8 Radiation Shielding

Concrete, both normal and high density, is effective for shielding radiation. The density of normal-weight concrete varies from 2500 kg/m^3 to 3800 kg/m^3. Concretes of density up to 5300 kg/m^3 have been produced. These are excellent shielding materials from harmful radiation such as x-rays and gamma rays. They have the necessary mechanical properties and can be made relatively cost-effective compared to lead which is commonly used in radiation protection. They are easy to maintain in nuclear plants. The heavy aggregates in such concretes occur naturally and can also be manufactured. Heavier aggregates occur in the form of smelted metal.

The commonly employed heavy aggregates are barite, magnetite, limonite, and hematite. Steel and iron aggregates in the form of shots and scrap shavings can also be used. The aggregates are chosen based on their availability.

The capacity of concrete to absorb gamma rays is proportional to its density; hence heavy aggregates are preferred for radiation shielding. Table 25.3 summarizes the properties of heavy aggregates.

Table 25.3 Properties of high-density aggregates

Aggregate type	Placement method	Density (N/m^3)	Compressive strength (MPa)	Modulus of elasticity (GPa)	Shrinkage strain (%)
Limonite	Conventional	30,000	40–45	32	0.020
Limonite + magnetite	Pre-placed	34,000	30–35	36	0.022
Barite	Conventional	36,000	40–45	31	0.030
Barite	Pre-placed	37,000	20–25	26	0.030
Magnetite	Conventional	36,000	40–45	31	0.018
Iron limonite	Pre-placed	44,000	20–25	40	0.015

Structural strength is required even at elevated temperatures for shielding concrete. This is achieved by controlling the water–cement ratio and the other techniques mentioned earlier.

High-density concrete is placed by pre-placing techniques. Coarse aggregate is placed first and then the grout is injected. All the procedures and methods of handling materials and the precautions discussed in Chapter 5 in connection with the production and control of conventional concrete should be followed when producing heavyweight concrete of required quality.

Exercises

Review Questions

1. Describe the effect of pores in concrete.
2. What are the factors that induce tensile zone cracking?
3. Explain the impact of temperature on the strength and deformation characteristics of concrete.
4. Explain the freezing and thawing mechanism of water in pores.
5. Write short notes on the following:
   a. Acid attack
   b. Sulphate attack
   c. Alkali attack
6. Describe the effect of coastal atmospheric conditions on a building.
7. List the various exposures experienced by a typical structure in a marine environment.
8. How does concrete behave under fire?
9. Explain how the radiation shielding property of concrete can be improved.

Multiple Choice Questions

1. The performance and service life of the structure is affected by
   a. Design specification
   b. Construction quality
   c. Maintenance of the structure
   d. All of the above

2. In concrete, capillary pores are of size
   a. 10^{-2} to 10^{-4}
   b. 10^{-4} to 10^{-7}
   c. 10^{-8} to 10^{-10}
   d. $<10^{-10}$

3. In concrete, water transport through pores depends on
   a. Evaporation
   b. Capillary suction
   c. Hydraulic gradient
   d. All of the above

4. Plastic shrinkage cracks in concrete occur within a period of _____ after concreting.
   a. One week
   b. 6 to 8 hours
   c. One month
   d. One year

5. Plastic shrinkage cracks can be avoided by proper
   a. Formwork design
   b. Early curing
   c. Reinforcement spacing
   d. Cover to rebars

6. 45 degree diagonal cracks in a concrete beam are generally caused by
   a. Corrosion
   b. Shrinkage
   c. Shear
   d. Flexure

7. Frost action on concrete can be reduced by adopting
   a. Air entrainment
   b. Larger cover
   c. Higher strength
   d. Painting

8. When chlorides are present in soil, it is advisable to use
   a. OPC cement
   b. PPC cement
   c. Sulphate-resisting cement
   d. Rapid hardening cement

9. The worst zone for corrosion in a marine structure is
   a. Splash zone
   b. Submerged zone
   c. Atmospheric zone
   d. Tidal zone

10. Modulus of elasticity of concrete
    a. Does not vary with temperature
    b. Increases with increasing temperature
    c. Decreases with increasing temperature
    d. Increases with increase in temperature up to 200°C and then decreases

Answers to Multiple Choice Questions

1. d	4. b	7. a	10. c
2. b	5. b	8. b	
3. d	6. c	9. a	

CHAPTER 26

Concreting under Special Circumstances

There are often situations in which concreting must be done under special circumstances such as constructions in basements, sewerage, and marine works. In such situations, concrete is placed underwater or in contact with earth or sand. Concrete placed in such difficult locations needs special care. If proper precautions are not taken, the cement may leach or the aggregates may segregate. It is also sometimes necessary to place concrete under extreme weather conditions—either very hot weather as in the summer in Rajasthan or in very cold weather as in Kashmir during the winter. Because of the continued demand of building and civil engineering structures in extreme hot and cold weather conditions, it is necessary to have complete knowledge of concreting in these extreme conditions. This chapter deals with concreting under such special circumstances.

26.1 Underground Construction

When concrete is to be laid below ground level in a deep excavation (which may be waterlogged), the space for working is temporarily created by excluding the soil and water. This technique is generally called *cofferdam construction*. A cofferdam is a retaining structure, usually temporary in nature, which is used to support the sites of deep excavations where water is present. These structures generally consist of

a. vertical sheet piling,
b. a bracing system composed of either wales or struts or prestressed tiebacks, and
c. a bottom seal course, if required, to seal out the water.

Cofferdams are used in situations where the adjacent ground must be supported against settlements or slides, and in the construction of bridge piers and abutments in relatively shallow water.

Cofferdams usually fall under the category of temporary structures necessary to construct an underground structure. As such, the site in-charge is responsible for the proper design, construction, maintenance, and removal of cofferdams. They are required to prepare working drawings ensuring safety at all times for the successful implementation of the project.

26.1.1 Sheet Piles and Bracing

There are three basic materials used for the construction of sheet piles: wood, concrete, and steel. Wood sheet piling consists of a single line of boards (Fig. 26.1)—'single-sheet piling'—but is suitable only for comparatively small excavations in which there is no serious groundwater problem.

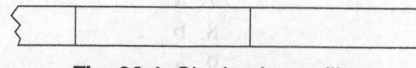

Fig. 26.1 Single-sheet piling

In saturated soils, particularly in sand and gravel, it is necessary to use a more elaborate form of sheet piling which can be made reasonably watertight with overlapping boards spiked or bolted together, such as the 'lapped-sheet piling' or 'Wakefield' system as shown in Fig. 26.2.

'Tongue-and-groove' sheet piling (Fig. 26.3) is also used. This is made from a single piece of timber cut at the mill with a tongue shape at one end and a groove shape at the other.

Precast concrete sheet piles are normally used in situations where the precast members are going to be incorporated into the final structure or are going to remain in place after they fulfil their purpose. One normally does not encounter precast sheet piling in minor structure works.

Precast concrete sheet piling is usually made in the form of a tongue-and-groove section. The piles vary in width from 45 to 60 cm and in thickness from 20 to 45 cm. They are reinforced with vertical bars and hoops in much the same way as precast concrete-bearing piles. This type of sheeting is not always perfectly watertight, but the spaces between the piles can be grouted (Fig. 26.4).

In order to provide a more watertight precast concrete sheet pile, two halves of a straight steel web sheet pile, which has been split in half longitudinally, are embedded in the pile (Fig. 26.5).

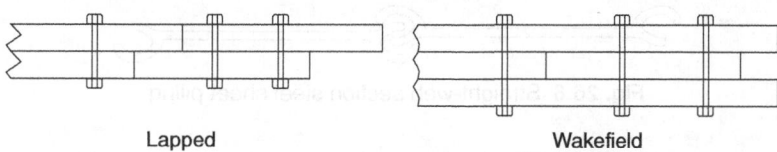

Lapped   Wakefield

Fig. 26.2 Lapped and Wakefield sheet piling

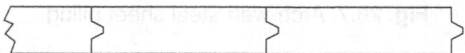

Fig. 26.3 Tongue-and-groove wood sheet piling

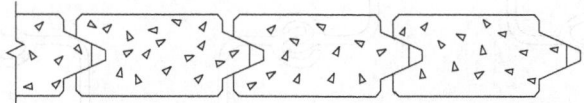

Fig. 26.4 Precast concrete sheet piling

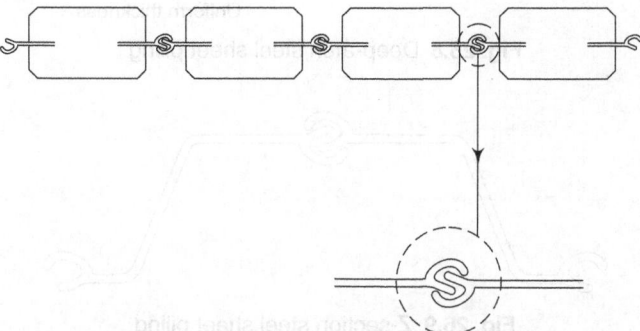

Fig. 26.5 Concrete sheet piling with steel interlocks

Steel sheet piling is most commonly used and is available in a number of different shapes. The shape provides for bending strength and the interlock (connection between the sheets) provides for the alignment. Each steel company that manufactures sheet piling has its own form of interlock. The simplest shape is known as the 'straight-web' type. These come in various widths ranging from 40 to 50 cm. The web thickness varies from 10 to 12 mm. Straight-web sheet piling is comparatively flexible and requires a considerable amount of bracing in areas where the horizontal thrusts are large (Fig. 26.6).

In order to provide a pile with a greater resistance to bending, steel companies have developed the 'arch-web' section (Fig. 26.7), in which the centre of the web is offset so as to provide a greater moment of inertia in the cross section. To provide for even greater stiffness, there is the 'deep-arch' section (Fig. 26.8). The deep-arch section is similar to the arch-web section except that the offset in the web is increased considerably. A type known as the Z-section (Fig. 26.9) has a stiffness considerably greater than that of the deep-arch section.

The choice of the type of steel sheet pile to be used for a given job depends largely on the capacity it is intended to be used in. The straight-web section is comparatively flexible so that it requires a considerable amount of bracing when subject to a large horizontal thrust. However, its size allows it to be used in closed quarters, where a deep-arch section or Z-section will not fit.

Fig. 26.6 Straight-web section steel sheet piling

Fig. 26.7 Arch-web steel sheet piling

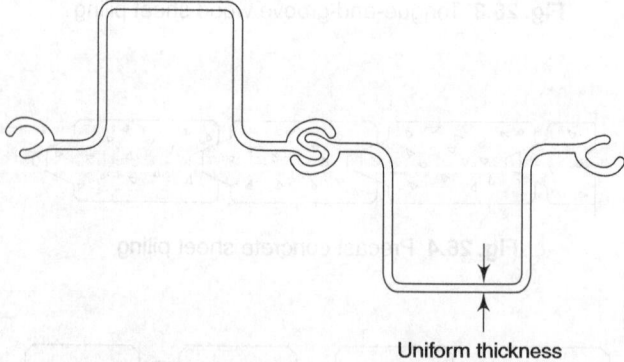

Uniform thickness

Fig. 26.8 Deep-arch steel sheet piling

Fig. 26.9 Z-section steel sheet piling

The composition of the bracing system inside a cofferdam depends upon the forces that will act on it, the availability of materials, and the costs connected with the system. Wood and steel are the materials used normally. Prestressed tiebacks are often used in large land cofferdams, where the system of cross-bracing is impractical.

It is required that excavations be completed at the bottom of the footings before driving the piles. As in many other areas of our work, there are times when engineering judgement should be used depending on the type of soil, amount of excavation required, type of pile, and depth below the water surface. Normally, excavation would be done by submerged clamshell, with the elevations being checked by sounding.

In the case of pile foundations, it is often advisable to over-excavate a predetermined amount to compensate for the heave of the material caused by pile driving displacement. This is done to eliminate the need for excavation after driving. If excavation is needed, one should ensure that none of the driven piles are damaged.

Following the installation of the cofferdam, the footing can be excavated and the piles driven. Usually the footing area must also be dewatered. Depending on the volume of water present, this can be achieved by pumping. Otherwise, a seal course may be necessary. A seal course is intended to control and remove water from the excavation.

As the name implies, a properly constructed seal course seals the entire bottom of a cofferdam and prevents subsurface water from entering the cofferdam. In so doing, it permits the construction of footings and columns or other facilities in the dry. The seal course is a concrete slab placed underwater and constructed thick enough so that its weight is sufficient to resist the uplift from hydrostatic forces. In terms of its importance to the designed structure, the seal course normally has no structural significance.

The decision as to the thickness of seal course required, or whether the seal course is to be eliminated, is based on the conditions encountered on the jobsite. The plans also contain the provision for adjusting the footing elevations if seal courses are eliminated. In usual field practice, this decision is not a difficult one to make. In most cases of absence of water, a seal course is clearly not needed.

Concrete directly placed underwater without creating a water-excluded zone using a cofferdam involves the tremie and other methods, which will be discussed later in underwater concreting techniques.

In addition to the usual pre-pour issues, such as access and suitability or adequacy of equipment, sufficient soundings should be taken to verify the elevations. Particular attention should be given to the perimeter of the cofferdam and the pile locations. Soundings can be accomplished using a flat plate of suitable size and weight at the end of a rod or rag tape. This device can be used not only to determine elevations but also, to some extent, to determine the nature of the material (soft or firm). During the pouring, soundings can again be used to verify the elevation of the top surface of the concrete. Because of the type of operation, surface irregularities can be expected, particularly in pile footings. The important thing is to check for proper thickness throughout and the absence of excessive low spots. A minimum curing period of 5 days is required before dewatering.

The thickness of the seal course is important. For example, seal courses in pile footings are constructed one foot thicker than required to allow for surface irregularities. The bond friction between sheet piling and concrete is disregarded. The minimum thickness of seal course concrete is 60 cm.

Dewatering can present some anxious moments since the cofferdam and the seal course are put to the test in this phase. Dewatering is sometimes conducted in stages for a moderately deep cofferdam. At each stage, intermediate bracing systems are installed before proceeding deeper. Depending on the particular design, these internal braces restore the stability of the system.

Sheet pilings are not watertight and minor leaks can be expected as the cofferdam is dewatered. These leaks occur along the joints between adjacent sheets. Sawdust, cement, or some other material can be used to plug these leaks, which are ordinarily not a problem. Dropping the blocking material into the water adjacent to the leaking sheets usually achieves this. Flow through the leak carries the fine material to the problem area and seals the crack or opening. A sump built into the surface of the seal would be helpful in keeping the work area reasonably dry. Obviously, sumps should be located outside the footing limits.

Prior to proceeding with footing work, all high spots in the seal have to be removed. All scum, laitance, and sediment must also be removed from the top of the seal. This can be very time-consuming and expensive. The effect can be reduced significantly if the seal is placed carefully.

Cofferdam work presents safety problems unique to this type of construction. Among them are limited access, limited work areas, damp or wet footing, and deep excavations. Provisions must be made for safe access in terms of adequate walkways, rails, ladders, or stairs into and out of the lower levels. The worksite may be within a waterway, in which case additional safety regulations may apply. These would include provisions for flotation devices, boats, warning signals, and suitable means for a rapid exit in case of accidents.

26.2 Concrete Construction in Marine Environment

Even as concrete is replacing structural steel for all types of construction on land and even for tall structures such as towers and chimneys, the use of concrete for the construction of on- and off-shore marine structures is also increasing. This has special significance in India with over 6000 km of coastline, over 15 major ports and dock installations, and a large number of navigational aids such as lighthouses. The search for alternative techniques and materials for the construction of such structures has assumed importance in the wake of off-shore oil exploration. Increasing transportation costs also calls for the construction of large-span bridges over lakes and backwaters to cut down transport distances and, therefore, save fuel and reduce the delivery time of goods. Building all this infrastructure calls for increasing construction activities of the types listed in Table 26.1, in or near the sea.

Table 26.1 Types of marine structures

Facility	Type I Freely floating structures	Type II Floating structures anchored to the seabed	Type III Structures supported at the seabed	Type IV Coastal or extended structures
Floating storage tank	✓			
Floating docks	✓			
Barges, boats and ships	✓			
Floating navigational aids		✓		
Underwater tunnels/bridges		✓		
Lighthouses off-shore/on-shore		✓	✓	
Off-shore oil platforms			✓	
Large span-bridges and piers			✓	
Coastal protection structures				✓
Fort and harbour structures				✓
Airports and runways (sea side)				✓

The use of materials and techniques for constructing such structures is discussed in the following section.

While many of the structures mentioned here could be claimed to have been built early in the nineteenth century, the scope for using concrete for all these structures has increased since 2000. To emphasize the experience gained by now, the world over, with these concrete structures, some examples of historic structures are quoted below:

Oil platforms built in 1960–70	Ekofisk, Ardyne point, and Tasmania
Boats built in 1848 by Lambot	France
Harbours built during 1910–20	Norway
Bridge piers 150–300 m² at the base built in 12–25 m deep water (1925–42)	Denmark
Carriers built in 1917	Norway
Underwater tunnel built in 1970 (40 m seawater)	Bay Area Rapid Transit (BART), USA
Floating docks, from 1929	Erstwhile USSR
100 boats during World War I and II	USA
100 boats weighing 6000 t dwt	Erstwhile USSR
Seaside airports extended to the sea	Hong Kong, New York, and Nice, France

The use of reinforced concrete in on- and off-shore structures started when Joseph-Louis Lambot constructed a boat using reinforced concrete in 1848. One such boat built in 1876 is still afloat. Concrete has been utilized for coastal structures, such as lighthouses, since 1920. An inspection done during 2010 of 67-year-old concrete blocks from the Los Angeles breakwater showed essentially no loss of strength and the form marks are still clearly visible.

Piles, piers, bridge piers, and floating drydocks can exhibit high durability performance extending to over 50 years in environmental conditions ranging from the subartic to the sub-tropical.

Concrete ships were built in substantial numbers during World War I and II. They demonstrated excellent performance as far as corrosion and fatigue resistance are concerned. One of the World War I vessels, the Selma, has been beached at Galveston, Texas, since 1922. It utilized expanded shale lightweight aggregate and a design cover of 2 cm with an actual cover of only 1 cm in many places and still is in a good condition.

26.2.1 Developments Since 2000

The availability of a set of new techniques and materials has led to the wider use of concrete structures in sea. Some of these are listed in Table 26.2.

Table 26.2 Materials and techniques used for building marine structures

Materials/techniques	Application
Ferro-cement	Thin watertight walls
Fibre-reinforced concrete	High-impact resistance
Polymer concrete	Impermeable, high-strength concrete
Epoxy coating	Water proofing even under high pressure in deep sea
Lightweight concrete	Self-weight reduction up to 75% compared to normal, dense concrete
Sacrificial anodes	Protection against corrosion
Prestressed concrete	Crack-free massive structures
Expanded polystyrene foams	Adds buoyancy to hollow concrete
Pumping of concrete	Continuous supply of uniform grade concrete in large quantities to inaccessible areas
Slipforms	Continuous concreting to a height of 6 m/day without joints and, therefore, speeding up construction and avoiding critical joints vulnerable to water penetration
Galvanized/stainless steel/FRP reinforcement bars	Reducing or eliminating corrosion in embedded steel

If we analyse these materials and techniques, it can be seen that fabricated steel had to be eliminated as a structural material in preference to other corrosion-resistant materials for oceanic structures. The main reasons for this are the following:

a. Large structures call for using a plate thickness of up to 100–125 mm at the nodes when welded. To obtain the notch durability needed, it is difficult to achieve good weldability in such situations.
b. The corrosion of steel and the resulting loss of metal at about 0.4 mm of thickness per year could reduce the strength of the structure substantially over a 50-year lifecycle. The maintenance required to counter-act this could become extremely expensive. Moreover, corrosion of steel causes cracks and these could substantially reduce the fatigue strength of steel under seawater.
c. In the case of storage-type oil platforms—producing 30,000 kL per day—it is not easy to construct large caissons in steel to store the oil.
d. Design parameters rule out steel for some structures. For example, a jacket-type steel platform may carry a 180-tonne load directly on one leg as well as 540 tonnes resulting from moments due to wave/current forces in 50-m-deep seawater. In other words, only 33% of the capacity is used for carrying the load. When this design is used for 150-m-deep water, the direct land payload will be roughly the same, whereas the component due to the moments will increase three-fold to 1620 tonnes, i.e., the direct land payload will be just about 10%. In such a situation, for the purpose of stability, a heavier base or a lowered centre of gravity is essential, which cannot be provided by steel.

26.2.2 Underwater versus Land Construction

The use of concrete for large structures has been quite common on land and, therefore, the technology and related parameters may be considered as well understood. However, there are special considerations for construction in or near the sea. Structures on land are exposed entirely to a uniform environment, whereas in the sea the following environments are common.

a. A temperature range of 20–40°C above water—in the ocean around India—even as the submerged zone has only a near uniform temperature, below 20°C. In cold regions, these variations could be much higher.
b. The atmosphere around is laden with salt and fully saturated. The salinity of water is around 3.5%. The salts which are present in significant quantities in most seas are sodium chloride, magnesium chloride, calcium sulphate, potassium chloride, and potassium sulphate. Concentrations vary from sea to sea although the average salt content is about 35 g/L.
c. Winds overseas are higher than over land. Structures are subjected to cyclones/hurricanes/typhoons of speeds up to 200 km/h.
d. A free-standing structure in sea could be subjected to a total lifecycle loading of small waves at 2×10^8 cycles with an estimated period of 9–18 s.
e. Lateral forces from waves could go to such values as 40,000 tonnes in extreme cases on free-standing large structures in sea.
f. Accidents caused by hits from floating objects such as a bulk carrier of 500,000 dwt are possible in the case of off-shore structures.
g. The temperature and oxygen content of surface water may change due to waves and wave-induced mixing. Water well below the sea level remains in uniform temperature, resulting in a temperature gradient. In the case of storage platforms, gradient could be quite steep.
h. Implosion due to pressure on the walls of supporting columns is also possible.
i. Organic growth swell of concrete and marine fouling can add to self-weight by 7–10% and therefore affect buoyancy.
j. Sulphate of magnesium present in seawater and its reaction with cement in concrete could lead to the expansion of concrete.
k. External pressure or loading on concrete at 20 m depth of seawater is around 2 t/m² in case the structure has cavities.

26.2.3 Durability of Structures in Marine Environment

Chapter 25 discussed in a detail the durability problems in a marine environment. Here, only the construction and repair of such structures are considered. The durability problem of reinforced and prestressed concrete structures for underwater service is more extensive than just prevention of corrosion. Rather, it might be defined as the continuance of the ability of the structure to perform its functions in accordance with the intended behaviour, as well as maintenance of the ability to resist extreme loads without collapse. Thus, durability is the freedom of the structure from having degrading time-dependent behaviour.

Reinforced or prestressed concrete is a composite material, and the durability considerations must reflect the effect on the structure as a whole, rather than on its individual components. We find considerable interaction between the time-dependent effects on steel and those on concrete, which makes it necessary to adopt procedures that will optimize the life of the structure as a whole.

Durability considerations primarily involve the concrete, the passive reinforcement, and the prestressing system, along with the embedded fittings. They must also encompass effects due to creep, fatigue, and cracking, since these interact with corrosion of steel in the seawater environment. Durability failure occurs when either the concrete or the steel has suffered physical or chemical change, which reduces its capacity to function.

26.2.4 Special Considerations

Cover

Experience has shown that durability is related to the amount of cover. The thicker the cover over the steel is, the longer it will take the chloride ions to reach the steel and reduce the pH and passivity provided by the cement. However, excessive cover can lead to the development of a few wide cracks under overstress, whereas a thinner cover results in many small cracks

As opposed to the above-mentioned facts, which would appear to justify the rigid rules on cover, are the following facts:

a. Ships built during World War I and II had covers of only about 20 mm, yet they did not suffer corrosion of steel.
b. In the erstwhile USSR, many floating drydocks were built with covers of 15 and 20 mm with highly successful durability over many years of adverse exposure.

It is confirmed opinion that the impermeability of the cover is of major importance. The thickness should be related to the steel bar diameter and the maximum size of the coarse aggregate.

The general factors affecting permeability, such as cement content, water–cement ratio, compaction and consolidation of the concrete, and curing, are important. While many feel that prestressed steel should have a greater cover than non-stressed steel, because of the more serious consequences of corrosion, prestressed concrete piling by hundreds of thousands are rendering completely successful service with only 4–6 cm of cover. Other factors which affect cover are the tolerances of placement of steel and forms, and the depths of honeycomb and bug holes and other surface defects.

Cracking

Existing recommendations assume a direct relation between crack width and corrosion. Yet, actual field experience with high-quality dense concrete that has been cracked indicates no definite cause-and-effect relationship with corrosion. Corrosion seems to be far more related to the permeability of concrete, whether cracked or not cracked.

Reinforced concrete theory assumes a model of perfect bond and no slip between steel reinforcement and concrete. Research undertaken in Japan during 2010–17 indicates that this model of crack should be revised because of the assumed bond (no slip), and this may better explain the general lack of corrosion at the root of small cracks. For example, a number of prestressed concrete piles, driven in San Francisco Bay in 1960, were badly cracked (up to 3 mm) due to rebound tensile stresses during driving. In some cases, the concrete popped off to expose 1 cm of prestressing strand. While all permanent piles were replaced, four of the cracked piles were left in place as a long-term test. All cracks occurred 3 to 6 m below low water. Subsequent inspections by divers have failed to reveal corrosion. A surface coating of marine growth and exuded lime from the adjoining concrete appears to have protected the steel.

Electrolytic corrosion sets in when cracking is severe by allowing oxygen to reach the steel. This has serious effects. The oxygen can be atmospheric or dissolved. The fact that cold water contains more dissolved oxygen than warm water should be kept in mind.

However, cracking does have a definite correlation with fatigue, and fatigue gets greatly accelerated in seawater. Hence, the present rules on cover and cracking as specified by IS: 456 are justified.

Electrochemical corrosion

This is by far the most prevalent type of corrosion and is directly related to permeability of concrete. Porosity (discontinuous pores) does not seem to be a factor, since lightweight aggregate and air-entrained concrete show no more corrosion than conventional concrete. Because of permeability, and consequent capillary attraction, salt water migrates to the surface of the steel, where it reduces the pH of the cement paste. If oxygen is available, electrolytic corrosion rapidly sets in.

Chloride concentrations in the mix itself appear to greatly accelerate the destruction of passivity; thus they should be avoided.

Permeable concrete is usually traceable to high water–cement ratios (above 0.50) and low cement content. Steel in very permeable concrete corrodes much more quickly than bare steel. Since a high water–cement ratio is generally prevalent throughout the structure, once such electrolytic corrosion sets in, piecemeal repairs serve only to shift the anodes to another location.

Electrolytic corrosion requires oxygen to carry on the process (in the complete absence of oxygen, it takes 30 times the salt concentration to initiate corrosion as and when oxygen becomes present).

The salt concentration, which reduces the passivity, builds up as a result of permeation and evaporation. When concrete is deeply submerged and surrounded on all sides by seawater, there is no evaporation; hence the chloride concentration is not critical. Furthermore, there is no substantial source of oxygen. However, in the case of a habitat or chamber located at depth, with air at atmospheric pressure on the inside, there is

a. substantial permeation,
b. evaporation, with salt cell concentration in the concrete, and
c. a plentiful supply of oxygen.

The inside of such chambers should be considered the same as the splash zone, with due consideration given to protective coatings.

Biological attack

Marine growth fouling can be of significance near a sewage outfall. As far as can be determined, attack by rock-boring and other marine biota affect only weak porous concrete and have no effect on the structure itself. However, marine growth adheres to the structure and increases the drag resistance. Such marine growth on concrete is usually substantially less than that on comparable steel structures, apparently due to the inhibiting effect of the alkaline cement paste.

Corrosion along interstices of wire strands

This type of corrosion caused a great deal of theoretical concern many years ago but is not borne out by wide experience with prestressed girders of over-water bridges and jetties. Usually all that is done to avoid such corrosion is the application of a cover of epoxy mortar. However, some accelerated tests in a tropical environment, e.g., in coastal India, have shown that corrosion still penetrates along the strand. It is, therefore, recommended that for important sea structures, pre-tensioned strands be cut back 2 to 3 cm and the hole plugged with epoxy mortar.

Corrosion of inserts

Metal inserts and exposed penetrations are required in many sub-sea structures. These should be protected with epoxy coating and by using sacrificial anodes. The points at which these inserts in turn are connected to piping, conductors, piles, or skirts, there should be no direct metal-to-metal contact with the reinforcement of the structure. It is believed that adequate insulation can be achieved in most cases by using anodes, or additional steel thickness and epoxy coating.

Ducts and anchorages for post-tensioning

Ducts should be semi-rigid or rigid, and watertight. There is a considerable debate as to whether they should be black, galvanized, or lead-coated. Lining and coating with lead (lead being non-reactive with steel) is a superlative solution, which reduces friction. While no direct connection with hydrogen embrittlement has been positively proven, one should guard against zinc in the vicinity of prestressing steel in a corrosive environment.

Grouting must be given special attention and special procedures and proper admixtures should be added to ensure that the ducts are completely filled. High vertical ducts are especially prone to develop excessive bleed, which, if not countered, could lead to unfilled or water-filled pockets at the top, with the probability of bursting due to freeze–thaw cycles and corrosion.

Anchorages should be, as far as practicable, of the recessed type, with epoxy mortar to give 4 cm cover over the steel.

Monitoring of corrosion

Monitoring of corrosion until 2015 consisted in only detailed visual examination for cracking and rust stains. During 2015–16 non-destructive monitoring devices have been developed using a copper–copper sulphate electric half-cell potential measurement, with one terminal connected to a voltmeter and the other connected to the steel. It has been found that when sodium chloride reaches the steel, the voltage differential increases (from 0.19 to 0.36 V) indicating the potential for corrosion.

When corrosion takes place, the voltage differential rises to between 0.40 and 0.70 V. It is recommended that half-cell potential measurements be made on a grid system and then all the locations of equal potential be interconnected so as to construct an equipotential contour to indicate corrosion potential and activity. It is very important to distinguish between chloride penetration into the steel, which is inevitable, and the actual situation of corrosion, which requires oxygen and may never occur.

Cathodic protection

It is considered generally impracticable to electrically connect all the reinforcement and prestressed tendons. Welding reduces fatigue strength drastically and thus might do more harm than good.

When adjoining steel structures are protected by an impressed current cathodic system, there is the danger that with substantial current flow, the non-connected reinforcement will liberate monotonic hydrogen, leading to hydrogen embrittlement. Thus, sacrificial anodes are recommended in lieu of impressed current systems.

Another approach is to shield the reinforcement in the concrete by a surface sheet of electrically welded wire mesh which is electrically connected.

Coating rebars

Coated reinforcement has been proposed and is now being used for bridge decks. Epoxy coating is apparently satisfactory if the coatings are complete. If there are abraded areas, corrosion may be locally intensified. Galvanized reinforcement has been much used in the past, but concern has been voiced over the liberation of hydrogen by reaction with the cement. This can be prevented by chromate treatment of the galvanized steel.

26.2.5 Repair Considerations

When cracking occurs without corrosion, the cracks should be injected with an epoxy selected so as to bond to the damp surfaces. When corrosion occurs, it should not just be repaired piecemeal but the cause carefully determined. In some cases, epoxy coatings may be sufficient to arrest further corrosion, in others, the removal and replacement of concrete and reinforcement may be needed.

26.2.6 Recommended Practices to Ensure Maximum Durability of Important Underwater Structures

Concrete placed underwater (or grout) may play a wider role in undersea structures in the future: it is already extensively used for reinforced bell footings under steel platforms in many off-shore structures. The rules and procedures advocated herein apply to concrete placed underwater as well. Proper procedures must be followed: The cover must be substantially increased to cover practical tolerances and the local entrapment of laitance, etc. The water–cement ratio should be kept to about 0.40. The practices recommended to ensure maximum durability are listed below:

1. The amount of chlorides and sulphates in the mix should be reduced to a minimum.
   Mixing water: Chlorides less than 500 ppm, sulphates less than 1000 ppm
   Aggregates: Chlorides less than 0.2% by weight
   Admixtures: not more than 0.25% by weight
   Curing water: Reduce chlorides as much as practicable, especially on the first day
2. The water–cement ratio should be less than 0.45; preferably 0.40 or less.
3. Cement content not less than 400 kg/m^3.
4. C_3A should be between 4% and 7%.
5. Cement should be low alkali, with less than 0.60% Na_2O plus K_2O.
6. Aggregates should be non-reactive, and sound, as tested. Those with a satisfactory past history of durability in seawater and, where applicable, under freeze–thaw exposure (less relevant in India) should be selected. Fine aggregates should be especially tested because of their large specific surface.
7. Air entrainment (5–6%) should be used in the freeze–thaw zone.
8. The cover over reinforcement steel in submerged zones should not be less than twice the maximum size of the coarse aggregate or 1.5 times the diameter of the reinforcing bar, whichever is larger, plus the allowable tolerance on reinforcement placement.
9. Reinforcement should be well distributed, with a maximum spacing of 30 cm in each direction on each face.
10. The concrete should be a workable mix, well compacted and consolidated by vibration, and free from honeycomb or air pockets. Air–water bleed holes or surface defects should be limited to a surface skin depth of one-sixth the cover.
11. Prestressed ducts should be semi-rigid or rigid, with not less than 0.6 mm wall thickness, and can be galvanized (passivity by chromate), lead-lined and coated, or black pitched.
12. Grouting of ducts should be done with neat cement and water, and admixtures (free from chlorides) should completely fill the ducts. Proper procedures (stage grouting or stand pipes) or thixotropic admixtures should be employed to overcome the problem of excessive bleed in vertical ducts.
13. Thorough curing should be carried out using fresh water where practicable or heavy membrane-curing compounds. Steam curing when followed by water curing is excellent. Doubling the water curing period gives 50% greater resistance to chloride penetration.

14. When the reinforcement has a cover which is less than 2-cm thick, it should be galvanized, with the galvanized steel passivity increased by chromate.

15. The ends of the exposed prestressed strands should be cut back 2 cm and the recess filled with epoxy mortar.

16. The post-tensioning anchorages should be recessed. The anchorage and recess should be painted with epoxy and the pocket filled with epoxy mortar where practicable. It can also be filled with well-compacted concrete that is tied to the structure by reinforcement (e.g., mesh). In the latter case, the joint between new and old concrete should also be epoxied.

17. Inserts should be coated with epoxy, galvanized, or coated with cadmium. The points at which they can pick up stray currents should have a separation of 6 cm of concrete between them and the reinforcing steel.

18. Construction joints should be cleaned of laitance and roughened to a depth of 6 mm. Prior to new pour, the surface should be sprayed with an epoxy bonding compound or else a rich mortar should be used to start the next lift. The external face of the joint should be painted with epoxy, up to 30 cm on each side.

19. Epoxy or special penetrating coatings should be used for the splash zone and the interior surface, which are subjected to air and atmospheric pressure. They may add 10 to 20 years to the time it takes the chloride to reach the steel.

20. The cathodic protection of adjoining steel structures should be done by sacrificial anodes only unless a thorough study is carried out and special steps are taken to protect the concrete otherwise.

21. A quality assurance programme should be adopted to ensure uniform application of the specifications adopted.

22. Important structures should be inspected thoroughly (once a year) by visual and electric potential methods so as to discover any corrosion at an early stage, when simple remedial steps may be adequate.

26.3 Underwater Construction

In some structures, such as basements, sewerage, and marine works, concrete is to be placed underwater. In diaphragm wall construction, concrete is placed in a trench filled with bentonite slurry. Concrete cast underwater should not fall freely through the water; otherwise it may be leached and become segregated. When concrete has to be deposited underwater, one of two courses may be adopted. Either the space can be enclosed and the water excluded temporarily using a cofferdam or the concrete can be placed directly in the water using one of the following underwater placement methods.

Tremie method

A tremie pipe is about 20 cm in diameter. It is a steel pipe strong enough to withstand the external pressure of water. A flanged steel pipe of adequate strength is used for the purpose. A funnel is fitted to the top of this pipe for easy pouring of concrete. The bottom end of the pipe is closed with a thick polyethylene cap or similar material to prevent the ingress of water.

The pipe is lowered into the water and made to rest at the location of the concreting as shown in Fig. 26.10(a). Concrete slump is kept high (150–250 mm) so that it can flow freely without the mix having a tendency to segregate or bleed. To obtain cohesive and easy-flowing concrete, a high percentage of sand (40–50% of the total weight of aggregate) is used. In addition, the cement content is kept higher. For 40-mm maximum size of aggregate, IS: 456-2000 recommends a minimum cement content of 350 kg/m^3 for underwater concreting. For 20-mm aggregate, the cement content is kept in the range of 380 to 390 kg/m^3. Water-reducing admixtures and pozzolans can also be used to make the concrete durable and workable.

The whole length of the tremie pipe is filled with concrete. It is then lifted up and given a slight jerk by a winch-and-pulley arrangement as shown in Fig. 26.10(b). This helps the concrete to dislodge the bottom cover and the concrete gets discharged in place. However, care must be taken to ensure that the bottom end of the pipe remains inside the discharged concrete, so that no water can enter the tremie pipe from the bottom. Concrete is again poured through the funnel, and when the whole length of the pipe is filled with concrete, it is lifted slightly and jerked taking care that the lower end remains embedded in concrete. This way the tremie pipe is filled with concrete and discharged sequentially till the concrete attains a level higher than the water level. In this way concrete does not get affected by water except at the top or bottom layer. This defective concrete is then removed and rectified. For good results, it is essential that a steady, uninterrupted flow of concrete is maintained through the tremie pipe.

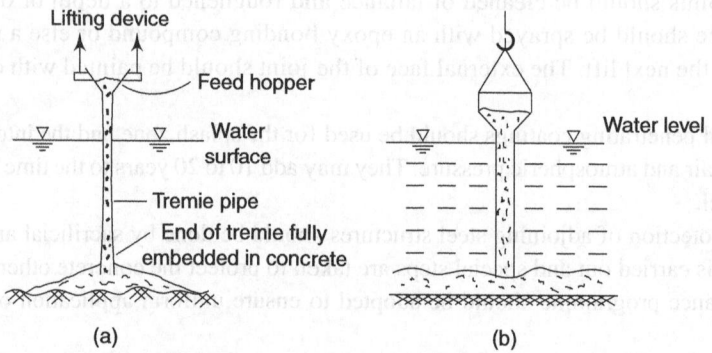

Fig. 26.10 Tremie method of concreting

Deep dump bucket method

The deep dump bucket is a specially made bucket [Fig. 26.11(a)] with its bottom opening downwards and outward when it is lowered down and tipped. When the bucket touches the bed, the skirts get lowered, the bucket is opened by a mechanism, and the concrete is discharged as shown in Fig. 26.11(b). The lowered skirts protect the concrete from the surrounding water. However, the results may not be fully satisfactory, as a certain amount of cement in the concrete may still be washed off by the water.

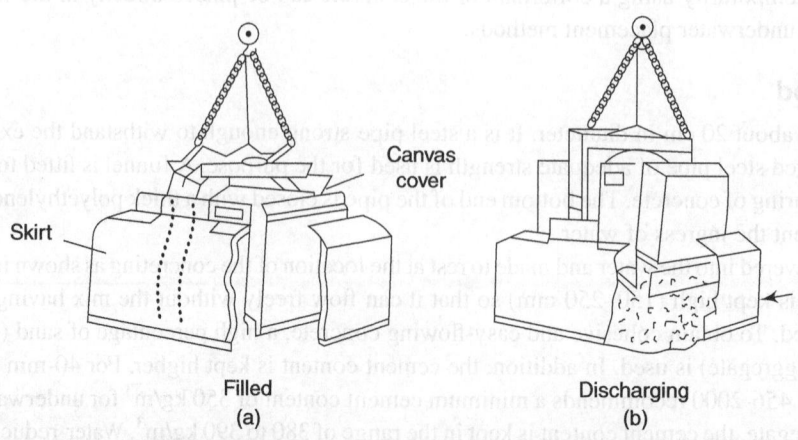

Fig. 26.11 Placement with bottom dump bucket

Instead of a bucket, gunny bags filled with dry or semi-dry mixtures of cement and fine and coarse aggregate can also be used and lowered down into water and opened out at the bottom without disturbing the water.

Grouting method

A series of round gauges made with 50-mm mesh of 6-mm steel and extending over the full height to be concreted are prepared and laid vertically over the area to be concreted, so that the distances between the casters of the cages and also between the faces of the concrete do not exceed 1 m. Stone aggregates of sizes between 50 and 200 mm are carefully deposited outside the steel cages over the full area and over the height to be concreted. This is done to prevent the displacement of the cages.

A cement–sand (1:2) grout with a water–cement ratio between 0.6 and 0.8 is prepared using a mechanical mixer. This is sent down under pressure (of about 20 kg/cm²) through a 38–50 mm diameter pipe terminating into steel cages. As the grouting proceeds, the pipe is raised gradually up to a height of not more than 60 cm above its starting level, after which it is withdrawn and placed into the next cage for further grouting by the same procedure. After grouting the whole area for a height of about 60 cm, the same operation is repeated, if necessary, for the next layer of 60 cm and so on.

The grout quantity sent down should be sufficient to fill all the voids, which may be either ascertained or assumed to be 55% of the volume to be concreted.

Using bags

Old cement polypropylene bags are filled up to about two-thirds with concrete and their open ends are securely tied or sewn to make the bags square-ended. These bags are deposited underwater in alternate header-and-stretcher courses so that all the bags are interlocked to form one solid mass. The placement of bags should be such that the mouths of the bags face away from the free surface. This enables a good bond between the placed bags. The course of bags may be held together by driving steel spikes through them after placing. In deep water, the services of a diver may be needed for a satisfactory job.

Pumping

Concrete can be placed underwater through direct pumping. The vertical end of the pipeline is inserted sufficiently deep into the previously poured concrete. The pipe should not move sideways during pumping. The direct pumping method is similar to the tremie method except that in the former case concrete is pumped through the pipe instead of depositing through buckets.

Pre-placed aggregate method

In this method, the formwork placed underwater is packed with well-graded stone aggregates and then grout is injected into the mass to fill the voids. This method is used particularly for repairing existing underwater structures such as jetties, wharfs, spillways, and piers.

26.3.1 Precautions Required during Underwater Concreting

The following precautions should be taken during underwater concreting:

a. Dry ingredients should not be dumped into water, nor should the concrete be allowed to fall through water from any height.
b. Pumping or bailing out of water should not be done while the concrete is being deposited, and within 24 hours of placing the concrete, as it may suck the cement particles of the laid concrete.
c. No tamping, ramming, or compaction of concrete should be done until the concrete surface rises above the water level.
d. The concrete mix should be rich and have 10–30% extra cement. The cement content of concrete should not be less than 380–390 kg/m³.

e. The aggregates should be properly graded to reduce voids to a minimum and produce a concrete mass of excellent plasticity and strength.

f. Concrete should not be placed in very cold water, as this causes hardening problems. Accelerators such as sodium chloride and sodium silicates can be used based on requirement.

g. No construction joint is allowed within 600 mm below the water level. Underwater concreting should be done in one continuous operation.

Underwater concreting is a specialized operation. It needs appropriate equipment, material, and operation skills. It should be done under good supervision.

26.4 Piles

Piles are used for deep foundations, usually described as column elements inserted in the earth for the purpose of transferring loads from the superstructure to the ground. The factors to be considered in selecting piles are the following:

- Structural capacity of the pile
- Durability and resistance to handling
- Feasibility of splicing
- Ground displacement during driving
- Penetrability

26.4.1 Drilled Shafts

Drilled shafts, also known as belled or under-reamed piles, are referred to as drilled sub-piers or caissons. They are installed by placing concrete in drilled holes. A typical drill shaft is shown in Fig. 26.12. The drilled shaft installation may be dry or casing-protected depending upon the soil condition and the presence of ground water.

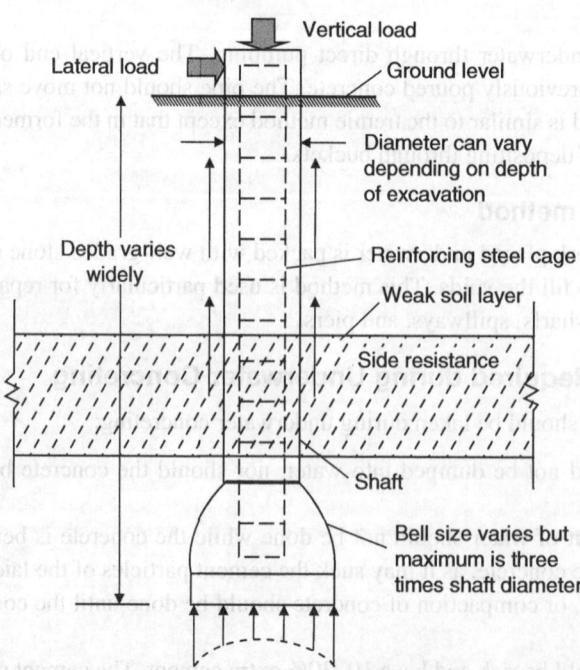

Fig. 26.12 A typical drilled shaft

Dry installation

This method is used in soil above the groundwater level. It may also be used below the natural water table in soils of low permeability and soils in which the seepage can be controlled. The excavation is carried out with suitable drilling equipment. Reinforcement is provided along the entire depth of the hole. Concrete is usually poured using tremie pipes, exercising care not to strike the side of the hole. Figure 26.13 shows the dry method.

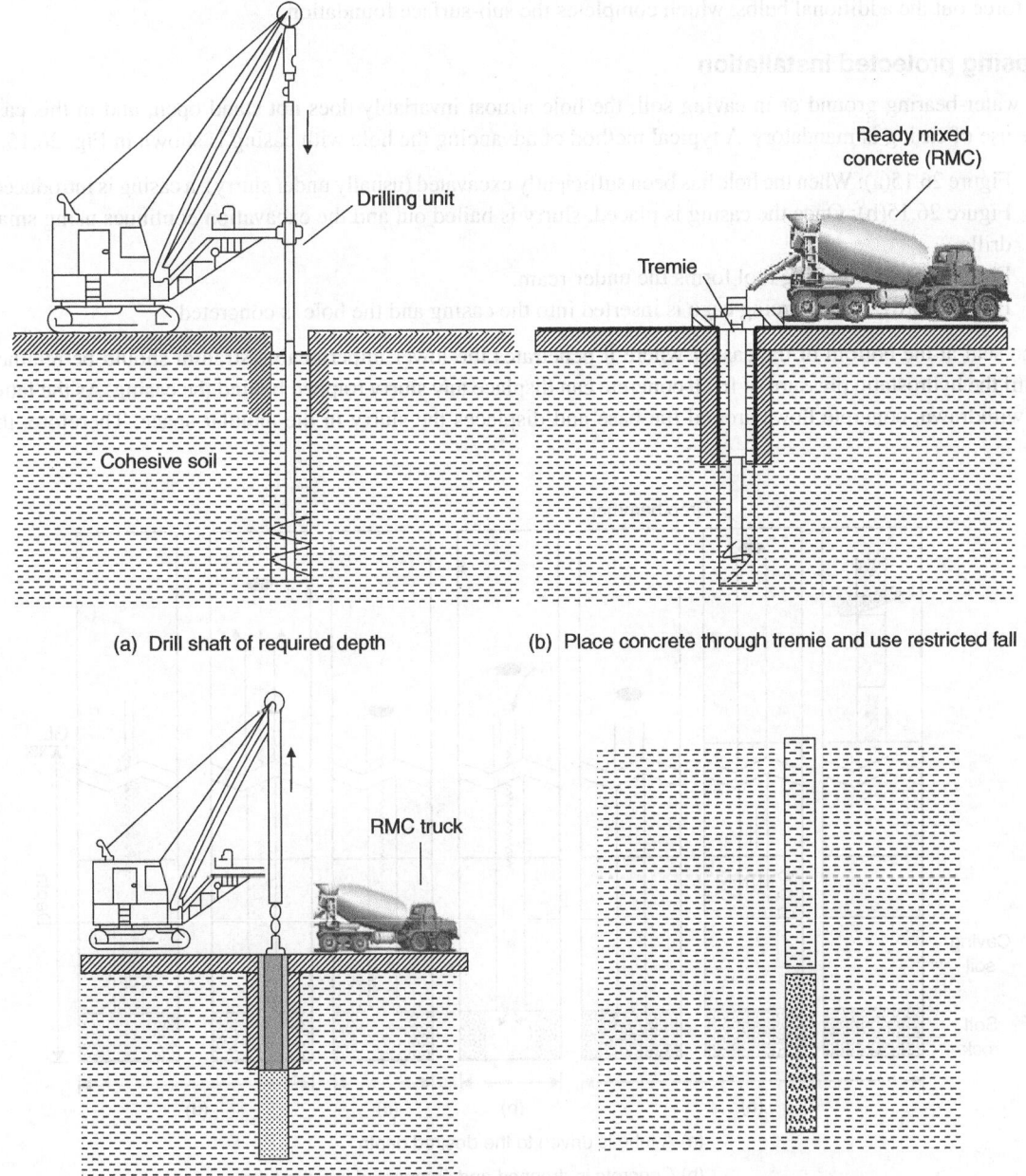

(a) Drill shaft of required depth

(b) Place concrete through tremie and use restricted fall

(c) Pull out and set reinforcement cage for required depth

(d) Completed shaft

Fig. 26.13 Dry method of drilled pier construction

Franki process

The Franki pressure-injected footing technique is illustrated in Fig. 26.14. A steel tube of desired diameter and length is fitted with an expandable boot to close the bottom end. The tube is first driven to the desired depth. After removing the driver, a charge of dry mix concrete is dropped and compacted by a hammer. This creates a watertight plug at the bottom. At this juncture, the tube is raised and held in position while the hammer drop forces the plug out from the lower end partially. This leaves a portion of concrete in the tube, which makes it sufficiently watertight preventing the sub-soil water outside from entering the tube. The operation is repeated to force out the additional bulbs, which completes the sub-surface foundation.

Casing protected installation

In water-bearing ground or in caving soil, the hole almost invariably does not stand open, and in this case the use of casing is mandatory. A typical method of advancing the hole with casing is shown in Fig. 26.15.

- Figure 26.15(a): When the hole has been sufficiently excavated (usually under slurry), a casing is introduced.
- Figure 26.15(b): Once the casing is placed, slurry is bailed out and the excavation continues using small drills.
- Fig. 26.15(c): A belling tool forms the under-ream.
- Fig. 26.15(d): A reinforcing cage is inserted into the casing and the hole is concreted.

The seal at the bottom of the casing where it penetrates the lower impermeable stratum should be retained until the hydrostatic pressure of fresh concrete can displace any slurry trapped behind the casing. As the latter is withdrawn, concrete flows around the base and displaces the slurry in the annular space. Therefore, the

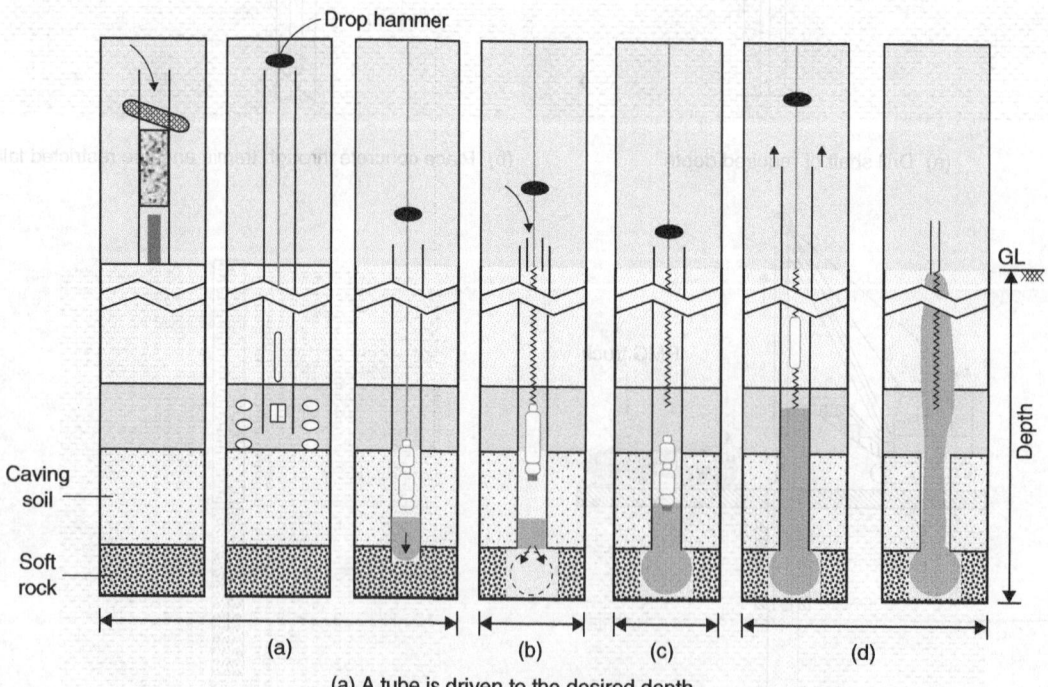

(a) A tube is driven to the desired depth

(b) Concrete is dropped and compacted into the tube

(c) Drop hammer expels concrete from the tube

(d) Additional steps complete the footing

Fig. 26.14 Steps for constructing the Franki pressure-injected pile system

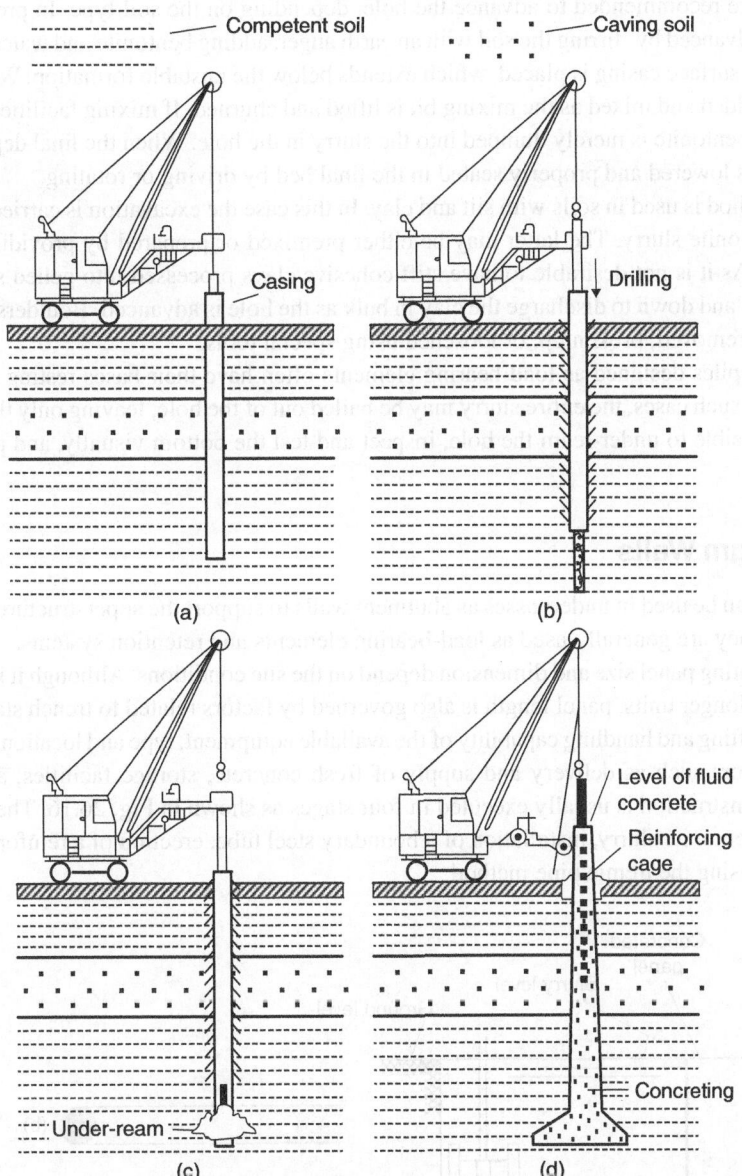

Fig. 26.15 Advancing the hole with casing: (a) introducing the casing and removing the slurry, (b) drilling below the casing in competent soil, (c) forming the under-ream, and (d) removing the casing and concreting the hole

withdrawal of the casing is a critical part of the operation. If concrete sets prematurely, it will move up with the casing, forming random cracks into which the slurry can flow. The same situation can result with unworkable concrete. If the casing is pulled out prematurely, the slurry is likely to flow in and mix with the concrete. Hence, the casing should be pulled out carefully and slowly at the most appropriate time.

Slurry method

Initially, slurry was used to remove sand from the excavation, but later it became evident that it could improve the stability of the face, lubricate the tools thereby avoiding stick-pipe problems, and prevent formation fluids from entering the excavation and keeping the cutting in suspension even if the pumping was stopped.

Two methods are recommended to advance the hole, depending on the soil type. In predominantly loose sands, the hole is advanced by stirring the soil with an earth auger, adding bentonite and water as the excavation becomes deeper. A surface casing is placed, which extends below the unstable formation. Water and bentonite are continuously added and mixed as the mixing bit is lifted and churned. If mixing facilities are not available at the site, the dry bentonite is merely dumped into the slurry in the hole. When the final depth is reached, the remaining casing is lowered and properly seated in the final bed by driving or rotating.

The second method is used in soils with silt and clay. In this case the excavation is carried out with drilling buckets under bentonite slurry. The latter may be either premixed or prepared by providing bentonite with water in the hole. As it is not desirable to have stiff cohesive clays processed into gelled slurry, the drilling bucket is moved up and down to discharge the clay in bulk as the hole is advanced. Boulders and heavy gravel if encountered are removed, broken up, or loosened using special tools.

Large-diameter piles designed as load-bearing elements often have their bases resting on rocks or other firm formations. In such cases, the entire slurry may be bailed out of the hole, leaving only the casing. In these conditions it is possible to under-ream the hole, inspect and test the bottom visually, and place the concrete in the dry state.

26.5 Diaphragm Walls

Diaphragm walls can be used in underpasses as abutment walls to support the superstructure load and transfer them to the soil. They are generally used as load-bearing elements and retention systems.

The factors dictating panel size and dimension depend on the site conditions. Although it is advantageous to construct a wall in longer units, panel length is also governed by factors related to trench stability, concreting requirements, the lifting and handling capability of the available equipment, type and location of lateral bracing, and incidental factors such as delivery and supply of fresh concrete, storage facilities, and water supply. Diaphragm wall construction is usually executed in four stages as shown in Fig. 26.16. The stages consist of excavation under bentonite slurry, installation of a boundary steel tube, erection of a reinforcement cage, and finally concreting using the tremie pipe method.

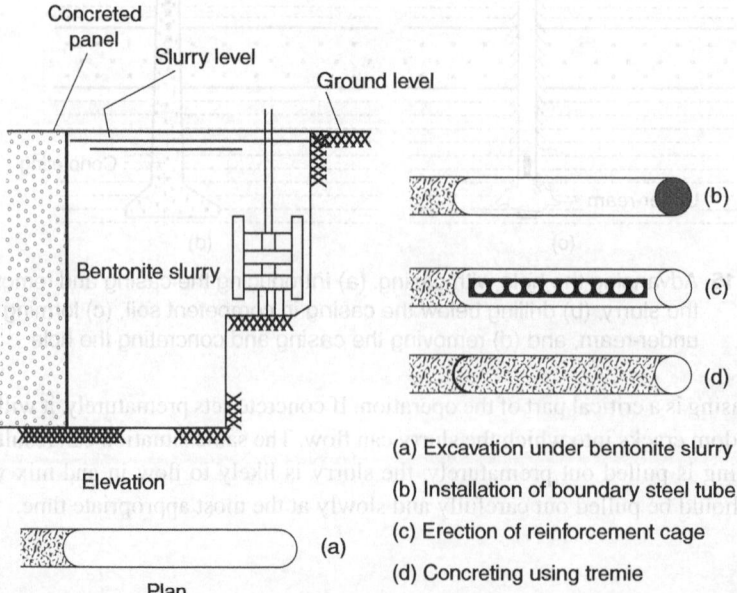

(a) Excavation under bentonite slurry
(b) Installation of boundary steel tube
(c) Erection of reinforcement cage
(d) Concreting using tremie

Fig. 26.16 Construction sequence of a diaphragm wall, executed in heat

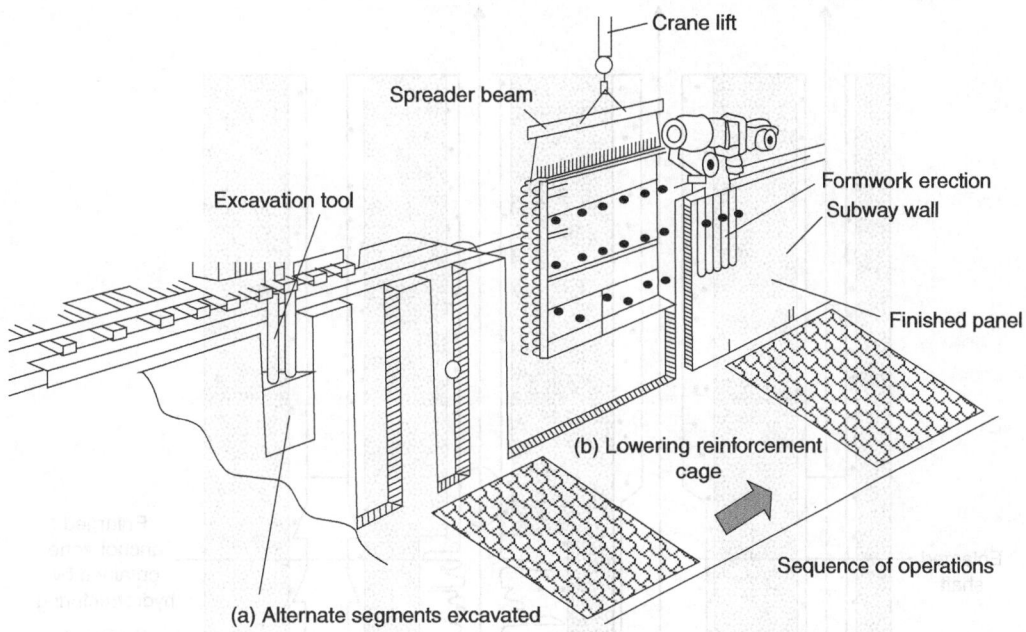

Fig. 26.17 Panel installation and construction sequence of a typical subway

A typical panel installation illustrating the construction sequence for a metro subway wall is shown in Fig. 26.17.

26.5.1 Anchored Walls

Anchored diaphragm walls are used in places where the depth of excavation exceeds the limit within which the diaphragm thickness range is applicable to the free cantilever. In this case, the anchor serves as a permanent bracing, and hence one of the design requirements is the protection of anchor against corrosion. Usually the anchor capacity ranges between 200 and 900 kN.

The anchor method can be considered a special application of prestressing in foundation and ground work. Thus, anchors are grouped into the following three main categories:

- soil anchors
- rock anchors
- Marine anchors, installed in alluvial environment or aggressive water

In a broad sense, anchors are used to mobilize shear strength, and often the passive resistance of the soil. Anchors can thus provide the required resistance to the earth pressure acting on the structure. Likewise, tensile forces may act on the foundation of a structure as a result of a particular combination of loads. When such tensile forces exist, they must be resisted by the careful provision of ground anchors.

Depending upon the method of load transfer, anchors may be classified into the following four types as shown in Fig. 26.18.

Type A

This type is characterized by a tremie-routed straight shaft cylindrical hole of a uniform diameter, which may be lined or unlined according to the requirement of hole stability. This type is suitable in rocks as well as in very stiff-to-hard cohesive layers, where it is most commonly used. The load transfer is by shear resistance mobilized along the ground–grout interface.

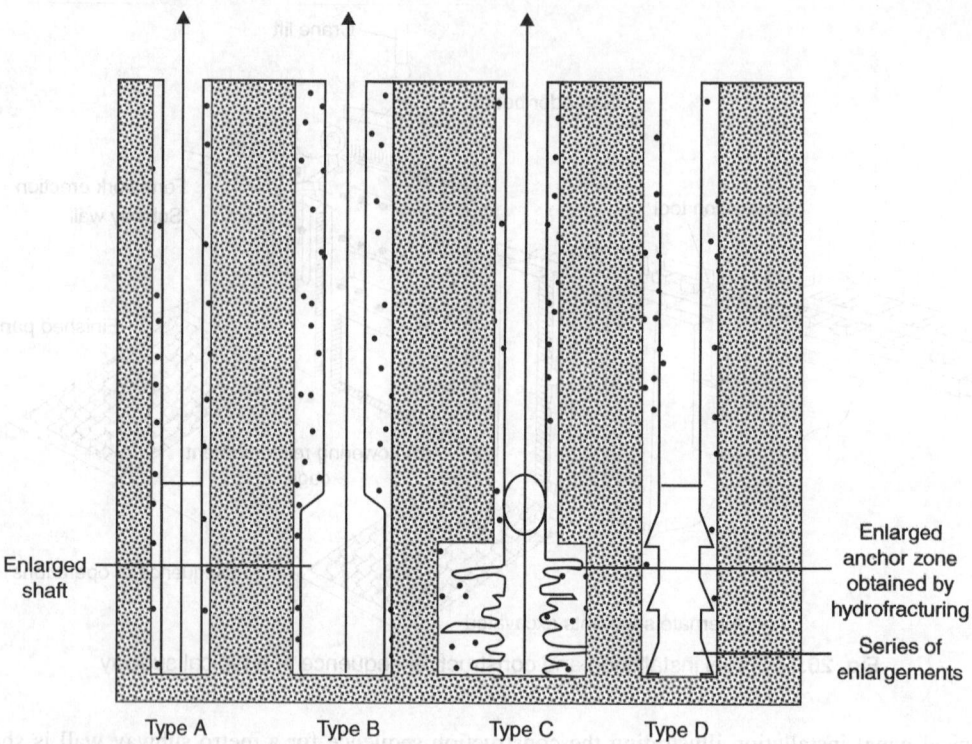

Fig. 26.18 Different types of anchors

Type B

With this type, the anchor zone is created as an enlarged cylinder formed in a grout borehole under low injection pressure. In this process, the actual effective diameter of the fixed zone is increased with minimum disruption to the surrounding earth materials as the grout permeates through the pores or natural fractures under injection pressure, which is normally less than the overburden pressure. The enlarged cylinder is suitable in soft fissure rocks and coarse alluvium, but may also be used in fine-grained soils. In the latter case, the cement particles will not always invade the small soil pores, but under pressure the grout will compact the soil locally to increase the effective diameter. For this type of anchor, resistance to withdrawal begins with side shear, but ends with bearing at the upper end of the enlarged shaft.

Type C

In this case, the grout under high pressure forces the cement particles to penetrate the soil irregularly and thereby enlarges the anchor zone through hydrofracturing. A grout root or fissure system is thus produced beyond the core diameter shown. This anchor is suitable primarily in cohesion-less soil, although it has been used successfully in stiff cohesive deposits. The design is based on an assumed uniform shear along an appropriate diameter at the designated fixed zone. It is not always clear whether injection produces type B or C, and in many instances a composite system results.

Type D

As in type A, the borehole is tremie-grouted, but it includes a series of enlargements formed mechanically in the fixed zone. This type is used in stiff-to-hard cohesive deposits. The pull-out capacity is derived primarily from side shear, but the plug and end bearing contribute to resistance to withdrawal.

Figure 26.19 shows the schematic arrangement of a typical ground anchor. Note the three main components of the anchor, i.e., the anchor head, the free length, and the bond length (fixed length).

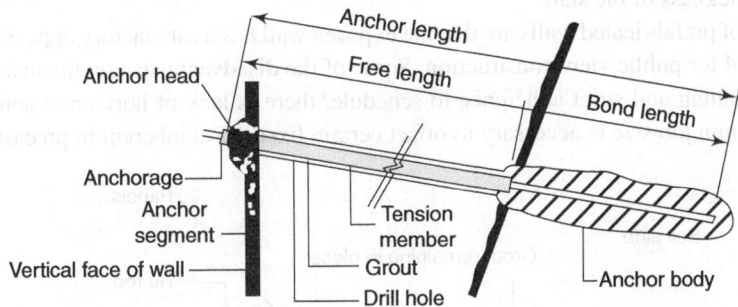

Fig. 26.19 Schematic presentation of a typical ground anchor

26.5.2 Post-tensioned Diaphragm Walls

The general principle of prestressing can be applied to diaphragm walls mainly to increase their effective depth. In this concept, internal stress of predetermined magnitude and distribution is introduced into the concrete, which partially or fully counteracts and balances the tensile stresses. The common procedure used for the construction of post-tensioned members is used in this method.

Figure 26.20 shows the following two categories of walls suitable for prestress:

a. vertical cantilever fully restrained at the base by sufficient embedment
b. walls supported at the top and bottom

Walls of the first type depend on adequate embedment below the excavation level to ensure stability and balance active and passive pressure, and also to limit lateral movement. Free cantilevers should therefore be used primarily in stiff or dense soil. Walls of the second type have their top braced, and are therefore beams simply supported at the top and bottom. Note the difference between the bending moment diagrams for the two cases.

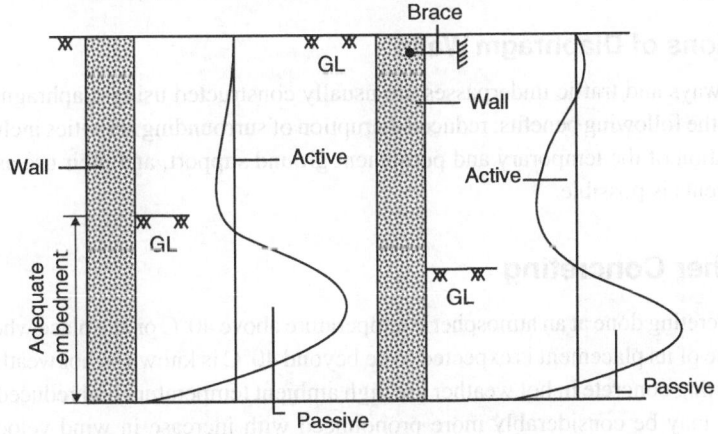

Fig. 26.20 Pattern of lateral earth pressure on walls

26.5.3 Prefabricated Diaphragm Walls

These walls can be imagined as the combination of prefabricated interlocking piles and sheet pile walls. The trench is likewise excavated under slurry, but the in situ tremie placement of concrete is replaced by inserting precast panels. Guide walls must be detailed according to the size and weight of the panels, since they must hold these sections until the wall becomes a self-supporting system. Figure 26.21 shows typical panels that slot together. Figure 26.22 shows an alternative type, which has a beam-and-slab section with the beam usually twice the thickness of the slab.

The advantages of prefabricated walls are that the exposed wall has a satisfactory appearance and a smooth face and can be used for public view construction. Some of the disadvantages are: the installation procedure requires careful planning and strict adherence to schedule, there is lack of horizontal continuity across the panels, and a minimum job size is necessary to offset certain fixed costs inherent in precasting.

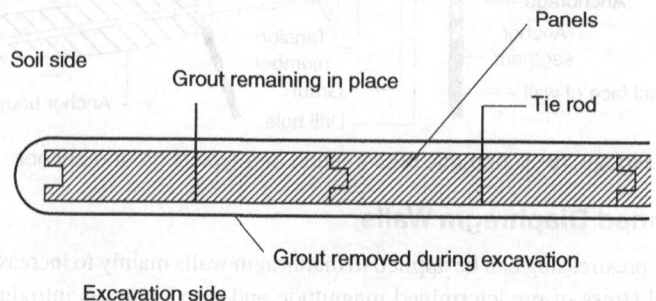

Fig. 26.21 Plan of prefabricated wall with identical panels

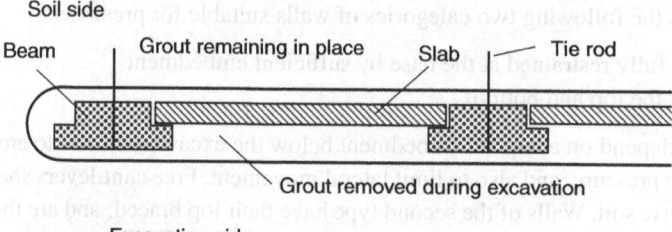

Fig. 26.22 Prefabricated wall with beam-and-slab panels

26.5.4 Applications of Diaphragm Walls

Underground motorways and traffic underpasses are usually constructed using diaphragm walls. The use of diaphragm walls has the following benefits: reduced disruption of surrounding activities including traffic during construction, integration of the temporary and permanent ground support, and their use as both substructure and foundation elements is possible.

26.6 Hot Weather Concreting

Any operation of concreting done at an atmospheric temperature above 40°C or at a place where the temperature of concrete at the time of its placement is expected to be beyond 40°C is known as hot weather concreting. The climatic factors affecting concrete in hot weather are high ambient temperature and reduced relative humidity, the effects of which may be considerably more pronounced with increase in wind velocity. Effects of hot weather are more critical during rising temperature, falling relative humidity, or both. These occur during the summer season even in normal climates. Precautionary measures are required during the summer season in such climates. The measures required on a calm, humid day are less stringent than those required on a dry, windy day, even if the air temperatures are identical.

Damage to concrete caused by hot weather can never be fully alleviated, since improvisation on site is rarely successful. Therefore, early preventive measures may be taken with the emphasis on materials, advanced planning, and coordination of all phases of work on site.

26.6.1 Effects of Hot Weather on Concrete

The effects of hot weather on concreting in the absence of special precautions are explained below.

Accelerated setting

High temperature increases the rate of setting of concrete. Sufficient duration of time during which the concrete can be handled is required for ease in placement. Quick stiffening may necessitate undesirable re-tempering by adding water.

Reduction in strength

High temperature increases the quantity of mixing water required for achieving the desired workability. This in turn reduces the strength of concrete. Concrete mixed and placed at a high temperature produces high strength at an early stage and generally lower strength after 28 days. This is shown in Fig. 26.23.

Increased tendency to crack

In hot weather conditions, the rate of evaporation is high, which results in plastic shrinkage cracks. High air temperature, high wind velocity, and low relative humidity also favour cracking. The graphical method for estimating the loss of surface moisture for various air temperatures, relative humidities, and wind velocities is given in Fig. 26.24. Drying shrinkage increases with the increase in water content in the mix and the decrease of relative humidity. Increase in concreting temperature increases the water demand, which may lead to increased drying shrinkage. Figure 26.25 shows the relationship between drying shrinkage and water content for various amounts of cement in the mix. It shows that shrinkage is a direct function of the unit water content of fresh concrete and cement content only has secondary importance.

Rapid evaporation of water during curing period

It is difficult to retain moisture for hydration and maintain reasonable uniform temperature conditions during the curing period.

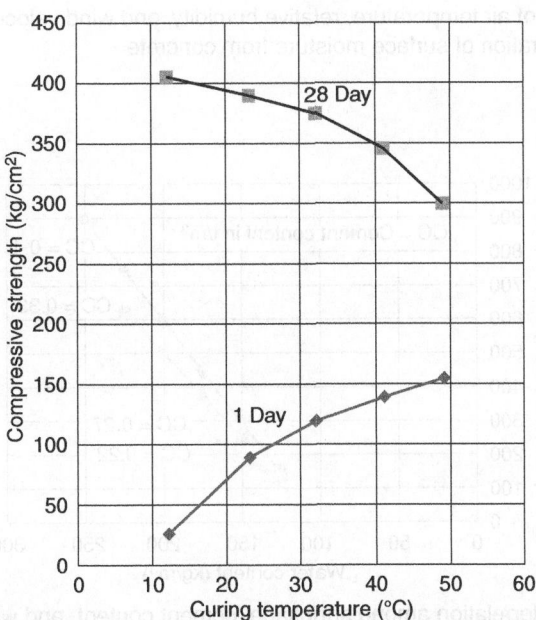

Fig. 26.23 Effect of curing temperatures on 1-day and 28-day strength of concrete

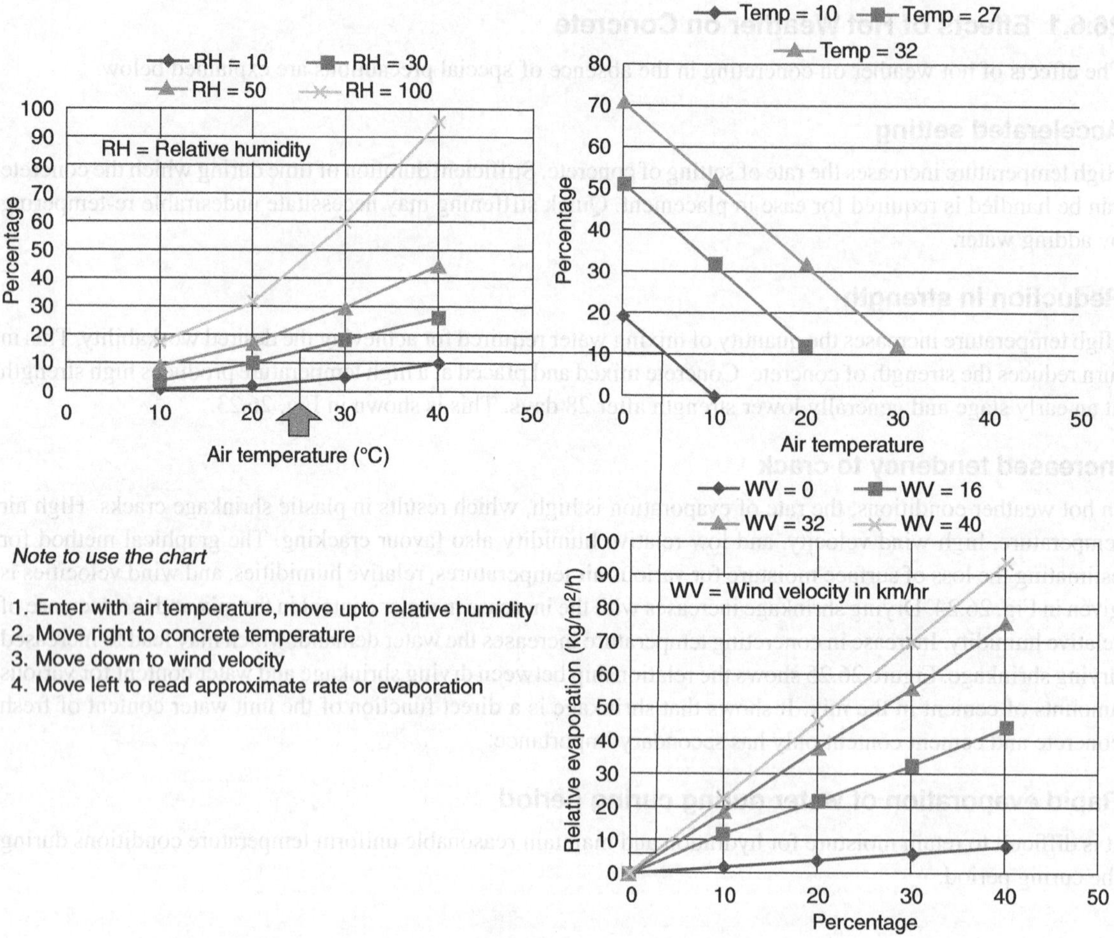

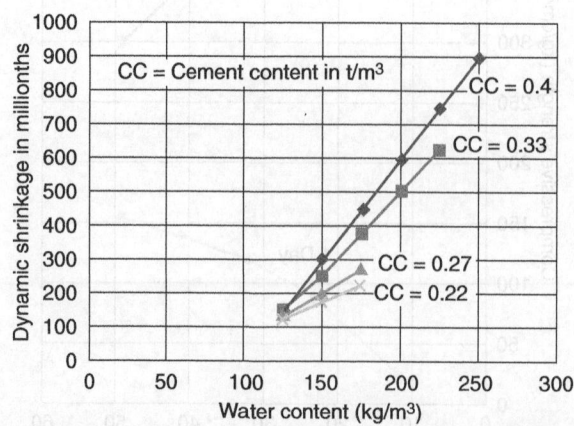

Fig. 26.24 Effect of air temperature, relative humidity, and wind velocity on the rate of evaporation of surface moisture from concrete

Note to use the chart

1. Enter with air temperature, move upto relative humidity
2. Move right to concrete temperature
3. Move down to wind velocity
4. Move left to read approximate rate or evaporation

Fig. 26.25 Interrelation among shrinkage, cement content, and water content

26.6.2 Temperature Control

The most direct way to keep the temperature of concrete down is to control the temperature of its ingredients. Some of the methods of controlling the temperature are as follows:

a. Shading stockpile of aggregates from the direct rays of the sun.
b. Sprinkling water on stockpiles and keeping them moist. This results in cooling by evaporation and is especially effective when relative humidity is very low. However, the addition of moisture has to be corrected in the mix design.
c. Cooling coarse aggregates by either spraying them with cold water or by circulating refrigerated air through pipes or by other suitable methods.
d. Using cold water for mixing to reduce the placing temperature. Figure 26.26 shows the effect of cold water mixing.
e. Using ice as a part of the mixing water, which will reduce the concrete temperature. Figure 26.27 shows the effect of ice in mixing water.
f. Designing the mix to have minimum cement content with other functional requirements, as the quantity of cement used in the mix affects the rate of increase in temperature. As far as possible, cement with lower heat of hydration should be used.

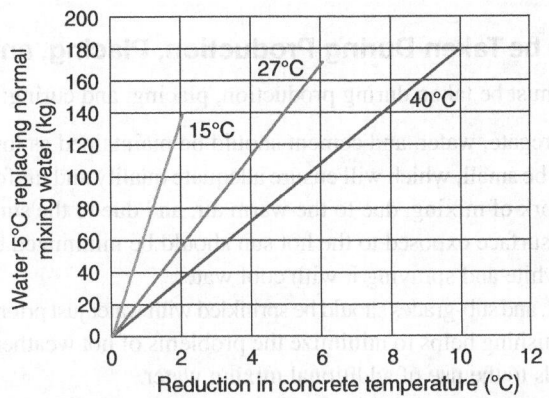

Fig. 26.26 Effect of cooled mixing water on concrete temperature

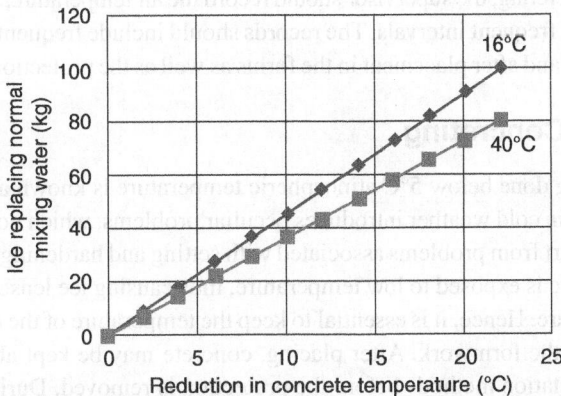

Fig. 26.27 Effect of ice in mixing water on concrete temperature

26.6.3 Temperature of Concrete as Placed

The temperature of concrete at the time of leaving the mixer or batch plant can be calculated using the following formulas. The actual change in temperature from the time it leaves the mixer to the time of placing should be estimated, but it may be assumed to be 2°C in the absence of any information in this respect.

Cold water for mixing

$$T = \frac{S(T_a W_a + T_c W_c) + T_w W_w + T_{wa} W_{wa}}{S(W_a + W_c) + W_w + W_{wa}}$$

With ice added to the mixing water

$$T = \frac{S(T_a W_a + T_c W_c)}{S(W_a + W_c) + W_w + W_i + W_{wa}} + \frac{(W_a - W_i)T_w + W_{wa} T_{wa} - 79.6 W_i}{S(W_a + W_c) + W_w + W_i + W_{wa}}$$

where T is the temperature of the freshly mixed concrete (°C); T_a, T_c, T_w, and T_{wa} are the temperature of the aggregate, cement, mixing water, and free water on aggregate, respectively (°C); W_a, W_c, W_w, W_{wa}, and W_i are the mass of the aggregate, cement, mixing water, free water on aggregate, and ice, respectively (kg); and S is the specific heat of cement and aggregate.

26.6.4 Precautions to be Taken During Production, Placing, and Curing

The following precautions must be taken during production, placing, and curing:

a. The temperature of aggregate, water, and cement should be maintained as low as possible.
b. The mixing time should be small, which will ensure adequate quality and uniformity, because the concrete is made warm by the work of mixing, due to the warm air, and due to the sun rays.
c. The effect of the mixer surface exposed to the hot sun should be minimized by painting and keeping the mixer drum yellow or white and spraying it with cool water.
d. The forms, reinforcement, and sub-grades should be sprinkled with water just prior to the placement of concrete.
e. Rapid placement and finishing helps to minimize the problems of hot weather concreting. Delay reduces the workability and leads to the use of additional mixing water.
f. Immediately after placing and surface finishing the concrete, it should be protected from evaporation of moisture, without allowing the ingress of external water. Once concrete has attained a certain degree of hardness, moist curing must be done.
g. During hot weather concreting, the supervisor should record the air temperature, general weather conditions, and relative humidity at frequent intervals. The records should include frequent checks on the temperature of concrete as delivered and after placement in the forms as well as the protection and curing times adopted.

26.7 Cold Weather Concreting

Any operation of concreting done below 5°C atmospheric temperature is known as cold weather concreting. The production of concrete in cold weather introduces peculiar problems, which do not arise while concreting at normal temperatures. Apart from problems associated with setting and hardening, severe damage may occur if concrete in the plastic state is exposed to low temperature, thus causing ice lenses to form and expansion to occur within the pore structure. Hence, it is essential to keep the temperature of the concrete above a minimum value before it is placed in the formwork. After placing, concrete may be kept above a certain temperature with the help of proper insulation methods before the protection is removed. During periods of low ambient temperature, special cold weather concrete techniques must be adopted.

26.7.1 Effects of Cold Weather on Concrete

The effects of cold weather concreting in the absence of special precautions are briefly explained below.

Delayed setting

When the temperature is below 5°C, the development of concrete strength is retarded compared with the strength development at normal temperature. The hardening period necessary before the removal of forms is thus increased and the experience gained from concreting at normal temperature cannot be used directly.

Freezing of concrete at an early stage

Fresh concrete is vulnerable to freezing temperature: if the water in fresh concrete is allowed to freeze, irreparable damage to the quality of concrete and permanent lowering of compressive strength can occur. During freezing, water expands in volume by 9%, which has a considerable impact on the strength of concrete. If concrete while still plastic is allowed to freeze, the expansion of the water cloud renders it useless and such damage would be obvious once the formwork is removed. The effect of early freezing of concrete is shown in Fig. 26.28.

Impact on strength

Concrete produced at lower temperatures normally develops strength at a slower rate. The effect of low temperature on strength is shown in Fig. 26.29. Even though the development is slow, long-term strength is not affected much.

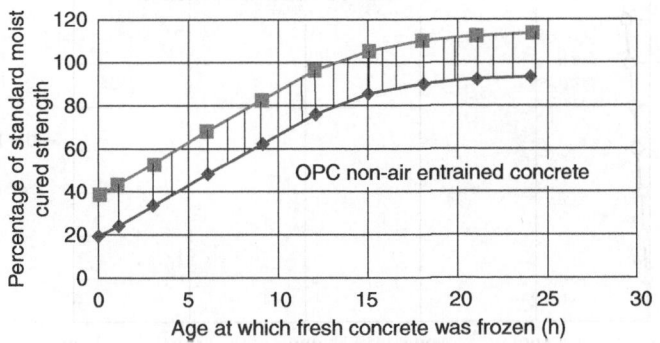

Fig. 26.28 Effect of early freezing on concrete

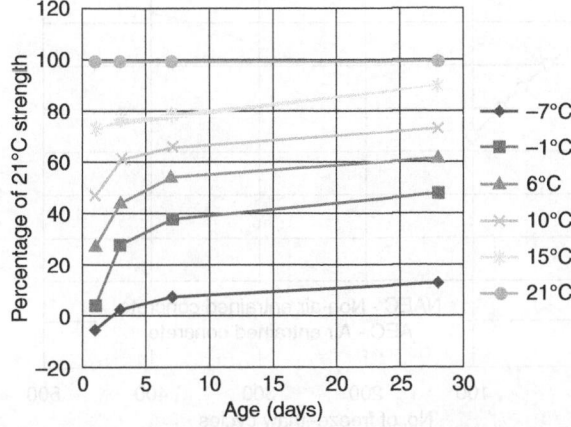

Fig. 26.29 Effect of low temperature on strength of concrete

Effects of freezing and thawing of concrete

Repeated freezing and thawing after the final setting and during the hardening period affects the final quality of concrete. Damage due to freezing and thawing is usually measured in terms of the lowering of the dynamic modulus of elasticity of concrete with the number of freezing and thawing cycles. Figure 26.30 shows the effect of repeated freezing and thawing on strength as well as stiffness. The entrainment of discrete air bubbles in the pore system of the concrete by means of air-entraining agents is the best remedy for this problem. Such voids do not contain water and behave like expansion chambers to accommodate the increase in volume of the ice formed. Properly air-entrained concrete is satisfactory even in severe winter conditions, provided the precautions required during mixing, placing, and curing have been taken.

Stress due to temperature differential

A large temperature differential within a concrete member may promote cracking and can cause harmful effects on the durability. Such differentials are likely to occur in cold weather at the time of removal of form insulations.

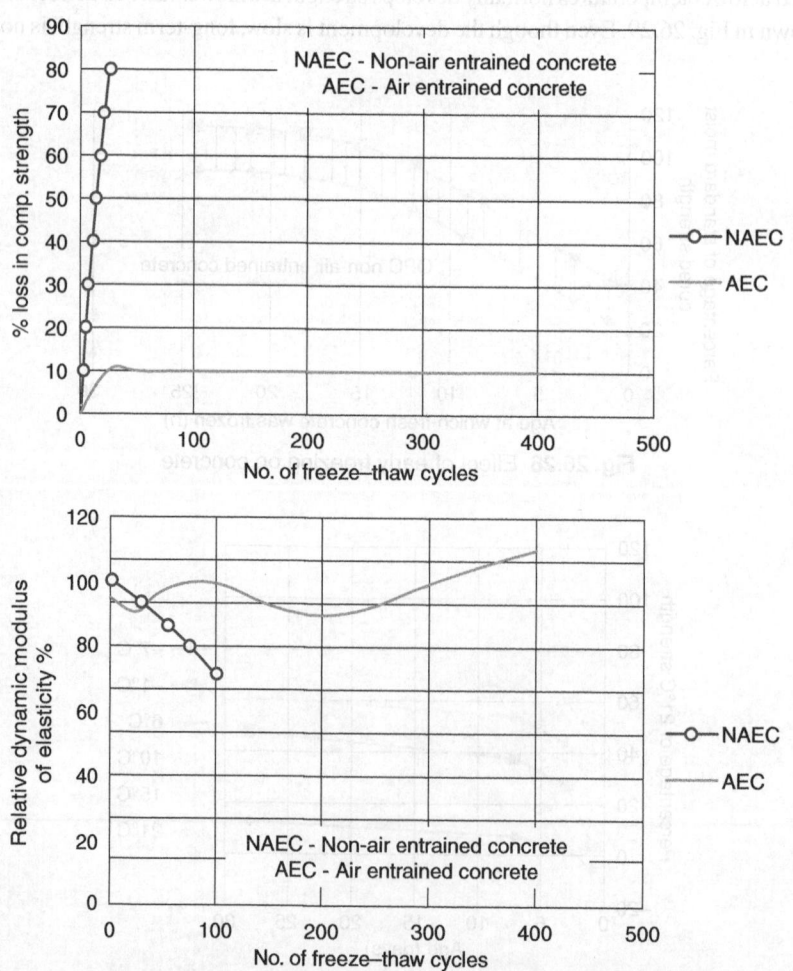

Fig. 26.30 Effect of repeated freezing and thawing on hardened concrete

26.7.2 Temperature Control

The most direct way to control the temperature of concrete is to control the temperature of its ingredients. Some of the methods of controlling the temperature of the materials are as follows:

a. The aggregates should be heated in such a way that frozen lumps, ice, and snow are eliminated and at the same time overheating is avoided. If coarse aggregate is dry and frost-free, the required temperature can be obtained just by heating the sand.
b. The mixing water should be heated to avoid appreciable fluctuations in temperature from batch to batch. Figure 26.31 shows the mixing water temperature requirement for producing heated concrete. Water at temperature up to boiling point can be used, provided the aggregate is cold enough to reduce the temperature of the mixing water.
c. The type and quantity of cement used in the concrete mix affect the rate of development of compressive strength and the rate of increase in the temperature of concrete. Additional quantities of ordinary Portland cement and rapid-hardening Portland cement or the use of accelerating admixtures helps in the development of the required strength in proper time by maintaining the temperature.

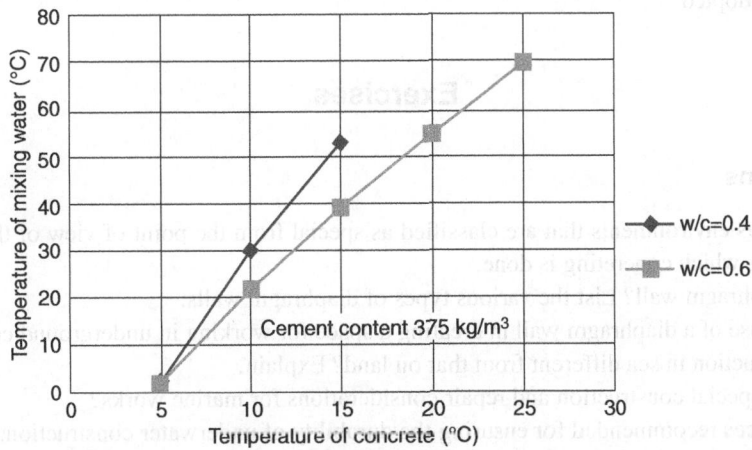

Fig. 26.31 Required temperature of mixing water to produce heated concrete

26.7.3 Temperature of Concrete as Placed

When the temperatures of all the constituents are known, the actual drop in temperature from the time the concrete leaves the mixing plant to the time of placing should be estimated; it may be taken as 2°C in the absence of proper information. The temperature of concrete at the time of leaving the mixer or batch plant may be calculated from the following formula:

$$T = \frac{S(T_a W_a + T_c W_c) + T_w W_w + T_{wa} W_{wa}}{S(W_a + W_c) + W_w + W_{wa}}$$

where T is the temperature of the freshly mixed concrete (°C); T_a, T_c, T_w, and T_{wa} are the temperature of the aggregate, cement, mixing water, and free water on aggregate, respectively (°C); W_a, W_c, W_w and W_{wa} are the mass of the aggregate, cement, mixing water, free water, and free moisture on aggregate, respectively (kg); and S is the specific heat of the cement and aggregate. In practice, the specific heat of cement and aggregate is taken as 0.22 and $T_a = T_{wa}$.

26.7.4 Precautions to be Taken During Production, Placing, and Curing

The following precautions must be taken during the production, placing, and curing of concrete in cold weather:

a. Wind breakers should be erected to shield the mixing and batching plants; tarpaulin, plastic sheets, and other covering materials should be made available at the plant.

b. Suitable protective clothing should be available to the site staff.

c. It is necessary to place the concrete quickly and cover the top of the concrete with insulated material.

d. Before placing the concrete, all ice, snow, and frost should be removed and the temperature of all surfaces that will be in contact with the new concrete should be raised to a temperature closer to that of the fresh concrete.

e. During freezing or near freezing conditions, water curing is not necessary, as loss of moisture from concrete by evaporation is greatly reduced in cold air conditions.

f. Low-pressure, wet-steam curing provides the best means of both heating the enclosure and curing the concrete. Early curing with liquid-membrane-forming compounds may be followed on the concrete surface with heated enclosures.

g. During cold weather concreting, a supervisor should record the air temperature, general weather conditions, and relative humidity at frequent intervals. The records should include frequent checks on the temperature of concrete as delivered and after placement in the forms as well as the protection and curing times adopted.

Exercises

Review Questions

1. List the various environments that are classified as special from the point of view of the precautions to be taken under which concreting is done.
2. What is a diaphragm wall? List the various types of diaphragm walls.
3. Describe the use of a diaphragm wall in creating a space for working in underground construction sites.
4. Why is construction in sea different from that on land? Explain.
5. What are the special construction and repair considerations for marine works?
6. List the practices recommended for ensuring the durability of underwater construction.
7. What are the various methods of underwater construction? Explain.
8. Define hot weather and cold weather concreting.
9. What are the effects of cold weather concreting?
10. What are the effects of hot weather concreting?
11. What precautions need to be taken for hot weather concreting?
12. How do you estimate the temperature of concrete that has been placed?

Multiple Choice Questions

1. When concrete is to be laid below ground level in deep excavation, space for working by excluding water is created by the temporary structure known as
   a. Scaffolding
   b. Formwork
   c. Shuttering
   d. Cofferdam
2. Cofferdam is generally constructed using
   a. Earthen mound
   b. Interlocked sheet piles anchored to the ground
   c. Formwork tied together
   d. Suitable shutters

3. In addition to the loads on ordinary land construction, underwater construction should withstand
   a. Wind load
   b. Earthquake load
   c. Live load
   d. Wave load

4. Slipform aids in
   a. Continuous concreting
   b. Achieving joint free structure
   c. Fast track construction achieving 6 m/day in height
   d. All of the above

5. The pipe used to pour concrete underwater without segregation is known as
   a. Tremie pipe
   b. Bucket pipe
   c. Insulating pipe
   d. Form pipe

6. Anchored walls are used when
   a. Excavation depth is large
   b. Rock for anchoring is available
   c. Excavation depth is waterlogged
   d. Excavation depth is shallow

7. Temperature control in hot weather concreting is done by
   a. Adding fly ash as admixture
   b. Using ice instead of water
   c. Storing the aggregates in shade
   d. All of the above

8. Cold weather concreting has the effect of
   a. Reducing setting time
   b. Increasing setting time
   c. Increasing the flow characteristics
   d. Decreasing the voids in concrete

9. The precaution to be taken for cold weather concreting is
   a. Placing the concrete quickly
   b. Removing all ice before placing
   c. Using wind shielding for batching plant
   d. All of the above

10. Underwater concreting is possible by
    a. Tremie method
    b. Dump bucket method
    c. Using bags filled with concrete
    d. All of the above

Answers to Multiple Choice Questions

1. d	4. d	7. d	10. d
2. b	5. a	8. b	
3. d	6. a	9. d	

CHAPTER 27

Tests on Concrete

Concrete structures are prone to wear and tear and damage due to many factors. These factors include undue loading due to earthquakes and damage due to environmental effects. Thus, repair and rehabilitation is needed to augment a weakened structure's architectural appearance and its functionality. The problems that afflict concrete structures are:

- moisture
- interior staining
- cracking
- spalling
- mortar deterioration
- loose components
- corrosion
- displacement
- blistering of coating
- efflorescence

Most of these problems occur even in new buildings. Some of these defects occur due to insufficient detailing, improper construction, disregard for quality control during construction, and overemphasis on reducing the construction cost at the expense of durability and safety. Even buildings that have served for a long lifespan have suffered damage because of changes in the environment caused by industrialization and heavy vehicular traffic. Therefore, it becomes necessary to evaluate the stability and safety of a structure periodically.

27.1 Evaluation of Existing Buildings

All structures are subjected to environmental impact and their condition deteriorates slowly. Therefore, the performance of any as-built structure after passage of time, say after 4 or 5 years, will be different. The evaluation of the current condition of any existing building thus becomes necessary both for gravity loads and for its resistance against lateral loads caused by cyclones and earthquakes. Such evaluation should be an integral part of the routine inspection or should be taken up as a step to measure the performance in light of signs of its deterioration. Investigation and subsequent evaluation become important to achieve an appropriate balance between the safe life of the structure and the protection of investment over a period of desired time. The evaluation of the current status of a building enables us to take decision on what to do with the structure, i.e., whether to repair and rehabilitate or to demolish and build anew.

27.1.1 Investigation Plan

There are several important factors which need consideration while investigating an existing building. The investigation evaluates the structure from stability and functionality points of view. The main factors that affect buildings are discussed below.

Strength

In a completed building the characteristic strength of the 'as built' structure, which is assumed to be constant throughout the lifespan at the time of designing, is likely to get weakened. For example, the concrete would have weakened through chloride contamination, ingress of moisture, and microcracking due to temperature stress. Contrary to this, in rare cases, strength enhancement due to the age factor is also possible in some structures. Therefore, testing of representative samples of materials in order to get a realistic statistical value of strength for use in the assessment becomes necessary. Material weakness is further aggravated by the loss of the cross section of elements due to corrosion or leaching effects. These factors need to be taken into account while assessment.

Structural form

The assessment of a building is done by analysing a theoretical model of the structure. It can be a frame, an arch, or a wall. The analysis model must be simple and should represent the structure in a 'safe' or a 'fail' situation. Based on the structural form, one should be capable of working out the extent of 'safety' based on a postulated collapse or failure condition. At each stage of investigation, the purpose and objective must be clear and, if needed, modified as new information requires fine-tuning of the original assumed criteria. Often, investigating the cause(s) of one problem may reveal other problems or deficiencies.

Loading

A schematic representation of the loads that a structure is subjected to during its lifespan is shown in Fig. 27.1. All these loads should be considered while chalking out an investigation plan. Initially the loads considered in

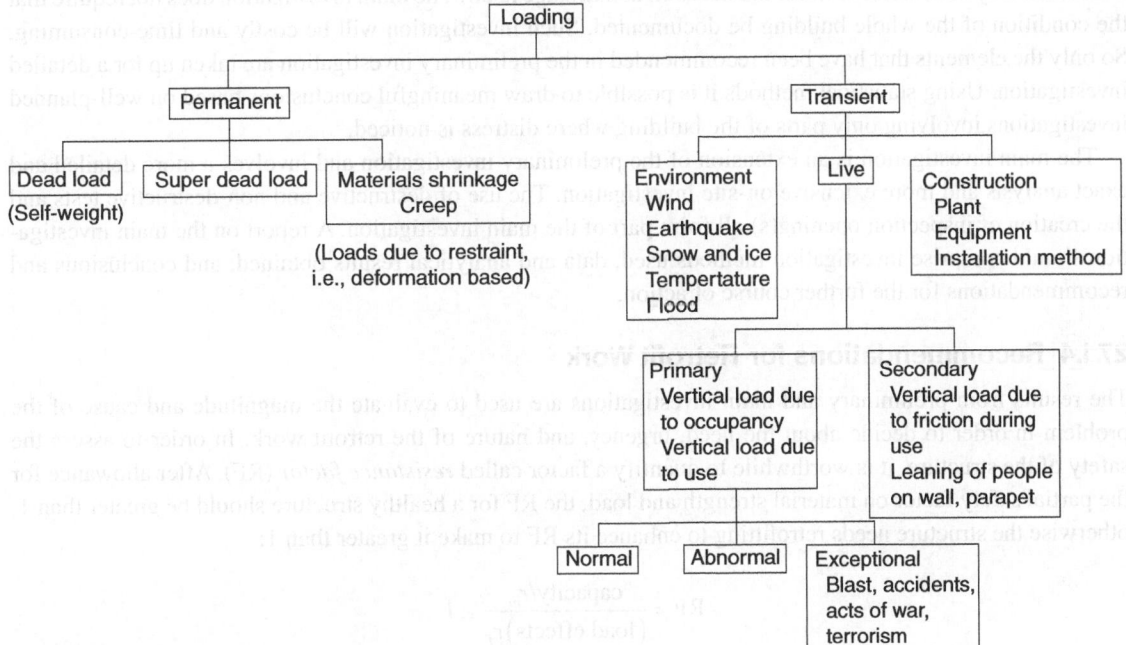

Fig. 27.1 Schematic representation of all loads

assessment are due to gravity which may be permanent (self-weight) or transient (live load due to occupancy). Permanent loads do not vary with time; these comprise the self-weight of the structure, superimposed fill, surfacing, slabs, parapets, weathering course, etc. Uncertainties in loading exist due to material density and construction dimensions, but these can be estimated with reasonable accuracy by prudent inspection and site measurement.

27.1.2 Preliminary Investigation

In the preliminary investigation a walk-around survey is the first step after collecting the following information:

- a set of building plans
- construction specifications
- inspection reports
- reports on previous investigation (if any)
- previous retrofit or repair works (if any)
- complaints by occupants (if any)
- other problems noticed

Collection of these data enables the engineer to have a better visual observation during the preliminary investigation stage. The preliminary investigation, though limited, should be thorough enough to have some preliminary analysis in order to define the likely repair options. Primarily, it should focus on the problem areas which can be easily repaired and should establish the need and scope for further investigation.

27.1.3 Main Investigation

The purpose of the main investigation is to analyse the structure thoroughly to provide sufficient documentation which forms the basis for the design of the remedial work. Only rarely would this sort of investigation conclude that the remedial work is not required. However, this could occur if the data from preliminary information was not sufficiently conclusive to make the decision at that stage itself. The main investigation does not require that the condition of the whole building be documented. Such investigation will be costly and time-consuming. So only the elements that have been recommended in the preliminary investigation are taken up for a detailed investigation. Using statistical methods it is possible to draw meaningful conclusions based on well-planned investigations involving only parts of the building where distress is noticed.

The main investigation is an extension of the preliminary investigation and involves a more detailed and exact analysis and more extensive on-site investigation. The use of destructive and non-destructive tests and the creation of inspection opening(s) all form part of the main investigation. A report on the main investigation should comprise investigation methods used, data and analytical results obtained, and conclusions and recommendations for the further course of action.

27.1.4 Recommendations for Retrofit Work

The results from preliminary and main investigations are used to evaluate the magnitude and cause of the problem in order to decide about the need, urgency, and nature of the retrofit work. In order to assess the safety of the structure, it is worthwhile to quantify a factor called *resistance factor* (RF). After allowance for the partial safety factor on material strength and load, the RF for a healthy structure should be greater than 1; otherwise the structure needs retrofitting to enhance its RF to make it greater than 1:

$$RF = \frac{\text{capacity}/r_m}{(\text{load effects})r_f} \geq 1$$

where r_m and r_f are partial safety factors for material strength and load effects, respectively.

After carrying out the investigation and analysis conservatively, the RF value is checked as shown in Fig. 27.2 to ascertain the need for any retrofit work. Once the problem is identified and the decision is taken for taking up retrofitting work, it is important to determine to what extent the problem needs to be addressed. For example: Should a weak defective parapet be simply removed or should it be retained with stitching it to the roof by providing additional reinforcements? In addition, local building regulations specifying minimum requirements such as fire safety should be considered compulsorily and adopted in the retrofit scheme.

Ensuring adequate structural capacity and fixing other problems that impose an immediate safety hazard have economic considerations also, and the details of retrofit and remedial work are significantly influenced by cost considerations. To effect utmost economy the retrofit design may also accept a larger extent of serviceability damaged than would normally be tolerated in the original design. Thus, the final retrofit scheme is based on an engineering judgement and on a sound database to ensure safety.

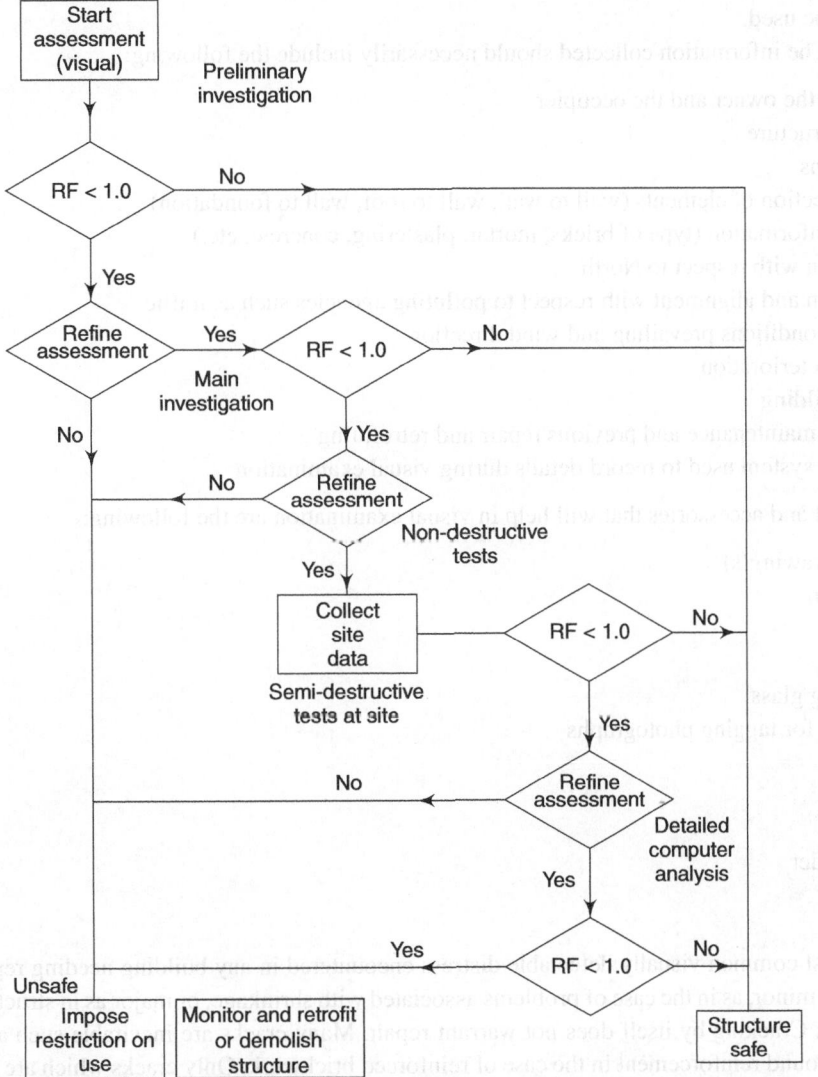

Fig. 27.2 Assessment flow diagram

27.2 Investigation Techniques

Investigation techniques can be broadly classified as destructive and non-destructive. But to decide on the technique to be used, visual investigation and a walk-around survey are to be undertaken first. At this stage, photography and notes should be taken. Also coordinate systems to identify the specific location of the damage(s) should be established. Digital photography of large areas is helpful in documenting overall conditions, and close-up views can be used to check details such as crack width and direction. During reconnaissance, binoculars and tele-photo lenses are used in addition to a photographic camera. These can be helpful to get more information and act as a record for further use.

27.2.1 Visual Examination

One of the most important components of any investigation plan is the visual examination. This involves looking for and analysing cracks, spalling, and other such structural defects which can be seen without digging into or removing the surface of the structure or its element(s). To aid visual examination, the following checklist can be used.

   Checklist The information collected should necessarily include the following:

1. Details of the owner and the occupier
2. Type of structure
3. Dimensions
4. Interconnection of elements (wall to wall, wall to roof, wall to foundation)
5. Material information (type of bricks, mortar, plastering, concrete, etc.)
6. Orientation with respect to North
7. Orientation and alignment with respect to polluting agencies such as traffic
8. Climatic conditions prevailing and wind direction
9. Signs of deterioration
10. Age of building
11. Details of maintenance and previous repair and retrofitting
12. Reference system used to record details during visual examination

The equipment and accessories that will help in visual examination are the following:

- Building drawing(s)
- Layout plan
- Camera
- Binoculars
- Magnifying glass
- Post-it pad for tagging photographs
- Flashlight
- Clipboard
- Compass
- Tape recorder

Cracking

This is the most common visually detectable distress encountered in any building needing repair or retrofit. Cracks may be minor, as in the case of problems associated with shrinkage, or major as in structures subjected to overloading. Cracking by itself does not warrant repair. Many cracks are inevitable such as those due to shrinkage or around reinforcement in the case of reinforced brickwork. Only cracks which are wider than the acceptable limit are to be considered as a sign of distress. Cracks of smaller widths can be of aesthetic concern and hence need cosmetic treatment such as putty work. Wide cracks (greater than 1 mm in width) signify structural problems and need to be investigated in greater detail. Cracks in a concrete member take the path of least resistance and form where the shear, tensile, or bond strength is the least. In strong brick units cracks may develop along the mortar joints. In case mortar joints are strengthened either due to the gravity load acting from top or due to grouting during repair, cracks form across the brick unit. Based on the crack direction as well as its stability with time on the affected element, cracks are classified as:

Dormant

- Do not increase in size once formed
- Typically caused by shrinkage, initial movement of supports, previous overload
- May or may not need repair

Active

- Change size under load
- Continuing movement and overload
- Difficult to repair—if the cause is not treated, a new crack next to the repaired one is common

On the basis of the width of cracking, cracks are classified as:

Fine

- Generally less than 1 mm

Medium

- Between 1 and 2 mm

Wide

- Greater than 2 mm

On the basis of the depth of cracking, cracks are classified as:

Surface cracks

- Local cracks on the top layer alone, mostly in non-structural components

Crack widths can be anywhere between 0.1 and 25 mm. Cracks smaller than 0.1 mm are not visible to the naked eye and, generally, do not affect the structural or functional utility of the building.

It is necessary to ascertain whether a crack is active or dormant. Active cracks propagate and hence a repeated inspection of cracks is necessary to classify cracks as active or dormant.

Cracks usually occur due to environmental effects. Excessive restraint within concrete to thermal movements may cause these cracks. Proper location of movement joints can help avoid cracks resulting from restrained movement, overloading, vibration, unintended interaction with other structural members such as beams and columns, differential settlement or settlement of supports. Figure 27.3 shows an example of crack commonly witnessed in structures.

Spalling

Delamination of the surface of concrete is called *spalling*. It can occur due to internal stress or due to external loads. Concentrated eccentric external load can cause a highly stressed narrow compression zone, which encourages spalling. Additionally, corrosion of steel embedded in concrete can also cause spalling (Fig. 27.4). Spalling may also occur due to the freeze–thaw effect of entrapped water. A careful observation of the location of spalling will throw some light on the agents responsible for spalling. Joint mortar can also suffer from

Fig. 27.3 Horizontal crack in a column

Fig. 27.4 Corrosion of steel embedded in concrete causes spalling

spalling especially if the mortar is weak or if the joint is too thick. Normally the mortar joint thickness should be less than 20 mm. Chemical reactions, efflorescence, and repeated wetting and drying in coastal areas, especially if concrete is not maintained, are some of the causes that can lead to spalling.

Staining

Staining of concrete is caused by absorption of water, which contains minerals/salts and results in leaching or drainage over other components. *Efflorescence* is the major cause of staining which is caused by deposition of water-soluble salts on the surface of masonry or concrete as the water carrying these salts evaporates. Efflorescence disfigures the building surfaces and can weaken members because of internal crystallization. However, the root cause of efflorescence is moisture penetration, which is a more troublesome problem.

In reinforced walls subjected to moisture penetration, rust straining may occur in localized zones. Because of increase in volume due to rust formation, spalling and cracking occur and, if unattended, can expose the reinforcement to faster corrosion and ultimate destruction of the wall. Therefore, for both corrosion and efflorescence staining, the moisture penetration problem has to be addressed.

Moisture problem

Moisture (due to rain or excessive humidity) causes wetness and encourages moulds and fungi. It can damage insulation and cause reduction in the quality of the performance of concrete elements. Moisture caused due to condensation resulting from the vapour pressure difference is difficult to detect. However, condensation and penetration of rain and its pathway can be detected through visual observation of wet areas and patches after a rainy day.

Construction and design defects

The visual inspection should record the following deficiencies, which occur primarily due to defective workmanship:

- Use of incorrect wall thickness
- Out-of-plumb walls
- Failure to connect inserting walls and columns
- Lack of movement joints
- Defective joints
- Defective bond
- Defective flashing

- Plugged weep holes
- Staining that defines drainage paths
- Misalignment of joints

The following problems are primarily due to ineffective design effort:

- Inadequate diaphragm stiffness for distributing lateral shear
- Poor layout that causes excessive torsion
- Careless and unaccounted provision of infills
- Excessive out-of-plane deflection
- Cracking below beams due to an insufficient gap for deflection
- Use of poor quality concrete
- Use of poor construction procedures
- Bowing due to dissimilar material used in masonry

Most of these defects are noticed and recorded during visual inspection.

27.3 Tests on Hardened Concrete

A good number of tests and techniques are available to evaluate the quality of concrete. Some tests conducted on hardened concrete do not affect the structure or element on which these tests are done. Such tests which do not cause any destruction of the structure or the element are classified as *non-destructive* tests. Some testing procedures and techniques cause negligible (but repairable) damage to the structural element. These tests are classified as *semi-destructive* tests. We will discuss in detail both the types of tests in the following sections.

27.4 Non-destructive Testing

There are a number of non-destructive tests and testing techniques that are being used to evaluate concrete. In this section we will discuss some of these.

27.4.1 Rebound Hammer Test

Schmidt's rebound hammer consists of a spring-controlled mass that slides on a plunger within a tubular housing. When the plunger is pressed against the surface of the concrete, the spring-controlled mass rebounds and the extent of rebound depends upon the surface hardness of concrete (Fig. 27.5). The surface hardness and,

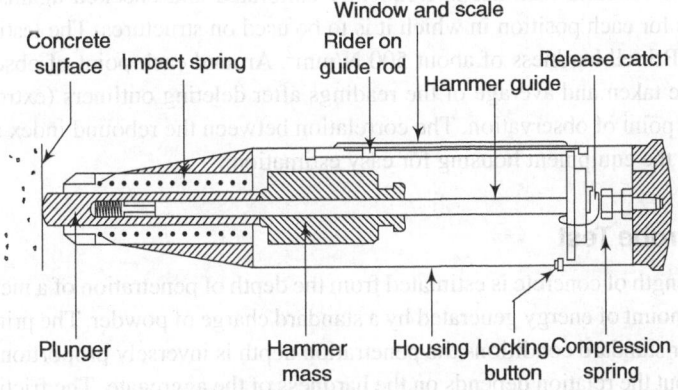

Fig. 27.5 Rebound hammer

therefore, the rebound are related to the compressive strength of the material. The rebound is read off along a graduated scale, and this value is designated as the 'rebound number' or 'rebound index'.

The rebound hammer is used for

- estimating the compressive strength of concrete with the help of suitable correlations between the rebound index and compressive strength,
- establishing the uniformity of concrete, and
- assessing the quality of one element of concrete in relation to another.

Depending upon the impact energy required for different applications, Schmidt's hammer test can be classified into four types as follows:

Type	Application	Approximate impact value required for rebound hammer
N	For testing normal grades of concrete in ordinary buildings and bridge construction	2.25
L	For testing lightweight concrete or small and impact-sensitive parts of concrete or artificial stones	0.75
M	For testing mass concrete in roads, air-field pavements, and hydraulic structures	30
P	For testing cement mortar and plaster, concrete of low strength (5–25 N/mm^2)	0.90

Limitations

The following are the limitations of the rebound hammer test:

1. The accuracy of the predicted strength of concrete is ±25%.
2. Results are affected by the angle of test, surface smoothness, and mix proportion.
3. It is only suitable for close-textured concrete.
4. All correlations assume full compaction; strength of partially compacted concrete bears no unique relationship to the rebound number.

Calibration

It is necessary that the rebound hammer is frequently calibrated and checked against the testing anvil to ensure reliable results for each position in which it is to be used on structures. The testing anvil should be of steel having an anvil Brinell hardness of about 500 N/mm^2. Around each point of observation, six readings of rebound indices are taken and average of the readings after deleting outliners (extreme values) gives the rebound index for the point of observation. The correlation between the rebound index and concrete strength is given as a graph on the equipment housing for easy estimation.

27.4.2 Windsor Probe Test

In this method the strength of concrete is estimated from the depth of penetration of a metal rod driven into the concrete by a given amount of energy generated by a standard charge of powder. The principle underlying this technique is that under standard conditions the penetration depth is inversely proportional to the compressive strength of concrete, but the relation depends on the hardness of the aggregate. The frictional resistance to the probe and the energy absorbed by cracking of concrete are assumed to be negligible.

The penetration depth (Fig. 27.6) is an inverse measure of the compressive strength of concrete. Charts of strength versus penetration for the normal probe used are available for aggregates with hardness between 3 and 7 on Mohs scale (Fig. 27.7). However, in practice, the penetration resistance should be correlated with the compressive strength of a standard test specimen or the core of the actual concrete used. The main advantage of this test is that hardness is measured up to a certain depth. Hence, it gives more realistic result than the rebound hammer test.

Specifications

- Diameter of probe = 7.94 mm
- Length of probe = 79.5 mm
- Velocity of cartridge = 183 m/s
- Variation in diameter = ±1%

Limitations

The following are the limitations of this technique:

1. This test basically measures hardness and cannot yield absolute reliable values of strength.
2. The type of aggregate affects the penetration depth, and hence a separate calibration chart for each type of aggregate is required.

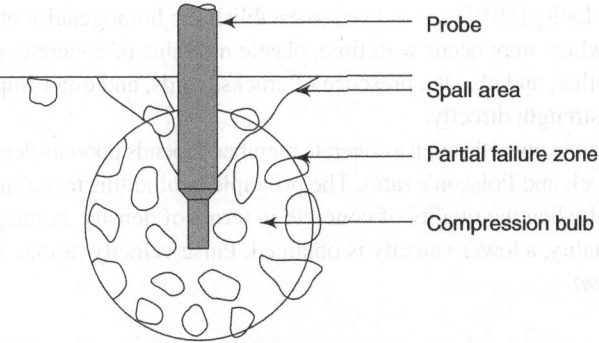

Probe

Spall area

Partial failure zone

Compression bulb

Fig. 27.6 Windsor probe test

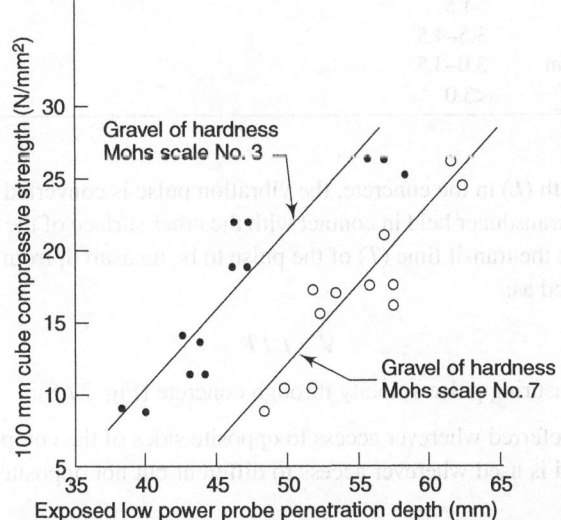

Fig. 27.7 Strength versus penetration depth relationship for Windsor probe

3. Reliability of results varies by 4%.
4. The shape and moisture content of aggregates also affects the results.

Calibration

Indentations are measured by a spring-loaded calibrated depth gauge to the accuracy of 0.5 mm. Alternatively, the probes in sets of three can be measured by using a triangle template system with probes located at 120° and using triangular measuring plates to get the average of three readings. While using this method, the following recommendations are followed:

- Low power drive for strength up to 26 N/mm^2
- Standard power drive for higher strength
- Edge distance should be minimum 100 mm

27.4.3 Ultrasonic Pulse Velocity Test

An ultrasonic pulse is generated by an electro-acoustical transducer. When the pulse is introduced into the concrete from the transducer, it undergoes multiple reflections at the boundaries of the different material phases within the concrete. A complex system of waves is developed, which includes longitudinal, transverse, and surface waves. The receiving transducer detects a set of longitudinal waves, which is the fastest.

The ultrasonic pulse velocity (UPV) test is used to establish the homogeneity of concrete, changes in the structure of the concrete which may occur with time, elastic modulus of concrete, quality of one element of concrete in relation to another, and also the presence of cracks, voids, and other imperfections. It may not be able to give a measure of strength directly.

The velocity of an ultrasonic pulse through a concrete member depends upon its density, modulus of elasticity, presence of reinforcing steel, and Poisson's ratio. The principle behind this technique is that a comparatively higher velocity is obtained when the quality of concrete in terms of density, homogeneity, and uniformity is good. In case of poorer quality, a lower velocity is obtained. Pulse velocity ratings for various quality grades of concrete are given below:

Quality	Pulse velocity—as per Bureau of Indian Standards (km/s)
Excellent	>4.5
Good	3.5–4.5
Fair/Medium	3.0–3.5
Poor	<3.0

After traversing a known path (L) in the concrete, the vibration pulse is converted into an electrical signal by a second electro-acoustical transducer held in contact with the other surface of the concrete member. An electronic timing circuit enables the transit time (T) of the pulse to be measured, from which the ultrasonic pulse velocity (V) can be calculated as:

$$V = L / T$$

There are three ways of measuring pulse velocity through concrete (Fig. 27.8):

1. The direct method is preferred wherever access to opposite sides of the component is possible.
2. The semi-direct method is used wherever access to different but not opposite sides of the component is possible.
3. The indirect method is least satisfactory and should be used when access to only the surface is possible.

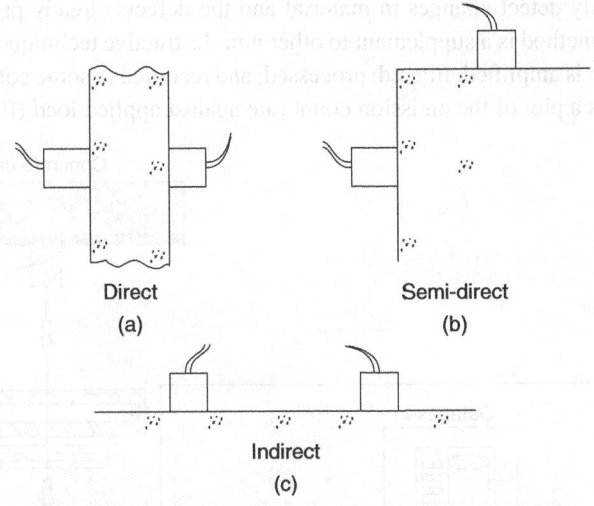

Fig. 27.8 Ultrasonic pulse velocity test—three types

Specifications

The apparatus required for UPV test are as follows:

- One electrical pulse generator
- A pair of transducers
- An amplifier
- An electronic timing device

Limitations

The following are the limitations of this testing procedure:

1. Any variation of concrete temperature between 5°C and 30°C does not affect the pulse velocity measurement. At temperatures between 30°C and 60°C, there can be reduction in the pulse velocity by 5%. Below the freezing temperature, free water freezes within concrete and the pulse velocity is increased by 7%.
2. The pulse velocity of concrete increases with an increase in the moisture content of concrete. The influence is pronounced for low-strength concrete than for high-strength concrete. The pulse velocity of saturated concrete may be increased by 2%.
3. The pulse velocity measured in reinforced concrete in the vicinity of reinforcing bars is usually higher than in plain concrete of the same composition. This is because the pulse velocity through steel is 1.2–1.9 times higher than the velocity through concrete.
4. When concrete is subjected to stress which is abnormally high for the quality of concrete, pulse velocity may be reduced due to development of micro-cracks.

27.4.4 Acoustic Emission Method

In certain materials the change of condition can occur, which causes emission of sound waves due to thermal, mechanical, or other effects. The change of condition can be a fracture, crack formation, crack growth, or metallurgical change such as plastic deformation, dislocation, and change in crystal structure. During such changes in the condition, energy released propagates in the surrounding material as elastic vibrations. These vibrations or sound waves can be detected by placing a sensor on the surface of the material. Acoustic emission (AE) sensors are in principle high-frequency microphones, which are glued or fixed in some manner to the clean surface so that the best possible acoustic contact is achieved.

The AE method can only detect changes in material and the defects already present cannot be revealed through this method. This method is a supplement to other non-destructive techniques. The signal detected by the piezoelectric transducer is amplified, filtered, processed, and recorded in some convenient form (Fig. 27.9).

Results are presented as a plot of the emission count rate against applied load (Fig. 27.10).

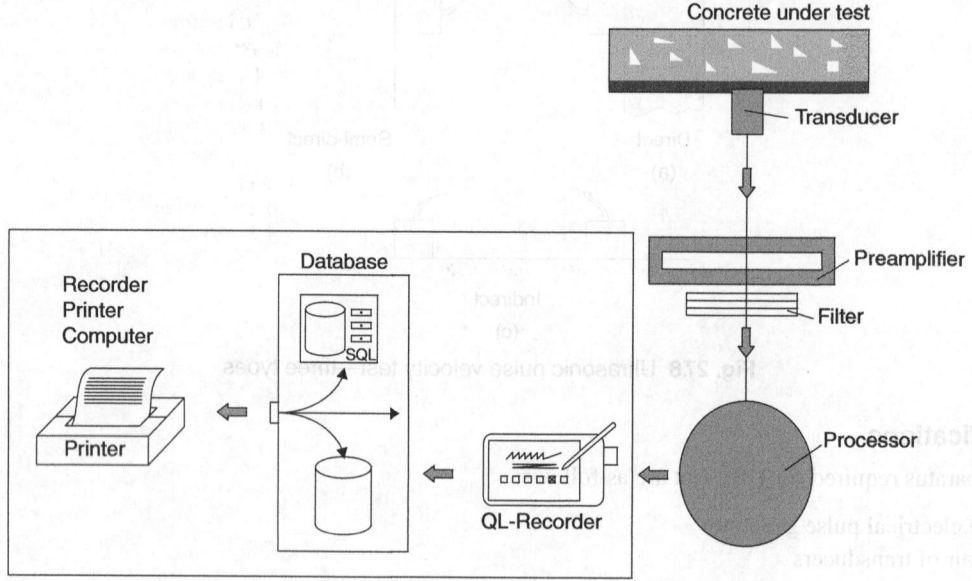

Fig. 27.9 Acoustic emission method

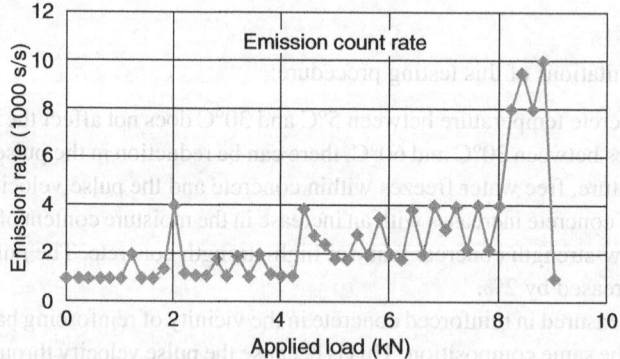

Fig. 27.10 Plot of emission count rate against applied load

27.4.5 Pulse-Echo Method

The pulse-echo method is used for locating defects such as voids, cracks, and zones of deterioration. Ultrasonic examination (Fig. 27.11) by means of the pulse-echo method (Fig. 27.12) is based on monitoring the interaction of acoustic waves with the internal structure of an object. A high-frequency pulse is transmitted through the material. The pulse has a character of sound oscillation, but it cannot be heard because of its high frequency. Pulses propagate in the material until they strike an interface or the opposite surface of the object and entirely or partly get reflected back to the transmitter, which now functions as the receiver. The time interval between transmission and reception of a pulse depends upon the distance travelled and the speed for a particular type of material. If one wants to localize internal defects in an object, one must know the starting point of the sound pulse and the direction in which it has travelled as well as the distance traversed. Larger defects are mapped by

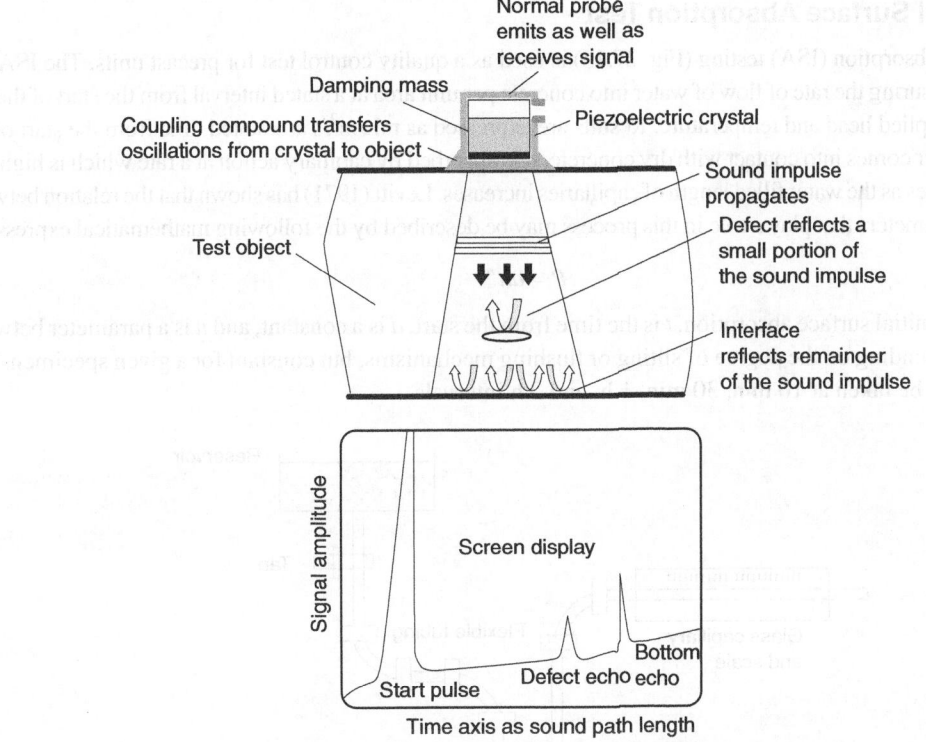

Fig. 27.11 Principle of pulse-echo examination

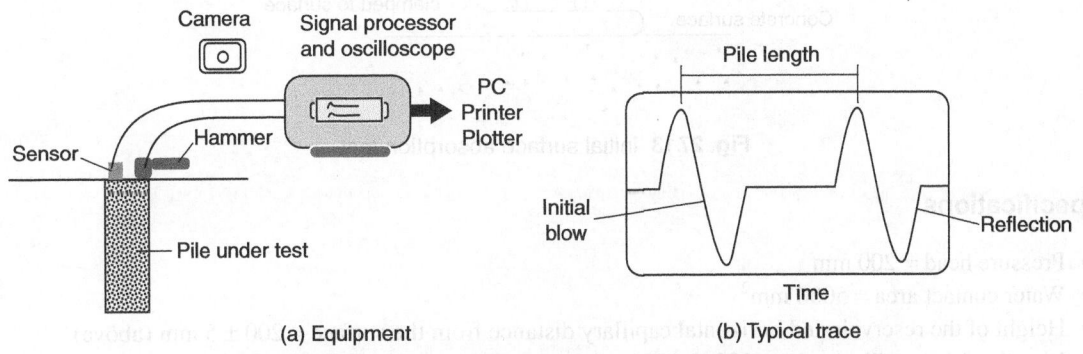

(a) Equipment   (b) Typical trace

Fig. 27.12 Pulse-echo method of pile integrity test

moving the probe or transducer which transmits and receives the sound pulse across the surface of the object. This search pattern is called *scanning* and often follows a certain pattern. Ultrasonic examination is mostly used in steel industry and also, to a lesser extent, to monitor concrete.

Limitations

This technique has the following two drawbacks:

1. The reliability of this technique appears to decrease with increase in the thickness of concrete.
2. Pulse-echo testing using a mechanically generated stress pulse is applicable only if the test object dimensions exceed certain minimum requirements. The specimen must be large enough such that the reflected pulse arrives at the receiver after the surface wave.

27.4.6 Initial Surface Absorption Test

Initial surface absorption (ISA) testing (Fig. 27.13) is used as a quality control test for precast units. The ISA test consists in measuring the rate of flow of water into concrete per unit area at a stated interval from the start of the test at a constant applied head and temperature. Results are expressed as mL/m²/s at a stated time from the start of the test. When water comes into contact with dry concrete, it is absorbed by capillary action at a rate which is high, but this rate decreases as the water filled length of capillaries increases. Levitt (1971) has shown that the relation between the various parameters that play a role in this process may be described by the following mathematical expression:

$$P = a/t^n$$

where P is the initial surface absorption, t is the time from the start, a is a constant, and n is a parameter between 0.3 and 0.7 depending on the degree of silting or flushing mechanisms, but constant for a given specimen. The readings are to be taken at 10 min, 30 min, 1 h, and 2 h intervals.

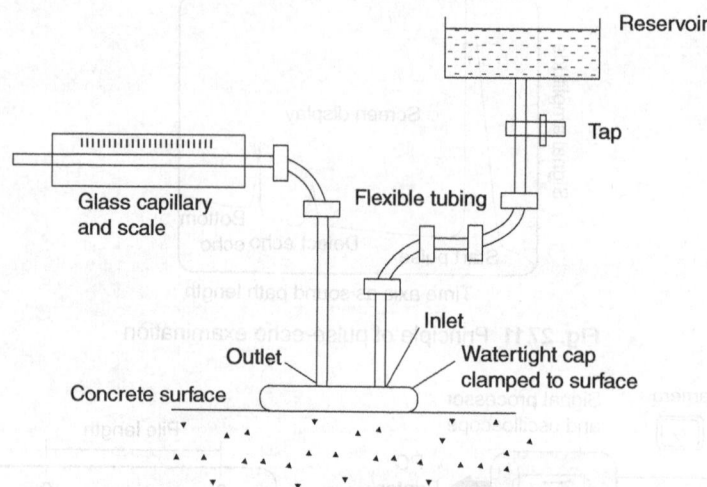

Fig. 27.13 Initial surface absorption test

Specifications

- Pressure head = 200 mm
- Water contact area = 5000 mm²
- Height of the reservoir and horizontal capillary distance from the surface = 200 ± 5 mm (above)
- Length of the capillary tube = 100 to 1000 mm
- Bore of the capillary tube = 0.4 to 1.0 mm radius
- Minimum edge distance proposed from the circumference of the cap = 30 mm

Limitations

ISA testing has the following limitations:

1. This test is very sensitive to changes in quality and is difficult to correlate with the observed weathering behaviour.
2. This method is not applicable for porous or honeycombed concrete.

27.4.7 Radar Technique

High-frequency pulse radars have been used to detect deterioration in concrete pavements and bridge decks. This procedure consists in using bituminous-surfaced bridge decks for measuring the thickness of the surface. The echo produced from the pavement surface and the interface with the bridge deck concrete is distinct such that the thickness can be determined accurately. A short-duration pulse of radio-frequency energy is directed onto the deck; a portion is reflected from any interface and the output is displaced on an oscilloscope is recorded and examined. Reinforcing bars, voids, and ducts can be identified as well as zones of varying moisture content, and the thickness of slab is estimated. A permanent record is usually stored on a magnetic tape. The unit is normally mounted on a vehicle and the large amount of data is collected as the vehicle moves slowly across the deck. Equipment consists of transmitting and receiving antennae together with a control unit and recorder, and is available in portable form (Fig. 27.14).

Application of radar technique for rebar location has advantages over other methods. The profile taken on parallel and perpendicular lines rapidly builds up a picture of the reinforcement pattern.

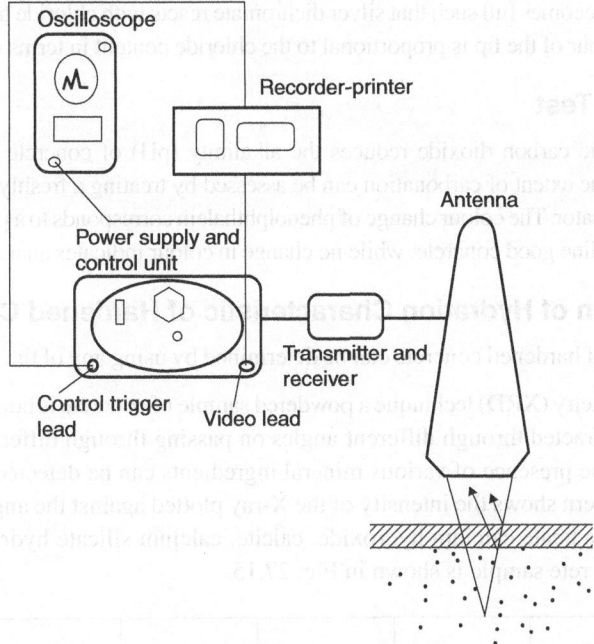

Fig. 27.14 Radar technique

Limitations

Some of the limitations of this testing technique are the following:

1. Analysis of large amounts of data collected and relating different radar signals to physical distress creates practical problems.
2. The testing does not work effectively if the deck surface is wet or if there is significant moisture on the surface.

27.4.8 Infrared Thermography

Infrared thermography has been found to be capable of detecting delamination in bridge decks. The difference between surface temperatures at various points can be measured using a sensitive infrared detection system. This method works on the principle that a discontinuity within concrete, such as de-lamination parallel to the surface, interrupts the heat transfer through concrete. When exposed to heat due to sunlight, the surface temperature of delamination is higher than the surrounding concrete. At night when there is loss of heat from concrete to the surrounding air, the surface of de-lamination is cooler than the average temperature of the solid concrete. The essential components of such a system are an infrared scanner, a control unit, a battery pack, and a display screen. The images can be recorded on photographic plates or videotape.

A number of operation modes have been used ranging from a hand-held scanner at the deck level to those mounted on a boom truck or a helicopter. A truck-mounted plate from 4 m to 6 m above the deck has been found to give best results with respect to definition, accuracy, and speed. This configuration also permits lane width to be scanned at a single pass and is very convenient in the field.

27.4.9 Quantab Test

This test is used to measure the chloride concentration in concrete. A solution containing 5 g of powdered concrete is prepared first. A plastic strip, 75 mm long and 15 mm wide, with a vertical capillary column-like arrangement impregnated with silver dichromate is used. At the top of the column is a horizontal air vent containing a yellow, moisture-sensitive indicator which changes to blue when the capillary is full. The lower end is placed in the chloride solution until the capillary becomes full such that silver dichromate reacts with chloride to form silver chloride. The degree of change in the colour of the tip is proportional to the chloride content in terms of ions/litre.

27.4.10 Carbonation Test

Carbonation by atmospheric carbon dioxide reduces the alkalinity (pH) of concrete and increases the risk of reinforcement corrosion. The extent of carbonation can be assessed by treating a freshly exposed concrete surface with a phenolphthalein indicator. The colour change of phenolphthalein corresponds to a pH of about 9. A purple-red colour indicates highly alkaline good concrete, while no change in colour indicates an acidic concrete with pH < 5.

27.4.11 Determination of Hydration Characteristic of Hardened Concrete

Hydration characteristics of hardened concrete can be determined by using any of the following three methods:

1. In the X-ray diffractometry (XRD) technique a powdered sample of concrete is bombarded with high-energy X-rays. X-rays get refracted through different angles on passing through different mineral constituents. Using this property, the presence of various mineral ingredients can be detected by examining the XRD pattern. The XRD pattern shows the intensity of the X-ray plotted against the angle of the reflected beam. The peaks generally indicate calcium hydroxide, calcite, calcium silicate hydrate, etc., in concrete. An XRD pattern of a concrete sample is shown in Fig. 27.15.

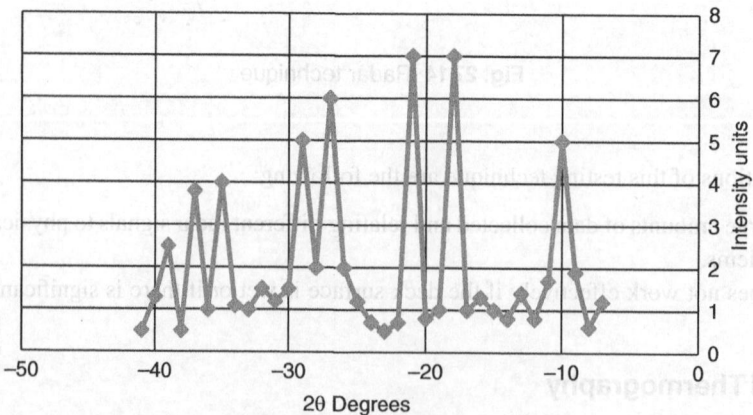

Fig. 27.15 X-ray diffraction pattern

2. In the X-ray fluorescence (XRF) spectroscopy technique, the sample of concrete is bombarded with high-energy X-rays and the fluorescent emission so caused is collimated into a parallel beam, directed on to the analysing crystal within a spectrometer and reflected into a detector. The wavelength and density of the fluorescent emission are measured and the constituent elements together with their properties can be calculated from this data. This method is a comparative one, and the results obtained are compared with samples of known properties.

3. Differential thermal analysis (DTA) technique is concerned with the rate of change of temperature of a sample when heated at a constant rate of heat input and involves heating a small sample of powdered concrete in a furnace together with a sample of inert material. The DTA graph shows a series of peaks at particular temperatures, which are characteristic of minerals in the concrete sample under test. Figure 27.16 shows a DTA graph for a concrete sample. DTA studies are useful for fire-damaged concrete structures for assessment of temperature to which the concrete was exposed during the fire and also for assessment of the depth of affected concrete.

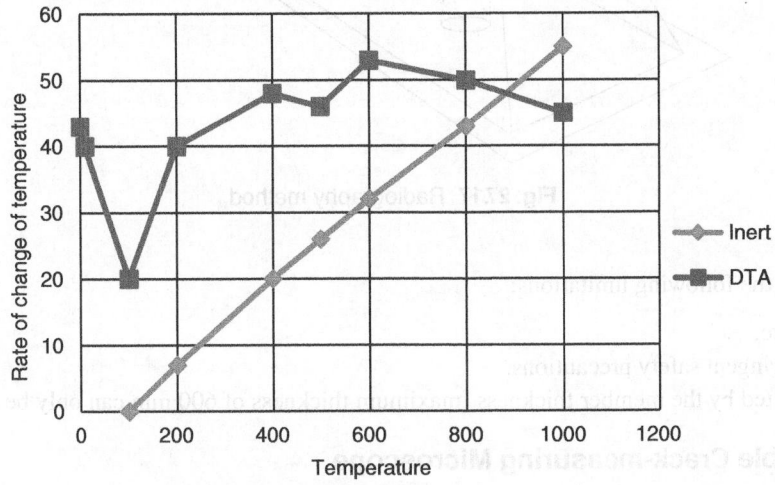

Fig. 27.16 Differential thermal analysis graph

27.4.12 Dye Penetration Examination

This is a simple method in which non-porous materials are examined with penetrants. Penetrants reveal defects when they reach the surface such as crack, porosity, and cleavage. Initially penetrants are applied to the cleaned surface of the component and allowed to act for a period of time on the component to be examined. Any excess penetrant is carefully removed, and a developing liquid (developer) is applied thereafter. The developer widens the defect so that it can be seen clearly. This method is used to find surface defects and leaks. Some commonly used developers are

- dry powder developer
- water-based wet developer
- non-water-based developer

Of these, wet developers are used most commonly.

27.4.13 Radiography

Radiography is used to determine the position of cables, voids in grouting, and in situ density of concrete. There are three methods used for testing concrete: X-ray radiography, α/β-ray radiography, and γ-ray radiometry. In order to observe an object, one of these rays is chosen and passed through the object. This causes radiation of varying degrees in the object, depending upon the thickness, composition, and wavelength of material. The portion which penetrates through the object can be recorded on a film (Fig. 27.17). By examining the film in different exposures, the thickness and composition of the member can be found. In reinforced concrete, rebars can easily be located.

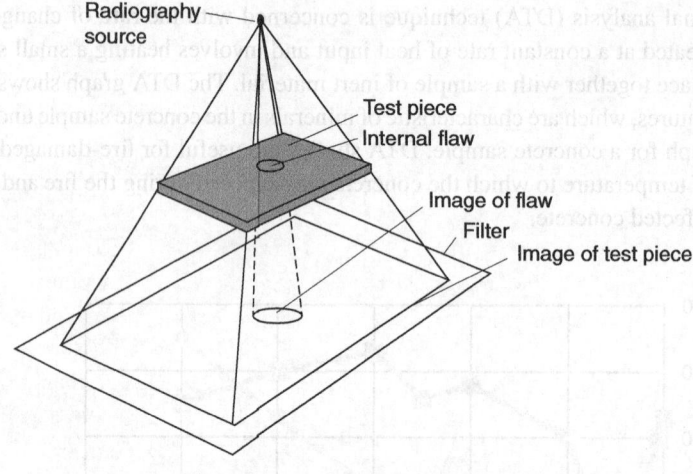

Fig. 27.17 Radiography method

Limitations

Radiography has the following limitations:

1. It is expensive.
2. It requires stringent safety precautions.
3. It is also limited by the member thickness (maximum thickness of 600 mm can only be used).

27.4.14 Portable Crack-measuring Microscope

The portable crack-measuring microscope (Fig. 27.18) is powered by battery and is used to measure the crack width. The concrete surface is illuminated by a small internal light bulb and the magnified crack width is

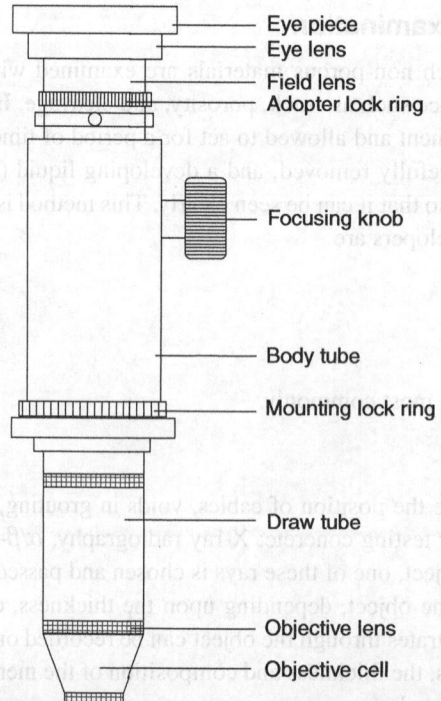

Fig. 27.18 Crack-measuring microscope

measured. The instrument consists of an objective cell containing objective lens, an ocular cell containing the eye and field lenses, and a reticle-mounting arrangement, all assembled in an optical body tube. The ocular cell is adjustable in the tube and allows the eyepiece to focus sharply on the reticle. The complete tube can also be adjusted by means of a rack and pinion for focusing the microscope. The least count of the instrument is usually 0.01 mm.

27.4.15 Cover Meter

When a metallic object is placed in the varying magnetic field of a coil, the field induces eddy currents in the object. These eddy currents in turn produce an additional magnetic field in the vicinity of the magnetic object. A magnetic field gets superimposed and the magnetic field near the coil also gets modified in the presence of metal. This modification has the same effect as would be obtained if the characteristic of the coil itself had been changed. The change depends upon the electrical conductivity, dimension, magnetic permeability, presence of discontinuity such as cracks or cavities, frequency of the field of the coil, size and shape of the coil, and the distance of the coil from the metallic object.

It is possible to measure the cover thickness for a known diameter by keeping all other parameters constant. By placing the coil at two different distances from the rebar, both the cover thickness and the diameter of the rebar can be found.

The electronic circuit of the arrangement consists of an oscillator, a probe coil, and an amplifier. The oscillator generates a stabilized sine wave of 1 kHz frequency. This is applied to the probe coil through a resistance. The probe coil has inductance and along with a condenser forms a parallel resonant circuit. The voltage across the probe coil is connected to a voltage follower which acts as a high-impendence buffer. The AC voltage is rectified and the DC amplifier stage follows.

If there is no metallic object in the vicinity of the probe coil, the voltage across the probe coil is maximum and the output of the amplifier stage is maximum. The cover thickness in such a case is infinity. When the probe is placed over a concrete cube containing a rebar, the voltage across the probe coil gets reduced and the output of the DC amplifier is also reduced. The reduction depends on the cover thickness and diameter of the rebar.

Portable battery-operated cover thickness meters based on eddy currents are available in the market. The instrument is useful in measuring the actual concrete cover provided over embedded steel reinforcement. The orientation and profile of prestressing members can also be ascertained with the help of this instrument. Generally, a cover meter consists of a unit containing a power source, amplifier, meter, and a separate scanner unit containing the electromagnetic unit, which is coupled to the main by a cable.

27.4.16 Linear Variable Differential Transformer

The linear variable differential transformer (LVDT) is the most commonly used transducer to measure deformations. It provides an AC voltage output proportional to the relative displacement of the transformer coil to the winding. Figure 27.19 shows the arrangement and the section of an LVDT.

The central coil is energized by an external AC power source, and the two end coils connected together in phase position are used as pick-up coils. The output amplitude depends on the relative coupling between the two pick-up coils and the power coil. Relative coupling in turn depends on the position of the core.

Theoretically, there should be a core position for which the voltage induced in each of the pick-up coils is of the same magnitude. Within limits on either side of the null position, the core displacement results in proportional output. Generally, the linear range is primarily dependent on the length of the secondary coil. Although output voltage magnitudes are ideally the same for equal core displacement, on either side of null balance, the phase relation existing between the power source and the output changes by 180° through null. It is therefore possible through phase determination or the use of phase-sensitive circuitry to distinguish between outputs resulting from displacements on either side of null. As displacement recording instrument, LVDTs require a stationary base.

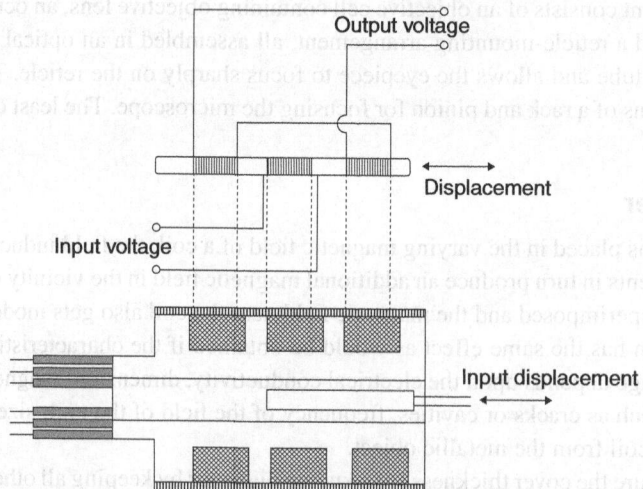

Fig. 27.19 Linear variable differential transformer

27.4.17 Vibrating Wire Strain Gauge

As the name indicates, this instrument is used to measure the strain of a member. Vibrating wire strain gauges are of two types.

Surface-mounted strain gauges are designed for welding them to steel structures. By modifying the end blocks, they can also be used for concrete members. In this type of strain gauge, a length of steel wire is tensioned between two end blocks which are welded to the steel surface being investigated. Any deformation in the structure makes the two end blocks to move, thus changing the tension in the steel wire. Tension is measured by exciting the wire and measuring the resonant frequency of vibration using an electromagnetic coil.

In *embedded strain gauge* a lengthy steel wire is tensioned between two end blocks which are embedded directly into concrete. The principle for these two types of strain gauges is the same. Figure 27.20 shows fixing details for the strain gauges.

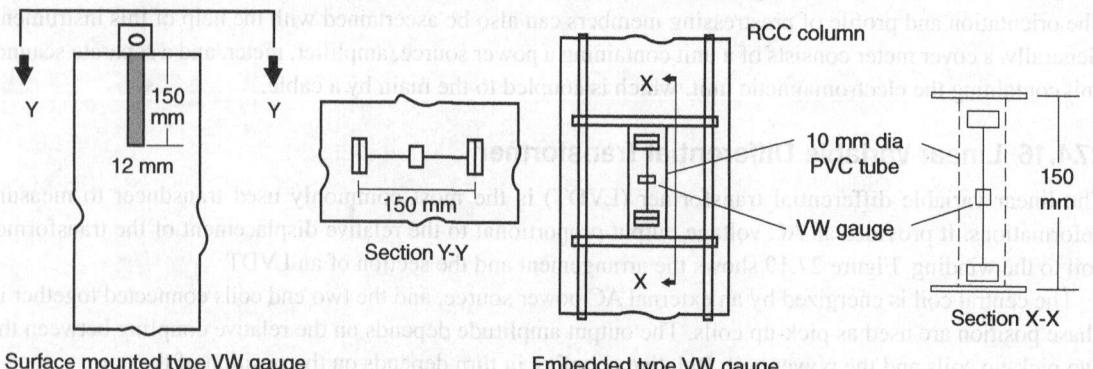

Surface mounted type VW gauge   Embedded type VW gauge

Fig. 27.20 Types of strain gauges

27.4.18 Contact Type Strain Gauge

Contact type strain gauge was developed by Pfender (1947). It is set in position with the help of bolts. One contact point is stationary and the other is movable. It has a dial gauge to show variation in the length. Due to loading, the length of measuring tract changes and the contact point also moves according to the variation in length. Thus, change in length can be measured, from which strain can be calculated.

Specifications

- Measuring length (variable) = 20–100 mm
- Maximum variation of length (L_{max}) = ±0.5 mm
- Minimum measurable extension (L_{min}) = ±0.001 mm

27.4.19 Tiltmeter

A tiltmeter measures tilt (i.e., rotation of the vertical plane) in structures. It has potential application for measuring effects of foundation settlement. The tiltmeter system consists of a tiltplate, a tiltmeter sensor, and an indicator. The tiltplate is bonded to surface of structure. The sensor is oriented on three pegs of the tiltplate and senses change in the tilt of the tiltplate in two orthogonal directions. The sensor provides an electrical signal proportional to the sine angle of inclination of the base plate from a reference plane which can be horizontal or vertical. The portable indicator displays the reading. The technique can be used to cross-check the supplementary deflection measuring system.

Specifications

- Range of inclination = ±53° from the horizontal
- Sensitivity (smallest change in tilt angle) = 1 in 10,000

27.4.20 Inclinometer

An inclinometer (Fig. 27.21) is a high-precision instrument for measuring subsurface displacement or deformation. Lateral movements within landslides, earth dams, and foundations can generally be measured more precisely and economically with an inclinometer than with any other type of instrument. The instrument is normally lowered down a grooved plastic or aluminium inclinometer casing installed in the structure near vertical. The grooves control the orientation of the instrument in a predetermined direction. Inclinometer readings are taken at frequent intervals of depth and are subsequently converted into displacements.

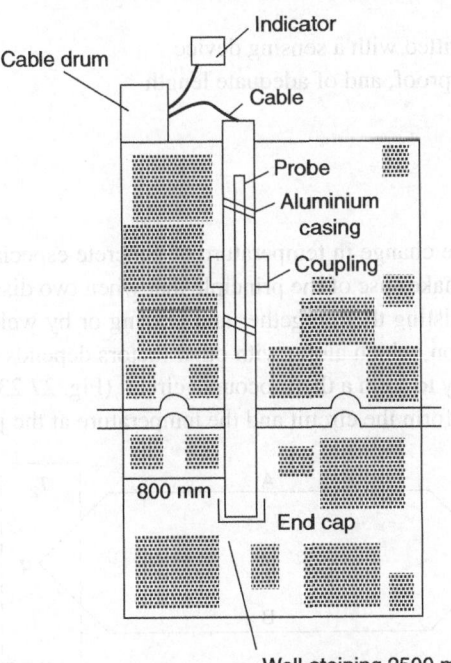

Fig. 27.21 Inclinometer system

Specifications

- Range of inclination = ±53° from the horizontal
- Sensitivity (smallest change in tilt angle) = 1 in 20,000
- Accuracy (error in deflection measurement) = ±7.5 mm

27.4.21 Pressure Transducer

A pressure transducer, usually a piezometer, is attached to a fluid-filled pressure cell. The soil pressure on the flat walls of the cell is converted into fluid pressure and measured by the piezometer. Pressure cells are fixed at the interface between the wall and the soil. Typical applications include the measurement of pressure at soil–concrete surface interface, foundation bearing pressure, and the orientation and magnitude of stress within dams and embankments. Figure 27.22 shows a pressure cell assembly.

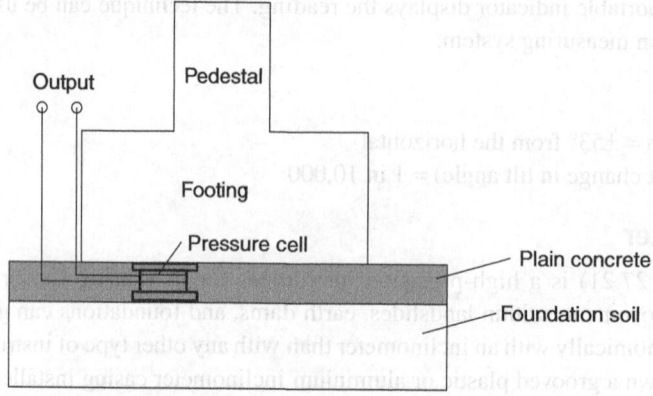

Fig. 27.22 Pressure cell mounting assembly

Specifications

- Pressure cell with a piezometer fitted with a sensing device
- Signal cable factory fitted, waterproof, and of adequate length
- Gauge body of stainless steel
- Compatible readout unit

27.4.22 Thermocouple

A thermocouple is used to determine change in temperature of concrete especially due to heat of hydration when concrete is in a fresh state. It makes use of the principle that when two dissimilar metals are brought in contact with each other either by twisting them together and brazing or by welding, an electromotive force (EMF) is produced across the junction, which along with other factors depends on the junction temperature. At least two conductors are necessary to form a thermocouple circuit (Fig. 27.23). The net EMF is a function of the two relative materials used to form the circuit and the temperature at the junction.

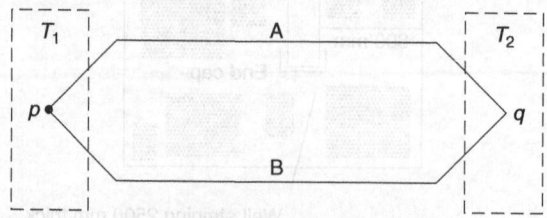

Fig. 27.23 Thermocouple circuit

27.4.23 Vibrating Wire Temperature Gauge

Vibrating wire temperature gauge is used to measure change in temperature on hardened concrete due to environmental effects. It consists of a stainless steel transducer body to which a wire element is attached. Different coefficients of thermal expansion of the body and the wire make it a simple temperature-sensitive device. The gauge is protected by a stainless steel housing, which allows it to survive in the most severe environments. The output signal from the vibrating wire temperature gauge is in the form of frequency. The stability and accuracy of the measurement is unaffected by any change in cable resistance caused by water penetration and temperature variation. These gauges can be used conveniently where other types of vibrating wire transducers are employed as all of them may require the same data logging device.

27.4.24 Open-circuit Potential Measuring Technique

The tendency of a metal to react with the environment is indicated by the potential it develops on contact with that environment. In a reinforced concrete structure, concrete acts as an electrolyte and the reinforcement develops a potential depending on the concrete environment, which may vary from place to place. The principle involved in this technique is essential for measuring the corrosion potential of the rebar with respect to a standard reference electrode. The open-circuit potential (OCP) measurement scheme is shown in Fig. 27.24. The steel rebar in the concrete structure should be accessible at a few locations to facilitate electric connection. The positive terminal of a high-impedance voltmeter is connected to the exposed rebar, and the negative terminal reference electrode is moved along the nodal points, and corresponding potentials are recorded. Note that the sponge in a wet condition at the tip of the calomel electrode (negative terminal) helps in making electrical contact.

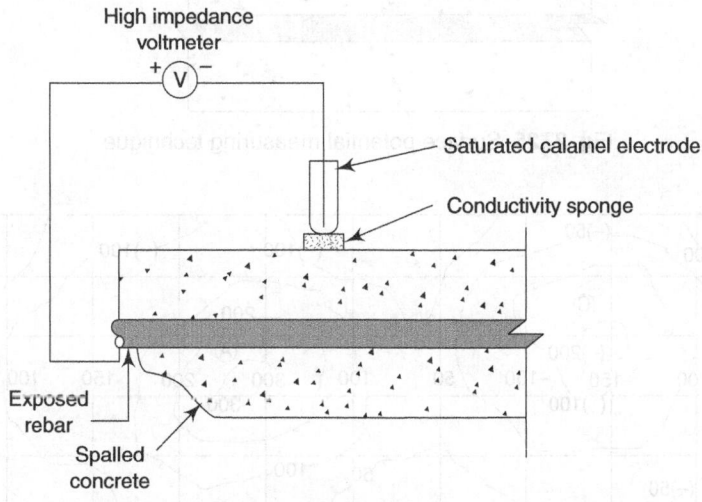

Fig. 27.24 Open-circuit potential measuring technique

Limitations

Some of the limitations of this technique are the following:

1. Monitoring of reinforcement in a submerged zone is not reliable.
2. Potential measurement on a coated rebar will not give real OCP values, and hence it will not reflect the real condition of rebar.
3. Delamination in concrete can also affect the potential measurement.

27.4.25 Surface Potential Measuring Technique

During corrosion an electric current flows between the cathodic and anodic sites through the concrete. This flow can be detected by measuring the potential drop in the concrete. Hence the surface potential measurement is used as a non-destructive technique for identifying anodic and cathodic regions in concrete structures and for indirectly detecting the corrosion state of rebar. The schematic diagram of the technique is shown in Fig. 27.25. No electric connection to the rebar is required in this method.

In this technique one electrode is kept fixed on the structure at a symmetrical point. The other electrode, called the moving electrode, is moved along the structure on the nodal points of the grid and OCP measurements are made. The potential of the movable electrode when placed at nodal points is measured against the fixed electrode with the help of a high-impedance voltmeter. Equipotential contours are plotted as shown in Fig. 27.26. A more positive potential reading represents anodic areas where corrosion is possible. Greater the potential difference between anodic and cathodic areas, greater is the probability of corrosion.

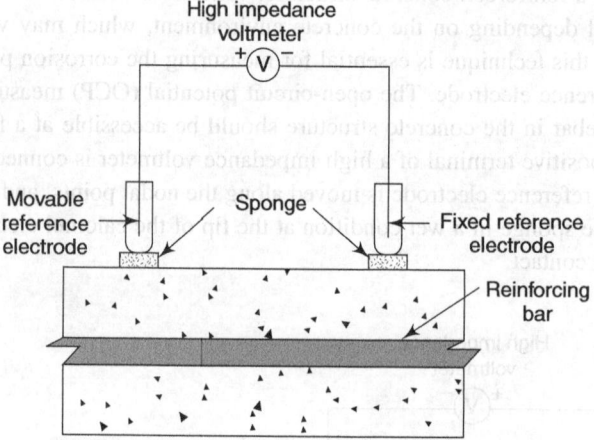

Fig. 27.25 Surface potential measuring technique

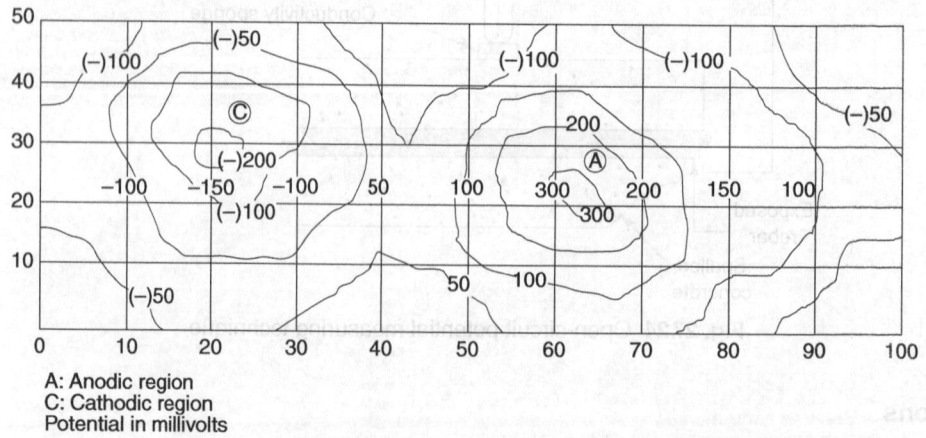

A: Anodic region
C: Cathodic region
Potential in millivolts

Fig. 27.26 Equipotential contours

27.4.26 Polarization Resistance Technique

Among the electrochemical techniques, the best technique for evaluation of instantaneous corrosion rate in the laboratory is the resistance polarization method. There is a linear relationship between potential and applied current at potentials only slightly shifted from the corrosion potential. Based on the kinetics of electrochemical reactions and the concept of the mixed potential theory postulated by Wagner and Traud (1938), an equation has been derived which relates quantitatively the slope of the polarization curve in the vicinity of the corrosion potential to the corrosion current density (i_{corr}) as follows:

$$i_{corr} = \frac{(b_a)(b_c)}{2.303(b_a + b_c)R_p} = \frac{K}{R_p} \qquad \text{where} \quad K = \frac{(b_a)(b_c)}{2.303(b_a + b_c)} \approx 11.2$$

b_a is the anodic slope constant, b_c is the cathodic slope constant, and $R_p = \Delta E/\Delta I$ is the polarization resistance.

A small amount of DC current is applied to the rebar and the corresponding potential (ΔE) is monitored. This is called *polarization voltage* and this can be carried from (−)10 mV to (+)10 mV in the vicinity of open-circuit potential. There are three methods to carry out this test.

Galvano-static method

By applying a small increment of current (ΔI), the change in potential is monitored. For each increment of current a waiting time of 10 min is necessary to obtain the corresponding stabilized (ΔE) value.

Potentio-static method

By applying a small increment of the potential, the change in the current is measured for each increment of potential (ΔE); and the current value (ΔI) is recorded after 30–60 s.

Potentio-dynamic method

By using a potentiostat coupled to a voltage scan generator, polarization can be carried out. Best results can be obtained at the scan rate of 5–10 mV/min.

27.4.27 Electrochemical Noise Analysis

Electrochemical noise analysis is a tool for monitoring corrosion. Information on corrosion mechanism and rate at areas identified in concrete structures can be obtained by using this technique. A low-amplitude variation of the corrosion potential (1–10 mV) of steel in concrete is measured to obtain a noise data as a record of potential fluctuation in the form of power spectra. The noise source is located within the probable corroding area. A time record of sufficient interval is monitored over the frequency range (10 μHz to 1 Hz). Noise data as a record of potential fluctuation is obtained. The noise signal is transformed from time domain to frequency domain, displaying in the form of amplitude and frequency based on either fast Fourier transform or maximum entropy method of spectral analysis. Figure 27.27 shows the block diagram of noise analysis.

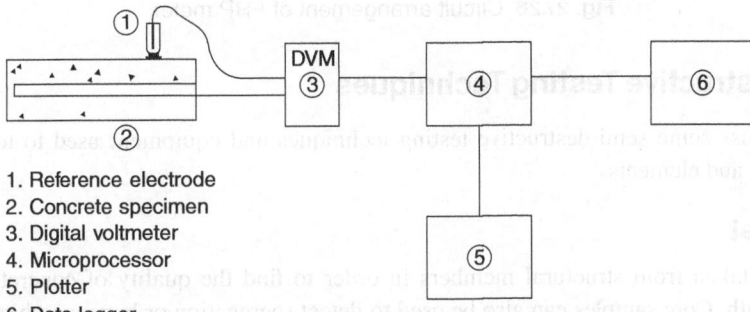

1. Reference electrode
2. Concrete specimen
3. Digital voltmeter
4. Microprocessor
5. Plotter
6. Data logger

Fig. 27.27 Block diagram for noise analysis

Specifications

- Calomel copper–copper sulphate electrode
- Digital voltmeter
- Data logger
- Microprocessor

Limitation

Since fluctuations are in microvolt range, highly sensitive equipment is necessary.

27.4.28 Resistivity of Concrete

Electrical resistivity of concrete is an important parameter which is related to various other aspects, such as strength, porosity, and deterioration. It is well known that the reinforcing steel embedded in concrete is protected by the concrete cover, and that this protection is mainly due to higher alkalinity and the fairly high electrical resistance of concrete. During corrosion, current has to flow from anode to cathode through the electrolyte, and the resistivity of concrete has an influence on the flow of this corrosion current. In the case of reinforced concrete structures, high electrical resistances can impede the flow of such currents. However, resistivity of concrete has been found to vary considerably depending on the moisture content and other soluble salts present in concrete. Hence, depending on the resistivity of concrete, the rate of corrosion process can be inhibited or accelerated.

Figure 27.28 shows the circuit arrangement for measuring resistivity. Four metallic probes are placed over concrete at equal spacing 'a'. On passing a known current I, potential drop V is measured. Resistance R is given by V/I and resistivity can be calculated using the following formula:

$$\text{Resistivity of concrete } (R_c) = 2\pi aR$$

where a is the inter-electrode distance (in cm) and R is the measured resistance (in ohm).

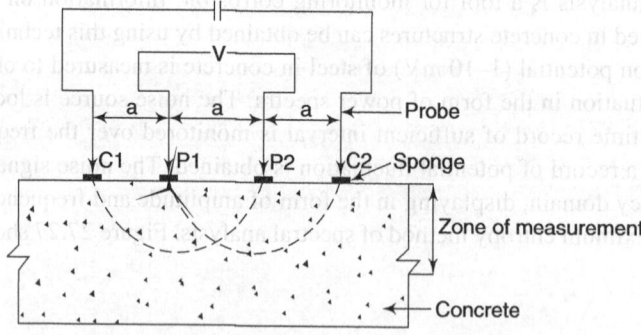

Fig. 27.28 Circuit arrangement of FRP meter

27.5 Semi-destructive Testing Techniques

Now we will discuss some semi-destructive testing techniques and equipment used to test the quality of concrete structures and elements.

27.5.1 Core Test

Core samples are taken from structural members in order to find the quality of concrete in terms of its compressive strength. Core samples can also be used to detect segregation or honeycombing or to check the bond at construction joints. Figure 27.29 shows some typical core samples taken for testing.

Fig. 27.29 Core samples

Specifications

- Maximum size of aggregate = 10–20 mm
- Core diameter = 50–150 mm
- Core diameter (as per ASTM C42-82) = 3 × (nominal size of aggregate)
- Core diameter (as per BS: 1881) = 100–150 mm

Limitation

The test is possible only if the quality of concrete is reasonably good in the structure being tested. If the quality is very poor, core samples themselves get damaged during either the coring operation or testing in the laboratory.

Calibration

1. As the slenderness ratio (*L/d*) decreases below 2, strength increases at an increasing rate; for the ratio between 2 and 3, strength remains sensibly constant; beyond 3, it shows reduction. In general, the slenderness ratio in core testing is limited to between 1 and 2, and the measured strength is expressed as the equivalent strength value for a specified *L/d*.
2. Because of inherent anisotropy of concrete, the strength of cores obtained by drilling in vertical and horizontal directions is different.
3. Strength of concrete varies in the vertical direction due to the water gain effect, and this produces weaker and even weaker concrete as the height increases. It has been observed that the actual strength is usually lower at top by 20%.
4. Concrete in the member represented by the core sample shall be considered acceptable if the average equivalent cube strength of the core is equal to at least 85% of the cube strength of the grade of concrete specified for the corresponding age and no individual core has strength less than 75% of the specified cube strength.
5. The measured compressive strength of the test specimen should be corrected using a suitable correction factor given in Fig. 27.30. The corrected compressive strength represents the equivalent strength of the cylinder having slenderness (height/diameter) ratio of 2. The equivalent cube strength of concrete shall be determined by multiplying the corrected cylinder strength by 5/4.

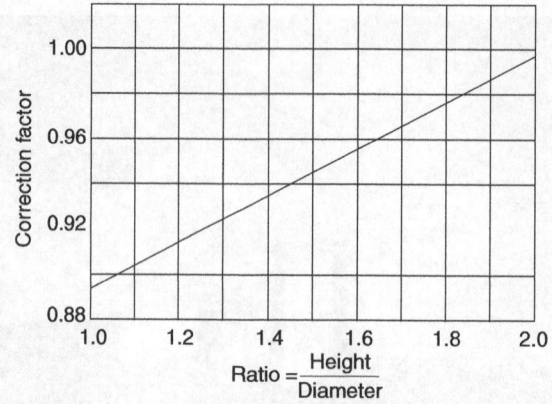

Fig. 27.30 Correction factor

27.5.2 LOK Test

LOK test is a patented standard pull-off method of assessing the strength of concrete. A cast-in unit is locked inside the concrete when the concrete is still fresh.

This test was developed at the Danish Technical University in the late 1960s. It consists in measuring the force required to pull out a previously cast-in steel rod with an embedded enlarged end from concrete (Fig. 27.31). This force is taken as a measure of the compressive strength of concrete. This test can also be used to determine the stressing time in post-tensioned construction and for in situ strength monitoring. The Construction Industry and Research Information Association (CIRIA) of Great Britain recommends the LOK test as a suitable method for assessing formwork stripping time.

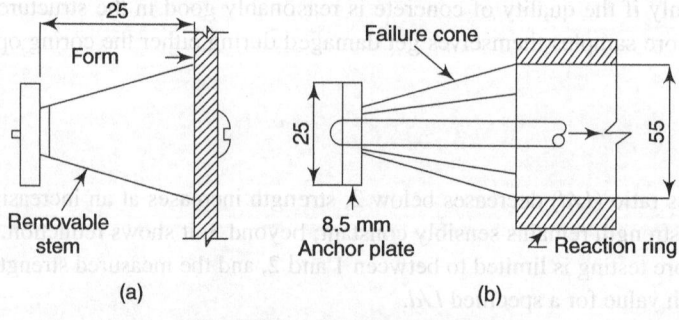

Fig. 27.31 LOK test assembly

Specifications

- Diameter of anchor plate = 25 mm
- Thickness of anchor plate = 8.5 mm
- Inner diameter of reaction ring = 55 mm
- Length of removable stem = 25 mm
- Rate of loading = 30 ± 10 kN/min

Calibration

Calibration charts for LOK testing are usually provided by the manufacturers. The reliability of this method is reported to be good with correlation coefficients for laboratory calibration of about 0.96 on straight line relationship and a corresponding coefficient of variation of about 7%. Strength prediction is more dependable when compared with other non-destructive or partial destructive methods.

27.5.3 CAPO Test

This test is based on the same principle as the LOK test, but it allows hardened concrete to be tested at random without having a pull-out disc embedded in fresh concrete before hand. The pulling-out unit almost looks like an inverted cap. Probably because of this reason it is named as 'CAPO' test. This test was developed in Denmark in the 1970s. The procedure involves drilling a hole in hardened concrete, cutting a groove at a specified depth of the hole by milling operation, and then expanding the CAPO insert in a similar way as the LOK test. The test results directly depend on the compressive strength. Figure 27.32 shows the internal state of cracking in the CAPO test.

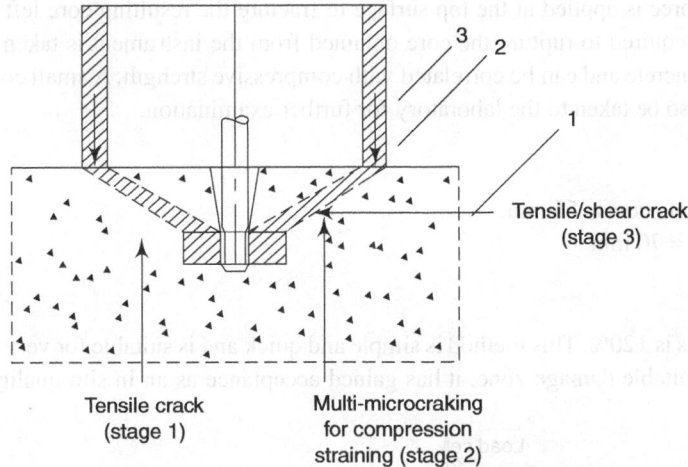

Fig. 27.32 Internal cracking stage in CAPO test

27.5.4 North American Pull-out Method

This test is also similar to the LOK test. The test assembly (Fig. 27.33) bears some simple modifications. The depth of penetration is greater in this method and the apex angle of failure is set at 64°.

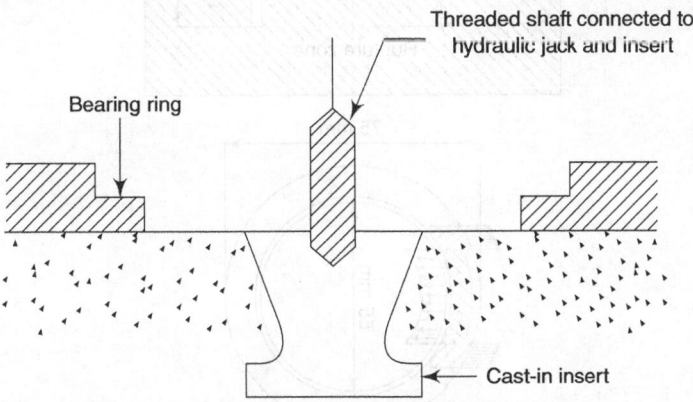

Fig. 27.33 North American pull-out method assembly

Calibration

The pull-out strength is calculated as follows:

$$f_P = F/A$$

where F is the force on the arm, A is the failure surface area ($A = (d_3 + d_2)(h/4) [4h^2 + (d_3 - d_2)^2]^{1/2}$), d_2 is the diameter of pull-out insert head, d_3 is the inside diameter of the reaction ring, and h is the distance from the insert head to the surface. Pull-out strength is approximately equal to 0.21 cube compressive strength.

27.5.5 Break-off Test

This test was developed as a method for in situ determination of the flexural strength of concrete by Norwegian Technical University and A/S NORCEM. A tubular disposable form is inserted into the fresh concrete. Alternatively, a shaped hole can be drilled to form a slot of the type shown in Fig. 27.34. Using a hydraulic testing device, a transverse force is applied at the top surface to fracture the resulting core left after the removal of the insert. The force required to rupture the core obtained from the instrument is taken as a measure of the flexural strength of concrete and can be correlated with compressive strength. A small concrete core obtained during the test may also be taken to the laboratory for further examination.

Specifications

- Diameter of concrete core = 55 mm
- Depth of concrete = 70 mm

Calibration

The accuracy of results is ±20%. This method is simple and quick and is suitable for very young concrete and, although it leaves a suitable damage zone, it has gained acceptance as an in situ quality control test where

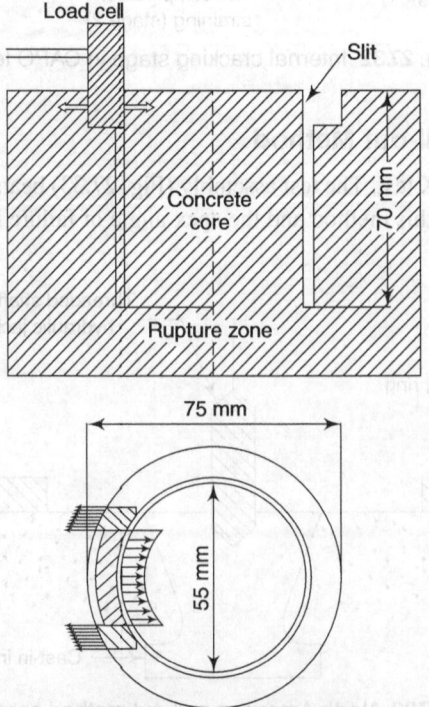

Fig. 27.34 Break-off test

tensile strength is important. Although quicker than compressive testing of cores, the use of results for the strength estimation of old concrete may be unreliable unless a specific calibration relation is available.

27.5.6 Pull-off Test

This method actually measures the nominal tensile strength of concrete, which is correlated with the compressive strength of concrete. A circular steel probe is bonded to the surface of concrete by means of an epoxy resin adhesive (Fig. 27.35). Before this operation, the surface is roughened with the help of sandpaper to remove laitance and then degreased with the help of a suitable solvent. After sufficient time has elapsed for the epoxy resin adhesive to cure, a slowly increasing tensile force is applied to the probe. And as the tensile strength of the bond is greater than that of concrete, the latter eventually fails in tension. The amount of over-break is usually small so that the area of failure can be taken as being equal to that of the probe. From this area and the force applied at failure, it is possible to calculate a nominal tensile strength of the concrete specimen.

Specifications

- Probe diameter = 50 mm (approx.)
- Minimum capacity of hydraulic apparatus = 10 kN
- Coefficient of variation between individual pull-off results = 8–20%
- Depth of partial core = 10 mm

Calibration

An inspection of the probe, after the test, reveals whether or not the concrete has failed. Results from unsatisfactory failures can, therefore, be discounted.

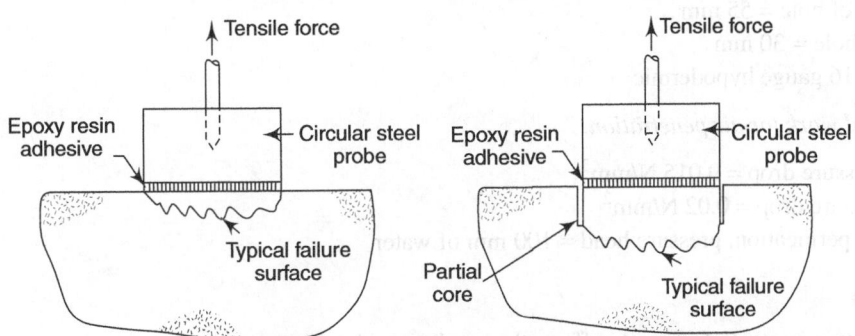

Fig. 27.35 Pull-off test: (a) arrangement for testing uncored specimens,
(b) arrangement for testing partially cored specimens

27.5.7 Figg's Air and Water Permeability Test

This test was developed by Building Research Establishment, UK, in the early 1970s. Later it was modified by Cather, Figg, Marsden, and O'Brien and is called as Figg's method. This method is an alternative to the ISA method for quality control checking.

In this method a hole is drilled into the concrete surface. After thorough cleaning, the hole is plugged from the outside surface by polyether foam and then sealed with a catalysed silicon rubber. After hardening of rubber, a hypodermic needle is pushed through the silicon plug (Fig. 27.36). For air permeation, a mercury-filled manometer and a hand pump is attached to the needle. The pressure within the system is first reduced to a standard value, and then, after isolating the pump, the time for the pressure to rise to a standard value is recorded. This unit time is a measure of the air permeation index of concrete.

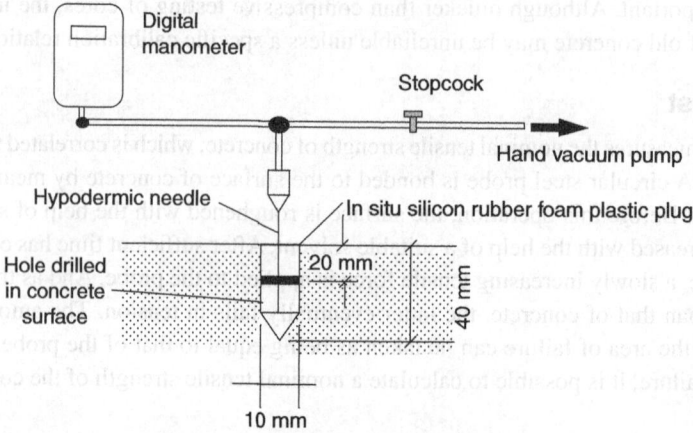

Fig. 27.36 Figg's air-permeability test

For water absorption, a water head of 100 mm is used. Water is forced into the assembly with a syringe for 1 min after first contact between water and concrete. The time for the meniscus in the capillary tube to travel 150 mm is recorded and taken as a measure of the water absorption index of concrete.

In the case of air permeability measurement, the higher the index, the less permeable the concrete. The advantage of using air rather than water is that readings may be repeated on the same specimen without delay, since the passage of air through it does not change the specimen in any way.

Specifications

- Diameter of hole = 55 mm
- Depth of hole = 30 mm
- Needle = 16 gauge hypodermic

In the original work for air penetration:

- Initial pressure drop = 0.015 N/mm^2
- Final pressure drop = 0.02 N/mm^2
- For water permeation, pressure head = 100 mm of water

Limitations

1. The moisture content of concrete affects the result considerably.
2. Tests on various specimens taken from the same batch and from different batches cast using the same mix proportion produced results with large variation.

27.5.8 Compression Test

The compression test is used to determine the strength of cubical and cylindrical specimens of concrete. The strength of a concrete specimen depends upon cement, aggregate, bond, water–cement ratio, curing temperature, and age and size of specimen. Mix design is the major factor controlling the strength of concrete.

Cubes of size 15 cm × 15 cm × 15 cm (as per IS: 10086-1982) should be cast. The specimen should be given sufficient time for hardening (approx. 24 h) and then it should be cured for 28 days. After 28 days, it should be loaded in the compression testing machine and tested for maximum load. Compressive strength should be calculated by dividing maximum load by the cross-sectional area. The behaviour of the specimen under load has been discussed in Chapter 7.

27.5.9 Direct Tensile Stress Test

Direct tensile stress testing of concrete is the most basic test for determining the tensile strength of concrete. The testing machine should conform to the applicable requirements of ASTM-C 39 (CRD-C 14) and ASTM-E 4 (CRD-C 512).

In the testing apparatus, cylindrical metal caps cemented to the specimen ends provide the means through which direct tensile load can be applied. The diameter of the metal cap should not be less than that of the test specimen, nor should it exceed the test specimen diameter by more than 1.6 mm. Caps should have a thickness of at least one-third the diameter of the specimen. Caps should be provided with a suitable linkage (robust chains above and below) system for load transfer from the testing machine to the test specimen. The linkage system should be so designed that the load is transmitted through the axis of the test specimen without the application of bending or torsional stress. The length of linkages at each end should be at least two times the diameter of the caps. One such system is shown in Fig. 27.37. Roller or link chains of suitable capacity have been found to perform quite well in this application. Because a roller chain flexes in one plane only, the upper and lower segments should be positioned at right angles to each other to effectively reduce bending in the specimen. Ball-and-socket, cable, or similar arrangements have been found to be generally unsuitable because their tendency to bend and twist makes the assembly unable to transmit a pure direct tensile stress to the test specimen.

Figure 27.37 also shows the specimen used for testing. The specimen should be attached to the testing machine. High-strength, rapid-setting epoxy should be used to glue specimens to end metal caps. Tensile load should be applied continuously at an approximately constant rate and without shock to failure. Loads at various intervals should be taken. The tensile strength of the specimen can be calculated by dividing the maximum load withstood by the specimen by its cross-sectional area.

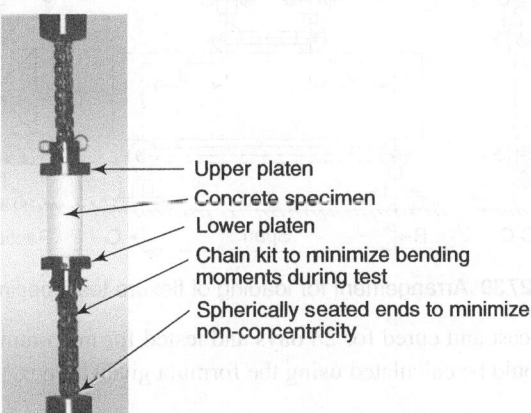

Upper platen
Concrete specimen
Lower platen
Chain kit to minimize bending moments during test
Spherically seated ends to minimize non-concentricity

Fig. 27.37 Concrete direct tension

27.5.10 Indirect Tensile Test

A cylindrical specimen is used in this test. This test is also known as split tensile test. Figure 27.38 shows the test arrangement. The specimen is loaded until failure occurs and failure load is noted. Tension can be calculated using the following formula:

$$T = 2p/\pi Ld$$

where p is the concentrated load, L is the length of cylinder, and d is the diameter of the cylinder.

Fig. 27.38 Split cylinder test

27.5.11 Flexure Test

A beam specimen can be cast for determining the flexural strength of concrete. The standard specimen sizes are 15 cm × 15 cm × 70 cm and 10 cm × 10 cm × 50 cm. A Universal Testing Machine (UTM) can be used for the test. The testing machine may be set to any reliable type of sufficient capacity for the test. Permissible errors should not be greater than ±0.5%. The bed of machine should be provided with two steel rollers, of 38 mm diameter, on which the specimen is supported. Rollers are placed at a centre-to-centre distance of 60 cm for the 15 cm specimen and at 40 cm for the 10 cm specimen. The suitable arrangement is shown in Fig. 27.39.

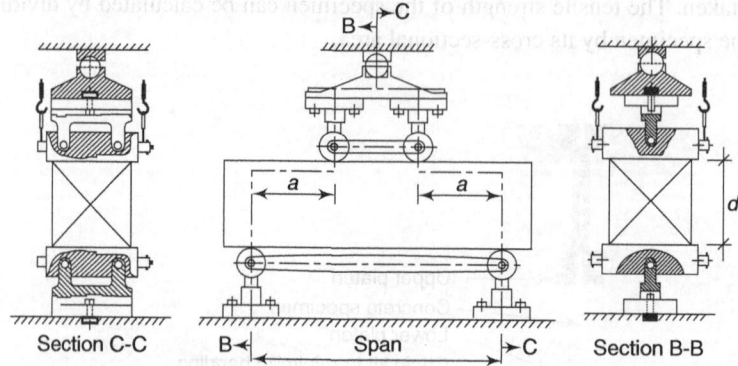

Section C-C   B   Span   C   Section B-B

Fig. 27.39 Arrangement for loading of flexure test specimen

The test specimen should be cast and cured for 28 days and tested for maximum load. The flexural strength or modulus of rupture (f_b) should be calculated using the formula given below:

$$f_b = (3pa) / (bd^2)$$

where b is the measured width (in cm) of the specimen, a is the distance as shown in Fig. 27.39, d is the measured depth (in cm) of the specimen at the point of failure, l is the length (in cm) of the span on which the specimen is supported, and p is the maximum total load (in kg) applied to the specimen. Modulus of rupture is about 60–100% higher than the direct tensile strength.

27.6 Tests on Fresh Concrete

Various tests can be carried out on fresh concrete. All these tests are used to determine the workability of concrete. Some important tests carried out on fresh concrete are slump cone test, Kelly ball test, and compaction factor test.

27.6.1 Slump Cone Test

This test is used to determine the workability of concrete. The apparatus is a cone of 10 cm top diameter, 20 cm bottom diameter, and 30 cm height (Fig. 27.40).

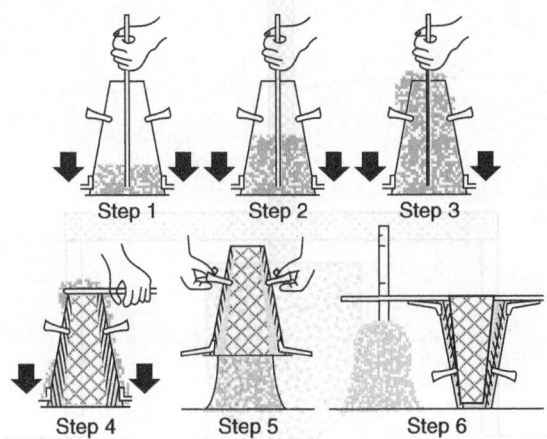

Fig. 27.40 Slump cone test

It has two handles for lifting purposes. Initially the cone is cleaned and oil is applied on the inner surface. Then the concrete to be tested is placed into the cone in three layers. Each layer is compacted 20 times by a standard tamping rod. After filling the cone, it is lifted slowly and carefully in the vertical direction. Concrete is allowed to subside and this subsidence is called *slump*.

Figure 27.41 represents the slump patterns. If the slump is even, then it is termed as *true slump*. If one half of cone slides, it is called *shear*. If entire concrete slides, it is called *collapse*. Shear slump indicates that concrete is non-cohesive and shows a tendency for segregation. Generally, the slump value is measured as the difference between the height of the mould and the average height after subsidence. Slump test is a simple test and is widely used.

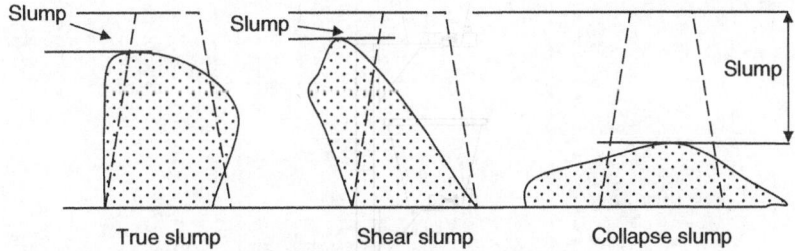

Fig. 27.41 Different patterns of slump

27.6.2 Kelly Ball Test

Kelly ball test is a simple test and can be conducted at the site itself. This test was introduced by Kelly and Polivka (1955). A Kelly ball (shown in Fig. 27.42) is freely placed on fresh concrete surface and the depth of its penetration, which indicates workability, is noted. This test is very simple and takes very less time.

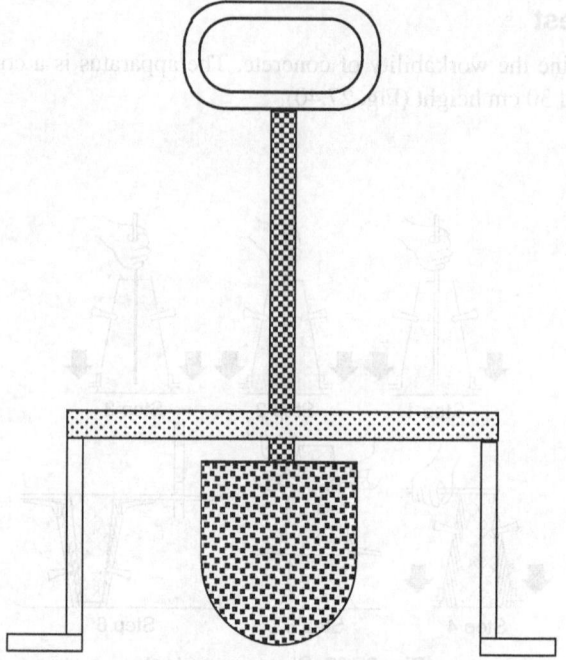

Fig. 27.42 Kelly ball apparatus

27.6.3 Compaction Factor Test

Compaction factor test was developed by Road Research Laboratory (UK) and is one of the most efficient tests for measuring workability. A standard amount of concrete is allowed to fall from a standard height and the compaction factor is measured. Figure 27.43 shows the apparatus. It has two hoppers and a cylinder. The first hopper has a top diameter of 25.4 cm, bottom diameter of 12.7 cm, and an internal height of 27.9 cm. The second hopper has a top diameter of 22.9 cm, bottom diameter of 12.7 cm, and an internal height of 22.9 cm. The cylinder has an internal diameter of 15.2 cm and height of 30.5 cm. The distance between the two hoppers and between the lower hopper and the cylinder is 20 cm.

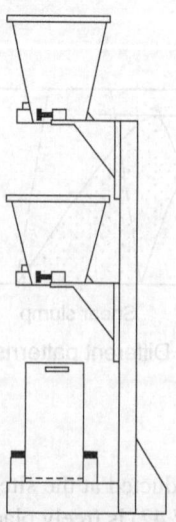

Fig. 27.43 Compaction factor apparatus

The concrete sample to be tested is first placed in the top hopper. The trap door is opened and concrete is allowed to fall to the lower hopper. Similarly, from the lower hopper, concrete is allowed to fall into the cylinder below. Now the cylinder is weighed and the weight is called as *weight of partially compacted concrete*. Now the cylinder is emptied and refilled with concrete from the same sample. It is rammed or vibrated heavily and its weight is taken. This weight is known as *weight of fully compacted concrete*. These values are used to calculate the compaction factor by using the following relation:

$$\text{Compaction factor} = \frac{\text{weight of partially compacted concrete}}{\text{weight of fully compacted concrete}}$$

27.7 Load Test on Structural Components

Load tests on components of a structure should be carried out as soon as possible after expiry of 28 days from the time of placing concrete. The component should be subjected to a load equal to its full dead load plus 1.25 times the imposed load for a period of 24 h and then the imposed load should be removed. Any deflection due to the imposed load should be recorded. If within 24 h of the removal of the imposed load the structure does not recover at least 75% of deflection under superimposed load, the test should be repeated after 72 h. If recovery is less than 80%, the component should not be used in the construction and the test is declared to be a 'failure'.

Exercises

Review Questions

1. Explain the need for evaluation of a structure under use.
2. Explain in detail what you understand by 'investigation plan'.
3. What is resistance factor?
4. Differentiate between cracking, spalling, and staining.
5. What is the difference between non-destructive and semi-destructive testing methods?
6. How will you classify the grade of concrete in UPV testing?
7. What is carbonation? Explain.
8. How will you determine the hydration of hardened concrete?
9. What test will you use to determine the chloride content in concrete?
10. How will you measure the cover thickness of concrete?
11. Explain polarization resistance technique.
12. Differentiate between pull-out and pull-off tests.
13. Explain Figg's test.
14. What tests are used to determine the workability of fresh concrete? Explain any one of them.
15. Describe how to conduct a load test on a component of a structure.

Multiple Choice Questions

1. One of the limitations of rebound hammer test is that
   a. It damages the structure being tested
   b. The accuracy is only ±25%
   c. It needs power supply to operate
   d. It is not suitable for all types of concrete

2. A pulse velocity of >4.5 km/s in UPV test indicates that the tested concrete is
   a. Excellent
   b. Good
   c. Fair
   d. Poor

3. Linear variable differential transformer is used to measure
   a. Force
   b. Displacement
   c. Stress
   d. Strain

4. In structures, tiltmeter is used to measure
   a. Lateral displacement
   b. Rotation of a vertical plane
   c. Strain
   d. Moment acting on a vertical plane

5. Thermocouple is an instrument used to record
   a. Permeability
   b. Young's modulus
   c. Poisson's ratio
   d. Heat of hydration

6. Surface potential measuring technique is used to assess
   a. Permeability
   b. Stiffness
   c. Thermal conductivity
   d. Corrosion susceptibility

7. Extraction of core to test the strength of concrete is classified as
   a. Destructive test
   b. Non-destructive test
   c. Semi-destructive test
   d. Field/lab test

8. LOK test on concrete sample is done to assess
   a. Permeability
   b. Strength
   c. Corrosion potential
   d. Modulus of elasticity

9. Kelly ball test on concrete is used to assess
   a. Workability
   b. Strength
   c. Corrosion potential
   d. Thermal conductivity

10. In a slump test, the shear slump of concrete indicates
    a. Flowing concrete
    b. Concrete liable to segregate
    c. Concrete with insufficient workability
    d. Good workable concrete

Answers to Multiple Choice Questions

1. b
2. a
3. b

4. b
5. d
6. d

7. c
8. b
9. a

10. b

CHAPTER 28

Special Materials in Construction

In this chapter we will study a few important materials used in construction. Normally, concrete is weak under tension. By distributing steel in a thin member, the element can be made flexible and ductile. The members are capable of carrying loads more due to their form than due to their strength. The forces developed in these members are known as *membrane forces*. These membranes are termed *ferro-cement*. This chapter will describe the various aspects of this modern material. One of the serious problems encountered in concreting is to make fresh concrete flow around reinforcements. Many problems arise because of poor compaction. Using a suitable technology, fresh concrete can be made to flow without any external effort. Such a flowing concrete which compacts itself due to its own flowability is known as *self-compacting concrete*.

A special type of concrete is used in pavements where one needs to get the rainwater to permeate and recharge groundwater below to ensure rainwater harvesting. This material is termed *pervious or porous concrete*. Another special type of concrete known as *reactive powder concrete* which has very high strength (>100 MPa) and ductility is used for structural applications requiring ultrahigh strength and ductility. *Geopolymer concrete* is an alternative to conventional concrete which utilizes alkaline liquid to react with silica or fly ash to produce the binder. It is to be noted that the intention is to eliminate cement and thus achieve less carbon footprint to save the environment in the long run.

In this chapter we will deal with details of these special types of concrete.

28.1 Ferro-cement

In the early 1940s, Pier Luigi Nervi, a noted Italian engineer, revived the concept of 'ferro-cement' by using it for building fishing boats. He pointed out that the distribution of reinforcing meshes in concrete produces a material with approximately homogeneous mechanical properties, capable of resisting high impacts. Following some preliminary tests on slabs, he showed that ferro-cement possesses exceptional elasticity, flexibility, strength, and resistance to cracking, and in the year 1943 the Italian Navy accepted ferro-cement as a vessel-building material. Ferro-cement gained wide acceptance in the 1960s for boat building in the United Kingdom, New Zealand, Canada, and Australia. In 1968, the Fisheries Department of the Food and Agricultural Organization (FAO) of the United Nations started ferro-cement boat-building projects in Asia, Africa, and Latin America and later in other countries. Today ferro-cement is widely accepted and utilized.

The definition of ferro-cement was first published by ACI Committee 549 in the year 1980: *Ferrocement is a type of thin wall reinforced concrete commonly constructed of hydraulic cement mortar reinforced with closely spaced layers of continuous and relatively small size wire mesh. The mesh may be made of metallic or other suitable materials.*

Based on experience and advancement in ferro-cement, Antoine E. Naaman extended the definition proposed by ACI Committee 549.

Ferrocement is a type of thin wall reinforced concrete commonly constructed of hydraulic cement mortar reinforced with closely spaced layers of continuous and relatively small size wire mesh. The mesh may be made of metallic or other suitable materials. The fineness of the mortar matrix and its composite should be compatible with the mesh and armature systems it is meant to encapsulate. The matrix may contain discontinuous fibres.

The last two sentences were added to ascertain the compatibility of the matrix with the reinforcement in order to build a sound composite material and to accommodate the use of fibres or microfibres to improve performance in hybrid composites when desirable.

28.1.1 Applications of Ferro-cement

Marine applications

Ferro-cement is used for constructing boats, fishing vessels, barrages, docks, floating buoys, and water or fuel tanks. Using ferro-cement in marine structures imparts desirable properties such as water-tightness, impact resistance, small thickness, and light weightness

Water supply and sanitation applications

Water tanks, sedimentation tanks, well casings, septic tanks, sanitary tanks, and linings of swimming pools

Agricultural applications

Grain storage bins, silos, water tanks, pipes, linings for underground pits, and irrigation channels

Housing applications

Shelters, sheds, domed structures, precast housing elements, wall panels, sandwich panels, corrugated roofing sheets, hollow-core slabs, permanent formwork, water tanks, and repair and rehabilitation of existing houses

Rural energy applications

Biogas digesters, biogas holders, incinerators, and panels for solar energy collectors

Permanent formwork

For reinforced or prestressed concrete columns, beams, slabs, etc.

Other structures

Bus shelters, sheds, pedestrian walkways, service core units for housing, slides for playgrounds, and industrial shelters

28.1.2 Materials used in Ferro-cement

Ferro-cement is basically made of a cement matrix and steel mesh reinforcement. Various materials are used for making ferro-cement. An overview of ferro-cement composition, the reinforcing materials, and their parameters and properties are given in Fig. 28.1.

Cement mortar mix

A cement mix is designed using standard methods only. Its components are Portland cement, fine aggregates, water, and admixtures (if required). Material should satisfy all requisite standards similar to reinforced concrete. Additives such as superplasticizers, silica fumes, and fly ash can also be used.

Skeletal steel

To form the skeleton of the structures, steel is often used in ferro-cement in the form of welded wires or a simple grid of steel wires, rods, or strands. Mesh layers are attached around this skeleton steel. The steel also acts as a spacer, leading to savings in the mesh layer. It helps in resisting tensile and punching shear. The properties of skeletal reinforcement are similar to those of reinforcement bars. Figure 28.2 shows skeletal steel details, and Fig. 28.1 gives information about skeletal reinforcement.

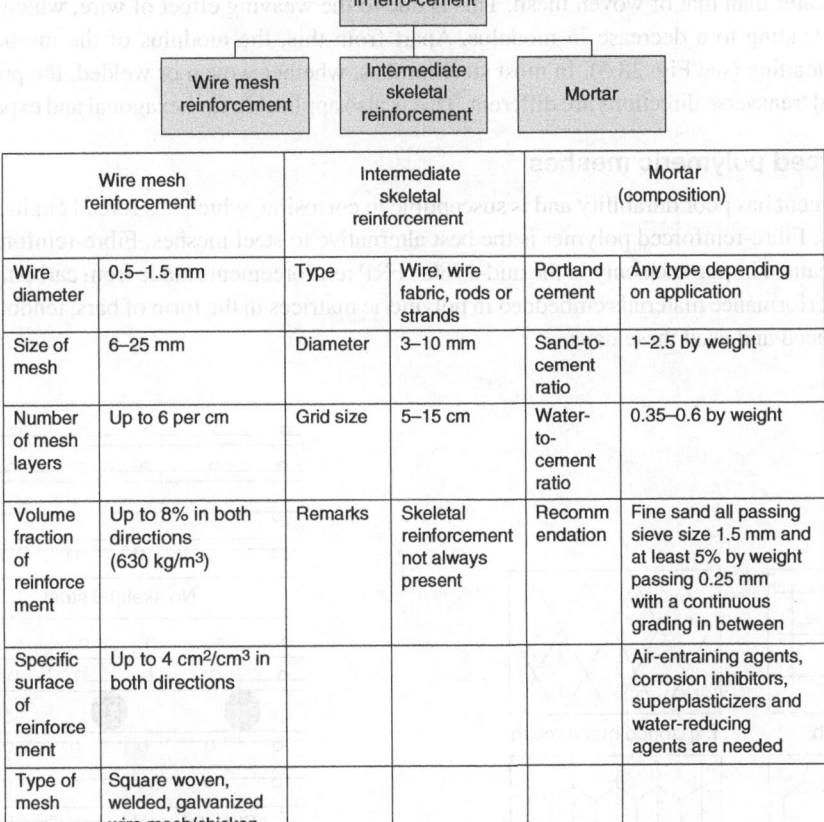

Wire mesh reinforcement		Intermediate skeletal reinforcement		Mortar (composition)	
Wire diameter	0.5–1.5 mm	Type	Wire, wire fabric, rods or strands	Portland cement	Any type depending on application
Size of mesh	6–25 mm	Diameter	3–10 mm	Sand-to-cement ratio	1–2.5 by weight
Number of mesh layers	Up to 6 per cm	Grid size	5–15 cm	Water-to-cement ratio	0.35–0.6 by weight
Volume fraction of reinforcement	Up to 8% in both directions (630 kg/m^3)	Remarks	Skeletal reinforcement not always present	Recommendation	Fine sand all passing sieve size 1.5 mm and at least 5% by weight passing 0.25 mm with a continuous grading in between
Specific surface of reinforcement	Up to 4 cm^2/cm^3 in both directions				Air-entraining agents, corrosion inhibitors, superplasticizers and water-reducing agents are needed
Type of mesh	Square woven, welded, galvanized wire mesh/chicken wire mesh or expanded wire mesh				

Fig. 28.1 Parameters and properties of materials used in ferro-cement

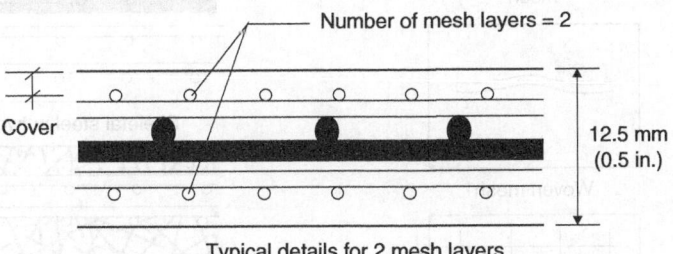

Typical details for 2 mesh layers

Number of mesh layers	Total thickness (mm)	Cover (mm)
2	12.5	2
4	12.5	2
5	12.5	1.5
6	12.5	1

Fig. 28.2 Skeletal steel details

Steel mesh reinforcement

Steel meshes are the primary reinforcement for ferro-cement. The meshes can be square woven or welded, or chicken wire meshes of hexagonal shape and sheet similar to those used in plaster and stucco applications. Figure 28.3 shows steel meshes used in ferro-cement, and Fig. 28.4 shows their cross-sectional details. The properties of wire meshes and reinforcing bars are similar to each other. Moreover, the elastic modulus of steel wire is greater than that of woven mesh. This is due to the weaving effect of wire, which under tension straightens up leading to a decrease in modulus. Apart from this, the modulus of the mesh changes with loading and unloading (see Fig. 28.5). In most steel meshes, whether woven or welded, the properties in the longitudinal and transverse directions are different. This is also applicable for hexagonal and expanded meshes.

Fibre-reinforced polymeric meshes

Steel reinforcement has poor durability and is susceptible to corrosion, which has forced engineers to think of new techniques. Fibre-reinforced polymer is the best alternative to steel meshes. Fibre-reinforced polymeric (FRP) meshes gained importance only in the mid-1980s. FRP reinforcements made from carbon, glass, aramid, or other high-performance materials embedded in polymeric matrices in the form of bars, tendons, and strands are being produced and used these days.

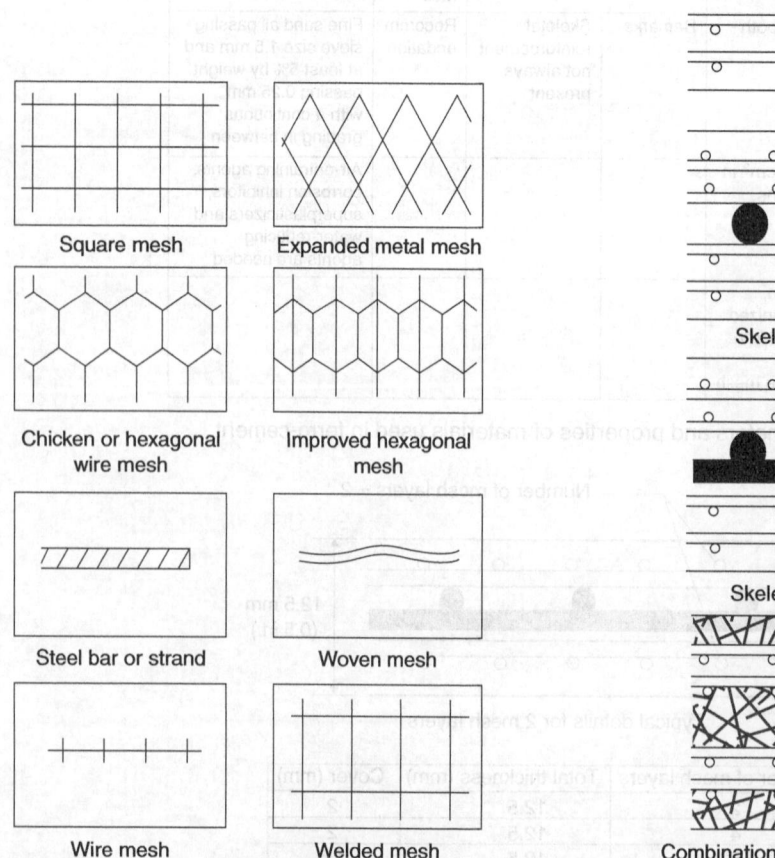

Square mesh

Expanded metal mesh

Chicken or hexagonal wire mesh

Improved hexagonal mesh

Steel bar or strand

Woven mesh

Wire mesh

Welded mesh

Fig. 28.3 Steel meshes used in ferro-cement

No skeletal steel

Skeletal steel in one direction

Skeletal steel in two directions

Combination of mesh and discontinuous fibres

Fig. 28.4 Cross-sectional details of steel meshes

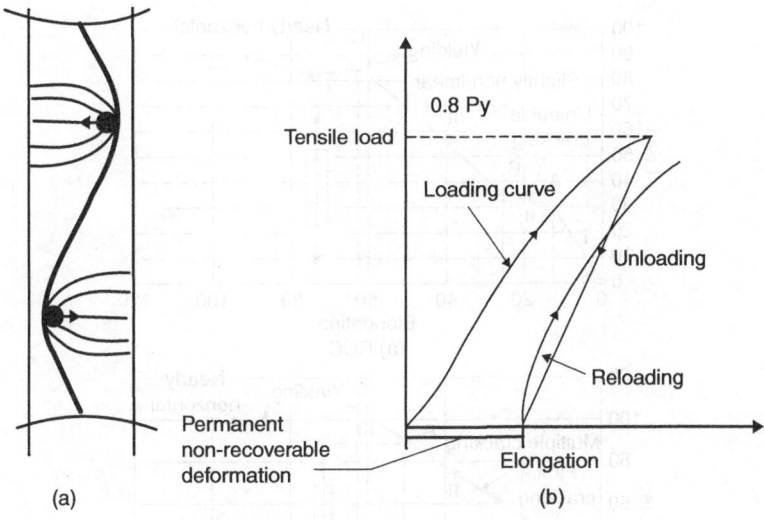

Fig. 28.5 Load-deformation curve

Advantages of FRP The following are the advantages of FRP:

- Good resistance to corrosion
- High strength
- Lower unit weight
- Easy to handle
- Good damping and fatigue behaviour
- Convenient to use for repairing structures

Disadvantages of FRP The following are the disadvantages of FRP:

- High cost
- Low shear strength
- Low ductility
- Susceptibility to rupture especially under compression due to buckling/delamination

28.1.3 Behaviour of Ferro-cement in Tension

The typical load elongation curve for reinforced concrete prism and ferro-cement prism is shown in Figs 28.6(a) and (b), respectively. The behaviour is mainly divided into three main stages. Stage I corresponds to the ascending linear elastic portion of the curve (*OA*). Stage II corresponds to the unstable portion (*AB*) where cracking starts and stabilizes. Stage III is where load elongation is almost linear elastic and the crack width increases with an increase in applied load because of cracking. This stage exists until the reinforcing steel yields. There will be only a few wide cracks across the steel.

Upon comparing these two figures, it can be seen that the main difference lies in unstable stage II (*AB*). The behaviour of ferro-cement can be seen clearly: it slowly adapts to the increasing load by increasing its extensibility. Many fine cracks form. When cracks form, the increase in crack width is small compared to reinforced concrete. Crack width in ferro-cement can be one to two orders of magnitude smaller than that of reinforced concrete. However, there are a number of fine cracks.

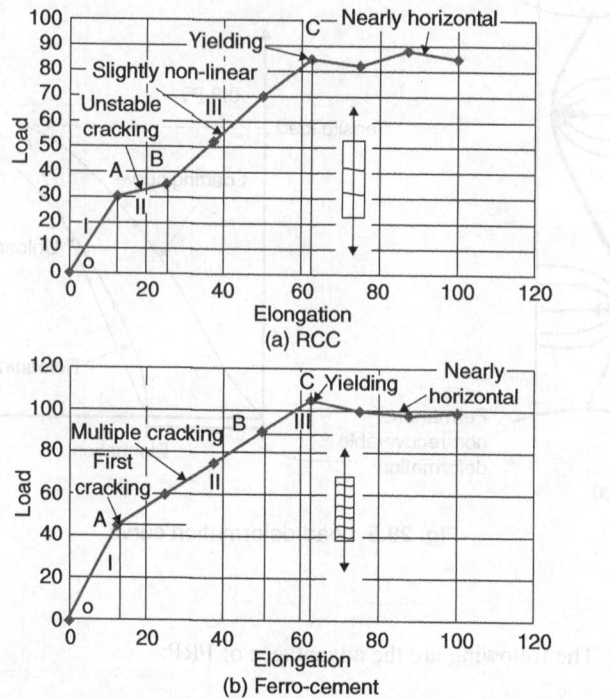

Fig. 28.6 Behaviour of (a) RCC compared with that of (b) ferro-cement

28.1.4 Advantages of Ferro-cement

The following are the advantages of ferro-cement:

• Favourable tensile property
• High ductility
• High resistance to cracking width and crack opening
• Ability to undergo large deflection
• Improved impact resistance and toughness
• Good fire resistance
• Good impermeability
• Low strength-to-weight ratio
• Low maintenance cost, owing to smaller width of cracks
• Environmental friendly, because of lower width of cracks

28.1.5 Differences between Ferro-cement and Reinforced Cement Concrete

Physical properties

The physical properties of ferro-cement and reinforced concrete members have been compared using a beam section shown in Fig. 28.7. Note the following in ferro-cement:

• reinforcement distribution is uniform throughout
• reinforcement is provided in both directions
• thinner section compared to RCC

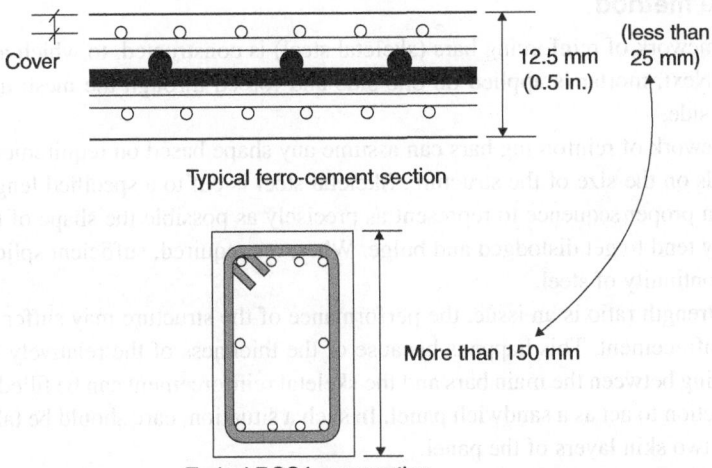

Typical ferro-cement section

More than 150 mm

Typical RCC beam section

Note: Beam has both flexural and shear reinforcement

Fig. 28.7 Comparison of ferro-cement section with RCC

Mechanical properties

The mechanical properties of ferro-cement vary from those of reinforced concrete as follows:

- Ferro-cement can have homogeneous, isotropic properties in two directions. Because of two-way action, there is a high level of redundancy in behaviour.
- Ferro-cement generally has high tensile strength and a high modulus of rupture. Its tensile strength can be of the same order as its compressive strength.
- Ferro-cement generally has a high reinforcement ratio in both tension and compression and in both directions.
- Ferro-cement has a large specific surface of reinforcement, which is one to two orders of magnitude that of reinforced concrete.
- Primarily due to the large specific surface of reinforcement, which is due to the presence of transverse reinforcement, the cracking and multiple cracking processes in ferro-cement under tension differ from those in reinforced concrete, which exhibit large, wide cracks.
- Extensibility of ferro-cement, that is, its elongation up to failure under tension, or its deflection at maximum load increases with an increase in the number of mesh layers used. Its ductility increases with the volume fraction and specific surface of reinforcement. Such behaviour is different from that of reinforced concrete in bending, where generally lower ductility is observed with an increase in tensile reinforcement ratio.
- Since the mesh in ferro-cement is galvanized and the crack widths are generally very small, it shows good durability under various kinds of environmental exposure.
- Ferro-cement has two-dimensional reinforcement and better resistance towards punching shear as well as resistance to impact compared to reinforced concrete.

28.1.6 Construction Methods

Various methods can be used for fabricating ferro-cement sections. Fabrication and finishing require good quality control in order to achieve a well-compacted mortar matrix without any voids or entrapped air. Some of the methods used for construction are described below:

Skeletal armature method

In this method a framework of reinforcing bars (skeletal steel) is constructed, to which a layer of meshes is applied (Fig. 28.8). Next, mortar is applied on one side and forced through the mesh until a slight excess appears on the other side.

The skeletal framework of reinforcing bars can assume any shape based on requirement. The diameter of the steel bars depends on the size of the structure. Skeletal steel is cut to a specified length and bent to suit the shape. It is tied in proper sequence to represent as precisely as possible the shape of the structure. If the bars are not tied, they tend to get dislodged and bulge. Wherever required, sufficient splice length should be provided to ensure continuity of steel.

If the weight-to-strength ratio is an issue, the performance of the structure may suffer from replacing the mesh by skeletal reinforcement. This happens because of the thickness of the relatively large skeletal bars. In this case, the opening between the main bars and the skeletal reinforcement can be filled with a lightweight core, allowing the section to act as a sandwich panel. In such a situation, care should be taken to ensure shear transfer between the two skin layers of the panel.

Advantages   The following are the advantages of the skeletal armature method:

- No elaborate form material required
- Easy to patch up (repair) the whole area from both sides
- Good penetration
- Easy to repair when damaged

Disadvantages   The following are the disadvantages of the skeletal armature method:

- Time consuming
- Application of mortar from one side may be difficult for a thick mesh system
- Galvanic corrosion may develop between the mesh and skeletal steel
- Embedment of skeletal reinforcement near the centre of the section leads to reduced performance in bending
- The steel provided is underutilized

Closed mould method

In the closed mould method, several mesh or mesh-and-rod combinations are held together in position against the surface of a mould (Fig. 28.9). Mortar is then applied from the open side. The mould either remains a permanent part of the structure or can be removed and reused. The selection of the closed mould eliminates the necessity of skeletal rods, providing mesh reinforcement only. This method is more suitable for the lay-up technique of mortar application. In this method, a thin layer of mortar is placed first and allowed to settle, over which the mesh is placed and the second layer of mortar poured. This procedure is repeated until the required number of layers are placed.

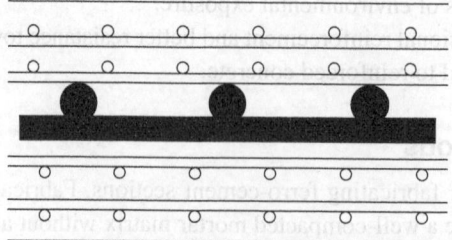

Fig. 28.8 Skeletal armature method

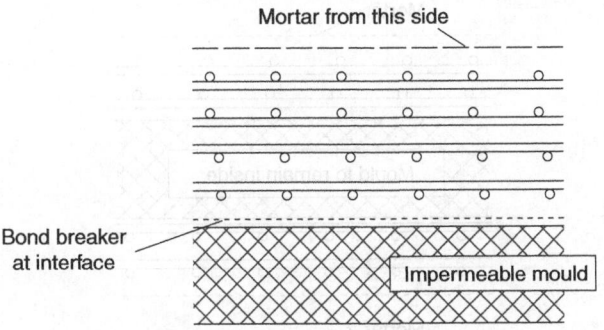

Fig. 28.9 Closed mould method

Advantages The following are the advantages of the closed mould method:

- Ideal for factory production since the reuse of moulds is permitted
- Skeletal reinforcement not required
- Suitable for patented lay-up method

Disadvantages The following are the disadvantages of the closed mould method:

- The mould is costly for one-time use
- Difficult to avoid internal voids, especially below reinforcement mesh
- Complete penetration of mortar from one side may not be possible

Integral mould method

This method involves a semi-rigid framework (Fig. 28.10). An integral mould may be formed using foam material such as polystyrene or polyurethane as the core. Mortar is poured from both sides of the mould. The mould is left inside the ferro-cement itself. This method is ideal for field operation. There are increased chances of variation due to disruption of the foam and therefore the construction should be carried out carefully.

Advantages The following are the advantages of the integral mould method:

- Provides good rigidity
- Provides good water-tightness
- Provides thermal insulation

Disadvantages The following are the disadvantages of the integral mould method:

- Special detailing is required for adequate shear connections between rigid ferro-cement layers, especially across insulating cores
- Both the sides need to be finished

Open mould method

The open mould method is a traditional method used for boat building. The open mould is made of lattice wood or some other suitable material and stiffened by ribs. The mortar is applied through one side only. To facilitate mould removal, the mould is covered with a release agent or entirely covered with polyethylene sheets. Figure 28.11 illustrates the open mould method.

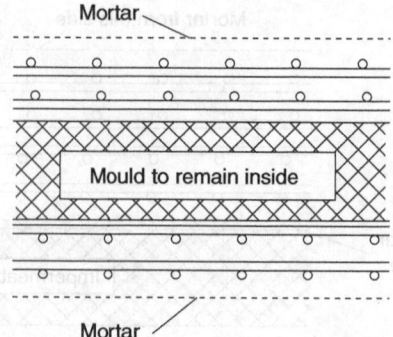

Fig. 28.10 Integral mould method

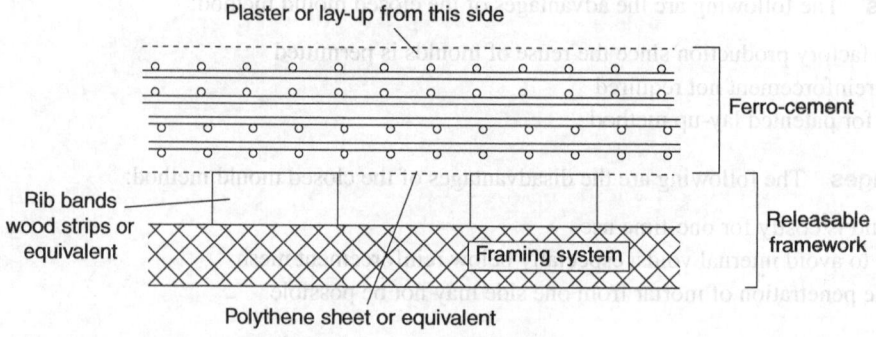

Fig. 28.11 Open mould method

Advantages The following are the advantages of the open mould method:

- No skeletal reinforcement is required
- Better control of finishes than the closed mould method
- Moulds can be reused
- Suitable for patented lay-up method
- Adoptable for traditional boat-building

Disadvantages The following are the disadvantages of the open mould method:

- Requires finishing on both sides
- Framing and shoring system is costly
- Complete penetration from one side is not guaranteed
- Amenable for any shape

Other special techniques

Funicular elements and sections may be produced using various other methods. Some of them are described below.

Rivas Wainshtok (1994) explained a technique for constructing U, V, and W-shaped panels from flat panels (Fig. 28.12). The technique involves pouring the mortar over a series of parallel panels, leaving strips of mesh between them without mortar. These strips act as joints traversed by mesh reinforcement. After hardening, the panels are lifted to the proper position and the joints are properly made with fresh in situ concrete.

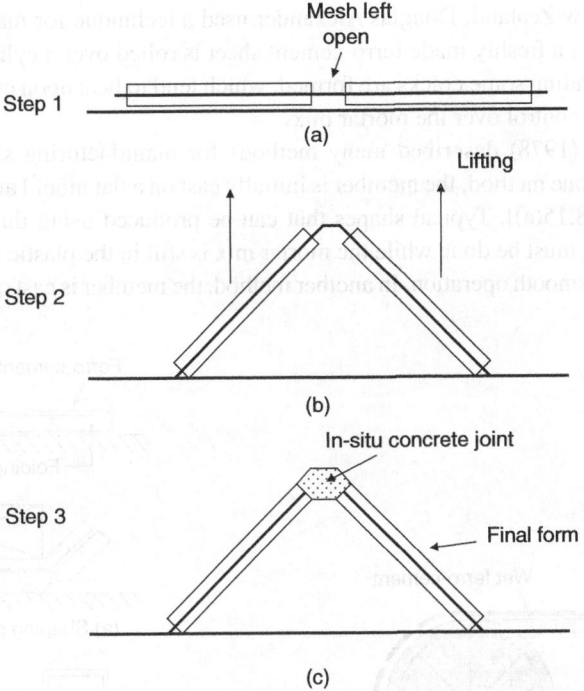

Fig. 28.12 V and W-shaped element production process

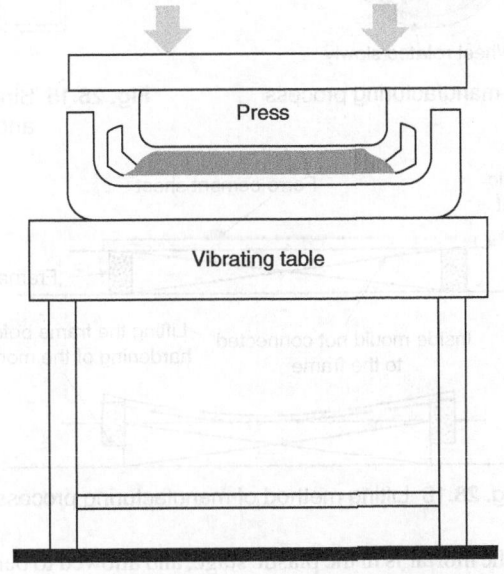

Fig. 28.13 Pressing technique

The pressing method is shown in Fig. 28.13. In this method, a strong mould press is used to shape the element. The mesh reinforcement can be constructed either as a flat sheet or roughly for the required shape of the final product. Good control over the mortar mix is required to use this method. After laying the mortar, it is pressed over the reinforcement mould to get the required shape. This method is suitable for making small ferro-cement tanks.

During the 1960s in New Zealand, Douglas Alexander used a technique for manufacturing ferro-cement members. In this technique, a freshly made ferro-cement sheet is rolled over a cylindrical drum (Fig. 28.14) or a large roller. During bending some cracks are formed, which tend to heal upon curing. In this method also, it is essential to have good control over the mortar mix.

Raichvarger and Tatsa (1978) described many methods for manufacturing singly and doubly curved ferro-cement members. In one method, the member is initially cast on a flat mould and then bent up to produce U-shaped sections [Fig. 28.15(a)]. Typical shapes that can be produced using this technique are shown in Fig. 28.15(b). The bending must be done while the mortar mix is still in the plastic stage. The mould must be properly designed to allow smooth operation. In another method, the member is cast on a flat shape (Fig. 28.16).

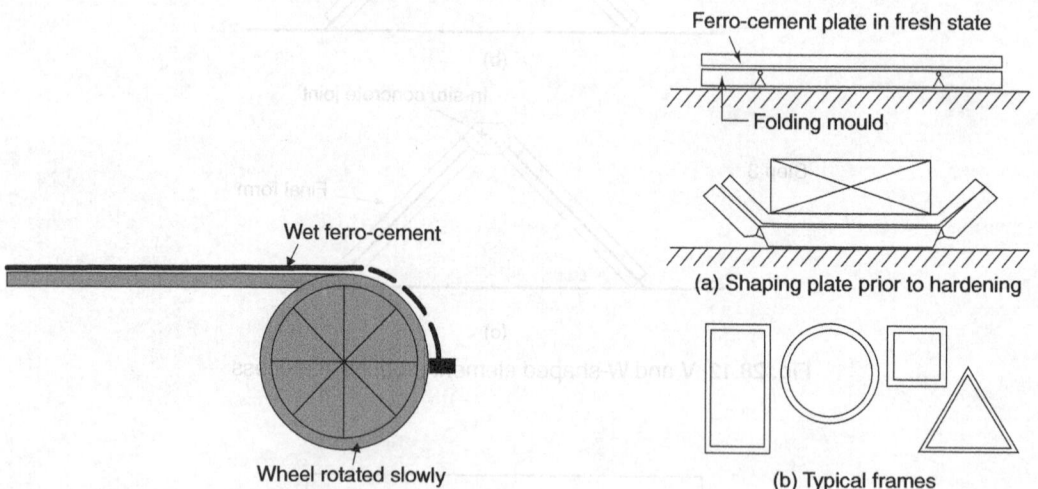

Fig. 28.14 Cylindrical element manufacturing process

(a) Shaping plate prior to hardening

(b) Typical frames

Fig. 28.15 Single and doubly curved panels and their manufacturing process

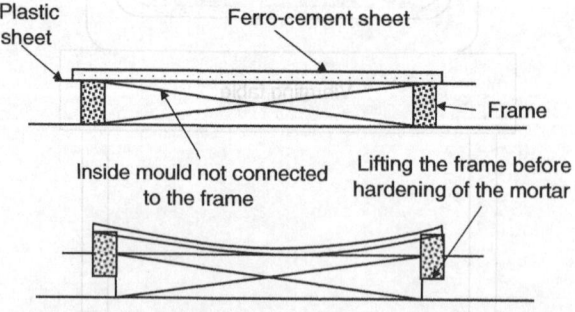

Fig. 28.16 Lifting method of manufacturing process

Later it is slightly lifted, while the mortar is in the plastic stage, and allowed to bend itself due to its self-weight. This way funicular shapes can be made easily.

28.1.7 Case Studies of Applications of Ferro-cement in Structures

Ferro-cement doors

Ferro-cement doors are an alternative to conventional wooden doors. Figure 28.17 shows a typical ferro-cement door. Special types of hinges are required for ferro-cement doors. In one type of hinge, pivots are fixed at the top and bottom, and a second type has forks and holdfasts; holdfasts are fixed in the wall to hold shutters in position.

Using ferro-cement to make shutter doors reduces the cost of joinery. Such doors are strong, durable, termite resistant, and less prone to fire and weathering. These are monolithic and provide better dimensional stability. Ferro-cement doors are more suitable for cost-effective buildings.

Roof shells

Ferro-cement can also be used to produce curved or pitched roofs. It is a good alternative to roofs made using asbestos sheets. Such roofs enhance the aesthetic appearance of buildings. Ferro-cement roofs also have a better provision for ventilation. Roof shells can be individual or continuous units (Fig. 28.18).

Rafters

Ferro-cement rafters are a good alternative to wooden or timber rafters. They display better performance and durability compared with wooden rafters. They are also cheap by 30% compared with timber rafters. Ferro-cement rafters can be fabricated by using cement mortar (1:1) and a steel skeleton.

Furniture

Ferro-cement can also be used for making furniture. Concrete benches and chairs used in parks and community centres are typical examples of ferro-cement furniture.

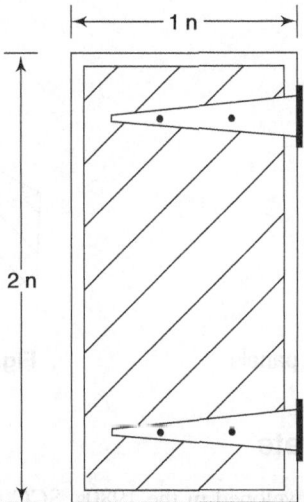

Fig. 28.17 Ferro-cement door

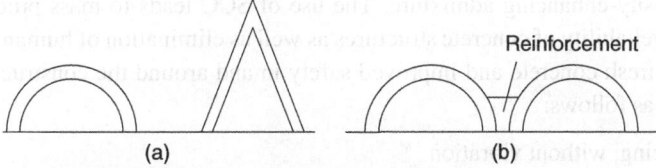

Fig. 28.18 (a) Individual roof units; (b) Combined roof unit

Ferro-cement housing panels

The panel unit shown in Fig. 28.19 can be manufactured using ferro-cement. These are used for dividing rooms or open spaces.

Ferro-cement water tanks

Water tanks of small capacity (Fig. 28.20) are one of the most common structures that can be made using ferro-cement. All the four sides and the bottom of the water tank can be made of wire meshes and mortar applied from both sides.

Ferro-cement manhole cover

Ferro-cement manhole covers are cheaper, lightweight, and corrosion free. Light, medium, and heavy-duty manhole covers can be produced using ferro-cement. Ferro-cement manhole covers can take all the loads specified for cast iron manhole covers (IS: 1726).

Ferro-cement meter box

A ferro-cement meter box is lightweight, simple to install, and cheap compared with conventional meter boxes.

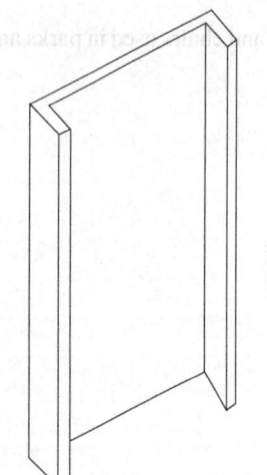

Fig. 28.19 Ferro-cement housing panels

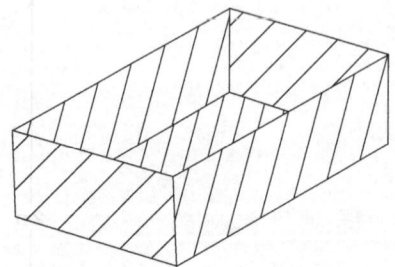

Fig. 28.20 Ferro-cement water tank

28.2 Self-compacting Concrete

Self-compacting concrete (SCC) was developed in the 1980s. SCC is a concrete which flows to a virtually uniform level under the influence of gravity without segregation, during which it de-aerates and completely fills the formwork and the spaces between the reinforcement without any need for induced compactions. SCC is obtained by limiting the water–cement ratio, adding an effective plasticizer, increasing sand–aggregate ratio, and adding some viscosity-enhancing admixture. The use of SCC leads to mass production and improved quality, durability, and reliability of concrete structures as well as elimination of human errors. It has replaced manual compaction of fresh concrete and improved safety in and around the construction site. Some of the advantages of SCC are as follows:

- A faster rate of placing, without vibration
- Uniform surface finish can be obtained with less remedial work

- Improved pumpability
- Improved consolidation around reinforcement
- Labour and cost saving
- Reduced construction period
- Improved overall construction quality
- Human errors (poor workmanship) can be avoided

28.2.1 Properties of Fresh SCC

Properties of SCC in a plastic state are an important factor. The flowability of SCC is obtained by using proper admixture. Stability or resistance to segregation is achieved by increasing the total fines content in concrete or modifying viscosity. Proper grading of materials helps in reducing cement and admixture quantities in concrete.

SCC in plastic state exhibits three basic characteristics, namely, flowability, self-levelling ability, and resistance against segregation. Various tests are performed to check these properties of fresh SCC. Some of these tests are discussed below.

Slump test

Slump test is a common test used to check workability. Figure 28.21 shows a slump cone. First this cone is placed on a horizontal plate. SCC is filled in the cone, without aided compaction. Then the cone is lifted and concrete is allowed to spread.

The time taken to reach a diameter of 50 cm and the final spread is noted. For good flowable concrete, 50 cm should be reached within 10 s and the final spread should be greater than 70 cm.

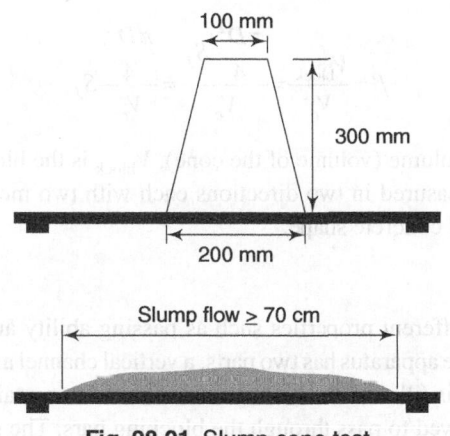

Fig. 28.21 Slump cone test

J-ring

The J-ring test is used to determine flowability. This test is similar to the slump cone test. Figure 28.22 shows the J-ring apparatus. J-rings are placed concentrically with the cone. After lifting the cone, concrete must flow under its own dead weight through steel rods, which simulates the flow through the maze of reinforcement inside the formwork. Depending upon the mixture composition and the number of steel rods, some amount of coarse aggregate will be left inside the ring. Blocking can be checked using this apparatus. According to the standard test method for passing ability of self-consolidating concrete using J-ring (ASTM C 1621/C 1621 M), slump flows with and without J-ring should not differ by more than 50 mm.

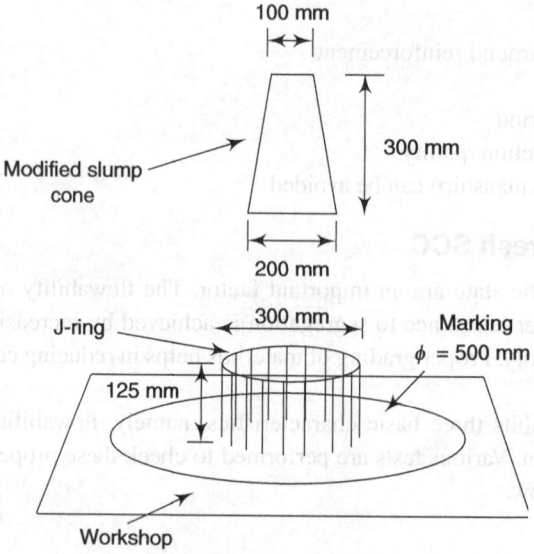

Fig. 28.22 J-ring test

The difference between heights of concrete mass inside and outside can also be used to measure blocking. Spread diameter S_J and flow time t_{500J} can be determined by measuring the height difference between the concrete behind and in front of the rods of the J-ring. A mean value can then be calculated. This mean value is defined as the step of blocking S_J. Using this value the blocking index (β) can be calculated:

$$\beta = \frac{V_{block}}{V_c} = \frac{\dfrac{\pi D^2}{4}S_J}{V_c} = \frac{\dfrac{\pi D^2}{4}}{V_c}S_J$$

where V_c is the whole concrete volume (volume of the cone), V_{block} is the blocked concrete volume, S_J is the step of blocking (mean value measured in two directions each with two measuring points at the ends), and D is the diameter of the idealized concrete shape.

L-box

The L-box is used to measure different properties such as passing ability and segregation of concrete. Figure 28.23 shows the apparatus. The apparatus has two parts, a vertical channel and a horizontal channel separated by a trap door. The vertical part is filled with concrete and allowed to remain therein for 1 min. The gate is then opened and concrete is allowed to pass through the blocking bars. The gap between reinforcement bars should be between 35 and 55 mm for 10 and 20 mm coarse aggregate, respectively. The height of concrete on either end of abstractions is measured and the blocking ratio (ratio of the height of concrete on either ends) p_s is calculated. A minimum blocking of 0.8 is expected for good flowable concrete.

V-funnel

This test for evaluating the flowability of SCC was developed at the University of Tokyo. Figure 28.24 shows a V-funnel. The funnel can be circular or rectangular in shape. Concrete is filled in the funnel. The funnel is opened and the time for flowing of concrete through the bottom aperture is noted. Normally the time for flow of concrete should be 5–20 s for good flowability.

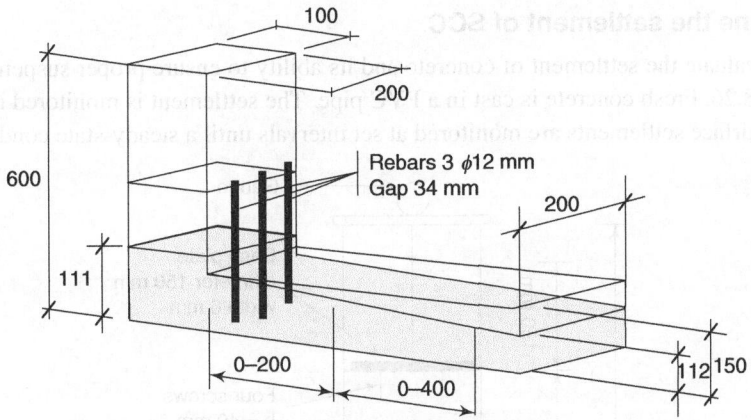

Fig. 28.23 L-box for measuring filling ability and passing ability

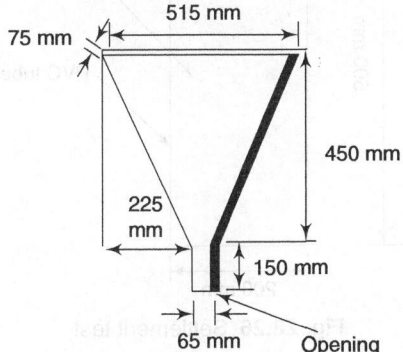

Fig. 28.24 V-funnel

U-box test

The U-box (Fig. 28.25) or U-flow test is used to determine the degree of compactability. It has a vertical partition at the centre, which has an opening at the bottom. This opening can be closed or opened with a trap door. A set of reinforcing bars are placed across the opening. Concrete is made to flow through the bars on opening the trap door under the influence of its own self-weight. Now the filling height of concrete should be noted. Normally SCC should have a minimum filling height of 300 mm. A U-box with straight base can also be used.

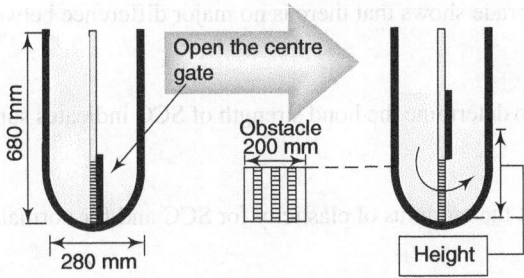

Fig. 28.25 U-box test

Test to determine the settlement of SCC

The test used to evaluate the settlement of concrete and its ability to ensure proper suspension of aggregate is shown in Fig. 28.26. Fresh concrete is cast in a PVC pipe. The settlement is monitored using a linear dial gauge or LVDT. Surface settlements are monitored at set intervals until a steady-state condition is reached.

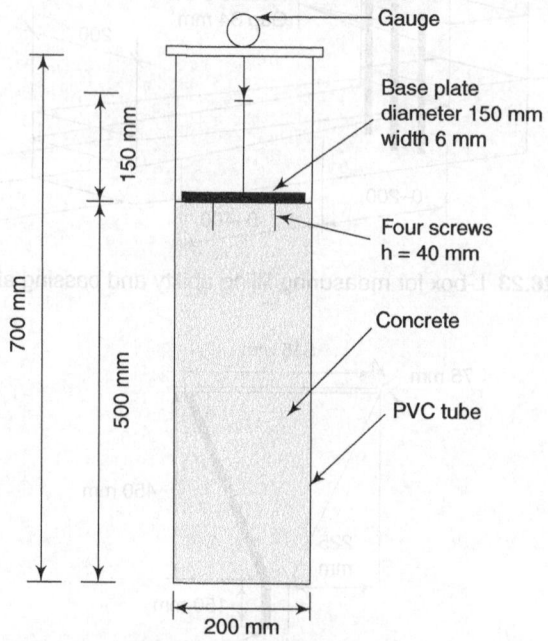

Fig. 28.26 Settlement test

28.2.2 Properties of Hardened SCC

Compressive strength

The compressive strength of SCC depends on the mix design. The compression test for SCC is the same test as is followed for normal concrete. Experimental investigation has proved that the compression strength of SCC when compared with normal concrete made for a particular strength is almost the same. The self-compacting property of SCC has very little effect on the strength of concrete.

Tensile strength

A splitting test on cylinders is followed for SCC also. A comparison between cylinders made of SCC and normal concrete of the same grade shows that there is no major difference between them.

Bond strength

The pull-out test carried out to determine the bond strength of SCC indicates superior bond strength of SCC.

Modulus of elasticity

All available results show that the modulus of elasticity for SCC and for normal concrete are the same.

Freeze–thaw resistance

This property is observed on the loss of ultrasonic pulse velocity (UPV). High-strength SCC and high-strength normal concrete show no loss of UPV. On the contrary, low-strength SCC and low-strength normal concrete show loss of UPV. The low-strength SCC has less resistance to freeze and thaw conditions as compared with low-strength normal concrete.

Creep

SCC normally is more pasty as compared with normal concrete. So its creep is slightly higher.

Durability

Various experiments carried out show that the durability of SCC is nearly equal to that of normal concrete. Moreover, durability is slightly higher in SCC because of the elimination of errors which may occur during placing and compaction of normal concrete. SCC is likely to have less voids.

Exposure to fire

SCC has a more compact microstructure. This can lead to high vapour pressure. So SCC has a higher risk of spalling when exposed to fire.

Property of hardened concrete as a full-size structural element

Many full-size beams and columns made with SCC have been loaded to failure and the test results have been compared with the standard ones. The results showed that the behaviours of the two concretes in terms of cracking moment, crack width, and load deflection were similar.

28.2.3 Mix Design

SCC should be designed in such a way that it has high fluidity, least or no segregation, and low risk of blocking. Generally SCC has high cement paste volume, low coarse aggregate and water content, and a proper dosage of superplasticizer. The sand–coarse aggregate ratio of SCC is normally about 1, which is slightly higher than that used for normal concrete. Most of the methods for designing SCC use empirical rules and differ from conventional concrete design methods. Some of the methods used for designing SCC are given below.

Petersson et al. (1996) described the Swedish Cement and Concrete Research Institute method of mix design. In this method, the minimum paste volume and aggregate proportions which guarantee free flow of concrete through reinforcement are determined. The superplasticizer quantity and water content are determined using a rheometer.

Okamura (1997) proposed a general method for designing SCC. In this method, the volumes of coarse and fine aggregates are fixed at 50% and 40%, respectively. On the basis of fluidity test, the superplasticizer quantity and water–fines ratio are determined. With these proportions, trial mixes are made and tests on fresh concrete are carried out. The mix proportions are modified based on the results to achieve the best result.

Gettu et al. (2001) developed the Universitat Politècnica de Catalunya method of mix design. This method is based on simple tests. Superplasticizer dosage is fixed on the basis of Marsh cone test. Using a mini-slump test, the filler dosage is fixed so that the concrete paste has good fluidity. Aggregate skeleton proportions are fixed by choosing a combination that has minimum voids in the dry, uncompacted stage. Finally, using fixed paste composition, a minimum paste volume that yields self-compacting mix with sufficient strength is chosen.

28.2.4 Mixing of Concrete

All common mixers can be used for mixing SCC. Truck mixers can be used, but these are less efficient and require more attention and longer mixing times. The mixing time for SCC is longer as compared with normal concrete.

Producing SCC will lead to smaller margins and more difficulties for the plant operator, who will need more training and experience. During the initial period, it is necessary to carry out more tests. But when the test results show a variation similar to traditional vibrated concrete, the test intervals can be reduced to normal levels.

Superplasticizers should be added towards the end of mixing to get better flowability.

28.2.5 Transporting

SCC should be transported in truck mixers to the construction site if it is not produced at the construction site. If it is produced at the construction site, concrete distributions can be carried out using a concrete pump, skip, or chute. The truck operator should be properly trained. After every trip, the truck drum should be cleaned. Checking the drum should be made a procedure before filling it. During transportation to the site and the waiting time, the drum should rotate at a low speed (not less than one rotation per minute). If it is not agitated properly, segregation takes place. Before delivery, the truck drum should be rotated at full speed (10–20 rotations per minute) for 3 min at the construction site.

Before using SCC at the site, slump test and visual investigation should be made. Placing of SCC is very easy as compared with that of normal concrete. The formwork should be properly checked before concreting. Errors in the formwork can lead to leaking of concrete.

28.3 Pervious or Porous Concrete

Pervious or porous concrete is a special type of concrete having high porosity. The porosity allows the water to pass through (see Fig. 28.27). When used in pavement, it reduces the run-off at the site, which helps in recharging the groundwater. The permeability is achieved by interconnected voids. It is made with very little or no fine aggregates so that cement paste coats the coarse aggregate particles without disturbing interconnected voids.

28.3.1 Need for Porous Concrete

Pervious concrete reduces the run-off from the paved areas and hence saves on the provision of a separate stormwater drainage system. This allows landowners to develop large areas of available property at an economical cost. It filters stormwater and reduces pollutant loads entering natural streams such as ponds or rivers. Pervious concrete also acts as a stormwater retention basin. It allows stormwater to enter into a large

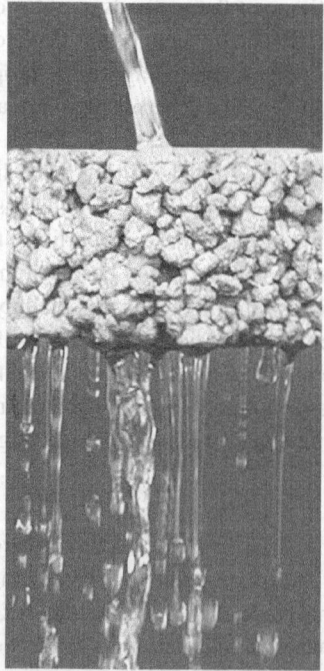

Fig. 28.27 Pervious pavement concrete

land area, thus leading to effective rainwater harvesting. Pervious concrete allows trees to grow unhindered in a paved area because it allows water to reach the root system. Its use is the best management practice for environmental protection because it permits stormwater to percolate.

28.3.2 Construction of Pervious Concrete Pavement

Quality control for pervious concrete pavement is important. Sub-grade has to be properly compacted to give uniform surface. When placed over gravelly or sandy soil, it should be compacted to about 95% density. With silty or clayey soil, a layer of graded stone may be required (Fig. 28.28). Fabric layering may be required between fine grained and stone layer. It is necessary to moisten the surface before laying the pervious concrete. Generally the water–cement ratio adopted is 0.35 to 0.45. Void content is about 20–25%. Three mix proportions used are shown in Table 28.1.

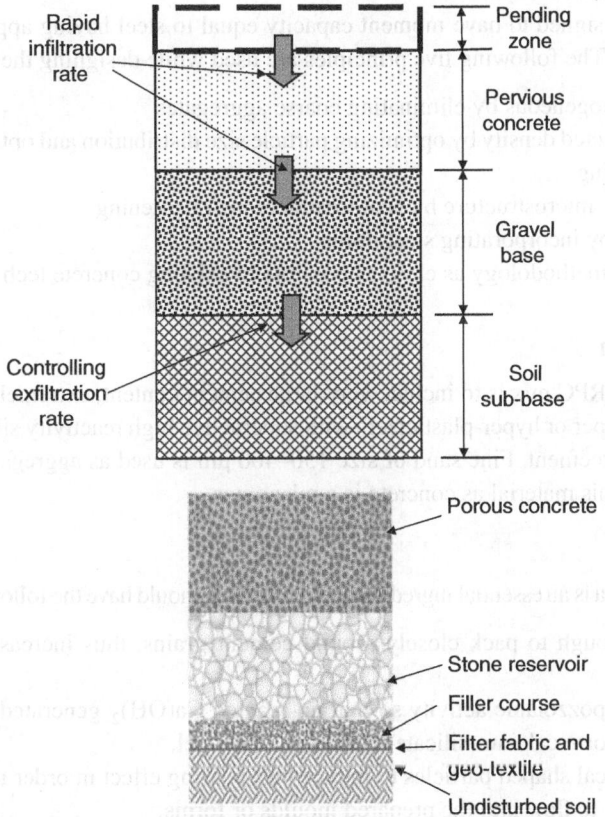

Fig. 28.28 Construction of pavement with pervious concrete

Table 28.1 Mix proportions used for pervious concrete (three different cases)

Materials	Case A	Case B	Case C
Cementitious material	355 kg/m^3	370 kg/m^3	360 kg/m^3
Water	75 to 90 kg/m^3	105 kg/m^3	80 kg/m^3
Water–cement ratio	0.21–0.25	0.28	0.22
Coarse aggregate	1540 kg/m^3	1600 kg/m^3	1365 kg/m^3
Voids content	22% to 25%	18%	35%

When pervious concrete is used for stormwater management, water that flows onto the pavement flows through the pavement in an open-graded aggregate base. From there, the water either infiltrates into the sub-grade or enters a conventional stormwater system. Any material that is picked up by the flowing water will also be taken into the concrete. The behaviour of particles within the system depends on the pore system, particle size, nature of the particles, and the flow rate. The designer should ensure that the pervious concrete has the required drainage characteristics by visual observation.

28.4 Reactive Powder Concrete

Reactive powder concrete (RPC) was developed by the technical division of Bouygues in France in 1990. It has excellent properties such as strength and ductility. However, RPC is more expensive to produce. Its isotropic properties and high ductility make it suitable for applications where only steel was considered previously. Here it has cost advantage over steel.

RPC beams can be designed to have moment capacity equal to steel having approximately same weight and cross-sectional area. The following five principles are used while designing the mix for RPC:

1. Making the mix homogeneous by eliminating coarse aggregates
2. Increasing the compacted density by optimizing particle size distribution and optionally applying pressure before or during setting
3. Obtaining favourable microstructure by heat treatment post hardening
4. Improving ductility by incorporating steel fibres
5. Maintaining casting methodology as close as possible to existing concrete technology practice

28.4.1 Composition

The important feature of RPC mix is to include very large cement content, extremely low water–binder ratio using latest generation super or hyper-plasticizers with superactive high reactivity silica fume and addition of fine steel fibres as reinforcement. Fine sand of size 150–400 μm is used as aggregates. No coarse aggregate is added. Hence calling this material as concrete is a misnomer.

Silica pozzolana

A highly reactive pozzolana is an essential ingredient in RPC mix. It should have the following three characteristics:

1. It should be fine enough to pack closely around cement grains, thus increasing density and thereby minimizing voids.
2. It should have good pozzolanic activity so that un-reacted $Ca(OH)_2$ generated by hydration of cement reacts with silica to form calcium-silicate-hydrate (C-S-H) gel.
3. It should have spherical shaped particles to induce ball-bearing effect in order to lubricate the fresh mix improving the ability to flow into the prepared moulds or forms.

These requirements will be satisfied if the surface area of silica fume is in the range of 15–22 m^2/g and bulk density of about 200–600 kg/m^3.

Cement

The cement should have high C_3S and C_2S and very small quantity of C_3A. This is because C_3A does not have good binding property. Commercially available sulphate-resisting cement will have these characteristics and hence is suitable.

Quartz Fine

For RPC mix cured at temperatures higher than 90 degrees in precast industries, additional silica is required to modify the ratio of CaO/SiO_2 binder. In these cases, quartz powder with a particle size of 10–15 µm is employed.

Sand

River sand with good grading and having a finesse modulus of 1 to 1.2 is suitable. Manufactured sand of similar properties can also be used.

Superplasticizer

The very low water–binder ratio adopted needs high-quality third-generation super-plasticizing agents. For this aqueous modified carboxylate is used.

Steel fibres

To impart ductility steel fibres of size 13 mm long and 0.2 mm in diameter ($l/d = 65$) with tensile strength 2000 MPa are suitable.

Mix ratio

The RPC mix that is generally used is given in Table 28.2.

28.4.2 Properties

Table 28.3 compares the properties of RPC with that of regular HPC.

28.4.3 Mechanical Performance and Durability

The compressive strength of RPC is significantly higher than that of HPC (see Fig. 28.29). RPC is highly durable. Fig 28.30 compares permeability of RPC with that of HPC.

Table 28.2 RPC mix ratio

Component	Mix for ordinary curing	Mix for steam curing
Cement	1.0	1.0
Sand	1.1	1.1
Silica fume	0.25	0.23
Powdered quartz	0	0.39
Steel fibres	0.175	0.175
Water–binder ratio	0.11–0.26	0.17–0.23
Superplasticizer (% of solids in cement)	0.6–1.6	1.9–2.5

Table 28.3 Properties of RPC compared with those of HPC

Property	HPC	RPC
Compressive strength (MPa)	60–100	180–200
Flexural strength (MPa)	6–10	40–50
Fracture energy (J/m^2)	140	1200–40000
Young's modulus (GPa)	23–37	50–60

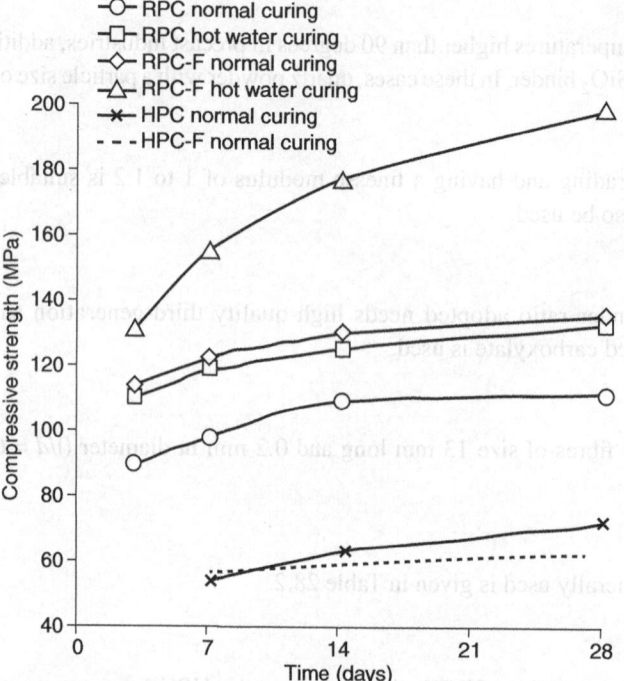

Fig. 28.29 Compressive strengths of RPC and HPC

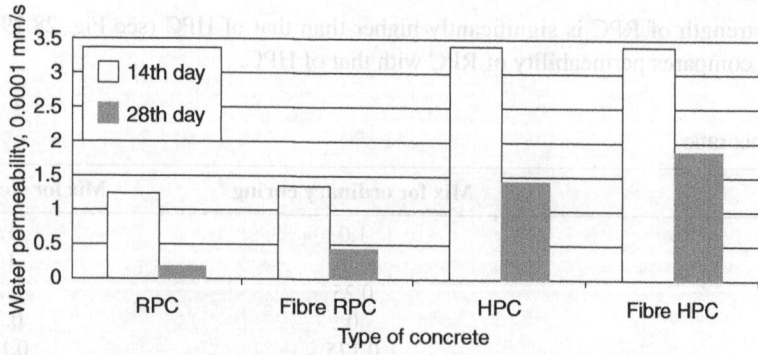

Fig. 28.30 Surface water permeability of RPC and HPC

28.4.4 Case Studies of Applications

RPC has the potential to replace steel in a few applications. Its superior strength results in lower dead loads. This leads to better seismic performance. In addition it can be made quite ductile. It has been used to construct Sherbrooke Pedestrian Bridge located in Quebec, Canada. Its applications include structures for isolation and containment of nuclear waste as well as blocking and stabilization of nuclear waste.

28.5 Geopolymer Concrete

Continuous demand for construction of infrastructure has led to the overuse of concrete and cement. Manufacture of cement causes release of carbon dioxide into the atmosphere. One tonne of cement production leads to a tonne of CO_2 being released. Thus, cement industry is held responsible for a sizable quantity of greenhouse emissions. Therefore, we have to look for alternative green options. Geopolymer concrete with much lower carbon footprint is considered to be appropriate for application in concrete industry.

Davidovits for the first time in 1988 suggested that an alkaline liquid could be used to react with Silicon (Si) and Aluminium (Al) in a source material of geological origin or an industrial byproduct such as fly ash, blast-furnace slag or rice husk ash to produce binders. There are two main constituents of a geopolymer—the source material and alkaline fluid. The most common alkaline liquid used is the combination of sodium hydroxide and potassium hydroxide or sodium silicate and potassium silicate (Rangan, 2015).

Fly ash available in India is of low calcium variety formed by burning bituminous coal in thermal stations. For applications similar to conventional concrete, the Si/Al atomic ratio is about 2.

28.5.1 Mix proportion

The binder in the geopolymer concrete is different from that of conventional concrete. The silicon and aluminium oxides in low-calcium fly ash react with alkaline liquid to form the geopolymer paste that acts as the binder and holds together aggregates and other unreacted materials and forms geopolymer concrete. A typical mix proportion developed by Rangan (2015) and used by Barber (2010) at Curtin University in Australia is shown below:

Constituent	kg/m^3
20 mm coarse aggregate	700
10 mm coarse aggregate	350
Fine sand	800
Fly ash (Class F)	380
Sodium silicate solution	110
Sodium hydroxide solution (8 M)	40

The workability of fresh mix obtained was 210. The tests showed that the strengths that could be reached are as follows:

Age (days)	Cylinder compressive strength (N/mm^2)
3	8
7	18
14	23
28	24
56	32

28.5.2 Properties

The behaviour of fly ash based geopolymer concrete is similar to conventional concrete. Tests show that with granite-type coarse aggregate, the unit weight is about 2400 kg/m^3. Available research data shows that the geopolymer concrete has lower drying shrinkage and low creep together with better sulphate and acid resistance compared with ordinary concrete. It also has better fire resistance.

28.5.3 Limitations

The use of this material is still in infancy and research in the area is ongoing. Most of the geopolymer systems proposed are difficult to work with practically in the field [CPTP Tech Brief (US Department of Transportation, Federal Highway Administration), 2010]. There is also a risk associated with high alkalinity of the activated solution. The polymerization reaction is very sensitive to temperature and usually requires geopolymer concrete to be cured at elevated controlled temperature regime (Rangan, 2009).

28.5.4 Economic Advantages

Geopolymer concrete is bound to be cheaper than conventional concrete as the cost of fly ash and slag are only a fraction of the cost of cement. Allowing for the cost of alkaline liquids, geopolymer concrete should be cost-effective compared to conventional concrete. It is also a low-carbon alternative to conventional concrete. Current research is focussing on finding user-friendly geopolymers which can be easily applied in infrastructure construction activities.

Exercises

Review Questions

1. Define ferro-cement.
2. What are the applications of ferro-cement?
3. Explain in detail the skeletal steel requirement for the manufacture of ferro-cement.
4. What are fibre-reinforced meshes?
5. Explain the behaviour of ferro-cement under tension.
6. What are the advantages of ferro-cement?
7. What are the major differences between ferro-cement and reinforced concrete?
8. Explain the following methods of manufacture of ferro-cement:
   (a) Closed mould method
   (b) Integral mould method
   (c) Open mould method
9. What is self-compacting concrete?
10. What are the tests carried out in the fresh stage of SCC to determine its properties? Explain any two of them.
11. Describe the J-ring test with a neat sketch.
12. Compare the hardened properties of normal concrete and SCC.
13. Explain the effect of creep in SCC.
14. What are the advantages of SCC?
15. What precautions are necessary when ready mix SCC is adopted?
16. What is the need to provide pervious concrete for pavements? Explain.
17. Distinguish between pervious concrete and porous concrete.
18. What is the material composition of reactive powder concrete?
19. Distinguish between RPC and HPC. Bring out the difference in behaviour of RPC when compared with HPC.
20. How adopting geopolymer concrete helps in reducing carbon footprint? Discuss.
21. What are the difficulties in adopting geopolymer concrete instead of conventional concrete in the field?

Multiple Choice Questions

1. Ferro-cement is made with
   a. Steel only
   b. Concrete only
   c. Concrete reinforced with steel rods
   d. Mortar reinforced with wire mesh
2. Ferro-cement reinforced with steel micro-fined fibres is a
   a. Composite material
   b. Two-phase material
   c. Homogeneous material
   d. Hybrid material

3. Ferro-cement is advantageous to use when the member is in
   a. Tension
   b. Compression
   c. Shear
   d. Torsion

4. The thickness of ferro-cement is usually
   a. <25 mm
   b. >150 mm
   c. <150 mm
   d. <5 mm

5. Ferro-cement is used for constructing
   a. Thin members predominantly in tension
   b. Thin members predominantly in compression
   c. Members in flexure
   d. Columns

6. Self-compacting concrete
   a. Requires hand vibration for compaction
   b. Requires precautions against honeycombs
   c. Requires jolting for consolidation
   d. Flows under the influence of gravity and gets consolidated without segregation

7. Using slump cone for SCC a record of _____ is made to assess the workability
   a. Slump value
   b. Time taken for a spread of 50 cm
   c. Spread achieved by concrete
   d. Spread achieved in 10 s

8. For SCC, J-Ring test is used for
   a. Flowability
   b. Slump
   c. Strength
   d. Passability

9. For SCC, passability can be assessed by
   a. Slump cone test
   b. J-Ring test
   c. Flow table test
   d. L-Box test

10. For SCC, the degree of compactability is assessed by
    a. Slump cone
    b. U-Box Test
    c. L-Box test
    d. J-Ring test

11. Pervious concrete is used in pavements because
    a. It is capable of reducing the discharge into stormwater drain during rainy seasons
    b. It is able to conserve sand
    c. It is more economical
    d. All of the above

12. The aggregate used in reactive powder concrete is
    a. Sand
    b. Granite
    c. Silica fume
    d. Fly ash

13. The binder in geopolymer concrete is
    a. Cement
    b. Epoxy
    c. Reaction product of fly ash and alkaline fluid
    d. Lime

14. Geopolymer concrete
    a. Is cheaper than conventional concrete
    b. Has lesser carbon footprint
    c. Has better sulphate resistance
    d. All of the above

Answers to Multiple Choice Questions

1. d	5. a	9. d	13. c
2. d	6. d	10. b	14. d
3. d	7. b	11. d	
4. a	8. a	12. a	

CHAPTER 29

Concreting Machinery and Equipment

Till few decades back, the construction industry mainly depended on manpower only. The rate, cost, and quality of construction were affected by the efficiency level of trained manpower. In order to improve these and save construction time, machinery is the inevitable choice. Various machines and tools are used to satisfy these needs. Even though the initial cost of the equipment is high, the improvement in construction quality and the reduction in construction time they bring about make them a must in modern age.

Infrastructure development involves projects such as tunnelling, rock blasting, mass excavation and pipeline laying, dam construction, bridge construction, and industrial parks and towns. In order to complete such massive projects in time and achieve world-class quality, modern equipment and machines are necessary.

Concreting generally involves various steps, namely batching and mixing, transporting, placing, and finishing. Various machines are used for these operations. Machines are preferred not only for concreting but also for other associated operations of concrete construction. Transporting precast concrete piles and using cranes for lifting precast concrete units are two such examples. This chapter discusses in detail the various machines used for the preparation of materials used in concrete, concrete construction, and erection of structures.

29.1 Concreting Equipment

In this section we will discuss various equipment used for manufacturing aggregates, batching, mixing, placing, consolidating, and curing concrete. It is not necessary that all these equipment are used at every construction site because some of these activities are still carried out manually without the use of machinery. The scale of the job and economics very often dictate whether to use machines or to execute the job manually.

29.1.1 Preparation of Aggregates

Concrete is made using a proper mix of cement, fine aggregates, coarse aggregates, and water. The preparation of aggregates is a fundamental operation in concreting. It can also be considered a pre-concreting operation.

Sand is the conventional fine aggregate used for construction. It is usually collected from riverbeds and may contain impurities; therefore cleaning is an important process for obtaining good quality sand. A simple method for cleaning sand is the *unit tank process*. In this process, sand and water are fed to classifier unit tanks. As the water flows to the outlet end of the tank, sand particles settle at the bottom of the tank—coarse particles at the bottom and fine particles at the top. When the layer of deposited sand reaches a predetermined level, a sensing paddle actuates a discharge valve at the bottom of the compartment to permit the material to flow into a splitter box, from which it can be removed and used.

A *screw-type classifier* can also be used to produce good quality sand. A sand screw is erected so that the material must move up the screw to be discharged. Sand and water are fed into the hopper. As the screw rotates, sand is moved up to the outlet by the motor. The undesirable material is flushed out of the tank with overflowing water.

Coarse aggregates are obtained in various (larger) sizes and are reduced to the required size in crushers. Crushers are classified as primary and secondary crushers. Stones obtained directly from blast are processed through the primary crusher and their sizes are reduced (large stones are reduced to medium-sized stones). The output of the primary crusher is fed to the secondary crusher and the required size of aggregate is obtained.

Crushers are classified mainly into three types based on their mechanical operation methods. They are jaw crusher, gyratory crusher, and roller crusher.

Jaw crusher

A jaw crusher consists of two jaws; stones are allowed into the space between the two jaws and crushed. Generally one jaw is fixed and the other is movable. The movable jaw can produce high pressure to crush even the hardest rock. These crushers are usually designed with a toggle as the weakest part. The *toggle* breaks when uncrushable parts are encountered, limiting the damage to the crusher. Figure 29.1 shows a single-toggle-type jaw crusher.

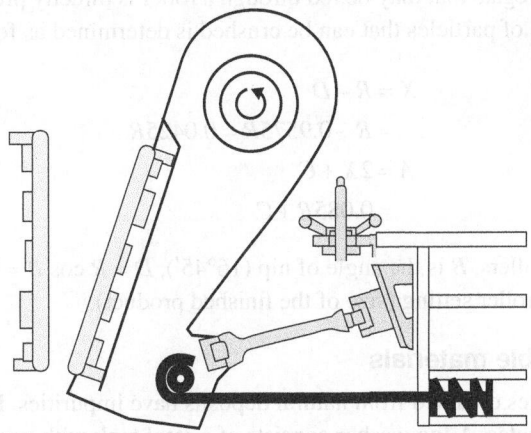

Fig. 29.1 Toggle-type jaw crusher

Gyratory crusher

A gyratory mantle mounted within a deep bowl characterizes gyratory crushers. These crushers provide continuous crushing action and are used for both primary and secondary crushing. To protect the crusher from uncrushable objects and overload, the outer crushing surface can be spring-loaded.

The gyratory crusher unit (see Fig. 29.2) consists of a cast-iron or steel frame with an eccentric shaft and driving gears in the lower parts of the unit. The upper part has a cone-shaped crushing chamber, lined with

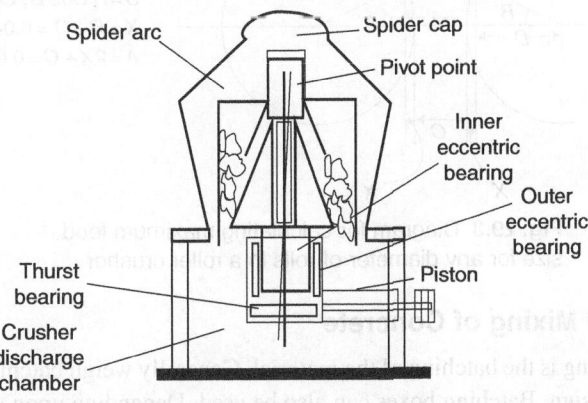

Fig. 29.2 Gyratory crusher

hard steel plates called *concaves*. The crushing member includes a hard steel crushing head mounted on a vertical steel shaft. The head suspended from the spider frame at the top is so constructed that some vertical adjustment of the shaft is possible. The eccentric support at the bottom causes the shaft and the crushing head to gyrate as the shaft rotates, thereby varying the width of the space between the concaves and the head. As a rock fed in at the top of the crushing chamber moves downward, it goes through a reduction in size until it finally passes through the opening at the bottom of the chamber.

Roller crusher

A roller crusher consists of heavy iron rollers, each mounted on a separate horizontal shaft. There may be a single roller or a double roller. In the case of a double roller, one roller is fixed and the other is allowed to roll. The movable roller is spring-loaded to provide safety against damage to the roller when non-crushable material passes through it.

The maximum size of aggregate that may be fed through a roller is directly proportional to the diameter of the roller. The maximum size of particles that can be crushed is determined as follows (see Fig. 29.3):

$$X = R - D$$
$$= R - 0.9575R = 0.0425R$$
$$A = 2X + C$$
$$= 0.085R + C$$

where R is the radius of the rollers, B is the angle of nip (16°45′), $D = R \cos B = 0.9575R$, A is the maximum size of the feed, and C is the roller setting (size of the finished product).

Elimination of undesirable materials

Both fine and coarse aggregates obtained from natural deposits have impurities. Log washers can be used for cleaning both types of aggregates. A log washer consists of a steel tank with two motor-driven shafts. When the washer is erected in the plant, the discharge end is raised. The aggregate to be processed is fed into the unit at the lower end, while water is supplied constantly into the elevated end. As the shaft rotates in the opposite direction, the paddle moves the aggregate to the upper end of the tank, producing a continuous scrubbing action between the particles. The stream of water removes the impurities from the lower end and clean aggregate is discharged from the upper end.

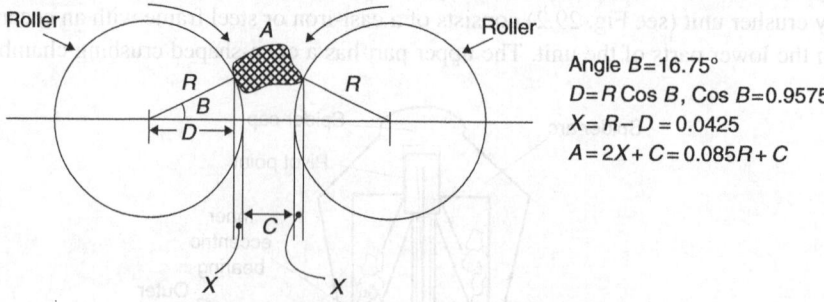

Fig. 29.3 Diagram for calculating maximum feed size for any diameter of rolls in a roller crusher

29.1.2 Batching and Mixing of Concrete

The initial step in concreting is the batching of the material. Generally weigh batching is adopted. Batching is done based on the mix design. Batching boxes can also be used. Depending upon the need, boxes of various volumes can be produced. After batching, the concrete should be mixed properly. Mixing can be done using two methods—*batch mixing* and *continuous mixing*. In the first method, after every cycle, the mixer should be cleaned and mixing done afresh. In the second method, the constituents are continuously entered and mixing done continuously.

Mixing up the various ingredients for any job is a laborious and time-consuming task. Mixers are used for mixing. A mixer has two main components—the drum and the propellers. The drum is a barrel-type unit through which water, aggregate, and cement are poured into the mixing bowl. There are three main types of drums: non-tilting drum, tilting drum, and reversing drum. The propellers are used to stir the cement and aggregates. Industrial mixing applications use left-hand propellers because they create a downward flow, resulting in better mixing. The truck mixer is the best example of this type. Another fairly common piece of equipment is a *suction strainer*. Suction strainers prevent small particles from entering and damaging the mixing system and are usually made of stainless steel wire cloth. Mixers are classified mainly into two types—*central mixer* and *transit mixer*.

Central mixers are transported to construction sites and mixing is done on the site itself. The mixer is transported in a truck. Fresh concrete is not transported. Depending upon the need, mixers of various sizes can be used. These mixers can be tilted to discharge the concrete.

A transit mixer is used to transport fresh concrete from the concrete manufacturing unit to the site. Generally, ready-mixed concrete (RMC) uses a transit mixer. Transit mixers are available in various sizes. They are capable of mixing concrete with about 100 revolutions of the mixing drum. The mixing speed is generally 8–12 rpm. This mixing during transit usually results in stiffening of the concrete, which can be prevented by adding suitable retarding admixtures.

In addition to these two types, concrete mixers are also classified as electric powered mixer, gas powered mixer, and diesel powered mixer.

29.1.3 Placing of Concrete

Though concrete can be either ready mixed or conventionally site made, its placing is very important. Concrete is required to be moved vertically and/or horizontally. The various equipment used for these operations are buckets, chutes and drop pipes, conveyor belts, and pumps. A properly designed bottom-dump bucket is a common equipment used for placing concrete. These buckets can be used for both vertical and horizontal movements. Care should be taken to prevent the concrete from getting segregated as a result of being discharged from too high above the surface or because of fresh concrete being allowed to fall against an obstruction.

Wheelbarrows are usually used to transport concrete in the horizontal direction, provided there is a smooth and rigid runway to operate on. Hand buggies are safer compared to wheelbarrows because they have two wheels while wheelbarrows have only one wheel. These means of horizontal transportation are preferred when the distance to be covered is less than 60 m.

Drop pipes and chutes are used to transfer concrete from higher elevations to lower elevations. The chutes used are round-bottomed and have sufficient slopes to allow the free flow of concrete. The chutes can be made of steel or aluminium. The drop pipes used are round pipes of diameter equal to at least eight times the maximum size of aggregate used for concreting. Drop pipes are used in wall or column constructions to avoid the segregation of concrete by allowing the free flow of concrete through the reinforcement.

Conveyor belts are used to transport low slump concrete. They provide rapid movement of fresh concrete. Particular attention should be given to points where the concrete leaves one conveyor or is discharged into another conveyor, where segregation can occur.

Pumps are commonly used simple equipment. By applying appropriate pressure, concrete can be moved through large distances in both horizontal and vertical directions provided a proper lubricating layer is provided within the pipe. RMC normally uses this technique.

Pumps are generally of three types:

1. Piston pump (Fig. 29.4)
2. Pneumatic pump (Fig. 29.5)
3. Squeeze pressure pump (Fig. 29.6)

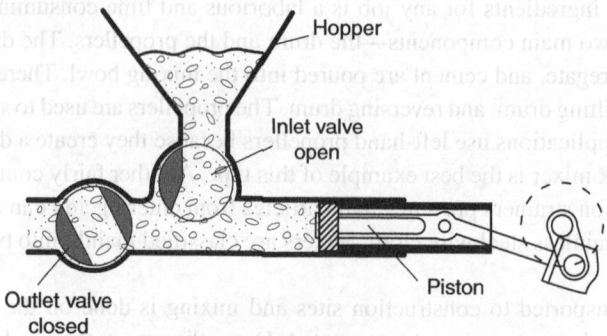

Fig. 29.4 Piston pump

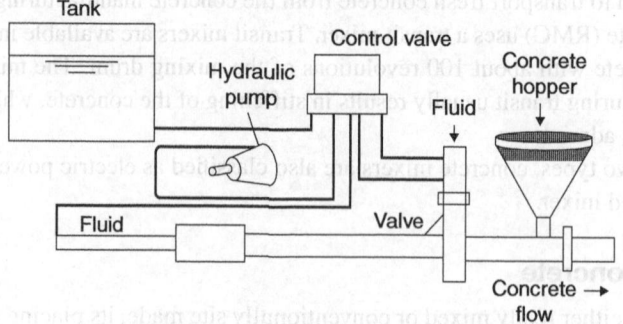

Fig. 29.5 Pneumatic concrete pump

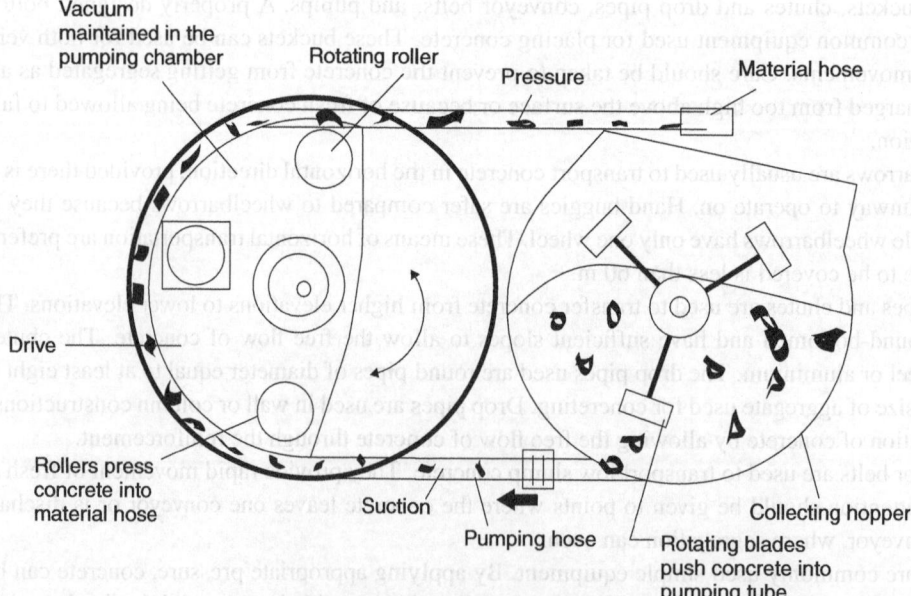

Fig. 29.6 Squeeze pressure pump

Most piston pumps normally contain two pistons, one retracting during the forward stroke of the other to provide continuous flow of concrete. Pneumatic pumps normally use a re-blending discharge box at the discharge end to bleed off the air and prevent segregation. In the case of squeeze pumps, hydraulic-powered rollers rotate on the flexible hose within the drum and squeeze the concrete out from the top. The vacuum maintains a steady supply of concrete in the tube from the receiving hopper. Pumps can be mounted on trucks or trailers and carried to the sites of construction.

For pumping concrete, the following rules must be followed:

a. Use a combined gradation of coarse and fine aggregates that will ensure that the gaps are eliminated so that the paste cannot squeeze through in between the coarser particles under the pressure induced in the line. This is the most often overlooked aspect of good pumping.

b. Use a minimum pipe diameter of 125 mm.

c. Always lubricate the line with cement paste or mortar before beginning the concrete pump operation.

d. Ensure a steady, uniform supply of concrete, with a slump between 50 and 125 mm as it enters the pump.

e. Always pre-soak the aggregates before mixing them into the concrete to prevent them from soaking up the mix water under imposed pressure. This is especially important when aggregates that have high absorption capacities are used.

29.1.4 Consolidation of Concrete

Concrete is consolidated to remove the entrapped air and improve the quality of concrete. This consolidation can be done by vibration, for which three types of vibrators are used. These are the following:

1. Internal or needle vibrator
2. Surface vibrator
3. Form or shutter vibrator

An internal vibrator, as the name indicates, is immersed into the concrete and vibrated. It has a vibrating casing and head which is immersed into the concrete and made to vibrate at high frequency (10,000–15,000 vibrations/min). The internal vibrator is normally preferred in all construction sites.

Surface vibrators are used in slab construction. These vibrators are used for consolidation only from the top surface. Their vibration speed is about 3000–6000 vibrations/min.

A form vibrator is an external vibrator. As the name indicates, these vibrators shake the entire form (vibrate the entire form). They are attached to the outer surface of the form. Form vibrators are used in large panel construction.

All the three types of vibrators can be powered by electricity, gas, or diesel. They should never be used to move concrete laterally. Judicious re-vibration (vibrating the concrete again after allowing it to remain undisturbed for some time) can be done to improve the quality of concrete.

After compaction, proper finishing of the surface is another important process. For this purpose a float is used, which can be of two types—power float and disc float. A power float has adjustable blades, which are used for the prefect finishing of a surface. A disc float has a circular disc and is used to remove undulations prior to finishing the surface.

After compaction and levelling, concrete must be allowed to set. In order to remove the excess water in concrete and allow for fast setting and finishing of concrete, a vacuum pump can be used to remove the excess water in concrete. This technology is known as *vacuum dewatering*.

29.1.5 Curing of Concrete

Curing is an important factor governing the strength of concrete. The main objective of curing is to avoid shrinkage cracks and enable full hydration to take place. The conventional method of curing concrete is ponding or spraying water on it and covering it with polyethylene sheets to prevent evaporation loss. All exposed surfaces of the concrete must be kept wet continuously for the required curing time. Another method of curing is steam curing. Steam curing is defined as the use of steam at temperatures above 40°C for curing. When concrete is steam-cured, the temperature inside the curing jacket at the surface of the concrete must not exceed 75°C for more than 1 h during the entire steam-curing period. Concrete exposed to temperatures exceeding 85°C is not acceptable. Sufficient moisture must be provided inside the curing jacket so that all the surfaces of the concrete are wet. An unobstructed air space of not less than 15 cm should be provided between all surfaces of the concrete and the curing jacket. The steam outlets should be positioned such that live steam is not applied directly on the concrete, reinforcing steel, or tendons. The location of steam lines, location of control points for the discharge of steam into the curing jacket, and the number and type of openings for steam distribution within the curing jacket should be so arranged that the temperature variation between any two points in the enclosure does not exceed 10°C. Steam curing should not be commenced until the concrete has been in place for a minimum period of 3 h. During the application of steam, the temperature inside the curing jacket should be raised uniformly at a rate not exceeding 20°C/h. The rate of temperature decrease at the end of the curing operation should not exceed the same rate. When elevated temperature curing is used, the members should remain protected until the difference between the temperature inside the curing jacket and that of the outside air is less than 12°C.

29.1.6 Slipform Paving

For the purpose of making concrete pavements, slipform pavers are used. They perform the function of spreading, vibrating, striking off, consolidating, and finishing the concrete pavements according to the prescribed cross section. This machine increases the speed of construction and is also quite economical.

The basic operating process involves moulding the plastic concrete to the desired cross section and profile under a relatively large screed. This is accomplished with a full-width screed that is maintained at a predetermined elevation and cross-section slope using hydraulic jacks that are actuated through an automatic control system. The control is referenced to offset the grade lines pre-erected parallel to the planned profile for each pavement edge. Concrete is delivered into the front of the paver. It is then internally vibrated as it flows under the machine. Slipform pavers also have automatic dowel inserters, which push or vibrate the dowels into fresh concrete in the exact position with minimum disturbance to the concreting operations. All pavers are equipped with final float finishers.

29.1.7 Shotcreting

Shotcreting can be defined as transporting mortar or concrete through a hose and pneumatically projecting it at high velocity onto a receptive surface. This method is used mostly in repair work. In this technique, high compressed air is used to pressurize and move the concrete. The cement, sand mix, and water are kept in separate containers, which are connected to a nose pipe. Compressed air is forced into these containers through a motor. Concrete flows through the nose and falls on the required surface with great pressure. Currently fibre-reinforced shotcreting equipment are available, which have the advantage of reducing the rebound of mortar. Shotcreting equipment is normally used in tunnel-lining operations.

29.2 Lifting Devices

The construction of heavy structures involves handling precast units. Precast units are cast at a factory and transported to the site and erected. Heavy lifting devices are required for handling these precast units. These lifting devices can also be used for erecting precast RCC, prestressed, or composite steel structures.

Cranes are the main lifting devices used. Cranes are designed and manufactured depending on the load to be carried and other specifications. Cranes are mainly classified into the following two types:

1. Mobile crane
2. Tower crane

29.2.1 Mobile Cranes

Some of the types of mobile cranes used are the following:

a. Crawler crane
b. Truck-mounted crane
c. Rough terrain crane
d. All-terrain crane
e. Heavy lift crane

Crawler cranes

A crawler crane has a full revolving superstructure mounted on a pair of continuous parallel crawler tracks. The crawler provides the crane with the ability to travel around the entire site. The crawler track has a large contact area with the soil and this sometimes causes soil failure under the crane. Before using this crane, the machine must be levelled (the ground surface should be levelled so that the machine stands level). This crane can be dismantled and reassembled. It has a low initial cost, but the reassembling and transporting costs are high. Hence it is preferred for long-duration projects only.

Truck-mounted cranes

Truck-mounted cranes are mounted on a truck and can also travel on public roads between project sites with minimum dismantling. Once the crane is levelled on the site, it is ready to operate. Its initial cost depends on the lifting capacity of the crane. If a job requires a crane for a few hours or days, the truck-mounted crane is the best option. Truck-mounted cranes can be further classified into two types—telescoping boom and lattice boom truck-mounted cranes.

The first type is a truck-mounted crane having a self-contained telescoping boom. The telescoping boom is a permanent part of the full revolving superstructure. These cranes have extendable outriggers for stability. In fact, many units cannot be operated safely with a full reach of the boom unless the outriggers are fully extended and the machine raised so that the tyres are clear off the ground. In the case of large structures, the operation must be carefully planned. Soil-bearing capacity should also be considered while using this crane. Large-sized steel or timber mats can be used to distribute the load evenly on the soil.

The second type of structure consists of a lattice boom, which is lightweight and hence has additional load-carrying capacity. A lattice boom is longer in assembly and requires more effort to dismantle and reassemble. During lifting, a second crane is often required to support this crane.

Rough terrain cranes

Rough terrain cranes are mounted on large wheels and closely spaced axles. They also have high ground clearance as well as the ability to move on slopes. They can travel on highways with a speed of 30 mph. In the case of long journeys, they should be transported by low bed trailers. They are primarily lift machines but capable of light, intermittent duty-cycle work.

All-terrain cranes

An all-terrain crane is designed with an undercarriage capable of long-distance highway travel. Additionally, the carrier has an axle drive, large tyres and high ground clearance. It has dual cabs—the lower cab for long highway travel and the superstructure cab for crane control and small-distance travel. This crane can be used for scattered projects located at moderate distances as well as multiple utilities for a single project. All terrain cranes are costly compared to other cranes.

Heavy lift cranes

As the name indicates, a heavy lift crane can carry large loads, in the range of 600–2000 tonnes. This crane consists of a boom and a counter-weight, each mounted on independent crawlers. Various systems have been introduced for increasing the load-carrying capacity of this type of crane. Some of them are the following:

a. Trailing counter-weight
b. Ring system
c. Guy derrick

Figure 29.7 shows the trailing counter-weight system crane. The counter-weight is provided at the back of the crane, which balances the load and increases the load-bearing capacity of the crane.

In the ring system, a large turn-table is created outside the base of the machine. A heavy counter-weight system is supported on this ring. Frames connected by axial pins are present at the front and rear of the base of the machine. These allow the counter-weight and the boom to move away from the machine.

The guy derrick method immobilizes the crane but increases the load-carrying capacity of the crane by nearly seven times. This can be achieved by tying the vertical mast with a guy cable mounted on the base crane.

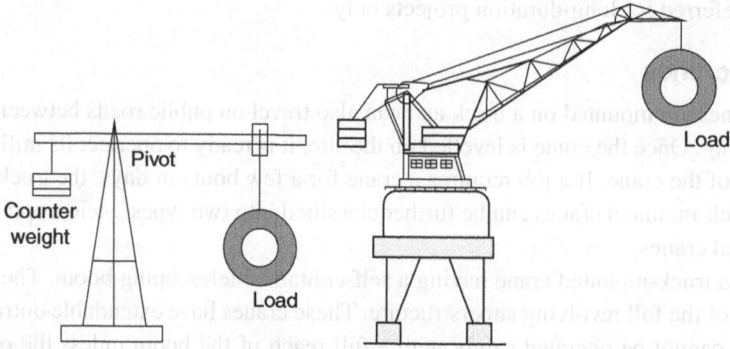

Fig. 29.7 Counter-weight crane

Figure 29.8 shows four types of crane booms depending upon the load-carrying capacity and reach required.

29.2.2 Tower Cranes

Tower cranes provide excellent working radii and lifting heights but have low load-carrying capacity compared to mobile cranes. The mobility of tower cranes is limited. Fixing or installing a tower crane on the site for construction requires a proper plan. Once it is installed, it should be able to support all the activities. So, a tower crane should be selected for the job based on the following criteria:

a. Weight, dimension, and lift radius of the heaviest load
b. Maximum free standing height
c. Machine climbing arrangement
d. Weight of machine support on structure
e. Area that must be reached
f. Speed of the machine

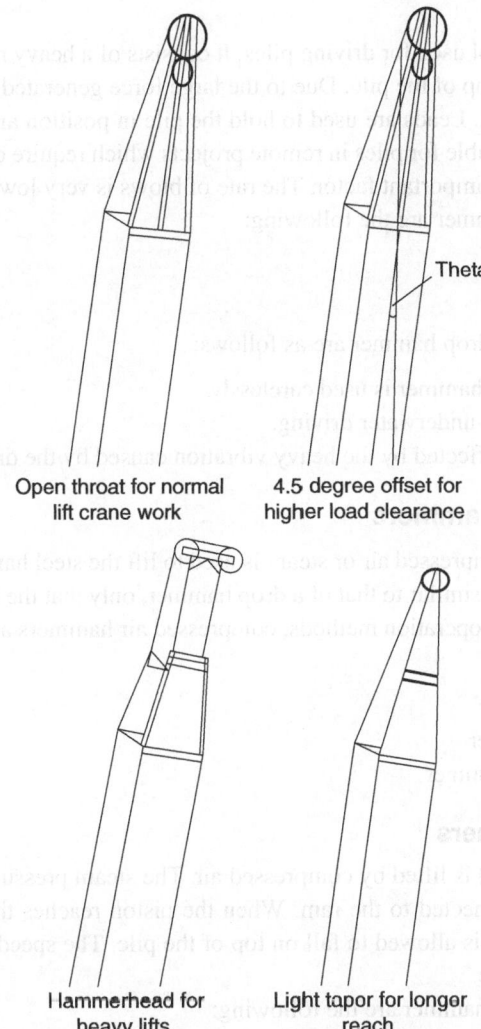

Theta

Open throat for normal
lift crane work

4.5 degree offset for
higher load clearance

Hammerhead for
heavy lifts

Light taper for longer
reach

Fig. 29.8 Optional crane booms

Tower cranes generally have two configurations—top slewing and bottom slewing. The first one has a fixed tower and a swing circle mounted on the top, allowing the jib, tower top, and operator cabin to rotate. The second one has a swing circle at the bottom, allowing both the tower assembly and the jib assembly to rotate.

29.3 Pile Driving Equipment

Piles are structures used to transfer the load of the structure they support to the rock formation below. Piles are made of concrete, steel, timber, etc., and should be inserted into the soil. For the purpose of driving a pile inside the soil, various types of hammers are used. Some of these are the following:

a. Drop hammer
b. Compressed air hammer
c. Diesel hammer
d. Hydraulic hammer
e. Vibratory driver

29.3.1 Drop Hammers

The drop hammer is a simple tool used for driving piles. It consists of a heavy metal, which is lifted and then released and allowed to fall on top of the pile. Due to the large force generated by dropping the hammer, the pile slowly moves inside the soil. Leads are used to hold the pile in position and guide the movement of the hammer. Drop hammers are suitable for piles in remote projects which require only a few piles and for which the time of completion is not an important factor. The rate of blows is very low, 4–8 blows per minute.

The advantages of a drop hammer are the following:

- small investment
- simplicity of operation

Some of the disadvantages of a drop hammer are as follows:

- Piles can get damaged if the hammer is used carelessly.
- It cannot be used directly for underwater driving.
- Adjacent buildings may be affected by the heavy vibration caused by the dropping hammer.

29.3.2 Compressed Air Hammers

In a compressed air hammer, compressed air or steam is used to lift the steel hammer which is allowed to fall on top of the pile. This system is similar to that of a drop hammer, only that the lifting is done by compressed air or steam. Depending on their operation methods, compressed air hammers are further classified into three types. These are as follows:

a. Single-acting steam hammer
b. Double-acting steam hammer
c. Differential-acting steam hammer

Single-acting steam hammers

In this hammer, the weight (ram) is lifted by compressed air. The steam pressure is applied from the bottom side of the piston, which is connected to the ram. When the piston reaches the required height, the steam pressure is released and the ram is allowed to fall on top of the pile. The speed of driving is high at the rate of about 40–60 blows per minute.

The advantages of using this hammer are the following:

- Faster driving
- The greater frequency of blows minimizes the increase in skin friction between blows
- The heavier ram falling at a lower velocity transmits a greater portion of the energy, driving the pile faster
- The closed-type mechanism makes the operation of this hammer feasible for underwater construction

The disadvantages of this type of hammer are the following:

- It is more complicated.
- It requires more time for the initial set-up.
- It requires a large crew for operating the equipment.
- The initial investment is high.

Double-acting steam hammers

The set-up and all other functions of the double-acting steam hammer are the same as that of the single-acting steam hammer. The only difference is that in the operation. In this method, the steam pressure is applied to the underside to raise the ram, and pressure is also applied to the top side to increase the speed of the blow. The ram is therefore of less weight and the striking velocity is increased.

The advantages of using this hammer are the following:

- Faster and easier driving of piles
- The greater frequency of blows minimizes the development of static skin friction between blows

The disadvantages of this type of hammer are the following:

- This hammer is more complicated.
- Its relative low weight and high velocity makes it less suitable for driving heavy piles.

Differential-acting steam hammers

This hammer incorporates the functions of both single and double-acting hammers. In this type, steam is used to lift the ram and the pressure used is not exhausted as in the single type, but is used to accelerate the ram in the downward direction similar to the double-acting hammer. This hammer requires the use of pile protection absorbers to avoid damage to piles.

29.3.3 Diesel Hammers

The diesel hammer is a self-contained driving unit. It is simpler and easier to move from place to place. The entire unit consists of a vertical cylinder, piston or ram, fuel tank, fuel pump, and mechanical lubricator. Its operation is similar to that of the compressed air hammer with the difference that it is powered by diesel. Figure 29.9 shows the operation of diesel hammers. There are two types of diesel hammers—the open type and the closed type. The open type delivers 40–55 blows per minute and the closed type delivers 75–85 blows per minute.

The advantages of using this hammer are the following:

- It does not need any external source of energy.
- It is economical to operate.
- It is convenient to operate in remote areas.
- It can be used in very cold areas.
- It is light in weight and easy to transport from one place to another.
- Its maintenance and servicing is simple.
- The energy per blow increases as the driving resistance of a pile increases.

The disadvantages of this type of hammer are the following:

- The length of the diesel hammer is greater compared to other types of hammers.
- This hammer may not work properly in soft ground unless the pile offers sufficient driving resistance.
- The number of strokes per minute is less compared to steam hammers.

29.3.4 Hydraulic Hammers

The hydraulic hammer functions on the basis of the differential pressure of hydraulic fluid. The dynamic pile driving equation is applicable to this hammer. A hydraulic hammer can be used to drive steel sheet pile and steel diaphragm piles. It can also be used in areas where the overhead space is restricted.

29.3.5 Vibratory Drivers

These drivers are preferred when piles are to be driven into water-saturated non-cohesive soils. Operation with these drivers may be difficult in dry or cohesive sand. The driver is equipped with a horizontal shaft to which eccentric weight is attached. The shaft rotates in the opposing direction and the force produced by the rotating concentric weights produces vibrations. These vibrations are transmitted to the piles and the pile vibration is transmitted into the soil. The agitation of the soil reduces the skin friction between the soil and the pile; this is more effective when the soil is saturated with water. The dead load of the pile and the vibrations drive the pile effectively. The relationship between exciting force and rotation is shown in Fig. 29.10.

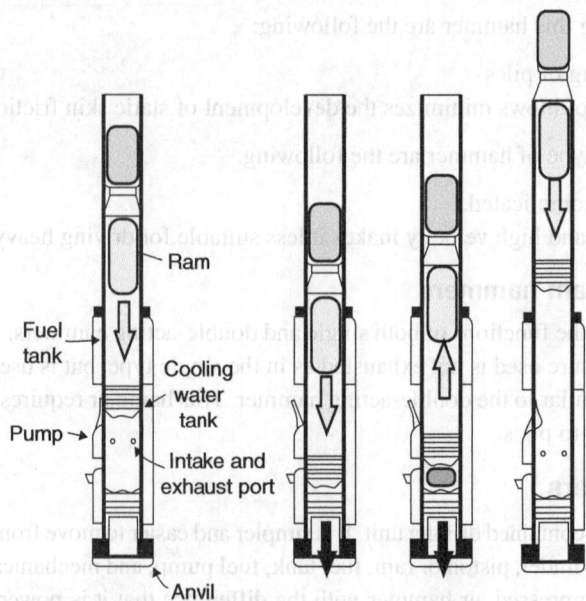

Fig. 29.9 Diesel hammer operation

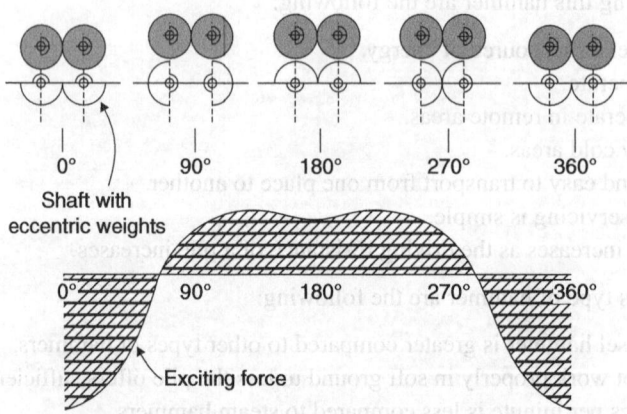

Fig. 29.10 Force rotation relationship for vibratory drivers

Exercises

Review Questions

1. Write a note on the equipment used in the production of aggregates.
2. What type of equipment is used for placing concrete? In what way does this equipment avoid segregation during placing?
3. Describe the various aspects of pumping concrete.
4. What are the methods used for consolidating concrete? What type of equipment is used for consolidation?
5. What are the precautions to be taken while adopting the steam curing method?
6. Describe the method of slipform paving and state its advantages.

7. What are the various lifting devices? Describe their use in construction.
8. How are piles driven into the ground? What equipment and machines are used for this operation?
9. What are the advantages of using construction equipment over the conventional manual method of construction?
10. Should a builder have all the equipment he needs for construction? Otherwise, how should he manage a good quality construction using appropriate equipment? Discuss.

Multiple Choice Questions

1. Construction machinery is employed in order to
   a. Speed up construction
   b. Improve quality
   c. Economize
   d. All of the above

2. Washers are used for
   a. Cleaning aggregates
   b. Manufacturing sand
   c. Manufacturing coarse aggregates
   d. Blending aggregates

3. The equipment used to move concrete horizontally is
   a. Wheelbarrow
   b. Lift
   c. Dump bucket
   d. Tower crane

4. The minimum pipe diameter for pumping concrete is
   a. 125 mm
   b. 100 mm
   c. 150 mm
   d. 200 mm

5. The equipment used to place concrete anywhere in a multistorey building construction is
   a. Wheelbarrow
   b. Belt conveyor
   c. Pump
   d. Boom placer

6. Pavement durability can be improved by
   a. Surface vibrator
   b. Slipform paving
   c. Vacuum dewatering
   d. Float finisher

7. Drop hammer method of pile driving is not preferred in urban areas because
   a. It introduces vibration in adjacent buildings
   b. It is costly
   c. It requires energy to drive
   d. It reduces skin friction

8. Diaphragm piles can be driven by a
   a. Drop hammer
   b. Compressed air hammer
   c. Diesel hammer
   d. Hydraulic hammer

9. Tower cranes are preferred because
   a. They have excellent working radius
   b. They are mobile
   c. They can handle large weight
   d. They are stable

10. Equipment that is preferred for driving piles in water-saturated non-cohesive soil is
    a. Hydraulic hammer
    b. Vibratory hammer
    c. Drop hammer
    d. Double-acting steam hammer

Answers to Multiple Choice Questions

1. d   4. a   7. a   10. b
2. a   5. d   8. d
3. a   6. c   9. a

CHAPTER 30

Performance and Maintenance of Concrete Structures

The whole-life performance of a concrete structure will be satisfactory only if all the processes of concreting—planning, analysis, design, and execution, involving fabrication and construction, regular inspection and maintenance, and repair when necessary, including the aspects of final demolition and disposal of debris and salvage or disposal of waste—are earnestly considered. If any of these aspects is ignored, the structural integrity of the concrete mass will be weakened; this may lead to a premature failure of the structure or may render it unsafe even before completing its service life. The service life of a structure can be broadly defined as *'the time period for which a structure in a specific environment under the accepted code prescribed load conditions will retain its desirable properties of service and provide security against collapse in addition to exhibiting an acceptable aesthetic appearance'*.

This chapter deals with the various aspects of the maintenance of concrete structures. It is a common belief that concrete once made will last forever. However, time and again we have seen that concrete structures deteriorate even immediately after their construction. This may happen due to the factors we have discussed in earlier chapters. So, at the time of design, due consideration must be given to ensure that the structure is serviceable and strong.

The concern for maintainability of concrete structures stems from the need to achieve durability through measures taken for preventing or slowing down the process of deterioration of concrete. The factors that are responsible for deterioration of concrete are the following:

- Presence of water or moisture in and around concrete
- Severity of the environment surrounding concrete
- Ineffective compactness (minimization of pores) of concrete
- Inadequate thickness of concrete cover
- Defective grouting compactness in case of prestressing
- Width of crack(s), if any, in concrete

During the service life of the structure, protection against overloading due to actions that were not anticipated in the design should be ensured. Regular inspections and maintenance by repair, restoration, and/or rehabilitation are essential. Serviceability of the structure has to be ensured at all times. Hence it is necessary to examine the performance profile of the structure at regular intervals. If proper maintenance and servicing of the structure is ensured, it may reach the limit state of collapse only rarely because of extraordinary events such as earthquakes.

A concrete structure may consist of structural components of varying strengths. For instance, the expansion joint is regarded as a weak item, whereas the beam is considered to be a strong item. While the strong items need less maintenance, the weak items should be monitored periodically and even replaced if necessary. From the point view of introducing a maintenance strategy, it is necessary to have a deterioration model for various components of a concrete structure.

30.1 Factors Affecting Whole-life Performance

In this section we will discuss some major factors that affect the performance of concrete structures during their service life.

30.1.1 Water or Moisture

Water ingress or moisture ingress is one of the major causes of deterioration of concrete structures. The detailing of a structure at the time of design should consider shapes which will easily facilitate drainage. Rainwater should not accumulate and stagnate in and around the structure; it should drain away quickly. Figure 30.1 shows some good shapes to facilitate easy and quick drainage of rainwater.

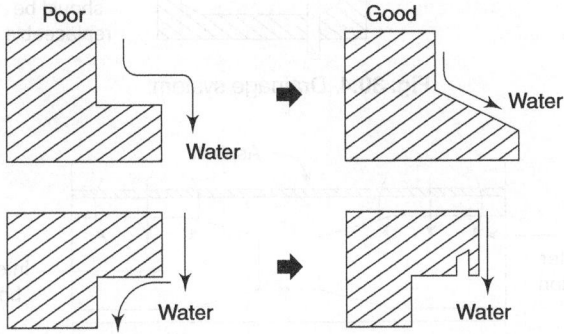

Fig. 30.1 Design detail to drain water

All structural components made of concrete should be designed in a way as to protect them from splash water. To avoid splash water, a drain/tube has to be provided (as shown in Fig. 30.2) so that water does not accumulate near the kerb. Drain pipes should never be embedded in concrete; these should be exposed to inspection as shown in Fig. 30.3. The leakage of water from damaged drains into the concrete mass is a serious problem and

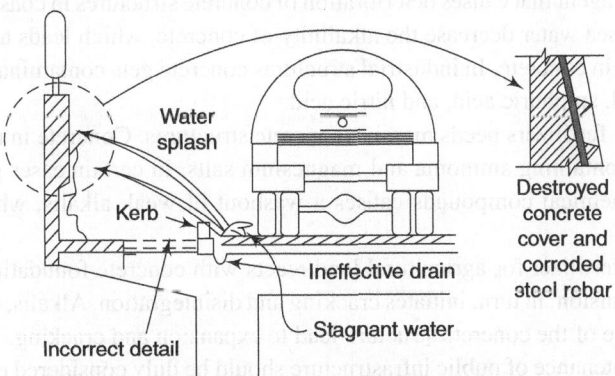

Fig. 30.2 Splashing due to stagnation of water and vehicle movement

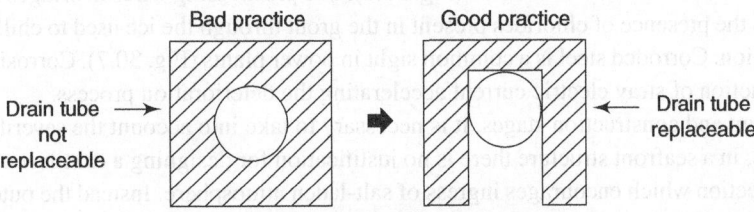

Fig. 30.3 Draining water

may cause early corrosion in the deck slab. The detail shown in Fig. 30.4 will help avoid premature service life problems.

Asphalt pavements on concrete bridges should not be assumed to be watertight. Water may penetrate the deck (Fig. 30.5). It is important to make the waterproofing between the deck and the pavement foolproof.

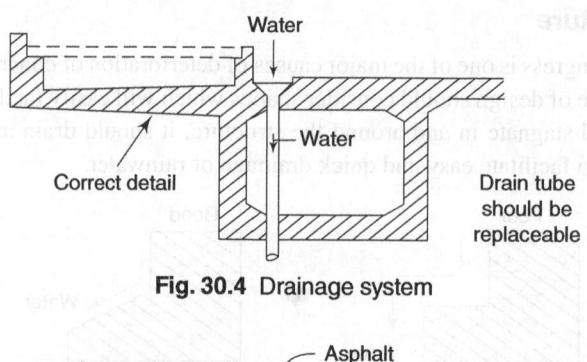

Fig. 30.4 Drainage system

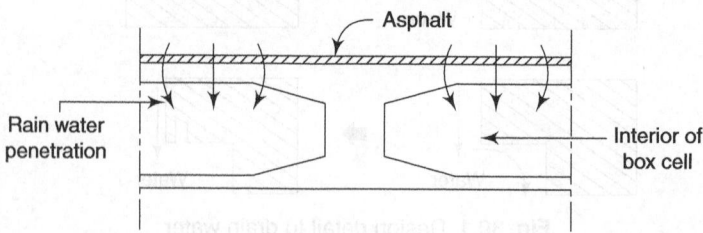

Fig. 30.5 Water penetration through asphalt into interior of a box girder

30.1.2 Severity of Environment

The deterioration and service life of a structure very much depend on the severity of the environment surrounding it. The most predominant agent that causes deterioration of concrete structures in coastal areas is the sea water. Chloride ions present in sea water decrease the alkalinity of concrete, which leads to depassivation and corrosion of steel embedded in concrete. In industrial structures concrete gets contaminated with inorganic acids such as hydrochloric acid, sulphuric acid, and nitric acid.

Production of urea for fertilizers needs massive concrete structures. Concrete in these structures are subjected to environments containing ammonia and magnesium salts. In certain cases groundwater containing magnesium and other chemical compounds causes a washout of weak alkalis, which leads to significant deterioration.

The sulphate used as fertilizer for agricultural land reacts with concrete foundation, forming compounds that expand, and this expansion, in turn, initiates cracking and disintegration. Alkalis, carbonates, and silicates in contact with the surface of the concrete structure lead to expansion and cracking.

The aspect of the maintenance of public infrastructure should be duly considered even during the planning stage. In box girder bridges, impurities such as bird droppings, eggs, and dirt can lead to severe deterioration, thus warranting replacement of the bottom slab (Fig. 30.6). The prestressing wires in bridges may be subjected to corrosion due to the presence of chlorides present in the grout through the ice used to chill the grout during hot weather execution. Corroded steel is a common sight in power plants (Fig. 30.7). Corrosion of steel in this case is due to the action of stray electric current accelerating the deterioration process.

During the design and construction stages, it is necessary to take into account the severity of the environment. For example, in a seafront structure there is no justification for designing a cantilever sunshade since it acts as a cracked section which encourages ingress of salt-laden atmosphere. Instead the outer surface should be designed as non-cracking section avoiding re-entrant corners on the outer surface.

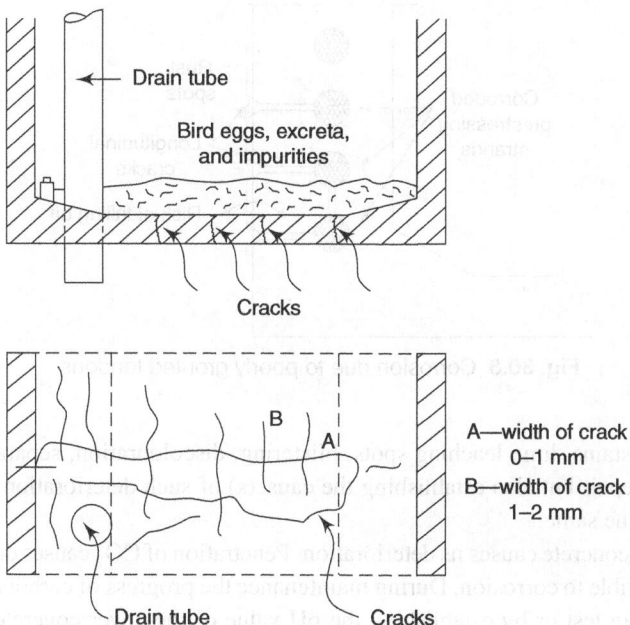

Fig. 30.6 Presence of impurities and cracking of lower chord of a box girder

Fig. 30.7 Corrosion due to stray currents

During the service life of the structure, suitable measures should be adopted to help decrease the severity of environment, which in turn will help in slowing down the deterioration process. Similarly, maintenance measures taken to clean and remove aggressive chemicals deposited on the structure will increase its service life and durability. The micro-climate surrounding the structure is of concern to both the designer and the maintenance engineer. It is in the hands of the designer and the maintenance engineer to change the micro-climate and to make it a favourable one.

30.1.3 Concrete Cover

The state of concrete cover determines the service life of a structure. Concrete cover which is porous, thin, poorly made, or made with porous cover blocks can cause corrosion of steel. It is necessary to have the best possible compactness and necessary thickness for the concrete cover. Grouting for prestressing tendon should be made with good grout of sufficient thickness (Fig. 30.8). During maintenance, it is important to check and control concrete cover quality, grout, and seal cracks and porosity when noticed.

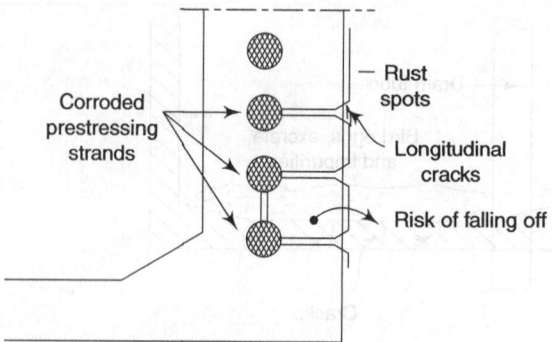

Fig. 30.8 Corrosion due to poorly grouted tendons

The appearance of rust stains, lime leaching spots, blistering, discolouration, separation, and falling off of cover require not only repair but also establishing the cause(s) of such deterioration so that necessary steps can be taken to remove the same.

Carbonation of cover concrete causes its deterioration. Penetration of CO_2 causes depassivation of concrete and renders steel susceptible to corrosion. During maintenance the progress of carbonation can be established using the phenolphthalein test or by establishing the pH value of the cover concrete to know the extent of carbonation and hence depassivation.

30.1.4 Cracks

Cracks can occur during either construction or operation of the structure which causes deterioration. Though concrete is designed to crack, the crack width must be limited. Penetration of aggressive chemicals occurs through cracked areas. Hence in environments of aggressive chemical activity, concrete should be designed as uncracked. For example, it is prudent to design the top roof of a cantilever as a non-cracking section. This will enhance the service life and durability of the structure.

At the design stage, it is important not only to assess the capacity of the section but also to check and limit crack widths as per the environment. During construction temperature and shrinkage cracks should be controlled by proper planning of the sequence of concreting. During maintenance it is necessary to monitor the crack widths and seal them with appropriate polymeric repair material if the width exceeds the safe limits. Thus control of crack width is an effort that starts from the planning stage and continues through the design, construction, operation, and maintenance stages of the structure.

30.1.5 Curing

After placement and compaction of concrete, adequate measures need to be taken to obtain expected properties from the hardened concrete. The main properties of concrete, which are desirable from the long-term behaviour point of view, are its strength, impermeability, and durability. Curing is a process that helps avoid premature drying of concrete and makes available adequate water, after placement, for a sufficiently long period of time for concrete to gain adequate degree of hydration within its mass and particularly in the surface cover layer. Curing protects concrete against drying due to sunshine and wind and effects due to early age shrinkage.

Careful and adequate curing ensures high-quality concrete, which is durable and lasts longer. It is recommended that curing and protection should be started immediately after compaction of fresh concrete. The time the concrete surface loses its sheen may be taken as an indicator for this. Concrete starts losing sheen when the concrete surface starts losing moisture. Any delay or interruption in curing at this stage cannot be compensated later on by extending the duration of curing.

Ponding is the most effective method of curing concrete. But this method is not always possible owing to site constraints and some other factors. Covering the concrete surface with wet gunny bags or straw is also effective. This can be used easily almost anywhere, both at the construction site and at the laboratory.

Keeping formwork in place for a longer period also prevents rapid drying of wet concrete. However, wooden formwork may absorb moisture from concrete. Such formwork must be kept moist externally.

Curing with plastic films is ideal for large concrete surfaces. However, films should be properly placed and secured to prevent them from getting dislodged due to either wind or other construction activity. Membrane-forming curing compounds can also be used. However, these may be useful only for a short period of time as these may be disturbed by the construction process used. In addition, their effect on the bond at the construction joints needs to be checked and made good. Curing by sprinkling of water may result in cracking of concrete due to the temperature differential between concrete and the sprinkled water. Therefore this method of curing should be discouraged and discontinued with.

30.2 Measures to Improve Safe Life and Durability

Construction of concrete structures involves a number of different activities and stages. These activities and stages need to be planned and executed in the most efficient way so as to achieve the desired results. A concrete structure is constructed to serve some particular purpose. This structure must serve the specified purpose for some specified minimum life (i.e., safe life) without compromising the safety of the users/occupants. Some of the measures that help engineers achieve these objectives are discussed briefly below.

30.2.1 Constituents of Concrete

Cement: High-strength cement with low C_3A content and moderate alkali content (5–8%) and of uniform quality. Pozzolana or slag would be an advantage.

Aggregate: Impurities need to be controlled. Fines less than 0.3 mm to be controlled to ensure slump and stability.

Admixture: Efficient water reducers (superplasticizers) and air entrainers at high slump and ensuring good batching process is desired. Particle packing should be efficient (0.2 mm).

30.2.2 Mix Proportion

A water–cement ratio of less than 0.45 is imperative for long-term performance. A cement content of more than 380 kg/m^3 imparts a self-healing ability to the concrete. A stable mix at high slump requires good grading of sand and efficient admixtures. A small dosage (<5%) of condensed silica fumes improves the strength and stability of concrete. A large dosage impairs constructability. Full-scale site trials are necessary before the correct mix is selected.

30.2.3 Batching Plants

Modern batching plants aid selection of the optimum batching procedure. Each batch should be checked using control tests. Adequate number of sample tests and uniform quality control are essential.

30.2.4 Compactness of Concrete

Concrete should be compact and free from any voids if it has to last long and remain stable. Re-vibration of the top layer in deep members helps minimize voids under embedded steel. Also the concrete cover quality should be maintained by proper compaction, especially below the bottom and top bars.

30.2.5 Control on Cover

The long-term performance of concrete depends on the durability of cover concrete. Steel reinforcement is covered and protected by the cover concrete. However, the basic environment surrounding reinforcement bars may be destroyed by carbonation, i.e., by the reaction of atmospheric CO_2 with $Ca(OH)$ of concrete. Two essential considerations to ensure protection of steel reinforcement against carbonation are

- the thickness of cover concrete and
- the density of cover concrete.

The concrete cover, in addition to the protection it gives to steel, serves to enhance the bond strength and provides protection against fire especially to steel. The thickness of the concrete cover is generally governed by the

- measurement of bent or suspended bars,
- clearance width between formwork and rebar,
- amount of dislocation of the rebar cage during concreting, and
- height/thickness of spacers.

To ensure proper concrete cover, the following checks are essential at the construction site:

- Inspection of rebar measurements, especially of bent or suspended bars
- Rejection of non-conforming bars
- Inspection of formwork for clearance of rebars prior to concreting

Dislocation of the reinforcement cage during concreting should be avoided completely. For this, a sufficient number of spacers and bridge planks should be provided. All factors that may produce deviation of the cage should be avoided.

Proper density of the cover concrete should be ensured by checking the resulting water–cement ratio in the cover region. Note that the water–cement ratio in the cover region should be more than the overall water–cement ratio adopted. However, one should plan to have less water–cement ratio in the cover region.

30.2.6 Construction Joints

Construction joints should be properly made and checked for durability. Any laitance should be removed and joints finished with rich mortar.

30.2.7 Temperature Effects

One of the major causes of cracking is the temperature effect. Concreting is a chemical process involving an exothermic reaction with liberation of heat. Hence, temperature effects should be taken care of by the procedure described earlier in Chapter 26 in Section 26.6.

30.2.8 Simple Design

Larger sections are easier to pour during concreting and do not deteriorate fast. The sections should have rounded smooth corners. Abrupt changes in the cross section should be avoided. Larger rebars take less space and make the section robust. Simplicity in design enables easy understanding of the bar schedule and hence fabrication mistakes at the site are reduced.

30.2.9 Training on Good Construction Practices

The quality and durability of concrete components and structures very much depend on the training, skill, and experience of human resources involved in the various stages of the construction process. For this, constant and periodic training of workers and construction supervisors is essential.

30.3 Deterioration Model

Reinforced concrete is deteriorated by carbonation and by chloride ingress. Generally, deterioration occurs in two stages. The first stage is *initiation* and the second stage is *progression of damage* (Fig. 30.9). During the initiation stage, the contaminants (chlorides and CO_2) from the atmosphere reach the steel reinforcement inside concrete. During the progression stage, the level of performance of the structure comes down.

The age at corrosion activation is dependent on the type of concrete (quality expressed in terms of strength) and its cover thickness (Fig. 30.10). The penetration rate of CO_2 from the atmosphere into concrete can be expressed as

$$d = k(t)^{0.5}$$

where d is the depth of penetration up to time t and k is the diffusion constant.

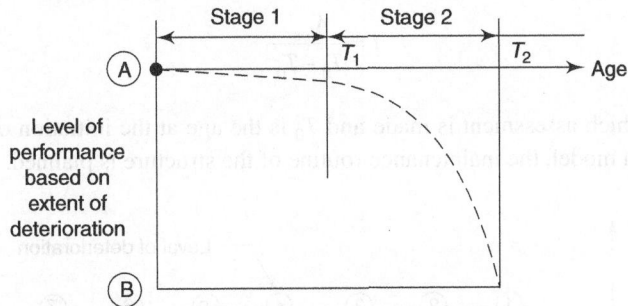

Fig. 30.9 Reduction in performance based on extent of deterioration

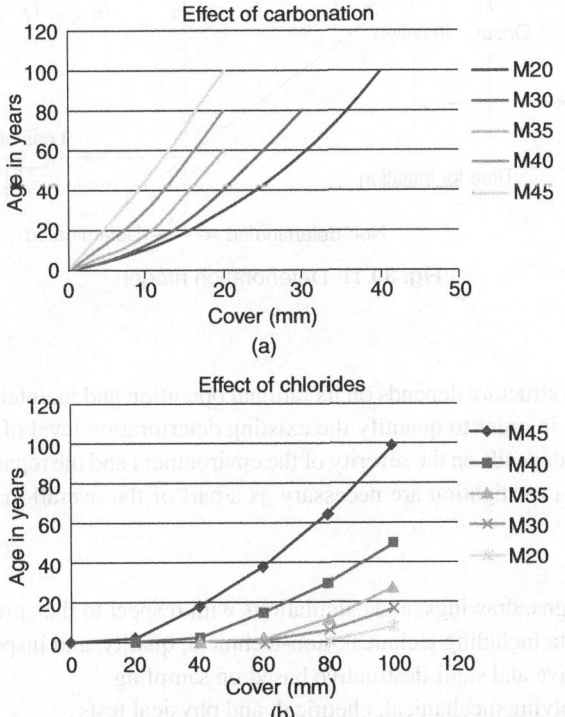

Fig. 30.10 Design charts for cover protection against corrosion

On the basis of Roberts and Atkins' observations (1997), it is possible to postulate deterioration model for reinforced concrete. The model identifies eight distinct levels of deterioration, after various ages (T_i). These are

1. Arrival of contaminants at the surface of concrete
2. Contaminant ingress
3. Onset of corrosion
4. Active corrosion
5. Delamination
6. Spalling
7. Reduction in steel area
8. Failure

Figure 30.11 shows these eight levels of deterioration. The damage can be assessed by the area loss of steel. The area loss per year, i, can be expressed as

$$i = \frac{A_{loss}}{T_n - T_0}$$

where T_n is the age at which assessment is made and T_0 is the age at the initiation of deterioration. On the basis of this deterioration model, the maintenance routine of the structure is planned.

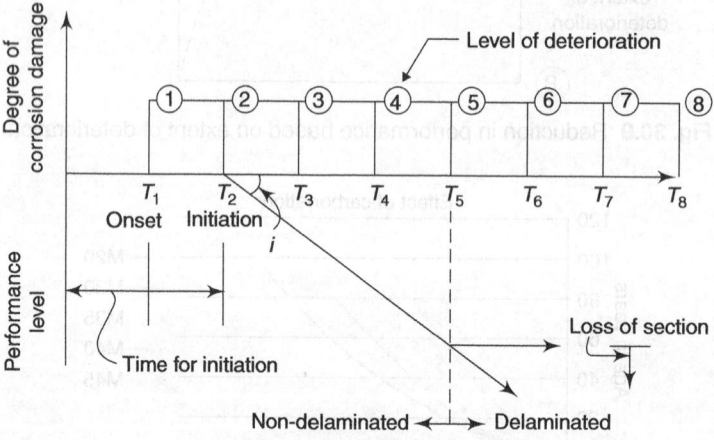

Fig. 30.11 Deterioration model

30.4 Inspection

The durability of a concrete structure depends on its rational operation and maintenance. Regular and systematic inspection is necessary in order to quantify the existing deterioration level of the structure. The way the structure behaves with time depends on the severity of the environment and the repair interventions undertaken. The following elements of investigation are necessary as a part of the overall maintenance during periodic inspection and reporting:

1. Visual inspection
2. Checking original designs, drawings, and calculations with respect to the current use of the structure
3. Checking execution data including technical, non-technical, quality, and inspection reports
4. In situ testing: destructive and semi-destructive based on sampling
5. Laboratory testing involving mechanical, chemical, and physical tests
6. Performing recalculation based on all the above data for assessing the current status of the structure

30.5 Tests and Monitoring

The whole-life performance of the concrete structure can be effectively monitored by periodic testing. Periodic testing is much more involved than a simple cube compression test conducted at the time of construction. These tests conducted at periodic intervals on the actual structure should concentrate on the environmental severity, structural detailing, loading conditions, existing cracks and their sizes, and the overall condition of the structure or member being investigated. The evaluation based on tests should include the condition of both concrete and reinforcing steel. The investigation procedure on concrete is shown in Fig. 30.12.

In situ load testing is required to assess the performance of the structural element under loading greater than the working load. A load test is undertaken either to clear a doubt regarding the acceptability of an element or to establish the behaviour of an existing structure for serviceability criteria. Where tests are undertaken to demonstrate the satisfactory behaviour under load, these are generally based on deflection measurements. The obtained results can be corroborated with the results of a computer analysis after accounting for the realistic properties of the structure under test. The results should show that at service load the deflections and strains are within acceptable limits. The deformations are almost completely recovered when the load is released.

Various non-destructive tests that can be used to monitor the behaviour of concrete structures have already been discussed in Chapter 27.

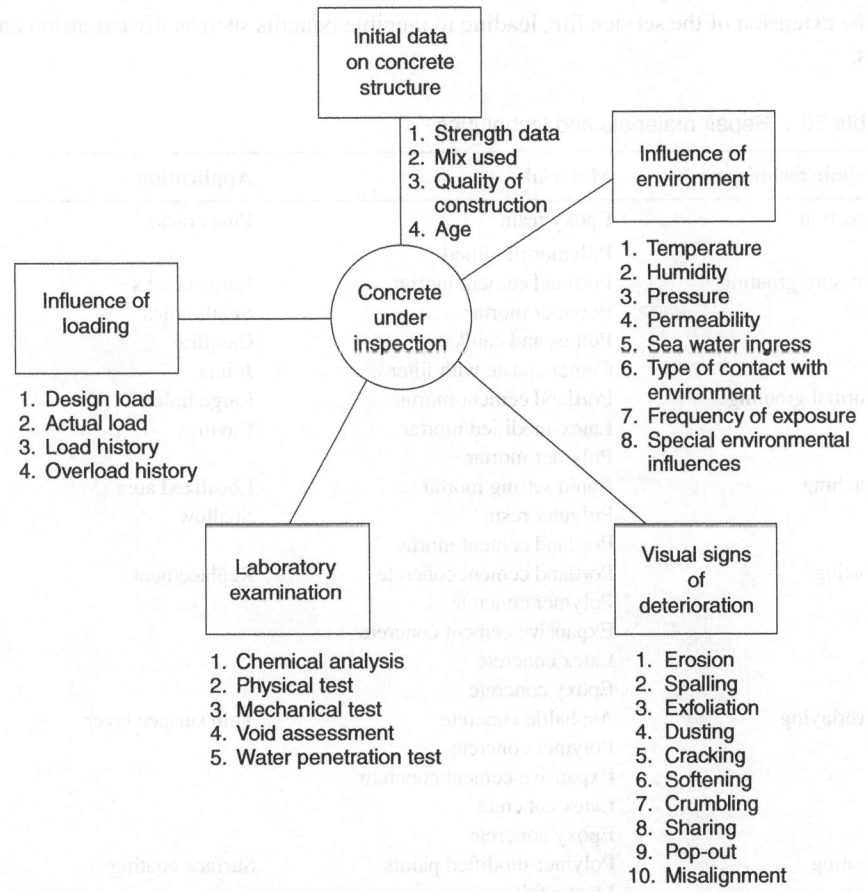

Fig. 30.12 Concrete investigation method

Long-term monitoring of a structure can be used to evaluate deterioration with load and response under service conditions. When a structure is continuously deteriorating, monitoring can be used to decide on either retrofitting or replacement of the deteriorating component of the structure at a particular instant of time. Since the measurements are made over a long period of time, the instruments used should have long-term stability. Allowance for daily variation in temperature and seasonal changes should be made while interpreting long-term test results.

30.6 Repair Materials and Techniques

After the structural evaluation is over, appropriate repair materials and techniques must be carefully chosen to suit the behaviour, field condition, and future whole-life performance of the evaluated structure. Various repair materials and techniques have already been discussed in Chapters 23 and 24. Table 30.1 shows various types of repair techniques involving different repair materials applicable to the different types of defects mentioned.

Prior to repair, all damaged and disintegrated portions should be removed until the undamaged parent material is exposed. This ensures good bonding of repair material with the parent material. Before finalizing the repair option, the whole-life cost model can be worked out with different alternatives. Figure 30.13 shows alternative costs over expected life using a particular discount rate. The whole-life cost is the sum of the first cost plus the discounted replacement cost in the future.

Rehabilitation is less costly than replacement. It is important to reduce its cost. It is well understood that one unit of maintenance cost will result in a reduction of four to five units of rehabilitation cost. This advantage results from the extension of the service life, leading to tangible benefits such as life extension and reduction in repair costs.

Table 30.1 Repair materials and techniques

Repair technique	Material	Application
Injection	Epoxy resin	Fine cracks
	Polymer-modified	
Pressure grouting	Portland cement mortar	Large cracks
	Polymer mortar	Small holes
	Putties and caulks	Cavities
	Cement paste with filler	Joints
Normal grouting	Portland cement mortar	Large holes
	Latex-modified mortar	Cavities
	Polymer mortar	
Patching	Rapid setting mortar	Localized area
	Polymer resin	Shallow
	Portland cement mortar	
Placing	Portland cement concrete	Replacement
	Polymer concrete	
	Expansive cement concrete	
	Latex concrete	
	Epoxy concrete	
Overlaying	Asphaltic concrete	Thin surface layer
	Polymer concrete	
	Expansive cement concrete	
	Latex concrete	
	Epoxy concrete	
Coating	Polymer-modified paints	Surface coating
	Mastic felt	

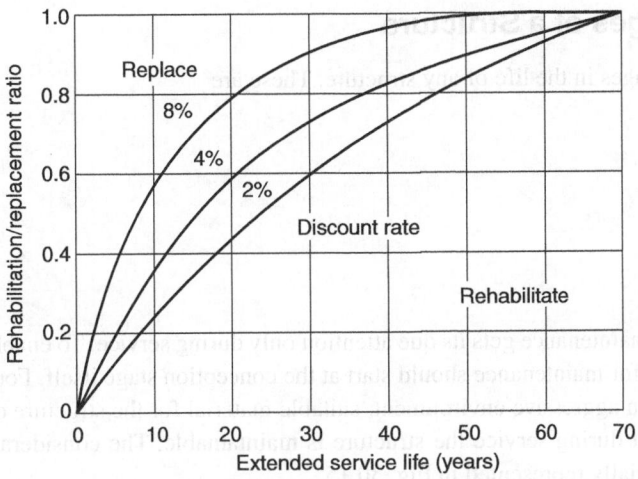

Fig. 30.13 Lifecycle cost comparisons

30.7 Maintenance Requirement

The purpose of the maintenance of a concrete structure is to ensure its safety and serviceability at acceptable performance levels. Structural safety and serviceability are essential. The performance level shows the level of satisfaction by the user with respect to the functional utility of the structure. This is illustrated with the help of a performance model in Fig. 30.14. As can be observed from the figure, the performance (A) of the structure deteriorates from the day the structure is opened for use and becomes unserviceable at stage B after, say, 100 years. However, the useful life of the structure may get terminated at stage C owing to the performance level going below the functionally acceptable limit. Hence, the realized service life is less than that for which the structure was designed. In the absence of any regular maintenance this reduction can range from 10 to 80 years depending on the severity of environment. To achieve the full designed service life, periodic maintenance thus becomes inevitable.

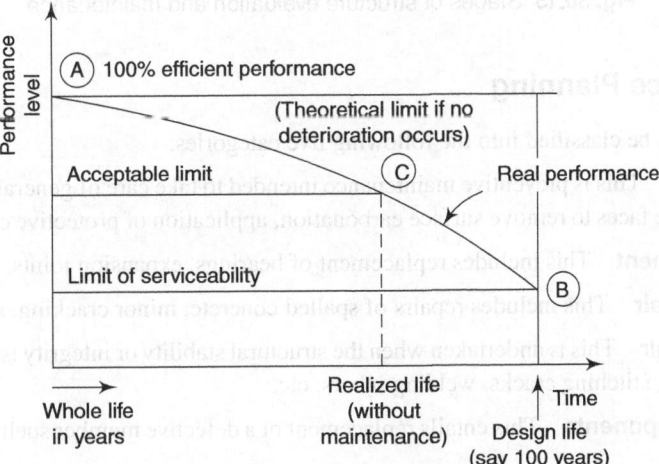

Fig. 30.14 Performance Model

30.8 Critical Stages of a Structure

There are six critical stages in the life of any structure. These are

1. conception
2. analysis
3. design
4. construction
5. service
6. failure/demolition

Generally, the issue of maintenance gets its due attention only during service. To enable good and proper performance, the planning for maintenance should start at the conception stage itself. For instance, if a structure is to be constructed in an aggressive environment, suitable material for the structure can be fixed even at the conception stage so that during service the structure is maintainable. The consideration of maintenance at different stages is pictorially represented in Fig. 30.15.

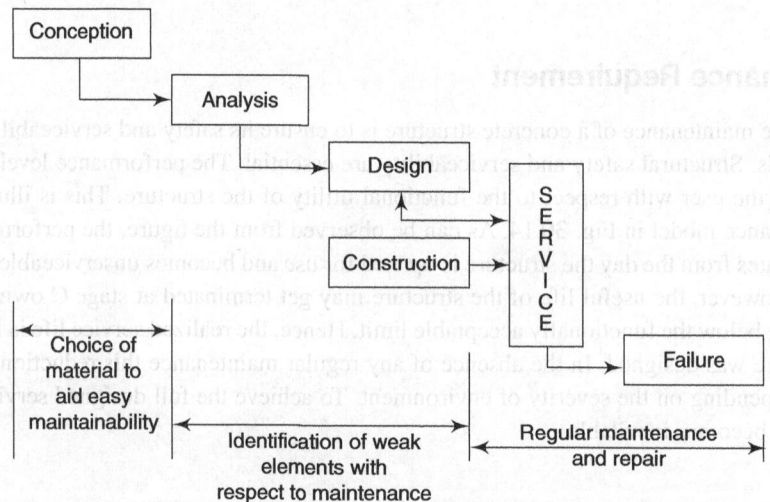

Fig. 30.15 Stages of structure evaluation and maintenance

30.9 Maintenance Planning

Maintenance work can be classified into the following five categories:

Cyclic maintenance   This is preventive maintenance intended to take care of general deterioration, such as flushing down concrete faces to remove surface carbonation, application of protective coat(s) to concrete, etc.

Equipment replacement   This includes replacement of bearings, expansion joints, etc.

Minor structural repair   This includes repairs of spalled concrete, minor cracking, and staining.

Major structural repair   This is undertaken when the structural stability or integrity is at stake. This includes major concrete repairs, stitching cracks, welding rebars, etc.

Replacement of components   This entails replacement of a defective member such as a beam or a column with a new member.

These five maintenance options should be judiciously undertaken during the lifespan of the structure. The effect of continuous maintenance on the lifespan is illustrated in Fig. 30.16. Note that due to material degradation, loading, and foundation settlement, the performance of the structure deteriorates progressively and becomes considerably less than that at the initial condition. Thus at a time interval T_1 repair becomes necessary. At intervals T_1 and T_2, repairs 1 and 2 are performed to improve the serviceability condition either to the initial condition or even above that. At T_d deterioration reaches such a level that it may not be worthwhile to repair the structure.

Generally, various parts of a structure are of varying quality and subject to different exposure conditions. So each element must be considered separately and its repair cost built into the lifecycle cost, which includes costs of all the repairs at T_1, T_2, T_3, etc. Each element must be assigned a life after which it must be either repaired or replaced.

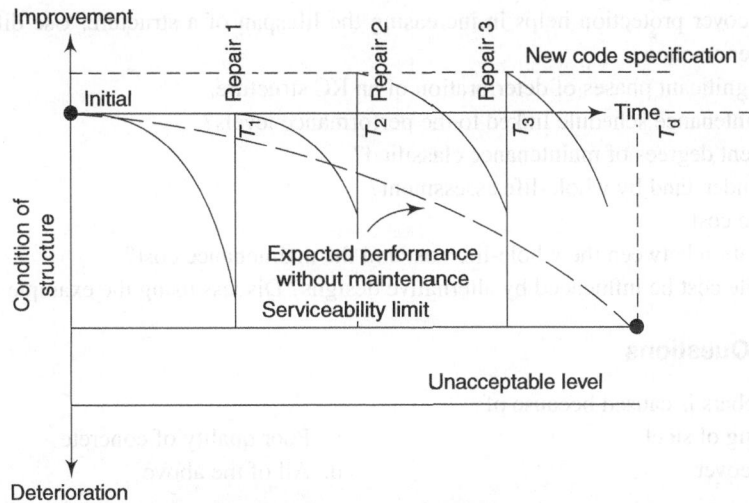

Fig. 30.16 Deterioration–Repair model

30.10 Whole-life Assessment

Whole-life assessment is a method of obtaining data for the whole-life performance of a structure. Both the whole-life assessment and whole-life performance profile of a structure are computed with a view to minimize or optimize the total expenditure on the whole and not in one segment alone.

The method of whole-life costing is considered for comparing different alternative designs initially so that the maintenance expenditure is the least during the lifespan. The service life and maintenance strategy are closely related to the whole-life cost. Many countries are now adopting some form of performance profiling for assessing the whole-life cost.

The whole-life assessment deals with specific structural elements and establishes their performance levels based on performance indicators. The indicators are then used to determine the type of maintenance action to be taken together with the cost considerations.

Though some deterioration models for structures have been proposed, we still do not have a model with acceptable detail. Various techniques used have been either statistical or stochastic. These models are highly theoretical. They use mathematical techniques such as Markov chain or Monte Carlo simulation. What is really needed is an expert system for assessment, choice between different technical solutions, products, and intervention frequencies, etc.

Some maintenance needs to be carried out as a routine to achieve the desired economic life for the structure. Principles of quality assurance and certification should also be considered while constructing structures to ensure their enhanced life and serviceability. The maintenance cost should be included as a sustenance fund in addition to allocating repair/rehabilitation/reconstruction costs.

Exercises

Review Questions

1. Discuss a deterioration model and its relation to the performance of a structure.
2. Use a deterioration model to indicate the level of performance of a typical structure after initiation and during progression stages of damage of a structure.
3. Describe how cover protection helps in increasing the lifespan of a structure. Use diffusion theory to explain this phenomenon.
4. Describe the significant phases of deterioration of an RC structure.
5. How is the maintenance schedule linked to the performance levels?
6. How are different degrees of maintenance classified?
7. What do you understand by whole-life assessment?
8. Define lifecycle cost.
9. What is the relation between the whole-life cost and the maintenance cost?
10. Can the lifecycle cost be influenced by alternative designs? Discuss using the example of an RC bridge.

Multiple Choice Questions

1. Corrosion of rebars is caused because of
   a. Poor detailing of steel
   b. Inadequate cover
   c. Poor quality of concrete
   d. All of the above

2. The design of a structure in general warrants that
   a. Cracking is not permitted
   b. Cracking is permitted but crack width is limited
   c. Cracking is permitted but maximum tensile stress is limited
   d. Cracking is permitted but crack length is limited

3. Corrosion protection of rebars can be effected by
   a. Provision of proper cover
   b. Coating the rebar with epoxy
   c. Using corrosion inhibitors in concrete mix
   d. All of the above

4. The chemical that gets formed during corrosion is
   a. Ferric chloride
   b. Ferric oxide
   c. Ferric sulphide
   d. Ferric sulphate

5. Visual sign of deterioration is
   a. Discoloration
   b. Cracking
   c. Spalling
   d. All of the above

6. When rebar inside starts to corrode, the concrete surface
   a. Bulges
   b. Shows signs of transverse cracking
   c. Shows signs of longitudinal cracking
   d. Shows signs of cover popping off

7. Maintenance is necessary
   a. At regular intervals
   b. When we notice distress
   c. When performance level goes down
   d. When there is fear of failure

8. Maintenance helps to
   a. Prolong life
   b. Keep up performance level
   c. Ensure availability of safety factors
   d. All of the above

9. Whole-life assessment is made to include
   a. Demolition cost
   b. Maintenance cost
   c. Initial conception cost
   d. All of the above

10. Test for locating voids in a concrete structure is
   a. Core test
   b. Rebound hammer test
   c. Thermograph test
   d. Permeability test

Answers to Multiple Choice Questions

1. d
2. b
3. d

4. b
5. d
6. a

7. a
8. d
9. d

10. c

Performance and Maintenance of Concrete Structures 653

8. Maintenance helps to
    a. Prolong life
    b. Keep up performance level
    c. Ensure availability of safety factors
    d. All of the above

9. Whole-life assessment is made to include
    a. Demolition cost
    b. Maintenance cost
    c. Inception cost
    d. All of the above

10. Test for leaching voids in a concrete structure is
    a. Rebound hammer test
    b. Permeability test

CHAPTER 31

Future Trends in Concrete Technology

All types of construction (small or large) use concrete. Though conventional concrete is a widely used construction material, there are other 'tailored' forms of concrete that are being increasingly used for special purposes. Depending upon the need, one may employ prestressed, self-compacting, and/or fibre-reinforced concrete. There are many varieties of modified concrete.

Developments in concrete industry can be classified into six major focus areas as shown in Fig. 31.1. Recycling of demolition waste, use of industrial wastes such as fly ash, and green building methods are developments that have stemmed from the need to conserve materials and obtain maximum output from them. High-strength and ultrahigh-strength concrete are, on the other hand, concentrating on improving the strength and ductility behaviour under challenging loading conditions. Currently, a good number of structures are built in very hostile environments. Such structures need to withstand severe and cyclic environmental changes. Therefore, durability of concrete has been an important consideration in the recent developments in concrete technology. Polymers and co-polymers are added to concrete to impart it new and improved properties that were thought impossible about 20 years ago. The use of special polymeric fibres has made it possible for a material technologist to produce concrete as flexible as rubber

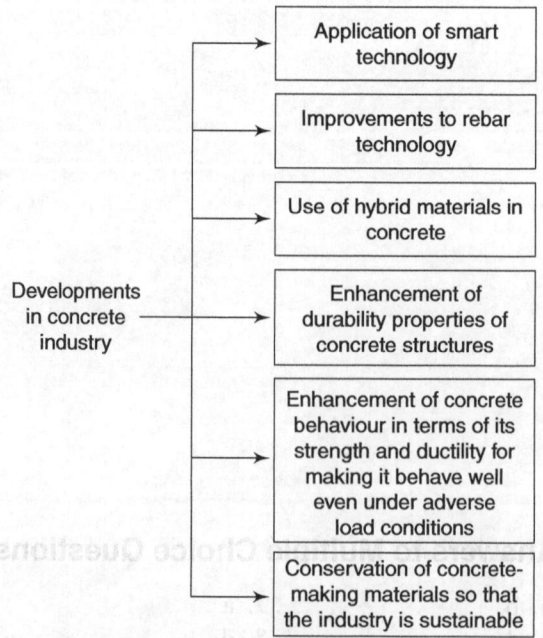

Fig. 31.1 Developments in concrete industry

and as strong as steel. Similarly, the introduction of glass into the concrete mix makes concrete look as beautiful as a jewel. Composites have enhanced the scope of the applicability of this widely used construction material.

31.1 Sustainability of Concrete Industry

Concrete industry is drawing upon enormous natural resources. Disposing of large quantities of construction and demolition wastes in landfills for building new and high-rise structures has become the order of the day. The cement manufacturing process, as well as the accumulation of considerably large volumes of debris and rubble resulting from the demolition of structures, is a serious threat to the environment.

The search for environment-friendly material to substitute cement in concrete becomes most important in the light of the world facing serious problems with CO_2 emissions. It is known that production of 1 tonne of Portland cement accounts for about 1 tonne of CO_2 being added to the atmosphere.

In such a situation the industry is faced with the task of finding ways and means to ease the pressure on the natural resources and to optimize the use of construction material. Efforts in this direction have led to the use of high-volume fly-ash concrete, which can help save about 60% of cement. Also, geopolymer, an inorganic aluminosilicate polymer synthesized from minerals, and by-products such as rice husk ash and fly ash are being used as a binder in concrete.

Rapid modernization and reconstruction of bridges, roads, and industrial structures involve demolition of old structures, which generates a lot of waste. The worldwide consumption of concrete is approximately 7 billion tonnes a year. It has been estimated that every year approximately 65 million tonnes of concrete is discarded in Europe, 70 million tonnes in the United States, and 20–25 million tonnes in Japan. It has been further estimated that by the year 2025 these figures will multiply several fold.

Apart from reconstruction, debris results from natural calamities and ravages of wars. For instance, the hurricane Katrina caused damages to more than 134,000 houses in New Orleans, and after the earthquake in Kobe, Japan, the amount of construction demolition waste was assessed to be of the order of 20 million tonnes. In India, the Gujarat earthquake resulted in the accumulation of large construction debris and wastes. Wars in Afghanistan and Iraq have resulted in large quantities of demolished structures without recycling. If the concrete industry is to sustain then the debris has to be recycled. The current trend is to use technology to recycle wastes so that available resources for making concrete do not get depleted. In addition, the structures that are built should be such that they do not generate waste and/or deplete the natural resources.

31.1.1 Recycled Aggregate Concrete

The depleting nature of concrete aggregates has led to recycling of waste/debris aggregates. In the long run, this will prove to be economically advantageous and sustainable. Such recycled aggregates were used in the reconstruction of European cities on a large scale after the Second World War. Demolished waste with partial substitution of natural aggregates by recycled aggregates originating from concrete provided a solution for the 200 million tonnes/year construction waste generated in European Union.

Both fine and coarse aggregates can be produced by recycling of waste. Physical properties of aggregates should be checked for compliance before using them in the concrete mix. Owing to variation in the composition of the debris at different locations, samples of debris from each location should be tested for basic properties. Thereafter, depending upon the requirement, design of the mix must be arrived at.

The demolition waste aggregates (Fig. 31.2) contain the following basic materials (the proportions are approximate):

a. Demolition concrete = 75%
b. Demolition brickwork = 20%
c. Mosaic tiles and other impurities = 5%

The typical recycled coarse aggregate is shown in Fig. 31.3.

Fig. 31.2 Demolition waste

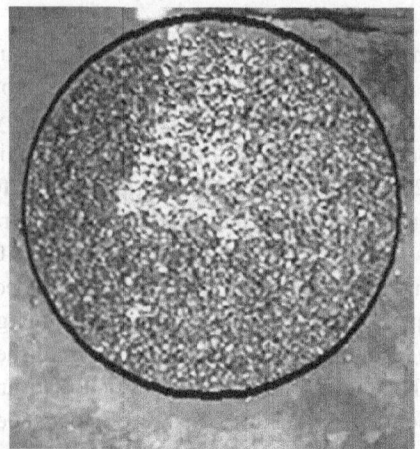

Fig. 31.3 Recycled coarse aggregate

Grading curves

A detailed sieve analysis should be done and various proportions of recycled aggregates should be identified. The proportion of recycled aggregates for which the fineness modulus is nearly the same as that for normal aggregates should be chosen and used. A typical grading curve drawn for recycled and conventional coarse aggregate used in a project is shown in Fig. 31.4.

Properties of recycled aggregate concrete

Recycled aggregate is a good substitute to conventional aggregate in concrete. Experimental investigations have shown that recycled aggregate concrete has compressive strength similar to conventional aggregate concrete. Moreover, demolition waste possesses relatively low bulk density, higher water absorption, and cause low workability in the fresh concrete. However, workability can be improved by using a suitable superplasticizer. Experimental investigations have also shown that recycled aggregate concrete can be effectively and successfully used for structural concrete work. From the point of view of sustainable growth of concrete industry, recycled aggregate concrete holds promise. However, it should be emphasized that proper mix design and quality control are essential for the successful use of recycled aggregate concrete.

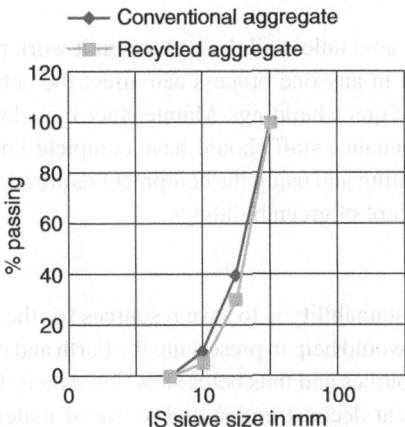

Fig. 31.4 Grading curves for conventional and recycled aggregates

31.1.2 Green Buildings

Green buildings are structures which are designed, built, operated, and/or reused in an ecological and resource-efficient manner. Green buildings are environment-friendly, protect occupants' health, and make efficient use of water, energy, and other resources. Such buildings also reduce the overall impact of concrete construction on the environment. Green buildings are also known as sustainable buildings.

Environmental benefits of green buildings are the following:

1. They sustain the quality of water and air.
2. They help conserve natural resources.
3. They help reduce quantity of waste produced.
4. They improve the ecosystem.

Economic benefits of such buildings include the following:

1. The initial cost of green building construction is a little high, but its operation cost is less.
2. Improved occupant health, comfort, productivity, and reduced pollution ensure that the higher initial cost incurred is justified and, in the long run, proves economical as well.
3. Green buildings need reduced lifecycle cost.

Material benefits from resorting to green building practices are as follows:

1. Reduction in building material needed and, hence, reduction in the cost of construction.
2. Recycling of construction waste.
3. Water is treated and reused (rainwater and water from washbasins can be treated and used for garden).
4. Solid waste management programme prevents waste generation and makes waste a resource.
5. Preserving environment.

Green building practices offer benefits for the occupants of the structure also. These include the following:

1. These buildings offer good quality environment and hence reduce the rate of respiratory diseases.
2. Adequate ventilation is provided.
3. Building material is chosen in such a way that it is non-toxic and helps protect environment.
4. Natural heating and cooling systems improve the room condition and comfort.
5. Prevents indoor microbial contamination by using materials resistant to microbial growth.
6. The overall life style and comfort of the occupant is improved.

Maintenance

A green building cannot achieve its goal unless all the systems in it work properly because all features in the building are interlinked. Stagnation in any one process can affect the entire system. Proper maintenance is important for realizing the benefits of green buildings. Maintenance includes mechanical, electrical, plumbing, and many other aspects. The maintenance staff should have complete knowledge about the green building system and have experience in operating and using the equipment required. Adequate and proper maintenance can ensure optimum use of the concept of green buildings.

Need for green buildings

One of the major components of sustainability is to save resources for the future generations. We should use natural resources judiciously which would help in preserving the Earth and its environment. The green building technology optimizes the use of resources and thus helps slow down their depletion. The concepts of multiple use, future modification, and efficient deconstruction and re-use of materials are intrinsic to green design. The green building construction technology offers real potential for increasing profitability for the society in an environment-friendly way.

31.1.3 Use of Supplementary Material for Sustainable Development

Conservation of Portland cement by partial substitution of supplementary cementitious products and thereby enhancement of service life of structures is essential for sustainable development. The supplementary materials that can be advantageously used (either singly or jointly) are

- fly ash,
- silica fume,
- rice husk ash, and
- metakaolin.

The typical physical and chemical characteristics of these materials are given in Table 31.1.

Table 31.1 Typical characteristics of supplementary cementitious materials

Characteristic	PFA	SF	RHA	GGBS	MK
			Chemical composition		
Silica (%)	48	85–95	85–95	35	52
Carbon	1–2	1–2	2–8	40	—
Alkali	20	10.5–2	2–3	10	<1%
			Physical properties		
Average particle size	0–1	0.1–0.2	7–10	0–1	0–1
Specific surface (m^2/kg)	300–600	15,000–30,000	700	400	12,000
Colour	Black/grey	Grey	Grey	Light grey	White
Specific gravity	2.3	1.4	2.4	2.8	2.5

Concrete containing fly-ash content of about 20% is used in India. Laboratory studies have shown excellent properties at low water–binder ratios. Elsewhere in the world, especially in Canada, concrete has been used successfully with a cement replacement of about 60%. The high-volume fly ash is an economical concrete mix component. It can be mixed, placed, and consolidated with conventional equipment. It is possible to use large quantities of fly ash and thereby make the construction industry sustainable, especially because it encourages bulk utilization of an industrial waste product which is actually a resource.

Silica fume and rice husk ash have been used in small amounts (10% by mass of binder) to enhance the desirable properties such as impermeability and durability of concrete. Though both the silica fume and rice husk ash have been successfully employed as admixtures to concrete, it is worthwhile to develop cement blends composed of fly ash, silica fumes, and/or rice husk ash in order to conserve cement clinker for the sustainability of cement industry in the long run. Silica as a product of agriculture, i.e., rice husk ash is a renewable resource. In India the use of rice husk ash blend with cement will not only help sustain the cement industry in an environment-friendly way but also ensure rural development in a sustainable manner.

Ground granulated blast-furnace slag, a by-product of steel industry, is a unique cementitious material which possesses an inherent ability to provide strength, stiffness, and durability to concrete structures. The most important contribution of slag is its ability to reduce the heat of hydration during the concreting process. Added to its ability to reduce the temperature rise due to heat of hydration and the resulting thermal strains and microcracking, it has delayed strength development and excellent durability. The mineralogy and chemistry of slag is able to mobilize a very fine pore structure. A proper use of this waste material will bring desired benefits to society in terms of energy resource conservation and environmental protection.

Metakaolin is a white pozzolana made by heating kaolin clay to the temperature of 600–800°C. It reacts rapidly with calcium hydroxide in the cement paste and converts it into a stable cementitious compound. It refines the microstructure of concrete and improves impermeability.

Supplementary cementitious materials are considered as alternative binders. These are cement replacement materials and have become a necessity for the construction industry because of the economic, environmental, and technological benefits resulting from their use. In the years to come, wider utilization of these materials and further search for such waste materials and industrial by-products will find application owing to their energy and cost-saving considerations. These materials may thus be 'unloading' the surcharged environment by providing a route to 'lock up' the waste, which is otherwise hazardous to environment. In the process, this also helps reduce CO_2 emissions attributed to cement/construction industry.

31.2 Enhancement in Strength/Ductility of Concrete

The current trend is to build tall and long-span structures. For such structures strength and ductility become important. Hence concrete with ultrahigh strength and which behaves in a significantly ductile manner is essential and needs to be researched upon. Developments in this domain will enable us to build tall and long-span structures, probably a building more than one kilometre tall or a bridge having span larger than the longest single-span steel bridge of the suspension type. For building tall or long-spanning structures, good ductility behaviour becomes necessary. Reinforced concrete unlike steel is a tailor-made material. Hence, the challenge is to produce concrete as strong and ductile as steel and at the same time retain all the advantages of a concrete structure such as durability, fire resistance, and mouldability. A few examples of initiatives towards this end are considered in the following section.

31.2.1 Use of High-strength Concrete

In the past high-strength concrete was mostly used in columns of tall buildings. As tall buildings began to be designed with larger heights, the need for high-strength concrete emerged. More recently high-strength concrete has been used in long-span bridges. It has been used in precast, prestressed components to achieve longer spans and wider beam-spacing shallower sections, and lighter superstructures. High-strength concrete has also become inevitable in offshore structures. Table 31.2 shows the progressive increasing application of high-strength concrete in deep sea structures.

The use of high-strength concrete with a higher modulus of elasticity reduces the axial shortening of tall columns and walls.

Table 31.2 Deep sea structures with high-strength concrete

Year installed	Water depth (m)	Concrete volume (m³)	Strength (MPa)	Project
1984	145	130,000	55	Statf jorf C (Norway)
1989	216	240,000	70	Gulfaks C (Sweden)
1993	251	80,000	75	Draugen (USA)
1995	303	235,000	80	Troll A (USA)

31.2.2 Use of Highly Ductile High-strength Concrete

Highly ductile high-strength concrete (Ductal®) is an ultrahigh-performance concrete product. Ductal concrete is a concrete proportioned with a maximum particle size of 600 μm and a minimum particle size of 0.1 μm, which provides a dense mix with minimum void spaces. Such a concrete has a poor ductile character. So, in order to improve ductility, fibres are mixed along with the concrete.

It has got the following advantages:

1. Compressive strength up to 200 MPa can be achieved.
2. Flexural strength exceeding 40 MPa can be achieved.
3. High abrasion resistance.
4. Good resistance to chemical agents.
5. Structures can be designed without passive and shear reinforcement. The properties of Ductal are tabulated in Tables 31.3 and 31.4.

Table 31.3 Mechanical properties of Ductal

Property	Without treatment	With heat treatment
Density	2500 kg/m³	2500 kg/m³
Compressive strength	150 MPa	180 MPa
Bending strength	30 MPa	33 MPa
Direct tensile strength	7 MPa	8 MPa
Shrinkage	500 μm	0
Creep	0.8	0.15
Young's modulus	55,000 MPa	50,000 MPa
Poisson ratio	0.2	0.2
Thermal expansion	12×10^{-6}	12×10^{-6}
Fatigue resistance	>Million cycles	>Million cycles

Table 31.4 Durability properties of Ductal

S. No.	Property	Value
1	Gas permeability (nitrogen)	1.5×10^{-20} m²
2	Water porosity	1.9%
3	Abrasion resistance	1.1
4	Chlorine ion diffusion	2×10^{-14} m²/s
5	Carbonation depth	Nil
6	Tritium water diffusion	10^{-5} cm²/J

Having a look at these tables it is clear that we can achieve many favourable behavioural parameters by use of Ductal. Figure 31.5 shows the behaviour of Ductal in compression, Fig. 31.6 in bending, and Fig. 31.7 in shrinkage.

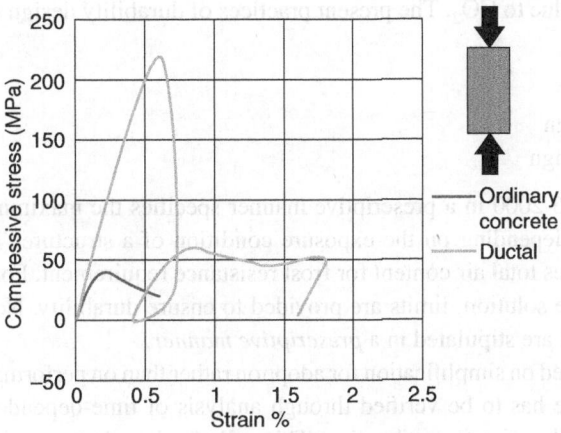

Fig. 31.5 Ductal behaviour in compression

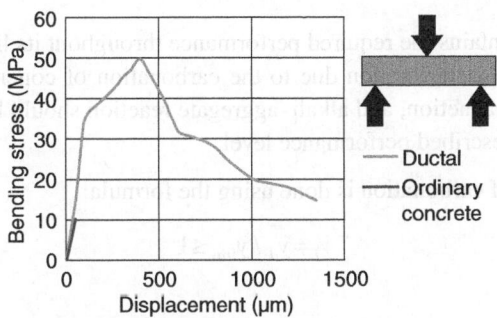

Fig. 31.6 Ductal behaviour in bending

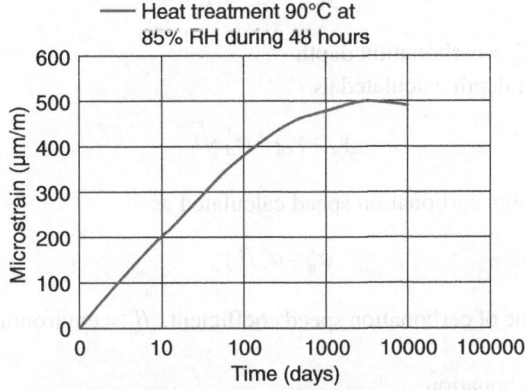

Fig. 31.7 Shrinkage behaviour of Ductal

31.3 Enhancement of Durability of Concrete Structures

Concrete is the most used construction material and concrete structures are generally assumed to have a very long life. However, durability problems have manifested in many cases especially for structures in severe environments. Design codes do not provide a comprehensive tool for evaluating either the chloride ion ingress or carbonation of concrete due to CO_2. The present practices of durability design can be divided into following categories:

1. Prescriptive design
2. Performance type design
3. Performance based design

As explained earlier, IS:456-2000 in a prescriptive manner specifies the maximum water–cement ratio and minimum cement content depending on the exposure condition of a structure. The ACI building code for structural concrete prescribes total air content for frost resistance requirement. For concrete exposed to deicing solution and/or sulphate solution, limits are provided to ensure durability. Note that all these clauses in IS:456-2000 or ACI:318-02 are stipulated in a *prescriptive manner*.

Prescriptive design is based on simplification for adoption rather than on performance. However, the required performance in a real sense has to be verified through analysis of time-dependent durability behaviour of concrete under the specified environmental action. This calls for lengthy numerical approach and currently many models for solution are being researched upon. Under these conditions, some of the codes such as Japan Society of Civil Engineers have introduced the *performance type* method for durability design. This calls for verification of the following:

- The concrete structure maintains the required performance throughout its life.
- The resistance to structural deterioration due to the carbonation of concrete, ingress of chloride ions, freeze–thaw attack, chemical action, and alkali–aggregate reaction should be assessed and the combined effect should satisfy the prescribed performance level.

For example, the verification of carbonation is done using the formula:

$$\gamma_i = y_d / y_{\lim} \leq 1$$

where

γ_i = structure factor

$y_{\lim}$ = limit depth of carbonation to induce corrosion initiation calculated as

$$y_{\lim} = C - C_k$$

where C = design cover and C_k = carbonation depth

y_d = designed carbonation depth calculated as

$$y_d = \gamma_{cd} + \alpha'_d \sqrt{t}$$

where α'_d = coefficient of design carbonation speed calculated as

$$\alpha'_d = \alpha'_k \beta_c \gamma_c$$

where α'_k = characteristic value of carbonation speed coefficient, β_c = environment factor, and γ_c = material factor for concrete

t = design life span for carbonation

γ_{cd} = coefficient to consider y_d

Approximately, α'_k is taken based on experiments and statistical evidence,

$$\alpha'_k = 3.57 \pm 9.0 \; w/b$$

where w/b is the water–binder ratio.

To consider the effect of environment, the environment factor β_c is introduced. The value of β_c varies from dry environment to wet environment (i.e., varies between 1.6 and 1.0). Thus, the framework of concrete technology is made to be performance based by integrating structural and durability design parameters in terms of performance parameters. Since detailed probabilistic analysis is not done and the analysis is based on semi-empirical factors such as α'_k and β_c, this approach is called performance type design method.

If considerable data on environmental factors are collected and are available, the values of α'_k and β_c can be refined using probabilistic methods. For such an approach considerable effort is needed to prepare the probabilistic models, which are under research study in various parts of the world. When these results are available, we may be able to postulate a design philosophy which will lead us to performance-based design approach.

31.4 Use of Hybrid Systems in Concrete

It is well known that mineral cementitious materials contribute to the durability of concrete structures. There are a number of cementitious materials, such as fly ash, ground granulated blast-furnace slag, and silica fume, which are widely available and are being used extensively. Similarly, the use of different types of steel and polypropylene fibres has enhanced the ductility performance of concrete. Composite mineral admixtures as well as various types of fibres in the same mix can be used to get the desirable performance. This section discusses the use of a few types of such composites and hybrids which are likely to find application in the construction industry.

31.4.1 Polypropylene Structural Fibres

Fibre-reinforced concrete (FRC) is gaining importance with growing demand for concrete. Glass, carbon, and steel fibres are being used to produce fibre-reinforced concrete. Polypropylene fibres can also be used for producing fibre-reinforced concrete. Using polypropylene fibres has the following advantages:

1. The mode of failure is changed from brittle to ductile.
2. It can be incorporated in specific amounts without causing balling and segregation.
3. It is corrosion resistant (when compared with steel fibres).
4. Its deadweight is less.
5. It can be used in magnetic areas as well.

31.4.2 Slurry Infiltrated Fibrous Concrete

Slurry infiltrated fibrous concrete (SIFCON) is a variety of fibre-reinforced concrete. In normal FRC, fibres are mixed with wet concrete and placed. As against this, cement slurry is infiltrated into the fibre-packed bed in the case of SIFCON. SIFCON has higher ductility and impact resistance when compared to normal FRC.

31.4.3 Types of Cementitious Materials

Portland cement is being used widely since 1824. Modernization and technology have led to the development of new cementitious materials. Developments in cementitious materials have been from the points of view of quality, environment friendliness, and utilization of industrial by-products. A brief discussion on these developments is given next.

Environment-friendly cement

The process of Portland cement manufacture releases considerable amount of CO_2 into the atmosphere. This causes environmental pollution. The need to reduce environmental pollution has led to the development of environment-friendly cement.

Geopolymer cement

Geopolymer cement has good durability. It is produced using oxides of Si and Al. When compared with the conventional Portland cement, CO_2 emission is reduced by 80% by resorting to geopolymer cement production. Geopolymer concrete undergoes low creep and very little shrinkage and is a good alternative to conventional concrete.

Sodium cement

Sodium is the major component in this cement. This is also an environment-friendly cement. In this cement the calcium content is reduced, which in turn reduces the CO_2 gas emission during production of cement.

Calcium sulpho-aluminate cement (CSA)

CSA cement uses $Ca_4(Al, Fe)$ $6O_{13}SO_3$ instead of C_3S. This cement has a good resistance to corrosive ions such as sulphates and chlorides. The strength character of CSA cement and its setting property permit it to be used for repair works.

Reactive belite cement

The manufacture of this variety of cement saves energy. Additionally, the quantity of limestone used is minimized. In this type the alite phase is restricted to a minimum and the belite phase is increased to a range of 55–60% with a low proportion of C_3A and C_4AF.

Belite cement has low carbonation and a capillary porosity of 7–14%. However, a major challenge in this cement is that it has a low hydration rate.

Magnesium phosphate cement (MPC)

MPC is suitable for quick repair of concrete pavements. MPC has excellent compressive strength. Its setting time is 8–10 min. It has good resistance to freezing and thawing conditions. The reaction responsible for bond is

$$MgO + NH_4H_2PO_4 \rightarrow NH_4MgPO_4 \cdot 6H_2O$$

MPC is corrosion resistant and does not require special care for handling.

Microdefect-free system

By adapting special rheological aids and special processing techniques, tensile and flexural strengths of cement paste can be improved with a reduced water–cement ratio.

Densified particle system (DPS)

This system was developed by Aalborg Portland in Denmark in the 1970s. It consists of low-porosity matrices produced from mixing OPC and condensed silica fumes. DPS concrete can also be considered to be an improved version of high-strength concrete with a lower water–cement ratio. It has more abrasion resistance, but low permeability to ions and gases.

Microfine cement

Microfine cement is produced by ultrafine grinding in a cement mill to a higher fineness of about 750–800 m^2/kg. This cement is suitable for repair works.

31.4.4 Fibre-reinforced Self-consolidating Concrete

Fibre-reinforced concrete has proved to be a good alternative to conventional concrete. Much research has been conducted to enhance the workability of FRC. In Taiwan a local mixture proportion method called densified mixture design algorithm (DMDA) has been developed to produce fibre-reinforced self-compacting concrete (FRSCC). An important advantage of the DMDA method is that fly ash can be utilized properly to fill the voids between aggregate and cement paste. The relationship between the unit weight and fibre content of concrete is shown in Fig. 31.8.

The following are the advantages of FRSCC:

a. Improved ductility
b. Least voids
c. FRSCC designed by DMDA also reduces the balling problems of fibres and improves the flowing and self-consolidating capacity

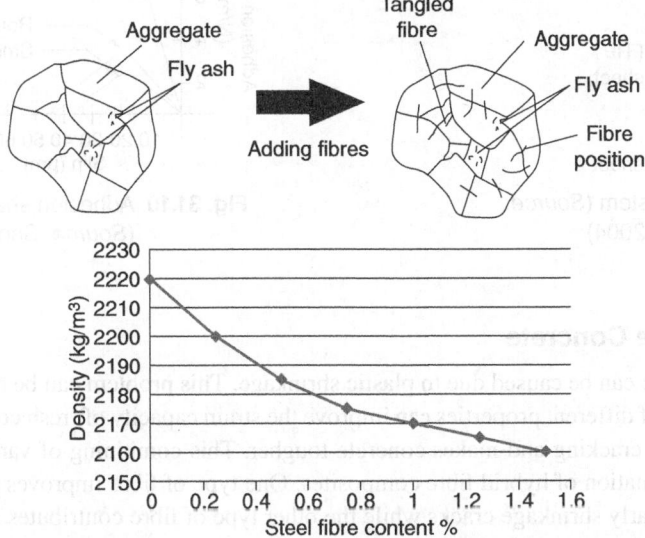

Fig. 31.8 Relationship between density and fibre content of FRC

31.4.5 High-performance Fibre-reinforced Cementitious Composites

The demand for improvement in the quality of structure and repairing them is increasing. To meet this demand, a special concrete or construction material with more durability is required. This has led to the development of high-performance fibre-reinforced cementitious composites (HPFRCC). The large ductility of HPFRCC makes it an excellent construction material. Engineered cementitious composite (ECC), a type of HPFRCC, has high tensile strain capacity and is a very good repair material.

31.4.6 Fibre-reinforced System for Seismic Strengthening

Earthquake is a natural phenomenon capable of exerting unspecified, extraordinary loads on a structure. So the likelihood that a structure might be subjected to devastating seismic loads during its service life has forced engineers and designers to search for ways and means to minimize the consequences due to seismic activity. These efforts have led to the development of FRC and FR construction systems.

Introduction of fibre into the construction system imparts more ductility to the construction system, which in turn helps a structural system to better dissipate the earthquake energy. Many structures constructed before 1990 need to be strengthened if they are to be made safe for the current specification requirements. This can be achieved by using fibre-reinforced systems. In this method, the columns and beams to be strengthened are clad with FRP sheets (Fig. 31.9). The adhesion between FRP structural members is provided by epoxy coating. The entire member is covered by concrete grouting. In order to carry out strengthening, adhesion shear versus load slip between old and new materials may have to be accounted for, by taking into account the value of slip. The typical load slip curves as a function of adhesion shear are shown in Fig. 31.10.

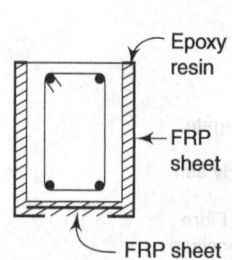

Fig. 31.9 FRP repair system (*Source:* Santhakumar 2004)

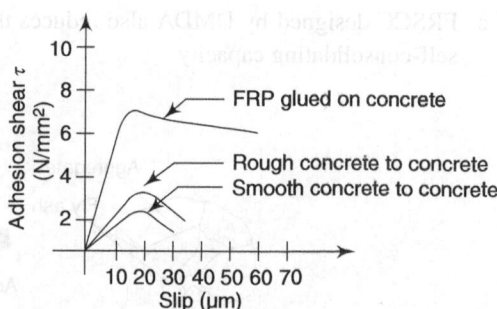

Fig. 31.10 Adhesion shear variation with slip (*Source:* Santhakumar 2004)

31.4.7 Hybrid Fibre Concrete

Cracks in fresh concrete can be caused due to plastic shrinkage. This problem can be avoided by using FRC. Addition of two fibres of different properties can improve the strain capacity of fresh concrete. A combination of fibres prevents early cracking and makes concrete tougher. This combining of various types of fibres in a mix results in the formation of hybrid fibre composites. One type of fibre improves the properties of fresh concrete and prevents early shrinkage cracks while the other type of fibre contributes to the improvement of ductility of hardened concrete.

Fibres used in hybrid concrete should be selected carefully. The volumes of both the fibre materials combined together should not exceed the permitted fibre volume fraction.

Hybrid steel fibres (long and short steel fibres) can also be used in concrete. This combination of hybrid steel fibres can decrease the shrinkage crack by 10–34% and also provide good ductility for seismic resistance. Researches have proved that hybrid fibres reduce the plastic shrinkage crack of concrete when compared with normal FRC.

31.4.8 Hybrid Micro Silica Concrete

Fly-ash concrete is used to produce high-performance/strength concrete. By adding a small percentage of micro silica, the structural performance of high-strength concrete is increased. Fly ash blended with micro silica or other such materials is used to produce hybrid admixture concrete. Hybrid fly-ash concrete has more compressive strength and flexural strength as compared to normal fly-ash concrete. The use of micro silica in concrete makes the transition zone denser. The transition zones of hybrid fly-ash concrete after 7 and 135 days are shown in Fig. 31.11. Table 31.5 shows hybrid admixture concrete mix proportions when two or more cementitious materials are mixed.

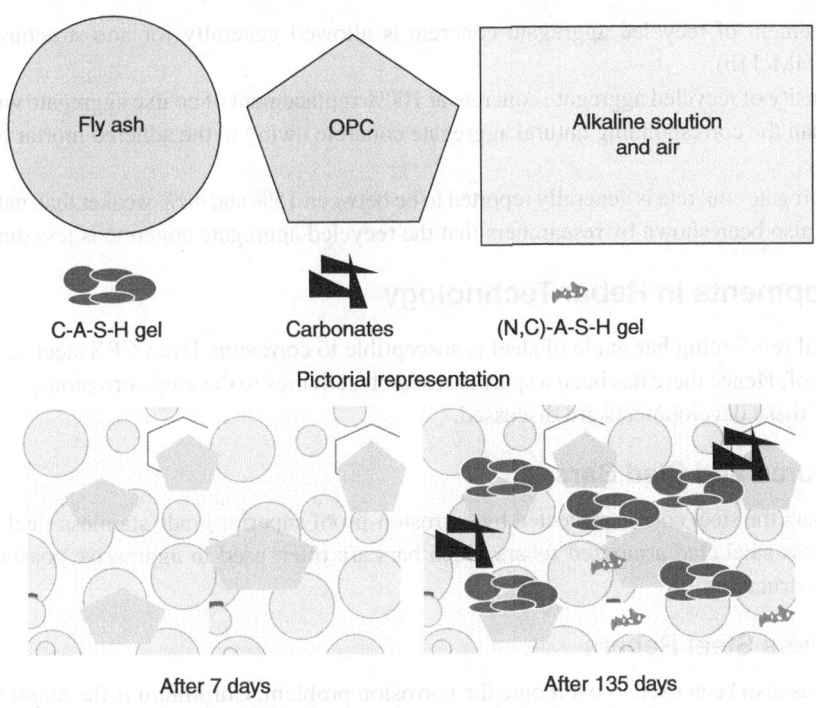

Fig. 31.11 Hybrid fly ash concrete

Table 31.5 Cementitious material proportions if three or more materials are used

	Percentage of total cementitious material		
	Conventional large aggregate mass concrete	Structural concrete	Underwater or Infill mass concrete
Portland cement	20–50	40–70	15–40
GGBFS	30–50	20–50	30–60
Fly ash	20–50	0–25	15–45
Silica fume	0	0	0–8

31.4.9 Recycled Aggregate Concrete

The reuse of broken concrete pieces as coarse aggregate is a proven technology. Old concrete can be crushed and used in fresh concrete as partial replacement for conventional natural aggregates. The old concrete can be from demolition waste (recycled concrete aggregate) or leftover concrete at a construction site (leftover concrete aggregate). The fresh concrete leftover at a site can also be washed free of cement paste and the aggregates recovered to be used subsequently (recovered concrete aggregates). Waste materials from other industries (e.g. broken glass pieces) can be used as secondary aggregates. Such action will save the environment by reducing the landfill load.

The quality of recycled aggregate concrete depends on externally verifiable quality certification standards and compliance system adopted in the field. There is need to develop quality specifications and guidelines for adoption for Indian conditions such as the one practiced in New Zealand (Park, 2001).

Research into new and innovative use of recycled waste is being pursued in many countries. Some of the materials tried are glass, rubber, and wood chips.

100% replacement of recycled aggregate concrete is allowed generally for non-structural applications (ASTM C94/C94M-11b).

The fresh density of recycled aggregate concrete at 100% replacement of coarse aggregate will generally be 5–10% lower than the corresponding natural aggregate concrete owing to the adhered mortar on the recycled coarse aggregate.

Recycled aggregate concrete is generally reported to be between 15% and 40% weaker than natural aggregate concrete. It has also been shown by researchers that the recycled aggregate concrete is less durable.

31.5 Developments in Rebar Technology

The conventional reinforcing bar made of steel is susceptible to corrosion. Even CRS steel cannot be said to be corrosion proof. Hence there has been a spurt in research activities to develop corrosion-proof steel. In this section a few of these developments are discussed.

31.5.1 Armoured and Clad Bars

In this type of bars the steel core is protected by corrosion-proof superior grade stainless steel. Such bars are known as stainless steel clad armoured rebars. Such bars are often used in aggressive coastal environment and in off-shore structures.

31.5.2 Stainless Steel Rebars

Stainless steel has also been used to overcome the corrosion problem. Chromium is the major metal in stainless steel alloy and imparts the corrosion resistance property to steel. The corrosion resistance property of steel can be increased by increasing the chromium content. Stainless steel can withstand chloride attacks. The advantages of using stainless steel rebars are the following:

- Stainless steel is environment friendly.
- It has good resistance against corrosion (no need of surface protection).
- It has good engineering properties.
- The cover of concrete can be reduced if stainless steel bars are used.
- It is easy to handle.
- It helps reduce maintenance.
- It is recyclable.
- Its design, detailing, and construction techniques are similar to that of ordinary steel bars.

31.5.3 FRP Rebars

Steel is the conventional material used for providing reinforcement. In order to overcome corrosion problem, FRP rebars provide an alternative. FRP rebars are made of fibres of glass, carbon, aramid, and basalt. Using FRP rebars in the construction has the following advantages:

- FRP rebars can be manufactured as per the required specification.
- They are more effective in the flexural zone.
- They offer resistance against corrosion.
- Using FRP bars helps reduce the dead load.
- They do not conduct current and can be used near high voltage and magnetic fields.

31.6 Smart Concrete

Before we summarize this chapter, imagine concrete which can think! Can such concrete be produced? The indications are that we can build smart concrete buildings by incorporating sensors and other biological or nanomaterials into concrete. Though some of the trends look to be a figment of imagination today, in reality the concepts that are reported are likely to be realized in the coming future. This section attempts at throwing light on some developments that may become reality in the coming generation.

31.6.1 Bacterial Concrete

The concept of bacterial concrete is considered one step towards the smart concrete technology. The concrete that has the ability to adapt itself to the environment is called smart concrete. This concept is based on the ability of bones to grow and repair. The bone has the ability to repair itself, provided nutrients are supplied properly.

Bacillus pasteurii is a common soil bacterium. It has the ability to produce calcite. When this bacterium is used in concrete, it produces a highly impermeable layer of calcite over the surface of already-existing mortar layer. Bacterial concrete is a self-repairing bio-material; therefore, it is called 'smart bio-material'.

This smart concrete has proved to be effective in successfully taking care of plastic shrinkage cracks. It also increases the durability of concrete. Higher the bacterial dosage, higher the performance. Bacteria can be suspended in water or phosphate buffer and mixed in concrete. Concrete made by suspending bacteria in phosphate buffer is more effective. There are many advantages of using bacterial concrete:

1. Increased resistance to alkali and sulphate attacks
2. More resistance to the freezing and thawing conditions
3. Reduces plastic shrinkage cracks
4. Additional calcite layer formed by bacteria increases the impermeability of material, which in turn prevents the penetration of harmful gases and chemicals into the concrete. This reduces the possibilities for rebar corrosion
5. The compressive strength of bacterial concrete is also increased by 5–10%

31.6.2 Nanocomposite Materials

Natural bone is a natural bio-nanocomposite material of hydroxyapatite $[Ca_{10} (PO_4)_6(OH)_2]$, mineral (HAP), and organic matter. This composite has shown improved recovery and smaller plastic strain as compared to other composites. HAP polymer composite was produced and its material properties tested. Such a composite can be used by soaking it in water and body fluids. Body fluid is prepared using $NaCl$, $NaHCO_3$, KCl, $K_2HPO_4 \cdot 3H_2O$, $MgCl_2 \cdot 6H_2O$, $CaCl_2$, and Na_2SO_4. It is now possible to produce composites of strengths of 80 MPa.

Another nanocomposite material is NACRE, which improves the physical properties such as strength and toughness of concrete. NACRE is a ceramic-laminated composite consisting of highly organized polygonal-shaped aragonitic platelet layers of a thickness of 0.5 µm separated by thin 10–30 nm layers of organic matter composed mainly of proteins and polysaccharides.

Further research is going on in this field to improve and optimize the nanocomposite concrete technology and its application in the construction industry.

31.6.3 Fibre-optic Sensors

Currently the need to develop infrastructure world over is being felt never than before. This has resulted in large-scale construction activities throughout the length and breadth of our country. More and more massive structures are being constructed to meet the increasing demand. This spree of construction activities has made it necessary to devise a technique for assessing the quality of the structure and its useful life. In order to meet this necessity, fibre-optic sensors are being made use of.

Using fibre-optic sensors is a non-destructive technique for evaluating the condition of a structure. Indeed it is a strain sensor for monitoring the behaviour of the concrete structures.

Extrinsic Fabry–Pèrot interferometric (EFPI) sensors (Fig. 31.12) have a cavity comprising two mirrors which are parallel to each other and perpendicular to the axis of optic fibre. Sensing and reference fibres are the same up to the first mirror. This is the starting of the sensing region. The cavity formed between the mirrors is known as the Fabry–Pèrot cavity. A change in the distance between the two fibre end faces (called air gap length) causes interferometric fringe variation. From this distance change, strain can be calculated using the following known relationship:

$$\text{Strain} = (\text{Change in the length of air gap})/(\text{gauge length})$$

Installation of such advanced sensing devices will enable us to monitor our structures live and assess the condition of our vital infrastructure.

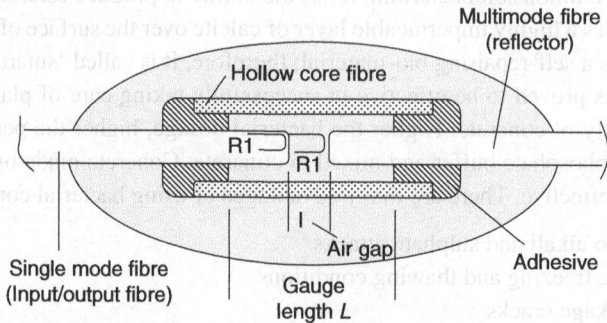

Fig. 31.12 Schematic of fibre optic sensor

31.6.4 Concrete with Manufactured Sand

Concrete is the most widely used man-made material. Sand plays an important role in concrete. River sand which has been used traditionally is in short supply. Mining of sand from rivers poses a threat to environment. In concrete whether it is site mixed or ready mixed, sand consumption is around 30% of the total volume.

Manufactured sand is defined as a purpose-made crushed fine aggregate produced from a suitable source material. Production generally involves crushing, screening, and possibly washing.

It is clear from the definition of manufactured sand that it was never acceptable for quarries to produce a crusher dust that results from the fine screenings of all quarry crushing and call this material manufactured sand. Particularly for quarries that have different rock types or that have differential weathering or alteration patterns across the extraction area, crushed fine aggregate needs to be produced to specification and controlled by quality programmes.

Instead of river sand, manufactured sand can be used to make concrete. The stone is crushed in a vertical shaft impact (VSI) crusher. It is possible to obtain the particle sizes in M sand similar to conventional river sand. The grading curve of M sand is shown in Fig. 31.13.

It has been reported (Syam Prakash, 2007) that on testing the sand had a specific gravity of 2.59 and the grading can be considered to fall under Zone II. Test results on concrete made with M sand reported by

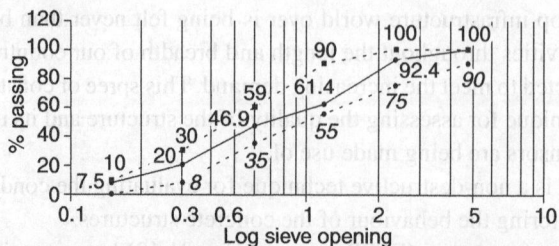

Fig. 31.13 Grading curve of fine aggregates

Syam Prakash (2007) shows that the characteristic compressive strength has been obtained for all mixes with the standard deviation within the permissible limit. It could also be noticed that the minimum quantity of cement has been used for all the mixes. This ensures that the use of M sand is quite appropriate for making RMC satisfying workability, transportability and pumpability of concrete in its fresh stage leading to a strong and durable concrete in its hardened stage.

Exercises

Review Questions

1. What is a green building?
2. What are the benefits of green buildings?
3. Why are green buildings needed?
4. What are the main advantages of recycling aggregates?
5. Describe the importance of the grading curve for recycled aggregates.
6. What are the environmental benefits of using recycled aggregates?
7. What is Ductal and where is it used?
8. What are the advantages of nanocomposite materials?
9. Differentiate between FRC and SIFCON.
10. What are the advantages of using stainless steel rebars?
11. What do you understand by hybrid concrete?
12. What is meant by smart concrete? Explain.
13. What is M Sand? Can Crusher dust be used as M Sand?

Multiple Choice Questions

1. Cement is considered to be an unfriendly material for sustainability because
   a. Manufacture of cement involves emission of CO_2
   b. It introduces dust in the atmosphere
   c. It affects skin of workers
   d. It causes allergic reaction

2. Recycling of aggregates is considered necessary because it
   a. Improves concrete quality
   b. Is economical
   c. Can be used without processing
   d. Enables the use of debris generated as waste resulting from natural calamities like earthquakes and floods

3. A green building is one which is
   a. Painted green in colour
   b. Built, operated and used in a resource-efficient manner
   c. Cheap
   d. Comfortable to live

4. Extensively used supplementary cementitious material in India is
   a. Silica fume
   b. Rice husk ash
   c. Metakaolin
   d. Fly ash

5. Ultrahigh strength concrete normally has
   a. High ductility and high strength
   b. High strength and required ductility
   c. Medium ductility
   d. Explosive failure

6. Ductal properties indicate that it has
   a. High strength and high ductility
   b. High strength and medium ductility
   c. High strength and low ductility
   d. Explosive failure

7. The design complying with IS:456 can be said to be
   a. Prescriptive design
   b. Performance type design
   c. Performance-based design
   d. Code-based design

8. Hybrid concrete consists of
   a. Fine aggregate only
   b. Coarse aggregate only
   c. Coarse and fine aggregates
   d. Coarse and fine aggregates with fibres and/or two or more cementitious materials

9. The approximate size of silica fume particle is
   a. 0.1 μm
   b. 1.0 μm
   c. 0.01 μm
   d. 0.001 μm

10. The size of nanoparticles is
   a. 0.1 μm
   b. 1.0 μm
   c. 10.0 μm
   d. 0.001 μm

Answers to Multiple Choice Questions

1. a	4. d	7. d	10. d
2. d	5. d	8. d	
3. b	6. a	9. a	

Bibliography

'Laminated process for ferrocement', *Journal of Ferrocement*, vol. 22, no. 3, 1992, p. 305.

'NEG Micon India's Green Building', *The Masterbuilder*, vol. 8, 2006, p.103.

'Recycled aggregate concrete for sustainable development', *Indian Concrete Journal*, August 1999, Editorial part.

Abdullah, Abang A. 1995, 'Application of ferrocement as a low cost construction material in Malaysia', *Journal of Ferrocement*, vol. 25, no. 2, pp.123–8.

Accelerators and chemicals for concrete construction, Literature from Dayton Sure Grip & Shore Company, Ohio, 1977.

ACI 1973, *ACI Monograph*, no. 6.

ACI 1973, 'Recommended practice for selecting proportions for normal-weight concrete', *ACI 211.1-70, ACI Manual of Concrete Practice*, 1973, Part 1, American Concrete Institute, Redford Station, Detroit, Michigan, pp. 211–1 to 211–16.

ACI Committee 209 (1990), 'Prediction of creep, shrinkage and temperature effects in concrete structures', *Manual of Concrete Practice*, part I.

ACI Committee 212 (IR-81) 1981, *Admixtures for concrete, Concrete International Design and Construction*, vol. 3, no. 5, pp. 24–52.

Ahmed, H.I. and L. Robles-Austriaco 1991, 'State-of-art report on rehabilitation and restrengthening of structures using ferrocement', *Journal of Ferrocement*, vol. 21, no. 3, pp. 243–58.

Ahmed, S.H. 1994, *Short Term Mechanical Properties, Higher Performance Concrete*, S.P. Shah and S.H. Ahamed (eds), McGraw Hill Book Company, pp. 28–64.

Ahmed, S.H. and S.P. Shah 1985, 'Structural properties of high strength concrete and its implication for present prestressed concrete', *PCI Journal*, Nov-Dec, pp. 91–119.

Aitcin, P. and P. Richard 1996, 'The pedestrian bike way bridge of Sherbrook', in *The 4th International Symposium on utilization of High strength/High Performance Concrete*, Paris.

Aitcin, P.C. and Neville Adam 1993, *High Performance Concrete Demystified*, vol. 15, no. 1, pp. 21–6.

Aitcin, P.C., S.C. Sankar, R. Ranc, and C. Levy 1991, 'A high silica modulus cement for HPC', in *Advances in Cementitious Materials Ceramic Transactions*, S. Minders (ed.), vol. 16, American Ceramic Society, pp. 102–21.

Alexander, M.G. and *Construction through Innovation*, ed. M.C. Limbachiya & HY Kew, Taylor and Francis, Oxford.

Alexanderson, J. A. 1971, 'Design of Concrete Mixes', *RILEM Bulletin*, Jul Aug, pp. 203 12.

American Concrete Institute 1993, 'High performance concrete construction materials and systems: An essential programme for America and it infrastructure', *Technical Report 93-5011*, Civil Engineering Foundation, Washington, D.C.

Antoine, E. Naaman 1987, 'Advances in high performance fibre reinforced cement based composites', in *Proceedings of International Symposium on Fibre Reinforced Concrete*, Madras, India, vol. pp. 7–87.

Anwar, A.W., P. Nimityongskul, R.P. Pama, and L. Robles-Austriaco, 'Methods of rehabilitation of structural beam elements using ferrocement', *Journal of Ferrocement*, vol. 21, no. 3, pp. 229–34.

Baker, L. 1956, 'Trial-mix method', *Journal of Institution of Engineers (Australia)*, April-May.

Barber, S. 2000. Final year Project Report, Curtin University, Australia.

Barfoot, R.J. 1974, 'Joseph, James and William—The Aspdin Jigsaw', *Concrete*, vol. 8, no. 8, pp. 18–26.

Barr, B. 1987, 'Compact shear test specimens for FRC materials', *Composites*, vol. 18, no. 1, pp. 54–60.

Bauer, E.E. 1949, *Mortar Void-Plain Concrete*, HMSO, London, pp. 111–12 and 368–74.

Beall, Christine 1984, *Masonry Design and Detailing for Architects, Engineers and Builders*, Prentice-Hall, Englewood Cliffs, NJ.

Berman H.A. 1975, *Sodium Chloride, Corrosion of Reinforcing Steel and the pH of Calcium Hydroxide Solution*, American Concrete Institute.

Berry, N.K. 1975, 'Controlled Concrete Mix Designs on the Beas Dam Project', *Indian Concrete Journal*, vol. 49, pp. 7 and 73.

Betancourt, G. Hill 1988, 'Admixtures, Workability, Vibration and Segregation', *Materials and Structures*, vol. 21, no. 124, pp. 286–8.

Bischoff, P.H. and Perry, S.H. 1991. 'Compression behaviour of concrete at high strain rates', *Materials and Structures*, Vol. 24(144), pp. 425–450.

Blanks, R.F. 1945, 'Recommended practice for the design of concrete mixes', in *ACI 613–44 Proceedings*, vol. 41, pp. 651–69.

Blanks, R.F. and H.L. Kennedy 1955, *The Technology of Cement and Concrete*, vol. I, Wiley Publications.

Browne, R.D. 1986, 'Practical considerations in producing durable concrete', in *Seminar on Improvement of Concrete Durability*, ICE, pp. 97–116.

Browne, R.D. and P. Domone 1974, 'How concrete stands us to a life in the ocean wave', *New Civil Engineer*, Special Supplement, London.

BRS 1971, 'Concrete mix proportioning and control', *BRS Digest*, no. 13.

BS: 1881-1970(part 4), Methods for testing concrete for strength', BSI, London.

BS: 1881-1983(part 120), Methods for determination of compressive strength concrete cores, BSI, London.

BS: 6098-1985, Guide to assessment of concrete strength in existing structures.

Buchanan, Andrew H. 2001, *Structural Design for Fire Safety*, John Wiley, England.

Cabahug, R.R. and L. Robles-Austrio 1996, 'Notable ferrocement structures', *Journal of Ferrocement*, vol. 26, no. 4, pp. 281–88.

Calder, Nigel 1983, *Time Scale*, Viking Press, p. 288.

Castillo, C. and A.J. Duranni 1990, 'Effect of Transient High Temperature on High Strength Concrete', *ACI Materials Journal*, vol. 87, no. 1, pp. 47–53.

CEB/FIB 1970, International Recommendations for the design and construction of concrete structure.

Chakravarty, S.M. 1996, 'Ready mixed concrete—Need of the hour', in *Proceedings of the Seminar on Ready Mixed Concrete*, Cement Manufacturers' Association, Chennai.

Chancellor, W.J. 1996, 'Ferrocement floating house: Thermal comfort', *Journal of Ferrocement*, vol. 26, no. 4, pp.251–57.

Chandrasekaran, S. 1988, Report on 'Corrosion in Civil Engineering', CECRI, Madras.

Chatterjee, A.K. 1997, 'Cement in high performance concrete—As seen from the manufacturer's position', in *Proceedings of TCDC International Workshop on Advances in HPC Technology and Application*, pp. III 35–41

Chemical Solutions for Concrete Problems, Literature from Non-chem, Inc., USA, 1975.

Collins, M.P., Mitchell, D., and MacGregor, J.G. 1993. 'Structural design: Considerations for high-strength concrete', *Concrete International*, 15(5), pp. 27–34.

Comite Euro-International Du Beton 1992, 'Durable Concrete Structures—Design Guide', Thomas Telford, London.

Comportment Des Betons Exposes al Eau de Mer, Collogue Rilem, Palerme, 24–26, Mai 1965.

Concrete Association of India 1979, *Concrete Mix Design*, A booklet published by the Concrete Association of India, Bombay, p. 36.

Concrete forming systems accessories and chemicals, Literature from Symons Corporation, USA, 1979.

Concrete Mix Design—Theory and Practice, Refresher coarse proceedings, vols I and II, Structural Engineering Division, College of Engineering, Guindy, Madras (India), 1976.

Concrete Society (London) 1973, *Technical Paper*, no. 101.

Cordon, W.A. and H.A. Gillespie 1963, 'Variations in Concrete Aggregates and Portland Cement Paste which Influence the Strength of Concrete', *Journal of the American Concrete Institute*, pp. 1029–52

Cowie, K.J. 1996, 'Setting up ready mixed concrete plant', *Proceedings of the Seminar on Ready Mixed Concrete*, Cement Manufacturers' Association, Chennai.

Davidovits, J. 1988. 'Soft mineralogy and geo-polymers', Proceedings of the Geopolymer International Conference, Compiegne, France, Universite de Technologie.

Davis, A.C. 1924, *A Hundred Years of Portland Cement*, Concrete Publications Ltd, p. 282.

Department of Environment (DOE) 1998, Design of normal concrete mixes (U.K.).

Dewar, J.D. 1994, 'The structure of fresh concrete', in *Cement and Concrete Science Conference of the Institute of Materials*, St. Avins College, Oxford.

Dhir, R.K. 1997, 'High durability performance concrete', in *Proceedings of TCDC International Workshop, Advances in HPC Technology and Application*, SERC, India, pp. V-13–V-34.

Dhir, R.K. 1997, 'Towards a holistic approach to material selection for HPC', in *Proceedings of TCDC International Workshop on Advances in HPC Technology and Application*, SERC, India, pp. I-1–I-18.

Dhir, R.K. and C.Y Lee 1996, *Permeation of High PPA Content Cement*, Internal Report, University of Dandee.

Dhir, R.K. and Y.N. Chan 1995, *Near Surface Characteristics and Durability of Concrete*, Internal Report, University of Dundee.

Dhir, R.K, M.R. Jones, and H.E.H. Ahmed 1991, 'Concrete durability—Estimation of chloride concentration changing design life', in *Concrete Research*, vol. 43, no. 154.

Do, Minh-Tan 1995, 'Fatigue does betons a hautes performances', PhD Thesis, University of Sherbrooke (Sherbrooke, Canada Lonan).

EFNARC (in co-operation with the European Federation of Concrete Admixture Associations). 2006. 'Guidelines for viscosity modifying admixtures for concrete', p. 3.

EFNARC 2002, Specification and Guidelines for Self-Compacting Concrete, Farnham.

Ekasit, Limsuwan 1997, 'Recent development and utilisation of HPC in Thailand', in *Proceedings of TCDC International Workshop on Advances in High Performance Concrete Technology and its Applications*, pp. III 21–41.

Euro-International Committee for Concrete (CEB) 1978, 'Trial and Comparison Calculations', *Bulletin d' Information no. 129*, Paris.

European Committee for Concrete (CEB) 1973, 'Bewehrungsfuehrungen Stahlbetontragwerken', *Bulletin d' Information no. 87, Drafting Manual* Paris, completed by an addendum *in Bulletin d' information no. 104*, pp. 95–135, Paris, 1974.

European Project Group (EPG) 2005, The European Guidelines for Self Compacting Concrete.

Evans, E.P. 1998, 'High performance concrete—From laboratory to construction site', in *Proceedings High Performance High Strength Concrete*, Perth, Australia 1998, pp. 35–49.

Fernandez, A. 1996, 'Ferrocement research and development in Oaxaca, Mexico', *Journal of Ferrocement*, vol. 26, no. 4, pp. 259–66.

Francis, Xavier A. 1988, 'Corrosion monitor for reinforced concrete structures', *Transactions of SAEST*, vol. 23, nos 2-3, pp.229–31.

Franz, G 1970, Konstrukionslehre des stahlbetons, 1. Band: Grundlagen und Bauelements, 3.Auflage, Springer Verlag, Berlin.

Fuller, W.B. and Thompson S.E. 1907. 'The laws of proportioning concrete', *Transactions of ASCE*, Vol. LIX(2), pp. 67–143.

Gambhir, M.L. 1995 , *Concrete Technology*, 2nd edn, Tata McGraw-Hill, New Delhi.

Garnett, J.B. 1959, The effect of vacuum processing on some properties of concrete, cement concrete. Also, test reports TRA/326 (London, Oct.1959).

Gerwick, B.C. Jr 1971, *Construction of Prestressed Concrete Structures*, Wiley Inter Science, New York.

Gerwick, B.C. Jr 1973, 'Practical methods of ensuring durability of prestressed concrete ocean structures', presented at 1973 ACI Convention.

Gerwick, B.C. Jr 1974, 'Chapter on marine concrete', *Handbook of Concrete Engineering* (Fintel), McGraw-Hill, 1974.

Gerwick, B.C. Jr 1999, *Construction of Marine and Offshore Structures*, 2nd edn, CRC Press, London.

Gettu, R., Gomes, P., Agullo, L., Bernard, C. 2001. 'Experimental optimisation of high strength self-compacting concrete', Proceedings of the 2nd International Symposium on Self Compacting Concrete, Kochi, Japan, pp.377–86,

GIMERT & Co. 1989, *Equipments of Non-Destructive Testing of Bridges*, New Delhi.

Gjorv, O. 1968, 'Durability of reinforced concrete wharves in Norwegian harbours', *Ingeniorforlaget* A/S, Oslo.

Glassville, W.H., A.R. Collins, and D.D. Matthews 1947, *The Grading of Aggregates and Workability of Concrete*, Road Research Tech. Paper no. 5, HMSO, London.

Glucklich, J. 1968, *Proceedings of International Conference on the Structure of Concrete, Cement and Concrete Association*, Wexham Springs Slough, UK, pp. 176–85.

Grasim Industries Ltd 2003, *Concrete Mix Design Manual*, Cement Business Marketing Publication # 4.

Green, H. 1964, 'Impact strength of concrete', in *Proceedings of Institution of Civil Engineers*, London, vol. 28, pp. 283–96.

Green, J.K. and Perkins, PH. 1980. *Concrete Liquid Retaining Structures, Design, Specification, and Construction*, Applied Sciences Publishers, London.

Griffith, A.A. 1921. 'The phenomena of rupture and flow in solids', *Transactions of the Royal Society of London*, Series A, 221 pp. 163–198.

Gunnar, I. 1997. *Concrete: Progress from Antiquity to the Third Millennium*. Thomas Telford, London, p. 359.

Hager, I. 2013. 'Behaviour of cement concrete at high temperature', *Bulletin of the Polish Academy of Technical Sciences*, Vol. 61(1). https://suw.biblos.pk.edu.pl/resources/i3/i8/i6/i5/i0/r38650/2013_Hager_Bulletin%20PAN.pdf

Hager, I. 2014. 'Colour change in heated concrete', *Fire Technology*, Vol. 50 (4), pp. 945–958.

Hanager, C.H. 1977, *Steel Fibrous Concrete Joint for Seismic Resistant Structures, Reinforced Concrete Structures in Seismic Zone*, ACI Publications, SP-53, pp. 371–86.

Hanna, E., K. Luke, D. Perraton, and P.C. Aitcin 1989, *Rheological behaviour of Portland Cement in presence of super-plastizer and other chemical admixture in concrete*, SP 119 ACI, Detroit, pp. 17–188.

Haynes, H.H. and P.C. Zuhiate 1973, 'Compressive strength of 67 year old concrete submerged in sea water', *Technical Note N-1308*, Naval Civil Engineering Laboratory, Port Huenema, CA.

Henager, C.H. 1977. 'Steel fibrous ductile concrete joints for seismic resistant structures', *Reinforced Concrete Structures in Seismic Zone*, ACI Publications, SP 53, pp. 371–386.

Henderson, N.A. and R.K. Dhir 1996, *Durability Studies in Multiaggressive Exposure Condition*, Report CTU/5/3 of Dept. of Civil Engg., University of Dundee, p. 40.

Hoff, George C. 2004, *Symposium on High Performance Concrete and Concrete for Marine Environment*, Las Vegas.

Holm, T.A. and T.W. Bremner 1994, 'High strength light weight aggregate concrete', in *High performance concrete properties and application*, S.P. Shah and A.H. Ahamed (eds), McGraw Hill, New York, pp. 341–71.

Hsu, T.C. 1971, *ACI Monograph*, no. 6, p. 100. http://www.fhwa.dot.gov/infrastructure/materialsgrp/admixture.html (visited on 26/07/2006)

ICI 2002, *Formwork*, ICI Technical Monograph TM03, Chennai.

Idom, Gunnar 1997, *Concrete Progress from Antiquity to the third Millennium*, Thomas Telford, p. 359.

IGNOU 1989, *Construction Methodology and Construction Technique* (ET-522, Block 3), School of Engineering and Technology.

IRC Highway Research Board 1996, *State of the Art Non-destructive Testing Technique of Concrete Bridge*, New Delhi.

Irons, M.E. 1986, 'Ferrocement in North America', *Journal of Ferrocement*, vol. 16, no. 2, pp. 150–56.

Irons, M.E. and L.L. Watson 1977, 'Ferrocement boats reinforced with expanded metal', *Journal of Ferrocement*, vol. 7, no. 1, pp. 9–16.

IRS: 87-1984, Guidelines for design and erection of false work for road bridges.

IS: 10262-1982, *Recommended Guidelines for Concrete Mix Design*, Bureau of Indian Standards, New Delhi.

IS: 1139-1977, *Specification for Hot Rolled Mild Steel, Medium Tensile Steel and High Yield Strength Steel Deformed Bars for Concrete Reinforcement*, Bureau of Indian Standards.

IS: 1150-1976, Trade names and abbreviation symbols for timber services (2nd edn).

IS: 1331-1971, Specification for cut sizes of timber (2nd edn).

IS: 13311-1992(part 1), Methods of non-destructive testing of concrete, Part 1: Ultrasonic pulse velocity', Bureau of Indian Standards, DOC-CED-2, pp. 1–9.

IS: 13311-1992(part 2), Methods of non-destructive testing of concrete, Part 2: Rebound hammer, Bureau of Indian Standards, DOC-CED-2, pp. 1–7.

IS: 13935-1993, Repair and Seismic Strengthening of Buildings—Guidelines.

IS: 14687-1999, False work for concrete structures—Guidelines.

IS: 1597-1992(part 1), Code of practice for construction of rubble stone masonry (1st revision).

IS: 1905-1987, Code of practice for structural safety of buildings, masonry walls (3rd revision).

IS: 2212-1991, Code of practice for brickwork (1st revision).

IS: 2366-1983, Code of practice for nail jointed timber construction (1st edn).

IS: 2750-1964, Specification for steel scaffolding (amendment 31).

IS: 3337-1978, Specification for ballies for general purpose (1st revision).

IS: 33 64-1976(part 1), Methods of measurement and evaluation of defects in timber logs (1st revision).

IS: 3364-1976(part 2), Methods of measurement and evaluation of defects in timber-converted timber (1st revision).

IS: 3629-1966, Specification for structural timber in buildings.

IS: 3696-1966(part 1), Safety code for scaffold and ladders.

IS: 383-1970, *Specification for Coarse and Fine Aggregate from Natural Source of Concrete*, Bureau of Indian Standards, New Delhi.

IS: 399-1963, Classification of commercial timber and their zonal distribution (revised).

IS: 4014-1967(part 1), Code of practice for steel tubular scaffolding.

IS: 456:2000, *Indian Standard Plain and Reinforced Concrete Code of Practice*, 4th revision, Bureau of Indian Standards, New Delhi.

IS: 4891-1968, Specification for preferred cut size of structural timber (1st revision).

IS: 4970-1973, Key for identification of commercial timber (1st revision).

IS: 4990-1969, Plywood of shuttering work.

IS: 6461-1972(part 5), Glossary of items relating to comment concrete form for concrete.

IS: 875-1987(part 1), Code of practice for design loads (other than earthquake) for building and structures, Part 1: dead loads—unit weight of building and stored materials (2nd revision).

IS: 875-1987(part 2), Code of practice for design loads (other than earthquake) for building and structures, Part 2: imposed loads (2nd revision).

IS: 875-1987(part 3), Code of practice for design loads (other than earthquake) for building and structures, Part 3: wind loads (2nd revision).

IS: 875-1987(part 4), Code of practice for design loads (other than earthquake) for building and structures, Part 4: snow loads (2nd revision).

IS: 875-1987(part 5), Code of practice for design loads (other than earthquake) for building and structures, Part 5: special loads and load combination (2nd revision).

IS: 883-1994, Code of practice for design of structural timber in building (4th revision).

Ismail, M.S. and A.M. Waliuddin 1996, 'Network of ferrocement drainage system', *Journal of Ferrocement*, vol. 26, no. 2, pp.113—19.

Jagadish, R. 1992, 'Structural Failure of Multi-storeyed Buildings—Case Hostories', *Training course notes on Damage Assessment and Repairs in Low Cost Housing*, Regional Housing Development Centre (RHDC), National Building Organisation (NBC), and Anna University, Chennai.

Jain, A.K. 2003, 'RMC-growth prospects in India', *Journal of Builder's Association of India* (Southern Center), Aug/Sept., pp. 33–7.

John, R.D., S. Ramalingam, and V.R. Rangaswamy 1987, 'Epoxy Application in Civil Engineering Structures in Sriharikota', *ICI Bulletin no. 18*, pp. 33–7.

Johnson, Sidney M. 1965, *Deterioration, maintenance and repair of structures*, McGraw Hill.

Jones, C.J.F.P. 1995, 'Performance profiles and the concept of zone defense', in *Seminar on Whole Life Costing—Concrete Bridges*, Concrete Bridge Development Group, pp. 62–76.

Jones, M.R., Dhir, R.K., Newlands, M.D., Abbas, A.M.O. 2000. 'A study of the CEN test method for measurement of the carbonation depth of hardened concrete', *Materials and Structures*, Vol. 33(2), pp. 135–142.

Jones, R., J.T.J. Hirsch, and H.K. Stephenson 1971, *ACI Monograph*, no. 6, p. 178.

Jones, R., J.T.J. Hirsch, and H.K. Stephenson 1971, *Texas Transportation Institute Report*, no. E52.

Jose M.V. Gomez-Sobern 2002, 'Porosity of recycled concrete with substitution of recycled concrete aggregate an experimental study', *Cement Concrete Research*, vol. 32. pp. 1301–11.

Kalyanasundaram, P. 1990, A report on 'Assessment of Concrete Deterioration and Corrosion Damage', IIT, Madras.

Kamiyama, S. 1972, 'Cracks and corrosion of reinforcing bars in concrete', *Cement & Concrete*, no. 308, Japan Concrete Engineering Association.

Kar, J.N. and H.C. Sharma 1992, 'Tests on epoxy bonded concrete and RC elements', *Indian Concrete Journal*, vol. 66, no. 7, pp. 327–31.

Keith Green, J. and P.H. Perkins 1980, *Concrete Liquid Retaining Structures*, Applied Science Publishers, London.

Kelly, A. 1970, 'Interface effects and the work of fracture of a fibrous concrete', in *Proceedings of Royal Society of London*, A319, pp. 95–116.

Kelly, J.W. and Polivka, M. 1955. 'Ball test for field control of concrete consistency', *Journal of the American Concrete Institute*, Vol. 51, pp. 881–888.

Klaus Holschemacher et al. 2005, 'Structural aspects of self-compacting concrete', *The Masterbuilder*, vol. 7, no. 8.

Kukreja, C.B. and Partap Singh 1991, 'Environmental effects on concrete', *ICI Bulletin*, no. 36, pp. 35–7.

Kulkarni, P.D., R.K. Ghosh, and Y.R. Phull 1998, *Textbook of Concrete Technology*, 2nd edn, New Age International Publishers, p. 320.

Kulkarni, V.K. 1977, 'Concrete mix design', in *Proceedings of Winter School in Concrete Technology*, Indian Society for Technical Education (ISTE), L.D. College of Engineering, Ahmedabad, pp. K/1 to K/26.

Kupfer, H., Hilsdorf, H.K., Rusch, H. 1969. 'Behavior of concrete under biaxial stresses', *ACI Materials Journal*, Vol. 66(8), pp. 656–666.

Lakshmipathy. M. 1983, 'Ductility of Reinforced Fibrous Concrete Structural Members', PhD Thesis, University of Madras.

Lakshmipathy, M. and A.R. Santhakumar 1987, 'Moment redistribution characteristics of a reinforced fibrous concrete hinging zone', in *Proceedings of International Symposium on Fibre Reinforced Concrete*, Madras, India, vol. I, pp. 2.143–2.152.

Lal, A.K. 1991, *Sulphate attack in buildings*, ICI Bulletin no. 36, pp. 33–4.

Lea, F.M. 1970, *The Chemistry of Cement and Concrete*, 2nd edn (rev.), Edward Arnold Ltd, London, p. 637.

Leonhardt, F. 1977, vorlesugen Ueber Massivbau, 3 Tail, 'Grundlagen Zum Bewehrenim Stahlietonbau:, 3. Auflage, Springer Verlag, Berlin; 'Das Bewehren von Stahibetontragweken' in *Betonkalander*, W. Ernst & Sohn, Berlin, 1979.

Lerch, W. 1957, 'Plastic Shrinkage', *Journal of American Concrete Institute*, vol. 53, pp. 797–802.

Levitt, H. 1971. 'The ISAT – A non-destructive test for durability of concrete', *British Journal of Non Destructive Testing*, pp. 106–112.

Lew, H.S. and Reichard, T.W. 1978. 'Prediction of strength of concrete from maturity', *Accelerated Strength Testing*, ACI SP 56, American Concrete Institute, Detroit MI, pp. 229–248.

Lewis, D.W. 1982. 'Resource conservation by use of iron and steel slags', Extending aggregate resources, ASTM special Technical Publication, ASTM, 1982.

Liew, S.C. and M.K. Cheong 1991, 'Flexural behaviour of jacketed RC beams', *Concrete International*, Dec., pp. 43–7.

Limaye, R.G., M.K. Kamat, and A.K. Sopori 1991, *Experimental Investigation on Strengthening of Reinforced Concrete Beams with Epoxy Grouting*, ICI Bulletin no. 37, pp. 36–40.

Lincoln, J.D. 1976, 'Admixtures for concrete in the ready mixed concrete environment', *Advances in Ready Mixed Concrete Technology*, Pergamon Press, pp. 112–22.

Liquin, G and Z. Guofan 1987, 'A study on post-cracking behaviour of SFRC working together with steel bar in uni-axial tension', in *Proceedings of the International Symposium on Fibre Reinforced Concrete*, Madras, India, vol. I, pp. 2.153–2.165.

Loov, Loov 1998, 'Concrete and Society—Past, Present and Future', in *Proceedings High Performance High Strength Concrete*, Perth, Australia, *1998*, pp. 75–96.

Lydon, F.D. 1972, *Concrete Mix Design*, Applied Science Publishers, London.

Malhotra, V.M. 1976, 'Testing hardened concrete: Non-destructive methods', American Concrete Institute Monography no. 9.

Maltone, R. 1992, 'Ferrocement and replica ships', *Journal of Ferrocement*, vol. 22, no. 3, pp. 273–81.

Mangat, P.S. 1976, 'Tensile strength of steel fibre reinforced concrete', *Cement and Concrete Research*, vol. VI, pp. 246–52.

Manjrekar, S.K. 1992, 'Polymers in concrete—Mechanism, properties and applications', *Indian Concrete Journal*, vol. 66, no. 3, pp. 127–31.

Manjrekar, S.K. 2000, 'Case study of the repair of a major bridge and some thoughts on repair materials', *Indian Concrete Journal*, vol. 74, no. 3, pp. 39–44.

Mattock, A.H. 1965. 'Rotation capacity of hinging regions in reinforced concrete beams', Portland Cement Association, Bulletin D101, Skokie, IL.

McIntosh, J.D. 1964, *Concrete Mix Design*, CACA.

McIntosh, J.D. and H.C. Erntroy 1955, *The Workability of Concrete Mixes with ¾ Aggregates*, CACA Research report no. 2.

McIntosh, R.H., Botton, J.D., and Muir, C. H. D. 1956. 'No-fines concrete as a structural material', *Proceedings of Institution of Civil Engineers*, Part I, 5(6), pp. 677–94.

McIntosh, R.H., J.D. Bottom, and C.H.P. Muir 1956, 'No fines concrete as a structural materials, in *Proceedings of Institution of Civil Engineers*, London, part 1, 5, no. 6, pp. 677–94.

Mehta, P.K. 1986, 'Concrete—Structure, Properties and Materials', Prentice-Hall International Series in Civil Engineering, Prentice Hall, Englewood Cliffs, NJ.

Mehta, P.K. 2005, 'Fundamental principle for radical enhancement of durability of concrete', *The Masterbuilder*, vol. 7, pp. 96–102.

Meininger, R.C. 1969, *Study of ASTM Limits on Delivery Time*, National Ready Mixed Concrete Association, Publ. no. 121, Washington D.C., p. 17.

Meyer, L.M. and Perenchio, W.F. 1980. 'Theory of concrete slump loss related to use of chemical admixtures', PCA Research and Development Bulletin RD069.0IT, pp. 1–8, Skokie, IL.

Mindess, S. 1994, 'Material selection, proportioning and quality control', in *High performance concrete properties and application*, S.P. Shah and A.H. Ahamed (eds), McGraw Hill, New York, pp. 1–25

Mindess, S. and J.F. Young 1981, *Concrete*, Prentice Hall, Englewood Cliffs, pp. 486, 501.

Modak, M.S. 2005, 'Factors for success of self compacting concrete', *The Masterbuilder*, vol. 7, no. 8.

Murashev, V.I, E.Y. Sigalove, and V.N. Baikov (English translation by G. Leib) 1968, *Design of Reinforced Concrete Structures*, Mir Publishers, Moscow.

Naaman, Antoine E. 2000, *Ferrocement & Laminated Cementitious Composites*, Techno Press.

Naik, Anil M 1989, 'The chemical connection', *Indian Architect and Builder*, April, pp. 24–5.

Narayan, Deepak and A.K. Sharma 1998, 'Thermo-mechanically treated bars: Advantages & limitations', in *Proceedings of National Seminar on Steel & Construction Industry Partnership*, Ministry of Steel & Mines (Govt of India), Bangalore.

National Geographic Society 1986, *Builders of the Ancient World*, p. 199.

National Ready Mixed Concrete Association 1993, *Outline and Tables for Proportioning Normal Weight Concrete*, Silver Spring, Maryland, p. 6.

Nawy, E.G. 1997, *Concrete Construction Engineering Handbook*, CRC Press, Boca Raton, New York.

Nawy, E.G. 2001, *Fundamentals of High Performance Concrete*, John Wiley, New York.

Neelamangalam, M., J.K. Dattatreya, and V.S. Parameshwaran 1992, *CIVCON-92*, CE & CR.

Nemati, K.M. 2015. '(a) Etched surface of concrete (b) Freeze–thaw scaling of a railroad bridge'. http://courses.washington.edu/cm425/durability.pdf

Nemati, K.M. 2015. 'Alkali–silica gel'. http://courses.washington.edu/cm425/durability.pdf

Neville, A.M. 1977, *Properties of Concrete*, Pitman Publishing, London, p. 687.

Neville, A.M. 1981, *Properties of Concrete*, Addison Wesley Longman Ltd, England.

Neville, A.M. 1981, *Properties of Concrete*, Pitman Publishing, Marshfield, MA.

Nishioka, K., N. Kakimi, S. Yamakawa, and K. Sirakawa 1975, 'Effective application of steel fibre reinforced concrete', *Fibre Reinforced Cement and Concrete Rielem Symposium*, Construction Press Ltd, Lancaster, pp. 409–18.

Okamura, H. 1997. 'Self-compacting high performance concrete - Ferguson Lecture for 1996', *Concrete International*, Vol. 19(7), pp. 50–54.

Okamura, Hajime and M. Ouchi 2003, 'Self-compacting concrete', *Journal of Advanced Concrete Technology*, vol. 1, no. 1, pp. 5–15.

Orchard, D.F. 1973, *Concrete Technology*, vol. 1, Properties of Materials, 3rd edn, Applied Science Publishers, London, p. 375.

Paillere, A.M., Buli M., and Serrano, J.J. 1987. 'The use of steel fibres to improve the durability of high strength silica fume concrete', Paper presented at ACI Annual Convention, San Antonio.

Paillere, A.M., Buli, M., and Serrano, J.J. 1989. 'Effect of fibre addition on the autogenous shrinkage of silica fume concrete', *ACI Materials Journal*, Vol. 86(2), pp. 139–144.

Pakvor, A. 1975, 'Rehabilitation and reconstruction of concrete structures', SP 128-87, pp. 1362–1422.

Palit, B.K. et al. 1987, 'New concrete products, precast concrete production techniques and light weight concrete', *Report on Roving Seminar in Modern Concrete Construction Practices*, ICI.

Palmer, Dennis 1977, *Concrete Mixes for General Purposes*, CACA, London, p. 8.

Park, R. and T. Paulay 1975, *Reinforced Concrete Structures*, John Wiley.

Park, S.G. 2001. 'Effect of recycled concrete aggregate on new concrete, Study Report no. 101. BRANZ, Wellington, New Zealand.

Patnaik, A.K. and S.K. Jain 1988, 'Ductility requirements in Indian codes for aseismic design of RC frame structures: A review', *Bulletin of the ISTE*, vol. 28, no. 2, pp. 11–40.

Perenchio, W.F. 1973, 'An evaluation of some of the factors involved in producing very high strength concrete', *Research & Development Bulletin no. RD 014, OIT, Portland Cement Association, Skokie, US.

Perkins, P.H. 1977, 'The durability of concrete in underground pipelines', Paper G1, in *Second International Conference on Internal and External Protection of Pipes*, Canterbury, p. 11.

Perkins, Philip H. 1979, Repair, protection and water proofing of concrete structures (private communication).

Petersson, O. Billberg, P., van B.K. 1996. 'A model for self-compacting concrete', RILEM Conference on Production Methods and Workability of Concrete, E&FN Spon, pp. 483–92.

Peurifoy, Robert L. and Garold D. Oberlender 1995, *Formwork for Concrete Structures*, McGraw Hill, New York.

Pfender, E. 1947. 'Contact type strain gauge for measuring mechanical deformation', *British Journal of Material Testing*, pp. 4–8.

Phan, L.T. 1996. 'Fire performance of high strength concrete: A report of the state of the art', *NISTIR 5934*, National Institute of Standards and Technology, Gaithersburg, Maryland, p. 81–82.

Plowman, J.M. 1956. 'Maturity and the strength of concrete', *Magazine of Concrete Research*, vol. 8, no. 22, pp. 13–22.

Portland Cement Association 1954, 'Design of concrete mixes', *PCA Concrete Information*, ST-56.

Portland Cement Association. 1976. *Handbook on Block masonry*.

Portland Cement Association 1987, *Concrete Information*.

Portland Cement Association 2000, *RCC News Letter*, vol. 16.

Powers, T.C. 1932, 'Studies of Workability of Concrete', *Journal of American Concrete Institute*, vol. 28, pp. 419–48.

Powers, T.C. 1958, 'Structure and physical properties of hardened Portland cement paste', *Journal of American Ceramic Society*, vol. 41, no. 1, pp. 1–6.

Powers, T.C. 1968. *The Properties of Fresh Concrete*. John Wiley, New York.

Powers, T.C. and Wiler, F.M. 1941. 'A device for studying the workability of concrete', Proceedings of the American Society for Testing Materials (1941), Philadelphia PA.

Powers, T.C., L.E. Copeland, and H.M. Mann 1959, 'Capillary continuity or discontinuity in cement pastes', J. Portland Cement Association, Res. Deve. Lab., pp. 38–45.

Prakash Rao, D.S. 1984, 'Detailing of reinforced concrete structure', *Indian Concrete Journal*, vol. 12, pp. 348–53; 1985, vol. 59, no. 1, pp. 22–5.

Price, W.H. 1951, 'Factors Influencing Concrete Strength', *Journal ACI*, vol. 47, no. 6, pp 417–32, 1951.

Proceedings of FIP Symposium on Concrete Sea Structures, Tbilisi, Georgia 1972, published by Federation of Internationale de Precontranite, London, 1973.

Proceedings of *ICFRC International Conference on Fiber Composites, High Performance Concretes and Smart Materials*, 2003, vols 1 and 2, International Center for Fiber Reinforced Concrete Composites, Chennai, India.

Proceedings of *the international symposium on Concrete Technology for Sustainable Development in the Twenty First Century*, February, 1999, Hyderabad, India.

Raichvargar, Z. and E.Z. Tatsa 1978, 'Manufacturing technologies of structural elements made of ferrocement', *Journal of Ferrocement*, vol. 8, no. 4, pp. 259–66.

Raju, M.K. 1974, *Design of Concrete Mixes*, Sehgal Educational Consultants & Publishers Pvt. Ltd, Faridabad (India), pp. 39–48.

Raju, N. Krishnan and N.S. Nadgir 1991, 'Limit state behaviour of reinforced concrete beams strengthened by epoxy-bonded steel plates', *Indian Concrete Journal*, vol. 62, pp. 124–9 .

Ramakrishnan, V. 1987, 'Material and property of fibre reinforced concrete', in *Proceedings of the International Symposium on Fibre Reinforced Concrete*, Madras, India, vol. I, pp. 2.3–2.23.

Randall, Frank A. and William C. Panarese 1976, *Concrete Masonry—Handbook for Architects, Engineers and Builders*, Portland Cement Association, Illinois, 1976.

Rangan, BV. 2015. 'Geopolymer concrete: State of the art report', *ISTE-SRM International Journal of Civil Engineering*, pp. 1–15.

Rehm, G., R. Eligehausen, and B. Neubert 1978, 'Rationalization of Reinforced Concrete Construction-Simplified Shear Reinforcement in Beams', *Betonwerk Fertigteil-Technik*, 44, pp. 147–55 and 222–7.

Rehm, G., R. Eligehausen, and B. Neubert 1979, Erlauterungen zu den Bewehrusrichtlinien nach DIN 1048 (December 1978), DAfstb. Heft 300, W. Ernst & Sohn, Berlin.

RKKR Steels Ltd and Tor-Steel Research Foundation of India 1987, *Torsteel-Grade TOR-50 in RC Structures—Economy and Structural Requirements*.

Road Research Laboratory, *Design of Concrete Mixes* (Road Note no. 4), 2nd edn, HMSO, London, p. 15

Roberts, M.B. and C. Atkins 1997, 'Deterioration modeling', *Concrete Engineering International*, vol. 1, no. 1, pp. 38–40.

Ronneburg, H. and M. Sandrik 1990, 'High strength concrete for North Sea platforms', *Concrete International*, vol. 12. no. 1, pp. 20–34

Ross, A.D. 1944, *Shape, Size and Shrinkage Concrete and Construction Engineering*, HMSO, London, pp. 193–9.

Rothfuchs, G. 1962. *Betonfibel, Band II* (Concrete Primer Vol. 1), Bauverlag, Wiesbaden-Berlin.

RRL 1950, 'Design of concrete mixes', RRL, Road Note no. 4.

Rusch, H., 1960. 'Research towards a general flexural theory for structural concrete', *Proceedings of the Journal of the American Concrete Institute*, Vol. 57, pp. 1–28.

Ryall, M.J. 2001, *Bridge management*, Butterworth Heinemann, Oxford.

Santhakumar, A.R. 1990, *Concrete Mix Design and Contribution Practices*, Short course notes for serving engineers of Tamil Nadu Housing Board, Nandanam Madras, ICI Local Chapter, Madras (India).

Santhakumar, A.R. 1996, 'Quality control of ready mixed concrete', in *Proceedings of the Seminar on Ready Mixed Concrete*, Cement Manufacturers' Association, Chennai.

Santhakumar, A.R. et al. 2004, 'Properties of concrete made with demolition waste as recycled coarse aggregate' in *the Seventh CANMET/ACI International Conference on Recent Advances in Concrete Technology*, Supplementary paper, Las Vegas, pp. 121–31.

Santhakumar, A.R. et al. 2005, 'Proportioning of recycled aggregate concrete,' *The Indian Concrete Journal'*, vol. 79, no. 10, pp. 46–50.

Santhakumar A.R. and K. Jagadeesan 1995, *Corrosion resistance of chemical coatings*, A short-term course on 'Corrosion damage, maintenance and Rehabilitation of Structures', Anna University, Madras.

Santhakumar, A.R., S. Sathayanarayan, C.K. Venkatasubramanian, and R.N. Reddy 1986, 'Semi-destructive methods of testing in-situ concrete', *Indian Concrete Journal*, May, pp. 127–33.

Sauzeat, K. et al. 1996, 'Textural analysis of reactive powder concretes', in *The 4th International Symposium on Utilization of High Strength/High Performance Concrete*, Paris.

Schneider, Robert R. and Walter L Dickey 1987, *Reinforced Masonry Design*, Prentice-Hall Englewood Cliffs, NJ.

Schrader, Ernest. K. 1997, 'Roller compacted concrete', in *Concrete and Construction Engineering Handbook*, Edward. G. Nawy (ed.), CRC Press, New York.

Scott, P. Ewing and R. Hutchinson 1953, A report on 'Corrosion', p.221.

Sekar, M. 1992, *Course Notes of Seminar on Recent Trends in Design & Construction of Concrete Structures*, ICI, Chennai.

Sengupta, A.K. 2003, 'Constitutive models for plain and reinforced concrete—An overview', *Indian Concrete Journal*, vol. 77, no. 8, pp. 1244–9.

Seshadri, S. and S.V. Ramana Kumar 1992, 'Practical applications of polymers in concrete', *Indian Concrete Journal*, vol. 66, no. 3, pp. 133–7.

Seventh CANMET/ACI International Conference on Recent Advances in Concrete Technology, Las Vegas, May, 2004.

Shacklock, B.W. 1974, *Concrete Constituents and Mix Proportions*, CACA, London, p. 102.

Shah, S.P. 1997, 'High performance concrete in 21st century, Parts I and II', in *Proceedings of TCDC International Workshop on High Performance concrete Technology and its Application* SERC, Madras, India.

Shah, S.P. and S.H. Ahamed 1994, *High Performance Concrete: Properties and Applications*, McGraw Hill, New York.

Sharma, R.C. 1968, 'Design of Concrete Mixes for use in the Tarapur Atomic Power Project', *Indian Concrete Journal*, vol. 42, pp. 155–62.

Shetty, M.S. 2005, *Concrete Technology—Theory and Practice*, S. Chand and Co.,

Shirley, D.D. 1975, *Mix Design—Introduction to Concrete*, CACA (no. 45.028), London, pp. 15–17.

Short, Andrew and William Kinniburgh 1963, *Lightweight Concrete-Mix Design*, Asia Publishing House, Bombay, chap. 3, pp. 20–32.

Shoshroda, T. 1990, 'Effects of bleeding and segregation on the interval structure of hardened concrete', in *Properties of Fresh Concrete*, H.J. Wieving (ed.), Chapman & Hall, London, pp. 253–60.

SHRP-C/FR-91-103 1999, *High Performance Concrete: A State of The Art Report*, Strategic Highway Research Council, Washington, D.C.

Spellman, D. and R. Stratfull 1968, 'Laboratory corrosion test of steel in concrete', *Interim Report M&R no. 635116-3*, California Division of Highways.

Spellman, D.L. and R.F. Stratfull 1984, 'Concrete Variables and Corrosion Testing', *Highway Research Board no. 423*.

Structural Concrete text book on Behaviour, Design and Performance, Updates knowledge of the CEB/FIP Model Code 1990, vol. 1: Introduction—Design Process, Materials Fib. Bulletin no. 1, 1999.

Subramanian, B.V. 1990, *Concrete Chemicals*, Course notes on concrete mix design and construction practice, ICI Tamil Nadu (India) Centre.

Subramanian, S. 1997, 'High strength concrete in high rise structures', in *Proceedings of TCDC International Workshop on High Performance concrete Technology and its Application*, SERC, Madras, India.

Surlakar, Samir 1992, 'Evaluation and testing of polymers for repair', *Indian Concrete Journal*, vol. 66, no. 3, pp. 145–9.

Surlakar, Samir 2000, 'Rehabilitation of structures—Criteria for material selection', *Indian Concrete Journal*, vol. 74, no. 3, pp. 31–7.

Sutherland, A. 1974, *Mix Design Air-entrained Concrete*, CACA (no. 45-022), London, 1974, pp. 6–7.

Syam Prakash, V. 2007. 'Ready mixed concrete using manufactured sand as fine aggregate', 32nd Conference on World in Concrete and Structures, 28–29 August, Singapore.

Tangtermsirikul, A. 1994, 'Methods for quality control of fluoride concrete in fresh state', in *Proceedings of Workshop on HPC* (in Thai), Engineering Institute of Thailand and ACI Thailand Chapter Bangkok.

Teychenne D.C., R.E. Franklin, and H.C. Erntroy 1975, *Design of Normal Concrete Mixes*, Department of the Environment, Building Research Establishment, Transport and Road Research Laboratory (TRRL), HMSO, London, p. 30.

The Associated Cement Companies Ltd. (Mumbai) 2001, Cement Users' Guide, 3rd rev. edn, p. 147.

The Concrete Association of India 1979, *Concrete Mix Design.*

The Concrete Society London 1976, 'Concrete core testing for strength', *Technical Report no. 11*, London, p. 44.

TISCON-50 & TISCON TMT-50, Tata Steel Information.

Troxell, G.E. et al. 1971, *ACI Monograph*, no. 6, p. 151.

Troxell, G.E. et al. 1971, *ASTM*, vol. 58, p. 128.

U.S. Department of Interior Water and Power 1998, *Concrete Manual.*

Ujhelyi, J. 1975. 'Industrial application of volcanic materials in Iceland', Technical report no ICE/73/008/11-02/03, Govt. of Iceland.

Vanikar, S.N. and C.H. Goodspeed 1997, 'High performance concrete bridge and pavements', in *Proceedings of TCDC International Workshop on Advances in High Performance Concrete Technology and its Applications*, pp. VI–21 to VI–35.

Verback, G.J. 1973, 'Field and laboratory studies of sulphate resistance of concrete', *Research Record no. 433*, National Council Washington, D.C.

Vollick, C.A. 1958. 'Effects of re-vibrating concrete', *Journal of the American Concrete Institute*, Vol. 54, pp. 721–732.

Wagner, C., and Traud, W.Z. 1938. *Electrochem.*, Vol. 44, pp. 391.

Wainshtok, Rivas H. 1994, 'Ferrocement swimming pools: A new and advantageous technique', *Journal of Ferrocement*, vol. 25, no. 2, pp. 153–9.

West, J.A. 1985, *The Traveler's key to Ancient Egypt*, Harrup Columbus, London, p. 480.

Weymouth, C.A.G. 1933. 'Effects of particle interference in mortars and concrete', *Rock Products*, Feb 25, p. 26.

Weymouth, C.A.G. 1938. 'A study of fine aggregate in freshly mixed mortars and concrete', Proceedings of the American Society for Testing Materials, vol. 38, part II, p. 354.

Weymouth, C.A.G. 1938. 'Designing workable concrete', *Engineering News Record*, Dec 29, p. 818.

Wright, P.J.F. 1964, *The Flexural Strength of Plain Concrete—Its Measurement and Use in Designing Concrete Mixes*, Road Research Technical paper no. 67, HMSO, London.

Wüstholz, Timo 2003, 'Fresh properties of self-compacting concrete (SCC)', *Otto-Graf-Journal*, vol. 14.

Zacharia, George 1988, 'Failure of concrete and masonry structures: Some case studies', *ICI Bulletin*, no. 35, pp. 17–25.

Index

About the Author

A.R. Santhakumar is a former Dean and Chairman of Faculty of Civil Engineering, Anna University, and a former Emeritus Professor, Indian Institute of Technology Madras. He obtained his B.E. and M.Sc.(Engg.) from College of Engineering Guindy, and later as a Commonwealth Scholar he received his Ph.D. from the University of Canterbury, Christchurch, New Zealand. Dr Santhakumar has more than 45 years of teaching experience and has guided a number of Ph.D. students during his long teaching tenure at Anna University.

He is a leading consultant to several organizations and has designed a number of infrastructure projects such as bridges, novel buildings, towers, dams, and plants for public sector corporations, Indian Railways, and the Government of Tamil Nadu. His work on diagonally reinforced coupling beams has been incorporated in all codes of practice including IS: 13920.

He has received several awards during his illustrious professional career, including the prestigious ICI–L&T Lifetime Achievement Award.

Notes